Biomarkers and Biosensors

Biomarkers and Biosensors

Detection and Binding to Biosensor Surfaces and Biomarkers Applications

Ajit Sadana
Chemical Engineering Department,
University of Mississippi, MS, USA

Neeti Sadana
MD, Department of Anesthesiology and Pain Management,
University of Texas Southwestern Medical School,
Dallas, TX, USA

ELSEVIER

AMSTERDAM • BOSTON • HEIDELBERG • LONDON • NEW YORK • OXFORD
PARIS • SAN DIEGO • SAN FRANCISCO • SINGAPORE • SYDNEY • TOKYO

Elsevier
Radarweg 29, PO Box 211, 1000 AE Amsterdam, Netherlands
The Boulevard, Langford Lane, Kidlington, Oxford OX5 1GB, UK
225 Wyman Street, Waltham, MA 02451, USA

Notices

Knowledge and best practice in this field are constantly changing. As new research and experience broaden our understanding, changes in research methods, professional practices, or medical treatment may become necessary.

Practitioners and researchers must always rely on their own experience and knowledge in evaluating and using any information, methods, compounds, or experiments described herein. In using such information or methods they should be mindful of their own safety and the safety of others, including parties for whom they have a professional responsibility.

To the fullest extent of the law, neither the Publisher nor the authors, contributors, or editors, assume any liability for any injury and/or damage to persons or property as a matter of products liability, negligence or otherwise, or from any use or operation of any methods, products, instructions, or ideas contained in the material herein.

ISBN: 978-0-444-53794-2

British Library Cataloguing in Publication Data
A catalogue record for this book is available from the British Library

Library of Congress Cataloging-in-Publication Data
A catalog record for this book is available from the Library of Congress

Printed in the United States of America

Contents

Preface

Biosensors and Biomarkers is the next installment in a series of books by the author (A.S) and co-author (N.S.). The series has transitioned from explaining the basics of biosensors to how they function and now in a more sophisticated manner, practical applications of the biosensors to make our human lives safer. Previous books have focused on the kinetics of biosensors and then on to the features of these tiny miracle workers. The primary focus of this book as mentioned is the practical application of this nanotechnology in medicine and technology. The amalgam of theoretic and applied science as well as the combination of a doctor of science with a doctor of medicine sets this particular book apart from all other textbooks on the subject.

The American Cancer Society estimates 1,660,290 new cancer diagnoses in 2013 and 580,350 deaths in the same year with approximately 1600 cancer-related deaths per day. Despite advances in medical care, cancer will still be responsible for 1 in 4 deaths in the United States. Chapter 3 in *Biomarkers and Biosensors* is devoted to the use of biosensors for the detection of biomarkers of cancer and their assistance in clinical decision-making. Nonscientists are aware that DNA and RNA make up the molecular structure of human cells, but biosensors have elucidated a concept of microRNA that has been found to be dysregulated in certain pathologic states such as breast cancer.

Chapter 8 addresses the binding and dissociation of biomarkers for systemic lupus erythematosus (SLE). SLE remains an illusive disease in diagnosis, symptomatology, and treatment. Its many variations have plagued physicians and patients alike for years and other illness of autoimmune nature such as rheumatoid arthritis and even Celiac disease are making front-page news for plaguing star athletes as well as the common man. A simple diagnosis of Celiac disease and resultant change in diet can change the life course of every patient afflicted with this painful and inconvenient disorder. The mathematical modeling and fractal analysis of the kinetics of chosen biomarkers has led to the detection of auto-antibodies that aid in the diagnosis and prognosis of the above disease states.

Whether this text is of interest to you as a scientist to aid in the understanding of what biosensors are and how they may work and be applied, or if you are a medical doctor searching for how translational research can be applied in the future toward clinical decision-making and patient care, *Biomarkers and Biosensors* is for you. Many chapters are devoted to the mathematical modeling of biosensors and biomarkers, while the meat of this book is discerning how to utilize this sophisticated technology. Without practical application, this information is largely theoretic and useful only in the classroom. This textbook can provide the tools to grasp the science and fully appreciate the application.

The author Dr Ajit Sadana would like to acknowledge Ms Emily Jane Pugh (EP), Ms Nichole Brasel Williams, and Ms Stephanie Amber Hall for being the graders for my courses, and freeing up a lot of time that facilitated the writing of this manuscript. The students in the classes very much appreciated the help sessions conducted by them. Ms Jaya Viswanathan, my undergraduate student at present, uploaded all the files and I do thank her for her assistance. The daily interaction with students and colleagues and quest for unearthing the unique is what drives the author to make biosensors approachable and understandable to all. The author (AS) wishes to express his appreciation to Ms Anne Buntin Pringle for all the assistance she has provided over the years.

The co-author Neeti Sadana MD, would like to thank her father Dr Ajit Sadana, the author, for introducing her to the world of nanotechnology for its academic interest and application for the treatment of disease states. She would also like to thank her husband, Dr Pooya Pouralifazel and her two fat bulldogs, Bentley and Stewie, for making her world such a fun place.

Neeti Sadana
Ajit Sadana

Chapter 1

Introduction

1.1 INTRODUCTION

A biomarker may refer to a protein whose concentration refers to the severity or presence of some disease state. These biomarkers may be detectable and measurable by a variety of methods including physical examination, laboratory assays, and medical imaging. College Hill indicates that "Biomarkers are valued tools used across the biological spectrum from research to diagnostics, as indicators of normal or disease processes to assess pharmacological responses." Biomarker (biological marker) is a characteristic that is objectively measured and evaluated as an indicator of normal biologic processes, pathogenic processes, or pharmacologic response to a therapeutic intervention. Biomarkers are proposed to measure the delivery of drugs to their intended targets, and to understand and predict pathophysiology, and how it is altered by therapy, through monitoring variables known to have chemical relevance. The intent is to use biomarkers for their predictive power to select compounds and design dosing regimens for meeting the pharmacokinetic criteria for a new drug.

Chabner (2008) has recently reviewed the challenges and advances in the use of biomarkers for clinical trials. Sheridan (2005) has indicated that protein chip companies have turned to biomarkers. Furthermore, they state that a new proteomics chip allows high-throughput protein interaction studies used in drug discovery. Baker (2005) further indicates that the "omics" revolution provides for quite a few decision-making tools. The author emphasizes that the trick lies on how to use them.

Illyin et al. (2004) indicate that the field of biomarkers has application in the diagnosis, prognosis, and in monitoring disease progression. They also assist in monitoring the responses to a therapeutic intervention, and in the delivery of personalized medicine. They further emphasize that biomarkers are involved in clinical, physiological, biochemical, developmental, morphological, and molecular measures. In drug discovery, biomarkers permit patient stratification as well as the efficacy determination of drugs. They conclude by indicating that the different technologies for data collection and analysis are critical in the different biomarker processes that involve identification, characterization, validation, and application.

Panteghini (2010) in a recent article on cardiac biomarkers emphasizes the need for a cardiac biomarker that detects the presence of myocardial damage

Biomarkers and Biosensors. http://dx.doi.org/10.1016/B978-0-444-53794-2.00001-X

even before the reversible myocardial damage is induced, and thereby help identify the vulnerable patients before major events occur, permitting prior treatment. This author indicates that the increase in the blood concentration of cardiac troponins is designated as a surrogate for cardiac necrosis and myocardial infarction (MI).

Labtechnologist (2010) defines a biomarker as a "biochemical feature that either directly or indirectly provides information about a disease and its remission, and the effects of a drug compound on the disease. These biomarkers may be used to help evaluate drug therapies in clinical trials, and also serve as 'surrogate endpoints' wherein the ultimate condition in clinical trials is the patient's death if not treated effectively". Furthermore, the author adds that biomarkers are particularly useful in the efficacy of drugs for neurological diseases, for example, Alzheimer's. He recognizes that the only way to know that if a patient has Alzheimer's is to open up the brain after the patient's death. Finally, the author adds that biomarkers may also be used to minimize adverse events by monitoring the patients' response to a drug.

Nagano et al. (2010) indicate that molecular biomarkers are keys to the development of new diagnostics, protocols, and therapies. They point out that recently significant research effort has been involved in the development of biomarkers using different approaches. One of these is disease proteomics. According to them disease proteomics involves analyzing and identifying the changes in the expression pattern in the disease-related condition, that is, in basically the disease-related proteins by using two-dimensional gel electrophoresis (2D-DIGE). They emphasize on being able to pickup the right proteins that are important and significant from a large collection of disease-related proteins that have been identified. They emphasize that on using antibody proteomics one is able to identify a wide variety of disease-related proteins by 2D-DIGE. Also, this technique permits the preparation of monoclonal antibodies to these proteins using a phage antibody library.

Ramachandran (2009) has recently provided some insights into the active and high growth biomarker market. The author indicates that the biomarker definitions working group at the National Institutes of Health (NIH) initially defined, in the year 2001, a biomarker as "a characteristic that is objectively measured and evaluated as an indicator of normal biological processes, pathogenic processes, or pharmacologic responses to therapeutic intervention". The author further indicates that biomarkers may be predictive, surrogate, and efficacy and safety. The different biomarkers may be simple molecules (glucose, cholesterol, triglycerides), macromolecules (insulin, prostate specific antigen, C-reactive protein (CRP), and hemoglobin), or complex molecules. Furthermore, the author states that the applications of biomarkers may be to (1) early disease identification, (2) help identify potential drug targets, (3) predict the response to the medication prescribed to patients, (4) help shorten the time for clinical trials, and (5) assist in the delivery of personalized medicine (since each individual is different, biomarkers will assist the physician to prescribe the right dose for a patient).

Biomarkers are especially valuable to help prioritize drug discovery resources by enabling early proof-of-concept studies for novel therapeutic targets. There is an increased acceptance and application of biomarkers in drug discovery by pharmaceutical companies. This is primarily due to how the application of biomarkers facilitates these pharmaceutical companies to overcome the challenges posed by conventional drug discovery processes which are time-consuming and expensive. Note that high drug attrition rates, safety and efficacy concerns, and time-consuming methods have convinced the pharmaceutical sector to use biomarkers in the drug discovery process on a large scale. The Food and Drug Administration (FDA) (FDA, 2004) indicates that biomarkers are 10% more efficient in predicting drug failures and hence can save approximately $100 million and 3–4 years in each drug discovery process. In order to place this in some sort of perspective, note that it takes about $800 million and about 12–15 years to get a drug from the bench scale to the market.

In June 2006, the FDA came out with a working paper to enhance the drug development process (Goldberg and Pitts, 2010). The intent was to provide the biopharmaceutical companies with guidelines for them to bring new medicines to the market in a safer, less expensive, faster, and streamline manner. The FDA task force comprised of 25 experts from the industry, government, and scientific community. This FDA task force suggested the use of new technologies (specifically the "omics" sciences), genetic tools, and faster computers. Some of the specific suggestions included:

1. the validation of biomarkers at every stage of the regulatory process,
2. the use of validated biomarkers to assess the safety and efficacy of these specific drugs,
3. articulate the importance of congressional appropriations for biomarker research.

The report emphasizes that "biomarkers are measures of the disease progression, pharmacology, or safety that can help identify unique disease mechanisms or responses to medicine." The report also indicates that how FDA can specify how biomarkers may be used to develop drugs, biologics, and companion diagnostics. Finally, the report also suggests that a strong collaboration is required amongst the biopharmaceutical community to help validate the biomarkers.

As early as November 2001, the Tufts Center for the study of drug development indicated that the cost for developing a drug was $800 million to $1.7 billion, and the estimated time for the development was 12–15 years. The biopharmaceutical companies have a limited time to recoup these very significant amounts of money spent on R&D. They are thus spending enormous resources according to Goldman and Pitts (2006) on the applications of biomarkers in the drug development process to help minimize the time and money spent. However, hurdles, still need to be overcome such as to predict hepatic injury (liver damage).

Among the various suggestions made by the FDA task force some are:

1. the testing and development of molecular and imaging biomarkers, and
2. specific directions for the use of biomarkers in clinical trials during drug development.

Finally, the report cautions that safety issues often come to light during clinical trials and even after marketing. These, of course, need to be avoided, or at least significantly minimized.

Turner (2012) indicates that 11% of all papers ever published on biosensors were published in 2011, and the total worldwide sales of biosensors exceeds 13 billion US dollars. He also indicates that the academic input has spawned extraordinarily.

Recently, the area of biosensor/biomarker research has expanded considerably. Recent presentations on biomarker detection have appeared in the literature including presentations at the 2010 Annual American Institute of Chemical Engineers Meeting held in Salt Lake City, Utah, November 7–12, 2010, and the 4th Biomarker Discovery and Development Conference held in San Francisco, October 20–22, 2010. Some of the presentations included:

1. Engineered knottin peptides: A new class of agents for noninvasive molecular imaging of tumor biomarkers (Moore et al., 2010). There is a critical need for noninvasive molecular imaging probes that specifically target receptors over-expressed on tumors, for earlier cancer detection and patient-specific treatment and disease management.
2. On-chip electrochemical detection of biomarkers for detection of water-borne toxins (Wilson et al., 2010).
3. Hybrid magnetic–plasmonic nanoplatelets for biomarkers (Sotiroub et al., 2010).
4. Identifying secreted biomarkers for immune evasion in cellular models of cancer (Kulkarni and Klinke, 2010). The authors attempt to analyze proteins in the bloodstream arising from the secretome of cancer cells. These proteins serve as potential noninvasive biomarkers.
5. Rational and combinatorial design of peptide affinity ligands for diagnostic assays (Chandra et al., 2010). Peptides are promising affinity ligands for the detection of proteins in biological samples. The authors used peptides to design single-step, high-sensitivity diagnostic assays for detecting s-protein as a model biomarker in human serum. The authors' intent was to develop a systematic approach for the design and discovery of peptide affinity ligands for the detection of protein biomarkers of interest.

Some of the biomarker presentations at the 4th Biomarker Discovery and Development Conference held recently in San Francisco, California from October 20–22, 2010 include:

1. Novel click chemistry-based tools for high resolution biomarker discovery (Agnew et al., 2010). These authors have developed for Molecular Probes-Life Technologies, a powerful click chemistry-based platform for applications in global biomarker discovery. For example, the authors presented a novel nascent RNA enrichment tool. This permitted global enrichment and identification of newly synthesized RNA transcripts.
2. Development of sandwich enzyme-linked immunosorbent assays (ELISAs) for potential biomarkers in pancreatic cancer (Brahmandan et al., 2010). The authors indicate that pancreatic cancer or pancreatic ductal adenocarcinoma (PDAC) is apparently one of the most deadly forms of cancer in the United States. It is also the fourth leading cause of death in men, and the fifth leading cause of death in women. The authors emphasize that biomarkers such as CA19-9 are widely used in clinics. The intent of the present study is the development of a sandwich type ELISA to help detect potential biomarkers for the early detection of PDAC.
3. Comprehensive analysis of serum peptidome using restricted access media and nanoliquid chromatography–tandem mass spectroscopy (Gil et al., 2010). These authors indicate that serum peptidome serves as a rich source of biomarkers for disease diagnosis and monitoring. These authors have developed a modular automated processing system (MAPS) for high-throughput analysis of complex biological samples.
4. The accelerated expansion of clinical use of plasma transforming growth factor beta-1 (TGFβ-1). The new paradigm of inflammation and fibrosis was given by Shoemaker et al., 2010. These authors indicate that the multifunction cytokine, TGFβ-1, regulates tissue morphogenesis and differentiation through its effects on cell proliferation, differentiation, and extracellular matrix production. Furthermore, these authors add that elevation or reduction in levels of TGFβ-1 may be connected with various disease states including cancer, diabetes, and fibrotic disease of the kidney, liver, and lung. The authors conclude by indicating that TGFβ-1 may be used as a biomarker for clinical diagnosis. It may also be used for gauging the efficacy of treatment for a diverse array of diseases and helps expand the knowledge of innate immune responses by bringing salutary new therapies to patients.
5. Antilipid response in systemic lupus erythematosus (SLE) (Jovanovic et al., 2010). These authors indicate that SLE is a chronic, multisystem, and autoimmune disorder with a broad range of clinical presentation. Several comprehensive computerized indices for measuring clinical disease activity in SLE are part of standard clinical practice. The authors emphasize that there is still a need for defining well-validated diagnostic and prognostic

biomarkers. These authors investigated lipid and autolipid antibody profiles in SLE patients using ELISA and gas chromatographic-mass spectrometry (GC–MS). The authors emphasize that reduction of lipid levels and anti-lipid IgGs at different time points would help to provide information as to the success of the treatment.

Other biomarker presentations or workshops of interest at different conferences include:

1. Translation of imaging biomarkers from research to clinical trials (van Bruggen et al., 2011).
2. Systematic biomarker data analysis workshop (Bio-IT World Conference, 2011).
3. Molecular diagnostic laboratories (MoDEL): a program to support cancer biomarker clinical assay development (Jessup et al., 2009).
4. Application of intraassay calibration curves to quantitate clinical biomarker assays (Rhyne, 2009).
5. Office of in vitro diagnostics (OIVD) outlook: biomarker-based in vitro diagnostic assays (Phillip, 2010).
6. Biomarker imaging: from molecules to man, discovery to diagnostics (Frank, 2010).
7. Circulating oncology biomarkers to guide targeted therapies: companion diagnostics for personalized medicine (Carney, 2010).
8. Establishing the interest for introducing new safety biomarkers into clinical trails (Furlong, 2011).
9. Using multiple omics principle in preclinical hepatotoxicity biomarker discovery (Salminen et al., 2011).
10. Development of biomarker panels: opportunities and challenges (Hanash, 2011).
11. Development of drug-induced vascular injury biomarkers (Brott, 2011).
12. Using mass spectrometry and immunoassays to discover and validate serum protein biomarkers for islet autoimmunity and type I diabetes (She, 2011).
13. Analyzing miR-122 as a biomarker for hepatotoxicity (Batheja, 2011).
14. Translating biomarkers from the lab to the clinic—short course (Furlong et al., 2011).

Case studies and examples are presented related to developed biomarkers.

15. Best practices in fit-for-purpose biomarker assay validation—short course (Batheja, 2011a,b).
16. Neutrophil gelatinase-associated lipocalin (NGAL) as a biomarker of acute kidney injury and nephrotoxicity (Devrajan, 2010).
17. Biomarkers of nephrotoxicity: uses and challenges in preclinical and clinical studies (Bonventre, 2010).
18. Discovering biomarkers to predict clinical outcome (2010).

19. Those biomarkers come from biospecimens: garbage in, garbage out (Compton, 2011).
20. Oncology biomarker discovery technologies today and tomorrow (Li, 2011).
21. Development of imaging agents (Nunn, 2006).
22. Implications of pharmacogenomics for drug development (Kirk et al., 2008).
23. Pretreatment circulating vascular endothelial growth factor (VEGF) levels as a predictive biomarker of efficacy in non-small-cell lung carcinoma (NSCLC) patients treated with vandetanib (Ryan, 2010).
24. The kristen rat sarcoma (KRAS) signaling pathway biomarker in oncology: from prognostic to predictive (Patterson, 2010).
25. Incorporation of biomarkers into tarceva clinical trials (Richardson, 2010).

DePalma et al. indicate that biosensors have the potential to revolutionize the field of in vitro diagnostics. This is for human, veterinary, and food applications. The authors emphasize the role of point-of-care (POC) applications of biosensors. They have developed a sensitive magnetic bead-sensing-platform for the detection of proteins. They have presented the optimization of their magnetic immunoassay that includes surface chemistry, the blocking procedure, and the type of magnetic particles for the highly specific detection of S100ββ (a diagnostic marker for stroke and minor head injury). These authors emphasize that the use of superparamagnetic particles is an appealing alternate for the commonly used fluorescent labels, especially since they have been successfully used for the isolation of cells, proteins, and nucleic acids.

Haes et al. (2005) have developed a nanoscale optical biosensor to detect a biomarker for Alzheimer's disease from synthetic and clinical samples. These authors used localized surface plasmon resonance spectroscopy to monitor the interactions between the antigen, amyloid-β-derived diffusible ligand (ADDLs), and specific anti-ADDL antibodies. Their technique permitted the authors to determine the ADDL concentration and provided physical insights into the aggregative mechanism of this Alzheimer disease pathogen at relevant monomer concentrations.

Cassiday (2010) indicates that carbon nanotubes have been used by Professor James Rusling of the University of Connecticut to stretch the boundaries of biomarker detection. In order to move cancer detection devices closer to the clinic these authors have developed an ultrasensitive electrochemical immunosensor for the oral cancer biomarker, interleukin-6 (IL-6). Cassiday (2010) indicates that although existing methods can detect cancer-related biomarkers, they are not suitable for POC applications. Liquid chromatography-tandem mass spectrometry (LC—MS) proteomics analyses are labor intensive, and require expensive instrumentation. Furthermore, ELISAs (the gold standard method) are not amenable to multiplexing. This time-consuming activity increases the patient's anxiety. Rusling indicates that "what is required is a device that could be used in a doctor's office routinely to screen people for the different types of cancer." Finally, Joseph Wang of the University of California, San Diego comments that "the excellent sensitivity of the device opens the door for the early diagnosis of cancer." This, of course, is essential for an early and timely intervention.

Cambridge Health Institute (2010) in a recent report entitled "Biomarkers in late stage clinical trials", indicates that biomarkers play a critical role in late stage clinical trials. They act as surrogate endpoints for patient monitoring and stratification. They emphasize that biomarkers may be used to lower drug attrition rates, and thereby increase the productivity and lower the cost and duration of clinical trials as indicated above. Furthermore, they help "speed up" the drug development process.

Grigsby et al. (2010) have recently designed and implemented a prototype software tool for the visualization and analysis of small molecule metabolite GC−MS and LC−MS data for biomarker discovery. The authors indicate that metabolomics is a growing field that helps characterize the metabolic profile of a specific tissue or biofluid. They further emphasize that metabolomics is an attractive approach to analyze and study the time-related quantitative multivariate metabolic responses to pathophysiological processes. These may be caused by drugs or indigenous metabolites involved in critical pathways (Reo, 2005). Reo (2005) indicates that the metabolic alterations may be expressed as a "fingerprint" of the biochemical perturbation which is characteristic of the type and target of a toxic insult or disease process. Dunn and Ellis (2005) indicate that biofluids may be obtained noninvasively (urine) or minimally invasively (blood), and they can be used in metabolomic studies. Furthermore, Wang et al. (2004) emphasize that if one were able to monitor a significant number of trace molecules, then this would be more predictive than a single biomarker. Grigsby et al. (2010) emphasize that environmental toxins and therapeutic interventions lead to nephrotoxicity. A list of metabolites indicating kidney damage would be helpful in monitoring renal conditions.

Qureshi et al. (2010) have recently developed a label-free capacitive biosensor for the sensitive detection of multiple biomarkers using gold interdigitated capacitor arrays. Their assay is able to detect a panel of disease biomarkers: CRP, tumor necrosis factor-α (TNF-α), and IL-6. Cardiovascular risk (CVR) is associated with more than one biomarker for its incidence. Hill and Martins (2006) have emphasized the development of a multianalyte immunoassay for panels of biomarkers for the diagnosis of a disease. Qureshi et al. (2010) indicate that CRP, TNF-α, and IL-6 have a strong and consistent relationship between markers of inflammation and future CVR. These authors emphasize that the early detection of a panel of biomarkers for a disease permits the prediction of the disease risk. They used a relative change in capacitive/detection properties for the detection of this panel of biomarkers. These authors also emphasize that multianalyte detection provides the advantages of shortened analysis time, simplified analytical procedure, minimal sampling volume, improved test efficiency, and cost effectiveness when compared with parallel single-analyte assays. Stoeva et al. (2006) have used optical immunosensor arrays for the detection of multianalyte protein biomarkers. Finally, Qureshi et al. (2010) emphasize that capacitive immunoassays may be used as an alternative to the existing immunochemical assay methods for the development of hand-held devices that may be used for POC applications.

Lin et al. (2010) have recently developed a biogenic nanoporous silica-based sensor for enhanced electrochemical detection of cardiovascular biomarker proteins. These authors indicate that proteomics research has been able to identify a number of biomarker proteins which exhibit the potential to improve disease diagnosis (Darain et al., 2004; Hahm and Lieber, 2004; Nam et al., 2003; Niwa et al., 1990). Lin et al. (2010) also indicate that the detection of multiple biomarkers provides the information to permit a robust diagnosis for a disease in any person (Abeloff et al., 2000; Chou et al., 2004; Danesh et al., 1998). They emphasize that the use of biomarkers will depend on the development of new techniques to permit the rapid and multiplexed detection of a wide range of biomarkers with high selectivity and sensitivity.

Chenevier-Gobeaux et al. (2010) have recently reviewed new biomarkers in emergency patients with cardiovascular conditions. These authors indicate that new biomarkers have changed the approach of diagnosis and treatment procedure in emergency medicine. This is especially true for cardiovascular disorders. They emphasize the integration of biomarkers in new strategies that help improve their effectiveness. Besides, the inclusion of biomarkers results in the development of tools that enhance safety and efficiency. They provide current knowledge on emergent biomarkers in emergency medicine in the field of cardiovascular diseases and infection.

Duffy et al. (2011) have recently analyzed the uses and limitations of cancer antigen 15-3 (CA15-3) as a disease biomarker for cancer. These authors indicate that CA15-3, which detects soluble forms of MUC-1 protein, is the most widely used serum marker in patients with breast cancer. It is used primarily for monitoring therapy in patients or metastatic disease. These authors emphasize that CA15-3 should be used in monitoring therapy in conjunction with diagnostic imaging, clinical history, and physical examination. According to these authors CA15-3 is particularly valuable for treatment monitoring in patients who have disease that may not be evaluated using the existing radiological procedures. They emphasize that CA15-3 may also be used in postoperative surveillance of asymptomatic women who have undergone surgery for invasive breast cancer.

Rubenstein (2007) in a report entitled “Disease-related biomarkers: their potential in patient screening, prognosis, and stratification”, indicates that disease-related biomarkers are not a new phenomenon as is evidenced by blood glucose for diabetes diagnosis and management and cholesterol for cardiovascular risk. The author indicates that oncology is the most active field for disease biomarker research and development because cancer therapy routinely provides autopsy and tissue, and also that pharmaceutical and biotechnology companies are heavily involved in cancer drug discovery. Rubenstein (2007) emphasizes that molecular biomarkers for neurological diseases such as Alzheimer's and schizophrenia are in focus since obtaining human tissue as a sample is difficult. The author emphasizes that this is essential for people who are involved in the discovery, development, validation, and commercialization of disease-related biomarkers.

Kinsinger (2010) of the National Cancer Institute (NCI)/NIH indicates the need for the development of diagnostics and for companion biomarkers. He emphasizes the importance of early detection. He indicates that one to two protein biomarkers are approved each year. He notes the three steps that are involved in the biomarker process: characterize, verify, and validate. One is able to whittle down the possible candidates from 1000s to 100s to 10s in this process. One needs to filter biomarkers before one moves them to the clinic. He emphasizes the need to develop a systematic and integrative approach.

Moore (2010) of the NCI/NIH indicates that the health costs related to cancer are around $189 million per year. She is presently the Director of the Biospecimen Research Network at NCI/NIH, and emphasizes that biospecimens are the foundations of translational research. They are at the center of the evolution of cancer research. There is a critical need for high quality specimens, and there is high emphasis on quality control.

Zheng (2010) has recently analyzed fit-for-purpose biomarker studies. He indicates that valuable biomarkers constructively aid in decision making. They emphasize that biomarker clinical trials balance right dose, right schedule, and the right patient. Two types of inputs are involved: (1) strategic input (what to test), and (2) operational input (how to test). Three simple steps are involved: sample, deliver, and assay/data. This is a multidisciplinary area which involves a multifunctional group. One needs to qualify the biomarker as a predictive or a prognostic biomarker. The authors indicate that BRAF (for melanoma trials) was examined in 500 patients. The study was still ongoing at that time. EGRF inhibitor is a predictive biomarker. The author emphasizes that informed consent is essential, and this is not a trivial process. Point-of-principle biomarkers are also involved.

Lakey (2010) has recently analyzed epigenetic biomarkers and their application in therapy selection. He indicates that 70% people benefit and 30% people do not benefit from epigenetic biomarkers. He emphasizes that people are different, but the treatments do not differ. One must take into account the molecular heterogeneity of the disease as well as the heterogeneity of the people. Lakey (2010) indicates that an epigenetic signal is a good biomarker if one can solve the spatial and temporal issues. Initially Crick had proposed the following sequence: DNA, coding, protein, metabolites. This is, however, not true now. The author is the President and Chief Financial Officer of Orion, a cancer diagnostic company. The company develops diagnostics and validation technologies based on epigenetic differences in man. They have developed a lead breast cancer biomarker which exhibits 90% sensitivity.

Lai-Goldman (2010) in a presentation on the translation of personalized cancer diagnostics emphasizes the need for the codevelopment of biomarkers and companion diagnostics. She mentions the three criteria that are required for the development of new diagnostics:

1. unmet medical need,
2. actionable result, and
3. resources to develop evidence of clinical utility.

Also, the test should be accessible. She estimates the time required for the development of a new biomarker to be around five years.

Lakey (2010) indicates that the biomarker insulin-like growth factor 2 (IGF2) assists in the screening of colorectal cancer (CRC). The company plans to market their colon cancer risk test to the 20–40 years age group. There are about 81.6 million people in this age group. For negative patients, one may delay the test for 5–10 years. For positive patients, the author suggests that they are on a "fast track" to get CRC by the time they are 50 years old. Loss of imprinting (LOI) of IGF2 increases the risk of CRC by 21.7 times the normal. The author indicates that LOI prevalence is stable. Also, the prevalence does not increase with age. Lakey (2010) further indicates that LOI prevalence of IGF2 trials are underway. About 75,000 patients are being screened for CRC for the IGF2 biomarker.

Huey (2010) of Cambridge Biomedical Research Group in Boston indicates that the elevation of TGFβ-1 is linked to cancer, diabetes, etc. There is a need to develop a diagnostic assay for TGFβ-1. This is a good therapeutic and diagnostic biomarker, and is elevated in chronic fatigue patients. The author indicates that there is no true normal for TGFβ-1, since there is a wide range; besides there is age variation in TGFβ-1 levels. Huey (2010) indicates that there is a reduction in TGFβ-1 levels with the drug, Cosartan.

Kavsak (2010) indicates that a functional biomarker may be classified into three categories: preanalytical, analytical, and post analytical. An iterative technique may be involved in developing an appropriate biomarker between these three categories. The author addresses the issue of which biomarkers are important to help identify people at risk for inflammation and fibrosis. One needs to adjust for sex, age, heart failure, etc. This would assist in the unbiased selection of patients.

Brahmandan et al. (2010) indicate that there is a need to distinguish between chronic pancreatic and PDAC. One may use proteomics and genomics to help discover biomarkers for PDAC. These biomarkers need to be verified and validated. The authors are developing a capture ELISA kit for DKK1. This needs to be validated before it can be tested on actual samples. They are also in the process of developing a sandwich ELISA kit for junction plakoglobin (JUP). JUP or plakoglobin is a common functional plaque protein. The membrane-associated plaques are architectural elements in an important strategic position to influence the arrangement and function of both the cytoskeleton and the cells within the tissue. The presence of plakoglobin in demosomes and the intermediate junction suggest that it plays a central role in the structure and function of submembranous plaques.

Presently, there are no antibodies available for this and they are trying to develop antibodies for this too.

Detmers (2010) has recently analyzed "difficult" proteins, peptides, and biomarkers. The author indicates that their "Impercicer" assay is an excellent sensitivity platform when compared with other platforms. The readout system is different from ELISA. The detection is amplified by polymerase chain reaction, and is about three orders of magnitude more sensitive than ELISA. It may be used on different types of human specimens. For example, it may be used to detect biomarkers (amyloid β and tau protein) for Alzheimer's disease. In these cases the biomarker concentration is low, as well as the sample volume may be limiting (2 μl). The detection device needs to be very sensitive since the biomarkers have to cross the blood−brain barrier. The author indicates that it may be used to detect cytokines in inflammation. Patients may be stratified into low, medium, and high levels of cytokines. Thus, one may be able to give cytokine scavenging drugs. Note that ultrasensitive patient screening enhances drug responder rates. Besides, their sensitive assay may be used for infectious testing. The author emphasizes that the assays should be "drug-tolerant" to help minimize the "immunogenicity effect". Another application of their technique which Detmers (2010) suggests is doping associated with muscle growth.

Aletta (2010) has recently analyzed the application of methylarginine proteins for the treatment of autoimmune diseases such as SLE, sclerodoma, and multiple sclerosis (MS). One may be able to stratify cancer patients using a difference in methylation status. Their company is trying to develop a diagnostic for protein arginine methyltransferases (PRMT) activity. PRMTs may be linked to DNA damage responses, and may be involved in neoplastic diseases such as hormone-dependent cancers. PRMT1 is linked, this author suggests, to breast cancer, and PRMT4 is linked to prostate cancer. Aletta (2010) also suggests that these methylarginine proteins may also be linked to cardiovascular disease, hypertension, artherosclerosis, and chronic lung disease.

Vasto et al. (2010) have recently analyzed the biomarkers of ageing. These authors state that ageing is a complex process that impacts in a detrimental way in the development of different systems, and their ability to function properly. They also indicate that the rate of ageing is not uniform because of genetic heterogeneity, and the influence of environmental factors. They define ageing rate as the decline of functional capacity and stress resistance. This, they claim, is different in every individual. They define age-related biomarkers as age-related changes in body function or composition that could serve as a measure of biological age. Furthermore, these age-related biomarkers predict the onset of age-related diseases and/or residual lifetime. The biomarkers that they put forward for analysis are based on immunosenescence, inflammatory responses, and oxidative stress. They indicate that their approach is to prevent infectious diseases and delay the onset of age-related diseases. Finally, their biomarkers help provide a better understanding of ageing as well as provide for new strategies to help counteract the ageing process.

1.2 BIOMARKER ECONOMICS AND MARKETS

Francis (2010) indicates that biomarkers are playing an important role in streamlining the drug development process. Biomarkers assist in preventing the wastage of time and money; critical resources which are in short supply, particularly for smaller companies. Biomarkers help in making decisions early of possible drug candidates that will never make it to the market by allowing research teams to find out about this as early as is possible. Francis (2010) emphasizes that biomarkers help speed up the process of getting the drug from the bench scale to the market. This very significantly assists in minimizing the costs for drug development. Thus, biomarkers assist in a faster progress for medical research.

Aarkstore enterprises (2010) emphasizes the growing recognition amongst pharmaceutical companies about the benefits of biomarkers and this has led to increasing commercial interest in this area. The authors emphasize that the potential exhibited by biomarkers in that rheumatologists may be able to predict responses of patients to expensive biological therapies based on biomarker profiles. The authors provide an identification of potentially promising biomarkers and assays. The authors emphasize that personalized medicine will eventually lead to market fragmentation. Also, an increased competition in marketed products, biomarkers, and increased market stratification will help ensure that new products get to the market and find a niche. They estimate that personalized medicine is about 5 years away for the treatment of rheumatology. They point out that cost effectiveness and health care economics are essential in this endeavor. They also emphasize the potential impact of personalized treatment approach to the treatment of rheumatology.

Pacific Biomarkers, Inc. (PBI) (2010), a provider of biomarker laboratory services to the pharmaceutical and diagnostic industries indicates that it has received an award of $244,000 for its ongoing organ injury biomarker initiative. The company indicates that their initiative met the guidelines for a qualified therapeutic discovery project and showed a reasonable potential to detect or treat chronic or acute diseases and conditions. Also, it exhibited potential to reduce the long-term care costs in the United States, and to significantly enhance the goal of curing cancer within 30 years.

PBI indicates that the aim of their program is to diagnose organ injury occurring because of specific toxic effects of drugs that are under development and thereby prevent their approval. They emphasize that presently the costs for developing a drug are around $1 billon, and approximately 30% of all drugs do not reach the market because of drug-induced toxicity effects. PBI emphasizes that their program exhibits potential to diagnose the toxic effects that may be missed in clinical trials.

Imminst (2010) indicates that one of the best ways to look for biomarkers is by mass spectrometry. They indicate that by the year 2010, mass spectrometry research for biomarkers will exceed $75 million. In their report they list the following types of biomarkers: screening, prognostic, staging, stratification, efficacy, target, toxicity, antecedent, translational, and surrogate. The authors further emphasize the value of the biomarkers in the industry, biomarker discovery, biomarker validation, and areas of application such as cancer, CVR, diabetes, etc. The report also addresses consortia and alliances, important company profiles, and challenges.

An Austin, Texas-based biomarker company, Rules-Based Medicine, Inc. is looking for capital that can help sustain the company's rapid growth in the foreseeable future (Austin Business Journal (ABJ) (ABJ entrepreneur, 2010) entrepreneur, Inc., 2010). In order that the company could accelerate its new products to the market, the company had planned to raise $90 million on an initial public offering (IPO) according to its filing with the Securities Exchange Commission. The company had recently launched a psychiatric biomarker based on a blood test for diagnosing schizophrenia and other mental illnesses. It is also working on oncology biomarkers as well as biomarkers for other psychiatric illnesses such as depression and bipolar disorder. In the year 2009, the company was able to generate $25–30 million revenue and had 120 employees.

Rules-Based Inc. indicates that the IPO market is presently challenging since (1) investors expect higher revenues, (2) more consistent profitability, and besides, (3) there are more legal and regulatory issues. Also, an IPO depends on the industry segment. A comparison can be made between biotech and software companies. The biotech companies require a lot more time and capital to get through clinical trials and to the market. Rules-Based Inc. further adds that investors presumably prefer that companies are further along in their development process. Previously, the public market was an effective way to raise equity in a business. Nowadays, there is an increasing trend towards raising private funds. However, investors are still interested in life sciences and biotechnology companies. Finally, Rules-Based Inc. adds that presently the global molecular diagnostic market is estimated at $3.7 billion, and is estimated to grow to $6.4 billion by the year 2015. This is an increase by a factor 1.73 in five years.

Levinson (2010), founder and chairman of US Biomarkers indicates that US Biomarkers, Inc. is a formation stage company that has developed biomarkers for the early detection and diagnosis of cancer. He estimates the overall market for biomarkers to be $12 billion per year with a compounded annual growth rate (CAGR) of 15%. This is much higher than the estimate presented at the end of the previous paragraph. He further indicates that his company is well positioned to take part in each of the following biomarker segments:

1. Biomarker discovery—$5.2 billion
2. Molecular diagnostics—$5.1 billion
3. Clinical trials—$2 billion

He indicates that the average time for the development of a biomarker should be 6–8 months presently. The company is seeking $1.1 million in initial investment for laboratory space and supplies, etc., and expects to have a revenue of $25 million in 5 years with a net income of $12 million.

Walker (2010) recently indicates that the Biomarker Factory is a company jointly owned by Duke University and Labcorp. The company coordinates medical, scientific, and commercial expertise related to biomarker development, biospecimen collection, clinical research, treatment practices, market analysis, business development, and manufacturing. He further indicates that their company is interested in diagnostics for all clinical applications, assay formats, and disease areas. For example, their work includes single biomarker assays, biomarker signatures, biomarker panels, as well as companion diagnostics for drugs.

Khetan (2007) indicates that the global biomarker market was expected to increase from $4.8 billion to $5.6 billon in the year 2007, and then to $12.8 billion by the year 2012. This is a CAGR of 18%. This author too divides the biomarker into three segments: biomarker discovery, clinical trials, and molecular diagnostics. These segments according to the author are expected to exhibit growth rates of 16.9, 23.5, and 17.5% respectively. In the year 2012, the market shares of these three segments are expected to be $5.843, 1.761, and 5.156 billion, respectively. The author emphasizes that the development of oncology biomarkers is in the forefront, followed by cardiovascular applications. This is indicated by their respective applications in clinical trials.

Markets and markets (2009) in a recent report entitled "Biomarkers-Advanced Technologies and Global Market (2009–2014)", indicates that there is a need to reduce the drug development time and cost. Besides, there are increasing concerns with regard to drug efficacy and drug safety, along with the trends towards personalized medicine. This has increased significantly the need to integrate biomarkers in the drug development process. The report emphasizes the need to obtain a better idea of the market dynamics, the competition, and the market size. Their report emphasizes biomarker tools (genomic markers and technologies, imaging biomarkers, etc.), biomarker services market (pre- and postclinical biomarker services, sample preparation), and biomarker application market (diagnosis, drug development, and discovery). The report analyzes the gaps and opportunities in the biomarker markets as well as what are the factors that are primarily responsible for the market growth.

Kirk (2008) indicates that the use of pharmacogenomics (PG_X) is almost ubiquitous in drug development especially since an increasing number of drugs are coming into the market with indicators that are related to the presence or absence of a biomarker. The authors provide a detailed review of pharmacogenomics and its tools in research, in clinical trails, and also in clinical medicine. The authors emphasize on the economic, regulatory, and technical driving forces for the adoption of pharmacogenomics in biomarker discovery. They also analyze the impediments to a more robust proliferation of these technologies in biomarker discovery and development. Nunn (2006) has also analyzed the cost of developing imaging agents for routine clinical use.

1.3 CHAPTER CONTENTS

Chapter 1 is introduction. Chapter 2 describes briefly the fractal analysis method to analyze the binding kinetics of biomarkers in solution to appropriate receptors immobilized on biosensor surfaces. The kinetics should help provide novel physical insights into the different binding and dissociation (if involved) reactions taking place on the biosensor surface.

Chapters 3 (Part I) and chapter 4 (Part II) examine and analyze the binding of the different cancer biomarkers on the different biosensor surfaces. Since this is a very active and prolific area of research, two chapters are devoted to this area of cancer biomarkers.

Chapter 5 analyzes the detection of biomarkers for MI. A very significant number of patients in the United States and worldwide suffer from this ailment. Chapter 6 analyzes the detection of biomarkers for arthritis. This includes both osteoarthritis as well as rheumatoid arthritis.

Chapter 7 discusses the detection of CVR biomarkers. The detection of glucose is a very important area of research for biosensors. Thus, two chapters, chapters 9 and 10, present the detection of glucose (biomarker for diabetes mellitus).

SLE is a very debilitating disease. Chapter 10 analyzes the detection of biomarkers for SLE using biosensors. Chapter 11 analyzes the detection of biomarkers for different nervous system diseases, such as Alzheimer's, Huntington, MS, and Parkinson's.

Chapter 12 discusses the detection of biomarkers for severe acute respiratory syndrome. Chapter 13 discusses the detection of biomarkers for different diseases such as HIV infection, hepatitis, stroke marker protein, etc.

Chapter 14 examines the markets and economics of disease-related biomarkers. This is a capstone chapter, and is very important. Especially so, since this information is scarcely presented in the open literature. One may obtain recent reports on this topic in the open market, but at a steep price of a few thousand dollars.

REFERENCES

Aarkstore enterprises, Stakeholders opinions biomarkers in rheumatology-awaiting the next step towards personalized medicine, posted November 18, 2010, downloaded November 24, 2010. http://www.articleshub.org/article/25326/stakeholder-opinions-biomarkers-in-rheumatoid.

Abeloff, M.D., Armitage, J.O., Lichter, A.S., Niederbuber, J.E., 2000. Clinical Oncology, second ed. Churchill Livingstone, New York.

Abj entrepreneur, Rules-Based Medicine Moves on, san IPO, 2010. http://abjentrepreneur.com/news2010/11/roles-based-medicine-moves-on-ipo.html.

Agnew, B.J., Qian, X.D., Nyberg, T., Hart, C., Huang, W., Kumar, K., October 20–22, 2010. Novel Click Chemistry-Based Tools for High Resolution Biomarker Discovery, 4th Biomarker Discovery and Development, San Francisco, California.

Aletta, J.M., October 20–22, 2010. Methylarginine Protein Biomarkers for Oncology Research and Development, 4th Biomarker Discovery and Development, San Francisco, California.

Batheja, A., January 31–February 2, 2011a. Analyzing MiR-122 as a Biomarker for Hepatotoxicity. Third Annual Biomarker Assay Development. Cambridge Healthtech Institute, San Diego.

Batheja, A., January 31–February 2, 2011b. Best Practices in Fit-for-Purpose Biomarker Assay Validation. Third Annual Biomarker Assay Development. Cambridge Healthtech Institute, San Diego, California.

Bio-IT World Conference & Expo 11, Systematic Biomarker Data Analysis, Workshop, April 12–14, 2011.

Bonventre, J., June 15–16, 2010. Biomarkers of Nephrotoxicity: Uses and Challenges in Pre-Clinical and Clinical Trials, Keynote Address, New Tools for Detecting Nephrotoxicity. CHI's Ninth Annual World Pharmaceutical Congress, Philadelphia.

Brahmandan, M., Holsti, M., Benesics, C., Johnson, M., Ciatto, C., Hua, P., Singh, A., Krothapallis, S., Hanash, S., Greinfried, E., DeCaprio, J.A., 2010. Development of sandwich ELISAs for potential biomarkers in pancreatic cancer, 4th Biomarker Discovery and Development Conference, San Francisco, California.

Brott, D., January 31–February 2, 2011. Development of Drug-Induced Vascular Injury (DIVI) Biomarkers. Third Annual Biomarker Assay Development. Cambridge Healthtech Institute, San Diego, California.

Cambridge Healthtech Institute, April 2010. Biomarkers in Late Stage Clinical Trials: Applications, Opportunities, and Activities of Leading Players.

Carney, W.P., 2010. Circulating Oncology Biomarkers to Guide Targeted Therapies: Companion Diagnostics for Personalized Medicine. Bridging diagnostic and drug development, Biomarker web symposia, symposia archive. Cambridge Healthtech Institute, 250 First Avenue, Suite 300, Needham, Massachussetts.

Cassiday, L., 2010. Carbon nanotubes stretch the boundaries of biomarker detection. Analytical Chemistry.

Chandra, U., Morrison, C.J., Cramer, S., Karande, P. Rational and Combinatorial Design of Peptide Affinity Ligands for Diagnostic Assays, 2010 Annual American Institute of Chemical Engineers Meeting, Wednesday, November 10, San Francisco, California.

Chenevier-Gobeaux,, C., Billemont,, M., Eche,, A., Trabattini,, E., Kiesek,, G., Pourriat,, J.L., Classens, Y.E, 2010. New biomarkers in emergency patients with sepsis and cardiovascular conditions. Reanimation.

Chou, S.F., Hsu, W.L., Hwang, J.M., Chen, C.Y., 2004. Biosensors & Bioelectronics 19 (9), 999–1005.

Compton, C., January 27–28, 2011. Those Biomarkers Come from Specimens: Garbage in, Garbage out, 3rd Annual Oncology Biomarkers, San Diego, California.

Danesh, J., Collins, R., Appleby, P., Pero, R., 1998. Journal of the American Medical Association 279 (18), 1477–1482.

Darain, F., Oark, D.S., Park, J.S., Shim, Y.B., 2004. Biosensors & Bioelectronics 19 (10), 1245–1252.

Detmers, J., October 20–22, 2010. Ligand-Binding Assays: Quantifying 'Difficult' Proteins, Peptides, and Biomarkers, 4th Biomarker Discovery and Development, San Francisco, California.

Devrajan, P., June 15–16, 2010. NGAL as a Biomarker of Acute Kidney Injury and Nephrotoxicity, New Tools for Detecting Nephrotoxicity. CHI's Ninth Annual World Pharmaceutical Congress, Philadelphia.

Duffy, M.J., Evoy, D., McDermott, E.W., December 14, 2010. CA 15–3: uses and limitations as a disease biomarker for breast cancer. Clinica Chimica Acta 23–24, 1869–1874.

Dunn, W., Ellis, D., 2005. Trends in Analytical Chemistry 24, 285–294.

FDA Critical Path Report, July 2004.

Francis, J. Biomarkers Becoming an Integral Part of Medical Research, articlesbase, downloaded November 24, 2010. http://www.articlesbase.com/causes-and-organizations-articles/biomarkers-becomingan-inregral.

Frank, R.A., 2010. Biomarker Imaging: From Molecule to Man, Discovery to Diagnostics, Bridging Diagnostic and Drug Development, Biomarker Web Symposia, Symposia Archive, Cambridge Healthtech Institute, 250 First Avenue, Suite 300, Needham, Massachussetts.

Furlong, S., June 15–16, 2010. How Successful will Biomarkers be in Predicting Renal Injury, New Tools for Detecting Nephrotoxicity. CHI's Ninth Annual World Pharmaceutical Congress, Philadelphia.

Furlong, S., January 31–February 2, 2011. Establishing the Context for Introducing New Safety Biomarkers into Clinical Trials. Third Annual Biomarker Assay Development. Cambridge Healthtech Institute, San Diego, California.

Furlong, S., Bocan, T., Mattes, W.B., Translating Biomarkers from the Lab to the Clinic, Pre-Conference Short Course, Third Annual Biomarker Assay Development, Cambridge Healthtech Institute, San Diego, California, January 1-February 2,011

Gil, G.C., Brennan, J., Throckmort, D., Branda, S., Chirica, G., 2010. Comprehensive Analysis of Serum Peptidome Using Restricted Access Media and Nanoliquid Chromatography-Tandem Mass Spectrometry. 4th A Biomarker Discovery and Development, San Francisco, California.

Goldberg, R., Pitts, P., A Working Paper of the 21st Century Task Force: Prescription for Progress, the Critical Path to Drug Development, downloaded November 24, 2010. http://www.manhattan–institute.org/fda-tak_1.htm

Grigsby, C.C., Rizki, M.M., Tamburino, L.A., Pitsch, R.H., Shlyanov, P.A., Cool, D.R., 2010. Metabolite differentiation and discovery lab (MeDDl): a new tool for biomarker discovery and mass spectral visualization. Analytical Chemistry 82, 4386–4395.

Haes, A., Chang, L., Klein, W.L., van Duyne, R.P., February 23, 2005. Detection of a biomarker for Alzheimer's disease from synthetic and clinical samples using a nanoscale optical biosensor. Journal of the American Chemical Society.

Hahm, J., Lieber, C.M., 2004. Nano Letters 4 (1), 51–54.

Hanash, S., January 31–February 2, 2011. Development of Biomarker Panels: Opportunities and Challenges. Third Annual Biomarker Assay Development. Cambridge Healthtech Institute, San Diego, California.

Hill, H.R., Martins, T.B., 2006. Methods 38 (4), 312–316.

Huey, J.G., October 20–22, 2010. The Accelerated Expression of Clinical Use of Plasma TGF-beta-1: The New Paradigm of Inflammation and Fibrosis, 4th Biomarker Discovery and Development, GTC bio, San Francisco.

Imminst, Mass Spectrometry Biomarker Research, downloaded November 25, 2010. http://www.iminst.org/forum/topic/13277-massspectrmetry-biomarker-research/.

Jessup, J.M., May 11, 2009. Molecular Diagnostic Laboratories (MoDEL): A Program to Support Biomarker Clinical Assay Development. Biomarker Web Symposia, Biomarker Assay Development. Cambridge Healthtech Institute.

Jovanovic, V., Poh, A.N.A., Pei, E.H.X., Lin, Y.T., Aziz, N.A., Lew, F.C., McKinney, E.F., Lyons, P.A., Kemeay, M.D., Chung Yung, J.L., Huang, S.H., Jenner, A.M., Halliwell, B., Smith, K.G.C., MacAry, P.A., Wenk, M., October 20–22, 2010. Anti-Lipid Response in Systemic Lupus Erythematosus (SLE), 4th Biomarker Discovery and Development, San Francisco, California.

Kavsak, P., October 20–22, 2010. Biomarker Selection: A Clinical Chemistry Perspective on Drug Efficacy and Toxicity, 4th Biomarker Discovery and Development, San Francisco, California.

Khetan, V. Biomarkers: The Expanding Global Market, BCC Research, Report code BIO061A, November 2007, downloaded April 15, 2010. http://www.bccresearch.com/report/BIO061A.html.

Kinsinger, C., October 20–22, 2010. Building a Reliable and Efficient Protein-Biomarker Development Pipeline, 4th Biomarker Discovery and Development, GTC bio, San Francisco.

Kirk, R.J., Hung, J.L., Horner, S.R., Perez, J.T., 2008. Implications of pharmacogenomics for drug development. Experimental Biology and Medicine 233 (2), 1484–1497.

Kulkarni, Y., Klinke, D.J., November 11, 2010. Identifying Secreted Biomarkers for Immune Evasion in Cellular Models of Cancer, 2010 Annual American Institute of Chemical Engineers Meeting, Thursday. San Francisco, California.

Labtechnologist, Biomarker Market Prepared to Explode, downloaded April 15, 2010. http://www.labtechnologist.com/industry-drivers/biomarker-market-oredicted-toexplode.

Lai-Goldman, M., October 20–22, 2010. A Novel Partnership Strategy for the Translation of Personalized Cancer Diagnostics, 4th Biomarker Discovery and Development, GTC bio, San Francisco.

Lakey, N., October 20–22, 2010. Discussing the Field of Epigenetic Biomarkers and Their Application in Therapy Selection, 4th Biomarker Discovery and Development, San Francisco, California.

Levenson, US Biomarkers-Development of Blood Based Biomarkers, downloaded November 25, 2010. http://www.usbiomarkers.com/about.html.

Li, T., January 2–28, 2011. Oncology Biomarker Discovery Technologies Today and Tomorrow, 3rd Annual Oncology Biomarkers, San Diego, California.

Lin, K.C., Kunduru, V., Bothara, M., Rege, K., Prasad, S., Ramakrishna, B.L., 2010. Biogenic nanoporous silica-based sensor for enhanced electrochemical detection of cardiovascular biomarkers proteins. Boisensors & Bioelectronics 25, 2336–2242.

Markets and Markets.com, Biomarkers—Advanced Technologies and Global Market (2009–2014). http://www.marketsandmarkets.com/Market-reports/biomarkers-advanced-technologies-and-globalmarket.

Moore, H., October 20–22, 2010. Biospecimen Science is Foundational to Biomarker Research and Development, 4th Biomarker Research & Development, GTC bio, San Francisco.

Moore, S.J., Apte, S., Graves, E.E., Cochran, J.E., Engineered Knottin Peptides: A New Class of Agents for Non-invasive Molecular Imaging of Tumor Markers, 2010 Annual American Institute of Chemical Engineers Meeting, Wednesday, November 10, San Francisco, California.

Nam, J.M., Thaxton, C.S., Mirkin, C.A., 2003. Science 301 (5641), 1884–1886.

Niwa, O., Morita, M., Tabei, H., 1990. Analytical Chemistry 62 (5), 447–452.

Nogano, K., Imai, S., Nakagawa, S., Tsunoda, S.I., Tsutsumi, Y., 2010. From disease proteomics to biomarker development. Establishment of antibody proteomics technology and exploration of cancer-related biomarkers. Yakuguku Zasslie 130 (4), 487–492.

Nunn, A.D., 2006. Investigative Radiology 41 (3), 206–212.

Pacific Biomarkers, Inc., 2010. Pacific biomarkers awarded $244K under healthcare reform program. Marketwatch. http://www.marketwatch.com/story/pacific_biomarkers_awarder_244K_under_healthcare.

Panteghini, M., 2010. Cardiac: is this biomarker ready for prime time? Scandinavian Journal of Clinical and Laboratory Investigation 70 (Suppl. 242), 66–72.

Philip, R., 2010. OIVD Outlook: Biomarker-Based in vitro Diagnostic Assays. Bridging diagnostic and drug development, Biomarker web symposia archive. Cambridge Healthtech Institute, 250 First Avenue, Suite 300, Needham, Massachusetts.

Qureshi, A., Nizi, J.H., Kallempudi, S., Gurbuz, Y., 2010. Label-free capacitive for sensitive detection of multiple biomarkers using gold interdigitated capacitor arrays. Biosensors & Bioelectronics 25, 2318–2323.

Ramachandran, R. An Insight into the Active and High Growth Biomarker Market, Frost and Sullivan Report, October 14, 2009. http://www.forst.com/prod/serlet/market-insight.

Reo, N., 2005. Drug Chemical Toxicology 25, 285–294.

Rhyne, P.W., May 11, 2009. Application of Intra-Assay Calibration Curves to Quantitate Clinical Biomarker Immunosensors. Biomarker Web Symposia, Biomarker Assay Development. Cambridge Healthtech Institute.

Richardson, F., November 29, 2010. Incorporation of Biomarkers into Traceva-Clinical Trials. Case Studies in Cancer Drug Development. Cambrige Healthtech Institute, Needham, MA. bethanyg, personal email communication.

Rubenstein, K., 2007. Disease-Related Biomarkers: Their Potential, in Patient Screening, Prognosis, and Stratification, Pharma Reports. http://www.insightpharmareports.com/2007/83_disease_biomarkers/overview.asp.

Ryan, A., November 29, 2010. Pretreatment Circulating VEGF Levels as a Predictive Biomarker of Efficacy in NSCLC Patients Treated with Vandetanib. Case Studies in Cancer Drug Development. Cambridge Healthtech, Boston, MS. personal email communication from bethanyg.

Saliminen, W., January 31–February 2, 2011. Using Multiple Omics Platforms in Pre-Clinical Hepatotoxicity Biomarker Discovery. Third Annual Biomarker Assay Development. Cambridge Healthtech Institute, San Diego, California.

Scott, November 29, 2010. The KRAS Signaling Pathway Biomarker in Oncology: From Prognostic to Predictive. Case Studies in Cancer Drug Development. Cambridge Healthtech Institute, Needham, MS. bethanyg, personal email communication.

She, J.X., January 31–February 2, 2011. Using Mass Spectrometry and Immunoassays to Discover and Validate Serum Protein Biomarkers for Islet Autoimmunity and Type 1 Diabetes. Third Annual Biomarker Assay Development. Cambridge Healthtech Institute, San Diego, California.

Shoemaker, R., Bronfin, B., Hoey, J.G., Han, R.C.T., Rundell, C.A., October 20–22, 2010. The Accelerated Expansion of Clinical Use of Plasma TGF Beta-1: The New Paradigm of Inflammation and Fibrosis, 4th Annual Biomarker Discovery and Development, San Francisco, California.

Sotiroub, G.A., Hirt, A.M., Taleki, A., Oratsins S.E. 2010 Annual American Institute of Chemical Engineers Meeting, Thursday, November 11, 2010, San Francisco, California.

Van Bruggen, N., February 23–25, 2011. Translation of Imaging Biomarkers from Research to Clinical Drug Trials, Cancer Biologics. San Francisco, California.

Vasto, S., Scapaguini, G., Bulati, M., Candore, G., Castaglia, L., Colonna-Romano, G., Lio, D., Caruso, C., 2010. Biomarkers I Ageing. Fontiers in Bioscience (Scholar edition) 2, 392–402.

Walker T., Biomarker Factory, downloaded November 25, 2010. http://www.dtmi.duke.edu/public-private-partnerships/biomarker-factory.

Wang, Y., Holmes, E., Nicholson, J., Cloarec, O., Chollet, J., Tanner, M., Singer, B., Utzinger, J., 2004. National Academy of Sciences 101, 12676–12681.

Wilson, R.C., Cropek, D.M., Banta, S.A., “On-chip Electrochemical Detection of Biomarkers for Detection of Water Borne Toxins,” 2010 Annual American Institute of Chemical Engineers Meeting, Monday, November 8, San Francisco, California.

Zheng, F., October 20–22, 2010. Integrating Various Fit-for-purpose Biomarker Studies in Oncology Early Phase Trials, Their Benefits and Challenges, 4th Biomarker Discovery and Development, GTC bio. San Francisco, California.

Chapter 2

Modeling and Theory

2.1 INTRODUCTION

In a biosensor-based assay, the molecule to be detected (analyte) is present in solution and the appropriate receptor is immobilized on a solid surface. The interaction between the analyte and the receptor on the solid biosensor surface is detected either by a change in the refractive index (in surface plasmon resonance (SPR) instruments) or by changes in the fluorometric intensity, ultraviolet light intensity, etc. The SPR biosensor protocol analyzes the binding (and dissociation where applicable) kinetic curves using classical saturation models involving analyte–receptor binding using 1:1, 1:2, etc. ratios, generally under diffusion-free conditions and assuming that the receptors are homogeneously distributed over the sensor surface. Computer programs and software that come with the equipment provide values of the binding (and the dissociation) rate coefficients. Though a careful analysis and experimental protocol may eliminate or minimize the influence of diffusional limitations; realistically speaking, it is more appropriate to include a heterogeneous distribution on the sensing surface. Heterogeneity on the sensing surface and in the biosensor systems itself may be due to other reasons, such as nonspecific binding, inherent irregularities on the sensing surface, mixture of receptors on the surface, and mixture of analytes in solution which includes the analyte of interest.

Two factors need to be addressed while analyzing the analyte–receptor binding and dissociation kinetics. The system by its design is heterogeneous. For example, as indicated above, the receptors immobilized on the biosensor surface may exhibit some heterogeneity, that is, surface roughness. No matter how careful one is in immobilizing the receptors on the biosensor surface, there will be some degree of heterogeneity on the surface.

Henke et al. (2002) have used the atomic force microscopy technique to determine the effects of cleaning fused silica and glass on surface roughness. This is for biosensor use. Note that prior to the immobilization of receptors on the surface, the surface needs to be cleaned to remove contaminants, and to create surface attachment sites for example, for hydroxyl groups. For the analyte–receptor binding (and dissociation) to take place the analyte, by the diffusion process, must come within the "proximity" of the active site on the receptor. Mass transport limitations may be minimized or eliminated if the

Biomarkers and Biosensors. http://dx.doi.org/10.1016/B978-0-444-53794-2.00002-1

system is either properly designed or properly operated or both. In most cases, however, both diffusional effects and heterogeneity aspects will be present in biosensor systems, and their influence on binding and dissociation kinetics need to be determined. Ideally, one would like to determine the influence of each of these separately on the binding and dissociation kinetics. In the theoretical analysis to be presented below, (the Havlin, 1989; analysis) the effects of diffusion and heterogeneity are presented coupled together. One possible way of accounting for the presence of diffusional limitations and the heterogeneity that exists on the surface is by using fractals. Ideally, and as indicated above, one would prefer to decouple the influence of diffusion and heterogeneity. Presumably, an approach other than fractal analysis is required to decouple these two effects.

A characteristic feature of fractals is self-similarity at different levels of the scale. Fractals exhibit dilatational symmetry. Fractals are disordered systems, and the disorder is described by nonintegral dimensions (Pfeifer and Obert, 1989). Fractals have nonintegral dimensions, and are smaller than the dimension they are embedded in. In other words, the highest value that a fractal can have is three. In our case, an increase in the degree of heterogeneity on the biosensor surface would lead to an increase in the value of the fractal dimension. Another way of looking at the fractal dimension is its "space filling" capacity. The more the space a surface fills, the higher is its fractal dimension. The fractal dimension cannot have a negative value, and very low values of the fractal dimension on the surface indicate that the surface exists as a Cantor-like dust.

Kopelman (1988) points out that surface-diffusion-controlled reactions that occur on clusters or islands are expected to exhibit anomalous and fractal-like kinetics. These kinetics exhibit anomalous reaction orders and time-dependent (e.g., binding) rate coefficients. As long as surface irregularities show scale invariance they can be characterized by a single number, the fractal dimension. Later on in this book we will characterize the surfaces of the biosensors used in different examples by a fractal dimension. More specifically, we will characterize the heterogeneity present on these biosensor surfaces by a fractal dimension.

The fractal dimension is a global property, and it is insensitive to structural or morphological details (Pajkossy and Nyikos, 1989). Markel et al. (1991) point out that fractals are scale self-similar mathematical objects that possess nontrivial geometrical properties. Furthermore, these authors state that rough surfaces, disordered layers on surfaces, and porous objects all possess fractal structure. A consequence of the fractal nature is a power-law dependence of a correlation function (in our case the analyte–receptor on the biosensor surface) on a coordinate (e.g., time).

Pfeifer (1987) shows that fractals may be used to track topographical features of a surface at different levels of scale. Lee and Lee (1995) point out that the fractal approach permits a predictive approach for transport (diffusion-related) and reaction processes occurring on catalytic surfaces. This approach may presumably be extended to diffusion-limited analyte—receptor reactions occurring on biosensor surfaces.

The binding of an analyte in solution to a receptor attached to a solid (albeit flow cell or biosensor surface) is a good example of a low dimension reaction system in which the distribution tends to be "less random" (Kopelman, 1988), and a fractal analysis would provide novel physical insights into the diffusion-controlled reactions occurring at the surface. Also, when too many parameters are involved in a reaction, which is the case for these analyte—receptor reactions on a solid (e.g., biosensor surface), a fractal analysis provides a useful lumped parameter. It is appropriate to pay particular care to the design of such systems and to explore new avenues by which further insight or knowledge may be obtained on these biosensor systems. The fractal approach is not new and has been used previously in analyzing different phenomena on lipid membranes.

Fatin-Rouge et al. (2004) have recently presented a summary of cases where the analysis of diffusion properties in random media has provoked significant theoretical and experimental interest. These cases include soils (Sahimi, 1993), gels (Starchev et al., 1997; Pluen et al., 1999), bacterial cytoplasm (Berland et al., 1995; Schwille et al., 1999), membranes (Saffman and Delbruck, 1975; Peters and Cherry, 1982; Ghosh and Webb, 1988), and channels (Wei et al., 2000). Coppens and Froment (1995) have analyzed the geometrical aspects of diffusion and the reaction occurring in a fractal catalyst pore. In this chapter, and in this book as a whole, we are extending the analysis to analyte—receptor binding (and dissociation) on biosensor surfaces. Fatin-Rouge et al. (2004) show that in most real systems disorder may exist over a finite range of distances. Harder et al. (1987) and Havlin (1989) point out that in this range the diffusion process cannot be characterized by the classical Fick's law. In this range, anomalous diffusion applies. Fatin-Rouge et al. (2004) emphasize that at larger distances than in the above window range, the effects of disorder on diffusion may be very small due to statistical effects, and may cancel each other.

Prior to presenting the Havlin (1989) analysis modified for the analyte—receptor binding occurring on biosensor surfaces, it is appropriate to discuss briefly the analysis presented by Fatin-Rouge et al. (2004) on size effects on diffusion processes within agarose gels, and apply it to analyte—receptor binding and dissociation for biosensor kinetics. This analysis provides some insights into general fractal-related processes. Fatin-Rogue et al. (2004) have considered diffusion within a fractal network of pores. They indicate that fractal networks such as percolating clusters may be characterized by a power-law distribution (Havlin, 1989):

$$M(L)^{D_{\mathrm{f}}} \tag{2.1}$$

Here M is the average number of empty holes in the (gel) space characterized by a linear size, L. The exponent, D_{f} is the mass fraction dimension.

Fatin-Rogue et al. (2004) emphasize that in the general case of fractals, D_f is smaller than the dimension of space of interest. Furthermore, the independence of D_f on scale is also referred to as self-similarity, and is an important property of rigorous fractals.

Havlin and Ben-Avraham (1987) point out that the diffusion behavior of a particle within a medium can be characterized by its mean square displacement, $r^2(t)$ versus time, Γt, which is written as:

$$r^2(t) = t^{(2/D_w)} \tag{2.2a}$$

Here Đ is the transport coefficient, and D_w is the fractal dimension for diffusion. Normal or regular diffusion occurs when D_w is equal to 2. In this case, $r^2(t)$ is equal to t. In other words,

$$r^2(t) = 2dDt \tag{2.2b}$$

Here d is the dimensionality of space, and D is the diffusion coefficient.

Harder et al. (1987) and Havlin (1989) describe *anomalous diffusion* where the particles sense obstructions to their movement. This is within the fractal matrix, or in our case due to heterogeneities on the biosensor surface, perhaps due to irregularities on the biosensor surface. Fatin-Rogue et al. (2004) are careful to point out that anomalous diffusion may also occur due to nonelastic interactions between the network and the diffusing particles in a gel matrix (Saxton, 2001). Furthermore, Fatin-Rouge et al. (2004) indicate that *anomalous diffusion* is different from *trapped diffusion* where the particles are permanently trapped in holes, and are unable to come out of these holes. When the particles (analytes in our case) are in these trapped holes, then as time $t \to \infty$, the mean square displacement, $r^2(t)$ tends to a constant value. Fatin-Rouge et al. (2004) emphasize that in real heterogeneous porous media anomalous diffusion of particles occurs over a limited length or timescale since the structure is only fractal over a limited size scale. In other words, there is a lower bound and an upper bound beyond which the fractal structure applies. Similarly, in our case, the anomalous diffusion of the analyte on the biosensor surface occurs over a limited range of length or timescale.

For anomalous diffusion, one may combine the right-hand side of Eqns (2.2a) and (2.2b). Then, the diffusion coefficient, D is given by (Fatin-Rouge et al., 2004):

$$D(t) = (1/4)t^{[(2/D_w)-1]} \tag{2.3}$$

Due to the temporal nature of $D(t)$, it is better to characterize the diffusion of the analyte in our case by D_w. If we were still talking about the medium and gels, then D_w would refer to the diffusing medium.

We will now develop the theory for the analyte−receptor binding and dissociation on biosensor surfaces. We will use the Havlin (1989) approach.

2.2 THEORY

We present now a method of estimating fractal dimension values for analyte–receptor binding and dissociation kinetics observed in biosensor applications. The following chapters will present the different examples of data that have been modeled using the fractal analysis. The selection of the binding and dissociation data to be analyzed in the later chapters is constrained by whatever is available in the literature.

2.2.1 Variable Rate Coefficient

Kopelman (1988) points out that classical reaction kinetics are sometimes unsatisfactory when the reactants are spatially constrained at the microscopic level by either walls, phase boundaries, or force fields. Such heterogeneous reactions, for example, bioenzymatic reactions, that occur at interfaces of different phases, exhibit fractal orders for elementary reactions and rate coefficients with temporal memories. In such reactions, the rate coefficient exhibits a form given by:

$$k_1 = k't^{-b} \quad 0 \leq b \leq 1 \ \ (t \geq 1) \tag{2.4}$$

In general, k_1 depends on time whereas $k' = k_1$ $(t = 1)$ does not. Kopelman (1988) points out that in three dimensions (homogeneous space) $b = 0$. This is in agreement with the results obtained in classical kinetics. Also, with vigorous stirring, the system is made homogeneous and b again equals zero. However, for diffusion-limited reactions occurring in fractal spaces, $b > 0$; this yields a time-dependent rate coefficient.

Antibodies may form fractal clusters on biosensor surfaces. These antibodies or receptors on the biosensor surface may consist of islands of highly organized or disorganized antibodies. This is similar to the growth of crystalline structures. It is quite possible that a cooperative effect may arise due to this tightly organized fractal structures. This is one possibility that could lead to an increase in the binding rate coefficient with an increase in the fractal dimension or the degree of heterogeneity on the biosensor surface.

The diffusion-limited binding kinetics of antigen (or antibody, or analyte, or substrate) in solution to antibody (or antigen, or receptor, or enzyme) immobilized on a biosensor surface has been analyzed within a fractal framework (Sadana and Beelaram, 1994; Sadana et al., 1995). One of the findings, for example, is that an increase in the surface roughness or fractal dimension leads to an increase in the binding rate coefficient. Furthermore, experimental data presented for the binding of HIV virus (antigen) to the antibody immobilized on a surface displays characteristic ordered "disorder" (Anderson, 1993). This indicates the possibility of a fractal-like surface.

A biosensor system (wherein either the antigen, antibody, analyte, or substrate is attached to the surface), along with its different complexities, which include heterogeneities on the surface and in solution, diffusion-coupled reaction, time-varying adsorption, or binding rate coefficients, etc., can be characterized as a fractal system.

The diffusion of reactants toward fractal surfaces has been analyzed (De Gennes, 1982; Pfeifer et al., 1984a,b; Nyikos and Pajkossy, 1986). Havlin (1989) has briefly reviewed and discussed these results. The diffusion is in the Euclidean space surrounding the fractal surface (Giona, 1992). Havlin (1989) presents an equation that may be utilized to describe the buildup of the analyte—receptor on a biosensor surface during the binding reaction. The receptor is immobilized on the biosensor surface. This equation is given below. In all fairness, at the outset, it is appropriate to indicate that the biosensor surface is assumed to be fractal, or possibly so.

Ideally, it is advisable to provide independent proof or physical evidence for the existence of fractals in the analysis of analyte—receptor reactions occurring on biosensor surfaces. Also, as indicated earlier, if the diffusion effects can be separated from the heterogeneity effects, then one may better understand the effects of each of these on analyte—receptors reactions occurring on biosensor surfaces. In general, diffusion effects may be minimized either by increasing flow rates or by immobilizing fewer receptors on the biosensor surface.

In general, to demonstrate fractal-like behavior log—log plots of distribution of molecules $M(r)$ as a function of the radial distance (r) from a given molecule are required. This plot should be close to a straight line. The slope of log $M(r)$ versus log (r) plot determines the fractal dimension. In our case, one could try to obtain a log—log plot of two variables, k and time, t and perform a least squares fit in this parameter space to find the slope of the curve. A regression coefficient at this stage could be beneficial in understanding the efficacy of this metric. However, an easier method, without the use of the required log—log plots, is presented below.

This is the equation developed by Havlin (1989) for diffusion of analytes toward fractal surfaces.

2.2.2 Single-Fractal Analysis

In the literature some authors refer to binding as comprising of two phases, an association phase and a dissociation phase. In this chapter and in this book, we will refer to binding as just binding. The dissociation phase is separate.

2.2.2.1 Binding Rate Coefficient

Havlin (1989) indicates that the diffusion of a particle (analyte) from a homogeneous solution to a solid surface (e.g., receptor-coated surface) on which it reacts to form a product (analyte—receptor complex) is given by:

$$(\text{Analyte}\cdot\text{Receptor}) \sim \begin{cases} t^{(3-D_{\mathrm{f},\mathrm{bind}})/2} = t^{p}, & t < t_{\mathrm{c}} \\ t^{1/2}, & t > t_{\mathrm{c}} \end{cases} \tag{2.5a}$$

where the analyte—receptor represents the association (or binding) complex formed on the surface. Here $p = -b$, and D_f is the fractal dimension of the surface. Havlin (1989) states that the crossover value may be determined by $r_c^2 \sim t_c$. Above the characteristic length, r_c, the self-similarity of the surface is lost and the surface may be considered homogeneous. Eqn (2.5a) indicates that the concentration of the product (analyte—receptor) on a solid fractal surface scale at short and intermediate times as analyte—receptor $\sim t^p$ with the coefficient $p = (3-D_f)/2$ at short timescales and $p = 1/2$ at intermediate timescales. Note that D_f, $D_{f,assoc}$, and $D_{f,bind}$ are used interchangeably. This equation is associated with the short-term diffusional properties of a random walk on a fractal surface. Note that, in perfectly stirred kinetics on a regular (nonfractal) structure (or surface), the binding rate coefficient, k_1, is a constant, that is, is independent of time. In other words, the limit of regular structures (or surfaces) and the absence of diffusion-limited kinetics leads to k_1 being independent of time. In all other situations, one would expect a scaling behavior given by $k_1 \sim k' t^{-b}$ with $-b = p < 0$. Also, the appearance of the coefficient, p different from $p = 0$ is the consequence of two different phenomena, that is, the heterogeneity (fractality) of the surface and the imperfect mixing (diffusion-limited) condition. Finally, for a homogeneous surface where D_f is equal to 2, and when only diffusional limitations are present, $p = 1/2$ as it should be. Another way of looking at the $p = 1/2$ case (where $D_{f,bind}$ is equal to 2) is that the analyte in solution views the fractal object, in our case the receptor-coated biosensor surface, from a "large distance". In essence, in the association process, the diffusion of the analyte from the solution to the receptor surface creates a depletion layer of width $(Dt)^{1/2}$ where Đ is the diffusion constant. This gives rise to the fractal power law, $(\text{Analyte} \cdot \text{Receptor}) \sim t^{(3-D_{f,bind})/2}$.

The values of the parameters k (binding rate coefficient), p, and D_f in Eqn (2.5a) may be obtained for analyte—receptor association kinetics data. This may be done by a regression analysis using, for example, Corel Quattro Pro (1997) along with Eqn (2.5a) where $(\text{analyte} \cdot \text{receptor}) = kt^p$ (Sadana and Beelaram, 1994; Sadana et al., 1995). The fractal dimension may be obtained from the parameter p. Since $p = (3 - D_{f,bind})/2$, $D_{f,bind}$ is equal to $(3 - 2p)$. In general, low values of p would lead to higher values of the fractal dimension, $D_{f,bind}$. Higher values of the fractal dimension would indicate higher degrees of "disorder" or heterogeneity or inhomogeneity on the surface.

Another way of looking at the diffusive process is that it inherently involves fluctuations at the molecular level that may be described by a random walk (Weiss, 1994). This author points out that the kinetics of transport on disordered (or heterogeneous) media needs to be described by a random-walk model. When both of these are present, that is the diffusion phenomena as well as a fractal surface, then one needs to analyze the interplay of both these fluctuations. In essence, the disorder on the surface (or a higher fractal dimension, D_f) tends to slow down the motion of a particle (analyte, in our case) moving in such a medium. Basically, according to Weiss (1994), the particle (random-walker analyte) is trapped in regions in space as it oscillates for a long time before resuming its motion.

Havlin (1989) indicates that the crossover value may be determined by $r_c^2 \sim t_c$. Above the characteristic length, r_c, the self-similarity of the surface is lost. Above t_c, the surface may be considered homogeneous, and "regular" diffusion is now present. One may consider the analysis to be presented as an intermediate "heuristic" approach in that in the future one may also be able to develop an autonomous (and not time-dependent) model of diffusion-limited kinetics in disordered media.

It is worthwhile commenting on the units of the association and the dissociation rate coefficient(s) obtained for the fractal analysis. In general, for SPR biosensor analysis, the unit for the analyte–receptor complex on the biosensor surface is RU (resonance unit). One thousand resonance units is generally 1 ng/(mm)^2 (of surface), or 1 RU is 1 pg/(mm)^2. Here, ng and pg are nanogram and picogram, respectively. Then, to help determine the units for the binding coefficient, k, from Eqn (2.5a):

$$(\text{Analyte}\cdot\text{Receptor}),\ \ \text{pg}(\text{mm})^{-2} = kt^p = kt^{(3-D_{\text{f,bind}})/2}$$

This yields a unit for the binding rate coefficient, k as $(\text{pg})(\text{mm})^{-2}(\text{s})^{(D_{\text{f,bind}}-3)/2}$. Note that the unit of dependence in time exhibited by the association (or binding) rate coefficient, k, changes slightly depending on the corresponding fractal dimension obtained in the binding phase, $D_{\text{f,bind}}$. The fractal dimension value is less than or equal to three. Three is the highest value of the fractal dimension, since the system is embedded in a three-dimensional system. k and k_{bind}, and D_{f}, and $D_{\text{f,bind}}$ are used interchangeably in this chapter and in this book.

It should be noted that different laboratories use different technologies or different experimental designs to analyze the binding affinity of ligands to target proteins (or analytes) of interest (or to determine the rate coefficients for association and dissociation kinetics for binding). The comparison of data between different technologies and experimental designs and conclusions thereof should be made with great caution. The fractal analysis is of value in that it provides the pros and cons of different in vitro technologies (or more precisely, in this case, analysis procedures). It makes the user of the technology aware of the quality of data generated and what can be done to improve the analysis.

One might very reasonably question the utility of the approach, considering the different dimensions, and subsequently the units one may obtain even for the same interactions. It would be difficult to compare this technique with other approaches for different interactions. Nevertheless, the inclusion of the surface effects is essential, albeit difficult. This is especially true, if the rate coefficients for association and dissociation for binding are very significantly dependent on the nature of the surface. Unless a simpler alternative approach that includes the surface effects is suggested, it is reasonable, for now, to follow this approach. Hopefully, modifications to this approach may be suggested that permit comparison for different interactions as well as with other approaches.

It would be useful to specify the carrier of fractal properties. It could either be the analyte surface, the receptor surface, or the immobilizing (in our case, the biosensor) surface. There is a considerable body of work on fractal surface properties of proteins (Li et al., 1990; Dewey and Bann, 1992; Le Brecque, 1992; Federov et al., 1999). Le Brecque (1992) points out that the active sites (in our case, the receptors on the biosensor surface) may themselves form a fractal surface. Furthermore, the inclusion of nonspecific association sites on the surface would increase the degree of heterogeneity on the surface, thereby leading to an increase in the fractal dimension of the surface. At present, we are unable to specify what the carrier of the fractal properties is. This is exacerbated by our reanalysis of kinetic data available in the literature. Presumably, it is due to a composite of some or all of the factors mentioned above. No evidence of fractality is presented.

2.2.2.2 Dissociation Rate Coefficient

The diffusion of the dissociated particle (receptor or analyte) from the solid surface (e.g., analyte–receptor complex coated surface) into the solution may be given as a first approximation by:

$$\begin{aligned}(\text{Analyte}\cdot\text{Receptor}) &\sim -t^{(3-D_{\text{f,diss}})/2} \\ &= -k_{\text{diss}}t^{(3-D_{\text{f,diss}})/2}, \quad t > t_{\text{diss}}\end{aligned} \tag{2.5b}$$

Here $D_{\text{f,diss}}$ is the fractal dimension of the surface for the dissociation step. t_{diss} represents the start of the dissociation step. This corresponds to the highest concentration of the analyte–receptor complex on the surface. Henceforth, its concentration only decreases. $D_{\text{f,bind}}$ may or may not be equal to $D_{\text{f,diss}}$. k_{d} and k_{diss}, and $D_{\text{f,d}}$ and $D_{\text{f,diss}}$ are used interchangeably in this chapter and in this book.

One may obtain a unit for the dissociation rate coefficient, k_{d}, in a similar manner as done for the binding rate coefficient. In this case, the units for the binding and the dissociation rate coefficient are the same. The unit for the dissociation rate coefficient, k_{d} is $(\text{pg})(\text{mm})^{-2}(\text{s})^{(D_{\text{f,diss}}-3)/2}$. Once again, note that the unit dependence on time exhibited by k_{d} changes slightly due to the dependence on $D_{\text{f,diss}}$.

2.2.3 Dual-Fractal Analysis

2.2.3.1 Binding Rate Coefficient

The single-fractal analysis we have just presented is extended to include two fractal dimensions. At present, the time $(t = t_1)$ at which the first fractal dimension "changes" to the second fractal dimension is arbitrary and empirical. For the most part it is dictated by the data analyzed and the experience gained by handling a single-fractal analysis. The r^2 (regression coefficient) value obtained is also used to determine if a single-fractal analysis is sufficient, or one needs to use a dual-fractal analysis to provide an adequate fit. Only if the r^2 value is less than 0.97 for a single-fractal analysis, do we use a dual-fractal model. In this case, the analyte–receptor complex is given by:

$$(\text{Analyte}\cdot\text{Receptor}) \sim \begin{cases} t^{(3-D_{f1,\text{bind}})/2} = t^{p1}, & t < t_1 \\ t^{(3-D_{f2,\text{bind}})/2} = t^{p2}, & t_1 < t < t_2 = t_c \\ t^{1/2}, & t > t_c \end{cases} \tag{2.5c}$$

In analyte–receptor binding, the analyte–receptor binds with the active site on the surface and the product is released. In this sense the catalytic surface exhibits an unchanging fractal surface to the reactant in the absence of fouling and other complications. In the case of analyte–receptor association, the biosensor surface exhibits a changing fractal surface to the analyte in solution. This occurs because as each association reaction takes place, smaller and smaller amounts of "association" sites or receptors are available on the biosensor surface to which the analyte may bind. Furthermore, as the reaction proceeds, there is an increasing degree of heterogeneity on the biosensor surface for some reaction systems. This is manifested by two degrees of heterogeneity or two fractal dimensions on the biosensor surface. In the theoretical limit one might envisage a temporal fractal dimension wherein there is a continuous change in the degree of heterogeneity on the surface; though of course, such situations would be very rare, if at all.

Surfaces exhibit roughness, or a degree of heterogeneity at some scale. This degree of heterogeneity on the surface may be due to fracture or erosion. In our case of biosensors, this may arise due to (1) the inherent roughness of the biosensor surface, or (2) due to the immobilization or deposition of the receptors on the biosensor surface. The method of deposition of the receptors on the surface would also lead to different degrees of heterogeneity on the surface. The binding reaction takes place between the analyte in solution and the receptors on the surface through chemical bond formation and subsequent molecular association. The geometric nature (or parameter) of the surface will significantly influence these reactions. The influence of surface morphology and structure has been analyzed (Lee and Lee, 1994; Chaudhari et al., 2002, 2003). It would be of interest to determine the scale of these roughness heterogeneities. Are these at the Angstrom level or lower? With the current emphasis on nanotechnology and nanobiotechnology these types of questions are becoming more and more relevant and of significance. The nature of surfaces in general, and of biosensors in particular (our case), should exhibit a fractal nature at the molecular level. Furthermore, one of the reasons for the emphasis on nanotechnology is that as one goes down in scale, the properties of some substances change, sometimes for the better. It is these beneficial changes that one wishes to exploit in nanotechnology and nanobiotechnology. Hopefully, similar parallels can be drawn on analyzing the fractal nature of biosensor surfaces. Do they exhibit self-similarity; and if they do what are their limits? In other words, what are their lower and upper bounds?

Furthermore, each binding event need not result in the formation of an analyte–receptor complex on the biosensor surface. All of the receptors on the biosensor surface are presumably not the results of binding events, and do not exhibit the same activity. In other words, their active sites should comprise of presumably a probability distribution in "activity". In lieu of any prior information, it is reasonable to assume a bell-shaped Gaussian (or normal) distribution of active sites on the surface. A probabilistic approach is more realistic here. Analyses of this sort have presumably not been performed (at least this author is unaware of them) for analyte–receptor reactions occurring on biosensor surfaces. Thus the fractal analysis is a convenient method of providing a lumped parameter analysis of analyte–receptor reactions occurring on biosensor surfaces.

Note that, at present, the dual-fractal analysis does not have a basis at the molecular level. This represents two different levels of heterogeneity on the biosensor surface. But, in some of the examples presented, a single-fractal analysis is clearly inadequate to model the data. Only in these cases does one resort to a dual-fractal analysis. The binding rate coefficients, k_1 and k_2 in the dual-fractal analysis have the same units $(\mathrm{pg})(\mathrm{mm})^{-2}(\mathrm{s})^{(D_{\mathrm{f1,diss}}-3)/2}$ and $(\mathrm{pg})(\mathrm{mm})^{-2}(\mathrm{s})^{(D_{\mathrm{f2,diss}}-3)/2}$, respectively, as the association rate coefficient, k, in the single-fractal analysis.

2.2.3.2 Dissociation Rate Coefficient

In this case the dissociation rate coefficient is given by:

$$(\mathrm{Ab}\cdot\mathrm{Ag}) \approx \begin{cases} -t^{(3-D_{\mathrm{f1,diss}})/2}, & t_{\mathrm{diss}} < t < t_{\mathrm{d1}} \\ -t^{(3-D_{\mathrm{f2,diss}})/2}, & t_{\mathrm{d1}} < t < t_{\mathrm{d2}} \end{cases} \tag{2.5d}$$

Here $D_{\mathrm{f,diss}}$ is the fractal dimension of the surface for the dissociation step. t_{diss} represents the start of the dissociation step. This corresponds to the highest concentration of the analyte–receptor on the surface. Henceforth, its concentration only decreases. $D_{\mathrm{f,bind}}$ or $D_{\mathrm{f,assoc}}$ may or may not be equal to $D_{\mathrm{f,diss}}$. The dissociation rate coefficients, k_{d1} and k_{d2} in the dual-fractal analysis have the same units $(\mathrm{pg})(\mathrm{mm})^{-2}(\mathrm{s})^{(D_{\mathrm{fd1,diss}}-3)/2}$ and $(\mathrm{pg})(\mathrm{mm})^{-2}(\mathrm{s})^{(D_{\mathrm{fd2,diss}}-3)/2}$, respectively, as the dissociation rate coefficient, k_{d}, in the single-fractal analysis.

2.2.4 Triple-Fractal Analysis

As will be shown later in this book, one resorts to a triple-fractal analysis when the dual-fractal analysis does not provide an adequate fit. The equation for fractal analysis is generic in nature, and one may easily extend the single- and the dual-fractal analysis equations (Eqns (2.5a) and (2.5c)) to describe the binding (and/or the dissociation) kinetics for a triple-fractal analysis. In fact, in the extreme case, n fractal dimensions may be present. In this case, the degree of heterogeneity, D_f, or the fractal dimension is continuously changing on the biosensor surface, and the surface needs to be represented by D_{fi} where i goes from 1 to n. Similarly, we have n binding rate coefficients on the biosensor surface. A similar representation may also be made for the dissociation phase.

It is perhaps appropriate here to at least mention one more approach that has been used to model the binding kinetics on surfaces.

2.2.5 Pfeifer's Fractal Binding Rate Theory

Pfeifer and Obert (1989) have suggested an alternate form of the binding rate theory. In the equation given in this reference, N is the number of complexes, N_0 is the number of receptors on the solid surface, D is the diffusion coefficient of the analyte, L is the receptor diameter, and λ is the mean distance between two neighboring receptors. This equation may also be used to analyze the analyte–receptor binding kinetics. The problem, however, is that it may not be possible in all instances to estimate a priori all the parameters described in the equation (not given here). In that case, one may have to approximate or assume certain values, and this will affect the accuracy and reliability of the analysis. The suggested equation does have an advantage compared to the fractal analysis described above in that it does include a prefactor necessary to convert the time interval over which fractal scaling is observed into a length interval. It also provides an expression for t_c ($=L^2/D$), which separates the short-term regime from the long-term regime. The short-term regime is the one in which the anomalous diffusion applies. At the end of the short-term interval ($t = t_c$), the self-similarity of the system is lost, the surface is homogeneous, and regular diffusion applies.

Pfeifer and Obert (1989) state that the application of the above equation is contingent on the:

1. analyte being uniformly distributed in the solution at time t equal to zero,
2. binding being irreversible and first-order (N equals the number of analyte particles that have reached the receptors), and
3. binding occurring whenever an incoming analyte particle hits a receptor surface for the first time. In other words, the "sticking" probability is one.

It is perhaps difficult to imagine any one or all of these conditions being satisfied for analyte–receptor binding interactions occurring in continuous flow reactors.

Given the extremely small volume of the flow channels there is a high probability of the mixing of the analyte not being proper. This in turn may lead to analyte depletion in the flow channel. Also, the binding cannot be assumed to be irreversible in all instances. There may be cases of extremely fast binding and dissociation, especially for analytes with low affinity, which can dissociate in the continuously flowing buffer without any regeneration reagent.

Condition (3) may be satisfied. However, it does not include the "sticking" probability in that each collision leads to a binding event. Also, the presence of nonspecific binding, avidity effects, and binding with reactions or binding of dissociated analytes may interfere with condition (3) being satisfied. Furthermore, the equation makes assumptions about the number of active sites, and the immobilized receptors. For example it states that the analyte binds to one specific active site. The receptor cannot bind to more than one analyte molecule at a time (1:1 binding). The equilibrium dissociation rate coefficient, $K_D = k_{diss}/k_{assoc}$ can be calculated using the above models. The K_D value is frequently used in analyte−receptor reactions occurring on biosensor surfaces. The ratio, besides providing physical insights into the analyte−receptor system, is of practical importance as it may be used to help determine (and possibly enhance) the regenerability, reusability, stability, and other biosensor performance parameters. K_D has the unit $(s)^{[D_{f,diss}-D_{f\,assoc}]/2}$. This applies to both the single- as well as the dual-fractal analysis. For example, for a single-fractal analysis, K_D has the units $(s)^{[D_{fd}-D_f]/2}$. Similarly, for a dual-fractal analysis, the affinity, K_{D1} has the units $(s)^{[D_{fd1}-D_{fassoc1}]/2}$ and K_{D2} has the units $(s)^{[D_{fd2}-D_{fassoc2}]1/2}$. Note the difference in the units of the equilibrium dissociation rate coefficient obtained for the classical as well as the fractal-type kinetics. Though the definition of the equilibrium dissociation rate coefficient is the same in both types of kinetics (ratio of the dissociation rate coefficient to the association rate coefficient), the difference(s) in the units of the different rate coefficients eventually leads to a different unit for the equilibrium dissociation rate coefficient in the two types of kinetics. This is not entirely unexpected as the classical kinetic analysis does not include the characteristics of the surface in the definition of the equilibrium dissociation rate coefficient whereas the present fractal analysis does. Thus, one may not be able to actually compare the equilibrium dissociation rate coefficient affinities in these two types of systems. This is a significant difference in the kinetic analysis of binding and dissociation reactions on biosensor surfaces from what is available in the literature.

It is perhaps appropriate to, at least, briefly present some of the other approaches that have recently appeared in the literature and to help model the binding and the dissociation kinetics of the different analytes (present in the liquid phase) on biosensor surfaces. The following kinetic modeling approaches will be presented and analyzed briefly:

1. application of the synthetic jet concept to low Reynolds number biosensor microfluidic flows for enhanced mixing (Mautner, 2004);
2. kinetics of analyte capture on nanoscale sensors (Solomon and Paul, 2006); and

3. probing the functional heterogeneity of surface binding sites along with the effect of mass transport limitation and its influence on binding and dissociation of analytes on biosensor surfaces (Svitel et al., 2007).

2.2.6 The Mautner Model

Mautner (2004) points out that microfluidic components used in biosensors may be effectively used for the analysis of chemicals and biologicals. He emphasizes that the microfluidic systems used in biosensors will operate at low Reynolds numbers. Reynolds number is a dimensionless number and is used to characterize flows in different types of systems. Reynolds number (Re) is equal to $(Dv(\rho)/\mu)$. ρ is a Greek letter and represents the density of the fluid, D is the diameter of the pipe or a characteristic dimension of the system, v is the velocity of the fluid in the system, and μ is the viscosity of the fluid. Low Reynolds number regime is characterized by Re less than 10. Furthermore, Mautner (2004) points out that these devices will have characteristic dimensions less than 100 μm. He also emphasizes that we are dealing here with small volumes of fluids (pico- to microliters). This type of slow Re laminar flow is characterized by diffusion-only mixing. However, Mautner (2004) emphasizes that rapid mixing is essential in immunoassays. Thus, enhanced mixing is essential to overcome the slow fluid mixing in low Re number flow.

Mautner (2004) points out that the following techniques have been used to enhance mixing in microfluidic networks:

1. Use of slanted wells to increase lateral flow transport (Johnson et al., 2002).
2. Flows over shallow grooves (Stroock et al., 2002).
3. Utilization of passive mixing in three-dimensional serpentine microchannels (Liu et al., 2000).
4. Application of surface layers creating hydrophobic or hydrophilic surface patterns to direct fluids (Zhao et al., 2002).

Mautner emphasizes that the application of readily available pumps can help the unsteady wall jets obtain the required time-dependent wall jet conditions which would mix both existing and merging flow stream conditions.

Mautner (2004) has proposed the application of macroscale jets to be applied to the low Reynolds ($Re = 10$) two-dimensional channel flows that may be found in biosensor microfluidic systems. The method includes a hybrid approach of the Lattice-Boltzmann (LB) method for flow field computations and a finite difference, convection–diffusion equation for passive scalar transport. This author emphasizes that the forced jet imparts momentum to the channel flow, thereby enhancing fluid mixing.

2.2.7 Kinetics of Analyte Capture on Nanoscale Sensors (Solomon and Paul, 2006)

Solomon and Paul (2006) point out that nanoscale electromechanical systems have been used to detect biomolecular targets with increasing sensitivity (Roukes et al., 2000; Paul and Cross, 2004; Ekinci and Roukes, 2005). Solomon and Paul (2006) call these devices, BioNems devices. These BioNems devices may be used to detect proteins, enzymes, viruses, and bacteria. They point out that the sensitivity of these devices is directly related to the binding kinetics of the analytes to the receptors immobilized on these BioNems devices. The BioNem device that these authors have developed and analyzed are different from the conventional devices that measure, for example, the binding rate coefficients, in that they identify the presence or absence of biological macromolecules such as large proteins, enzymes, viruses, and enzymes.

The Solomon and Paul (2006) system uses a single (or multiple) cantilever(s). The multiple cantilever configuration improves detection probabilities. The cantilever is located in a small channel (or "via"). Fluid containing the target analyte flows through these channels at some fixed velocity. The detection of the analyte by the cantilever is based on either (1) the mass-loading effects or (2) by a change in the effective damping constant of the cantilever that alters the mean square displacement of the cantilever tip. Solomon and Paul (2006) point out that a small region of the cantilever tip is functionalized with immobilized receptors specific to the target analyte. Self-assembling monolayers constructed with alkanethiols permits the functionalization by a linkage to analyte-specific receptors. These emphasize that their device is able to detect specific biomolecules down to concentration levels of 1 nM or less.

Solomon and Paul (2006) also attempted to analyze the influence of surface-diffusion enhancement on the analyte capturer. Their quantitative results indicate that under certain circumstances this mechanism might be advantageous. This will depend, as the authors indicate, on the particular parameters that characterize a specific device implementation.

Solomon and Paul (2006) also attempted to analyze the influence of convection on these types of BioNems devices. Using simple fluid dynamics arguments these authors show that at the bulk flow velocities used in these devices and for Damkohler (Da) numbers less than unity, capture efficiencies would be dominated by reaction-diffusion mechanisms. In other words, convection will only play a minor role. They emphasize that for Damkohler numbers $>>1$, the reaction-diffusion–convection coupled equations need to be solved to obtain a better perspective of analyte binding kinetics to receptors immobilized on these nanoscale sensors.

2.2.8 Probing the Functional Heterogeneity of Surface Binding Sites Along with the Effect of Mass Transport Limitation and Its Influence on Binding and Dissociation of Analytes on Biosensor Surfaces (Svitel et al., 2007)

Svitel et al. (2007) have recently probed the functional heterogeneity of surface binding sites under the influence of external mass transport limitations. They did this by analyzing experimental binding traces. These authors point out the need to analyze the binding of macromolecules to surfaces assuming that the surface binding sites are heterogeneous (Sips, 1948; Koopal and Vos, 1993; Vijayendran and Leckband, 2001; Lebedev et al., 2006). Svitel et al. (2007) suggest that there may be two (or more) reasons for inhomogeneity on the surface during the interactions of analyte–receptors on the sensor surface: (1) the surface sites may be intrinsically inhomogeneous in their binding properties, and (2) the receptors may be rendered heterogeneous by attaching them to the surface. Yeung and Leckband (1997), Rabbany et al. (1997), and Kloss et al. (2000) have pointed out that the immobilization of chemically homogeneous species frequently results in functionally impaired subpopulations due to (1) constraints in orientation, (2) variable cross-linking, and (3) the influence of the microenvironment of the surface. This heterogeneity, Svitel et al. (2007) explain, will influence the application of antibody-based affinity biosensors (Wilson and Nock, 2002; Taitt et al., 2005), and the analysis of protein interactions by SPR biosensors (O'Shannessy, 1994; O'Shannessy and Winzor, 1996; Schuck, 1997).

Svitel et al. (2007) explain that evanescent field biosensors have been used to characterize protein–protein, protein–small molecule, protein–nucleic acid interactions, and DNA hybridization reactions (Cooper, 2002; Livache et al., 2003). These biosensors have allowed the kinetic binding traces to be measured with high sensitivity. Svitel et al. (2007), however, also point out that these binding traces when analyzed kinetically are apparently not consistent with a simple 1:1 interaction (Karlsson et al., 1994; Glaser and Hausdorf, 1996; Schuck, 1997; Schuck et al., 1998). Svitel et al. (2003) attempt to interpret these deviations from a simple 1:1 interaction as a source of information on the homogeneity of the surface immobilized sites. They have proposed a computational model that assumes that the binding signal is a superposition of independent parallel reactions occurring on the biosensor surface. These parallel binding reactions result from a continuous distribution of thermodynamic and kinetic binding constants. In a more recent publication, Svitel et al. (2007) have expanded their computational model and their approach to include a compartment-like transport step, which describes the competitive binding to different surface sites in a zone of depleted analyte close to the biosensor surface. Just as in the fractal analysis approach presented in different chapters in this book to analyze the binding and the dissociation phase, the approach presented by Svitel et al. (2007) helps to analyze surface binding when both inhomogeneity on the biosensor surface and transport limitations are present simultaneously. Their approach, the authors claim, permits the evaluation of both the kinetic binding parameters as well as the effective transport rate coefficients.

REFERENCES

Anderson, J., July 1993. NIH Panel Review Meeting. Case Western Reserve University, Cleveland, OH.

Berland, K.M., So, P.T.C., Gratton, E., 1995. Two-photon fluorescence correlation spectroscopy: method and application to the intracellular environment. Biophysical Journal 68, 694–701.

Chaudhari, A., Yan, C.C., Lee, S.L., 2002. Multifractal analysis of diffusion-limited reactions over surfaces of diffusion-limited aggregates. Chemical Physics Letters 207, 220–226.

Chaudhari, A., Yan, C.C., Lee, S.L., 2003. Mathematical and general. Journal of Physics A 36, 3757.

Cooper, M.A., 2002. Optical biosensors in drug discovery. National Reviews in Drug Discovery 1, 515–528.

Coppens, M.O., Froment, G.F., 1995. Diffusion and reaction in a fractal catalyst pore. Geometrical aspects. Chemical Engineering Science 50 (6), 1013–1026.

Corel Quattro Pro, 8.0, 1997. Corel Corporation Limited, Ottawa, Canada.

De Gennes, P.G., 1982. Diffusion-controlled reactions. Polymer Melts Radiation and Physical Chemistry 22, 193.

Dewey, T.G., Bann, J.G., 1992. Diffusion-controlled reaction in polymer melts. Biophysical Journal 63, 594.

Ekinci, K.L., Roukes, M.N., 2005. Nanoelectrical systems. Reviews of Scientific Instrumentation 76, 1–12.

Fatin-Rouge, N., Starchev, K., Buffle, J., 2004. Size effects on diffusion process with agarose gels. Biophysical Journal 86, 2710–2719.

Federov, B.A., Federov, B.B., Schmidt, P.W., 1999. An analysis of the fractal properties of globular proteins. Journal of Chemical Physics 99, 4076–4083.

Giona, M., 1992. First-order reaction-diffusion in complex fractal media. Chemical Engineering Science 47, 1503–1515.

Ghosh, R.N., Webb, W.W., 1988. Results of automated tracking of low density lipoprotein receptors on cell surfaces. Biophysical Journal 53, A352.

Glaser, R.W., Hausdorf, X., 1996. Binding kinetics of an antibody against HIV p24 core protein measured with real-time biomolecular interaction analysis suggest a slow conformational exchange in antigen p24. Journal of Immunological Methods 189, 1–14.

Harder, F.H., Havlin, S., Bunde, A., 1987. Diffusion in fractals with singular waiting-time distribution. Physics Reviews B 36, 3874–3879.

Havlin, S., 1989. Molecular diffusion and reaction. In: Avnir, D. (Ed.), The Fractal Approach to Heterogeneous Chemistry: Surfaces, Colloids, Polymers. Wiley, New York, pp. 251–269.

Havlin, S., Ben-Avraham, D., 1987. Diffusion in disordered media. Advance in Physics 36, 695–798.

Henke, L., Nagy, N., Krull, U.J., 2002. An AFM determination of the effects of surface roughness caused by cleaning of fused silica and glass substrates in the process of optical biosensor preparation. Biosensors & Bioelectronics 17, 547–555.

Johnson, T., Ross, D., Locascio, L., 2002. Rapid microfluidic mixing. Analytical Chemistry 74, 45–51.

Karlsson, R., Roos, H., Fagerstam, L., Persson, B., 1994. Kinetic and concentration analysis using BIA technology. Methods in Computation and Methods in Enzymology 6, 99–110.

Kloss, A.A., Lavrik, N., Yeung, C., Leckband, D.E., 2000. Effect of the microenvironment on the recognition of immobilized cytochrome by soluble redox proteins. Langmuir 16, 3414–3421.

Koopal, L.K., Vos, C.H.W., 1993. Adsorption on heterogeneous surfaces. Calculation of the adsorption energy distribution function or the affinity spectrum. Langmuir 9, 2593–2605.

Kopelman, R., 1988. Fractal reaction kinetics. Science 241, 1620–1624.

Lebedev, K., Mafe, S., Stroeve, P., 2006. Convection, diffusion and reaction in a surface-based biosensor: modeling of cooperativities and binding site competition on the surface and in the hydrogel. Journal of Colloid and Interface Science 296, 527–537.

Le Brecque, M., 1992. Mosaic 23, 12–15.

Lee, C.K., Lee, S.L., 1994. Multifractal scaling analysis of autocatalytic and autopoisoning reactions over DLA surfaces. Chemical Physics Letters 228, 539–546.

Lee, C.K., Lee, S.L., 1995. Multifractal scaling analysis of reactions over fractal surfaces. Surface Science 325, 294–310.

Li, H.L., Chen, S., Zhao, H., 1990. Fractal mechanisms for the allosteric effects of proteins and enzymes. Biophysical Journal 58, 1313–1320.

Liu, R., Stremler, M., Sharp, K., Olsen, M., Santiago, J., Adrian, R., Aref, H., Beebe, D., 2000. Passive mixing in a three-dimensional serpentine microchannel. Journal of MEMS 9, 190–197.

Livache, T., Maillart, E., Lasalle, N., Mailley, P., Corso, B., Guiedon, P., Roget, A., Levy, Y., 2003. Polypyrrole based DNA hybridization assays: study of label free detection processes versus fluorescence on microchips. Journal of Pharmaceutical & Biomedical Analysis 32, 687–696.

Markel, V.A., Muratov, L.S., Stockman, M.I., George, T.F., 1991. Theory and numerical simulation of optical properties of fractal clusters. Physics Reviews B 43 (10), 8183.

Mautner, T., 2004. Application of the synthetic jet concept to low Reynolds number, biosensor microfluidic flows for enhanced mixing: a numerical study using the lattice Boltzmann method. Biosensors & Bioelectronics 19, 14009–14019.

Nyikos, L., Pajkossy, T., 1986. Diffusion to fractal surfaces. Electrochimica Acta 31, 1347.

O'Shannessy, D.J., 1994. Determination of kinetic rate and equilibrium binding constants for macromolecular interactions: a critique of the surface plasmon resonance literature. Current Opinions in Biotechnology 5, 65–71.

O'Shannessy, D.J., Winzor, D.J., 1996. Interpretations of deviations from pseudo-first-order kinetic behavior in the characterization of ligand binding by biosensor technology. Analytical Biochemistry 236, 275–283.

Pajkossy, T., Nyikos, L., 1989. Diffusion to fractal surfaces. II. Verification of theory. Electrochimica Acta 34, 171.

Paul, M.R., Cross, M.C., 2004. Stochastic dynamics of nanoscale mechanical oscillators immersed in a viscous fluid. Physics Reviews Letters 92, 235501–235504.

Peters, R., Cherry, R.J., 1982. Lateral and rotational diffusion of bacteriorhodopsin in lipid bilayers: experimental test of the Saffman-Delbruck equations. Proceedings of the National Academy of Sciences, USA 79, 4317–4321.

Pfeifer, P., Obert, M., 1989. In: Avnir, D. (Ed.), The Fractal Approach to Heterogeneous Chemistry: Surfaces, Colloids, Polymers. Wiley, New York, pp. 251–269.

Pfeifer, P., 1987. Characterization of surface irregularity, chapter 2. In: Preparative Chemistry Using Supported Reagents. Academic Press, San Diego.

Pfeifer, P., Avnir, D., Farin, D.J., 1984a. Molecular fractal surfaces. Nature (London) 308, 261.
Pfeifer, P., Avnir, D., Farin, D.J., 1984b. Surface geometric irregularity of particulate materials. The fractal approach. Journal of Colloid and Interface Science 103 (1), 112.
Pluen, A., Netti, P.A., Rakesh, K.J., Berk, D.A., 1999. Diffusion of macromolecules in agarose gels: comparison of linear and globular configurations. Biophysical Journal 77, 542−552.
Rabbany, S.Y., Piervincenzi, R., Judd, L., Kusterbeck, A.W., Brederhorst, R., Hakansson, K., Ligler, F.S., 1997. Assessment of heterogeneity in antibody-antigen displacement reactions. Analytical Chemistry 69, 175−183.
Roukesm, M., Fraser, S., Cross, M., Solomon, T., 2000. U.S Patent Application Number 60/224, vol. 109.
Sadana, A., Beelaram, A., 1994. Fractal analysis of antigen-antibody binding kinetics biosensor applications. Biotechnology Progress 9, 45.
Sadana, A., Alarie, J.P., Vo-Dinh, T., 1995. A β-cyclodextrin based fiber-optic chemical sensor: a fractal analysis. Talanta 42, 1567.
Saffman, P.G., Delbruck, M., 1975. Brownian motion in biological membranes. Proceedings of the National Academy of Sciences, USA 72, 3111−3113.
Sahimi, M., 1993. Flow phenomena in rocks: from continuum models to fractals, percolation, cellular automata and simulated annealing. Reviews in Modern Physics 65, 1393−1534.
Saxton, M.J., 2001. Anomalous diffusion due to binding: a Monte Carlo study. Biophysical Journal 77, 2251−2265.
Schwille, P., Haupts, U., Maiti, S., Webb, W.W., 1999. Molecular dynamics in living cells observed by fluorescence correlation spectroscopy with one- and two-photon excitation. Biophysical Journal 77, 2251−2265.
Schuck, P., 1997. Use of surface plasmon resonance to probe the equilibrium and dynamic aspects of interactions between biological macromolecules. Annual Reviews in Biophysical and Biomolecular Structures 26, 541−566.
Schuck, P., 1997. Reliable determination of binding affinity and kinetics using surface plasmon resonance biosensors. Current Opinion in Biotechnology 8, 498−502.
Schuck, P., Millar, D.B., Kortt, A.A., 1998. Determination of binding constants by equilibrium titration with circulating sample in a surface plasmon resonance biosensor. Analytical Biochemistry 265, 79−91.
Sips, R., 1948. On the structure of a catalyst surface. Journal of Chemical Physics 16, 490−495.
Solomon, J.E., Paul, M.R., March 2006. The kinetics of analyte capture on nanoscale sensors. Biophysical Journal 90, 1842−1852.
Stroock, A., Dertinger, S., Whitesides, G., Ajdari, A., 2002. Patterning flows used grooved surfaces. Analytical Chemistry 74, 5306−5312.
Svitel, J., Balbo, A., Mariuzza, R.A., Gonzales, N.R., Schuck, P., 2003. Combined affinity and rate constant distributions of analyte or ligand populations from experimental surface binding and kinetics and equilibria. Biophysics Journal 84, 4062−4077.
Svitel, J., Boukari, H., van Ryk, D., Willson, R.C., Schuck, P., March 2007. Probing the functional heterogeneity of surface binding sites by analysis of experimental binding traces and the effect of mass transfer limitation. Biophysical Journal 92, 1742−1758.
Starchev, K., Sturm, J., Weill, G., Brogen, C.H., 1997. Brownian motion and electrophoretic transport in agarose gels studied by epifluorescence microscopy and simple particle tracking analysis. Journal of Physical Chemistry 101, 5659−5663.
Taitt, C.R., Anderson, G.P., Ligler, F.S., 2005. Evanescent wave fluorescence biosensors. Biosensors & Bioelectronics 20, 2470−2487.

Vijayendran, R.A., Leckband, D.E., 2001. A quantitative assessment of heterogeneity for surface-immobilized proteins. Analytical Chemistry 73, 471–480.

Wei, Q.H., Bechinger, C., Leiderer, P., 2000. Single-file diffusion of colloids in one-dimensional channels. Science 287, 625–627.

Weiss, G.H., 1994. Fractals in Science. Springer-Verlag, Berlin, p. 119.

Wilson, D.S., Nock, S., 2002. Functional protein microarrays. Current Opinion in Chemical Biology 6, 81–85.

Yeung, C., Leckband, D.E., 1997. Molecular level characterization of microenvironmental influences on the properties of immobilized proteins. Langmuir 13, 6746–6754.

Zhao, B., Moore, J., Beebe, D., 2002. Principles of surface-directed liquid flow in microfluidic channels. Analytical Chemistry 74, 4259–4268.

Chapter 4

Detection of Cancer Biomarkers by Biosensors: Part II

4.1 THEORY

4.1.1 Single-Fractal Analysis

4.1.1.1 Binding Rate Coefficient

Havlin (1989) indicates that the diffusion of a particle (analyte [Ag]) from a homogeneous solution to a solid surface (e.g., receptor [Ab]-coated surface) on which it reacts to form a product (analyte–receptor complex; (Ab·Ag)) is given by

$$(\mathrm{Ab\cdot Ag}) \approx \begin{cases} t^{(3-D_{\mathrm{f,bind}})/2} = t^{\mathrm{p}} & t < t_{\mathrm{c}} \\ t^{1/2} & t > t_{\mathrm{c}} \end{cases} \tag{4.1}$$

Here $D_{\mathrm{f,bind}}$ or D_{f} is the fractal dimension of the surface during the binding step and t_{c} is the crossover value. Havlin (1989) indicates that the crossover value may be determined by $r_{\mathrm{c}}^2 \sim t_{\mathrm{c}}$. Above the characteristic length, r_{c}, the self-similarity of the surface is lost and the surface may be considered homogeneous. Above time t_{c}, the surface may be considered homogeneous, since the self-similarity property disappears and "regular" diffusion is now present. For a homogeneous surface where $D_{\mathrm{f}} = 2$ and when only diffusional limitations are present, $p = 1/2$ as it should be. Another way of looking at the $p = 1/2$ case (where $D_{\mathrm{f,bind}} = 2$) is that the analyte in solution views the fractal object, in our case, the receptor-coated biosensor surface, from a "large distance." In essence, in the association process, the diffusion of the analyte from the solution to the receptor surface creates a depletion layer of width $(Đt)^{1/2}$, where Đ is the diffusion constant. This gives rise to the fractal power law, (analyte receptor) $\sim t^{(3-D_{\mathrm{f,bind}})/2}$. For the present analysis, t_{c} is arbitrarily chosen and we assume that the value of the t_{c} is not reached. One may consider the approach as an intermediate "heuristic" approach that may be used in the future to develop an autonomous (and not time-dependent) model for diffusion-controlled kinetics.

Biomarkers and Biosensors. http://dx.doi.org/10.1016/B978-0-444-53794-2.00004-5

4.1.1.2 *Dissociation Rate Coefficient*

The diffusion of the dissociated particle (receptor [Ab] or analyte [Ag]) from the solid surface (e.g., (analyte [Ag]—receptor [Ab]) complex-coated surface) into solution may be given, as a first approximation, by

$$(\mathrm{Ab}\cdot\mathrm{Ag}) \approx -t^{(3-D_{\mathrm{f,diss}})/2} = -t^{\mathrm{p}} \quad (t > t_{\mathrm{diss}}) \tag{4.2}$$

Here $D_{\mathrm{f,diss}}$ is the fractal dimension of the surface for the dissociation step. This corresponds to the highest concentration of the analyte—receptor complex on the surface. Henceforth, its concentration only decreases. The dissociation kinetics may be analyzed in a manner "similar" to the binding kinetics.

4.1.2 Dual-Fractal Analysis

4.1.2.1 *Binding Rate Coefficient*

Sometimes, the binding curve exhibits complexities and two parameters (k and D_{f}) are not sufficient to adequately describe the binding kinetics. This is further corroborated by low values of r^2 factor (goodness-of-fit). In that case, one resorts to a dual-fractal analysis (four parameters: k_1, k_2, D_{f1}, and D_{f2}) to adequately describe the binding kinetics. The single-fractal analysis presented above is thus extended to include two fractal dimensions. At present, the time ($t = t_1$) at which the "first" fractal dimension "changes" to the "second" fractal dimension is arbitrary and empirical. For the most part, it is dictated by the data analyzed and experience gained by handling a single-fractal analysis. A smoother curve is obtained in the "transition" region if care is taken to select the correct number of points for the two regions. In this case, the product (antibody—antigen or analyte—receptor complex; Ab·Ag or analyte.receptor) is given by

$$(\mathrm{Ab}\cdot\mathrm{Ag}) \approx \begin{cases} t^{(3-\,D_{\mathrm{f1,bind}})/2} = t^{\mathrm{p1}} & (t < t_1) \\ t^{(3-\,D_{\mathrm{f2,bind}})/2} = t^{\mathrm{p2}} & (t_1 < t < t_2) = t_{\mathrm{c}} \\ t^{1/2} & (t > t_{\mathrm{c}}) \end{cases} \tag{4.3}$$

In some cases, as mentioned above, a triple-fractal analysis with six parameters (k_1, k_2, k_3, D_{f1}, D_{f2}, and D_{f3}) may be required to adequately model the binding kinetics. This is when the binding curve exhibits convolutions and complexities in its shape due to perhaps the very dilute nature of the analyte (in some of the cases to be presented) or for some other reasons. Also, in some cases, a dual-fractal analysis may be required to describe the dissociation kinetics.

4.2 RESULTS

Teker (2008) analyzed the binding of cancer cells to antibody-conjugated carbon nanotube field-effect transistor (CNTFET). Both the insulin-like growth factor 1 receptor (IGF1R) and the nonspecific monoclonal antibodies (mAbs) reduced the electrical conductance of the CNTFET devices. The application of the HBT474 and the MCF7 breast cancer cells increased the conductance for the IGF1R mAb CNTFET devices. Figure 4.1(a) shows the binding of the B7474 breast cancer cells to the IGF1R antibodies immobilized on the CNTFETs. Table 4.1 shows the modeling for comparisons sake:

1. first using a binding phase only and
2. then using a binding and a dissociation phase.

The values of the binding rate and the dissociation rate coefficients (as the case may be) and the corresponding fractal dimension for the binding and the dissociation phase, are given in Table 4.1 respectively. In the second case, the affinity, $K = (k/k_d)$, value is 0.05789.

Figure 4.1(b) shows the binding of the MCF7 breast cancer cells to the IGF1R antibodies on the CNTFETs. Table 4.1 shows the modeling once again, for comparisons sake:

1. first using a binding phase only and
2. then using a binding and a dissociation phase.

The values of the binding rate and the dissociation rate coefficients (as the case may be), and the corresponding fractal dimension for the binding and the dissociation phase, are given in Table 4.1 respectively. In the second case, the affinity, K $(= k/k_d)$, value is 0.01727.

Kim et al. (2010) have recently analyzed the detection of a prostate cancer marker (PSA-ACT complex) through a label-free protein biosensor based on the CNTFET. The authors functionalized the CNTFET with a solution

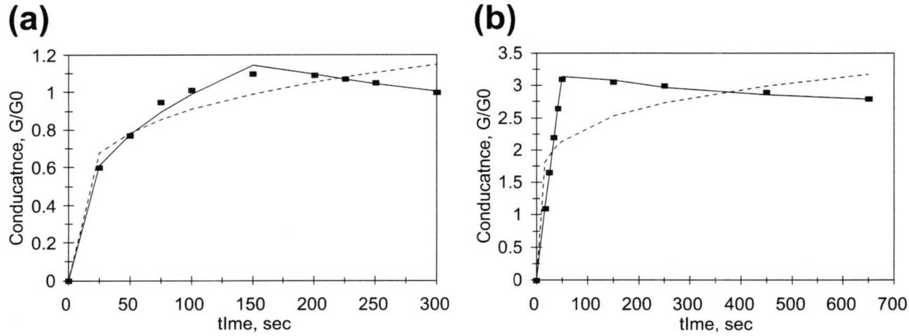

FIGURE 4.1 Binding of cancer cells to the monoclonal antibodies on the carbon nanotube field-effect transistor (Teker, 2008): (a) B7474 breast cancer cells (b) MCF7 breast cancer cells.

TABLE 4.1 Binding and Dissociation Rate Coefficients and Fractal Dimensions for the Binding and the Dissociation Phase for (a) B474 Cancer Cells Applied to IGF1R-specific mAb Carbon Nanotube Field-Effect Transistor (CNTFET) and (b) MCF7 Breast Caner Cells Applied to IGF1R-Specifc mAb CNTFET (Teker, 2008)

Analyte in Solution	k	k_d	D_f	D_{fd}
B7474 breast cancer cells	0.3395 ± 0.0359	NA	2.5720 ± 0.08532	NA
B7474 breast cancer cells	0.1969 ± 0.0086	3.3921 ± 0.0207	2.2970 ± 0.06254	3 − 0.040
MCF7 breast cancer cells	1.622 ± 0.3556	NA	2.6898 ± 0.1242	NA
MCF7 breast cancer cells	0.07436 ± 0.00039	4.3034 ± 0.055	1.0864 ± 0.04292	3 − 0.0927

containing various linker-to-spacer ratios. Prostate-specific antigen (PSA) is a glycoprotein that contains 93% peptide and 7% sugar. It is produced by the prostate tissue. This according to is the best serum marker for diagnosing and monitoring prostate cancer (Loeb and Catalona, 2007). The major forms comprise a combination of PSA and protein C inhibitor, α_1-antitrypsin, and α-trypsin (Armbuster, 1993; Lilja et al., 1991; Savage & Waxman, 1990). PSA-ACT and f-PSA are two molecules that may be measured to determine prostate cancer. The normal range is 4 ng/ml, whereas higher than 20 ng/ml is indicative of the presence of cancer.

Kim et al. (2010) indicate that present-day methods suffer from several disadvantages that include the requirement of multiple reagents and signal amplification, relatively large sample size, etc. A highly sensitive and specific method is required, which does not require labels, permits real-time monitoring, is rapid, and is also portable, disposable, and of low cost.

CNTFETs whereby the conductance changes by the charge transfer from the molecules adsorbed on carbon nanotube (CNT) surface satisfy the abovementioned requirements (Allen et al., 2007; Bradley et al., 2004; Chen et al., 2003; Guo et al., 2005; Lin, 2008; Merkoci et al., 2008; Qi et al., 2003; Star et al., 2003; Robertson, 2004; and Trojanowicz, 2006). For biodetection, Kim et al. (2008) state that the CNTFET surface is functionalized with the specific receptor molecule that binds to the target biomolecules. The charges of the biomolecules affect the conductance of the CNTFETs.

Aptamers are used to alleviate the size problems in these types of biosensors (Machashi et al., 2007; So et al., 2005). Peptides are also used (Kim et al., 2008).

have used a CNTFET biosensor using 1-pyrenebutanoic acid, succinimidyl ester, and 1-pyrenebutanol as spacers. The CNTFET was functionalized with five kinds of solutions containing various linker-to-spacer ratios.

Figure 4.2(a) shows the binding of 1 ng/ml of PSA-ACT complex in solution to a functionalized CNTFET biosensor. A single-fractal analysis is adequate to describe the binding kinetics. The values of the binding rate coefficient, k, and the fractal dimension, D_f, for a single-fractal analysis are given in Table 4.2.

Figure 4.2(b) shows the binding of 10 ng/ml of PSA-ACT complex in solution to a functionalized CNTFET biosensor. A single-fractal analysis is adequate to describe the binding kinetics. The values of the binding rate coefficient, k, and the fractal dimension, D_f, for a single-fractal analysis are given in Table 4.2.

Figure 4.2(c) shows the binding of 50 ng/ml of PSA-ACT complex in solution to a functionalized CNTFET biosensor. A single-fractal analysis is adequate to describe the binding kinetics. The values of the binding rate coefficient, k, and the fractal dimension, D_f, for a single-fractal analysis are given in Table 4.2.

Figure 4.2(d) shows the binding of 100 ng/ml of PSA-ACT complex in solution to a functionalized CNTFET biosensor. A single-fractal analysis is

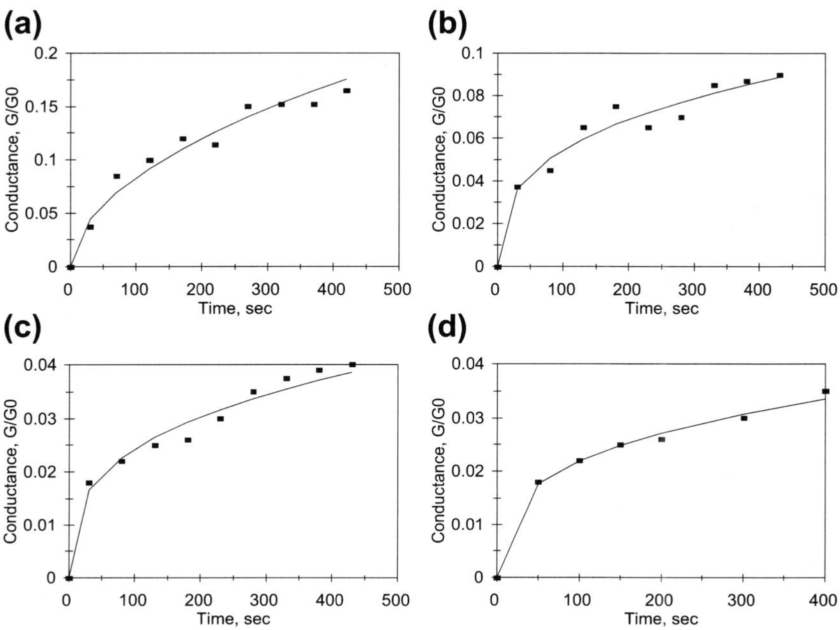

FIGURE 4.2 Electrical signal (binding) of carbon nanotube field-effect transistor modified with surface C following the immune reaction of PSA-ACT (monoclonal antibody): (a) 1 ng/ml (b) 10 ng/ml (c) 50 ng/ml (d) 100 mg/ml (Kim et al., 2010).

TABLE 4.2 Binding Rate Coefficients (k) and Fractal Dimensions (D_f) for PSA-ACT Complex in Solution to a Functionalized Carbon Nanotube Field-Effect Transistor biosensor (Kim et al., 2010)

Analyte in Solution	k	D_f
1 ng/ml	0.00519 ± 0.00018	2.3902 ± 0.03928
10 ng/ml	0.005642 ± 0.000418	2.3656 ± 0.05876
50 ng/ml	0.11166 ± 0.0012	2.3302 ± 0.0374
100 ng/ml	0.005913 ± 0.000591	1.8602 ± 0.07754

adequate to describe the binding kinetics (Figure 4.3). The values of the binding rate coefficient, k, and the fractal dimension, D_f, for a single-fractal analysis are given in Table 4.2.

Figure 4.3(a) and Table 4.3 show the decrease in the binding rate coefficient, k, with an increase in the PSA-ACT mAb in solution. For the data shown in Figure 4.3(a), the binding rate coefficient, k, is given by

$$k = (0.024118 \pm 0.0187)[\text{PSA} - \text{ACT mAb}]^{-0.1578-0.1621} \quad (4.4a)$$

There is scatter in the data. Only four data points are available. The availability of more data points would lead to a more reliable fit. The binding rate coefficient, k, exhibits only a mild negative order (equal to −0.1578) on the PSA-ACT mAb in solution.

Figure 4.3(b) and Table 4.3 show the increase in the binding rate coefficient, k, with an increase in the fractal dimension, D_f. For the data shown in Figure 4.3(b), the binding rate coefficient, k, is given by

$$k = (8.5\text{E} - 08 \pm 5.5\text{E} - 08)D_f^{7.872 \pm 3.76} \quad (4.4b)$$

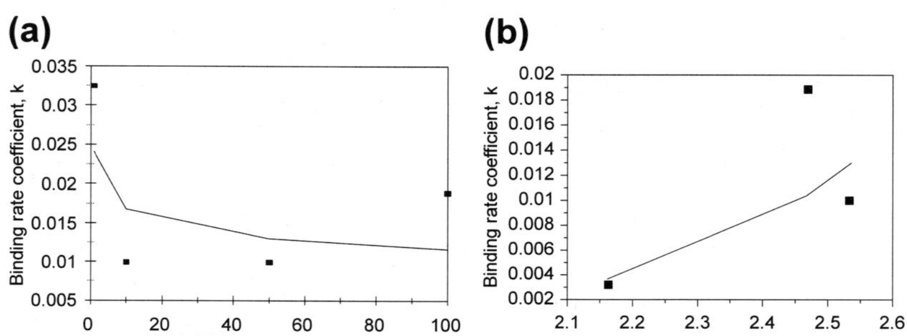

FIGURE 4.3 (a) Decrease in the binding rate coefficient, k, with an increase in the PSA-ACT concentration in solution (b) Increase in the binding rate coefficient, k, with an increase in the fractal dimension, D_f.

TABLE 4.3 Binding Rate Coefficients, *k*, and Fractal Dimensions, D_f, for PSA-ACT Monoclonal Antibodies, mAb, and the PSA-ACT Na CNTFET Biosensor Modified with Surface D Following the Immune Reaction of the PSA-ACT mAb and the PSA-ACT Complex (Kim et al., 2010)

Analyte in Solution	k	D_f
1 ng/ml	0.003258 ± 0.000488	2.1618 ± 0.2048
10 ng/ml	0.0100 ± 0.011	2.5350 ± 0.1117
50 ng/ml	0.00999 ± 0.0007	2.5322 ± 0.1103
100 ng/ml	0.0189 ± 0.0007	2.4674 ± 0.1600

There is scatter in the data. Only four data points are available. The availability of more data points would lead to a more reliable fit. Two of the points are almost the same. The binding rate coefficient, *k*, is very sensitive to the fractal dimension or the degree of heterogeneity on the surface, since it exhibits close to an eighth order of dependence (equal to 7.872) on the fractal dimension or the degree of heterogeneity that exists on the surface.

Wee et al. (2005) have presented a novel electrical detection technique for the detection of PSA based on piezoelectric self-sensing microcantilevers. Basically, the piezoelectric sensor measures the film resistance variation with respect to surface stress caused by biomolecular-specific binding. These authors used the cantilever biosensors for the detection of PSA and C-reactive protein (CRP), which are specific markers for prostate cancer and cardiac diseases.

Wee et al. (2005) indicate that microcantilevers find applications in the sensing of biological analytes. These authors indicate that cantilever fabrication has benefited from nanomechanics (Raiteri et al., 1999, 2001; Fritz et al., 2000; Cherian et al., 2003). The advantages of the microcantilever are that it does not require any label or reporter molecule. Baselt et al. (1996) and Alvarez et al. (2003) indicate the use of microcantilevers permits the real-time and label-free measurement.

Wee et al. (2005) indicate that these microcantilevers can operate in two modes: (1) cantilever bending (surface stress method) and (2) resonance response variation (microbalance method). The first method is based on static deflection originating from the stress generated (Jenessius et al., 2000; Ji and Thundat, 2002; Kooser et al., 2003). The second method detects the change in the cantilever using resonance energy shift. The first method is three orders of magnitude more sensitive than the second method.

Wee et al. (2005) have analyzed a static (DC) mode for the electrical detection of label-free disease-marker proteins (PSA and CRP) using microresistive self-sensing microcantilevers. PSA is an extremely useful biomarker for the monitoring of and detection of prostate cancer. Prostate cancer is currently the most prevalent form of cancer in men and the second leading cause of cancer deaths in the United States.

Wee et al. (2005) analyzed the piezoelectric response curves of the cantilevers (due to static deflection) with respect to the PSA concentration (without antigen, 1 μg/ml, 100 ng/ml, 10 mg/ml) in solution. All solutions contained 0.1 mg/ml BSA, which was used to estimate the nonspecific adsorption on the cantilever surface. In the case of no PSA, there was no significant variation of piezoelectric-induced voltage. Figure 4.4(a) shows PSA antigen—antibody (Ag-Ab) binding of 1 μg/ml to the microcantilever surface, as denoted by the microcantilever deflection. The binding kinetics (deflection) is adequately described by a single-fractal kinetics model. The values of the binding rate coefficient, k, and the fractal dimension, D_f, are given in Table 4.4.

Figure 4.4(b) shows PSA antigen—antibody (Ag-Ab) binding of 100 ng/ml to the microcantilever surface, as denoted by the microcantilever deflection. The binding kinetics (deflection) is, once again, adequately described by a

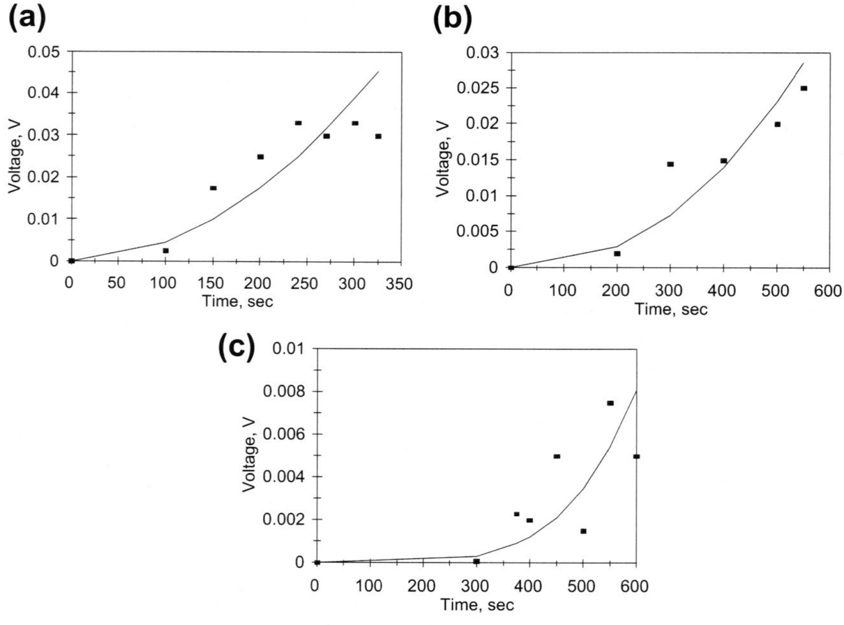

FIGURE 4.4 Real-time monitoring of the microcantilever deflection signal for different prostrate-specific antigen (PSA) concentrations. Binding of PSA concentrations to a functionalized Au surface (Wee et al., 2005): (a) 1 μg/ml (b) 100 ng/ml (c) 10 ng/ml.

TABLE 4.4 Binding Rate Coefficients, *k*, and Fractal Dimensions, D_f, for Different Concentrations of prostrate-specific antigen in Solution During Microcantilever detection (Wee et al., 2005)

Analyte in Solution	*k*	D_f
1 μg/ml	5E-07 ± 2.9E-07	0 ± 0.8894
100 ng/ml	1.9E-08 ± 1.2 E−08	0 ± 1.1394
10 ng/ml	7.8E-16 ± 7.2 E−16	0 ± 1.9846

single-fractal kinetics model. The values of the binding rate coefficient, *k*, and the fractal dimension, D_f, are given in Table 4.4.

Figure 4.4(c) shows PSA antigen—antibody (Ag-Ab) binding of 10 ng/ml to the microcantilever surface, as denoted by the microcantilever deflection. The binding kinetics (deflection) is, once again, adequately described by a single-fractal kinetics model. The values of the binding rate coefficient, *k*, and the fractal dimension, D_f, are given in Table 4.4.

Figure 4.5 shows the increase in the binding rate coefficient, *k*, with an increase in the PSA concentration in solution. For the data shown in Figure 4.5, the binding rate coefficient, *k*, is given by

$$k = (3\mathrm{E} - 19 + 8.3\mathrm{E} - 17)[\mathrm{PSA}]^{4.403 \pm 1.722} \quad (4.4c)$$

Only three data points are available. There is scatter in the data. This is reflected in the error in the estimated value of the binding rate coefficient, *k*. The binding rate coefficient, *k*, is very sensitive to PSA concentration in solution, as is noted by the close to four and one-half order of dependence (equal to 4.403) exhibited.

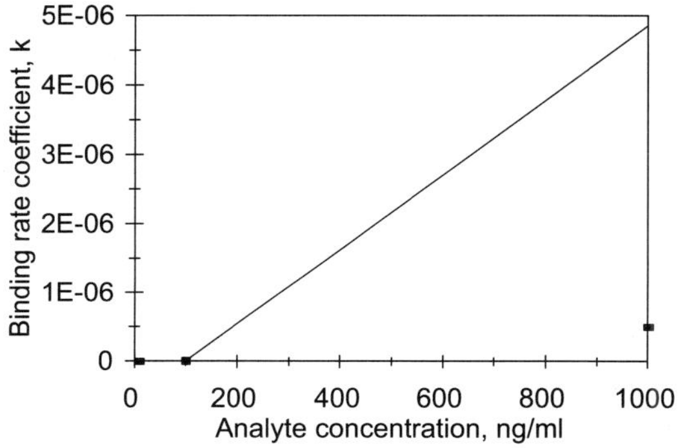

FIGURE 4.5 Increase in the binding rate coefficient, *k* with an increase in the analyte concentration.

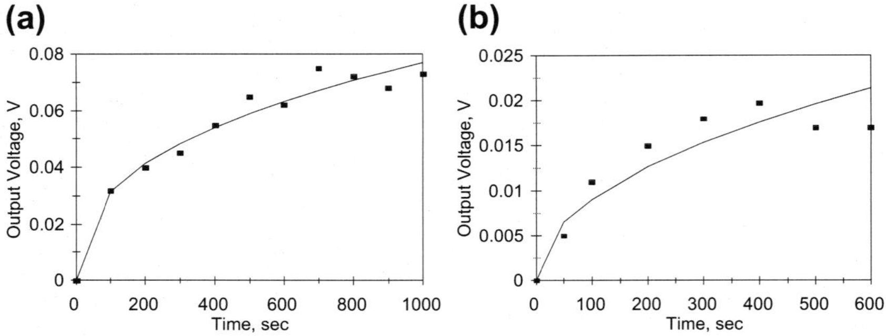

FIGURE 4.6 Binding of C-reactive protein (CRP) in solution to anti-CRP antibody immobilized on a microcantilever surface (Wee et al., 2005): (a) 1 μg/ml (b) 100 ng/ml.

Figure 4.6(a) shows the binding of 1 μg/ml of CRP concentration in solution to the anti-CRP antibody immobilized on a cantilever surface (Wee et al., 2005). The signal is output voltage due to CRP concentration. The binding may be described by a single-fractal analysis. The values of the binding rate coefficient, *k*, and the fractal dimension, D_f, are given in Table 4.5.

Figure 4.6(b) shows the binding of 100 ng/ml of CRP concentration in solution to the anti-CRP antibody immobilized on a cantilever surface (Wee et al., 2005). The signal is output voltage due to CRP concentration. The binding may be described by a single-fractal analysis. The values of the binding rate coefficient, *k*, and the fractal dimension, D_f, are given in Table 4.5.

Uludag and Tothill (2010) indicate that the WHO (2008) states that 13% of all the deaths worldwide in the year 2007 were caused by cancer. Ferlay et al. (1989) indicate that biosensor technology exhibits the potential to produce point-of-care devices to test for biomarkers. Brawer and Lange (1989) indicate that prostate cancer is multifactorial and complex disease and was responsible for 24.1% of all deaths to cancer in Europe. PSA is a good prognostic and diagnostic biomarker for this disease (Diamandis, 1998). Uludag and Tothill (2010) indicate that different types of biosensors have been applied for the

TABLE 4.5 Binding Rate Coefficients, *k*, and Fractal Dimensions, D_f, during the Real-Time Monitoring of the Cantilever Output Voltage at Different C-reactive protein (CRP) (Antigen) Concentration (Wee et al., 2005)

Analyte in Solution/Receptor on Cantilever Surface	k	D_f
1 μg/ml/anti-CRP	0.005329 ± 0.000388	2.268 ± 0.06382
100 ng/ml/anti-CRP	0.00101 ± 0.00025	0.9540 ± 0.1998

detection of PSA (Okuno et al., 2007; Panini et al., 2008; Sarkar et al., 2002; Cao et al., 2006; Choi et al., 2008; Schweitzer et al., 2000; Soukka et al., 2003; Ye et al., 2004; Hwang et al., 2004; Wee et al., 2005; Ding et al., 2008; Zhang et al., 2007).

Uludag and Tothill (2010) describe a simple and rapid detection method for cancer biomarkers using a quartz crystal microbalance (QCM) biosensor. The immunoassay method used was optimized to detect the cancer biomarkers in human serum. They minimized the matrix effects by analyzing the addition of detergent, salt, and other additives to the buffer solution. The authors used PSA and PSA-ACT complexes for the detection of cancer.

Figure 4.7(a) shows the binding and dissociation of 5000 ng/ml PSA in solution to monoclonal PSA detection antibody immobilized on a QCM biosensor surface (Uludag and Tothill, 2010). A dual-fractal analysis is required to adequately describe the binding kinetics. A single-fractal analysis is adequate to describe the dissociation kinetics. The values of (1) the binding rate coefficient, k, and the fractal dimension, D_f, for a single-fractal analysis and (2) the binding rate coefficients, k_1 and k_2, and the fractal dimensions, D_{f1} and D_{f2}, for a dual-fractal analysis are given in Table 4.6a and b. It is of interest to note that as the fractal dimension increases by a factor of 1.4281 from a value of $D_{f1} = 2.0464$ to $D_{f2} = 2.9226$, the binding rate coefficient increases by a factor of

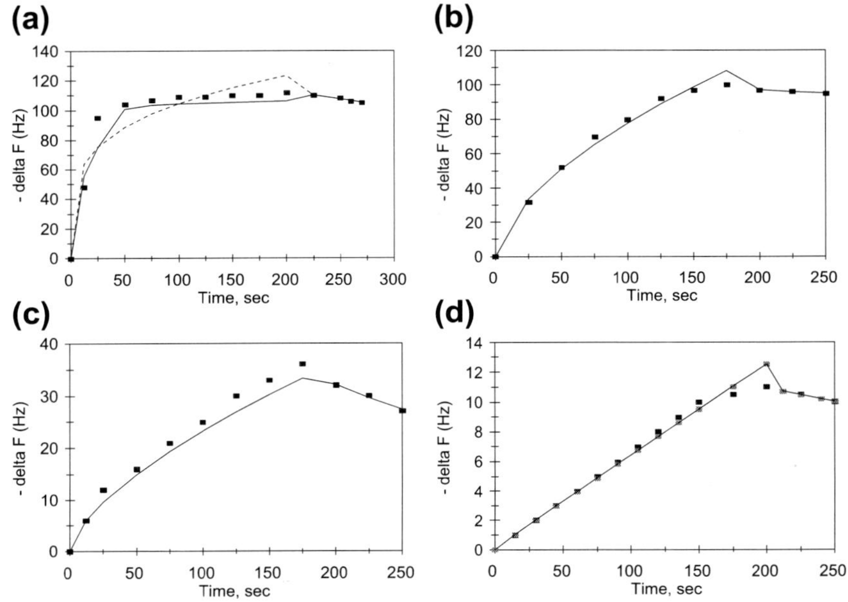

FIGURE 4.7 Binding of different concentrations of prostrate-specific antigen (PSA) in solution to PSA detection monoclonal antibody immobilized on a QCM biosensor surface. *Uludag and Tothill (2010):* (a) 5000 ng/ml (b) 312 ng/ml (c) 78 ng/ml (d) 18.8 ng/ml.

TABLE 4.6a Binding Rate Coefficients and Dissociation Rate Coefficients for Prostrate-Specific Antigen (PSA) in Solution to Monoclonal PSA Detection Antibody Immobilized on a QCM Biosensor Surface (Uludag and Tothill, 2010)

Analyte in Solution/Antibody on QCM Biosensor Surface	k	k_1	k_2	k_d
5000 ng/ml PSA/monoclonal PSA detection antibody	35.046 ± 6261	18.997 ± 4.554	90.684 ± 0.513	0.03843 ± 0.00403
312 ng/ml PSA/monoclonal PSA detection antibody	4.913 ± 0.293	–	–	0.6776 ± 0.0192
78 ng/ml PSA/monoclonal PSA detection antibody	1.3205 ± 0.0924	–	–	0.3820 ± 0.0342
18.8 ng/ml PSA/monoclonal PSA detection antibody	0.07724 ± 0.00440	–	–	0.03567 ± 0.0023

TABLE 4.6b Fractal Dimensions for Prostrate-Specific Antigen (PSA) in Solution to Monoclonal PSA Detection Antibody Immobilized on a QCM Biosensor Surface (Uludag and Tothill, 2010)

Analyte in Solution/Antibody on QCM Biosensor Surface	D_f	D_{f1}	D_{f2}	D_{fd}
5000 ng/ml PSA/monoclonal PSA detection antibody	2.5250 ± 0.12346261	2.0464 ± 0.3146	2.0226 ± 0.01382	0.5636 ± 0.2588
312 ng/ml PSA/monoclonal PSA detection antibody	1.7020 ± 0.0702	–	–	2.0810 ± 0.07114
78 ng/ml PSA/monoclonal PSA detection antibody	1.7140 ± 0.1004	–	–	1.5568 ± 0.2182
18.8 ng/ml PSA/monoclonal PSA detection antibody	1.0786 ± 0.04322	–	–	1.3146 ± 0.1145

4.77 from a value of $k_1 = 18.997$ to $k_2 = 90.684$. An increase in the fractal dimension or the degree of heterogeneity on the biosensor surface leads to an increase in the binding rate coefficient. In this case, the affinity values of K_1 ($=k_1/k_d$) and K_2 ($= k_2/k_d$) are 494.33 and 2359.7, respectively.

Figure 4.7(b) shows the binding of 312 ng/ml PSA in solution to monoclonal PSA detection antibody immobilized on a QCM biosensor surface (Uludag and Tothill, 2010). A single-fractal analysis is adequate to describe the binding kinetics. The values of the binding rate coefficient, k, and the fractal dimension, D_f, for a single-fractal analysis are given in Table 4.6a and b.

Figure 4.7(c) shows the binding of 78 ng/ml PSA in solution to monoclonal PSA detection antibody immobilized on a QCM biosensor surface (Uludag and Tothill, 2010). Once again, a single-fractal analysis is adequate to describe the binding kinetics. The values of the binding rate coefficient, k, and the fractal dimension, D_f, for a single-fractal analysis are given in Table 4.6a and b.

Figure 4.7(d) shows the binding of 18.8 ng/ml PSA in solution to monoclonal PSA detection antibody immobilized on a QCM biosensor surface (Uludag and Tothill, 2010). Once again, a single-fractal analysis is adequate to describe the binding kinetics. The values of the binding rate coefficient, k, and the fractal dimension, D_f, for a single-fractal analysis are given in Table 4.6a and b.

Figure 4.8(a) and Eqn (4.5a) show the increase in the binding rate coefficient, k, for a single-fractal analysis with an increase in the PSA concentration in solution in nanograms per milliliter. For the data shown in Figure 4.8(a), the binding rate coefficient, k, is given by

$$k = (0.001278 \pm 0.00105)[\text{PSA}]^{1.481\pm0.302} \tag{4.5a}$$

Only three data points are available. The availability of more data points would lead to a more reliable fit. The binding rate coefficient, k, exhibits close to a one and one-half (equal to 1.481) order of dependence on the PSA concentration in solution in the range studied.

Figure 4.8(b) and Eqn (4.5b) show the increase in the dissociation rate coefficient, k_d, for a single-fractal analysis with an increase in the PSA concentration in solution in nanograms per milliliter. For the data shown in Figure 4.8(b), the dissociation rate coefficient, k_d, is given by

$$k_d = (0.002183 \pm 0.00295)[\text{PSA}]^{1.050\pm0.361} \tag{4.5b}$$

Only three data points are available. The availability of more data points would lead to a more reliable fit. The dissociation rate coefficient, k_d, exhibits close to a first (equal to 1.050) order of dependence on the PSA concentration in solution in the range studied.

Figure 4.8(c) and Eqn (4.5c) show the increase in the binding rate coefficient, k, for a single-fractal analysis with an increase in the fractal dimension, D_f. For the data shown in Figure 4.8(c), the binding rate coefficient, k, is given by

$$k = (0.04401 + 0.0717)\left[D_f\right]^{7.57 \pm 2.576} \quad (4.5c)$$

There is scatter in the data. Only three data points are available. The availability of more data points would lead to a more reliable fit. The binding

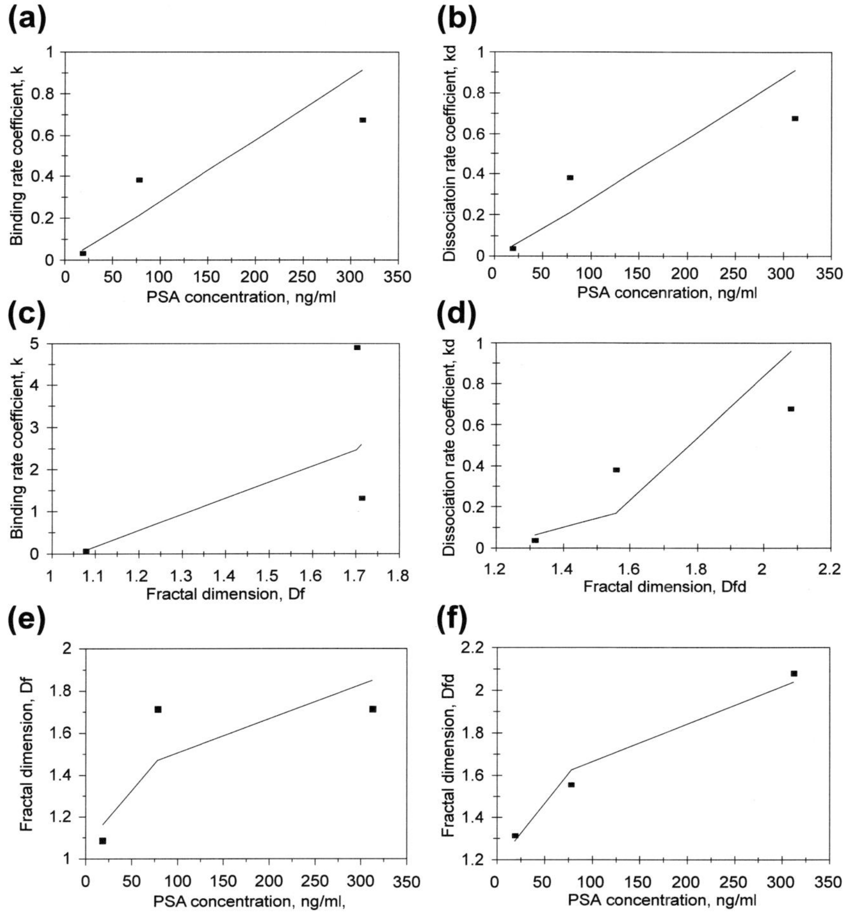

FIGURE 4.8 (a) Increase in the binding rate coefficient, k, with an increase in the prostrate-specific antigen (PSA) concentration in solution (b) Increase in the dissociation rate coefficient, k_d, with an increase in the PSA concentration in solution (c) Increase in the binding rate coefficient, k, with an increase in the fractal dimension, D_f (d) Increase in the dissociation rate coefficient, k_d, with an increase in the fractal dimension for dissociation, D_{fd} (e) Increase in the fractal dimension for binding, D_f, with an increase in the PSA concentration in solution (f) Increase in the fractal dimension for dissociation, D_{fd}, with an increase in the PSA concentration in solution.

rate coefficient, k, is very sensitive to the degree of heterogeneity on the biosensor surface, since it exhibits close to a seven and one-half (equal to 7.57) order of dependence on the fractal dimension, D_f, or the degree of heterogeneity on the surface.

Figure 4.8(d) and Eqn (4.5d) show the increase in the dissociation rate coefficient, k_d, for a single-fractal analysis with an increase in the fractal dimension, D_{fd}. For the data shown in Figure 4.8(d), the dissociation rate coefficient, k_d, is given by

$$k_d = (0.01197 + 0.02187)\left[D_f\right]^{5.928 \pm 3.162} \quad (4.5d)$$

Only three data points are available. The fit is good. The availability of more data points would lead to a more reliable fit. The dissociation rate coefficient, k_d, is very sensitive to the degree of heterogeneity on the biosensor surface, since it exhibits close to a sixth (equal to 5.928) order of dependence on the fractal dimension, D_f, or the degree of heterogeneity on the biosensor surface.

Figure 4.8(e) and Eqn (4.5e) show the increase in the fractal dimension, D_f, for a single-fractal analysis with an increase in the PSA concentration in solution. For the data shown in Figure 4.8(e), the fractal dimension, D_f, is given by

$$D_f = (0.7214 \pm 0.1506)[\mathrm{PSA}]^{0.163 \pm 0.093} \quad (4.5e)$$

Only three data points are available. The fit is quite good. The availability of more data points would lead to a more reliable fit. The fractal dimension, D_f, exhibits close to a zero (equal to 0.163) order of dependence on the PSA concentration in solution in the concentration range studied.

Figure 4.8(f) and Eqn (4.5f) show the increase in the fractal dimension in the dissociation phase, D_{fd}, for a single-fractal analysis with an increase in the PSA concentration in solution. For the data shown in Figure 4.8(f), the fractal dimension, D_{fd}, is given by

$$D_{fd} = (0.7973 \pm 0.0424)[\mathrm{PSA}]^{0.1633 \pm 0.0261} \quad (4.5f)$$

Only three data points are available. The fit is good. The availability of more data points would lead to a more reliable fit. The fractal dimension, D_{fd}, exhibits close to a zero (equal to 0.1633) order of dependence on the PSA concentration in solution in the concentration range studied.

Uludag and Tothill (2010) analyzed the effect of changing the buffer solution by using different additives to analyze the adsorption of sera proteins on the mouse IgG immobilized surface. Initially, a PBS/T buffer was injected on the surface. They noted that a significant amount of nonspecific binding occurred. Figure 4.9(a) shows the binding of human serum in PBS/T. A single-fractal analysis is adequate to describe the binding kinetics. The values of the binding rate coefficient, k, and the fractal dimension, D_f, for a single-fractal analysis are given in Table 4.6c and d.

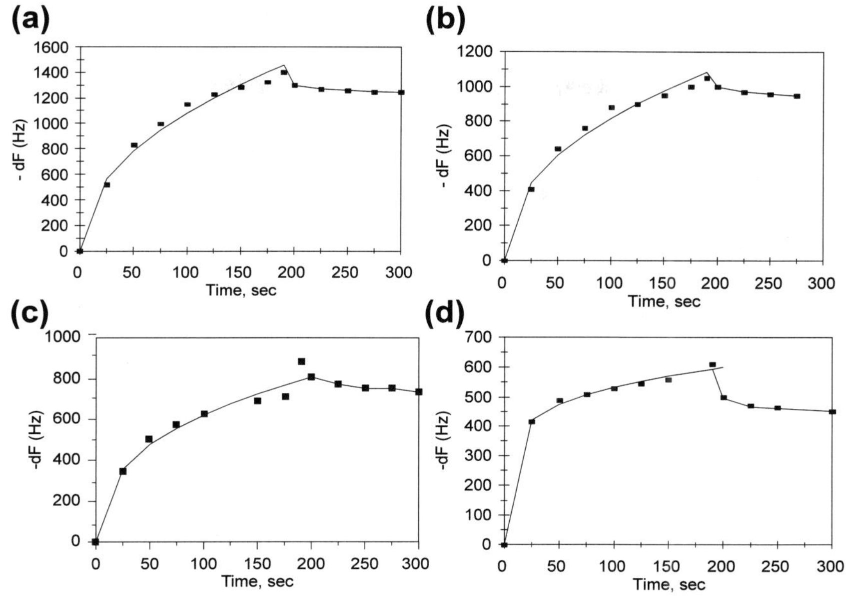

FIGURE 4.9 Binding and dissociation of the effect of different buffer solutions on the human serum signal (Uludag and Tothill, 2010): (a) PBS/T (b) PBS/T containing 100 μg/ml BSA (c) PBS/T containing 200 μg/ml BSA (d) PBS/T containing 200 μg/ml BSA and 0.5% Tween.

Uludag and Tothill (2010) also used different additives to minimize the influence of different additives. Figure 4.9(b) shows the influence of PBS/T containing 100 μg/ml BSA on the adsorption of sera proteins on the mouse IgG immobilized surface. Once again, a single-fractal analysis is adequate to describe the binding kinetics. The values of the binding rate coefficient, *k*,

TABLE 4.6c Binding and Dissociation Rate Coefficients for the Effect of Changing Buffer Solution on the Human Serum Signal (Uludag and Tothill, 2010)

Buffer Solution	k	k_d
PBS/T	125.91 ± 8.15	67.825 ± 1.849
PBS/T containing 100 μg/ml BSA	108.29 ± 6.79	24.117 ± 0.935
PBS/T containing 200 μg/ml BSA	105.11 ± 6.93	45.01 ± 1.80
PBS/T containing 200 μg/ml BSA and 0.5% Tween	243.19 ± 5.70	78.16 ± 2.262

TABLE 4.6d Fractal Dimensions for the Binding and Dissociation Phase for the Effect of Changing Buffer Solution on the Human Serum Signal (Uludag and Tothill, 2010)

Buffer Solution	D_f	D_{fd}
PBS/T	2.0662 ± 0.06814	2.6492 ± 0.02830
PBS/T containing 100 μg/ml BSA	2.1222 ± 0.0660	2.3516 ± 0.04684
PBS/T containing 200 μg/ml BSA	2.2288 ± 0.0650	2.4980 ± 0.0414
PBS/T containing 200 μg/ml BSA and 0.5% Tween	2.6588 ± 0.02702	2.6918 ± 0.02322

and the fractal dimension, D_f, for a single-fractal analysis are given in Table 4.6c and d.

Figure 4.9(c) shows the influence of PBS/T containing 200 μg/ml on the adsorption of sera proteins on the mouse IgG immobilized surface. Once again, a single-fractal analysis is adequate to describe the binding kinetics. The values of the binding rate coefficient, k, and the fractal dimension, D_f, for a single-fractal analysis are given in Table 4.6c and d.

Figure 4.9(d) shows the influence of PBS/T containing 200 μg/ml and 0.5% Tween on the adsorption of sera proteins on the mouse IgG immobilized surface. Once again, a single-fractal analysis is adequate to describe the binding kinetics. The values of the binding rate coefficient, k, and the fractal dimension, D_f, for a single-fractal analysis are given in Table 4.6c and d.

Figure 4.10(a) shows the increase in the binding rate coefficient, k, with an increase in the fractal dimension, D_f. For the data shown in Figure 4.10(a) and Table 4.6a, b, the binding rate coefficient, k, is given by

$$k = 11.02 \pm 2.66 D_f^{3.09 \pm 1.098} \tag{4.6a}$$

The fit is good. Only four data points are available. The availability of more data points would lead to a more reliable fit. The binding rate coefficient, k, is sensitive to the degree of heterogeneity on the biosensor surface or the fractal dimension, D_f, as it exhibits an order of dependence greater than three (equal to 3.09) on the fractal dimension on the biosensor surface.

Figure 4.10(b) shows the increase in the dissociation rate coefficient, k_d, with an increase in the fractal dimension, D_{fd}. For the data shown in Figure 4.10(b) and Table 4.6a, b, the dissociation rate coefficient, k_d, is given by

$$k_d = (0.01696 \pm 0.002) D_{fd}^{8.53 \pm 0.558} \tag{4.6b}$$

The fit is very good. Only four data points are available. The availability of more data points would lead to a more reliable fit. The dissociation rate coefficient, k_d, is very sensitive to the degree of heterogeneity on the biosensor

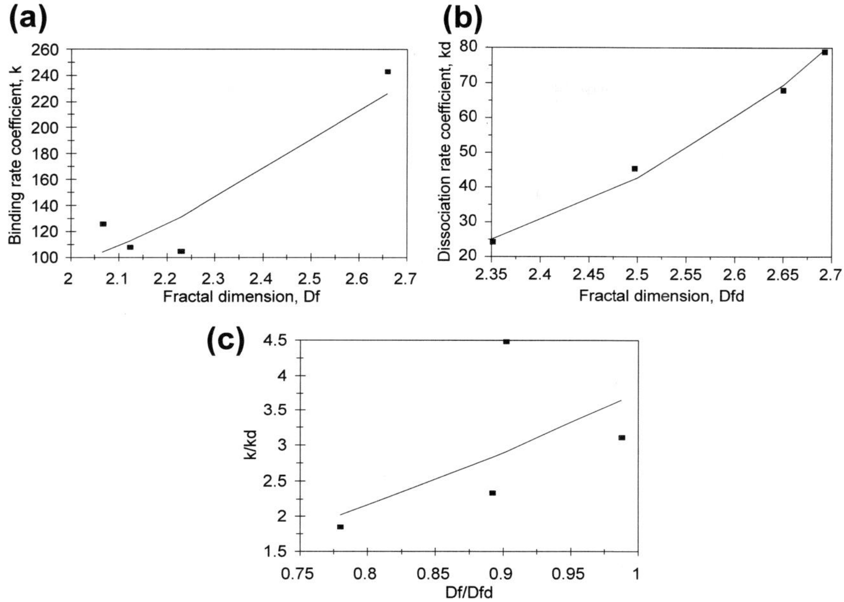

FIGURE 4.10 (a) Increase in the binding rate coefficient, k, with an increase in the fractal dimension, D_f (b) Increase in the dissociation rate coefficient, k_d, with an increase in the fractal dimension, D_{fd} (c) Increase in the affinity K (= k/k_d) with an increase in the fractal dimension ratio, D_f/D_{fd}.

surface or the fractal dimension, D_{fd}, as it exhibits an order of dependence greater than eight and one-half (equal to 8.53) on the fractal dimension in the dissociation phase, D_{fd} on the biosensor surface.

Figure 4.10(c) shows the increase in the affinity, K (= k/k_d), with an increase in the fractal dimension ratio, D_f/D_{fd}. For the data shown in Figure 4.10(c) and Table 4.6a, b, the affinity, K, is given by

$$K = (3.759 \pm 1.623)\left(D_f/D_{fd}\right)^{2.49 \pm 2.12} \tag{4.6c}$$

The fit is poor. Only four data points are available. The availability of more data points would lead to a more reliable fit. The affinity, K, is sensitive to the degree of heterogeneity on the biosensor surface or the ratio of the fractal dimensions, D_f/D_{fd}, as it exhibits an order of dependence close to two and one-half (equal to 2.49) on the ratio of fractal dimension on the biosensor surface.

Figure 4.11 shows the binding of 78 ng/ml spiked tPSA serum sample and tPSA antibody Au-nanoparticle on the surface using a direct assay (Uludag and Tothill, 2010). A dual-fractal analysis is required to adequately describe the binding kinetics. The values of the binding rate coefficient, k, and the fractal dimension, D_f, are given in Table 4.7a and b. It is of interest to note that as the

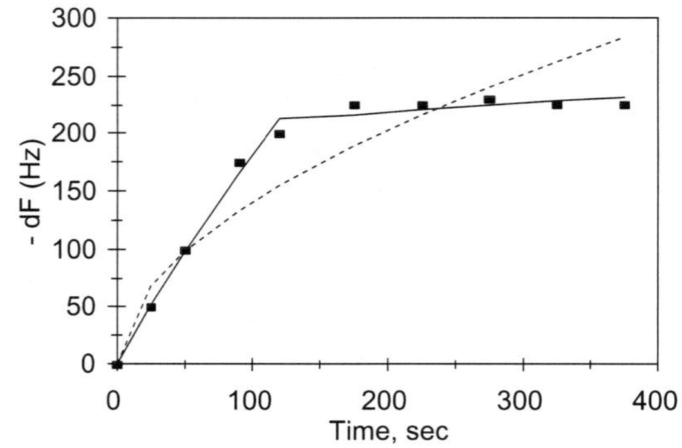

FIGURE 4.11 Binding of (Uludag and Tothill, 2010) spiked tPSA serum sample and tPSA antibody Au-nanoparticles on the surface using a direct assay.

fractal dimension increases by a factor of 1.91, and the binding rate coefficient, k, increases by a factor of 28.77 from a value of $k_1 = 3.649$ to $k_2 = 105$.

The binding of 9.4 μg/ml spiked tPSA serum sample and tPSA antibody Au-nanoparticle on the surface using a sandwich assay (Uludag and Tothill, 2010). No figure is shown. A dual-fractal analysis is required to adequately describe the binding kinetics. The values of the binding rate coefficient, k, and the fractal dimension, D_f, are given in Table 4.7 and b. It is of interest to note that as the fractal dimension increases by a factor of 1.32 from a value of $D_{f1} = 2.18$ to $D_{f2} = 2.88$, the binding rate coefficient, k, increases by a factor of 48.09 from a value of $k_1 = 2.79$ to $k_2 = 134.19$. An increase in the degree of heterogeneity on the sensor surface leads to an increase in the binding rate coefficient.

Chen et al. (2011) very recently indicate that biological processes are of importance in biomedical applications. They do point out that it is a challenge to convert the biological information into an electrical signal.

TABLE 4.7a Binding Rate Coefficient for (a) Spiked tPSA Serum Sample and tPSA Antibody Au-Nanoparticles on the Surface Using a Direct Assay and (b) Spiked tPSA Serum Sample and tPSA Antibody Au-Nanoparticle on an Active Surface Using a Sandwich Assay (Uludag and Tothill, 2010)

Analyte in Solution/ Type of Assay	k	k_1	k_2
78 ng/ml spiked 10% human serum on active surface + 1.5 μg/ml anti-PSA antibody/ direct assay	10.579 ± 3.437	3.649 ± 1.303	105
Injection of 9.4 μg/ml of spiked tPSA serum and tPSA-modified Au-nanoparticle on active surface/ sandwich assay	12.313 ± 3.044	2.7900 ± 0.2067	134.19 ± 4.98

TABLE 4.7b Fractal Dimensions for the Binding of (a) Spiked tPSA Serum Sample and tPSA Antibody Au-Nanoparticles on the Surface Using a Direct Assay and (b) Spiked tPSA Serum Sample and tPSA Antibody Au-Nanoparticle on an active Surface Using a Sandwich Assay (Uludag and Tothill, 2010)

Analyte in Solution/Type of Assay	D_f	D_{f1}	D_{f2}
78 ng/ml spiked 10% human serum on active surface + 1.5 μg/ml anti-PSA antibody/direct assay	2.1412 ± 0.2114	1.5680 ± 0.4080	3.0
Injection of 9.4 μg/ml of spiked tPSA serum and tPSA-modified Au-nanoparticle on active surface/sandwich assay	1.9404 ± 0.1712	2.1880 ± 0.1189	2.88162 ± 0.0774

Electrochemical biosensors have been used frequently to do this (Farre and Barceio, 2003; Wang et al., 2008). They have been used for medical diagnosis (Viswanathan et al., 2009; Rodriguez-Mozaz et al., 2008). indicate that the devices like the field-effect transistors (FETs) are effective for this application due to their ability to directly translate the interaction of the biological molecules with FET surface into a readable surface (Poghossian et al., 2007). IUPAC defines an "electrochemical sensor" as a self-contained device that allows for the specific analytical detection by using a biological element in direct contact with a transducer element (Thevenot et al., 2001; Grieshaber et al., 2008). Chen et al. (2011) mention that the electrochemical biosensor provides for an attractive platform, since it analyzes the contents of a biological sample by the direct conversion of the biological event to an electronic signal.

Chen et al. (2011) indicate that FETs are suitable candidates for designated sensors due to their ability to translate the interaction with target molecules taking place on the FET into a readable signal (Poghossian et al., 2007). Recently, semiconducting nanomaterials such as silicon nanowires and CNTs, configured with FETs (otherwise known as SINW-FET (Patolsky et al., 2006; Park et al., 2007) and CNTFET (Gruner, 2006; Katz and Willner, 2004; Chen et al., 2004)) have attracted much attention, since when they are used as a biosensor, they exhibit high selectivity, sensitivity, and real-time response time, and permit label-free detection.

Figure 4.12 shows the binding of CaM (calmodulin)-GST (glutathione-S-transferase, his-tag) in solution to GST-SINW-FET (silicon nanowire

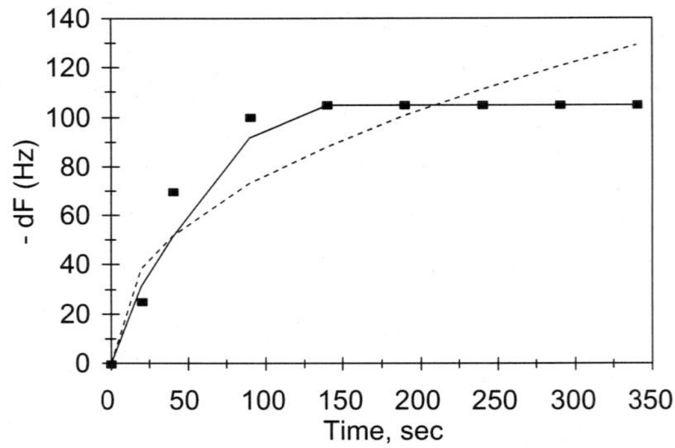

FIGURE 4.12 Binding of CaM-GST in solution to GST-SINW-FET.

field-effect transistors). A dual-fractal analysis is required to adequately describe the binding kinetics. The values of (1) the binding rate coefficient, k, and the fractal dimension, D_f, for a single-fractal analysis and (2) the binding rate coefficients, k_1 and k_2, and the fractal dimensions, D_{f1} and D_{f2}, are given in Table 4.8a and b. It is of interest to note that as the fractal dimension increases by a factor of 1.91 from a value of D_{f1} =1.5680 to D_{f2} = 3.0, the binding rate coefficient increases by a factor of 1.7 from $k_1 = 6.357$ to $k_2 = 10.81$.

Figure 4.13 shows the binding of 30 μg/ml cancer biomarker CA15-3 to anti-CA15-3-functionalized SINW-FET array. A single-fractal analysis is

TABLE 4.8a Binding (Electrical Conductance (nS)) of (a) CaM-GST with a GSH/SINW-FET in 0.1 M PSA Solution Supplemented with 0.5 mM EDTA, pH 7.4, and (b) Detection of Multiple Cancer Biomarkers (PSA, CEA, Mucin-Q!) with SINW-FET Arrays (Chen et al., 2011)

Analyte in Solution/Receptor on Surface	k	k_1	k_2
CaM-GST/GSH-SINW-FET	6.6554 ± 2.074	6.357 ± 2.352	10.81 ± 1.654
30 μg/ml cancer marker CA15-3/ anti-CA15-3-functionalized SINW-FET in an MPC-purified blood sample	0.000737 ± 0.000077	–	–
2.5 ng/ml PSA blood sample/anti-PSA-functionalized SINW-FET	6.6 E−05 ±0.7 E−05	–	–

TABLE 4.8b Fractal Dimensions for Binding (Electrical Conductance (nS)) of (a) CaM-GST with a GSH/SINW-FET in 0.1 M PSA Solution Supplemented with 0.5 mM EDTA, pH 7.4, and (b) Detection of Multiple Cancer Biomarkers (PSA, CEA, Mucin-Q!) with SINW-FET Arrays (Chen et al., 2011)

Analyte in Solution/Receptor on Surface	D_f	D_{f1}	D_{f2}
CaM-GST/GSH-SINW-FET	2.1412 ± 0.2114	1.5608 ± 0.4080	3.0 − 2.2E−14
30 μg/ml cancer marker CA15-3/anti-CA15-3-functionalized SINW-FET in an MPC-purified blood sample	0.000737 ± 0.000077	−	−
2.5 ng/ml PSA blood sample/ anti-PSA-functionalized SINW-FET	6.6 E−05 ± 0.7 E−05	−	−

adequate to describe the binding kinetics. The values of the binding rate coefficient, k, and the fractal dimension, D_f, for a single-fractal analysis are given in Table 4.8a and b.

Figure 4.14 shows the binding of 2.5 ng/ml PSA blood sample to anti-PSA-functionalized SINW-FET array. A single-fractal analysis is adequate to describe the binding kinetics. The values of the binding rate coefficient, k, and the fractal dimension, D_f, for a single-fractal analysis are given in Table 4.8a and b.

Figure 4.15(a) shows the binding of 0.9 ng/ml PSA in solution to anti-PSA antibody immobilized on a SINW-FET array. A dual-fractal analysis is required to adequately describe the binding kinetics. The values of (1) the binding rate coefficient, k, and the fractal dimension, D_f, for a single-fractal

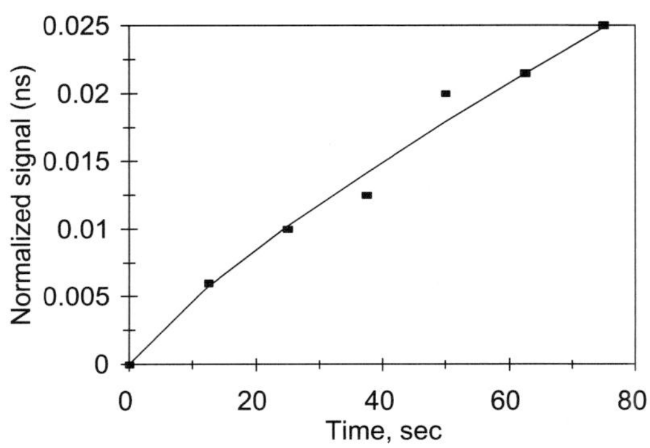

FIGURE 4.13 Binding of 30 μg/ml cancer biomarker CA15-3 to anti-CA15-3-functionalized SINW-FET array in an MPC-purified sample.

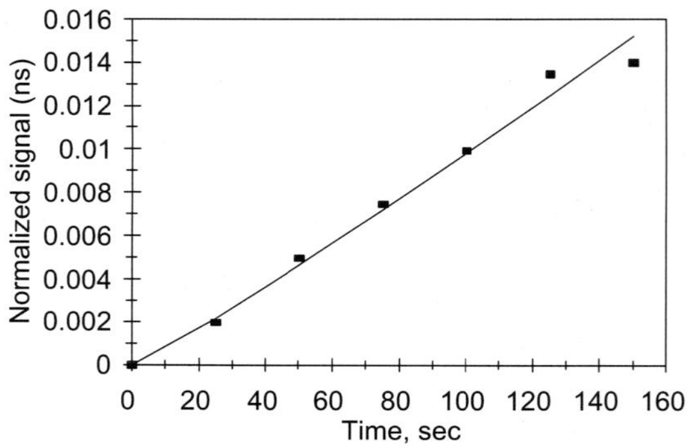

FIGURE 4.14 Binding of 2.5 ng/ml prostrate-specific antigen (PSA) blood sample to anti-PSA-functionalized SINW-FET array (Chen et al., 2011).

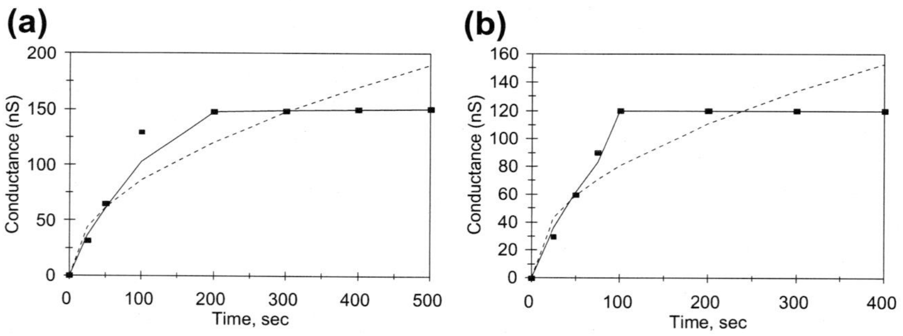

FIGURE 4.15 Binding of (a) 0.9 ng/ml of prostrate-specific antigen (PSA) in solution to anti-PSA immobilized on a SINW-FET array (b) CEA in solution to anti-CEA immobilized on a SINW-FET array (Chen et al., 2011).

analysis and (b) the binding rate coefficients, k_1 and k_2, and the fractal dimensions, D_{f1} and D_{f2}, for a dual-fractal analysis are given in Table 4.9a and b. It is of interest to note that as the fractal dimension increases by a factor of 2.0 from 1.4786 to 2.9567 for a dual-fractal analysis, the binding rate coefficient increases by a factor of 42.7 from $k_1 = 3.088$ to $k_2 = 131.85$. An increase in the fractal dimension on the surface or an increase in the degree of heterogeneity on the surface leads to an increase in the binding rate coefficient.

Figure 4.15(b) shows the binding of CEA in solution to anti-CEA antibody immobilized on a SINW-FET array. A dual-fractal analysis is required to adequately describe the binding kinetics. The values of (1) the binding rate coefficient, k, and the fractal dimension, D_f, for a single-fractal analysis and (2) the binding rate coefficients, k_1 and k_2, and the fractal dimensions,

TABLE 4.9a Binding Rate Coefficients for (a) 0.9 ng/ml Prostrate-Specific Antigen (PSA) to Anti-PSA Immobilized on SINW-FET Array, and (b) CEA to Anti-CEA Immobilized on SINW-FET Array (Chen et al., 2011)

Analyte in Solution/Receptor SINW-FET	k	k_1	k_2
0.9 ng/ml PSA/anti-PSA antibody	9.069 ± 2.867	3.088 ± 0.768	131.85 ± 0.019
CEA/anti-CEA antibody	9.706 ± 3.317	1.2 ± 0	120 ± 0

TABLE 4.9b Fractal Dimensions for the Binding of (a) 0.9 ng/ml Prostrate-Specific Antigen (PSA) to Anti-PSA Immobilized on SINW-FET Array, and (b) CEA to Anti-CEA Immobilized on SINW-FET Array (Chen et al., 2011)

Analyte in Solution/Receptor SINW-FET	D_f	D_{f1}	D_{f2}
0.9 ng/ml PSA/anti-PSA antibody	1.012 ± 0.19992	1.4786 ± 0.2868	2.9561 ± 0.0041
CEA/anti-CEA antibody	2.0804 ± 0.2354	$1.0 \pm 3.4E-14$	3.0 ± 0

D_{f1} and D_{f2}, for a dual-fractal analysis are given in Table 4.9a and b. It is of interest to note that as the fractal dimension increases by factor of 3.0 from a value of $D_{f1} = 1.0$ to $D_{f2} = 3.0$ for a dual-fractal analysis, the binding rate coefficient increases by a factor of 100 from $k_1 = 1.2$ to $k_2 = 120$. An increase in the fractal dimension on the surface or an increase in the degree of heterogeneity on the surface leads to an increase in the binding rate coefficient.

Chen et al. (2011) conclude by stating that SINW-FET sensors should find useful applications in monitoring protein–protein interactions, for example, antibody–antigen interactions. To make these types of biosensors more practical, they should be reusable. These authors suggest that one should apply a GSH/GST-tag or Ni ++/His-tag reversible system to the SINW-FET surface, which would make the SINW-FET biosensor reusable.

Chen et al. (2011) indicate that SINW-FETs may be applied for the detection of specific DNA by PNA-DNA hybridization. Finally, several advantages of SINW-FETs are that these are direct, label-free, and real-time electrical devices that exhibit ultrahigh sensitivity, excellent selectivity, and potential for use in multiple sensing arrays.

Ding et al. (2012) recently indicate that during cell proliferation, telomeres are progressively shortened, reaching a critical state wherein cell division stops and gets into senescence. Ding et al. (2012) indicate that telomeres are nucleic acids of constant repeat sequences that cap the ends of chromosomes. In most malignant cancer cells, the ribonuclease telomerase is accumulated. This results in the continuous elongation of the telomeres and the generation of immortal cells (Morin et al., 1989; Cohen et al., 2007; Stone et al., 2007; Harley and Villeponteau, 1995). Telomerase has been observed in 85% of all known tumors and is regarded as a biomarker for early-stage cancer diagnosis as well as a therapeutic target (Shay and Bachetti, 1997; Hiyama and Hiyama, 2002). Ding et al. (2012) indicate that a variety of techniques have been proposed to analyze telomerase activity (Counter et al., 1992; Gelmini et al., 1998; Hou et al., 2001; Tatematsu et al., 1996; Nemos et al., 2003).

Figure 4.16 shows the binding of telomerase activity in cancer cells by a displacement polymerization reaction to a biosensor surface (Ding et al., 2012). A dual-fractal analysis is required to adequately describe the binding kinetics. The values of (1) the binding rate coefficient, k, and the fractal dimension, D_f, for a single-fractal analysis and (2) the binding rate coefficients, k_1 and k_2, and the fractal dimensions, D_{f1} and D_{f2}, for dual-fractal analysis are given in Table 4.10a and b. It is of interest to note for a dual-fractal analysis, as the fractal dimension increases by a factor of 1.91 from a value of $D_{f1} = 1.5046$ to $D_{f2} = 2.8765$, the binding rate coefficient increases by a factor of 1.58 from a value of $k_1 = 1568.88$ to $k_2 = 2475.49$. An increase in the degree of heterogeneity on the biosensor surface, once again, leads to an increase in the binding rate coefficient.

Figure 4.17 shows the binding of PSA in solution to anti-PSA antibody immobilized on a biosensor surface (Li et al., 2005). The signal is measured in

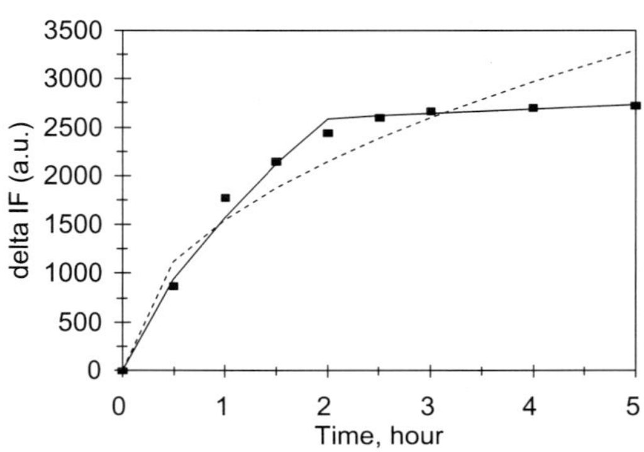

FIGURE 4.16 Binding of telomerase activity in cancer cells by a displacement polymerization reaction. *Ding et al. (2012).*

TABLE 4.10a Binding Rate Coefficients for (a) Telomerase Activity in Cancer Cells by Displacement Polymerization Reaction (Ding et al., 2012), (b) Prostate-Specific Antigen (PSA) in Solution to In_2O_3 (Li et al., 2005), (c) Interleukin-2 (IL-2) to Antibody IL-2 in a Sandwich Immunoassay (Luchansky and Bailey, 2011), and (d) IL-2 to Antibody Using a Sandwich Assay Using a Secondary Antibody for Added Specificity and Amplify the Primary Signal (Luchansky and Bailey, 2011)

Analyte in Solution/ Receptor on Surface	k	k_1	k_2	k_d	References
(a) Telomerase activity in cancer cells/ displacement polymerization reaction	1549.71 ± 284.37	1568.88 ± 138.3	2475.49 ± 72.42	NA	Ding et al. (2012)
(b) PSA/$In_2O_3^-$ nanowires	0.2049 ± 0.0287	NA	NA	0.396 ± 0.0278	Li et al. (2005)
(c) IL-2/IL-2 antibody	0.4738 ± 0.0186	NA	NA	NA	Luchansky and Bailey (2011)
(d) IL-2/IL-2/secondary antibody for added specificity	7.864 ± 0.843	NA	NA	NA	Luchansky and Bailey (2011)

TABLE 4.10b Fractal Dimensions for the Binding of (a) Telomerase Activity in Cancer Cells by Displacement Polymerization Reaction (Ding et al., 2012), (b) Prostate-Specific Antigen (PSA) in Solution to In_2O_3 (Li et al., 2005), (c) Interleukin (IL-2) to Antibody IL-2 in a Sandwich Immunoassay (Luchansky and Bailey, 2011), and (d) IL-2 to Antibody Using a Sandwich Assay Using a Secondary Antibody for Added Specificity and Amplify the Primary Signal (Luchansky and Bailey, 2011)

Analyte in Solution/Receptor on Surface	D_f	D_{f1}	D_{f2}	D_{fd}	References
(a) Telomerase activity in cancer cells/displacement polymerization reaction	2.0632 ± 0.680	1.5046 ± 0.2176	2.8765 ± 0.03394	NA	Ding et al. (2012)
(b) PSA/$In_2O_3^-$ nanowire	1.5512 ± 0.1918	NA	NA	1.784 ± 0.620	Li et al. (2005)
(c) IL-2/IL-2 antibody	2.0264 ± 0.0430	NA	NA	NA	Luchansky and Bailey (2011)
(d) IL-2/IL-2/secondary antibody for added specificity	1.8618 ± 0.1189	NA	NA	NA	Luchansky and Bailey (2011)

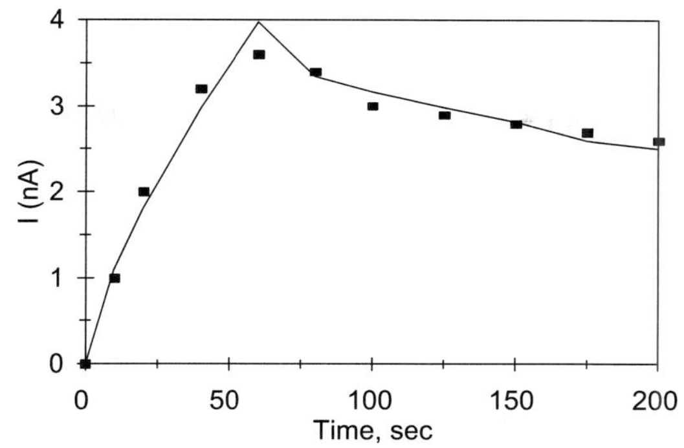

FIGURE 4.17 Binding of prostate-specific antigen (PSA) in solution to anti-PSA antibody immobilized on an In_2O_3 nanowire (biosensor) surface. *Li et al. (2005).*

nanoamps. A single-fractal analysis is required to adequately describe the binding kinetics. The values of (1) the binding rate coefficient, *k*, and the fractal dimension, D_f, for single-fractal analysis and (b) the dissociation rate coefficients, k_d, and the fractal dimension, for dissociation, D_{fd}, are given in Table 4.10a and b.

Luchansky and Bailey (2011) indicate that the incorporation of multiple sensors onto a single device permits the detection of multiple analytes by multiplexing sensors. These authors indicate that microring resonators permit sensor multiplexing and examples have appeared in the literature (Washburn et al., 2009). These can be monitored simultaneously using separate access waveguide to each ring. However, due to practical reasons, configurations with one sensor per waveguide are most common. Luchansky and Bailey (2011) emphasize that the integration of optical components, high-Q sensors, and microfluidics into a low-cost and automate device is a reasonable goal to achieve for medical devices, environmental monitoring, and homeland security applications (Jokerst et al., 2009; Monat et al., 2007).

Figure 4.18(a) shows the binding of IL-2 in solution to anti-IL-2 antibody immobilized on a biosensor surface (Luchansky and Bailey, 2011). A single-fractal analysis is adequate to describe the binding kinetics. The values of the binding rate coefficient, *k*, and the fractal dimension, D_f, for a single-fractal analysis are given in Table 4.10a and b.

Figure 4.18(b) shows the binding of IL-2 in solution to anti-IL-2 antibody including a secondary antibody for added specificity immobilized on a biosensor surface (Luchansky and Bailey, 2011). A single-fractal analysis is once again adequate to describe the binding kinetics. The values of the binding rate coefficient, *k*, and the fractal dimension, D_f, for a single-fractal analysis are given in Table 4.10a and b. The binding rate coefficient increases by a factor of 16.6 due to the added secondary antibody compared to when it is not added.

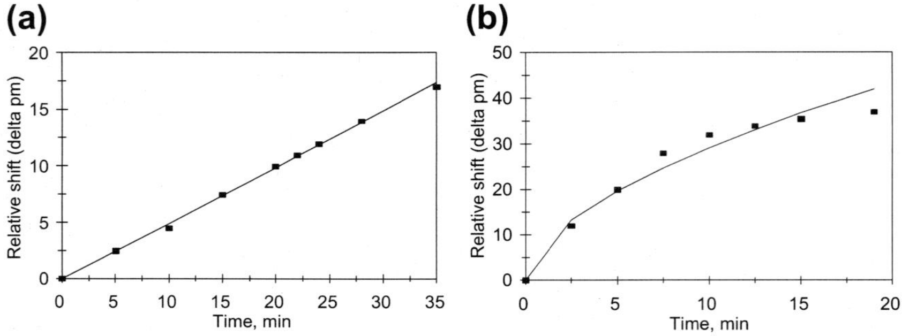

FIGURE 4.18 (a) Binding of interleukin-2 (IL-2) in solution to IL-2 antibody immobilized on a biosensor surface (Luchansky and Bailey, 2011) (b) Binding of IL-2 in solution to anti-IL-2 antibody and a secondary antibody for added specificity. *Luchansky and Bailey (2011).*

Washburn et al. (2010) attempted to detect different concentrations of PSA in solution to anti-PSA immobilized on a microring resonator array platform. The antibodies screened in parallel PSA from multiple vendors. In previous detection studies work, Washburn et al. (2010) have detected the five cancer biomarkers simultaneously—CEA, α-fetoprotein (AFP), PSA, tumor necrosis factor, and interleukin-8 (IL-8)—from the same solution using microring resonators. They demonstrated the ability to design a useful assay for multiple protein detection with limited cross-reactivity and no loss of sensitivity or precision as compared to single-parameter assays.

Figure 4.19(a) shows the binding and dissociation of 20 ng/ml PSA in solution to anti-PSA antibody immobilized to the microring array platform (Luchansky and Bailey, 2011). A single-fractal analysis is adequate to describe the binding and the dissociation kinetics. The values of (1) the binding rate coefficient, k, and the fractal dimension, D_f, for a single-fractal analysis and (2) the dissociation rate coefficient, k_d, and the fractal dimension for dissociation, D_{fd}, for a single-fractal analysis are given in Table 4.11. The affinity, K ($=k/k_d$), value is equal to 4.18.

Figure 4.19(b) shows the binding and dissociation of 50 ng/ml PSA in solution to anti-PSA antibody immobilized to the microring array platform (Luchansky and Bailey, 2011). A single-fractal analysis is, once again, adequate to describe the binding and the dissociation kinetics. The values of (1) the binding rate coefficient, k, and the fractal dimension, D_f, for a single-fractal analysis and (2) the dissociation rate coefficient, k_d, and the fractal dimension for dissociation, D_{fd}, for a single-fractal analysis are given in Table 4.11. The affinity, K ($=k/k_d$), value is equal to 1.644.

Figure 4.19(c) shows the binding and dissociation of 100 ng/ml PSA in solution to anti-PSA antibody immobilized to the microring array platform (Luchansky and Bailey, 2011). A single-fractal analysis is, once again, adequate to describe the binding and the dissociation kinetics. The values of (1) the binding rate coefficient, k, and the fractal dimension, D_f, for a

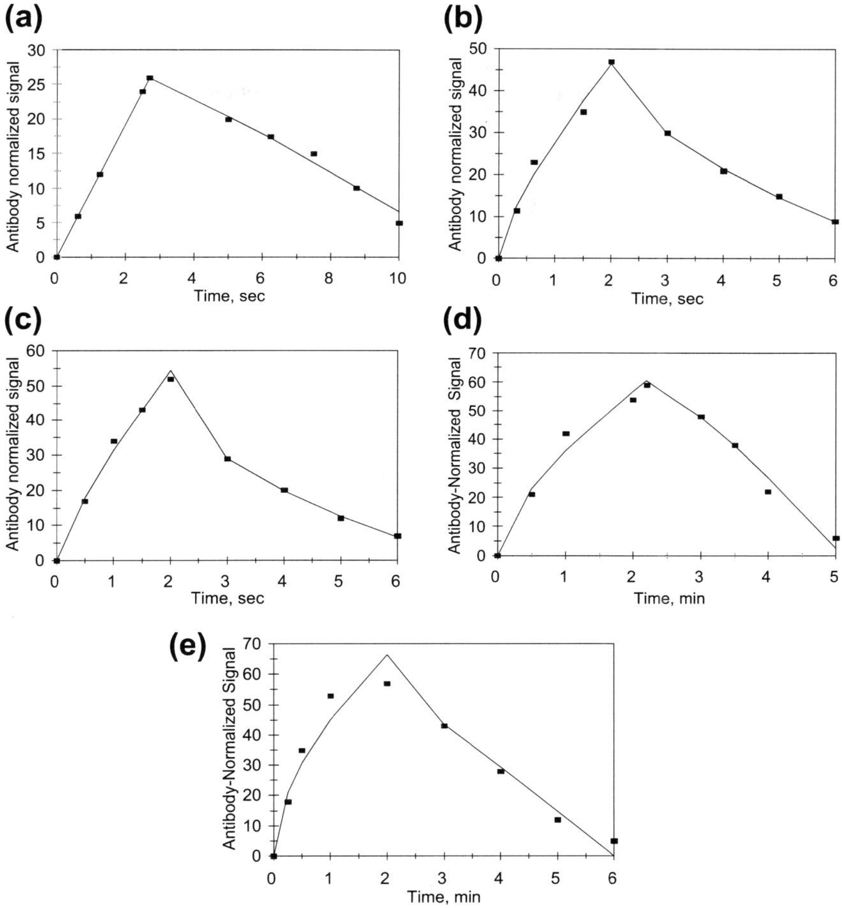

FIGURE 4.19 Binding of different concentrations of prostrate-specific antigen (PSA) in solution (in ng/ml) to anti-PSA on a microarray platform (Luchansky and Bailey, 2011): (a) 20 (b) 50 (c) 100 (d) 500 (e) 1000.

single-fractal analysis and (2) the dissociation rate coefficient, k_d, and the fractal dimension for dissociation, D_{fd}, for a single-fractal analysis are given in Table 4.11. The affinity, K ($=k/k_d$), value is equal to 1.35.

Figure 4.19(d) shows the binding and dissociation of 500 ng/ml PSA in solution to anti-PSA antibody immobilized to the microring array platform (Luchansky and Bailey, 2011). A single-fractal analysis is, once again, adequate to describe the binding and the dissociation kinetics. The values of (1) the binding rate coefficient, k, and the fractal dimension, D_f, for a single-fractal analysis and (2) the dissociation rate coefficient, k_d, and the fractal dimension for dissociation, D_{fd}, for a single-fractal analysis are given in Table 4.11. The affinity, K ($=k/k_d$), value is equal to 2.37.

TABLE 4.11 Binding Rate Coefficient for Different Concentrations of Prostrate-Specific Antigen (PSA) in Solution to Anti-PSA Immobilized on a Microring Resonator Array Platform. Antibodies Screened in Parallel for Antibodies from Multiple Vendors (Washburn et al., 2010)

Analyte in Solution/Receptor on Surface	k	k_d	D_f	D_{fd}
20 ng/ml/anti-PSA	9.641 ± 0.032	2.305 ± 0.224	1.008 ± 0.0056	0.850 ± 0.026
50 ng/ml/anti-PSA	28.16 ± 3.62	17.123	1.5594 ± 0.1686	1.8466 ± 0.02872
100 ng/ml/anti-PSA	31.073 ± 2.502	22.973 ± 0.319	1.3886 ± 0.1488	2.0188 ± 0.06644
500 ng/ml/anti-PSA	35.941 ± 4.930	15.144 ± 1.523	1.6788 ± 0.2176	0.7734 ± 0.2092
1000 ng/ml/anti-PSA	45.078 ± 10.528	13.481 ± 1.820	1.8826 ± 0.2708	0.92332 ± 0.3212

Figure 4.19(e) shows the binding and dissociation of 1000 ng/ml PSA in solution to anti-PSA antibody immobilized to the microring array platform (Luchansky and Bailey, 2011). A single-fractal analysis is, once again, adequate to describe the binding and the dissociation kinetics. The values of (1) the binding rate coefficient, k, and the fractal dimension, D_f, for a single-fractal analysis and (2) the dissociation rate coefficient, k_d, and the fractal dimension for dissociation, D_{fd}, for a single-fractal analysis are given in Table 4.11. The affinity, K ($=k/k_d$), value is equal to 3.34.

Figure 4.20(a) shows the increase in the binding rate coefficient, k, with an increase in the PSA in solution in the concentration range 10−1000 ng/ml. For the data presented in Figure 4.20(a), the binding rate coefficient, k, is given by

$$k = (6.264 \pm 2.078)[\text{PSA}]^{0.3117 \pm 0.0871} \qquad (4.7a)$$

The fit is good. Only five data points are available. The availability of more data points would lead to a more reliable fit. The binding rate coefficient, k, exhibits less than one-half (equal to 0.3117) order of dependence on the PSA concentration in solution.

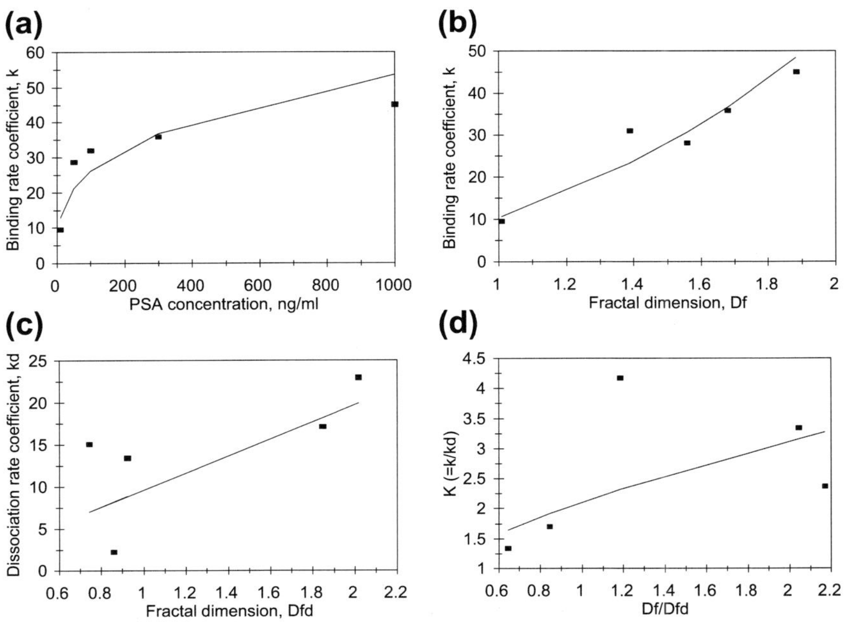

FIGURE 4.20 (a) Increase in the binding rate coefficient, k, with an increase in the prostrate-specific antigen (PSA) concentration in solution (b) Increase in the binding rate coefficient, k, with an increase in the fractal dimension, D_f (c) Increase in the dissociation rate coefficient, k_d, with an increase in the PSA concentration in solution (d) Increase in affinity, K ($= k/k_d$) with an increase in the ratio of fractal dimensions, D_f/D_{fd}.

Figure 4.20(b) shows the increase in the binding rate coefficient, k, with an increase in the fractal dimension, D_f, for the PSA in solution in the concentration range 10−1000 ng/ml. For the data presented in Figure 4.20(b), the binding rate coefficient, k, is given by

$$k = (10.565 \pm 2.221)D_f^{2.403\pm0.3995} \tag{4.7b}$$

The fit is good. Only five data points are available. The availability of more data points would lead to a more reliable fit. The binding rate coefficient, k, exhibits an order of dependence close to two and one-half (equal to 2.403).

Figure 4.20(c) shows the increase in the dissociation rate coefficient, k_d, with an increase in the fractal dimension, D_{fd}, for the PSA in solution in the concentration range 10−1000 ng/ml. For the data presented in Figure 4.20(c), the dissociation rate coefficient, k_d, is given by

$$k_d = (9.610 + 13.684)[\text{PSA}]^{1.036\pm0.961} \tag{4.7c}$$

There is scatter in the data. This is reflected in the error in the dissociation rate coefficient. Only five data points are available. The availability of more data points would lead to a more reliable fit. The binding rate coefficient, k, exhibits close to first order (equal to 1.036) order of dependence on fractal dimension for dissociation, D_{fd}.

Figure 4.20(d) shows the increase in the affinity, K (=k/k_d), with an increase in the ratio of the fractal dimensions, D_f/D_{fd}. For the data shown in Figure 4.20(d), the affinity, K, is given by

$$K(k/k_d) = (2.114 \pm 1.071)\left(D_f/D_{fd}\right)^{0.562\pm0.384} \tag{4.8d}$$

There is scatter in the data. Only five points are available. The availability of more data points would lead to a more reliable fit. The affinity, K, exhibits close to one-half (equal to 0.562) order of dependence on the ratio of fractal dimensions for the binding and the dissociation phase.

Sipova et al. (2012) indicate that nanotechnology advances have facilitated the probing of molecular events at the micro/nanoscale (Cui et al., 2001; Fritz et al., 2000). The detection of individual binding and dissociation is now feasible (Mayer et al., 2010; Sanomiya et al., 2008). An important aspect of the analysis is the transport of the analyte to the sensor surface (Sheehan and Whitman, 2005; Squires et al., 2008). Sipova et al. (2012) have recently analyzed bioanalytical applications. They compared the bioanalytical capabilities of flow-through surface plasmon resonance (SPR) sensors based on (1) localized surface plasmon registering a single-binding event and (2) propagating surface plasmon producing the sensor response by integrating the signal from a number of molecular interactions.

Asati et al. (2011) have been able to detect cancer biomarkers at neutral pH by the pH-tunable oxidase activity of cerium oxide nanoparticles and have achieved sensitive fluorogenic detection. These authors indicate that conventional biomarker assay methods such as ELISA use labile components such as hydrogen peroxide and horseradish peroxidase, which is a limitation. They indicate that nanoceria can oxidize the nonfluorescent substrate ampliflu either to a very stable fluorescent product resorcin at pH 7.0 or to the nonfluorescent resazurin at pH 4.0. They conjugated Protein G to immobilized antibodies on the surface of nanoceria and were able to detect the expression of prototypic cancer biomarkers at pH 7.0 such as folate receptor and EpCAM. They noted that within 3 h, they were able to identify the expressions of the folate receptor and EpCAM on lung cancer and breast adenocarcinoma cells, respectively. ELISA, on the other hand, had a detection time of 15 h, a higher threshold and required multiple washing steps.

In recent years, nanomaterials have been used to create sensitive and robust biosensors (Asati et al., 2011; Mani et al., 2007; Smith et al., 2007; Taton et al., 2000; Nam et al., 2003; Alivisatos, 2004; Ambrosi et al., 2010; Chikkaveeraiah et al., 2009). Asati et al. (2011) indicate that one benefit of using nanoceria in ELISA applications is that no hydrogen peroxide is required to facilitate the oxidation of dye (Asati et al., 2009). There is a drawback here in that the nanoceria-based ELISA method for the oxidation of the dye (TMB) is optimal in acidic conditions. This limits the use of antibodies and other pH-labile biomolecules as targeting ligands. Asati et al. (2011) emphasize that the application and detection at neutral pH would significantly enhance the applicability of the method.

Asati et al. (2011) analyzed the binding of ampliflu by fluorometric analysis of nanoceria-mediated oxidation at different pH values. Figure 4.21(a) shows the binding of ampliflu and its analysis by fluorometry at pH 6.0. A single-fractal analysis is adequate to describe the binding kinetics. The values of the binding rate coefficient, k, and the fractal dimension, D_f, are given in Table 4.12a and b.

Figure 4.21(b) shows the binding of ampliflu and its analysis by fluorometry at pH 7.0. A single-fractal analysis is, once again, adequate to describe the binding kinetics. The values of the binding rate coefficient, k, and the fractal dimension, D_f, are given in Table 4.12a and b. It is of interest to note that as the pH changes from acidic (pH 6.0) to neutral (pH 7.0), the fractal dimension increases by a factor of 1.41 from a value of $D_f = 0.526$ to $D_f = 0.7412$ and the binding rate coefficient, k, increases by a factor of 1.92 from a value of $k = 2868.82$ to $k = 5495.66$.

Figure 4.21(c) shows the binding of ampliflu and its analysis by fluorometry at pH 8.0. A dual-fractal analysis is now required to describe the binding kinetics. The values of (1) the binding rate coefficient, k, and the fractal dimension, D_f, for a single-fractal analysis and (2) the values of the binding rate coefficients, k_1 and k_2, and the fractal dimensions, D_{f1} and D_{f2}, are given in Table 4.12a and b. It is of interest to note that the estimated fractal

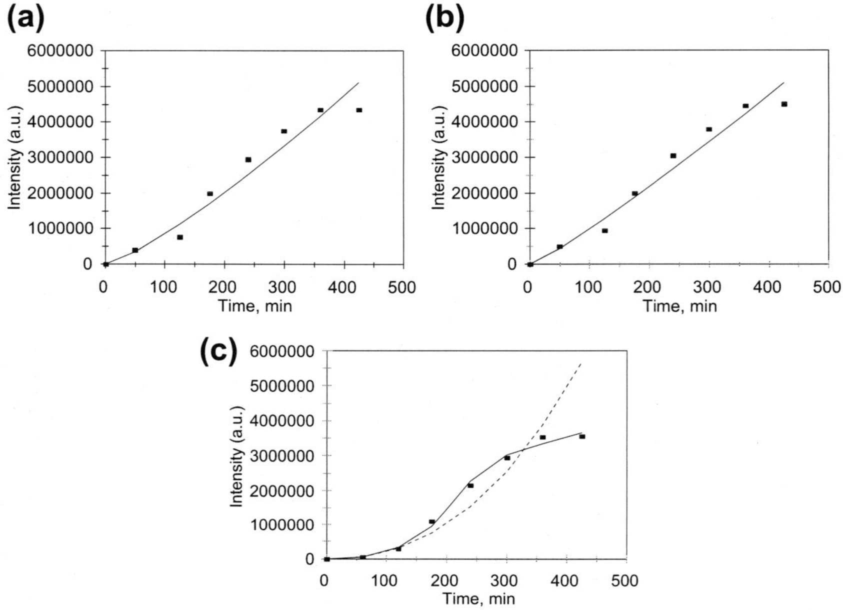

FIGURE 4.21 Analysis of binding of ampliflu by fluorometric analysis of nanoceria-mediated oxidation at different pH values (Asati et al., 2011): (a) pH 6.0 (b) pH 7.0 (c) pH 8.0.

dimension for the first phase is zero. This indicates that the surface acts like a Cantor-like dust (Vicsek, 1989).

Wilson and Nie (2006) indicate that sensitive, precise, and accurate multianalyte assays will considerably assist in measuring protein biomarkers. Furthermore, the early detection of tumor markers will assist in the response to

TABLE 4.12a Binding Rate Coefficients during Fluorometric Analysis of Nanoceria-Mediated Oxidation of Ampliflu at Different pH. Influence of pH 6, 7, and 8 (Sipova et al., 2012)

Analyte in Solution/ Receptor on Surface	pH	k	k_1	k_2
Ampliflu/fluorometric analysis	6	2868.82 ± 711.25	NA	NA
Ampliflu/fluorometric analysis	7	5495.66 ± 1030.39	NA	NA
Ampliflu/fluorometric analysis	8	5.584 ± 2.252	0.5803 ± 0.0843	127632.2 ± 9201.4

TABLE 4.12b Fractal Dimensions during Fluorometric Analysis of Nanoceria-Mediated Oxidation of Ampliflu at Different pH. Influence of pH 6, 7, and 8 (Sipova et al., 2012)

Analyte in Solution/ Receptor on Surface	pH	D_f	D_{f1}	D_{f2}
Ampliflu/fluorometric analysis	6	0.526 ± 0.245	NA	NA
Ampliflu/fluorometric Analysis	7	0.7412 ± 0.1902	NA	NA
Ampliflu/fluorometric analysis	8	0 ± 0.4022	0 ± 0.262	1.8894 ± 0.5650

therapy and predicting recurrence of diseases. Wilson and Nie (2006) indicate that due to their limited specificity, the measurement of a single marker is insufficient in the diagnosis of cancer. However, most cancers have more that one marker associated with their incidence. For example, CEA, AFB, hCG-β, and CA can be associated with lung, pancreatic, breast, and colorectal cancer (Niklinski and Furman, 1995; Beason et al., 2002; Gregory and Finlay, 1999). Thus, a panel that detects multiple biomarkers will improve the diagnosis of cancer (Carpelan-Holmstrom et al., 2002; Leuhimo et al., 2002; Hayakawa et al., 1999; Tsao et al., 2006). Quantitative microarray assays for a limited number of tumor markers have been described (Song et al., 2008). Nyholm et al. (2005) indicate that electrochemical immunosensors (EIS) can provide advantages in performing multianalyte assays. These are alternative tools, since they are easily quantifiable and produce excellent detection limits with small analyte samples.

Wilson and Nie (2006) developed multiplex EIS that can detect a limited number of analytes but were simple to use, robust, sensitive, precise, and accurate.

Figure 4.22(a) shows the binding of the antibody anti-CA15-3 to the corresponding tumor marker CA15-3 immobilized on an EIS chip (Wilson and Nie (2006)). A single-fractal analysis is required to adequately describe the binding kinetics. The values of the binding rate coefficient, k, and the fractal dimension, D_f, are given in Table 4.13.

Figure 4.22(b) shows the binding of the antibody anti-ferritin to the corresponding tumor marker ferritin immobilized on an EIS chip (Wilson and Nie (2006)). A single-fractal analysis is required to adequately describe the binding kinetics. The values of the binding rate coefficient, k, and the fractal dimension, D_f, are given in Table 4.13.

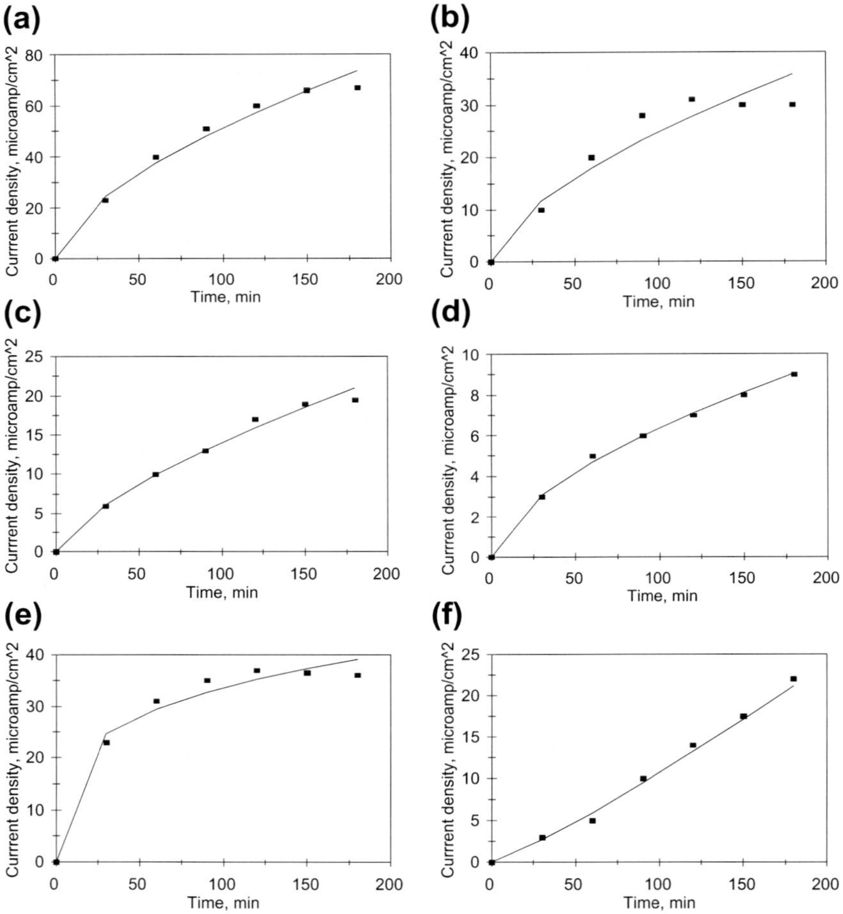

FIGURE 4.22 Binding of different tumor markers during multiplex measurement using an electrochemical chip (Wilson and Nie, 2006): (a) anti-CA15-3/CA15-3 (b) anti-ferritin/ferritin (c) anti-CEA/CEA (d) anti-CA15/CA15 (e) anti-AFB/AFB (f) anti-hCG-β/hCG-β.

Figure 4.22(c) shows the binding of the antibody anti-CEA to the corresponding tumor marker CEA immobilized on an EIS chip (Wilson and Nie (2006)). A single-fractal analysis is required to adequately describe the binding kinetics. The values of the binding rate coefficient, k, and the fractal dimension, D_f, are given in Table 4.13.

Figure 4.22(d) shows the binding of the antibody anti-CA15 to the corresponding tumor marker CA15 immobilized on an EIS chip (Wilson and Nie (2006)). A single-fractal analysis is required to adequately describe the binding kinetics. The values of the binding rate coefficient, k, and the fractal dimension, D_f, are given in Table 4.13.

TABLE 4.13 Binding Rate Coefficients and Fractal Dimensions for Six Tumor Markers during Multiplex Measurement Using an Electrochemical Chip (Wilson and Nie, 2006)

Analyte in Solution/ Receptor on Surface	k	D_f
Anti-CA15-3/CA15-3	3.090 ± 0.220	1.7892 ± 0.1010
Anti-ferritin/ferritin	1.3802 ± 0.2580	1.746 ± 0.2316
Anti-CEA/CEA	0.5937 ± 0.0324	1.6364 ± 0.07122
Anti-CA15/CA15	0.4098 ± 0.0156	1.8088 ± 0.0503
Anti-AFB/AFB	10.231 ± 0.787	2.4840 ± 0.0952
Anti-hCG-β/hCG-β	0.0538 ± 0.0064	0.6924 ± 0.01498

Figure 4.22(e) shows the binding of the antibody anti-AFB to the corresponding tumor marker AFB immobilized on an EIS chip (Wilson and Nie (2006)). A single-fractal analysis is required to adequately describe the binding kinetics. The values of the binding rate coefficient, k, and the fractal dimension, D_f, are given in Table 4.13.

Figure 4.22(f) shows the binding of the antibody anti-hCG-β to the corresponding tumor marker hCG-β immobilized on an EIS chip (Wilson and Nie (2006)). A single-fractal analysis is required to adequately describe the binding kinetics. The values of the binding rate coefficient, k, and the fractal dimension, D_f, are given in Table 4.13.

Figure 4.23 shows the increase in the binding rate coefficient, k, with an increase in the fractal dimension, D_f. For the data shown in Figure 4.23, the binding rate coefficient, k, is given by

$$k = (0.1602 + 0.2301)D_f^{3.764 \pm 0.9297} \quad (4.9)$$

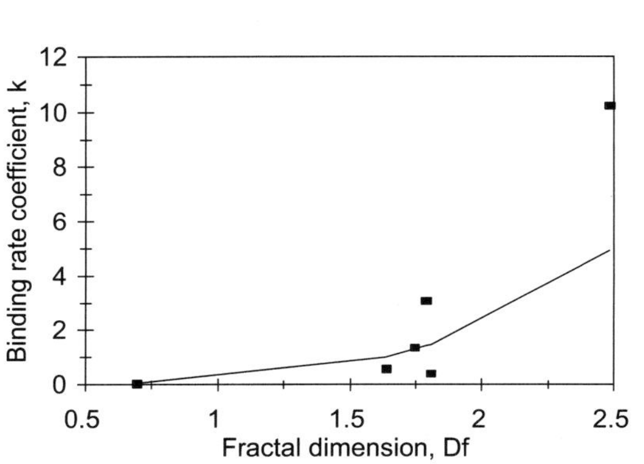

FIGURE 4.23 Increase in the binding rate coefficient, k, with an increase in the fractal dimension, D_f.

The fit is reasonable. There is scatter in the data. This is reflected in the error in the binding rate coefficient, k. The availability of more data points would lead to a more reliable fit. For the data presented in Figure 4.23, the binding rate coefficient, k, exhibits an order of dependence between three and one-half and four (equal to 3.764) on the fractal dimension, D_f, or the degree of heterogeneity that exists on the biosensor surface.

Wilson and Nie (2006) have recently indicated that the sensitive detection of disease-related proteins is critical in many areas of medical research. Kitano (2002) and Srinivas and Nie (2006) indicate that the detection of cancer biomarkers shows great promise for the early diagnosis of cancer. The detection of biomarkers also assists in following the disease progression and in monitoring Wilson and Nie (2006). indicate that the increasing emphasis on the early detection of cancer is pushing the enhancement of detection sensitivity by signal amplification or novel detection techniques (Polsky et al., 2006; Bao et al., 2006; GD Liu et al., 2006; Daniel and Anstruc, 2004; Liu and Lin, 2007).

Du et al. (2006) indicate that nanomaterials have drawn considerable attention in electrochemical biosensors due to their physical properties and conductivities (McCreey, 2008; Katz and Willner, 2004). Shang et al. (2008) and Tang et al. (2009) indicate that graphene has emerged as a useful material due to its electronic properties and large surface area.

Du et al. (2006) report on an electrochemical biosensor for the sensitive detection of biomarkers based on a dual amplification mechanism that is result of multienzyme antibody-functionalized graphene sheets as the sensor platform. The authors have used AFP as a model cancer biomarker and demonstrated the amplification process in sandwich detection.

Figure 4.24(a) shows the binding of AFP to HRP-Ab_2-CNSs. A dual-fractal analysis is required to adequately describe the binding kinetics. The values of (1) the binding rate coefficient, k, and the fractal dimension, D_f, for a single-fractal analysis and (2) the values of the binding rate coefficients, k_1 and k_2, and the fractal dimensions, D_{f1} and D_{f2}, are given in Table 4.14a and b. An

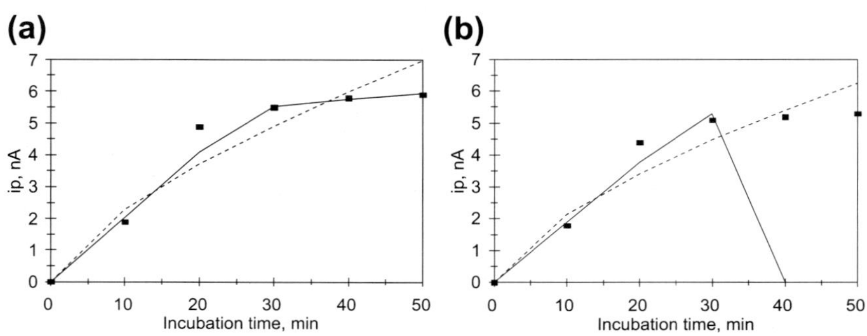

FIGURE 4.24 Binding of α-fetoprotein at (a) HRP-Ab_2-CNSs and (b) Ab_l/GS-CHI/SPCE (Du et al., 2006): (dependence of SWV current.)

TABLE 4.14a Binding of AFP at (a) HRP-Ab_2-CNSs and (b) Ab_l/GS-CHI/SPCE (Du et al., 2006). Dependence of SWV Current

Analyte in Solution/ Receptor on Surface	k	k_1	k_2
HFP/HRP-Ab_2-CNSs	0.4676 ± 0.1199	0.1982 ± 0.0496	2.434 ± 1.031
Ab_l/GS-CHI/SPCE	0.4697 ± 0.1152	0.1973 ± 0.0416	3.9500 ± 0.0072

TABLE 4.14b Fractal Dimensions for the Binding of AFP at (a) HRP-Ab_2-CNSs and (b) Ab_l/GS-CHI/SPCE (Du et al., 2006). Dependence of SWV Current

Analyte in Solution/ Receptor on Surface	D_f	D_{f1}	D_{f2}
HFP/HRP-Ab_2-CNSs	1.6174 ± 0.3588	0.980 ± 0.5686	2.7208 ± 0.0610
Ab_l/GS-CHI/SPCE	1.6758 ± 0.3450	1.0306 ± 0.4864	2.801 ± 0.0101

increase in the degree of heterogeneity on the biosensor surface leads to an increase in the binding rate coefficient.

Figure 4.24(b) shows the binding of AFP to Ab_1/GS-CHI/SPCE. A dual-fractal analysis is required to adequately describe the binding kinetics. The values of (1) the binding rate coefficient, k, and the fractal dimension, D_f, and (2) the binding rate coefficients, k_1 and k_2, and the fractal dimensions, D_{f1} and D_{f2}, are given in Table 4.14a and b. Once again, an increase in the degree of heterogeneity on the biosensor surface leads to an increase in the binding rate coefficient.

Figure 4.25 shows the increase in the binding rate coefficient, k, for a single-fractal analysis with an increase in the fractal dimension, D_f. For the data shown in Figure 4.25, the binding rate coefficient, k, is given by

$$k = (0.1602 + 0.232)D_f^{3.764 \pm 0.929} \tag{4.10}$$

The fit is reasonable. Only six data points are available. The availability of more data points would lead to a more reliable fit. The binding rate coefficient, k, exhibits between three and one-half and fourth (equal to 3.764) order of dependence on the fractal dimension or the degree of heterogeneity that exists on the biosensor surface.

Figure 4.26 shows the binding of antibody-immobilized $PEMC_a$ (Piezo-electric-excited millimeter-sized) sensor to a 1-ml sample of control urine at

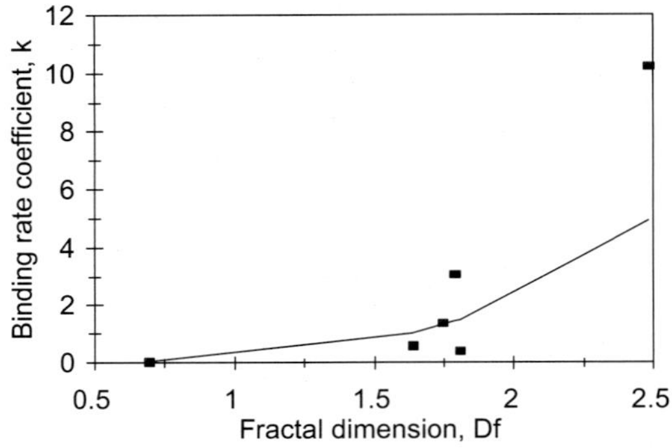

FIGURE 4.25 Increase in the binding rate coefficient, k, with an increase in the fractal dimension, D_f.

0.5 ml/min followed by a PBS flush (Maraldo et al., 2007). The intent was to quantitatively relate to AMACR (cancer biomarker, α-methylacyl-CoA racemase) concentration in the control urine. A dual-fractal analysis is adequate to describe the binding kinetics. The values of (1) the binding rate coefficient, k, and the fractal dimension, D_f, for a single-fractal analysis and (2) the binding rate coefficients, k_1 and k_2, and the fractal dimensions, D_{f1} and D_{f2}, for a dual-fractal analysis are given in Table 4.15a and b.

Figure 4.27 shows the $PEMC_a$ sensor response to the binding of AMACR. This was the detection of AMACR in a patient's sample. The binding is described by a single-fractal analysis. The values of the binding rate coefficient, k, and the fractal dimension, D_f, are given in Table 4.15a and b.

Figure 4.28 shows the typical resonance frequency shift of $PEMC_a$ upon binding to the antibody-immobilized sensor surface. The binding and the dissociation kinetics may be described by a single-fractal analysis. The values

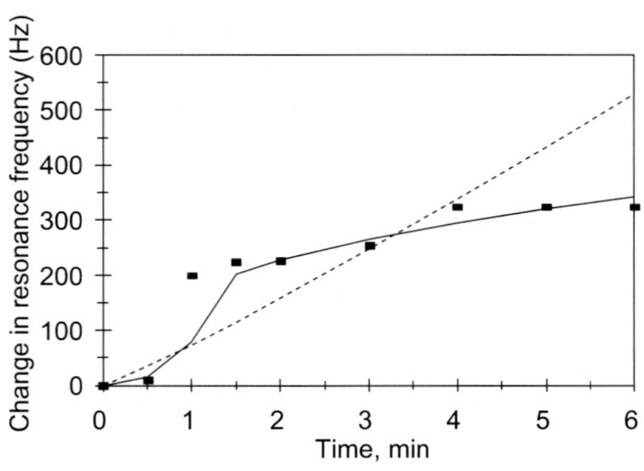

FIGURE 4.26 Binding of antibody-immobilized $PEMC_a$ sensor to AMACR in a urine sample. *Maraldo et al. (2007).*

TABLE 4.15a Binding Rate Coefficients for (a) Antibody-Immobilized $PEMC_a$ Sensor and Urine Sample, (b) $PEMC_a$ Response to AMACR, and (c) Resonance Frequency Shift of $PEMC_a$ upon Binding to Antibody-Immobilized Surface (Maraldo et al., 2007)

	k	k_1	k_2	k_d
(a)	73.64 ± 85.04	79.58 ± 103.51	176.79 ± 12.52	NA
(b)	240.81 ± 41.25	NA	NA	NA
(c)	478.55 ± 10.71	NA	NA	130.35 ± 25.42

TABLE 4.15b Fractal Dimensions for (a) Antibody-Immobilized $PEMC_a$ Sensor and Urine Sample, (b) $PEMC_a$ Response to AMACR, and (c) Resonance Frequency Shift of $PEMC_a$ upon Binding to Antibody-Immobilized Surface (Maraldo et al., 2007)

	D_f	D_{f1}	D_{f2}	D_{fd}
(a)	0.7964 ± 0.6082	0 + 0.1604	2.2588 ± 0.1576	NA
(b)	0.350 ± 0.2062	NA	NA	NA
(c)	2.7748 ± 0.0898	NA	NA	0 ± 0.342

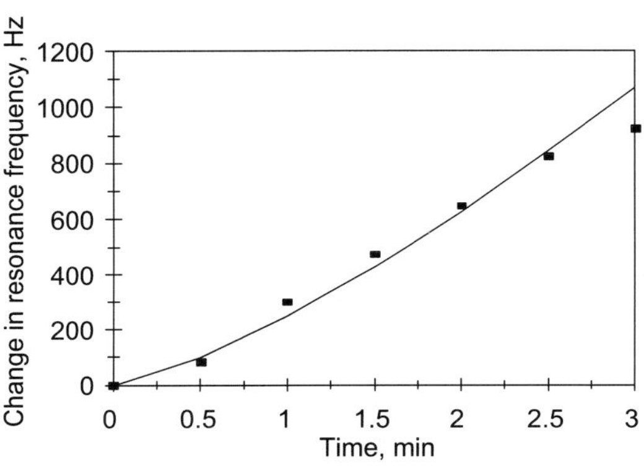

FIGURE 4.27 Binding of $PEMC_a$ sensor (response) to AMACR. *Maraldo et al. (2007).*

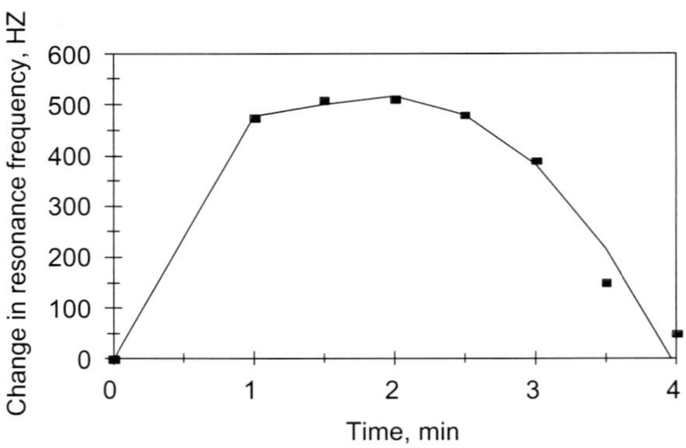

FIGURE 4.28 Resonance frequency shift of $PEMC_a$ upon binding of AMACR to antibody-immobilized surface. *Maraldo et al. (2007).*

of the binding rate coefficient, k, and the dissociation rate coefficient, k_d, are given in Table 4.15a and b. In this case, the affinity, K ($= k/k_d$), value is 3.67.

Luchansky and Bailey (2010) indicate that refractive index optical biosensors are useful tools, since they do not need fluorescent or enzymatic labels, which are costly and increase complexity and heterogeneity (Qavi et al., 2009; Fan et al., 2008). Microcavity resonators have been employed successfully to detect compounds of interest. These include microtoroids (Armani et al., 2007), microspheres (Arnold et al., 2003; Vollmer et al., 2008), liquid-core capillaries (White et al., 2009), and microrings (Xu et al., 2008; Chao et al., 2006). Bailey and coworkers (Washburn et al., 2009; Bailey et al., 2009; Iqbal et al., 2010) have described the operating principles of the microring detection platform. Luchansky and Bailey (2010) indicate that ring resonators have been used to detect proteins (de Vos et al., 2009; Zhu et al., 2009), nucleic acids (Suter et al., 2008), phage particles (Zhu et al., 2008), and whole cells (Ramachandaran et al., 2008).

Luchansky and Bailey (2010) indicate that cytokines, which are cell-signaling proteins and are secreted by lymphocytes and epithelial cells, present a challenge for detection in complex samples due to their small size. Luchansky and Bailey (2010) indicate that cytokines mediate human immune response and are involved in cell proliferation and inflammation via a complex network of cytokine secretion and cellular recognition (Young et al., 2009). They are prospective biomarkers for prostate (Fujita et al., 2008), breast (Chavez et al., 2007), and throat (Young et al., 1991) cancers, as well as different autoimmune and inflammatory diseases.

Luchansky and Bailey (2010) indicate that IL-2 is a T-cell growth factor and is a cytokine produced by T-lymphocytes that are responsible for T-cell proliferation. IL-2 levels have been used to gauge immune system health following chemotherapy (Mazur et al., 2004). Luchansky and Bailey (2010) have used a secondary antibody in a sandwich assay format for the sensitive

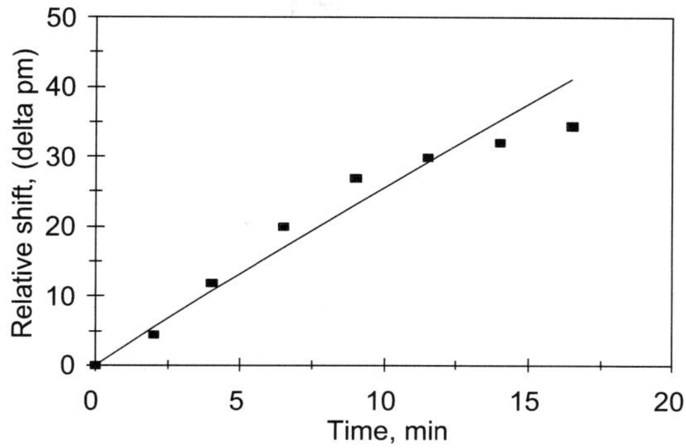

FIGURE 4.29 Binding of interleukin-2 (IL-2) to secondary antibody (IL-2) in a sandwich-type immunoassay. *Luchansky and Bailey (2010).*

detection of IL-2 in complex media. Previously, it has been demonstrated that using SPR, the use of a secondary antibody increases both the assay sensitivity and specificity to detect proteins present in complex media in low concentrations (Arima et al., 2009).

Figure 4.29 shows the binding of IL-2 in solution to anti-IL-2 antibody in a sandwich-type assay on a single microring optical resonator functionalized with the capture antibody. Quantitative signal enhancement is achieved by introducing an anti-IL-2 detection antibody. Figure 4.29 shows the binding of the IL-2 in solution to the functionalized biosensor. A single-fractal analysis is adequate to describe the binding kinetics. The values of the binding rate coefficient, k, and the fractal dimension, D_f, for a single-fractal analysis are given in Table 4.16.

Figure 4.30(a) shows the binding of 200 mM H_2O_2 by aliquot addition and the sensitive detection of IL-6 based on a dual amplification resulting from Au-nanoparticle dopamine-functionalized (Au-NP-PDOP 2) CNTs (Wang et al., 2002). This is Run#1 of a sequence of three consecutive runs. A dual-fractal analysis is required to adequately describe the binding kinetics. The values of (1) the binding rate coefficient, k, and the fractal dimension, D_f, for a single-fractal analysis and (2) the binding rate coefficients, k_1 and k_2, and the fractal dimensions, D_{f1} and D_{f2}, are given in Table 4.17a and b. It is of interest to note that as the fractal dimension increases by a factor of 1.63 from

TABLE 4.16 Binding Rate Coefficient and Fractal Dimension for the Interleukin-2 (IL-2) in Solution to an IL-2 Antibody Sandwich-Type-Functionalized Immunosensor (Luchansky and Bailey, 2010)

Analyte in Solution/ Antibody on Surface	k	D_f
IL-2/Anti-IL-2 antibody	2.825 ± 0.530	1.0874 ± 0.1868

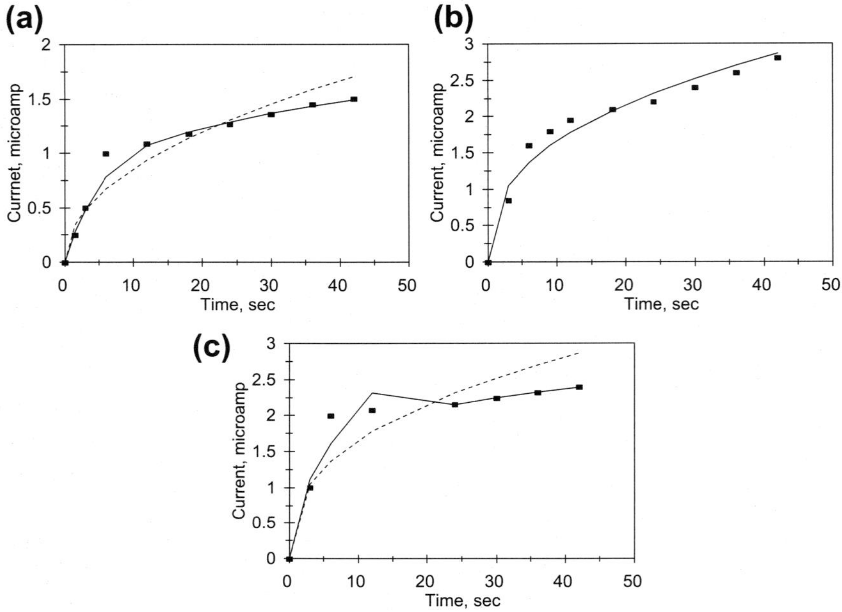

FIGURE 4.30 Binding of 200 nM H_2O_2 by aliquot addition and sensitive detection of Interleukin-6 (IL-6) (Wang et al., 2002). Influence of repeat runs: (a) Run#1 (b) Run#2 (c) Run#3.

a value of $D_{f1} = 1.5256$ to $D_{f2} = 2.4852$, the binding rate coefficient increases by a factor of 2.65 from a value of $k_1 = 0.2093$ to $k_2 = 0.5549$. An increase in the degree of heterogeneity on the biosensor surface leads to an increase in the binding rate coefficient.

Figure 4.30(b) shows the binding of 200 mM H_2O_2 by aliquot addition and the sensitive detection of IL-6 based on a dual amplification resulting from Au-NP-PDOP 2 CNTs (Wang et al., 2002). This is Run#2 of a sequence of

TABLE 4.17a Binding of 200 nM H_2O_2 by Aliquot Addition and the Sensitive Detection of Interleukin-6 (IL-6) Based on a Dual Amplification Resulting from Au-Nanoparticles Dopamine-Functionalized @ Carbon Nanotubes. Influence of Consecutive Runs (Wang et al., 2002)

Run#	k	k_1	k_2
Run#1	0.2827 ± 0.773	0.2093 ± 0.0555	0.5549 ± 0.0121
Run#2	0.6992 ± 0.0877	NA	NA
Run#3	0.9647 ± 0.1888	0.6241 ± 0.1910	1.1869 ± 0.0049

TABLE 4.17b Fractal Dimensions for the Binding of 200 nM H_2O_2 by Aliquot Addition and the Sensitive Detection of Interleukin-6 (IL-6) Based on a Dual Amplification Resulting from Au-Anoparticles Dopamine-Functionalized @ Carbon Nanotubes. Influence of Consecutive Runs (Wang et al., 2002)

Run#	D_f	D_{f1}	D_{f2}
Run#1	2.4844 ± 0.1438	1.5256 ± 0.3032	2.4852 ± 0.262
Run#2	2.2392 ± 0.09516	NA	NA
Run#3	2.4844 ± 0.1438	1.9434 ± 0.5446	2.6248 ± 0.0178

three consecutive runs. A single-fractal analysis is adequate to describe the binding kinetics. The values of the binding rate coefficient, k, and the fractal dimension, D_f, for a single-fractal analysis are given in Table 4.17a and b.

Figure 4.30(c) shows the binding of 200 mM H_2O_2 by aliquot addition and the sensitive detection of IL-6 based on a dual amplification resulting from Au-NP-PDOP 2 CNTs (Wang et al., 2002). This is Run#3 of a sequence of three consecutive runs. Once again, a dual-fractal analysis is required to adequately describe the binding kinetics. The values of (1) the binding rate coefficient, k, and the fractal dimension, D_f, for a single-fractal analysis and (2) the binding rate coefficients, k_1 and k_2, and the fractal dimensions, D_{f1} and D_{f2}, are given in Table 4.17a and b. It is of interest to note that as the fractal dimension increases by a factor of 1.35 from a value of $D_{f1} = 1.9434$ to $D_{f2} = 2.6248$, the binding rate coefficient increases by a factor of 2.65 from a value of $k_1 = 0.6241$ to $k_2 = 1.1869$. An increase in the degree of heterogeneity on the biosensor surface leads to an increase in the binding rate coefficient.

Figure 4.31 shows the binding and dissociation of standard antigen (IL-6) in solution to Au-NP-PDOP @ CNTs biosensor (Wang et al., 2011).

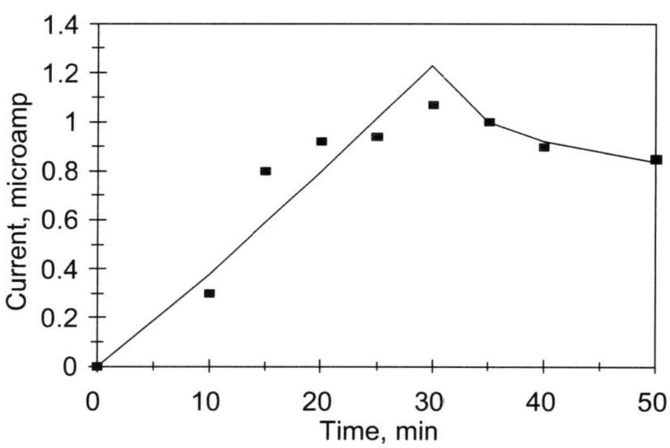

FIGURE 4.31 Binding of standard antigen (IL-6) in solution to Au-NP-PDOP @ carbon nanotubes *(Wang et al., 2011).*

TABLE 4.18 Binding and Dissociation Rate Coefficients and Corresponding Fractal Dimensions for Interleukin-6 (IL-6) in Solution to Au-Nanoparticles Dopamine-Functionalized (Au-NP-PDOP) @ Carbon Nanotubes (CNT) Biosensor (Wang et al., 2011)

Analyte in Solution/ Receptor on Surface	k	k_d	$K = k/k_d$
Standard antigen (IL-6)/Au-NP-PDOP @ CNT	0.0322 ± 0.0094	0.01306 ± 0.00185	2.374

A single-fractal analysis is required to adequately describe the binding kinetics. The values of the binding rate coefficient, k, and the fractal dimension, D_f, for a single-fractal analysis are given in Table 4.18.

Fang et al. (2009) have recently profiled cancer biomarkers in tumor tissue using a multiplexed nanostructured microelectrode integrated circuit. They indicate that biomarker analysis based on electronic reactant for chip-based devices is appropriate for clinical testing with regard to cost and sensitivity (Bell, 2004; Ludwig and Weinstein, 2005; Drummond et al., 2003; Westar and Hood, 2004). Fang et al. (2009) indicate that for biological samples during assay, direct electronic detection of biomarkers in clinical samples has yet to be shown. This is due to the high background noise levels present in biological samples during assay procedures, and they pose a challenge of generating multiplexed systems that are highly sensitive and specific.

Fang and Kelley (2009) emphasize that silicon nanowires that are used as FETs (Cui et al., 2001; Hahm and Lieber, 2001; Gao et al., 2007) display good dielectric ability for a wide range of analytes, including nucleic acids. They further indicate that, however, a low signal-to-background has their use to heterogeneous samples. Fang et al. (2009) have recently demonstrated the direct detection of cancer biomarkers using electrodes based on templates gold nanowires (Fang and Kelley, 2009). Fang and Kelley (2009) indicate that there is an urgent need for a more versatile and robust system to help expand the range of clinical diagnostic tools.

Fang and Kelley (2009) indicate that previous research has demonstrated the advantages of nanostructures for biomolecular sensors (Gasparac et al., 2004; Lapierre-Devlin et al., 2005). Fang and Kelley (2009) indicate that the nanostructured microelectrodes (NMEs) may function as nucleic acids biosensors by modifying then with thiolated peptide nucleic acid (PNA) probes. PNA has been used as probes previously that help increase the sensitivity of biosensing assays (Kerman et al., 2008; Liu et al., 2006).

Fang and Kelley (2009) selected the cancer biomarkers for analysis that were a group of gene fusions specific to prostate cancer. These gene fusions resulted from a chromosomal translocation that joins the ERG and TMPRSS2 genes. These genes have recently been discovered and appear in at least 50% of prostate tumors (Kumar-Sinha et al., 2008). There NME chip is designed for the direct multiplexed analysis of a panel of cancer biomarkers in relevant patient samples. In summary, they describe a new multiplexed electrode platform to read directly a panel of cancer biomarkers in clinically relevant samples using electronic signals.

Figure 4.32(a) shows the binding of 100 fM of the cancer biomarker TMPRSSR-ERG (DNA target) to DNA-modified multiplexed NMEs with complement DNA (Fang et al., 2009) by hybridization (Fang and Kelley, 2009). A dual-fractal analysis is required to adequately describe the binding kinetics. The values of (1) the binding rate coefficient, k, and the fractal dimension, D_f, for a single-fractal analysis and (2) the binding rate coefficients, k_1 and k_2, and the fractal dimensions, D_{f1} and D_{f2}, are given in Table 4.19a and b.

Figure 4.32(b) shows the binding of 100 fM TMPRSSR (DNA target) to DNA-modified multiplexed NMEs with complement DNA by hybridization (Fang and Kelley, 2009). A single-fractal analysis is required to describe the binding and the dissociation kinetics. The values of (1) the binding rate coefficient, k, and the fractal dimension, D_f, for a single-fractal analysis and

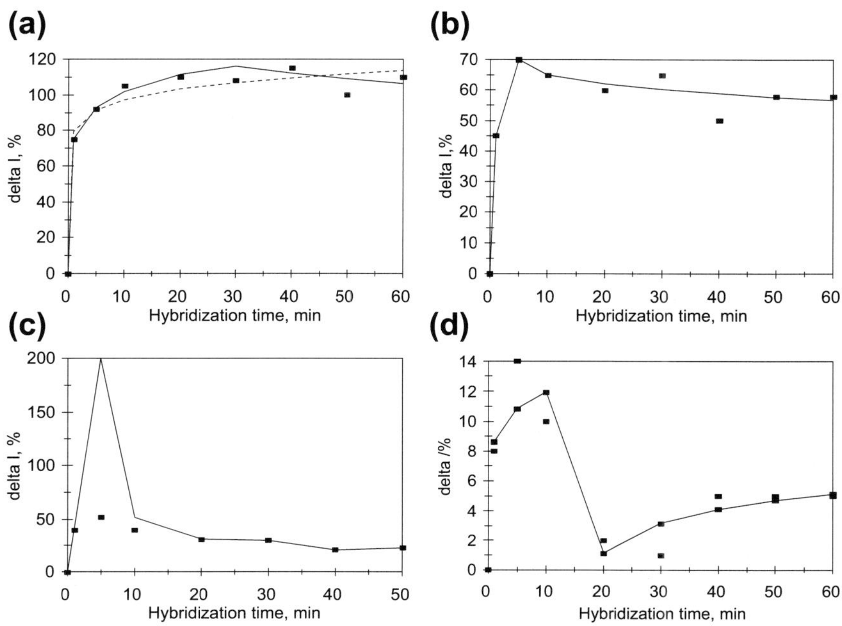

FIGURE 4.32 Binding of cancer biomarkers to multiplexed nanostructures microelectrodes by hybridization (Fang and Kelley, 2009): (a) 100 fM TMPRSSR-ERG (DNA target) (b) 100 fM TMPRSSR9 (DNA target) (c) 100 fM wt ERG (DNA target) (d) 100 fM non-comp (DNA target).

TABLE 4.19a Binding Rate Coefficients for Cancer Biomarkers to Multiplexed Nanostructure Microelectrodes (NMEs) by Hybridization (Fang and Kelley, 2009)

Analyte in Solution/ Receptor on Surface	k	k_1	k_2	k_d
100 fM TMPRSSR-ERG (DNA target)/DNA-modified NMEs with complement DNA	79.316 ± 5.742	75.122 ± 1.894	178.586 ± 17.598	NA
100 fM TMPRSSR (DNA target)/ DNA-modified NMEs with complement DNA	45 ± 0	NA	NA	2.650 ± 1.555
100 fM wt ERG (DNA target)/ DNA-modified NMEs with complement DNA	40 ± 0	NA	NA	6.258 ± 0.665
100 fM non-comp (DNA target)/ DNA-modified NMEs with complement DNA	8.639 ± 3.213	NA	NA	20.962 ± 4.549

(2) the dissociation rate coefficient, k_d, and the fractal dimension for dissociation, D_{fd}, are given in Table 4.19a and b.

Figure 4.32(c) shows the binding of 100 fM wt ERG (DNA target) to DNA-modified multiplexed NMEs with complement DNA by hybridization (Fang and Kelley, 2009). A single-fractal analysis is required to describe the binding and the dissociation kinetics. The values of (1) the binding rate coefficient, k, and the fractal dimension, D_f, for a single-fractal analysis and (2) the dissociation rate coefficient, k_d, and the fractal dimension for dissociation, D_{fd}, are given in Table 4.19a and b.

Figure 4.32(d) shows the binding of 100 fM non-comp (DNA target) to DNA-modified multiplexed NMEs with complement DNA by hybridization (Fang and Kelley, 2009). A single-fractal analysis is required to describe the binding and the dissociation kinetics. The values of (1) the binding rate coefficient, k, and the fractal dimension, D_f, for a single-fractal analysis and (2) the dissociation rate coefficient, k_d, and the fractal dimension for dissociation, D_{fd}, are given in Table 4.19a and b.

Figure 4.33(a) shows the decrease in the binding rate coefficient, k, with an increase in the fractal dimension, D_f. For the data shown in Figure 4.33(a), the binding rate coefficient, k, is given by

$$k = (0.9014 + 3.0)(\mathrm{E} + 07)D_f^{-13.457 \pm 15.26}$$

There is scatter in the data. Only three data points are available. The availability of more data points would lead to a more reliable fit. The binding

TABLE 4.19b Binding Rate Coefficients for Cancer Biomarkers to Multiplexed Nanostructures Microelectrodes (NMEs) by Hybridization (Fang and Kelley, 2009)

Analyte in Solution/Receptor on Surface	D_f	D_{f1}	D_{f2}	D_{fd}
100 fM TMPRSSR-ERG (DNA target)/DNA-modified NMEs with complement DNA	1.2736 ± 0.03792	2.7346 ± 0.2242	$0 + 0.65448$	NA
100 fM TMPRSSR (DNA target)/ DNA-modified NMEs with complement DNA	2.4510 ± 0	NA	NA	2.1920 ± 0.4672
100 fM wt ERG (DNA target)/ DNA-modified NMEs with complement DNA	2.674 ± 0	NA	NA	2.1606 ± 0.1163
100 fM non-comp (DNA target)/ DNA-modified NMEs with complement DNA	2.7178 ± 0.3846	NA	NA	0 ± 0.3090

rate coefficient, k, exhibits close to a negative 13 and one-half (equal to -13.45) order of dependence on the fractal dimension or the degree of heterogeneity on the surface.

Figure 4.33(b) shows the decrease in the affinity, K ($= k/k_d$), with an increase in the fractal dimension ratio, D_f/D_{fd}. For the data shown in Figure 4.33(b), the affinity, K, is given by

$$K = (13.563 \pm 8.823)\left(D_f/D_{fd}\right)^{-1.608 \pm 0.357}$$

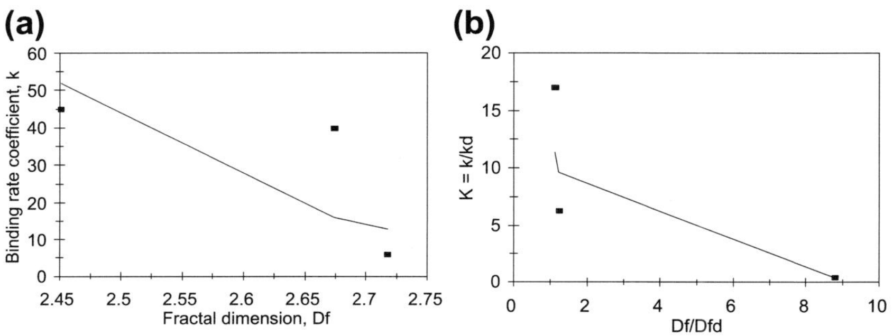

FIGURE 4.33 (a) Decrease in the binding rate coefficient, k, with an increase in the fractal dimension, D_f (b) Decrease in the affinity, K ($= k/k_d$), with an increase in the fractal dimension ratio, D_f/D_{fd}.

There is scatter in the data. Only three data points are available. The availability of more data points would lead to a more reliable fit. The affinity, K, exhibits close to a negative one and one-half (equal to -1.608) order of dependence on the ratio of fractal dimensions (D_f/D_{fd}) present on the biosensor surface (Figure 4.33b).

4.3 CONCLUSIONS

This is the second chapter in a series on the detection of cancer biomarkers. There is a wealth of information available in the literature (kinetic data to be specific) on the detection of the various types of cancer. Here we have attempted to present some of the more recent data available in the literature. As usual, we present a fractal analysis of the kinetics. As mentioned previously and throughout the book, the fractal analysis includes the presence of external diffusion limitations and provides for the binding and dissociation rate coefficients and the degree of heterogeneity that exists on the surface.

Some of the data from cancer biomarkers include

1. binding of cancer cells to antibody-conjugated CNTFET (Teker, 2008). The binding of B7474 breast cancer cells and MCF7 breast cancer cells was analyzed.
2. detection of prostate cancer marker (PSA-ACT complex) through a label-free protein biosensor based on CNTFET.
3. detection of PSA and CRP based on piezoelectric self-sensing micro-cantilevers (Wee et al., 2005).
4. binding and detection of different concentrations of PSA to an mAb immobilized on a QCM biosensor surface (Uludag and Tothill, 2010).
5. binding of 2.5 ng/ml PSA blood sample to anti-PSA-functionalized SINW-FET.
6. binding of CaM-GST in solution to GST-SINW-FET.
7. binding of carcinoembryonic antigen (CEA) in solution to anti-CEA antibody immobilized on a SINW-FET array.
8. binding of telomerase activity in cancer cells by a displacement polymerization reaction on a biosensor surface (Ding et al., 2008).
9. binding of IL-2 to anti-IL-2 immobilized on a biosensor surface.
10. binding and dissociation of different concentrations of PSA in solution to anti-PSA immobilized on a microring resonator platform (Washburn et al., 2010).
11. binding of ampliflu and its analysis by fluorometry at pH 7 (Asati et al., 2011).
12. binding of the antibody anti-CA15-3 to the corresponding tumor marker CA15-3 (Wilson and Nie, 2006).
13. binding on the antibody anti-ferritin to the corresponding tumor marker, ferritin (Wilson and Nie (2006)).

14. binding of antibody anti-CA15 to the corresponding tumor marker CA15 immobilized on an EIS chip (Wilson and Nie (2006)).
15. binding of the antibody anti-AFB immobilized on an EIS chip (Wilson and Nie (2006)).
16. binding of the antibody anti-hCGβ to the corresponding tumor marker hCG-β immobilized on an EIS chip (Wilson and Nie (2006)).
17. binding of AFP to HRP-Ab_2-CNS (Du et al., 2010).
18. binding of antibody-immobilized $PEMC_a$ in sample of control urine (Maraldo et al., 2007).
19. binding of IL-2 in solution to anti-IL-2 antibody in a sandwich-type assay on a single microring optical resonator functionalized with a capture antibody (Luchansky and Bailey, 2010).
20. binding of 200 mM H_2O_2 by aliquot addition and the sensitive detection of IL-2 based on a dual amplification resulting from Au-NP-PDOP CNTs (Wang et al., 2002).
21. binding of the cancer biomarkers (DNA target) to their corresponding DNA probes by hybridization: TMPRSSR-ERG; TMPRSSR; wt ERG, and non-comp (DNA target) immobilized on NMEs (Fang and Kelley, 2009).

REFERENCES

Allen, B.L., Kichambare, P.D., Star, A., 2007. Advances in Materials 19, 1439–1451.

Alivisatos, P., 2004. Nature Biotechnology 32, 47–52.

Ambrosi, A., Airo, F., Merkoci, A., 2010. Analytical Chemistry 82, 151–1156.

Arima, Y., Teramura, Y., Takiguchi, H., Kawano, K., Kotera, H., Iwata, H., 2009. Surface plasmon resonance enhanced fluorescence spectroscopy for sensitive detection if tumor markers. In: Rasooly, A., Herold, I.C.E. (Eds.), Methods in Molecular Biology: Biosensors and Biodetection, vol 503. Humana Press, New York, NY, pp. 3–20.

Armbuster, D.A., 1993. Clinical Chemistry 39, 181–195.

Alvarez, M., Calle, A., Tamayo, J., Lechuga, L.M., Abad, A., Montoya, A., 2003. Development of nanomechanical biosensors for detection of pesticide DDT. Biosensors & Bioelectronics 18, 649–653.

Armani, A.M., Kulkami, R.P., Fraser, S.E., Flagan, R.C., Vahala, K.J., 2007. Science 783–787.

Arnold, S., Khoshima, M., Teraoka, I., Holler, S., Vollmer, F., 2003. Optics Letters 28, 272–274.

Asati, A., Kaittanis, C., Santra, S., Perez, J.M., 2011. pH-tunable oxidase-like activity of cerium oxide nanoparticles achieving sensitive fluorogenic detection of cancer biomarkers at neutral pH. Analytical Chemistry 83, 2547–25553.

Asati, A., Santra, S., Kaittanis, C., Nath, S., Perez, J.M., 2009. Angewandte Chemie International Edition 48, 2308–2312.

Bailey, R.C., Washburn, A.L., Qavi, A.J., Iqbal, M., Gleeson, M., Tybor, F., Gun, L.C., 2009. Proceedings SPIE 7220, 72200N.

Bao, Y.P., Wei, T.F., Lefebvre, P.A., An, H., He, L.X., Kunkel, G.T., Muller, U.R., 2006. Analytical Chemistry 78, 2055–2059.

Baselt, D.R., Lee, G.U., Golton, R.J., 1996. Biosensor based on force microscope technology. Journal of Vacuum Science Technology B 14, 789–793.

Beason, K.L., Clinton, S.R., Bryant, S., Johnson, J.T., Jackson, M., Schultze, H., Fortenbeny, D., Bright, C., Hua, H., Ying, J., Sykes, P., Wilson, C., Holfiled, K., Vincent, C., Hall, M., 2002. Mississippi Academy of Sciences 47, 175–183.

Bell, J., 2004. Predicting disease using genomics. Nature 429, 453–456.

Bradley, K., Briman, M., Star, A., Gruiner, G., 2004. Nano Letters 4, 253–256.

Brawer, M.K., Lange, P.H., 1989. World Journal of Urology 7. 7–1.

Cao, C., Kim, J.P., Kim, B.W., Chae, H., Yoon, H.C., Yang, S.S., Sim, S.J., 2006. Biosensors & Bioelectronics 21, 2106–2113.

Carpelan-Holmstrom, M., Louhimo, J., Stenman, U.H., Alfhan, H., Haglund, C., 2002. Anticancer Research 22, 2311–2316.

Chao, C.Y., Fung, W., Guo, L.J., 2006. IEEE Selected Topics in Quantum Electronics 12 (134).

Chen, R.J., Bangsaruntip, S., Drouvvalakis, K.A., Kam, N.W.S., Shim, M., Li Y,kim, W., Ulz, P.J., Dai, H., 2003. Proceedings of the National Academy of Sciences 100, 4984–4989.

Chen, R.J., Choi, H.C., Bangsarunip, S., Yenimez, S., Tang, X.W., Wang, Q., Chang, Y.L., Dai, H.J., 2004. Journal of the American Chemical Society 126, 1563.

Chen, S., Bomer, J.G., Carlen, E.T., van den Berg, A., 2011. Al_2O_3/silicon nanoISFET with ideal Nennstian response. Nanoletters 11 (6), 2334–2341.

Cherian, S., Gupta, R.K., Mullin, B.C., Thundat, T., 2003. Detection of heavy metal ions using protein-functionalized micro-cantilever sensors. Biosensors & Bioelectronics 19, 411–416.

Chikkaveeraiah, B.V., Bhirde, A., Mahotra, R., Patel, V., Gutkind, J.S., Rusling, J.F., 2009. Analytical Chemistry 81, 9129–9134.

Choi, J.W., Kang, D.Y., Jang, Y.H., Kim, H.H., Min, J., Oh, B.K., 2008. Colloids and Surfaces A: Physicochemical and Engineering Aspects 313–314, 655–659.

Cohen, S.B., Graham, M.E., Lovrecz, G.O., Bache, N., Robinson, P.J., Reddel, R.R., 2007. Science 315, 1850–1853.

Counter, C.M., Avilon, A.A., LeFeuvre, C.E., Stewart, N.G., Greider, C.W., Harley, C.B., Bachetti, S., 1992. EMBO Journal 11, 1921–1929.

Cui, Y., Wei, Q.Q., Park, H.K., Leber, C.M., 2000. Science 293, 1289–1293.

Daniel, M.C., Anstruc, D., 2004. Chemical Reviews 104, 293–346.

Diamandis, E.P., 1998. Trends in Endocrinology and Metabolism 9, 310–316.

Ding, Y.J., Lu, H.X., Shi, G.R., Liu, J., Shen, G.L., Yu, R.Q., 2008. Biosensors & Bioelectronics 24, 228–232.

Ding, Z., Wu, C.J., Jaskeloff, M., Ivanova, E., Alimova, M.K., et al., March 2, 2012. Telomerase reactivation following telomere dysfunction yields murine prostate tumors with bone metastases. Cell 148 (5), 896–907.

Drummond, T.G., Hill, M.G., Barton, J.I.C., 2003. Biotechnology 21, 1192–1199.

Du, N., Liu, X.Y., Hew, C.L., 2006. Journal of Physical Chemistry 110, 20562–20567.

Fan, X., White, I.M., Shopova, S.I., Zhu, H., Suter, J.D., Sun, Y., 2008. Analytica Chimica Acta 620, 8–26.

Fang, Z., Kelley, S.O., 2009. Direct electrocatalytic mRNA detection using PNA-nanowire sensors. Analytical Chemistry 81, 612–617.

Farre, M., Barceio, D., 2003. TrAC Trends in Analytical Chemistry 22, 299.

Ferlay, J., Autier, P., Bonol, M., Heanue, M., Colombet, M., Boyle, P., 1989. Annals of Oncology 7, 7–11.

Fritz, J., Bailer, M.K., Lang, H.P., Rothulzen, H., Vettiger, P., Meyer, R., Guntherodt, H.J., Gerber, C., Gimzewski, J.K., 2000. Science 288, 316–318.

Fujita, K., Ewing, C.M., Sokoll, L.J., Elliott, D.J., Cunningham, M., Marzo, A.M.D., Isaac, W.B., Pavlovich, C.P., 2008. The Prostate 68, 872–882.

Gao, Z., Agarwal, A., Trigg, A.D., Singh, N., Fang, C., Tung, C.H., Fan, Y., Buddharaju, K.D., Kong, J., 2007. Silicon nanowire arrays for label-free detection of DNA. Analytical Chemistry 79, 3291–3297.

Gasparac, R., Taft, B.J., Lapierre-Devlin, M.A., Lazareck, A.D., XU, J.M., Kelley, S.O., 2004. Ultrasensitive electrocatalytic DNA detection at two- and three-dimensional nanoelectrodes. Journal of the American Chemical Society 126, 12270–12271.

Gelmini, S., Caldini, A., Becheni, L., Capaccioli, S., Pazzagli, M., Orlando, C., 1998. Clinical Chemistry 44, 2133–2138.

Gireshaber, D., MacKenzie, R., Voros, J., Reimhult, E., 2008. Sensors 8, 1400.

Gregory Jr., J.J., Finlay, J.L., 1999. Drugs 57, 463–467.

Gruner, G., 2006. Analytical and Bioanalytical Chemistry 384, 322.

Guo, X., Huang, L., O'Brien, S., Kim, P., Nuckolis, C., 2005. Journal of the American Chemical Society 127, 15045–15047.

Hahm, J., Lieber, C.M., 2001. Direct ultrasensitive electrocatalytic DNA detection at two- and three-dimensional nanoelectrodes. Journal of the American Chemical Society 126, 1289–1292.

Harley, C.B., Villeponteau, B., 1995. Current Opinions in Genetics 5, 249–255.

Havlin, S., 1989. Molecular diffusion and reaction. In The Fractal Approach to Heterogeneous Chemistry: Surfaces, Colloids, Polymers. In: Avnin, D. (Ed.). Wiley, New York, pp. 251–269.

Hayakawa, T., Naruse, S., Kitagawa, M., Ishiguro, H., Kondo, T., Kurimoto, K., Fukushima, M., Takayama, T., Horiguchi, Y., Kuno, N., Noda, A., Furukawa, T., 1999. International Journal of Pancreatology 25, 23–29.

Hiyama, E., Hiyama, K., 2002. Oncogene 21, 643–649.

Hou, M., Xu, D., Gruber, A., 2001. Clinical Chemistry 47, 519–524.

Hwang, K.S., Lee, J.H., Park, J., Yoon, D.S., Park, J.H., Kim, T.S., 2004. Lab on a Chip-Miniaturization for Chemistry and Biology 4, 547–552.

Iqbal M, Gleeson M.A, Spaugh B, Tybor F, Gunn W.G, Hochberg M, Baehr-Jones T, Bailey R.C and Gunn L.C, 2010. IEEE Journal Selected Topics in Quantum Electronics, (in press).

Iqbal, J., Weisenberger, D.D., Greiner, T.C., Vose, J.M., McKeithan, T., Kucik, C., Geng, H., Deffenbacker, K., Smith, L., Dybaker, K., Nakemura, S., Sato, M., Delabie, J., Berger, F., Loong, F., An, W.Y., Sing, I., Armitgae, J.O., Chan, W.C., 2010. International peripheral T-cell lymphoma and prognostication in angioimmunoblastic T-cell. Blood, Feb 4 115 (5), 1026–1036.

Jensenius, H., Thaysen, J., Rasmussen, A.A., Veje, L.H., Hansen, O., Boisen, A., 2000. A microcantilever based alcohol vapor sensor application and response model. Applied Physics Letters 76, 2615–2617.

Ji, H.F., Thundat, T., 2002. In situ detection of calcium ions with chemically modified micro-cantilevers. Biosensors & Bioelectronics 17, 337–343.

Jokerst, N., Royal, M., Palit, S., Luan, L., Dhar, S., Tyler, T., 2009. Journal of Biophotonics 2, 212–226.

Katz, E., Willner, I., 2004. Chemical and Physical Chemistry 5, 1085.

Kerman, K., Saito, M., Tamiya, E., 2008. Electroactive chitosan nanoparticles for the detection of single-nucleotide polymorphisms using peptide nucleic acids. Analytical and Bioanalytical Chemistry 391, 2759–2767.

Kim, J.P., Lee, B.Y., Hong, S., Sim, S.J., 2008. Analytical Biochemistry 381, 193–198.

Kim, S., James, D.B., Mohammed, S., Back, J., 2010. DC modelling and the course of flicker noise in passivated carbon nanotube transistors. Nanotechnology 38, 21.

Kitano, H., 2002. Science 295, 1662–1664.

Kooser, A., Manygoats, K., Eastman, M.P., Porter, T.L., 2003. Investigation of the antigen antibody reaction between anti-bovine serum (a-BSA) and bovine serum albumin (BSA) using piezoresistive microcantilever based sensors. Biosensors & Bioelectronics 19, 503–508.

Kumar-Sinha, C., Tomlins, S.A., Chinnalayan, A.M., 2008. Recurrent gene fusions in prostate cancer. Nature Reviews in Cancer 8, 497–511.

Lapierre-Devlin, M.A., Asher, C.L., Taft, B.J., Gasparic, R., Roberts, M.A., Kelley, S.O., 2005. Amplified electrocatalysis at DNA-modified nanowires. Nano Letters 5, 1051–1055.

Leuhimo, J., Finne, P., Alfthan, H., Stenman, U.H., Haglund, C., 2002. Anticancer Research 22, 1759–1764.

Li, C., Currell, M., Lin, H., Lei, B., Ishikawa, F.X., Datar, R., Cote, R.J., Thompson, R.E., Zhou, C., 2005. Complementary detection of prostate-specific anatigen using In2O3 nanowires and carbon nanotubes. Journal of the American Chemical Society 127 (36), 12484–12485.

Lilija, H.C.A., Dahlen, U., Matkainen, M., Nilsson, O., Petterson, K., Lowgren, T., 1991. Clinical Chemistry 37, 1818–1825.

Lin, A., 2008. Biosensors and Bioelectronics 24, 167–177.

Liu, G., Lin, Y., 2007. Talanta 74, 308–317.

Liu, G.D., Wang, J., Wu, H., Lin, Y.H., 2006. Analytical Chemistry 73, 7417–7423.

Liu, J., Telenauer, L., Tian, S., Nielsen, P.E., Knoll, W., 2006. PNA-DNA hybridization study using labeled streptavidin by voltammetry and surface plasmon fluorescence spectroscopy. Analytical Chemistry 78, 470–476.

Loeb, S., Catalona, W.J., 2007. Cancer Letters 249, 30–39.

Luchansky, M.S., Bailey, R.C., 2011. Rapid multiparameter profiling of cellular secretion using silicon photonic microring resonator arrays. Journal of the American Chemical Society 133, 20500–20506.

Luchansky, M.S., Bailey, R.C., 2010. Silicon photonic microring resonators for quantitative cytokine detection and T-cell secretion analysis. Analytical Chemistry 82, 1975–1981.

Ludwig, J.A., Weinstein, J.N., 2005. Biomarkers, in Cancer staging, prognosis, and treatment selection. Nature Reviews on Cancer 5, 84–856.

Mani, V., Chikkaveeraiah, B.V., Patl, V., Gtkind, J.S., Rusling, J.F., 2007. Nano 79, 3075–3082.

Maraldo, D., Rajal, K., Campbell, G., Mutharasan, R., 2007. Method for label-free detection of femtogram quantities of biologics in flowing liquid samples. Analytical Chemistry 79 (7), 2762–2770.

Mayer, K.M., Hao, F., Lee, S., Nordlander, P., Hafher, J.H., 2010. Nanotechnology 21.

Mazur, B., Mertas, A., Sonta-Jaykimczyk, D., Szcepanski, T., Janik, Moszant, A., 2004. Hematology and Oncology 22, 27–34.

McCreery, R.L., 2008. Chemical Reviews 108, 2646.

Merkoci, A., Pumera, M., Liopis, B., Perez, B., Valle, M.D., Allegret, S., 2008. Trends in. Analytical Chemistry 24, 826–828.

Monat, C., Domachuk, P., Eggleton, B.J., 2007. Nature Photonics 1, 106–114.

Morales, A.M., Lieber, C.M., 1998. A laser ablation technique for the synthesis of crystalline semiconductor nanowires. Science 208–211.

Morin, G.B., 1995. Cell 59, 521–529.

Nam, J.M., Thaxton, C.S., Mirkin, C.A., 2003. Science 301, 184–1886.

Nemos, C., Remy-Martin, J.P., Adami, P., Arbez-Gindre, F., Schaaal, J.P., Jouvenot, M., Delange-Mourroux, R., 2003. Clinical Biochemistry 36, 621–628.
Niklinski, J., Furman, M., 1995. European Journal of Cancer Prevention 4, 129–138.
Nyholm, L., 2005. Analyst 130, 599–605.
Okuno, J., Maehashi, K., Kerman, K., Takamura, Y., Matsumoto, K., Tamiya, E., 2007. Biosensors & Bioelectronics 22, 2377–2381.
Park, I.Y., Li, Z.Y., Li, X.M., Pisano, A.P., Williams, R.S., 2007. Biosensors & Bioelectronics 22, 2065.
Panini, N.V., Messina, G.A., Salinas, E., Fernandez, H., Raba, J., 2008. Biosensors & Bioelectronics 23, 1145–1151.
Patolsky, P., Zheng, G.F., Lieber, C.M., 2006. Nanomedicine 1, 51.
Poghossian, A., Ingebrandt, S., Abouzer, M.H., Schoning, M.J., 2007. Applied Physics A, Materials Science Process 87, 517.
Polsky, R., Gill, R., Kaganovsky, L., Willner, I., 2006. Analytical Chemistry 78, 2268–2271.
Qavi, A.J., Wasburn, A.L., Bycon, J.Y., Bailey, R.C., 2009. Analytical and Bioanalytical Chemistry 394, 121–135.
Qi, P., Vermesh, O., Green, M., Javey, A., Wang, Q., Dai, H., Peng, S., Cho, K.J., 2003. Nano Letters 3, 347–351.
Raiteri, R., Graltaola, M., Butt, H.J., Skaal, P., 2001. Sensing of biological substance based on the bending of micro fabricated cantilevers. Sensors & Actuators B 79, 115–125.
Raiteri, R., Grattarola, M., Butt, H.J., Skaldal, P., 1999. Micromechanical cantilever-based biosensors. Sensors & Actuators B 61, 213–217.
Ramachandaran, A., Wang, S., Clarke, J., Ja, S.J., Goad, D., Wld, L., Flood, E.M., Knobbe, E., Hryniwicz, J.V., Chu, S.T., Gill, D., Chen, W., King, O., Little, B.E., 2008. Biosensors & Bioelectronics 23, 939.
Robertson, J., 2004. Materials Today 7, 46–52.
Rodriguez-Mozaz, M., de Alda, M.J.L., Barcelo, D., 2008. Analytical and. Bioanalytical Chemistry 386, 1025.
Rosi, N.L., Mirkin, C.A., 2005. Chemical Reviews 105, 1547–1562.
Sanomiya, T., Hafner, C., Voros, J., 2008. Nano Letters 8, 3450–3455.
Sarkar, P., Pal, P.S., Ghosh, D., Setford, S.J., Tothill, I.E., 2002. International Journal of Pharmaceutics 238, 1–9.
Savage, P., Waxman, J., 1990. European Journal of Cancer 32A, 1087–1089.
Schwietzer, B., Wilshire, S., Lambert, J., O'Malley, S.O., Khaanskis, K., Zhu, Z., Kingsmore, S.F., Lizardi, P.M., Ward, D.C., 2000. Proceedings of the National Academy of Sciences of the United States of America 97, 10113–10119.
Shang, N.G., Papakonstantantinou, P., McMullan, M., Chu, M., Samboulis, A., Potenza, A., Dhesi, S.S., Marchetto, H., 2008. Advances Functional Materials 18, 3506.
Shay, J.W., Bachetti, S., 1997. European Journal of Cancer 33, 787–791.
Sheehan, T.M., Whitman, L.J., 2005. Nano Letters 5, 803–807.
Sipova, H., Adam, P., Pillarik, M., Springer, T., Vala, M., Homola, J., 2012. In: Bock, W.J., Gaozhi, J.X. (Eds.), Surface Plasmons for Biodetection in Photonic Sensing Principles & Applications for safety and monitoring. J Wiley & Sons, pp. 1–58.
Smith, J.E., Medley, C.D., Tang, Z.W., Shangguan, D., Lofton, C., Tan, W.H., 2007. Analytical Chemistry 79, 3075–3082.
So, H., Won, K., Kim, Y.H., Kim, B., Ryu, B.H., Na, P.S., Kim, H., Lee, J., 2005. Journal of the American Chemical Society 127, 11906–11907.
Song, S., Wang, L., Li, J., Zhao, J., Fan, C., 2008. Trends in Analytical Chemistry 27, 108–117.

Soukka, T., Antonen, K., Hama, H., Peikikangas, A.M., Huhinen, P., Lovgren, T., 2003. Clinica Chimica Acta 328, 45–58.

Squires, T.M., Messinger, R.J., Manalis, S.R., 2008. Nature Biotechnology 26, 417–426.

Srinivas, P.R., Nie, W.Y., 2006. Analytical Chemistry 78, 6476–6483.

Star, A., Gabriel, J.P., Bradley, K., Gruner, G., 2003. Nano Letters 3, 459.

Stone, M.D., Mihalusova, M., O'Connor, C.M., Prathapam, R., Collins, K., Zhuang, X., 2007. Nature 446, 458–461.

Suter, J.D., White, I.M., Zhu, H., Shi, H., Caldwell, C.W., Fan, X., 2008. Biosensors & Bioelelectronics 23, 1003.

Tang, L.H., Wang, Y., Li, Y.M., Feng, H.B., Lu, J., Li, J.H., 2009. Advances in Functional Materials 19, 1–8.

Tatematsu, K., Nakayama, J., Danbara, M., Sliionoya, S., Sao, H., Online, M., Ishikawa, F., 1996. Oncogene 13, 2265–2274.

Taton, T.A., Mirkin, C.A., Letsinger, R.L., 2000. Science 289, 1757–1760.

Teker, K., 25 September 2008. Biconjugated carbon nanotubes for targeting cancer biomarkers, Materials Science and Engineering. B 153, 83–87.

Thevenot, D.R., Toth, K., Durst, R.A., Wilson, G.S., 2001. Biosensors & Bioelectronics 16, 121.

Trojanowicz, M., 2006. Trends in Analytical Chemistry 25, 480–489.

Tsao, K.C., Wu, T.L., Chang, P.Y., Hong, J.H., Wu, J.T., 2006. Journal of Clinical Laboratory Analysis 20, 42–46.

Uludag, Y., Tothill, I.E., June 2010. Development of a sensitive detection method of cancer biomarkers in human serum (75%) using a quartz crystal microbalance sensor and nanoparticles amplification system. Talanta 82 (1), 277–282.

De Vos, K., Girones, J., Popelika, S., Schacht, E., Bacts, R., Bienstam, P., 2009. Biosensors & Bioelectronics 24, 2528.

Vicsek, T., 1989. Fractal Growth Phenomena. World Scientific, Singapore.

Viswanathan, S., Radecka, H., Radecki, J., 2009. Monatsch Chemistry 140, 891.

Vollmer, F., Arnold, S., Keng, D., 2008. Proceedings National Academic of Sciences USA 105, 20701–20704.

Wang, G., Huang, H., Zhang, G., Zhang, X., Fang, B., Wang, L., 2011. Dual-amplification strategy for fabrication of highly sensitive interleukin-6 amperometric immunosensor based on polydopamine. Langmuir 27 (3), 1224–1231.

Wang, H.T., Holmberg, B.A., Yan, Y.S., 2002. Journal of. Materials Chemistry 12, 3640–3643.

Wang, Y., Xu, H., Zhang, J.M., Li, G., 2008. Sensors B 2043.

Washburn, A.L., Luchansky, M.S., Bowman, A.L., Bailey, R.C., 2010. Analytical Chemistry 82, 69–72.

Washburn, A.L., Gunn, L.C., Bailey, R.C., 2009. Analytical Chemistry 81, 9499–9506.

Wee, K.W., Kang, G.Y., Park, J., Kang, J.Y., Yoon, D.S., Park, J.H., Kim, T.S., 2005. Biosensors & Bioelectronics 20, 1932–1938.

Weston, A.D., Hood, L., 2004. Systems biology, proteomics, and the future of health care; toward predictive preventative, and personalized medicine. Journal of Proteomic Research 3, 179–196.

White, I.M., Zhu, H., Suter, J.D., Fan, X., Zourob, M., 2009. Label-free detection with the liquid core optical ring resonator sensing platform. In: Rasooly, A., Herold, K.E. (Eds.), Methods in Molecular Biology; Biosensors and Biodetection, vol 503. Humana Press, New York, NY, pp. 139–165.

WHO, http://who.int/mediacentre/factsheets/fs297/en/index.html 2008.

Wilson, M.S., Nie, W., September 15, 2006. Multiplex measurement of seven tumor markers using an electrochemical protein chip. Analytical chemistry 78 (18), 6476–6483.

Xu, D.X., Densmoe, A., Delage, A., Waldron, P., McKinnon, R., Janz, S., Lapointe, J., Lopinski, G., Mischki, T., Post, E., Cheben, P., Sclimid, J.H., 2008. Optics Express 16, 15137.

Ye, Z., Tan, M., Wang, G., Yuan, J., 2004. Analytical Chemistry 76, 513–518.

Young, H.A., 2009. Cytokine multiplex analysis. In: Kozlov, S.V. (Ed.), Methods in Molecular Biology: Inflammation and Cancer, vol 511. Humana Press, New York, NY, pp. 85–105.

Young, B., Glesson, M., Crippe, A., 1991. C-reactive protein: a critical review. Pathology 23 (2), 118–124.

Zhang, B., Zhang, X., Yan, H.H., Xu, S.J., Tang, T.H., Fu, W.I., 2007. Biosensors & Bioelectronics 23, 19–25.

Zhu, H., Dale, P.S., Caldwell, C.W., Fan, X., 2009. Analytical Chemistry 81, 9858–9865.

Zhu, H., White, I.M., Suter, J.D., Zourob, M., Fan, X., 2008. Analyst 133, 356.

Chapter 5

Detection of Biomarkers for Myocardial Infarction

5.1 INTRODUCTION

Cardiovascular deaths account for a substantial proportion of deaths in developed as well as underdeveloped countries. In order to decrease the death rates due to these types of disease, the validation and the early detection of cardiac biomarkers assumes a predominant role in medicine. Thus, there is a lot of emphasis and effort involved in this direction. For example, one of the cardiac biomarkers that have gained a lot of attention is C-reactive protein (CRP).

Qureshi et al. (2009) have developed a novel interdigitated capacitor-based biosensor for the detection of CRP. These authors indicate that CRP is one of the plasma proteins known as acute phase proteins, and its level rises dramatically during inflammation in the body. They emphasize that CRP levels may increase by three orders of magnitude during inflammation. Pepys and Hirschfield (2003) and Casas et al. (2008) indicated that CRP is the only marker of inflammation that independently predicts the risk of a heart attack. Yang et al. (2009) emphasize that according to the American Heart Association and the Center of Diseases Control the risks for cardiovascular disease (CVD) are low for a CRP level below 1.0 mg/l, moderate for a CRP level of 1.0−3.0 mg/l, and high for a CRP concentration of 3.0 mg/l.

Panteghini (2010) very recently indicates that the blood concentration of troponin is considered as a surrogate of cardiac necrosis or myocardial infarction when a clinical situation is present. The author indicates that the presence of troponin indicates myocyte death. Thus, biomarkers of reversible myocardial damage are still required to help identify "vulnerable" patients before irreversible injury may be induced. This would facilitate an early treatment of these types of patients. The authors emphasize that the availability of biomarker information would prove to be invaluable in emergency-type situations to help provide additional independent diagnostic information.

Lin et al. (2010) have recently developed a biogenic nanoporous silica-based sensor for the enhanced electrochemical detection of two cardiovascular biomarker proteins, CRP and myeloperoxidase. These authors emphasize that the detection of multiple biomarkers is a robust method for the diagnosis of a disease (Abeloff et al., 2000; Chou et al., 2004; Danesh et al., 1998).

Biomarkers and Biosensors. http://dx.doi.org/10.1016/B978-0-444-53794-2.00005-7

Albrecht et al. (2010) have recently developed a new assay design for the clinical diagnosis of CRP levels based on alternative recognition elements. The authors indicate that in clinical settings, the sandwich-type assay format is most frequently used. However, there is still a loss of function, stability, and sensitivity in spite of the different immobilization procedures developed. They emphasize that an ideal recognition element for immobilization should possess a high affinity for the ligand, also exhibit a high specificity toward the ligand, and these properties should not be reduced upon immobilization. These authors have thus developed a new assay design assay format for CRP that uses a tailored binder as a capture element on the sensor surface, as well as an antibody for the detection element. They emphasize that their method minimizes the loss of the immobilized antibodies but still uses their evolved abilities by keeping them as the secondary detecting element.

In this chapter, we present a fractal kinetic analysis of the binding and the dissociation of the following cardiac biomarkers to appropriate receptors immobilized on biosensor surfaces: myoglobin (Mb) (Suprun et al., 2010), hFABP (fatty acid binding protein) (Ayyar et al., 2010), FABP, cTNT F11 and cTNT 7G7 (cardiac troponin) (Andersson et al., 2010), and CRP (Huang et al., 2008). It is hoped that the kinetic analysis of the different cardiac biomarkers will help provide novel physical insights into the pathological cardiovascular processes.

5.2 THEORY

5.2.1 Single-Fractal Analysis

5.2.1.1 Binding Rate Coefficient

Havlin (1989) indicates that the diffusion of a particle (analyte [Ag]) from a homogeneous solution to a solid surface (e.g., receptor [Ab]-coated surface) on which it reacts to form a product (analyte–receptor complex, (Ab.Ag)) is given by

$$(\mathrm{Ab.Ag}) \approx \begin{cases} t^{(3-D_{f,\mathrm{bind}})/2} = t^{p} & t < t_c \\ t^{1/2} & t > t_c \end{cases} \tag{5.1}$$

Here $D_{f,\mathrm{bind}}$ or D_f is the fractal dimension of the surface during the binding step and t_c is the crossover value. Havlin (1989) indicates that the crossover value may be determined by $r_c^2 \sim t_c$. Above the characteristic length, r_c, the self-similarity of the surface is lost and the surface may be considered homogeneous. Above time t_c, the surface may be considered homogeneous, since the self-similarity property disappears and "regular" diffusion is now present. For a homogeneous surface where D_f is equal to 2 and when only diffusional limitations are present, $p = ½$ as it should be. Another way of looking at the $p = ½$ case (where $D_{f,\mathrm{bind}} = 2$) is that the analyte in solution views the fractal object, in our case, the receptor-coated biosensor surface, from a "large distance." In essence, in the association process, the diffusion of the analyte from the solution to the receptor surface creates a depletion layer of width $(Đt)^{½}$, where Đ is the diffusion constant. This gives rise to the fractal power law, $(\text{Analyte.Receptor}) \sim t^{(3-D_{f,bind})/2}$. For the present analysis, t_c is arbitrarily chosen and we assume that the value of the t_c is not reached. One may consider the approach as an intermediate "heuristic" approach that may be used in the future to develop an autonomous (and not time-dependent) model for diffusion-controlled kinetics.

5.2.1.2 Dissociation Rate Coefficient

The diffusion of the dissociated particle (receptor [Ab] or analyte [Ag]) from the solid surface (e.g., (analyte [Ag]—receptor [Ab]) complex-coated surface) into solution may be given, as a first approximation by

$$(\text{Ab.Ag}) \approx -t^{(3-D_{f,diss})/2} = -t^{p} \quad (t > t_{diss}) \tag{5.2}$$

Here $D_{f,diss}$ is the fractal dimension of the surface for the dissociation step. This corresponds to the highest concentration of the analyte—receptor complex on the surface.

Henceforth, its concentration only decreases. The dissociation kinetics may be analyzed in a manner "similar" to the binding kinetics.

5.2.2 Dual-Fractal Analysis

5.2.2.1 Binding Rate Coefficient

Sometimes, the binding curve exhibits complexities, and two parameters (k, D_f) are not sufficient to adequately describe the binding kinetics. This is further corroborated by low values of r^2 factor (goodness-of-fit). In that case, one resorts to a dual-fractal analysis (four parameters: k_1, k_2, D_{f1}, and D_{f2}) to adequately describe the binding kinetics. The single-fractal analysis presented above is thus extended to include two fractal dimensions. At present, the time $(t = t_1)$ at which the "first" fractal dimension "changes" to the "second" fractal dimension is arbitrary and empirical. For the most part, it is dictated by the data analyzed and experience gained by handling a single-fractal analysis. A smoother curve is obtained in the "transition" region if care is taken to select the correct number of points for the two regions. In this case, the product (antibody—antigen or analyte—receptor complex, Ab.Ag or analyte.receptor) is given by

$$(\text{Ab.Ag}) \approx \begin{cases} t^{(3-D_{f1,blind})/2} = t^{p1} & (t < t_1) \\ t^{(3-D_{f2,blind})/2} = t^{p2} & (t_1 < t < t_2) = t_c \\ t^{1/2} & (t > t_c) \end{cases} \tag{5.3}$$

In some cases, as mentioned above, a triple-fractal analysis with six parameters (k_1, k_2, k_3, D_{f1}, D_{f2}, and D_{f3}) may be required to adequately model the binding kinetics. This is when the binding curve exhibits convolutions and complexities in its shape due to perhaps the very dilute nature of the analyte (in some of the cases to be presented) or for some other reasons. Also, in some cases, a dual-fractal analysis may be required to describe the dissociation kinetics.

5.3 RESULTS

A fractal analysis is presented for

1. the binding of cardiac Mb in solution to anti-Mb immobilized on an electrochemical nanobiosensor (Suprun et al., 2010),
2. the binding and dissociation of hFABP (human fatty acid binding protein) in solution to anti-FABP immobilized on an ELISA surface (Ayyar et al., 2010),
3. the binding of cardiac biomarkers FABP, cTNT 1F1, and cTNT 7G7 in solution to appropriate receptors (antibodies) immobilized on a CM5 sensor chip (Andersson et al., 2010), and
4. the binding and dissociation of CRP in solution to anti-CRP immobilized on different locations on the sensor surface (Huang et al., 2008).

Suprun et al. (2010) indicate that CVD causes nearly half of all deaths in the Western world (Allender et al., 2008) and over half of all the deaths in Russia. Suprun et al. (2010) indicate that myocardial infraction arises from a reduction in blood flow due to necrosis of the myocardium. These authors emphasize that the application of cardiac biomarkers assumes importance, since the levels are always elevated during heart occlusion and in CVD. They further indicate that due to the small size of Mb (17.8 kDa), it is released into the circulation in 1−3 h upon symptom onset. It reaches a maximum level in 6−12 h (Melanson and Tanasijevic, 2005). This, according to McDonnell et al. (2009), allows using Mb as a characteristic of early confirmation. These authors emphasize that this is one of the early acute myocardial infarction (AMI) biomarkers to increase after AMI onset. It returns to its baseline level within 24−48 h. Christenson and Azzazy (2009) indicate that there is a twofold search for cardiac biomarkers: (1) search for new cardiac biomarkers and (2) better and more sensitive and selective point-of-care (POC) detection devices for proven cardiac biomarkers.

Figure 5.1(a) shows the binding and dissociation of 2 μl of 14 μM Mb in solution to anti-Mb immobilized on an electrochemical nanosensor. Two microliters of Mb at pH 7.4 was dried overnight at 4 °C. A single-fractal analysis is adequate to describe the binding and the dissociation kinetics. The values of the binding rate coefficient, k, and the fractal dimension, D_f, for a single-fractal analysis are given in Table 5.1. In this case, the affinity, K (= k/k_d), value is 1.181.

Figure 5.1(b) shows the binding and dissociation of 2 μl of 14 μM Mb in solution to anti-Mb immobilized on an electrochemical nanosensor. Two

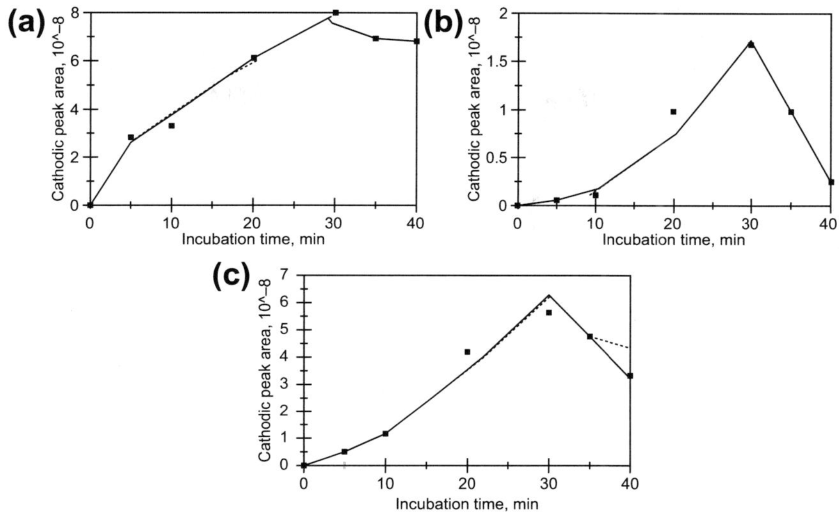

FIGURE 5.1 Binding of cardiac myoglobin in solution to anti-myoglobin immobilized on an electrochemical nanobiosensor at different temperatures (Suprun et al., 2010): (a) 4 °C (b) 25 °C (c) 37 °C.

microliters of Mb at pH 7.4 was dried for 15 min at room temperature (25 °C). A single-fractal analysis is, once again, adequate to describe the binding and the dissociation kinetics. The values of the binding rate coefficient, k, and the fractal dimension, D_f, for a single-fractal analysis are given in Table 5.1. In this case, the affinity, K (= k/k_d), value is 0.0098.

Figure 5.1(c) shows the binding and dissociation of 2 μl of 14 μM Mb in solution to anti-Mb immobilized on an electrochemical nanosensor. Two microliters of Mb at pH 7.4 was dried for 15 min at 37° C. A single-fractal analysis is, once again, adequate to describe the binding and the dissociation kinetics. The values of the binding rate coefficient, k, and the fractal dimension, D_f, for a single-fractal analysis are given in Table 5.1. In this case, the affinity, K (= k/k_d), value is 0.1297.

Table 5.1 and Figure 5.2(a) show the increase in the binding rate coefficient, k, with an increase in the fractal dimension, D_f, for a single-fractal analysis. For the data shown in Figure 5.2(a), the binding rate coefficient, k, is given by

$$k = (0.4248 \pm 0.1706) D_f^{0.9363 \pm 0.09137} \qquad (5.3a)$$

The fit is good. Only three data points at the different temperatures are available. The availability of more data points would lead to a more reliable fit. The binding rate coefficient, k, exhibits close to a first order of dependence (equal to 0.9363) on the fractal dimension or the degree of heterogeneity that exists on the electrochemical nanosensor surface in the binding phase.

TABLE 5.1 Binding and Dissociation Rate Coefficients (k and k_d, respectively) and Affinity Values (K), and Fractal Dimensions for the for the Binding and Dissociation Phase (D_f and D_{fd}, respectively) for 14 μM Myoglobin (Mb) in Solution to Anti-Mb Immobilized on an Electrochemical Nanosensor at Different Temperatures (Suprun et al., 2010)

Conditions and Temperature	k	k_d	D_f	D_{fd}	K
2 μl Mb at pH 7.4 Dried Overnight at 4 °C	0.9469 ± 0.1423	0.8014	1.7720 ± 0.2048	2.7250	1.181
2 μl Mb at pH 7.4 Dried for 15 min at Room Temperature (25 °C)	0.00126 ± 0.00075	0.1290	0 + 0.682	0.8988	0.0098
2 μl Mb at pH 7.4 Dried for 15 min at 37 °C	0.0497 ± 0.0084	0.3832	0.1542 ± 0.2284	1.9390	0.1297

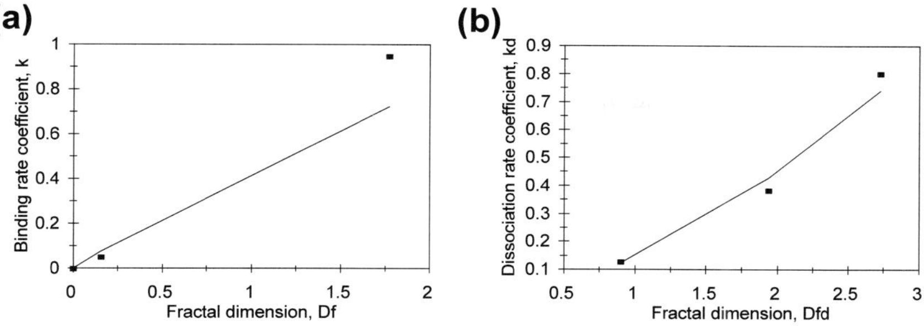

FIGURE 5.2 (a) Increase in the binding rate coefficient, k, with an increase in the fractal dimension, D_f (b) Increase in the dissociation rate coefficient, k_d, with an increase in the fractal dimension for dissociation, D_{fd}.

Figure 5.2(b) and Table 5.1 show the increase in the dissociation rate coefficient, k_d, with an increase in the fractal dimension, D_{fd}, for a single-fractal analysis. For the data shown in Figure 5.2(b), the dissociation rate coefficient, k_d, is given by

$$k = (0.1476 \pm 0.03624)D_{fd}^{1.609 \pm 0.1755} \qquad (5.3b)$$

The fit is good. Only three data points at the different temperatures are available. The availability of more data points would lead to a more reliable fit. The dissociation rate coefficient, k_d, exhibits an order of dependence slightly higher than one and one-half (equal to 1.609) on the fractal dimension or the degree of heterogeneity that exists on the electrochemical nanosensor surface in the dissociation phase.

Ayyar et al. (2010) have recently presented an analysis using ELISA to differentiate between heart-type fatty acid binding protein from noncardiac isoforms. These authors indicate that by the year 2020, it is estimated that the death rate from CVD in both developed and developing countries will reach 23.6 million, thus the need for identifying, validating, and deploying cardiac biomarkers for POC applications (McDonnell et al., 2009). Heart-type binding acid protein is involved in maintaining myocardial lipid homeostasis (Pasaolu et al., 2007). Kleine et al. (1992) and Glatz et al. (1998) indicate that hFABP is a promising early biomarker for myocardial infarction. Morrow et al. (2007) indicate that hFABP has similar rapid release and clearance kinetics to Mb. However, Van Nieuwenhoven et al. (1995) and Kragten et al. (1996) indicate that since hFABP has a greater relative myocardial/skeletal tissue content ratio than Mb, it may be considered as a more "cardiac-specific" marker than Mb. Ayyar et al. (2010) have developed a diagnostic assay specific to the hFABP isoform using a high-affinity recombinant antibody.

Figure 5.3(a) shows the binding and dissociation of 12.5 nM heart-type fatty acid binding in solution to anti-heart-type binding protein immobilized to an ELISA surface (Ayyar et al., 2010). A single-fractal analysis is adequate

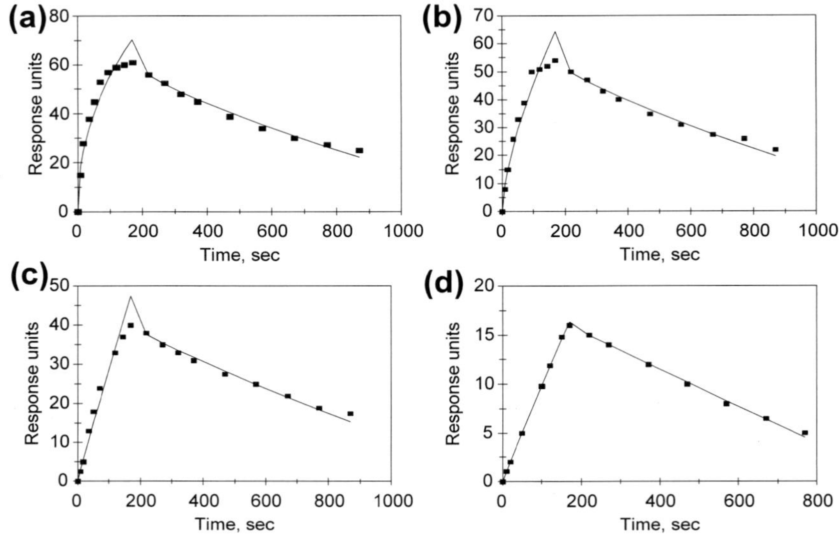

FIGURE 5.3 Binding and dissociation of different concentrations of human fatty acid binding protein (hFABP) (in nM) in solution to anti-hFABP immobilized on an ELISA surface (Ayyar et al., 2010): (a) 12.5 (b) 25 (c) 50 (d) 100.

to describe the binding and the dissociation kinetics. The values of (1) the binding rate coefficient, k, and the fractal dimension, D_f, for a single-fractal analysis and (2) the dissociation rate coefficient, k_d, and the fractal dimension, D_{fd}, for a single-fractal analysis are given in Table 5.2. In this case, the affinity, K, value is 5.290.

Figure 5.3(b) shows the binding and dissociation of 25 nM heart-type fatty acid binding in solution to anti-heart-type binding protein immobilized to an ELISA surface (Ayyar et al., 2010). Once again, a single-fractal analysis is adequate to describe the binding and the dissociation kinetics. The values of (1) the binding rate coefficient, k, and the fractal dimension, D_f, for a single-fractal analysis and (2) the dissociation rate coefficient, k_d, and the fractal dimension, D_{fd}, for a single-fractal analysis are given in Table 5.2. In this case, the affinity, K, value is 4.77.

Figure 5.3(c) shows the binding and dissociation of 50 nM heart-type fatty acid binding in solution to anti-heart-type binding protein immobilized to an ELISA surface (Ayyar et al., 2010). Once again, a single-fractal analysis is adequate to describe the binding and the dissociation kinetics. The values of (1) the binding rate coefficient, k, and the fractal dimension, D_f, for a single-fractal analysis, and (2) the dissociation rate coefficient, k_d, and the fractal dimension, D_{fd}, for a single-fractal analysis are given in Table 5.2. In this case, the affinity, K, value is 11.77.

Figure 5.3(d) shows the binding and dissociation of 100 nM heart-type fatty acid binding in solution to anti-heart-type binding protein immobilized

TABLE 5.2 Binding and Dissociation Rate Coefficients (k and k_d, respectively), Affinity Values (K) and Fractal Dimensions for the Binding and the Dissociation Phase (D_f and D_{fd}, respectively) for Different Concentrations of Human Fatty Acid Binding Protein (hFABP) in Solution to Anti-hFABP immobilized on an ELISA Surface (Suprun et al., 2010)

hFABP Concentration	k	k_d	D_f	D_{fd}	K
12.5	0.1185 ± 0.0125	0.02243 ± 0.0078	1.0816 ± 0.1569	1.1494 ± 0.2684	5.290
25	0.3677 ± 0.0665	0.0771 ± 0.0087	1.1180 ± 0.1196	1.2374 ± 0.0858	4.77
50	2.306 ± 0.329	0.1970 ± 0.0157	1.6042 ± 0.0940	1.4244 ± 0.0610	11.7
100	6.926 ± 0.993	0.2706 ± 0.0173	2.0966 ± 0.0944	1.4914 ± 0.4936	25.6

to an ELISA surface (Ayyar et al., 2010). Once again, a single-fractal analysis is adequate to describe the binding and the dissociation kinetics. The values of (1) the binding rate coefficient, k, and the fractal dimension, D_f, for a single-fractal analysis, and (2) the dissociation rate coefficient, k_d, and the fractal dimension, D_{fd}, for a single-fractal analysis are given in Table 5.2. In this case the affinity, K, value is 25.6.

Figure 5.4(a) and Table 5.2 show the increase in the binding rate coefficient, k, with an increase in the hFABP concentration in solution in nanomolarity. For the data shown in Figure 5.4(a), the binding rate coefficient, k, is given by

$$k = (0.000667 \pm 0.000171)[\text{hFABP}]^{2.0256 \pm 0.144} \tag{5.4a}$$

The fit is good. Only four data points are available. The availability of more data points would lead to a more reliable fit. The binding rate coefficient, k, exhibits an order of dependence slightly greater than second (equal to 2.0256) on the hFABP concentration in solution in nanomolarity. This indicates that the binding rate coefficient, k, is quite sensitive to the hFABP concentration in solution at least in this range.

Figure 5.4(b) and Table 5.2 show the increase in the dissociation rate coefficient, k_d, with an increase in the hFABP concentration in solution in nanomolarity. For the data shown in Figure 5.4(b), the dissociation rate coefficient, k_d, is given by

$$k_d = (0.001293 \pm 0.000503)[\text{hFABP}]^{1.2137 \pm 0.212} \tag{5.4b}$$

The fit is good. Only four data points are available. The availability of more data points would lead to a more reliable fit. The dissociation binding rate coefficient, k_d, exhibits an order of dependence greater than first (equal to 1.2137) on the hFABP concentration in solution in nanomolarity.

Figure 5.4(c) and Table 5.2 show the increase in the affinity, K, with an increase in the hFABP concentration in solution in nanomolarity. For the data shown in Figure 5.4(c) the affinity, K, is given by

$$K = (0.5157 \pm 0.2232)[\text{hFABP}]^{0.8118 \pm 0.232} \tag{5.4c}$$

The fit is good. Only four data points are available. The availability of more data points would lead to a more reliable fit. The affinity, K, exhibits less than a first (equal to 0.8118) order of dependence on the hFABP concentration in solution in nanomolarity.

Figure 5.4(d) and Table 5.2 show the increase in the binding rate coefficient, k, with an increase in the fractal dimension, D_f, or the degree of heterogeneity that exists on the surface plasmon resonance (SPR) biosensor surface. For the data shown in Figure 5.4(d), the binding rate coefficient, k, is given by

$$k = (0.1271 \pm 0.0860) D_f^{5.624 \pm 0.948} \tag{5.4d}$$

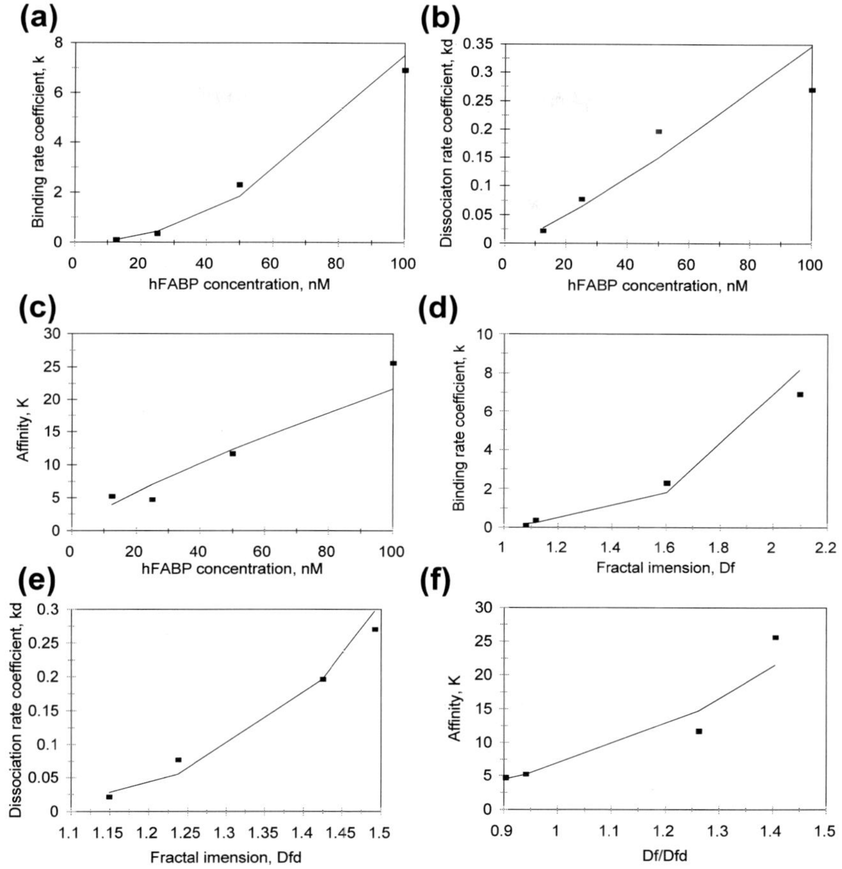

FIGURE 5.4 (a) Increase in the binding rate coefficient, k, with an increase in the human fatty acid binding protein (hFABP) concentration in solution (b) Increase in the dissociation rate coefficient, k_d, with an increase in the hFABP concentration in solution (c) Increase in the affinity, K, with an increase in the hFABP concentration in solution (d) Increase in the binding rate coefficient, k, with an increase in the fractal dimension, D_f (e) Increase in the dissociation rate coefficient, k_d, with an increase in the fractal dimension, D_{fd} (f) Increase in the affinity, K, with an increase in the fractal dimension ratio, D_f/D_{fd} (g) Increase in the fractal dimension, D_f with an increase in the hFABO concentration (in nM) in solution (h) Increase in the fractal dimension for dissociation, D_{fd} with an increase in the hFABO concentration in solution (i) Decrease in the fractal dimension ratio, D_f/D_{fd}, K with an increase in the hFABO concentration on solution.

The fit is good. Only four data points are available. The availability of more data points would lead to a more reliable fit. The binding rate coefficient, k, is very sensitive to the fractal dimension, D_f, or the degree of heterogeneity that exists on the SPR biosensor surface, since it exhibits greater than a fifth (equal to 5.624) order of dependence on the fractal dimension.

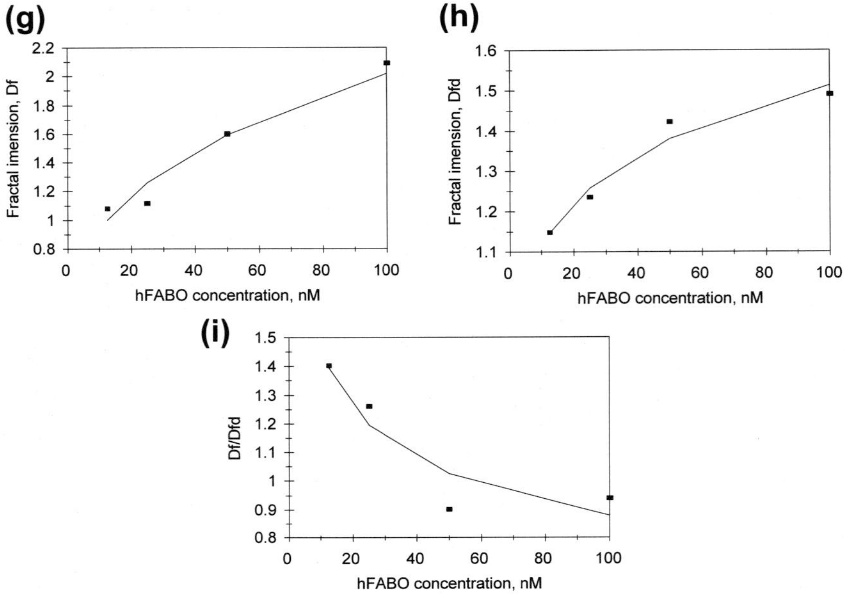

FIGURE 5.4 *Continued.*

Figure 5.4(e) and Table 5.2 show the increase in the dissociation rate coefficient, k_d, with an increase in the fractal dimension, D_{fd}, or the degree of heterogeneity that exists on the SPR biosensor surface in the dissociation phase. For the data shown in Figure 5.4(e), the dissociation rate coefficient, k_d, is given by

$$k_d = (0.01092 \pm 0.001810)D_{fd}^{9.011 \pm 1.413} \tag{5.4e}$$

The fit is good. Only four data points are available. The availability of more data points would lead to a more reliable fit. The dissociation rate coefficient, k_d, is extremely sensitive to the fractal dimension, D_{fd}, or the degree of heterogeneity that exists on the SPR biosensor surface in the dissociation phase, since it exhibits greater than a ninth (equal to 9.011) order of dependence on the fractal dimension.

Figure 5.4(f) and Table 5.2 show the increase in the affinity, K, with an increase in the fractal dimension ratio, D_f/D_{fd}, that exists on the SPR biosensor surface. For the data shown in Figure 5.4(f), the affinity, K, is given by

$$K = (6.488 \pm 2.493)D_f/D_{fd}^{3.525 \pm 0.5504} \tag{5.4f}$$

The fit is good. Only four data points are available. The availability of more data points would lead to a more reliable fit. The affinity, K, is sensitive to the fractal dimension ratio, D_f/D_{fd}, that exists on the SPR biosensor surface, since it exhibits slightly greater than a three and one-half (equal to 3.525) order of dependence on the fractal dimension ratio.

Figure 5.4(g) and Table 5.2 show the increase in the fractal dimension, D_f, with an increase in the hFABP concentration in nanomolarity in solution. For the data shown in Figure 5.4(g), the fractal dimension, D_f, is given by

$$D_f = (0.4246 \pm 0.0477)[\text{hFABP}]^{0.3385 \pm 0.171} \quad (5.4g)$$

The fit is good. Only four data points are available. The availability of more data points would lead to a more reliable fit. The fractal dimension exhibits less than one-half (equal to 0.3385) order of dependence on the hFABP concentration in solution in the range studied.

Figure 5.4(h) and Table 5.2 show the increase in the fractal dimension, D_{fd}, with an increase in the hFABP concentration in nanomolarity in solution. For the data shown in Figure 5.4(h), the fractal dimension, D_{fd}, is given by

$$D_{fd} = (0.8204 \pm 0.0228)[\text{hFABP}]^{0.1330 \pm 0.01768} \quad (5.4h)$$

The fit is good. Only four data points are available. The availability of more data points would lead to a more reliable fit. The fractal dimension exhibits close to a zero (equal to 0.1330) order of dependence on the hFABP concentration in solution in the range studied.

Figure 5.4(i) and Table 5.2 show the decrease in the fractal dimension ratio, D_f/D_{fd} with an increase in the hFABP concentration in nanomolarity in solution. For the data shown in Figure 5.4(i), the fractal dimension ratio, D_f/D_{fd}, is given by

$$K = (0.8204 \pm 0.0228)[\text{hFABP}]^{-0.2217 \pm 0.01767} \quad (5.4i)$$

The fit is good. Only four data points are available. The availability of more data points would lead to a more reliable fit. The fractal dimension ratio, D_f/D_{fd} exhibits close to a zero (equal to −0.2217) order of dependence on the hFABP concentration in solution in the range studied.

Andersson et al. (2010) have recently used an SPR biosensor to assay the cardiac biomarkers, FABP 3, and cardiac troponin. This is for the case when cardiac cells are exposed to the cardioactive drug, doxorubicin. These authors indicate that there is a need for cardiomyocytes for use in toxicity testing. However, access of primary cells for the human heart is limited. The human embryonic stem cells (hESCs) differentiated into cardiomyocyte-like cells may be potentially used to replace cell lines currently used in cardiotoxicity testing. Strumann et al. (2009) emphasize that the use of cells of human origin should improve the predictability of human cardiotoxicity tests by avoiding interspecies differences.

Andersson et al. (2010) emphasize that a detailed understanding of the physiology of hESC-derived cardiomyocytes is important in order that one may use them for the cardiotoxicity of drugs. They further indicate that Troponin T (cTNT) and FABP 3 are important biomarkers for assaying adverse cardiotoxic responses of drugs. Different authors have indicated that these biomarkers are also considered as sensitive translational biomarkers for drug-induced toxicity

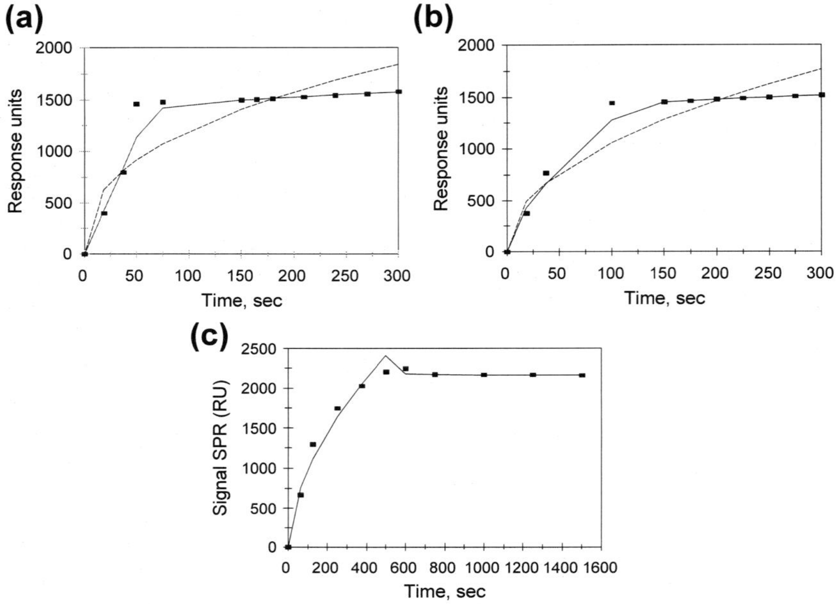

FIGURE 5.5 Binding and dissociation of different cardiac biomarkers to appropriate receptors immobilized on a CM5 sensor surface (Andersson et al., 2010): (a) FABP 3 (b) cTNT 1F11 (c) cTNT 7G7.

(Key et al., 1999; Nakata et al., 2003; Okamoto et al., 2000; O'Brien, 2008). Andersson et al. (2010) emphasize the need for sensitive and accurate bioanalytical tools for in vitro assays of biomarkers. They indicate that the following biosensor techniques are potential candidates for use as possible detection techniques for biomarkers: acoustic wave surface plasmon resonance and total internal reflection fluorescence methods (Kunz et al., 1996; Spangler et al., 2001; Dimitriev et al., 2003).

Figure 5.5(a) shows the binding of the cardiac biomarker FABP 3 to the appropriate antibody immobilized on a CM5 sensor chip on an SPR biosensor (Biacore 2000™, GE Healthcare AB). A dual-fractal analysis is required to adequately describe the binding kinetics. The values of (1) the binding rate coefficient, k, and the fractal dimension, D_f, for a single-fractal analysis, and (2) the binding rate coefficients, k_1 and k_2, and the fractal dimensions, D_{f1} and D_{f2}, for a dual-fractal analysis are given in Table 5.3. For a dual-fractal analysis, it is of interest to note that as the fractal dimension increases by a factor of 3.132 from a value of $D_{f1} = 0.9432$ to $D_{f2} = 2.9548$, the binding rate coefficient increases by a factor of 51.22 from a value of $k_1 = 20.302$ to $k_2 = 1039.91$. Changes in the fractal dimension or the degree of heterogeneity on the CM5 sensor surface and in the binding rate coefficient are in the same direction.

TABLE 5.3 Binding Rate Coefficients (k, k_1, k_2) and Fractal Dimensions (D_f, D_{f1}, D_{f2}) for Cardiac Biomarkers FABP, cTNT 1F1, and cTNT 7G7 to Appropriate Receptors (Antibodies) Immobilized on a CM5 Sensor Chip (Andersson et al., 2010)

Analyte in Solution	k	k_1	k_2	D_f	D_{f1}	D_{f2}
FABP 3	199.71 ± 57.70	20.302 ± 4.844	1039.91 ± 2.72	2.2212 ± 0.1758	0.9432 ± 0.4274	2.9548 ± 0.0822
cTNT 1F11	125.19 ± 13.86	60.835 ± 12.27	1056.18 ± 1.16	2.072 ± 0.126	1.6764 ± 0.2262	2.7728 ± 0.0036
cTNT 7G7	129.68 ± 23.98	63.983 ± 12.278	1056.16 ± 1.18	2.086 ± 0.1232	1.798 ± 0.2162	2.7228 ± 0.0037

Figure 5.5(b) shows the binding of the cardiac biomarker cTNT 1F11 (troponin) to the appropriate antibody immobilized on a CM5 sensor chip on an SPR biosensor (Biacore 2000™, GE Healthcare AB). Once again, a dual-fractal analysis is required to adequately describe the binding kinetics. The values of (1) the binding rate coefficient, k, and the fractal dimension, D_f, for a single-fractal analysis and (2) the binding rate coefficients, k_1 and k_2, and the fractal dimensions, D_{f1} and D_{f2}, for a dual-fractal analysis are given in Table 5.3. For a dual-fractal analysis, it is of interest to note that as the fractal dimension increases by a factor of 1.654 from a value of $D_{f1} = 1.6764$ to $D_{f2} = 2.7728$, the binding rate coefficient increases by a factor of 17.36 from a value of $k_1 = 60.835$ to $k_2 = 1056.18$. Changes in the fractal dimension or the degree of heterogeneity on the CM5 sensor surface and in the binding rate coefficient are, once again, in the same direction.

Figure 5.5(c) shows the binding of the cardiac biomarker cTNT 7G7 (troponin) to the appropriate antibody immobilized on a CM5 sensor chip on an SPR biosensor (Biacore 2000™, GE Healthcare AB). Once again, a dual-fractal analysis is required to adequately describe the binding kinetics. The values of (1) the binding rate coefficient, k, and the fractal dimension, D_f, for a single-fractal analysis and (2) the binding rate coefficients, k_1 and k_2, and the fractal dimensions, D_{f1} and D_{f2}, for a dual-fractal analysis are given in Table 5.3. For a dual-fractal analysis, it is of interest to note that as the fractal dimension increases by a factor of 1.542 from a value of $D_{f1} = 1.798$ to $D_{f2} = 2.7728$, the binding rate coefficient increases by a factor of 16.51 from a value of $k_1 = 63.983$ to $k_2 = 1056.16$. Once again, changes in the fractal dimension or the degree of heterogeneity on the CM5 sensor surface and in the binding rate coefficient are in the same direction.

Figure 5.6(a) and Table 5.3 show for a dual-fractal analysis the increase in the binding rate coefficient, k_1, with an increase in the fractal dimension, D_{f1}. For the data shown in Figure 5.6(a), the binding rate coefficient, k_1, is given by

$$k_1 = (22.695 \pm 1.294)D_{f1}^{1.829 \pm 0.110} \tag{5.5a}$$

The fit is good. Only three data points are available. The availability of more data points would lead to a more reliable fit. The binding rate coefficient, k_1, is sensitive to the fractal dimension, D_{f1}, or the degree of heterogeneity that exists on the biosensor surface, as k_1 exhibits an order of dependence between one and one-half and two (equal to 1.829) on D_{f1}.

Figure 5.6(b) and Table 5.3 show for a dual-fractal analysis the decrease in the binding rate coefficient, k_2, with an increase in the fractal dimension, D_{f2}. For the data shown in Figure 5.6(b), the binding rate coefficient, k_2, is given by

$$k_2 = (1354.65 \pm 0.02)D_{f2}^{-0.244 \pm 0.00026} \tag{5.5b}$$

The fit is very good. Only three data points are available. The availability of more data points would lead to a more reliable fit. The binding rate

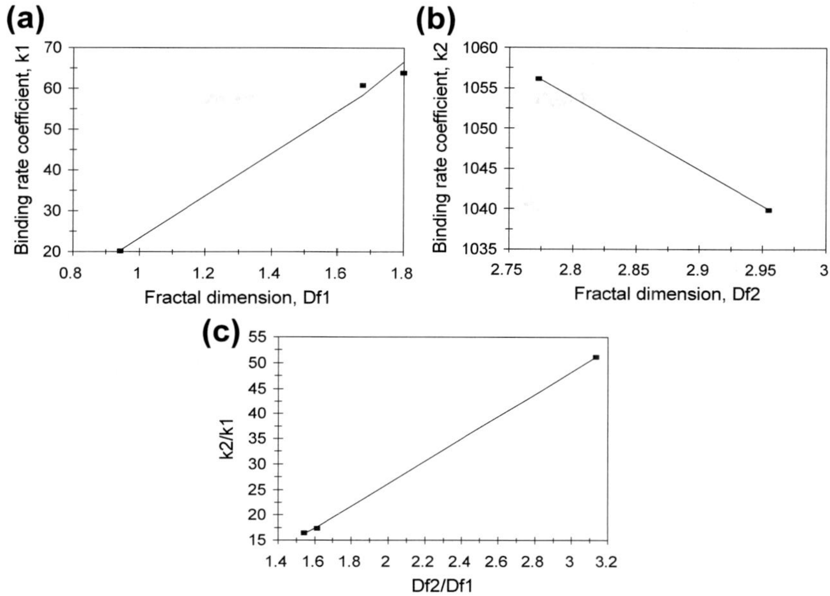

FIGURE 5.6 (a) Increase in the binding rate coefficient, k_1, with an increase in the fractal dimension, D_{f1} (b) Decrease in the binding rate coefficient, k_2, with an increase in the fractal dimension, D_{f2} (c) Increase in the binding rate coefficient ratio, k_2/k_1, with an increase in the fractal dimension ratio, D_{f2}/D_{f1}.

coefficient, k_2, exhibits a slight dependence on the fractal dimension, D_{f2}, or the degree of heterogeneity that exists on the biosensor surface, as k_2 exhibits an order of dependence less than one-half (equal to -0.244) on D_{f2}.

Figure 5.6(c) and Table 5.3 show for a dual-fractal analysis the increase in the ratio of the binding and dissociation rate coefficients, k_2/k_1, with an increase in the fractal dimension ratio, D_{f2}/D_{f1}. For the data shown in Figure 5.6(c), the binding rate coefficient ratio, k_2/k_1, is given by

$$k_2/k_1 = (8.134 \pm 0.123)\left(D_{f2}/D_{f1}\right)^{1.611 \pm 0.0268} \tag{5.5c}$$

The fit is very good. Only three data points are available. The availability of more data points would lead to a more reliable fit. The binding rate coefficient ratio, k_2/k_1, is sensitive to the fractal dimension ratio, D_{f2}/D_{f1}, since it exhibits an order of dependence between one and one-half and second (equal to 1.611) on the fractal dimension ratio, D_{f2}/D_{f1}, that exists on the biosensor surface.

Huang et al. (2008) have proposed and analyzed an electrothermal effect to help enhance the binding efficiency for an immunoassay. They used CRP and immunoglobulin G (IgG) as model analytes in physiological environments. The authors emphasize that in this diffusion reaction-type system, the diffusion boundary layer does hinder the binding and the dissociation reaction. This, in turn, restricts the response time as well as the overall performance of the biosensor. Mixing is always a concern in these types of systems. The authors have proposed and applied a nonuniform AC electric field to the flow microchannel of the biosensor. This, the authors indicate, generates a pair of vortices to help stir the flow field. By selecting an appropriate place for the biosensor, the flow field is accelerated fast enough to help depress the growth of the boundary layer. They emphasize that the interference patterns of the flow field thus created at different locations in the microchannel help enhance the binding and the dissociation kinetics. They indicate that by changing the location of the biosensor they found that the highest enhancement was found to be at a position near the negative electrode. They also noted that a change in the boundary temperature conditions also leads to an enhancement of the binding reaction. Their theoretical analysis is briefly summarized below.

Huang et al. (2008) indicate that the time required for a specific biomolecular reaction (such as that occurring between an analyte in solution and a receptor immobilized on a sensor surface) depends on the Damkohler (Da) number. The Da number (which is a dimensionless number) measures whether a reaction is diffusion limited or reaction limited. They further indicate that the Da number is the ratio of the reaction velocity to the diffusion velocity. The reaction velocity is the product of the association rate constant and the initial concentration of the ligand, and the diffusion velocity is the ratio of the diffusion coefficient of the analyte in the flow to the height of the microchannel (Deen, 1998). The reaction is diffusion limited when Da is greater than unity. In other words, the reaction at the biosensor surface is limited by the diffusion of the analyte in solution to the receptor (ligand) immobilized on the biosensor surface.

Hibbert and Gordon (2002) indicate that as the analyte takes a longer time to reach the receptor immobilized on the biosensor surface by a convection and diffusion mechanism, this leads to the formation of a diffusion boundary layer. It is this diffusion boundary layer that restricts the response time as well as the overall performance of the biosensor. Sigurdon et al. (2005) had initially indicated that the electrothermal microstirring effect may be used in a microchannel to help enhance the binding efficiency for diffusion-limited molecules (large-Da proteins).

A two-dimensional full time-scale process has been used to simulate the binding (association) and dissociation of CRP and IgG as model proteins to a biosensor surface. The authors also presented an "improved" design for their biosensor by introducing a "neck" design. This, the authors claim, reduces the distance between the diffusion distance of the analyte in the bulk solution, besides increasing the flow velocity of the solution over the reaction surface. This may be compared with the general case where the reaction surface is at the bottom position of the microchannel.

CRP is a protein produced by the liver, whose level rises in response to inflammation throughout the body (Ridker and Libby). It is extensively used as a marker for inflammation and diseases associated with it. Patients with a higher basal level of CRP are at an increased risk of diabetes, hypertension, and other CVD. Monitoring CRP levels in the human body can be very useful in determining the progress of a disease or evaluating the effectiveness of a treatment. A CRP test is a general test carried out to check for inflammation in any area of the body. If it exceeds a critical value, there is strong indication of the presence of an inflammation, though it may be difficult to give the specific location. On the other hand, a low CRP value does not always translate to the absence of inflammation, as observed in people with rheumatoid arthritis or lupus (systemic lupus erythematosus). In such cases, a higher sensitive test, hs-CRP test, could help determine an individual's vulnerability to heart disease.

Figure 5.7(a) shows the binding and the dissociation of CRP in solution to anti-CRP immobilized on the immunoassay surface at location (250, 1.51) and under the influence of the electrothermal effect (Huang et al., 2008). A dual-fractal analysis is required to adequately describe the binding kinetics. A single-fractal analysis is adequate to describe the dissociation kinetics.

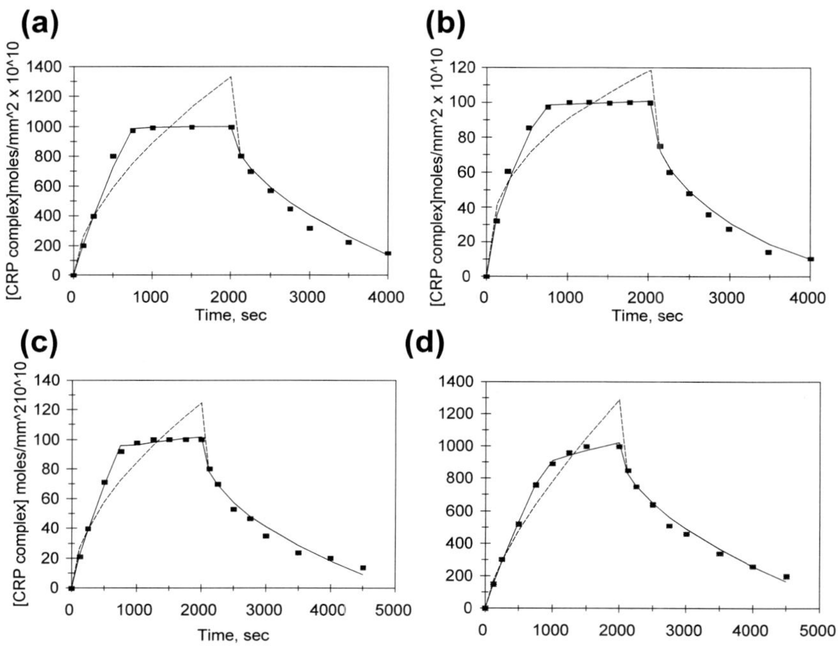

FIGURE 5.7 Binding and dissociation of C-reactive protein (CRP) in solution to anti-CRP immobilized at different locations on a biosensor surface (Huang et al., 2008): (a) (250,1.51) (b) (250,75) (c) (277.5,148.5) (d) (281,1.33).

The values of (1) the binding rate coefficient, k, and the fractal dimension, D_f, for a single-fractal analysis, (2) the binding rate coefficients, k_1 and k_2, and the fractal dimensions, D_{f1} and D_{f2}, for a dual-fractal analysis, and (3) the dissociation rate coefficient, k_d, and the fractal dimension in the dissociation phase, D_{fd}, for a single-fractal analysis are given in Tables 5.4 and 5.5. It is of interest to note that as the fractal dimension increases by a factor of 2.50 from a value of $D_{f1} = 1.1834$ to $D_{f2} = 2.9615$, the binding rate coefficient increases by a factor of 333.1 from a value of $k_1 = 2.584$ to $k_2 = 860.74$. An increase in the fractal dimension on the immunoassay surface leads to a substantial increase in the binding rate coefficient.

Figure 5.7(b) shows the binding and the dissociation of CRP in solution to anti-CRP immobilized on the immunoassay surface at location (250, 0.75) and under the influence of the electrothermal effect (Huang et al., 2008). A dual-fractal analysis is required to adequately describe the binding kinetics. A single-fractal analysis is adequate to describe the dissociation kinetics. The values of (1) the binding rate coefficient, k, and the fractal dimension, D_f, for a single-fractal analysis, (2) the binding rate coefficients, k_1 and k_2, and the fractal dimensions, D_{f1} and D_{f2}, for a dual-fractal analysis, and (3) the dissociation rate coefficient, k_d, and the fractal dimension in the dissociation phase, D_{fd}, for a single-fractal analysis are given in Tables 5.4 and 5.5. It is of interest to note that as the fractal dimension increases by a factor of 1.674 from a value of $D_{f1} = 1.7676$ to $D_{f2} = 2.9595$, the binding rate coefficient increases by a factor of 46.8 from a value of $k_1 = 1.7234$ to $k_2 = 83.006$. Once again, an increase in the fractal dimension on immunoassay surface leads to a substantial increase in the binding rate coefficient.

Figure 5.7(c) shows the binding and the dissociation of CRP in solution to anti-CRP immobilized on the immunoassay surface at location (277.5, 148.5) and under the influence of the electrothermal effect (Huang et al., 2008). A dual-fractal analysis is required to adequately describe the binding kinetics. A single-fractal analysis is adequate to describe the dissociation kinetics. The values of (1) the binding rate coefficient, k, and the fractal dimension, D_f, for a single-fractal analysis, (2) the binding rate coefficients, k_1 and k_2, and the fractal dimensions, D_{f1} and D_{f2}, for a dual-fractal analysis, and (3) the dissociation rate coefficient, k_d, and the fractal dimension in the dissociation phase, D_{fd}, for a single-fractal analysis are given in Tables 5.4 and 5.5. It is of interest to note that as the fractal dimension increases by a factor of 2.125 from a value of $D_{f1} = 1.3388$ to $D_{f2} = 2.8456$, the binding rate coefficient increases by a factor of 143.9 from a value of $k_1 = 0.3926$ to $k_2 = 56.501$. An increase in the fractal dimension on the immunoassay surface leads to an increase in the binding rate coefficient.

Figure 5.7(d) shows the binding and the dissociation of CRP in solution to anti-CRP immobilized on the immunoassay surface at location (281, 1.33) and under the influence of the electrothermal effect (Huang et al., 2008). A dual-fractal analysis is required to adequately describe the binding kinetics.

TABLE 5.4 Binding (k, k_1, k_2) and Dissociation Rate (k_d) Coefficients for the Binding and the Dissociation Phase for C-reactive protein (CRP) in Solution to Anti-CRP Immobilized on Different Locations on the Sensor Surface (Huang et al., 2008)

Location	k	k_1	k_2	k_d
(250,1.51)	15.377 ± 4.632	2.584 ± 0.247	860.74 ± 4.85	13.515 ± 1.813
(250,75)	6.095 ± 1.234	1.7234 ± 0.2252	83.006 ± 0.775	3.648 ± 0.327
(277.5,148.5)	1.8118 ± 0.3814	0.3926 ± 0.0021	56.501 ± 1.1605	2.2404 ± 0.1921
(281,1.33)	5.556 ± 0.953	2.0605 ± 0.0872	288.49 ± 9.089	12.542 ± 0.027

TABLE 5.5 Fractal Dimensions for the Binding (D_f, D_{f1}, D_{f2}) and the Dissociation (D_{fd}) Phase for C-reactive protein (CRP) in Solution to Anti-CRP Immobilized on Different Locations on the Sensor Surface (Huang et al., 2008)

Location	D_f	D_{f1}	D_{f2}	D_{fd}
(250,1.51)	1.8256 ± 0.2176	1.1834 ± 0.1334	2.9615 ± 0.01496	2.9082 ± 0.08887
(250,75)	2.2454 ± 0.1267	1.7676 ± 0.1250	2.9595 ± 0.02282	2.154 ± 0.0604
(277.5,148.5)	1.8970 ± 0.1435	1.3388 ± 0.0760	2.8456 ± 0.0489	2.0534 ± 0.05394
(281,1.33)	1.5672 ± 0.1262	1.2136 ± 0.0606	2.6674 ± 0.1221	1.9268 ± 0.05158

A single-fractal analysis is adequate to describe the dissociation kinetics. The values of (1) the binding rate coefficient, k, and the fractal dimension, D_f, for a single-fractal analysis, (2) the binding rate coefficients, k_1 and k_2, and the fractal dimensions, D_{f1} and D_{f2}, for a dual-fractal analysis, and (3) the dissociation rate coefficient, k_d, and the fractal dimension in the dissociation phase, D_{fd}, for a single-fractal analysis are given in Tables 5.4 and 5.5. It is of interest to note that as the fractal dimension increases by a factor of 2.198 from a value of $D_{f1} = 1.2136$ to $D_{f2} = 2.6674$, the binding rate coefficient increases by a factor of 140 from a value of $k_1 = 2.0605$ to $k_2 = 288.49$. An increase in the fractal dimension on the immunoassay surface once again leads to a substantial increase in the binding rate coefficient on the immunoassay surface.

Figure 5.8(a) and Tables 5.4 and 5.5 show the decrease in the affinity, K_1 ($= k_1/k_d$), with an increase in the ratio of the fractal dimensions, (D_{f1}/D_{fd}). For the data show in Figure 5.8(a), the affinity, K_1, is given by

$$K_1 = (1.98 + 3.2)E - 10\left(D_{f1}/D_{fd}\right)^{-41.97 \pm 13.07} \tag{5.6a}$$

The fit is good. Only four data points are available. The availability of more data points would lead to a more reliable fit. The affinity, K_1, is extremely sensitive to the ratio of fractal dimensions present in the binding and in the dissociation phase, since it exhibits an order of dependence greater than negative 41 and one-half (equal to -41.97) on the ratio of fractal dimensions, D_{f1}/D_{fd}. No explanation is presently offered for this extremely high negative order of dependence exhibited.

Figure 5.8(b) and Tables 5.4 and 5.5 show the decrease in the affinity, K_2 ($= k_2/k_d$), with an increase in the ratio of the fractal dimensions, (D_{f2}/D_{fd}). For the data show in Figure 5.8(b), the affinity, K_2, is given by

$$K_2 = (1061.57 \pm 238.43)\left(D_{f2}/D_{fd}\right)^{-11.29 \pm 0.769} \tag{5.6b}$$

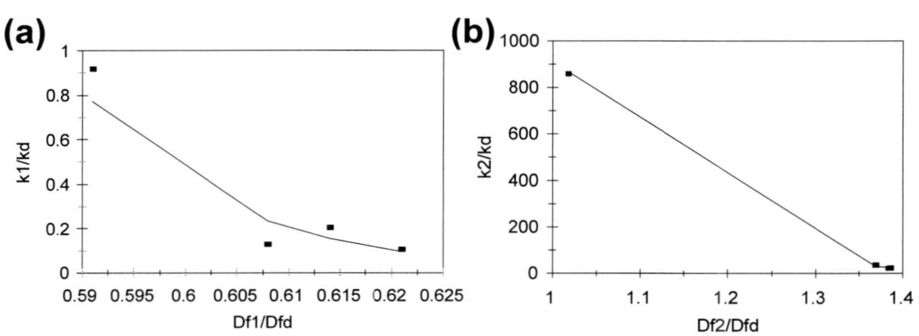

FIGURE 5.8 (a) Decrease in the affinity, K_1 ($= k_1/k_d$), with an increase in the fractal dimension ratio, (D_{f1}/D_{fd}), (b) Decrease in the affinity, K_2 (k_2/k_d) with an increase in the fractal dimension ratio (D_{f2}/D_{fd}).

The fit is good. Only four data points are available. The availability of more data points would lead to a more reliable fit. The affinity, K_2, is extremely sensitive to the ratio of fractal dimensions present in the binding and in the dissociation phase, since it exhibits an order of dependence greater than negative 11 (equal to −11.29) on the ratio of fractal dimensions, D_{f2}/D_{fd}. No explanation is, once again, presently offered to help explain this extremely high negative order of dependence exhibited.

5.4 CONCLUSIONS

A fractal analysis is presented for the binding and the dissociation kinetics of different cardiac biomarkers on biosensor surfaces. The biomarkers analyzed include (1) the binding of cardiac Mb to the anti-Mb immobilized on an electrochemical nanobiosensor at different temperatures (Suprun et al., 2010), (2) binding and dissociation of different concentrations of hFABP (in nanomolarity) to anti-hFABP immobilized on an ELISA surface (Ayyar et al., 2010), (3) binding and dissociation of FABP 3, cTNT 1F11, and cTNT 7G7 to appropriate receptors immobilized on a CM5 sensor chip surface (Andersson et al., 2010), and (4) binding and dissociation of CRP in solution to anti-CRP immobilized at different locations on a biosensor surface (Huang et al., 2008).

Both a single- and a dual-fractal analysis are utilized to model the binding and the dissociation (if applicable) kinetics. The dual-fractal analysis is used only if the single-fractal analysis did not provide an adequate fit. This was done using Corel Quattro pro (8.0) (1989) and only when the regression coefficient for the sum of squares was less than 0.97. The fractal dimension values provide a quantitative measure of the degree of heterogeneity present on the sensor chip surface. Binding and dissociation rate coefficient values and the affinity values are provided. The fractal dimension for the binding and the dissociation phase is not typical independent variable such as analyte concentration that may be directly manipulated. It is estimated from Eqn (5.1), and one may consider it as a derived variable.

An increase in the fractal dimension value or the degree of heterogeneity on the sensor surface leads, in general, to an increase in the binding and in the dissociation rate coefficient. Counterexamples do, however, exist. For example, (1) for the binding of Mb to anti-Mb immobilized on an electrochemical nanobiosensor, the binding rate coefficient exhibits close to a first order of dependence (equal to 0.9363) on the Mb concentration in solution (Suprun et al., 2010); (2) for the binding of hFABP in solution to anti-hFABP immobilized on an ELISA surface, the binding rate coefficient, k, exhibits slightly higher than second (equal to 2.0256) order of dependence on the hFABP concentration in solution in the range studied (Ayyar et al., 2010); (3) the dissociation rate coefficient, k_d, exhibits an order of dependence between first and one and one-half (equal to 1.2137) order on the hFABP concentration in solution (Ayyar et al., 2010); and (4) the affinity, K_1, exhibits an extremely high negative order of dependence (equal to −41.97) on the ratio of fractal dimensions, D_{f1}/D_{fd}, present on the ELISA surface (Ayyar et al., 2010).

The predictive-type relationships developed for the binding and the dissociation rate coefficients are of considerable value because they directly link the binding rate coefficient, for example, to the degree of heterogeneity that exists on the sensor surface and provides a means by which these rate coefficients may be manipulated by changing the degree of heterogeneity on the biosensor surface. Note that a change in the degree of heterogeneity on the biosensor surface would, in general, lead to changes in the binding and in the dissociation rate coefficient. Thus, this may require some thought and manipulation, especially if one is trying to improve, for example, the affinity, K (= k/k_d), value. The binding and the dissociation rate coefficient values are rather sensitive to their respective fractal dimension values or the degree of heterogeneity that exists on the biosensor surface. It is suggested that the fractal surface (roughness) leads to turbulence, which enhances mixing, decreases diffusional limitations, and leads to an increase in the binding (and dissociation) rate coefficient(s) (Martin et al., 1991).

For a better prognosis of the different diseases, it is imperative that one is able to detect cardiac as well as other validated biomarkers using biosensors as early as is possible, and at as low a concentration as is possible. The above relationships developed and presented is a step in that direction. Hopefully, by modifying the surface of biosensors in appropriate directions, one is able to fine-tune or enhance a biosensor parameter in a desired direction. It is quite possible that on modifying a biosensor surface, one is able to increase the sensitivity of the biosensor. However, simultaneously the selectivity and stability may be decreased. Perhaps, even the detection time required for the assay process may then become practically unfeasible. This is where the ingenuity of practicing biosensorists comes into play. This also depends on which of the biosensor parameters are important and critical for a particular application.

It is of interest to compare the binding of FABP with the two different biosensor surfaces: (1) hFABP immobilized on an ELISA surface (Ayyar et al., 2010) and (2) an appropriate receptor (antibody) immobilized on a CM 5 sensor chip surface (SPR) (Andersson et al., 2010). A comparison is useful even though (1) the analysis in each case is performed for different purposes and (2) different biosensor surfaces are used for the binding and dissociation kinetics. For the ELISA case (Ayyar et al., 2010), a single-fractal analysis is adequate to describe the binding kinetics, whereas for the CM5 chip (of an SPR biosensor) case, a dual-fractal analysis is required to adequately describe the binding kinetics. This clearly indicates that a different binding mechanism is involved in the two different biosensors (the concentrations of the FABP in solution notwithstanding).

More such kinetic studies are required to determine whether the binding and dissociation rate coefficient(s) are sensitive to their respective fractal dimensions or the degree of heterogeneity that exists on the biosensor surface.

If this is correct, and as indicated above, experimentalists may find it worth their effort to pay a little more attention to the nature of the surface and how it may be manipulated to control the relevant parameters and biosensor performance in desired directions. This is particularly true for the early detection of biomarkers for the different diseases, which should go a long way to help better manage these often insidious and intractable diseases.

REFERENCES

Abeloff, M.D., Armitage, J.O., Lichter, A.S., Niederbuber, J.E., 2000. Clinical Oncology, second ed. Churchill Livingstone, New York.

Albrecht, C., Fechner, P., Honcharenko, D., Baltzer, L., Gauglitz, G., 2010. A new assay for clinical diagnostics based on alternative recognition elements. Biosensors & Bioelectronics 25, 2302–2308.

Allender, S., Scarborough, P., Peto, V., Rayner, M., Leal, J., Luenso-Fernandez, R., Gray, A., 2008. European Cardiovascular Disease Stabilities. European Heart Netwrok, Brussels.

Andersson, H., Steel, D., Asp, J., Dahlenberg, K., Jonsson, M., Jeppson, A., Lindahl, A., Kagedal, B., Satripy, P., Mandenius, C.F., 2010. Assay cardiac biomarkers for toxicity testing using biosensing and cardiomyocytes derived from human stem cells. Journal of Biotechnology 150, 175–181.

Ayyar, V., Hearty, S., Kennedy, R.O., 2010. Heart sensitive recombinant antibodies capable of differentiating fatty acid binding protein from noncardiac forms. Analytical Biochemistry 407, 165–171.

Casas, J.P., Shah, T., Hingorani, A.D., Danesh, J., Pepys, M.B., 2008. Journal of Internal Medicine 264, 295–314.

Christenson, R.H., Azzazy, H.M.E., 2009. Clinical Biochemistry 42 (3), 150–157.

Chou, S.F., Hsu, W.L., Hwang, J.M., Chen, C.Y., 2004. Biosensors & Bioelectronics 19 (9), 999–1005.

Corel Quattro Pro (8.0), 1989. Corel Corporation, Ottawa, Canada.

Danesh, J., Collins, R., Appleby, P., Peto, R., 1998. Journal of the American Medical Association 279 (18), 1477–1482.

Deen, W.M., 1998. Analysis of Transport Phenomena. Oxford University Press, New York.

Dimitriev, D.A., Massino, Y.S., Segal, O.L., 2003. Kinetic analysis of interactions between biospecific monoclonal antibodies and immobilized antigens using a resonant mirror biosensor. Journal of Immunological Methods 280, 183–202.

Glatz, J.F.C., van der Vusse, G.J., Simmons, M.I., Kragten, J.A., van Diejien Vasser, M.P., Hermens, W.T., 1998. Fatty-acid binding protein and the early detection of acute myocardial infarction. Clinica Chimica Acta 272, 87–92.

Havlin, S., 1989. Molecular diffusion and reactions. In: Avnir, D. (Ed.), The Fractal Approach to Heterogeneous Chemistry: Surfaces Colloids, Polymers. Wiley, New York.

Hibbert, B., Gordon, J.J., 2002. Langmuir 18, 1770.

Huang, K.R., Chang, J.S., Chao, S.D., Wu, K.C., Yang, C.K., Lai, C.U., Chen, S.H., 2008. Simulation on binding efficiency of immunoassay for a biosensor with applying electrothermal effect. Journal of Applied Physics 104, 064702.

Kleine, K.H., Glatz, J.F.C., Vannieuvenhoeven, F.A., Vandervusse, G.J., 1992. Release of heart fatty-acid binding protein into plasma after acute myocardial infarction in man. Molecular and Cellular Biochemistry 116, 155–162.

Key, G., Schreiber, A., Foldbrugge, R., McNeil, C.J., Jorgensen, P., Pelsers, M.A.L., Glantz, J.R.C., Spener, F., 1999. Multicenter evaluation of amperometric immunosensor for plasma fatty acid binding protein: an early marker for acute myocardial infarction. Clinical Biochemistry 32, 229–231.

Kragten, J., van Nieuwenhoven, F., van Dieijen-Vasser, M., Hermens, W., Glatz, J., 1996. Distribution of myoglobin and fatty-aid binding protein in human cardiac autopsies. Clinical Chemistry 42, 337–338.

Kunz, U., Katerlamp, A., Renneberg, R., Spener, F., Cammann, K., 1996. Sensing fatty acid binding protein with planar and other fiber-optical surface plasmon resonance spectroscopic devices. Sensor & Actuators 32, 129–155.

Lin, K.C., Kunduru, V., Bothara, M., Rege, K., Prasad, S., Ramkrishna, B.L., 2010. Biogenic nanoporous silica-based sensor for enhanced electrochemical detection of cardiovascular biomarkers proteins. Biosensors & Bioelectronics 25, 2336–2342.

Martin, S.J., Granstaff, V.E., Frye, G.C., 1991. Effect of surface roughness on the response of thickness-shear mode resonators in liquids. Analytical Chemistry 65, 2910–2922.

McDonnell, B., Hearty, S., Leonard, P., Kennedy, R.O., 2009. Cardiac biomarkers and the case for point-of-care testing. Clinical Biochemistry 42, 549–561.

Melanson, S.F., Tanasijevic, M.J., 2005. Cardiovascular Pathology 14 (3), 156–161.

Morrow, D.A., Cannon, C.P., Jesse, R.L., Newby, I.K., Ravkilde, J., Storrow, A.B., Wu, A.H.B., Christenson, R.H., 2007. For national academy of clinical biochemistry, national academy of clinical biochemistry laboratory medicine practice guidelines: clinical characteristics and utilization of biochemical biomarkers in acute coronary syndromes. Circulation 115, e356–e375.

Nakata, T., Hashimoto, A., Hase, M., Tsuchihashi, C., Shimamoto, K., 2003. Human heart-type fatty acid-binding protein as an early diagnostic and prognostic marker in acute coronary syndrome. Cardiology 99, 96–104.

O'Brien, P.J., 2008. Cardiac troponin is the most effective translational safety biomarker for myocardial injury in cardiotoxicity. Toxicology 245, 206–218.

Okamoto, F., Sohmiya, K., Ohkaru, Y., Kawamura, K., Asayama, K., Kimura, H., Nishimura, H., Ishii, H., Sumahara, N., Tanaka, T., 2000. Human-heart-type cytoplasmic fatty acid binding protein (H-FABP) in comparison with myoglobin and creatine kinase isoenzyme –MB. Clinical Chemsitry and Laboratory Medicine 38, 231–238.

Pasaolu, R., Ofluouglu, E., Bhan, M.N., Cengel, A., Odemir, M., Duraklugil, E., Erden, M., 2007. The role of heart-type binding acid protein (H-FABP) in acute myocardial infarction (AMI) compared to conventional cardiac biomarkers. Turkish Journal of Medical Science 37, 61–67.

Panteghini, M., 2010. Cardiac: Is this biomarker ready for primetime? Scandinavian Journal of Clinical and Laboratory Investigation 242 (Suppl.), 66–72.

Pepys, M.B., Hirschfield, G.M., 2003. Journal Clinical Investigation 111, 1805–1812.

Qureshi, A., Gurbuz, Y., Keng, W.P., Davidson, J.L., 2009. A novel interdigitated capacitor based biosensor for detection of cardiovascular risk marker. Biosensors & Bioelectronics 25, 877–882.

Ridker, P.M., Libby, P. Risk factors for atherothrombotic disease. In: Libby, P., Bonow, R.O., Mann, D.L., Zipes, D.P. (Eds.), Lilly-Braunwald's Disease: A Textbook of Cardiovascular Medicine, eighth ed., Saunders, (Chapter 39).

Sigurdon, M., Wang, D., Meinhart, C.D., 2005. Lab on a Chip 5, 1366.

Spangler, B.D., Wilkinson, E.A., Murphy, J.T., Tyler, B.J., 2001. Comparison of the Spreeta ® surface plasmon resonance sensor and a quartz crystal microbalance for detection of *Escherichia coli* heat-labile enterotoxin. Analytica Chimica Acta 444, 149–161.

Strumann, T.C., Beilmann, M., Duker, C., Dumontier, B., Fredikson, J.M., Jones, R.L., Hasiwa, M., Kang, J., Mendenius, C.F., Mayer, T., Minotti, C., Valentin, V.I.P., Zunckler, B.J., Bremer, S., 2009. Report and recommendations of the workshop of the European Centre for the validation of alternative methods for drug induced cardiotoxicity. Cardiovascular Toxicology 9, 107–125.

Suprun, E., Bulko, T., Lisitsa, A., Gnedenko, O., Ivanov, A., Shumyantseva, V., Archakov, A., 2010. Electrochemical nanobiosensor for express diagnosis of acute myocardial infarction in undiluted plasma. Biosensors & Bioelectronics 25, 1694–1698.

Van Nieuwenhoven, F.A., Kleine, A.H., Wodzig, W.H., Hermens, W.T., Kragten, H.A., Maessen, J.G., Punt, C.D., van Dieijen, M.P., Vandervusse, G.J., Glatz, J.F.C., 1995. Discrimination between myocardial and skeletal muscle injury by assessment of the plasma ratio of myoglobin over fatty-acid binding protein. Circulation 92, 2848–2854.

Yang, Y.N., Lin, H.I., Wang, J.H., Shiesh, S.C., Lee, G.B., 2009. Biosensors & Bioelectronics 24, 3091–3096.

Chapter 6

A Fractal Analysis of the Binding and Dissociation Kinetics of Glucose to Different Biosensor Surfaces

6.1 INTRODUCTION

Diabetes has reached epidemic proportions, and it is estimated that over 350 million people worldwide are afflicted by this disease (Baxter, 2012). The author estimates that 3.5 million people die every year from this disease. There is an urgent need to develop a noninvasive method for the monitoring of glucose to increase compliance of the invasive (pin-prick) method of glucose detection in blood. Various companies and research laboratories are working furiously to develop such a method.

In this chapter, we analyze the binding and dissociation kinetics of

1. intracellular glucose in a single cell to the GIPi−Thr sensor using ratiometric FRET (fluorescence resonance energy transfer) measurement. The influence of repeat measurement is also presented (Jin et al., 2011).
2. binding of glucose during monitoring through FRET measurement with the GIPi-Thr sensor (Jin et al., 2011).
3. binding and dissociation of glucose during continuous monitoring of glucose by fluorophore-labeled genetically engineered glucose/galactose-binding protein (GGBP) immobilized on a hydrogel matrix (Weidamaier et al., 2011).
4. binding of different concentrations of glucose in solution in microliter to a boron-doped carbon-coated nickel (BCNi)-modified electrode (Yang et al., 2011a).
5. binding of glucose and its interferents, acetaminophen (AP), uric acid (UA), and ascorbic acid (AA), to glucose oxidase (GOx) immobilized on a biosensor surface.

Kuila et al. (2011) have recently reviewed graphene-based biosensors. They emphasize that the large surface area and excellent electrical conductivity of graphene allow it to act as an “electron wire” between the redox

Biomarkers and Biosensors. http://dx.doi.org/10.1016/B978-0-444-53794-2.00006-9

centers of an enzyme or protein and an electrode's surface. They emphasize that the rapid electron transfer permits the accurate and selective detection of the biomolecules.

Wang (2008) has indicated that since diabetes mellitus (DM) is a leading cause of death and disability, the monitoring of glucose to keep it in the normal range (80–120 mg/dl) is critical. Shao et al. (2010) further emphasize that graphene-based biosensors can lead to cost-effective biosensors for the effective management of DM. For example, Yang et al. (2009a) have used electrochemically single layers of graphene to detect glucose. Furthermore, Wang et al. (2011b) have shown that the presence of CdS nanocrystals with graphene has exhibited a low dielectric limit for glucose (0.7 mM). Zeng et al. (2010) have used organically modified graphene for enzyme-based glucose and maltose biosensors. They noted that the CV response of the modified electrode increased with increasing glucose in buffer solution.

Chen et al. (2011) have developed a novel microfluidic immunoassay system for the specific detection of hemoglobin A1c (HbA1c) based on a three-component shell/shell/core-structured magnetic nanocomposite Au/chitosan/Fe_3SO_4. They indicate that HbA1c is a stable minor hemoglobin derivative formed by a nonenzymatic reaction of glucose with the amino-terminal valine of Hb β-chains (Bunn et al., 1976). This is a good measure of long-term blood glucose levels (Nathan et al., 2007). This has also become an established procedure for the management of DM in both diagnosis and monitoring (Goodall, 2005; Koenig et al., 1976; Sacks et al., 2002).

Some of the more recent publications that have appeared in the literature or have been presented for the detection of glucose by biosensors include the following:

1. Nonenzymatic glucose sensing using novel metal oxide–metal composite nanofibers (Ding et al., 2010a)
2. Fluorescent nanosensors for glucose detection in vivo (Cash et al., 2010)
3. Development of model-based noninvasive continuous-time glucose monitoring device for non-insulin-dependent people (Rollins et al., 2011)
4. Ultrasensitive glucose detection using copper nanowires (Su et al., 2011)
5. Bioactive electroactive hydrogels (Guiseppi-Elie, 2011)
6. Glucose biosensor based on encapsulated enzyme with peptide nanotubes (Park et al., 2011)
7. Characteristics of a multisensory system for noninvasive glucose monitoring with external validation and prospective evaluation (Caduff et al., 2011)
8. Electrospun CO_3O_4 nanofibers for sensitive and selective glucose detection (Wang et al., 2010)
9. A novel enzyme entrapment in SU-8 microfabricated films for glucose microbiosensors (Psoma et al., 2010)

10. Glucose sensing through diffraction grating of hydrogel-bearing phenylboronic acid groups (Ye and Wang, 2010)
11. Single-walled carbon nanotube (CNT)-based chemiresistive affinity biosensors for small molecules: ultrasensitive glucose detection (Cella et al., 2011)
12. Measurement of glucose levels in the newborn (Anonymous, 2007)
13. A fluorescence lifetime-based fiber-optic glucose sensor using GGBP (Saxi et al., 2011)
14. Superior long-term stability of a glucose biosensor based on inserted barrel plating gold electrodes (Schwartz et al., 2009)
15. Electrochemical glucose biosensor of platinum nanospheres connected by CNTs (Claussen et al., 2010)
16. A self-referencing platinum nanoparticle-decorated enzyme-based microbiosensor for real-time measurement of physiological glucose transport (McLamore et al., 2011)
17. Noninvasive ultrasonic glucose sensing with large pigs (approximately 200 lbs) using a lightweight cymbal transducer array and biosensors (Park et al., 2010)
18. Glucose biosensors: an overview of use in clinical practice (Lee, 2010)
19. Self-monitoring of blood glucose in type I diabetes patients with insufficient metabolic control: focused self-monitoring of blood glucose intervention can lower glycated HbA1C (Skeie et al., 2009)
20. A novel nonenzymatic electrochemiluminescence (ECL) sensor for glucose using palladium nanoparticle supported in functional CNTs (Chen et al., 2009)
21. An electrochemical luminescent biosensor for glucose based on the ECL of luminol on the Nafion/GOx poly(nickel (II) tetrasulphothalocyanine) multiwalled CNTs-modified electrode (2009)
22. Nonenzymatic glucose sensor based on ultrasonic electrodeposition of bimetallic (M = Ru, Pd, and Au) nanoparticles on CNTs—ionic liquid composite film (Xiao et al., 2009)
23. Nonenzymatic glucose sensor based on flower-shaped Au@Pd core—shell nanoparticles—ionic liquids composite film-modified glassy carbon electrodes (GCEs) (2010)
24. Synthesis of functional SiO_2-coated graphene oxide nanosheets decorated with Ag nanoparticles for H_2O_2 detection (Lu et al., 2011)
25. Elimination of electrochemical interferences in glucose biosensors (Jia et al., 2010)
26. A novel glucose biosensor based on direct electrochemistry of GOx incorporated in bioremediated gold nanoparticles composite film
27. Preliminary investigations on a glucose biosensor based on the potentiometric principle (Liao et al., 2007)

28. Nano flake-like structure SnS_2 matrix for glucose biosensing based on direct electrochemistry of GOx (Yang et al., 2011a)
29. Direct electrochemistry of GOx immobilized on a hexagonal mesoporous silica-MCM-41 matrix (Dai et al., 2007)
30. Improved selectivity and stability of glucose biosensor based on in situ, electropolymerized polyaniline—polyacrylonitrile composite film (Xie et al., 2005)
31. Fluorescence-based glucose sensors (a review) (Pickup et al., 2005)
32. A novel glucose biosensor based on direct electrochemistry of GOx incorporated in biomediated gold nanoparticles—CNTs composite film (Zhang et al., 2011b)
33. A novel glucose sensor based on ordered carbon—gold nanoparticle nanocomposites (Wang et al., 2011a)
34. Glucose biosensors based on CNT nanoelectrode ensembles (NEEs) (Lin et al., 2004)
35. Ultrasensitive electrochemical immunosensor employing GOx reduction-catalyzed deposition on gold nanoparticles for signal amplification (Zhang et al., 2011a)
36. Electrochemical Glucose Biosensors, (Wang, 2008)
37. Glucose biosensor using GOx immobilized on multilayer films immobilized on a polyaniline (Parente et al., 1994)
38. Electrochemical contrast and kinetic analysis of water-dispersible carbon black nanoparticles and MWNTs in glucose biosensors (Yang et al., 2009b)
39. Amperometric glucose sensor based on layer-by-layer self-assembly by multiwall CNTs, gold particles, and GOx in Pt electrode
40. Biocompatible graphene oxide-based glucose biosensors (Lu et al., 2011)
41. Electrochemical glucose biosensors (Wang, 2008)
42. Transporting, fabrication and bifunctionalization of gold nanoelectrode arrays into versatile electrochemical glucose biosensors (Claussen et al., 2011)
43. Nanometal-decorated exfoliated graphite nanoplatelets-based glucose biosensors with high sensitivity and fast response (Lu et al., 2008a)
44. Adsorption of GOx onto single-walled CNTs and its application in layer-by-layer biosensors (Tsai et al., 2009)
45. Pt nanoparticles inserting in CNTs arrays: nanocomposite for glucose biosensors

A fractal analysis is presented for the binding and dissociation (if applicable) kinetics for

1. construction of a panel of glucose-indicator proteins (GIP) for continuous glucose monitoring (CGM) (Jin et al., 2011).

2. multiday preclinical demonstration of GGBP-based fiber-optic sensor (Weidamaier et al., 2011).
3. direct electrochemistry of GOx and biosensing for glucose-by-glucose-based BCNi-modified electrode (Yang et al., 2011a).
4. edge-plane microwire electrodes for highly sensitive hydrogen peroxide and glucose detection (Qiang et al., 2011).

6.2 THEORY

Havlin (1987) has reviewed and analyzed the diffusion of reactants toward fractal surfaces. The details of the theory and the equations involved for the binding and the dissociation phases for analyte—receptor binding are available (Sadana, 2001). The details are not repeated here, except that just the equations are given to permit an easier reading. These equations have been applied to other biosensor systems (Sadana, 2001; Ramakrishnan and Sadana, 2001; Sadana, 2005). For most applications, a single- or dual-fractal analysis is often adequate to describe the binding and the dissociation kinetics. Peculiarities in the values of the binding and the dissociation rate coefficients, as well as of the fractal dimensions with regard to the dilute analyte systems being analyzed will be carefully noted, if applicable.

6.2.1 Single-Fractal Analysis

6.2.1.1 Binding Rate Coefficient

Havlin (1989) indicates that the diffusion of a particle (analyte [Ag]) from a homogeneous solution to a solid surface (e.g., receptor [Ab]-coated surface) on which it reacts to form a product (analyte—receptor complex, (Ab.Ag)) is given by

$$(\mathrm{Ab\cdot Ag}) \approx \begin{cases} t^{(3-D_{f,bind})/2} = t^{p} & t < t_c \\ t^{1/2} & t > t_c \end{cases} \tag{6.1}$$

Here $D_{f,bind}$ or D_f is the fractal dimension of the surface during the binding step and t_c is the crossover value. Havlin (1989) indicates that the crossover value may be determined by $r_c^2 \sim t_c$. Above the characteristic length, r_c, the self-similarity of the surface is lost and the surface may be considered homogeneous. Above time t_c, the surface may be considered homogeneous, since the self-similarity property disappears, and "regular" diffusion is now present. For a homogeneous surface where $D_f = 2$ and when only diffusional limitations are present, $p = ½$ as it should be. Another way of looking at the $p = ½$ case (where $D_{f,bind} = 2$) is that the analyte in solution views the fractal object, in our case, the receptor-coated biosensor surface, from a "large distance." In essence, in the association process, the diffusion of the analyte from the solution to the receptor surface creates a depletion layer of width $(Đt)^{½}$, where

Đ is the diffusion constant. This gives rise to the fractal power law, (Analyte.Receptor) $\sim t^{(3-D_{f,bind}/2)}$. For the present analysis, t_c is arbitrarily chosen and we assume that the value of the t_c is not reached. One may consider the approach as an intermediate "heuristic" approach that may be used in the future to develop an autonomous (and not time-dependent) model for diffusion-controlled kinetics.

6.2.1.2 Dissociation Rate Coefficient

The diffusion of the dissociated particle (receptor [Ab] or analyte [Ag]) from the solid surface (e.g., (analyte [Ag]−receptor [Ab]) complex-coated surface) into solution may be given as a first approximation by

$$(\text{Ab.Ag}) \approx -t^{(3-D_{f,diss})/2} = -t^{p} \quad (t > t_{diss}) \tag{6.2}$$

Here $D_{f,diss}$ is the fractal dimension of the surface for the dissociation step. This corresponds to the highest concentration of the analyte−receptor complex on the surface. Henceforth, its concentration only decreases. The dissociation kinetics may be analyzed in a manner "similar" to the binding kinetics.

6.2.2 Dual-Fractal Analysis

6.2.2.1 Binding Rate Coefficient

Sometimes, the binding curve exhibits complexities and two parameters (k and D_f) are not sufficient to adequately describe the binding kinetics. This is further corroborated by low values of r^2 factor (goodness-of-fit). In that case, one resorts to a dual-fractal analysis (four parameters: k_1, k_2, D_{f1}, and D_{f2}) to adequately describe the binding kinetics. The single-fractal analysis presented above is thus extended to include two fractal dimensions. At present, the time ($t = t_1$) at which the "first" fractal dimension "changes" to the "second" fractal dimension is arbitrary and empirical. For the most part, it is dictated by the data analyzed and experience gained by handling a single-fractal analysis. A smoother curve is obtained in the "transition" region, if care is taken to select the correct number of points for the two regions. In this case, the product (antibody−antigen or analyte−receptor complex; Ab.Ag or analyte.receptor) is given by

$$(\text{Ab.Ag}) \approx \begin{cases} t^{(3-D_{f1,bind})/2} = t^{p1} & (t < t_1) \\ t^{(3-D_{f2,bind})/2} = t^{p2} & (t_1 < t < t_2) = t_c \\ t^{1/2} & (t > t_c) \end{cases} \tag{6.3}$$

In some cases, as mentioned above, a triple-fractal analysis with six parameters (k_1, k_2, k_3, D_{f1}, D_{f2}, and D_{f3}) may be required to adequately model the binding kinetics. This is when the binding curve exhibits convolutions and complexities in its shape due to perhaps the very dilute nature of the analyte (in some of the cases to be presented) or for some other reasons. Also, in some cases, a dual-fractal analysis may be required to describe the dissociation kinetics.

6.3 RESULTS

Jin et al. (2011) have recently constructed a panel of GIP for CGM. These authors emphasize the need for in vivo glucose monitoring and add that enzyme-based sensors often fail in vivo. They indicate that GIP can sense glucose without the need of enzymes. By site-directed mutagenesis, Jin et al. (2011) indicate that these GIP with their higher dissociation rate constant are more suitable for in vivo glucose monitoring.

Jin et al. (2011) emphasize the challenges involved in CGM. They indicate that though electrochemical biosensors are effective for use in vitro, however, they are unsuccessful for in vivo glucose sensing. To help circumvent problems encountered during in vivo applications, Campbell et al. (2009) suggested the use of FRET for a signal transduction mechanism, wherein the efficiency of energy transferred from one fluorescent donor molecule to its receptor molecule depends on glucose binding. Furthermore, Ballerstadt et al. (2006) and Schultz et al. (1982) have suggested the use of glucose-binding concanavalin A as receptors for the CGM biosensor.

Jin et al. (2011) emphasize the need for generating new molecules suitable for CGM and have selected a GIP that changes conformation upon glucose binding (Schultz, 2003). This is a 32-kDa periplasmic protein that is a glucose transporter in *Escherichia coli* (Careega and Falke, 1992). The protein adopts a "closed" form in the absence of glucose and an "open" form in the presence of glucose. Jin et al. (2011) have constructed a biosensor based on this GIP for CGM. They emphasize that similar biosensors have been constructed by other authors (Fehr et al., 2003, 2004: Weiss (2007); Khan et al., 2008; Takanaga et al., 2007). However, Jin et al. (2011) indicate that these sensors have a narrow detection range. Ye and Schultz (2003) and Miller et al. (1983) indicate that the glucose dissociation rate constant of wild-type GIP is around 5 μM. Jin et al. (2011) have used site-directed mutagenesis to increase this dissociation rate constant from 5 to 589 μM. These authors have used site-directed mutagenesis of GIP to help construct a panel of GIP for the effective detection of in vivo glucose.

Figure 6.1(a) shows the binding and dissociation of intracellular glucose to the intracellular protein to the GIP immobilized on the GIPi-Thr glucose sensor (Jin et al., 2011). GIPi is the site-directed mutagenesis form of GIP. FRET imaging measurement was used to visualize intracellular glucose (0–200 μM). This is the first run in a sequence of four runs. A single-fractal analysis is adequate to describe the binding and the dissociation kinetics. The values of (1) the binding rate coefficient, k, and the fractal dimension, D_f, for a single-fractal analysis and (2) the dissociation rate coefficient, k_d, and the

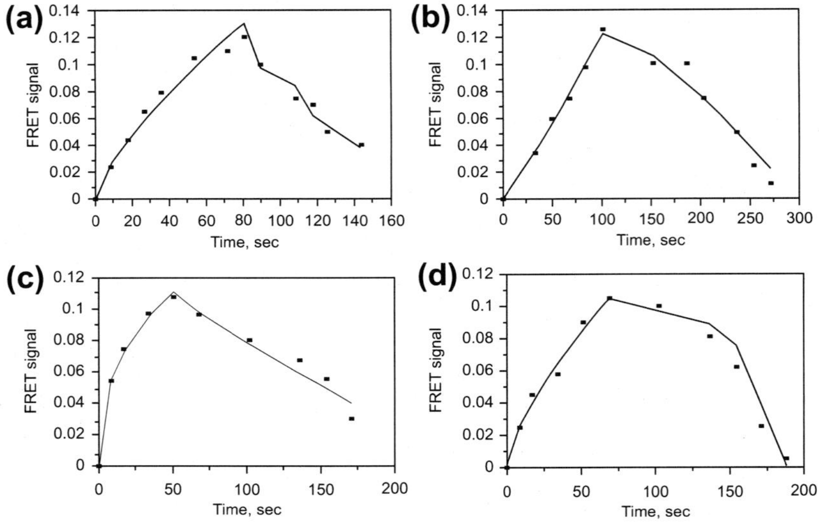

FIGURE 6.1 Binding and dissociation of intracellular glucose in single cell to the GIPi-Thr sensor. Measurement by ratiometric fluorescence resonance energy transfer (FRET) (Jin et al., 2011). Influence of repeat measurements (a), (b), (c), and (d).

fractal dimension for dissociation, D_{fd}, for a single-fractal analysis are given in Table 6.1. The affinity, K (=k/k_d), value is 1.106.

Figure 6.1(b) shows the binding and dissociation of intracellular glucose to the intracellular protein to the GIP immobilized on the GIPi-Thr glucose sensor (Jin et al., 2011). This is the second run in a sequence of four runs. A single-fractal analysis is once again adequate to describe the binding and the dissociation kinetics. The values of (1) the binding rate coefficient, k, and the fractal dimension, D_f, for a single-fractal analysis and (2) the dissociation rate coefficient, k_d, and the fractal dimension for dissociation, D_{fd}, for a single-fractal analysis are given in Table 6.1. The affinity, K (=k/k_d), value is 8.3.

Figure 6.1(c) shows the binding and dissociation of intracellular glucose to the intracellular protein to the GIP immobilized on the GIPi-Thr glucose sensor (Jin et al., 2011). This is the third run in a sequence of four runs. A single-fractal analysis is once again adequate to describe the binding and the dissociation kinetics. The values of (1) the binding rate coefficient, k, and the fractal dimension, D_f, for a single-fractal analysis and (2) the dissociation rate coefficient, k_d, and the fractal dimension for dissociation, D_{fd}, for a single-fractal analysis are given in Table 6.1. The affinity, K (=k/k_d), value in this case is 36.64.

Figure 6.1(d) shows the binding and dissociation of intracellular glucose to the intracellular protein to the GIP immobilized on the GIPi-Thr glucose sensor (Jin et al., 2011). This is the fourth and last run in a sequence of four runs. A single-fractal analysis is once again adequate to describe the binding and the dissociation kinetics. The values of (1) the binding rate coefficient, k, and the fractal dimension, D_f, for a single-fractal analysis and (2) the dissociation rate coefficient, k_d, and the fractal dimension for dissociation, D_{fd}, for a single-fractal analysis are given in Table 6.1. The estimated affinity, K, value is 43.210, since the k_d value is extremely small.

Figure 6.2(a) shows the increase in the binding rate coefficient k_1 for a dual-fractal analysis with an increase in the fractal dimension D_{f1}. For the data shown in Figure 6.2(a), the binding rate coefficient k_2 is given by

$$k_2 = (0.00150 \pm 0.00041)D_f^{3.191 \pm 0.304} \tag{6.4a}$$

The fit is good. Only four data points are available. The availability of more data points would lead to a more reliable fit. The binding rate coefficient k is sensitive to the fractal dimension D_f or the degree of heterogeneity that exists on the GIPi-Thr biosensor surface as noted by the order of dependence between three and three and one-half (equal to 3.191) exhibited.

Figure 6.2(b) shows the increase in the dissociation rate coefficient k_d for a dual-fractal analysis with an increase in the fractal dimension D_{f2}. For the data shown in Figure 6.2(b), the binding rate coefficient k_2 is given by

$$k_d = (0.00117 + 0.00168)D_{fd}^{1.887 \pm 0.602} \tag{6.4b}$$

The fit is reasonable. Only three data points are available. The availability of more data points would lead to a more reliable fit. The dissociation rate coefficient k_d is sensitive to the fractal dimension D_{fd} or the degree of heterogeneity that exists on the GIPi-Thr biosensor surface as noted by the order of dependence between one and one-half and two (equal to 1.887) exhibited.

Jin et al. (2011) also analyzed the influence of the increase in the glucose concentration in solution in millimolarity on the binding and dissociation kinetics to the GIPi-Thr biosensor. Figure 6.3(a) shows the binding and the dissociation kinetics of 1.25 mM glucose in solution to the GIPi-Thr biosensor. A single-fractal analysis is adequate to describe the binding and the dissociation kinetics. The values of (1) the binding rate coefficient, k, and the fractal dimension, D_f, for a single-fractal analysis and (2) the dissociation rate coefficient, k_d, and the fractal dimension in the dissociation phase, D_{fd}, are given in Table 6.2. Note that the value of the fractal dimension for dissociation, D_{fd}, is equal to zero. No explanation is offered, at present, to help explain the zero value obtained for D_{fd}. This would indicate that the surface acts like a "Cantor-like" dust in the dissociation phase. In this case, the affinity, K ($=k/k_d$), value is 30.91.

Figure 6.3(b) shows the binding and the dissociation kinetics of 2.5 mM glucose in solution to the GIPi-Thr biosensor. A single-fractal analysis is, once again, adequate to describe the binding and the dissociation kinetics. The

TABLE 6.1 Binding and Dissociation of Glucose in Solution During Continuous Glucose Monitoring of Intracellular Glucose in Single Live Cell with GIPi-Cys Through Radiometric FRET Measurement of Glucose Concentration (Jin et al., 2011)

Analyte in Solution/ Receptor on Surface	Run Number	k	k_d	K (=k/k_d)	D_f	D_{fd}
Intracellular glucose concentration/ GIPi-Cys sensor	1	0.005772 ± 0.0005776	0.00522 + 0.0010	1.106	1.5814 ± 0.0982	1.6686 ± 0.2248
Intracellular glucose concentration/ GIPi-Cys sensor	2	0.000688 ± 0.00034	0.0000829 ± 0.0000261	8.3	0.7594 ± 0.1092	0.8023 ± 0.5406
Intracellular glucose concentration/ GIPi-Cys sensor	3	0.02424 ± 0.00109	0.000644 ± 0.000591	36.64	2.2238 + 0.0662	1.0536 ± 0.2942
Intracellular glucose concentration/ GIPi-Cys sensor	4	0.00547 ± 0.000591	0.000000133 ± 0.000000075	43210	1.6238 ± 0.1187	0

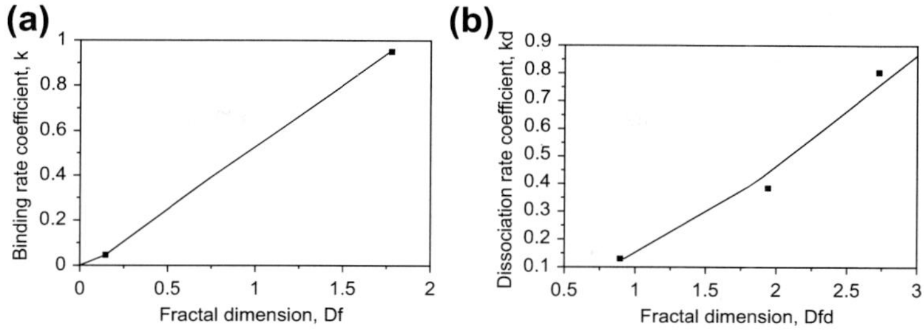

FIGURE 6.2 (a) Increase in the binding rate coefficient k_1 for a dual-fractal analysis with an increase in the fractal dimension D_{f1} (b) Increase in the dissociation rate coefficient k_d with an increase in the fractal dimension D_{f2}.

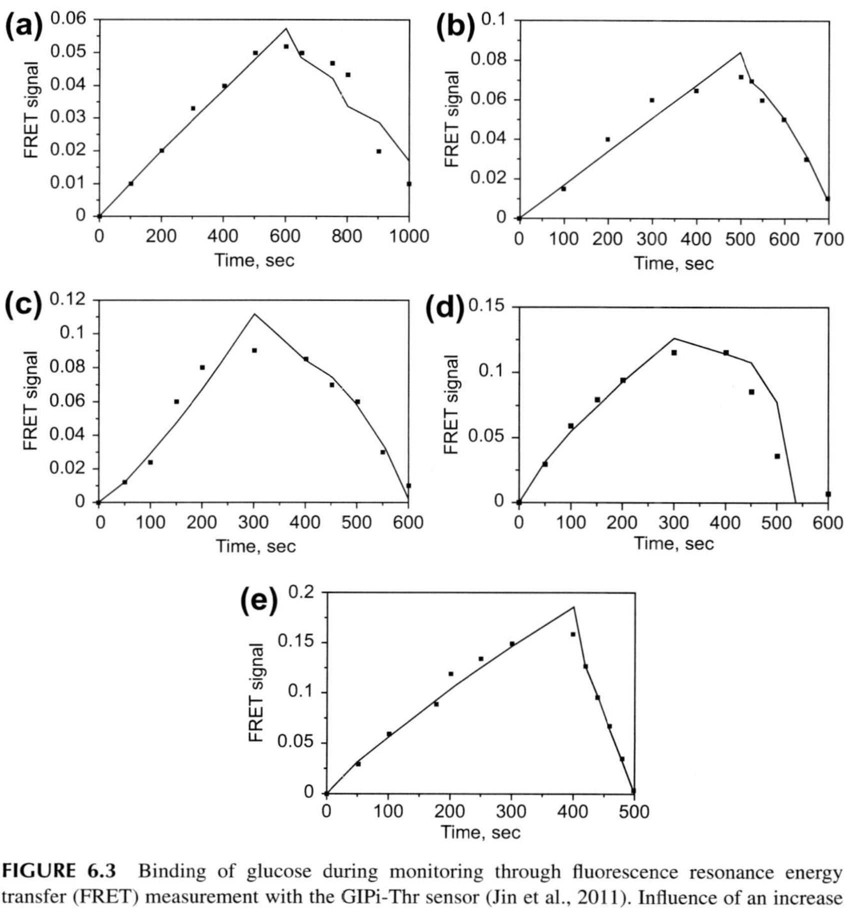

FIGURE 6.3 Binding of glucose during monitoring through fluorescence resonance energy transfer (FRET) measurement with the GIPi-Thr sensor (Jin et al., 2011). Influence of an increase in the glucose concentration (in mM): (a) 1.25 (b) 2.5 (c) 5 (d) 10 (e) 20.

TABLE 6.2 Binding and Dissociation Rate Coefficients and Fractal Dimensions for the Binding and Dissociation Phase During Continuous Monitoring of Glucose Through Radiometric FRET Measurement of Various Glucose Concentrations (in mM) (Jin et al., 2011)

Intracellular Concentration, mM	k	k_d	$K(=k/k_d)$	D_f	D_{fd}
1.25	0.000128 ± 0.000011	4.14E-06 ± 3.33E-06	30.91	1.8804 ± 0.1060	0
2.5	0.000187 ± 0.000037	1.78E-05 ± 0.81E-05	10.51	0.9828 ± 0.1419	0
5	9.97E-05 ± 2.63E-06	6E-08 ± 1.17E-08	1661.7	0.5406 ± 0.3394	0
10	0.001691 ± 0.000169	3.2E-13 + 9.2E-13	5.2	0.7564 ± 0.0695	0
20	0.001186 ± 0.000135	0.001688 ± 0.000027	0.703	1.3104 ± 0.1238	2.8309 + 0.0254

values of (1) the binding rate coefficient, k, and the fractal dimension, D_f, for a single-fractal analysis and (2) the dissociation rate coefficient, k_d, and the fractal dimension in the dissociation phase, D_{fd}, are given in Table 6.2. Note that the value of the fractal dimension for dissociation, D_{fd}, is, once again, equal to zero. In this case, the affinity, K (=k/k_d), value is 10.51.

Figure 6.3(c) shows the binding and the dissociation kinetics of 5 mM glucose in solution to the GIPi-Thr biosensor. A single-fractal analysis is, once again, adequate to describe the binding and the dissociation kinetics. The values of (1) the binding rate coefficient, k, and the fractal dimension, D_f, for a single-fractal analysis and (2) the dissociation rate coefficient, k_d, and the fractal dimension in the dissociation phase, D_{fd}, are given in Table 6.2. Note that the value of the fractal dimension for dissociation, D_{fd}, is, once again, equal to zero. In this case, the affinity, K (=k/k_d), value is 166.17.

Figure 6.3(d) shows the binding and the dissociation kinetics of 10 mM glucose in solution to the GIPi-Thr biosensor. A single-fractal analysis is, once again, adequate to describe the binding and the dissociation kinetics. The values of (1) the binding rate coefficient, k, and the fractal dimension, D_f, for a single-fractal analysis and (2) the dissociation rate coefficient, k_d, and the fractal dimension in the dissociation phase, D_{fd}, are given in Table 6.2. Note that the value of the fractal dimension for dissociation, D_{fd}, is, once again, equal to zero. In this case, the affinity, K (=k/k_d), value is 5.2.

Figure 6.3(e) shows the binding and the dissociation kinetics of 20 mM glucose in solution to the GIPi-Thr biosensor. A single-fractal analysis is, once again, adequate to describe the binding and the dissociation kinetics. The values of (1) the binding rate coefficient, k, and the fractal dimension, D_f, for a single-fractal analysis and (2) the dissociation rate coefficient, k_d, and the fractal dimension in the dissociation phase, D_{fd}, are given in Table 6.2. Note that the value of the fractal dimension for dissociation, D_{fd}, in this case is equal to 2.8309. In this case, the affinity, K (=k/k_d), value is 0.703.

Figure 6.4(a) shows the increase in the binding rate coefficient, k, for a single-fractal analysis with an increase in the glucose concentration in solution

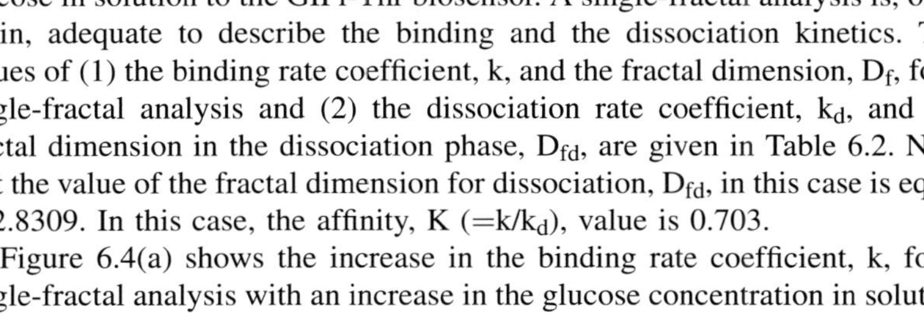

FIGURE 6.4 (a) Increase in the binding rate coefficient, k, for a single-fractal analysis with an increase in the glucose concentration (in mM). (b) Increase in the binding rate coefficient, k, for a single-fractal analysis with an increase in the fractal dimension, D_f.

in the 1.25−20 mM range. For the data shown in Figure 6.4(a), the binding rate coefficient, k, is given by

$$k = (0.0000733 + 0.00011)[\mathrm{Glu}]^{0.96 \pm 0.23} \tag{6.5a}$$

The fit is good except for one point. This is reflected in the estimated value of the binding rate coefficient. Only the positive error is presented, since the binding rate coefficient, k, cannot have a negative value. Only five data points are available. The availability of more data points would lead to a more reliable fit. The binding rate coefficient, k, for a single-fractal analysis exhibits close to a first (equal to 0.96) order of dependence on the glucose concentration in solution in the 1.25−20 mM range.

Figure 6.4(b) shows the increase in the binding rate coefficient, k, for a single-fractal analysis with an increase in the fractal dimension, D_f, in the 1.25−20 mM glucose concentration range. For the data shown in Figure 6.4(b), the binding rate coefficient, k, is given by

$$k = (0.000313 + 0.000491)D_f^{2.645 \pm 1.206} \tag{6.5b}$$

There is scatter in the data. This is reflected in the estimated value of the binding rate coefficient. Only the positive error is presented, since the binding rate coefficient, k, cannot have a negative value. Only five data points are available. The availability of more data points would lead to a more reliable fit. The binding rate coefficient, k, for a single-fractal analysis is sensitive to the fractal dimension, D_f, or the degree of heterogeneity that exists on the biosensor surface, as it exhibits an order of dependence between two and one-half and three (equal to 2.645) on the fractal dimension, D_f.

Weidamaier et al. (2011) have recently developed a GGBP-based fiber-optic biosensor. These authors indicate that the effective treatment of type I diabetes requires primarily the use of insulin to manage the glucose concentration within normal concentrations (Heinemann and Koschinsky, 2002; Ginsberg and Mazze, 1994; Giradin et al., 2009). Even with monitoring glucose concentrations three to four times a day abnormal concentrations of glucose can occur, and Weidamaier et al. (2011) indicate that dangerous excursions of hypoglycemic and hyperglycemic may occur. To avoid or help minimize such excursions, there is emphasis to develop continuous monitoring devices for the detection of glucose concentrations (Brauker, 2009; Moschou et al., 2004; Buckingham et al., 2007; Hirsch, 2009; De Brock et al., 2008; Giradin et al., 2009; Hirsch, 2009; Tubiana-Rufi et al., 2007; Piper et al., 2006; Klonoff, 2005). All of the currently commercially available glucose sensors are minimally invasive subcutaneous sensors. These sensors have limitations such as interferences by, for example, AP. Besides, Weidamaier et al. (2011) emphasize that for CGM, GOx-based sensors do have the accuracy in the hypoglycemic range (Klonoff, 2004, 2005). The inaccuracies could lead to dangerous clinical outcomes.

Weidamaier et al. (2011) emphasize that to avoid the above constraints nonenzymatic methods of monitoring glucose are being extensively explored. Some of these methods include fluorescent detection of boronate ester complexes formed between glucose and boronic acids (Fang et al., 2004; Capuccio et al., 2004). Other fluorescent approaches include binding of glucose to (1) concanavalin A (Ballerstadt and Schultz, 2000) and (2) *E. coli* GGBP (Amiss et al., 2007; Khan et al., 2008; Marvin and Hellinga, 2001), to name a few.

Weidamaier et al. (2011) indicate that GGBP immobilized on a surface plasmon resonance biosensor surface responds rapidly to glucose (Hsieh et al., 2004). GGBP generates a robust fluorescent signal (Weidamaier et al., 2011), and Zhou and Cass (1991) have termed GGBP as "reagentless sensors."

Figure 6.5(a) shows the binding and the dissociation of glucose in a swine to a GGBP-based fiber-optic biosensor (Weidamaier et al., 2011) during continuous glucose tracking by fluorophore-labeled and genetically engineered GGBP (Run #1). This is a 12-h experiment using a 31-gauge, 100-μm fiber with a PEG matrix and acrylodan-labeled SM4 GGBP placed subcutaneously.

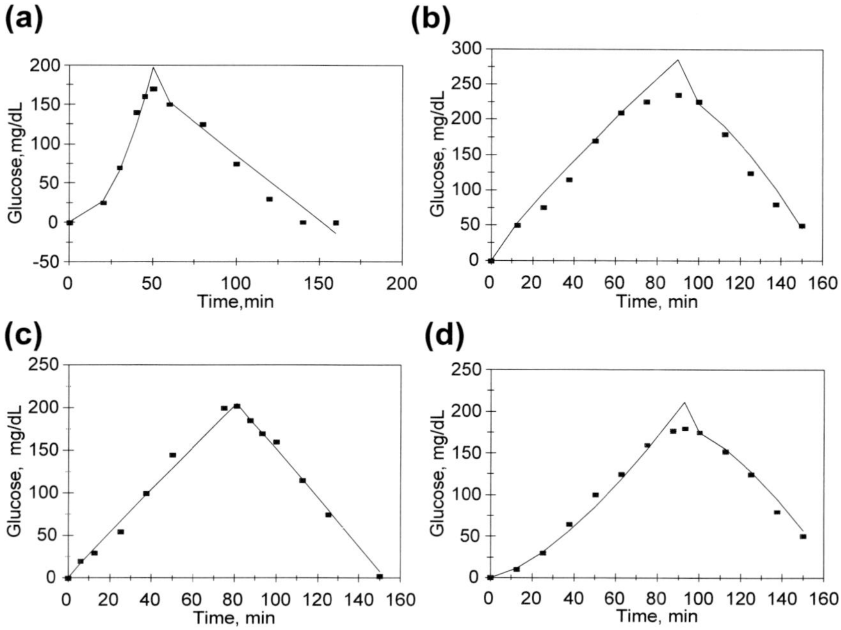

FIGURE 6.5 Binding and dissociation of glucose during continuous monitoring by fluorophore-labeled genetically engineered GGBP immobilized on a hydrogel matrix (Weidamaier et al., 2011): (a) Run# 1 (b) Run #2 (c) Run #3 (d) Run #4 (e) Increase in the binding rate coefficient, k, with an increase in the fractal dimension, D_f (f) Increase in the dissociation rate coefficient, k_d, with an increase in the fractal dimension for dissociation, D_{fd} (g) Increase in the affinity, K (= k/k_d) with an increase in the fractal dimension ratio, D_f/D_{fd}.

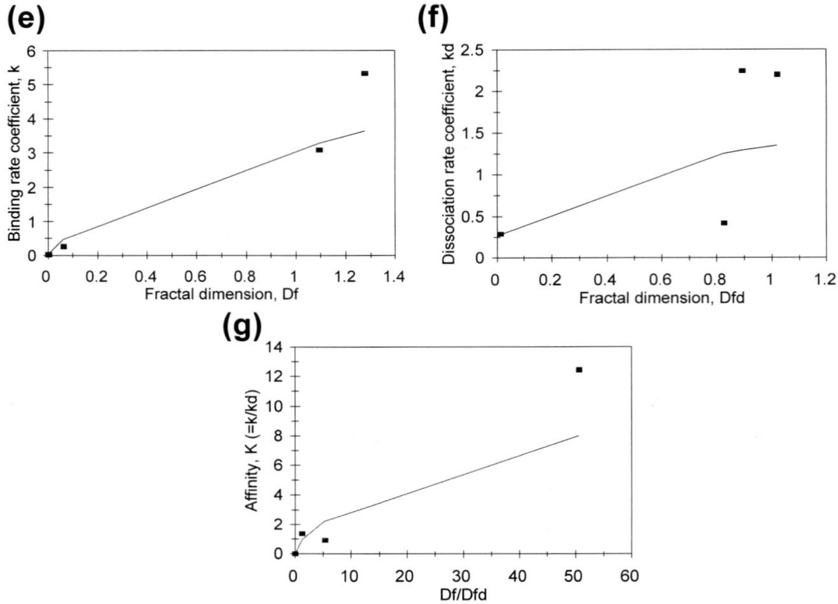

FIGURE 6.5 cont'd

This GGBP is immobilized on a hydrogel matrix at the tip of a small-diameter optical fiber contained in a stainless-steel needle. The binding and the dissociation kinetics may be described by a single-fractal analysis. The values of (1) the binding rate coefficient, k, and the fractal dimension, D_f, for a single-fractal analysis and (2) the dissociation rate coefficient, k_d, and the fractal dimension for dissociation, D_{fd}, are given in Table 6.3. In this case, the affinity, K, value is 0.0399.

Figure 6.5(b) shows the binding and the dissociation of glucose subcutaneously to a GGBP-based fiber-optic biosensor in a swine (Weidamaier et al., 2011). This is a 168-h experiment. This is Run #2. The binding and the dissociation kinetics may be described by a single-fractal analysis. The values of (1) the binding rate coefficient, k, and the fractal dimension, D_f, for a single-fractal analysis and (2) the dissociation rate coefficient, k_d, and the fractal dimension for dissociation, D_{fd}, are given in Table 6.3. In this case, the affinity, K, value is 12.56. No explanation is offered, at present, to help explain the almost two orders of magnitude difference in the affinity, K, values observed in Run #1 and 2.

Figure 6.5(c) shows the binding and the dissociation of glucose subcutaneously to a GGBP-based fiber-optic biosensor in a swine (Weidamaier et al., 2011). This is a 168-h experiment. This is Run #3. The binding and the dissociation kinetics may be described by a single-fractal analysis. The values

TABLE 6.3 Binding and dissociation rate coefficients and fractal dimensions for the binding and the dissociation phase for glucose subcutaneously to a GGBP fiber-optic sensor in a swine. The sensor is a 31-gauge, 100 micron fiber with a PEG matrix and acrylodan labeled SM4 GGBP placed subcutaneously (Weidamaier et al., 2011)

Type of experiment	k	k_d	D_f	D_{fd}
Single-day swine, 12 hours	0.0399 ± 0.0057	2.2025 ± 1.7483	0 + 0.3622	1.020 ± 0.2336
Multi-day swine, 168 hours, sensor 1	5.3355 ± 0.5768	0.4248 ± 0.1390	1.2761 ± 0.1215	0.8252 ± 0.3270
Multi-day swine, 168 hours, sensor 2	3.095 ± 0.462	2.2469 ± 0.2284	1.0926 ± 0.1186	0.8926 ± 0.0995
Multi-day swine, 168 hours, sensor 3	0.2739 ± 0.0377	0.2929 ± 0.0544	0.0606 ± 0.1408	0.0122 ± 0.1769

(Weidamaier et al., 2011)

of (1) the binding rate coefficient, k, and the fractal dimension, D_f, for a single-fractal analysis and (2) the dissociation rate coefficient, k_d, and the fractal dimension for dissociation, D_{fd}, are given in Table 6.3. In this case, the affinity, K, value is 1.377.

Figure 6.5(d) shows the binding and the dissociation of glucose subcutaneously to a GGBP-based fiber-optic biosensor in a swine (Weidamaier et al., 2011). This is a 168-h experiment. This is Run #4. The binding and the dissociation kinetics may be described by a single-fractal analysis. The values of (1) the binding rate coefficient, k, and the fractal dimension, D_f, for a single-fractal analysis and (2) the dissociation rate coefficient, k_d, and the fractal dimension for dissociation, D_{fd}, are given in Table 6.3. In this case, the affinity, K, value is 0.935.

Figure 6.5(e) shows the increase in the binding rate coefficient, k, with an increase in the fractal dimension, D_f, for a single-fractal analysis. For the data shown in Figure 6.5(e), the binding rate coefficient, k, is given by

$$k = (3.0987 \pm 2.0679) D_f^{0.6649 \pm 0.0881} \tag{6.6a}$$

The fit is good. Only four data points are available. The availability of more data points would lead to a more reliable fit. Since only a few data points are available and analyzed, the 12-h experiment and the 168-h experiment for subcutaneous detection in swine are presented together. The binding rate coefficient, k, is quite sensitive to the fractal dimension, D_f, or the degree of heterogeneity present on the biosensor surface as noted by the order of dependence between one-half and first (equal to 0.6649) exhibited.

Figure 6.5(f) shows the increase in the dissociation rate coefficient, k_d, with an increase in the fractal dimension for dissociation, D_{fd}, for a single-fractal analysis. For the data shown in Figure 6.5(f), the dissociation rate coefficient, k_d, is given by

$$k_d = (1.3404 + 2.0480) D_{fd}^{0.3531 \pm 0.2482} \tag{6.6b}$$

The fit is reasonable. There is scatter in the data. Only four data points are available. The availability of more data points would lead to a more reliable fit. Since only a few data points are available and analyzed, the 12-h experiment and the 168-h experiment for subcutaneous detection in swine, once again, are presented together. The dissociation rate coefficient, k_d, is only mildly sensitive to the fractal dimension, D_{fd}, or the degree of heterogeneity present on the biosensor surface as noted by the order of dependence between zero and one-half (equal to 0.3531) exhibited.

Figure 6.5(g) shows the increase in the affinity, K (=k/k_d), with an increase in the fractal dimension ratio, D_f/D_{fd}, for a single-fractal analysis. For the data shown in Figure 6.5(g), the affinity, K, is given by

$$K(=k/k_d) = (0.8623 + 0.9121)(D_f/D_{fd})^{0.5676 \pm 0.0891} \tag{6.6c}$$

The fit is reasonable. There is scatter in the data. Only four data points are available. The availability of more data points would lead to a more reliable fit. Since only a few data points are available and analyzed, the 12-h experiment and the 168-h experiment for subcutaneous detection in swine are presented together. The affinity, K, is mildly sensitive to the fractal dimension ratio, D_f/D_{fd}, or the degree of heterogeneity present on the biosensor surface as noted by the close to one-half order of dependence (equal to 0.5676) exhibited.

Yang et al. (2011a) have recently developed a BCNi-modified electrode for the direct electrochemistry of GOx for glucose biosensing. These authors indicate that electrochemistry of redox enzymes has attracted attention for the development of biosensors without the necessity of using a mediator (Zuo et al., 2008; Fu et al., 2009). Yang et al. (2011a) do, however, point out the difficulty involved in using redox enzymes and electrode surfaces. This is due to (1) deep immersion of the redox center of the enzyme, (2) passivation of the enzyme upon adsorption to the electrode surface, and (3) improper orientation of the adsorbed enzyme on the electrode surface (Li et al., 2007a; Salimi et al., 2006).

Yang et al. (2011a) indicate the different strategies that have been involved in the development of GOx biosensors in order to improve the electron transfer between the GOx and the biosensor surface. These include gold nanoparticle-functionalized multiwalled CNT composites (Li et al., 2009), carbon ionic fluid (Shangguan et al., 2008), ordered mesoporous silica SBA-15 (Wang et al., 2009a), and layer-by-layer self-assembling CNT and cationic polymer (Ding et al., 2010).

Yang et al. (2011a) indicate the use of nanostructured materials due to their novel properties (Huang et al., 2003; Josephson et al., 1990) for biotechnology applications such as immobilization of proteins (Rossi et al., 2004), detection of DNA hybridization (Pumera et al., 2005), and drug delivery (Huang et al., 2003). Carbon-coated nickel nanoparticle-modified electrodes have been shown to accelerate the electron transfer rate (Wang et al., 2007a, 2007b). Furthermore, properties of nanomaterials may be improved using boron doping (Deng et al., 2008; Wang et al., 2007c; Zhang et al., 2006). Thus, Yang et al. (2011a) have used a BCNi nanoparticle GCE to analyze the detection of glucose by immobilized GOx.

Yang et al. (2011b) analyzed the influence of binding of different amounts (in μL) of 0.1 M glucose in solution at a stirred pH of 7.0 to the GOx-chitosan BCNi/GCE. Figure 6.6(a) shows the binding of 10 μl of 0.1 M glucose. A dual-fractal analysis is required to adequately describe the binding kinetics. The values of (1) the binding rate coefficient, k, and the fractal dimension, D_f, for a single-fractal analysis and (2) the binding rate coefficients, k_1 and k_2, and the fractal dimensions, D_{f1} and D_{f2}, for a dual-fractal analysis are given in Table 6.4. For a dual-fractal analysis, an increase in the fractal dimension by a factor of 1.3862 from a value of $D_{f1} = 2$ to $D_{f2} = 2.7724$ leads to an increase in the binding rate coefficient value by a factor of 7.35 from a value of $k_1 = 0.0065$ to $k_2 = 0.04780$. In other words, an increase in the degree of heterogeneity on the biosensor surface leads to an increase in the binding rate coefficient.

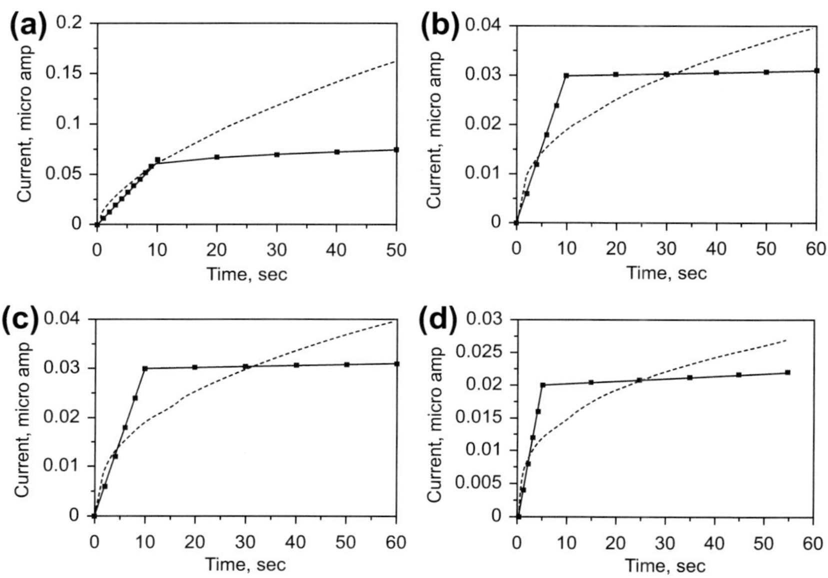

FIGURE 6.6 Binding of different concentrations of glucose in solution in microliter to a boron-doped carbon-coated nickel-modified electrode (Yang et al., 2011b): (a) 10 (b) 5 (c) 5 (d) 1.

TABLE 6.4 Binding Rate Coefficients and Fractal Dimensions for Different Amounts (in Micro-Liter) of Glucose (0.1 M) to a Born-Doped Carbon-Coated Nickel-Modified Electrode (Yang et al., 2011a)

Analyte in Microliter	k	k_1	k_2	D_f	D_{f1}	D_{f2}
10	0.0151 ± 0.0220	0.0065	0.04780 ± 0.00032	1.7842 ± 0.4430	2 ± 6.38 E-16	2.7724 ± 0.0192
5	0.00412 ± 0.00165	0.0025	0.01753 ± 08016	1.8736 ± 0.157	2 ± 2.12 E-15	2.7418 ± 0.0210
5	0.00743 ± 0.00260	0.003	0.02810 ± 0.00005	2.1804 ± 0.1729	2 ± 5.1 E-15	2.9529 ± 0.0043
1	0.00711 ± 0.00271	0.004	0.01738 ± 0.00011	2.3422 ± 0.1511	2 ± 8.06 E-16	2.8855 ± 0.01189

Figure 6.6(b) shows the binding of 5 μl of 0.1 M glucose. A dual-fractal analysis is required to adequately describe the binding kinetics. The values of (1) the binding rate coefficient, k, and the fractal dimension, D_f, for a single-fractal analysis and (2) the binding rate coefficients, k_1 and k_2, and the fractal dimensions, D_{f1} and D_{f2}, for a dual-fractal analysis are given in Table 6.4. For a dual-fractal analysis, an increase in the fractal dimension by a factor of 1.3709 from a value of $D_{f1} = 2$ to $D_{f2} = 2.7418$ leads to an increase in the binding rate coefficient value by a factor of 7.01 from a value of $k_1 = 0.0025$ to $k_2 = 0.01753$. Once again, an increase in the degree of heterogeneity on the biosensor surface leads to an increase in the binding rate coefficient.

Figure 6.6(c) shows the binding of 5 μl of 0.1 M glucose. A dual-fractal analysis is required to adequately describe the binding kinetics. The values of (1) the binding rate coefficient, k, and the fractal dimension, D_f, for a single-fractal analysis and (2) the binding rate coefficients, k_1 and k_2, and the fractal dimensions, D_{f1} and D_{f2}, for a dual-fractal analysis are given in Table 6.4. For a dual-fractal analysis, an increase in the fractal dimension by a factor of 1.476 from a value of $D_{f1} = 2$ to $D_{f2} = 2.9529$ leads to an increase in the binding rate coefficient value by a factor of 9.37 from a value of $k_1 = 0.003$ to $k_2 = 0.02810$. Once again, an increase in the degree of heterogeneity on the biosensor surface leads to an increase in the binding rate coefficient.

Figure 6.6(d) shows the binding of 1 μl of 0.1 M glucose. A dual-fractal analysis is required to adequately describe the binding kinetics. The values of (1) the binding rate coefficient, k, and the fractal dimension, D_f, for a single-fractal analysis and (2) the binding rate coefficients, k_1 and k_2, and the fractal dimensions, D_{f1} and D_{f2}, for a dual-fractal analysis are given in Table 6.4. For a dual-fractal analysis, an increase in the fractal dimension by a factor of 1.443

from a value of $D_{f1} = 2$ to $D_{f2} = 2.8855$ leads to an increase in the binding rate coefficient value by a factor of 4.345 from a value of $k_1 = 0.004$ to $k_2 = 0.01738$. Once again, and as above, an increase in the degree of heterogeneity on the biosensor surface leads to an increase in the binding rate coefficient.

Figure 6.7(a) shows the increase in the binding rate coefficient, k_2, with an increase in the glucose concentration in microliter of the 0.1 M glucose solution. For the data shown in Figure 6.7(a), the binding rate coefficient, k_2, is given by

$$k_2 = (0.01528 \pm 0.00736)[\text{Gl amount in microliter}]^{0.3853 \pm 0.2327} \tag{6.7a}$$

The fit is reasonable. Only four data points are available. The availability of more data points would lead to a more reliable fit. The binding rate coefficient, k_2, for a dual-fractal analysis exhibits a mild (equal to 0.3853) order of dependence on the glucose concentration in microliter of the 0.1 M solution.

Figure 6.7(b) shows the increase in the binding rate coefficient ratio, k_2/k_1, with an increase in the glucose concentration in microliter of the 0.1 M glucose solution. For the data shown in Figure 6.7(b), the binding rate coefficient ratio, k_2/k_1, is given by

$$k_2/k_1 = (0.662 \pm 0.199)[\text{Gl amount in microliter}]^{1.156 \pm 06234} \tag{6.7b}$$

There is a lot of scatter in the data. This is reflected in the estimated value of the binding rate coefficient ratio, k_2/k_1. Only four data points are available. The availability of more data points would lead to a more reliable fit. The binding rate coefficient ratio, k_2/k_1, for a dual-fractal analysis exhibits (equal to 1.156) an order of dependence between first and one and one half on the glucose concentration in solution.

Qiang et al. (2011) have developed an edge-plane microwire electrode for the sensitive detection of H_2O_2 and glucose. Wilson and Gifford (2005) have indicated the importance of continuous metabolic monitoring for the detection and care of different chronic diseases. Klonoff (2005) emphasizes the

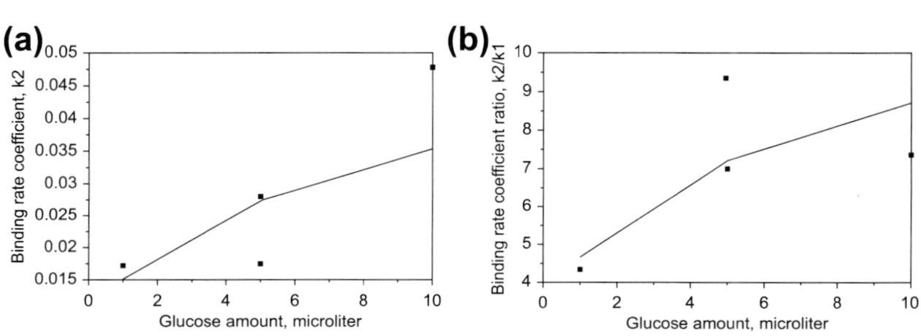

FIGURE 6.7 (a) Increase in the binding rate coefficient, k_2, for a dual-fractal analysis with an increase in the glucose concentration in microliter (b) Increase in the binding rate coefficient ratio, k_2/k_1 for a dual-fractal analysis with an increase in the glucose amount in microliter.

importance of CGM for the management of diabetes. Qiang et al. (2011) indicate the importance of enzymatic-based electrochemical method for transposing the concentration of a metabolite such as H_2O_2, O_2, etc. to an electroactive species (Vaddiraju et al., 2010; Wang et al., 2007; Wilson and Gifford, 2005). This permits the amperometric determination of the metabolite concentration (Vaddiraju et al., 2010; Wilson and Gifford, 2005). One cannot, however, emphasize the importance of device miniaturization (Li et al., 2007b; Vaddiraju et al., 2010). This is a strong constraint in amperometric devices for implantation purposes, since a decrease in the area of the working electrode leads to weaker signals.

Qiang et al. (2011) indicate the different techniques available to enhance the surface area of implantable electrodes:

1. *Increase porosity*: Etching and decorating techniques may be used to increase nanoporosity for (Ehlebacher et al., 2001; Lin et al., 2003), conducting polymers (Singh et al., 2009), CNTs (Yu et al., 2006), and nanostructured semiconductors (Singh et al., 2009).
2. *Increase enzyme loading*: Enzyme loading may be increased by layer-by-layer assembly (Wu et al., 2007), hydrogel (Ahmad et al., 2007), solgel (Chen and Dong, 2007), nanotubes (Yu et al., 2006), and nanoplatelets (Lu et al., 2008b; Kumar and Chaudhari, 2003; Patolsky et al., 2004).

Qiang et al. (2011) have fabricated and characterized a miniaturized biosensor geometry that provides for extreme sensitivity as well as flexibility and ease of fabrication. These authors emphasize that their technique of electrochemical rebuilding of the silver (Ag) working electrode along with Pt nanoparticle deposition helps increase the surface area and the electrocatalytic activity for H_2O_2 detection (Qiang et al., 2011).

Qiang et al. (2011) have analyzed the binding kinetics of the sequential addition of (1) 0.1 mM of AP, (2) 0.1 mM of AA, (3) 0.1 mM of UA, and (4) 5 mM of glucose (Glu). Figure 6.8a shows the binding of glucose + AP to the glucose sensor. A dual-fractal analysis is required to adequately describe the binding kinetics. The values of (1) the binding rate coefficient, k, and the fractal dimension, D_f, for a single-fractal analysis and (2) the binding rate coefficients, k_1 and k_2, and the fractal dimensions, D_{f1} and D_{f2}, for a dual-fractal analysis are given in Table 6.5. It is of interest to note that for a dual-fractal analysis as the fractal dimension increases by a factor of 1.1717 from a value of $D_{f1} = 2$ to $D_{f2} = 2.7432$, the binding rate coefficient increases by a factor of 6.795 from a value of $k_1 = 0.3$ to $k_2 = 2.0384$.

Figure 6.8(b) shows the binding of glucose + AA to the glucose sensor. A dual-fractal analysis is, once again, required to adequately describe the binding kinetics. The values of (1) the binding rate coefficient, k, and the fractal dimension, D_f, for a single-fractal analysis and (2) the binding rate coefficients, k_1 and k_2, and the fractal dimensions, D_{f1} and D_{f2}, for a dual-fractal analysis are given in Table 6.5. It is of interest to note that for a dual-fractal

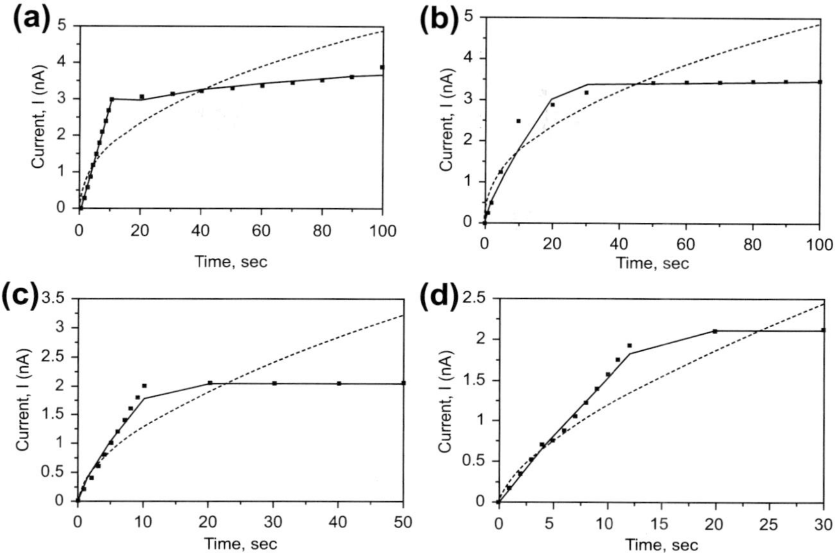

FIGURE 6.8 Binding of glucose and its interferents acetaminophen (AP), uric acid (UA), and ascorbic acid (AA) to glucose oxidase immobilized on a biosensor surface. Sequential addition (Qiang et al., 2011): (a) glucose + AP (b) glucose + AA (c) glucose + UA (d) glucose.

analysis as the fractal dimension increases by a factor of 2.04 from a value of $D_{f1} = 1.4504$ to $D_{f2} = 2.9586$, the binding rate coefficient increases by a factor of 10.8 from a value of $k_1 = 0.2943$ to $k_2 = 3.179$.

Figure 6.8(c) shows the binding of glucose + UA to the glucose sensor. A dual-fractal analysis is, once again, required to adequately describe the binding kinetics. The values of (1) the binding rate coefficient, k, and the fractal dimension, D_f, for a single-fractal analysis and (2) the binding rate coefficients, k_1 and k_2, and the fractal dimensions, D_{f1} and D_{f2}, for a dual-fractal analysis are given in Table 6.5. It is of interest to note that for a dual-fractal analysis as the fractal dimension increases by a factor of 1.5 from a value of $D_{f1} = 2.0$ to $D_{f2} = 3.0$, the binding rate coefficient increases by a factor of 10 from a value of $k_1 = 0.2$ to $k_2 = 2.0$.

Figure 6.8(d) shows the binding of glucose to the glucose sensor. A dual-fractal analysis is, once again, required to adequately describe the binding kinetics. The values of (1) the binding rate coefficient, k, and the fractal dimension, D_f, for a single-fractal analysis and (2) the binding rate coefficients, k_1 and k_2, and the fractal dimensions, D_{f1} and D_{f2}, for a dual-fractal analysis are given in Table 6.5. It is of interest to note that for a dual-fractal analysis as the fractal dimension increases by a factor of 1.214 from a value of $D_{f1} = 1.7054$ to $D_{f2} = 2.0702$, the binding rate coefficient increases by a factor of 11.64 from a value of $k_1 = 0.1778$ to $k_2 = 2.0702$.

TABLE 6.5 Binding Rate Coefficients and Fractal Dimensions for Glucose and its Interferents (Acetoaminophen (AP), Uric Acid (UA), and Ascorbic Acid (AA)) to Glucose Oxidase (GO) Immobilized on a Biosensor Surface. Sequential Addition (Qiang et al., 2011)

Sequential Addition of Analytes/Receptor on Surface	k	k_1	k_2	D_f	D_{f1}	D_{f2}
Glucose + AP/GO	0.6448 ± 0.2545	0.3 + 0	2.0384 ± 0.0556	2.12 ± 0.1108	2.0 ± 2.68 E-16	2.74332 ± 0.03550
Glucose + AA/GO	0.4108 ± 0.1603	0.2943 ± 0.0858	3.179 ± 0.002	1.9422 ± 0.1204	1.4504 ± 0.1705	2.9586 ± 0.00206
Glucose + UA/GO	0.3492 ± 0.1383	0.2 ± 0	2.0 ± 0	1.8588 ± 1.43 E-16	2.0 ± 1.428 E-16	3 − 5.12 E-16
Glucose/GO	0.2696 ± 0.0821	0.1778 ± 0.0097	2.0702 ± 0.0032	1.7054 ± 0.1295	1.122 ± 0.04256	2.9898 ± 0.004488

Figure 6.9(a) shows the increase in the binding rate coefficient, k_1, for a dual-fractal analysis with an increase in the fractal dimension, D_{f1}. For the data shown in Figure 6.9(a), the binding rate coefficient, k_1, is given by

$$k_1 = (0.1973 + 0.0698)D_{f1}^{0.3987+0.6234} \qquad (6.8a)$$

There is a lot of scatter in the data. This is reflected in the estimated value of the binding rate coefficient, k_1. Only the positive value is given, since the binding rate coefficient, k_1, cannot have a negative value. The availability of more data points would lead to a more reliable fit. The binding rate coefficient, k_1, exhibits only a mild and less than one-half (equal to 0.3987) order of dependence on the fractal dimension, D_{f1}, or the degree of heterogeneity that exists on the biosensor surface.

Figure 6.9(b) shows the increase in the binding rate coefficient, k_2, for a dual-fractal analysis with an increase in the fractal dimension, D_{f2}. For the data shown in Figure 6.9(b), the binding rate coefficient, k_2, is given by

$$k_2 = (0.6621 \pm 0.199)D_{f2}^{1.156+3.567} \qquad (6.8b)$$

There is a lot of scatter in the data. This is reflected in the estimated value of the binding rate coefficient, k_2. The availability of more data points would lead to a more reliable fit. The binding rate coefficient, k_2, exhibits greater than first (equal to 1.156) order of dependence on the fractal dimension, D_{f2}, or the degree of heterogeneity that exists on the biosensor surface.

McLamore et al. (2011) have recently developed a self-referencing platinum nanoparticle-decorated enzyme-based microbiosensor for real-time measurement of glucose transport. These authors indicate that glucose plays a significant role in metabolism, and glucose transport is analyzed in diabetes (Gulliam et al., 2000), cancer research (Busk et al., 2008), and cell/tissue culture applications (Morten et al., 2008). Some of the techniques used to measure glucose levels include wet-chemistry approaches (Lanevschi and Kramer, 2009), hollow fiber hybridoma cell bioreactors radiolabeling (Gulliam

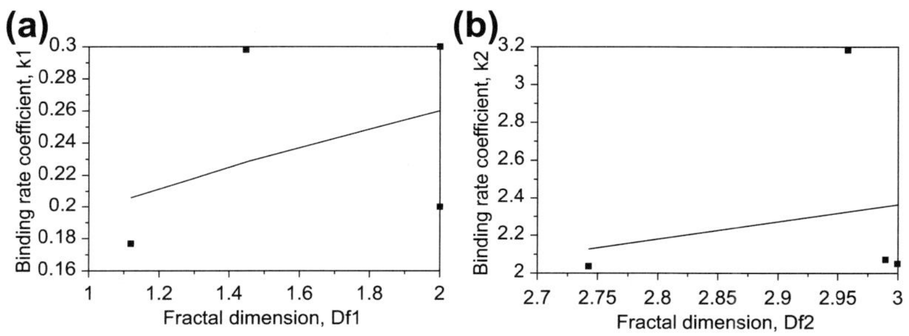

FIGURE 6.9 (a) Increase in the binding rate coefficient, k_1, for a dual-fractal analysis with an increase in the fractal dimension, D_{f1} (b) Increase in the binding rate coefficient, k_2, for a dual-fractal analysis with an increase in the fractal dimension, D_{f2}.

et al., 2000; Hellman et al., 1974; Sweet et al., 1996; Zawalich and Mitschinsky, 1977), Fourier transform infrared spectrometry (Zhang and Fang, 2005), and electrochemical biosensors (Wang, 2005). Electrochemical biosensors have been used in point-of-care diagnostics (Lowry et al., 1998) and preventative disease research. Hrapovic et al. (2004) and Wang (2005) indicate that enzymatic recognition of glucose is obtained by GO_x. In order that biosensor performance may be enhanced, CNTs (Wang et al., 2003; Gong et al., 2005; Gooding, 2005; Wang, 2005; Claussen et al., 2009, 2010), Nafion (Ni, 1999), and amorphous platinum clusters (Jaffe and Nuccitelli, 1974) have been used.

McLamore et al. (2011) have used very selective and sensitive GO_x biosensors for measuring glucose flux. Their microbiosensor used platinum black, Nafion, multiwalled CNTs, and GO_x. They emphasize that their microbiosensor permits direct quantitative glucose transport measurement under physiological conditions.

Figure 6.10(a) shows the binding and the dissociation kinetics for 5 mM glucose with 0.5% agar immersed in PBS during CGM and manual oscillation of the sensor at the surface of cultured pancreatic β-cells. This is Run #1 of the sequence of repeat runs #s 1 and 2. A single-fractal analysis is adequate to describe the binding kinetics. A dual-fractal analysis is required to adequately describe the dissociation kinetics. The values of (1) the binding rate coefficient, k, and the fractal dimension, D_f, for a single-fractal analysis, (2) the dissociation rate coefficient, k_d, and the fractal dimension for dissociation, D_{fd}, for a single-fractal analysis, and (3) the dissociation rate coefficients, k_{d1} and k_{d2}, and the fractal dimensions, D_{fd1} and D_{fd2}, for a dual-fractal analysis are given in Table 6.6. It is of interest to note that a decrease in the fractal dimension or the degree of heterogeneity on the sensor surface in the dissociation phase from D_{fd1} to D_{fd2} leads to a decrease in the dissociation rate

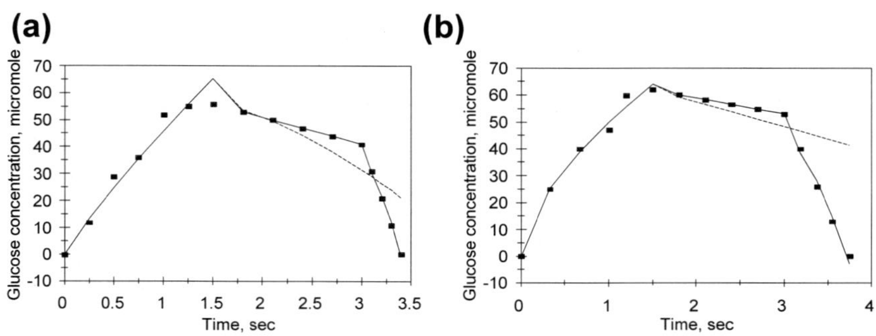

FIGURE 6.10 Binding and dissociation of 5 mM glucose with 0.5% agar immersed in PBS during continuous glucose monitoring and manual oscillation of the sensor at the surface of cultured pancreatic β-cells. Influence of repeat runs (McLamore et al., 2011): (a) Run #1 (b) Run #2.

TABLE 6.6a Binding and Dissociation Rate Coefficients for 5 mM Glucose with 0.5%Agar Immersed in PBS During Continuous Monitoring and Manual Oscillation of the Sensor at the Surface of the Cultured Pancreatic β cells. Influence of Repeat Runs (McLamore et al., 2011)

Run Number	k	k_d	k_{d1}	k_{d2}
One	45.77 ± 6.88	13.456 ± 5.764	10.0 ± 0	2.985 ± 0.114
Two	49.86 ± 3.047	9.140 ± 5.777	6.0 ± 0	3.7377 ± 0.2979

TABLE 6.6b Fractal Dimensions for the Binding and the Dissociation Phase for 5 mM glucose with 0.5% Agar Immersed in PBS During Continuous Monitoring and Manual Oscillation of the Sensor at the Surface of the Cultured Pancreatic β Cells. Influence of Repeat Runs (McLamore et al., 2011)

Run Number	D_f	D_{fd}	D_{fd1}	D_{fd2}
One	1.2484 ± 0.188	0 + 0.4071	2.0 ± 7.0 E-16	0 + 0.5996
Two	1.7706 ± 0.0996	0 + 0.5218	2.0 ± 2.4 E-16	0 + 0.7076

coefficient from k_{d1} to k_{d2}. In this case, the affinity K_1 (=k/k_{d1}) value is 4.577 and K_2 (=k/k_{d2}) value is 15.33.

Figure 6.10(b) shows, once again, the binding and the dissociation kinetics for 5 mM glucose with 0.5% agar immersed in PBS during CGM and manual oscillation of the sensor at the surface of cultured pancreatic β-cells. This is Run #2 of the sequence of repeat runs #s 1 and 2. A single-fractal analysis is, once again, adequate to describe the binding kinetics. A dual-fractal analysis is, once again, required to adequately describe the dissociation kinetics. The values of (1) the binding rate coefficient, k, and the fractal dimension, D_f, for a single-fractal analysis, (2) the dissociation rate coefficient, k_d, and the fractal dimension, D_{fd}, for a single-fractal analysis, and (3) the dissociation rate coefficients, k_{d1} and k_{d2}, and the fractal dimensions, D_{fd1} and D_{fd2}, for a dual-fractal analysis are given in Table 6.6. Once again, it is of interest to note that a decrease in the fractal dimension or the degree of heterogeneity on the sensor surface in the dissociation phase from D_{fd1} to D_{fd2} leads to a decrease in the dissociation rate coefficient from k_{d1} to k_{d2}. In this case, the affinity K_1 (=k/k_{d1}) value is 8.31 and K_2 (=k/k_{d2}) is 13.34.

Chen et al. (2010) have recently developed a nonenzymatic glucose sensor based on Au@Pd core–shell nanoparticles–ionic liquids composite film-modified GCE. These authors indicate that DM is a metabolic disorder that results from insulin deficiency. The normal range of glucose is 4.4–6.6 mM (Wang, 2008). Park et al. (2006) indicate that enzymatic biosensors exhibit long-term instability and unsatisfactory reproducibility. Furthermore, Feng et al. (2009) emphasize that electrooxidizable species act as interferents during the measurement and detection of glucose. Also, Zhu et al. (2009) indicate that different endogenous species such as AA and UA are also oxidized along with glucose in the potential range.

Chen et al. (2010) indicate that Pt-based metals have been used to develop potentiometric biosensors (Xiao et al., 2009; Song et al., 2005). Furthermore, Pt-surfaces modified by heavy metals such as Tl, Pb, Bi, or WO_3 have exhibited catalytic activity for GOx and have been analyzed (Sakamoto and Takamura, 1982; Wang et al., 2008; Wittstock et al., 1998; Zhang et al., 1997).

Chen et al. (2010) indicate that multimetallic nanoparticles with alloy or core–shell structure materials exhibit composition-dependent optical, catalytic, electronic, and magnetic properties (Toshima and Yonezawa, 1998). Rodriguez-Gonzalez (2006) and Park et al. (2004) indicate that there has been intense research activity in the controlled shapes of nanoparticles, since specific geometries lead to unusual physical and chemical properties. For example, Lu et al. (2004) indicate that nonspherical bimetallic clusters increase the surface area, but they also facilitate charge accumulation or valence change on the surface. This enhances the catalytic effect. Chen et al. (2010) have analyzed the synthesis and characterization of Au@Pd core–shell nanoparticles with a flower shape. These authors emphasize that the Pd provides a low resistance to the analyte. The Au provides the biocompatibility necessary for biofunctionalization. This facilitates the enhancement of response current during the detection of glucose.

Figure 6.11 shows the binding of 1.0 mM glucose in solution to the Au@Pd-ILs–Au@Pd/GCE biosensor (Chen et al., 2010). A dual-fractal

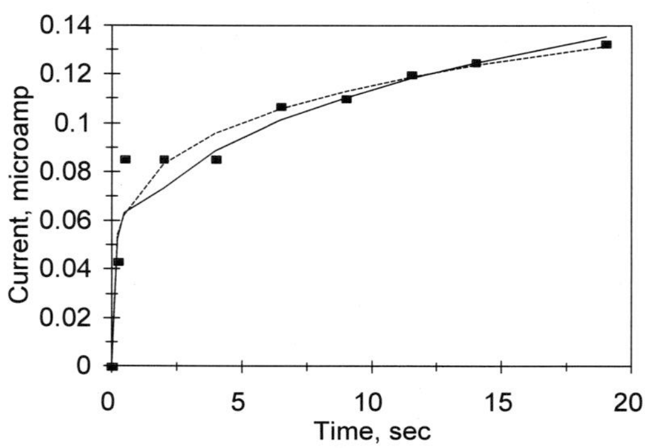

FIGURE 6.11 Binding of 1.0 mM glucose in solution to the Au@Pd-ILs–Au@Pd/GCE biosensor (Chen et al., 2010).

analysis is required to adequately describe the binding kinetics. The values of (1) the binding rate coefficient, k, and the fractal dimension, D_f, for a single-fractal analysis and (2) the binding rate coefficients, k_1 and k_2, and the fractal dimensions, D_{f1} and D_{f2}, for a dual-fractal analysis are given in Table 6.7.

It is of interest to note that as the fractal dimension increases slightly from a value of $D_{f1} = 2.4394$ to $D_{f2} = 2.4506$, the binding rate coefficient decreases from a value of $k_1 = 0.0771$ to $k_2 = 0.0604$. In this case, a slight increase in the degree of heterogeneity or the fractal dimension on the biosensor surface leads to a decrease in the binding rate coefficient.

Liao et al. (2007) have developed a glucose biosensor based on the potentiometric principle. These authors indicate that the enzyme and the electron mediator are immobilized on the SnO_2/indium tin oxide (ITO) glass using a covalent bond. These authors indicate that researchers have attempted for 4 years to develop suitable glucose monitoring devices for continuous monitoring of glucose in patients with DM (Patel et al., 2003). Mousty (2004) has indicated that interferences by AA and UA lead to nonspecific signals. Electron mediators have been used to minimize the potential applied to the working electrode to help minimize these nonspecific signals (Nagata et al., 1995; Harwod and Pouton, 1996; Saito and Watanabe, 1998; Liu et al., 2005). Other authors have suggested methods to improve the performance of glucose biosensors (Pan et al., 2005; Guan et al., 2005; Zhang et al., 2005; Cui et al., 2005; Yang and Zhu, 2006). Liao et al. (2007) indicate that polarizing voltage is the key factor that leads to interference.

Liao et al. (2007) indicate that a platinum electrode actuator on an ISFET sensitive gate has been used to electrolyze hydrogen peroxide (Seo et al., 1997; Lee et al., 2000). Liao et al. (2007) have fabricated their SnO_2/ITO glass pH sensor by depositing SnO_2 thin films onto ITO glass. They emphasize that their approach exhibits high sensitivity, ease of fabrication, and low cost (Yin et al., 2000, 2001a).

Figure 6.12 shows the binding of 360 mg/dL glucose in 5mM, pH 7.0 buffer solution to a disposable potentiometric glucose biosensor (Liao et al., 2007). A dual-fractal analysis is required to adequately describe the binding

TABLE 6.7 Binding and Dissociation Rate Coefficients and Fractal Dimensions for 1.0 mM Glucose in Solution to the Au@Pd-1Ls-Au@Pd/GCE Biosensors (Chen et al., 2010)

Analyte/Receptor	k	k_1	k_2	D_f	D_{f1}	D_{f2}
Glucose/GCE biosensor	0.07195 ± 0.01190	0.0771 ± 0.039	0.0604 ± 0.00024	2.5898 ± 0.07122	2.4394 ± 0.4866	2.4506 ± 0.0660

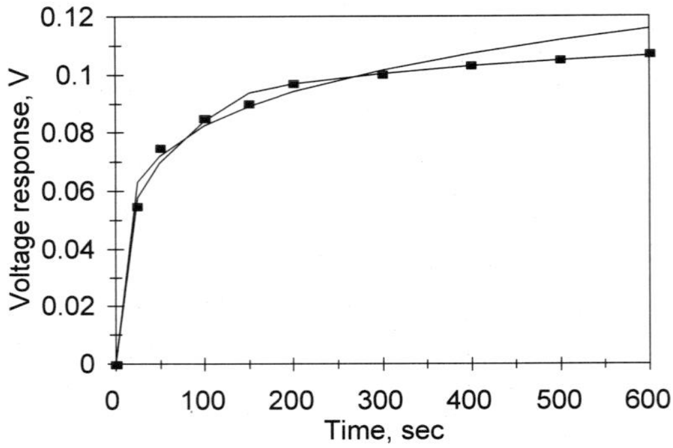

FIGURE 6.12 Binding of 360 mg/dl glucose in 5 mM, pH 7 buffer solution to a disposable glucose biosensor (Liao et al., 2007).

kinetics. The values of (1) the binding rate coefficient, k, and the fractal dimension, D_f, for a single-fractal analysis and (2) the binding rate coefficients, k_1 and k_2, and the fractal dimensions, D_{f1} and D_{f2}, for a dual-fractal analysis are given in Table 6.8. It is of interest to note that as the fractal dimension increases by a factor of 1.077 from a value of $D_{f1} = 2.6194$ to $D_{f2} = 2.8208$, the binding rate coefficient increases by a factor of 1.798 from a value of $k_1 = 0.02426$ to $k_2 = 0.0601$. Note that, once again, an increase in the fractal dimension or the degree of heterogeneity on the biosensor surface leads to an increase in the binding rate coefficient.

Dai et al. (2007) have immobilized GOx on a hexagonal mesoporous silica-MCM-41 matrix biosensor for the direct electrochemistry detection of glucose. These authors indicate that GOx has been used extensively to monitor blood glucose levels for diabetes. They further indicate that direct electron transfer of immobilized GOx has been obtained by platinum (Lu et al., 1994), gold (Du Poet et al., 1990), glassy carbon (Narsimhan and Wimgard, 1986), carbon paste (Savitri and Mitra, 1998; Liu and Ju, 2003), graphite surface (Lanniello et al., 1962), CNTs (Cai and Chen, 2004), CNTs/chisol (Liu et al., 2005), and CdS nanoparticles (Huang et al., 2005).

TABLE 6.8 Binding rate Coefficients and Fractal Dimensions for glucose in 5 mM Buffer Solution to a Disposable Potentiometric glucose Biosensor (Liao et al., 2007)

Analyte in Solution	k	k_1	k_2	D_f	D_{f1}	D_{f2}
360 mg/dL glucose	0.03343 ± 0.00225	0.02426 ± 0.00179	0.0601 ± 0.0002	2.6194 ± 0.04272	2.4612 ± 0.1016	2.8208 ± 0.0006

Dai et al. (2007) further indicate that inorganic porous materials have exhibited potential as immobilization matrices due to their high mechanical, thermal, and chemical stability. They also exhibit good adsorption and penetrability characteristics. These materials include clay (Lei et al., 1999), montmorillonite (Lei and Deng, 1996; Fan et al., 2000; Sallez et al., 2000), porous alumina (Ikeda et al., 1998), and solgel matrix (Yu and Ju, 2002). Dai et al. (2007) emphasize that a mesoporous molecular sieve with appropriate dimensions does exhibit the following two properties:

1. acts as a current nanocollector and
2. acts as an electron relay to a macroelectrode.

Dai et al. (2007) have applied MCM-41 to GOx immobilization and have demonstrated direct electron transfer and effective biosensing for glucose detection.

Figure 6.13(a) shows the binding and dissociation of 5 μl of 0.32 M glucose to the GOx/MCM-41/Nafion/GCE biosensor (Dai et al., 2007). This is the first (Run #1) of four consecutive runs. A dual-fractal analysis is required to adequately describe the binding kinetics. A single-fractal analysis is adequate to describe the dissociation kinetics. The values of (1) the binding rate coefficient, k, and the fractal dimension, D_f, for a single-fractal analysis,

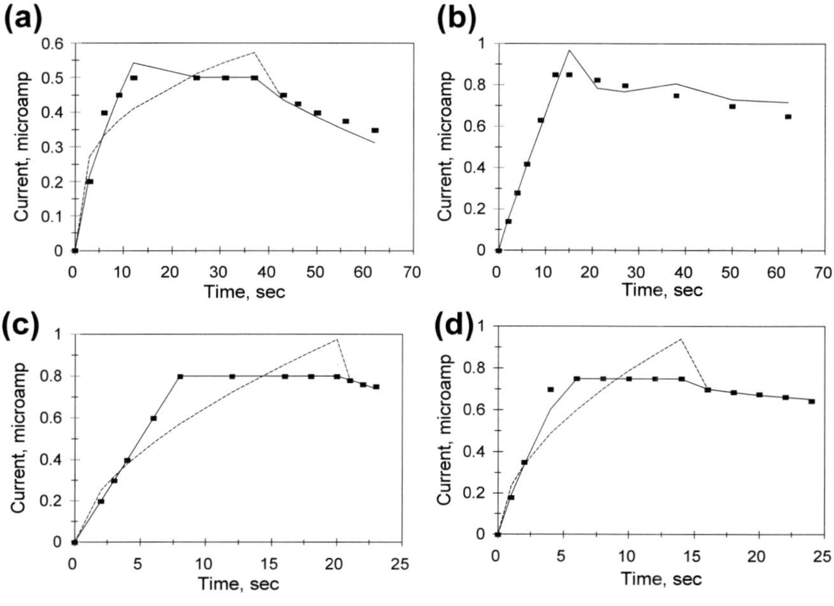

FIGURE 6.13 Binding and dissociation of 5 μl 0f 0.32 M glucose to the GOx/MCM-14/Nafion/GCE biosensor (Dai et al., 2007). Influence of consecutive runs: (a) Run #1 (b) Run #2 (c) Run #3 (d) Run #4.

(2) the binding rate coefficients, k_1 and k_2, and the fractal dimensions, D_{f1} and D_{f2}, for a dual-fractal analysis, and (3) the dissociation rate coefficient, k_d, and the fractal dimension for dissociation, D_{fd}, for a single-fractal analysis are given in Table 6.9. It is of interest to note that as the fractal dimension increases by a factor of 1.79 from a value of $D_{f1} = 1.6744$ to $D_{f2} = 3.0$ (the maximum value), the binding rate coefficient increases by a factor of 4.79 from a value of $k_1 = 0.1044$ to $k_2 = 0.5$. An increase in the degree of heterogeneity (fractal dimension) on the biosensor surface leads to an increase in the binding rate coefficient.

Figure 6.13(b) shows the binding and dissociation of 5 μl of 0.32 M glucose to the GOx/MCM-41/Nafion/GCE biosensor (Dai et al., 2007). This is the second (Run #2) of four consecutive runs. A single-fractal analysis is adequate to describe the binding and the dissociation kinetics. The values of

TABLE 6.9a Binding and Dissociation Rate Coefficients for 5 μL of 0.32 M Glucose to GOD/MCM-14/Nafion GCE (Dai et al., 2007). Influence of Consecutive Runs

Run Number	k	k_1	k_2	k_d
One	0.1963 ± 0.0463	0.1044 ± 0.0152	0.5 ± 0	0.01364 ± 0.0078
Two	0.07577 ± 0.00661	NA	NA	0.0473 + 0.04455
Three	0.1689 ± 0.0388	0.1 ±0	0.8 ± 0.0	0.2 ± 0.0
Four	0.2352 ± 0.0623	0.1897 ± 0.0290	0.75 ± 0.0	0.0608 ± 0.00211

TABLE 6.9b Fractal Dimensions for the Binding and Dissociation Phase for 5 μL of 0.32 M Glucose to GOD/MCM-14/Nafion GCE (Dai et al., 2007). Influence of Consecutive Runs

Run Number	D_f	D_{f1}	D_{f2}	D_{fd}
One	2.4060 ± 0.1861	1.6744 ± 0.2606	3.0 ± 0	0.01364 ± 0.0078
Two	1.1198 ± 0.0999	NA	NA	0.0473 + 0.04455
Three	1.8298 ± 0.1751	1.0 ± 0	3 − 1.6 E-16	1.0 ± 2.4 E-15
Four	1.9510 ± 0.1917	1.3264 ± 0.2084	3.0 − 3.0 E-16	1.8654 ± 0.08956

(1) the binding rate coefficient, k, and the fractal dimension, D_f, for a single-fractal analysis and (2) the dissociation rate coefficient, k_d, and the fractal dimension for dissociation, D_{fd}, for a single-fractal analysis are given in Table 6.9.

Figure 6.13(c) shows the binding and dissociation of 5 μL of 0.32 M glucose to the GOx/MCM-41/Nafion/GCE biosensor (Dai et al., 2007). This is the third (Run #3) of four consecutive runs. A dual-fractal analysis is required to adequately describe the binding kinetics. A single-fractal analysis is adequate to describe the dissociation kinetics. The values of (1) the binding rate coefficient, k, and the fractal dimension, D_f, for a single-fractal analysis, (2) the binding rate coefficients, k_1 and k_2, and the fractal dimensions, D_{f1} and D_{f2}, for a dual-fractal analysis, and (3) the dissociation rate coefficient, k_d, and the fractal dimension for dissociation, D_{fd}, for a single-fractal analysis are given in Table 6.9. It is of interest to note that as the fractal dimension increases by a factor of 3.0 from a value of $D_{f1} = 1.0$ to $D_{f2} = 3.0$ (the maximum value), the binding rate coefficient increases by a factor of 8.0 from a value of $k_1 = 0.1$ to $k_2 = 0.8$. An increase in the degree of heterogeneity (fractal dimension) on the biosensor surface, once again, leads to an increase in the binding rate coefficient.

Figure 6.13(d) shows the binding and dissociation of 5 μl of 0.32 M glucose to the GOx/MCM-41/Nafion/GCE biosensor (Dai et al., 2007). This is the fourth (Run #4) of four consecutive runs. A dual-fractal analysis is required to adequately describe the binding kinetics. A single-fractal analysis is adequate to describe the dissociation kinetics. The values of (1) the binding rate coefficient, k, and the fractal dimension, D_f, for a single-fractal analysis, (2) the binding rate coefficients, k_1 and k_2, and the fractal dimensions, D_{f1} and D_{f2}, for a dual-fractal analysis, and (3) the dissociation rate coefficient, k_d, and the fractal dimension for dissociation, D_{fd}, for a single-fractal analysis are given in Table 6.9. It is of interest to note that as the fractal dimension increases by a factor of 2.26 from a value of $D_{f1} = 1.3264$ to $D_{f2} = 3.0$ (the maximum value), the binding rate coefficient increases by a factor of 3.95 from a value of $k_1 = 0.1897$ to $k_2 = 0.75$. An increase in the degree of heterogeneity (fractal dimension) on the biosensor surface, once again, leads to an increase in the binding rate coefficient.

Figure 6.14(a) shows the increase in the dissociation rate coefficient, k_d, with an increase in the fractal dimension for dissociation, D_{fd}. For the data shown in Figure 6.14(a), the dissociation rate coefficient, k_d, is given by

$$k_d = (0.01685 \pm 0.00580) D_{fd}^{0.8184 \pm 0.4467} \tag{6.9a}$$

The fit is reasonable. Only four data points are available. The availability of more data points would lead to a more reliable fit. The dissociation rate coefficient, k_d, exhibits an order of dependence between one-half and first (equal to 0.8184) on the fractal dimension in the dissociation phase, D_{fd}.

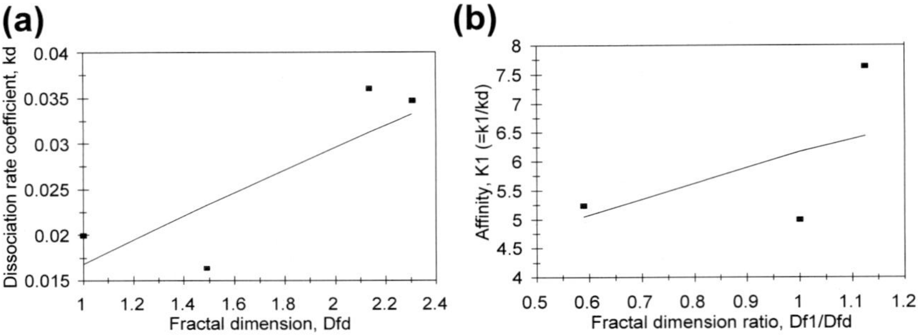

FIGURE 6.14 (a) Increase in the dissociation rate coefficient, k_d, with an increase in the fractal dimension for dissociation, D_{fd} (b) Increase in the affinity, K_1 ($=k_1/k_d$), with an increase in the fractal dimension ratio, D_{f1}/D_{fd}.

Figure 6.14(b) shows the increase in the affinity, K_1 ($=k_1/k_d$), with an increase in the fractal dimension ratio, D_{f1}/D_{fd}. For the data shown in Figure 6.14(b), the affinity, K_1 is given by

$$K_1 = (6.1657 \pm 1.9391)(D_{f1}/D_{fd})^{0.377+0.563} \quad (6.9b)$$

The fit is reasonable. There is scatter in the data. Only three data points are available. The availability of more data points would lead to a more reliable fit. This is reflected in the dependence exhibited by the affinity, K_1, on the ratio of fractal dimensions, D_{f1}/D_{fd}.

Lin et al. (2004) have developed a glucose biosensor based on CNT-NEEs. These authors emphasize the need for continuous ongoing research to help develop glucose biosensors for in vivo and in vitro applications (Free and Free, 1984). Lin et al. (2004) indicate that artificial mediators are used to increase the sensitivity and selectivity of amperometric biosensors (Few and Hill, 1987; Morris et al., 1992; Green and Hilditch, 1991). Also, to eliminate interference, permselective membranes have been used (Wang, 1991; Wang and Wu, 1994). Wang and coworkers (Wang et al., 1994, 1995) have also described mediator-free and membrane-free biosensors.

Lin et al. (2004) indicate the emergence of CNTs for their application in electrochemical biosensors (Musameh et al., 2002; Wang and Musameh, 2003; Li et al., 2003; Yu et al., 2006). Lin et al. (2004) emphasize that NEEs have practical application as electrochemical sensors, since their use results in an improvement in the signal-to-noise ratio as well as detection limits (Menon and Martin, 1995; Weber, 1995; Morf and de Rooij, 1997; Ugo et al., 2002). Lin et al. (2004) have used NEEs based on aligned CNTs for the development of a membrane-free and mediator-free glucose biosensor.

Figure 6.15(a) shows the amperometric response of NEE to glucose and a potential of −0.2 V in 0.1 M phosphate buffer. A singe-fractal analysis is

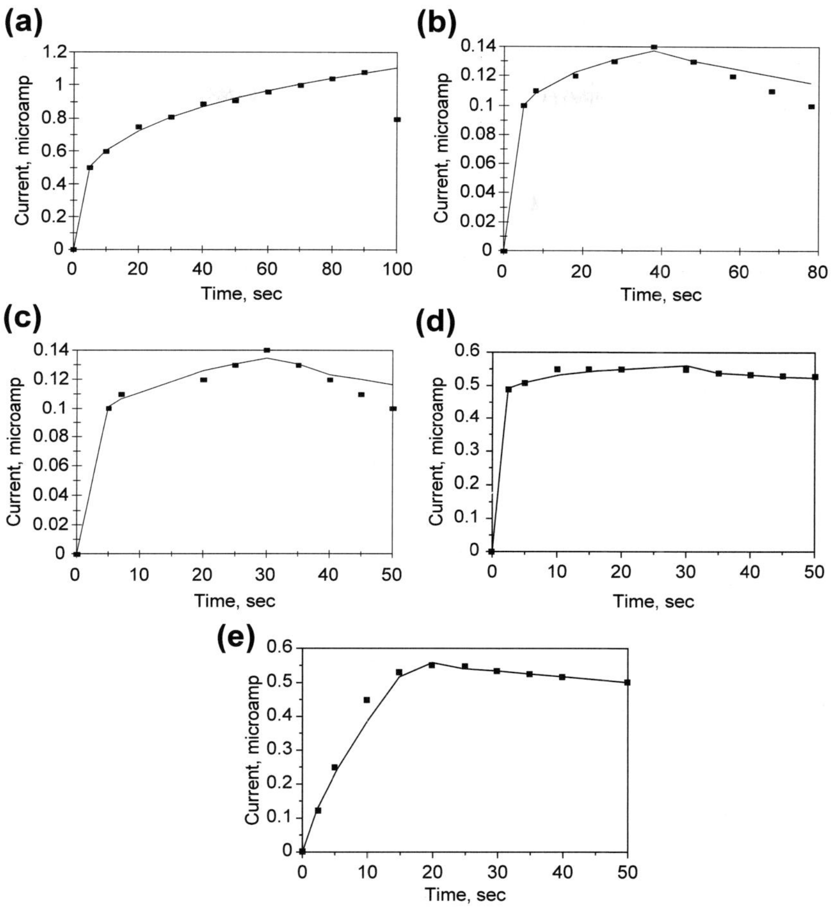

FIGURE 6.15 Amperometric response to (a) glucose and a potential of −0.2 V in 0.1 M phosphate buffer (b) glucose and a potential of 0.4 V in 0.1 M phosphate buffer (c) ascorbic acid and at a potential of 0.4 V in 0.1 M phosphate buffer (d) uric acid and at a potential of 0.4 V (e) acetaminophen (Lin et al., 2004).

adequate to describe the binding kinetics. The values of the binding rate coefficient, k, and the fractal dimension, D_f, for a singe-fractal analysis are given in Table 6.10.

Figure 6.15(b) shows the amperometric response of NEE to glucose and a potential of 0.4 V in 0.1 M phosphate buffer. A singe-fractal analysis is adequate to describe the binding and the dissociation kinetics. The values of the binding rate coefficient, k, and the fractal dimension, D_f, for a singe-fractal analysis, and the dissociation rate coefficient, k_d, and the fractal dimension for dissociation, D_{fd}, for a single-fractal analysis are given in Table 6.10.

TABLE 6.10 Binding and Dissociation Rate Coefficients for Glucose (−0.2 V and 0.4 V), Ascorbic Acid (AA: 0.4 V), Uric acid (UA: 0.4 V), and Acetaminophen (AC: 0.4 V)) to the NEE Biosensor (Lin et al., 2004)

Analyte in Solution/ Receptor on Surface	k	k_d	D_f	D_{fd}
Glucose (−0.2 V)	0.3321 ± 0.0052	NA	2.4776 ± 0.0102	NA
Glucose (0.4 V)	0.07825 ± 0.00156	0.001 ± 0.0	2.6896 ± 0.0232	2.0 ± 1.38E-15
Ascorbic Acid (0.4 V)	0.07832 ± 0.00331	0.002562 ± 0.000218	2.6814 ± 0.05148	1.3600 ± 0.15644
Uric Acid (UA)	0.4734 ± 0.0006	0.003414 ± 0.00001	2.8984 ± 0.0214	1.6864
Acetaminophen (AC)	0.07107 ± 0.01130	0.001571 ± 0.00036	1.5334 ± 0.17478	0.9549 ± 0.03256

Figure 6.15(c) shows the amperometric response of NEE to AA and at a potential of 0.4 V in 0.1 M phosphate buffer. A single-fractal analysis is adequate to describe the binding and the dissociation kinetics. The values of the binding rate coefficient, k, and the fractal dimension, D_f, for a singe-fractal analysis, and the dissociation rate coefficient, k_d, and the fractal dimension for dissociation, D_{fd}, for a single-fractal analysis are given in Table 6.10.

Figure 6.15(d) shows the amperometric response of NEE to UA and at a potential of 0.4 V 0.1 M in phosphate buffer. A single-fractal analysis is adequate to describe the binding and the dissociation kinetics. The values of the binding rate coefficient, k, and the fractal dimension, D_f, for a single-fractal analysis, and the dissociation rate coefficient, k_d, and the fractal dimension for dissociation, D_{fd}, for a single-fractal analysis are given in Table 6.10.

Figure 6.15(e) shows the amperometric response of NEE to AP and at a potential in 0.1 M phosphate buffer. A single-fractal analysis is adequate to describe the binding and the dissociation kinetics. The values of the binding rate coefficient, k, and the fractal dimension, D_f, for a single-fractal analysis, and the dissociation rate coefficient, k_d, and the fractal dimension for dissociation, D_{fd}, for a single-fractal analysis are given in Table 6.10.

Figure 6.16(a) shows the increase in the binding rate coefficient, k, with an increase in the fractal dimension, D_f, for a single-fractal analysis. For the data shown in Figure 6.16(a), the binding rate coefficient, k, is given by

$$k = \left(2.8 \pm 0.9\right)E - 14\Big) D_f^{2.881 \pm 2.545} \quad (6.10a)$$

The fit is good. Only three data points are available. The availability of more data points would provide a more reliable fit. The binding rate coefficient, k, exhibits an high order of dependence on the fractal dependence, D_f, between two and one-half and three (equal to 2.881).

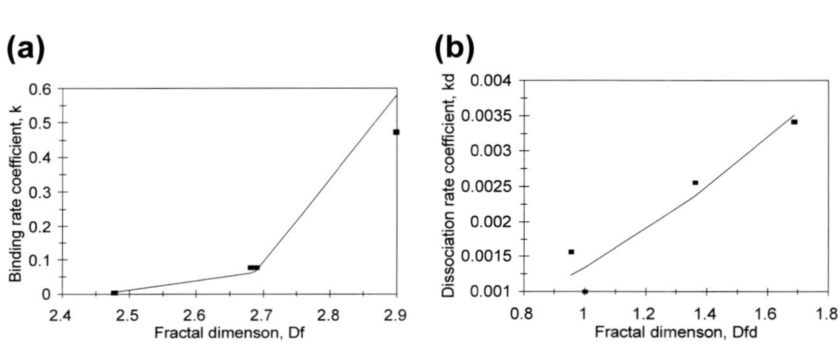

FIGURE 6.16 (a) Increase in the binding rate coefficient, k, with an increase in the fractal dimension, D_f (b) Increase in the dissociation rate coefficient, k_d, with an increase in the fractal dimension for dissociation, D_{fd}.

Figure 6.16(b) shows the increase in the dissociation rate coefficient, k_d, with an increase in the fractal dimension for dissociation, D_{fd}. For the data shown in Figure 6.16(b), the dissociation rate coefficient, k_d, is given by

$$k_d = (0.001344 \pm 0.000427) D_{fd}^{1.833 \pm 0.593} \quad (6.10b)$$

The fit is good. Only four data points are available. The availability of more data points would lead to a more reliable fit. The dissociation rate coefficient, k_d, exhibits a dependence between one and one-half and two (equal to 1.833) on the fractal dimension in the dissociation phase, D_{fd}.

6.4 CONCLUSIONS

A fractal analysis is presented for the binding and the dissociation (if applicable) of glucose to different biosensor surfaces. Both a single- and a dual-fractal analysis are used. The dual-fractal analysis is used only if the single-fractal analysis did not provide a reasonable and better fit. This is done using Corel Quattro Pro 8.0 (1997). The fractal dimension provides a quantitative measure of the degree of heterogeneity present on the biosensor surface. The fractal dimension for the binding and the dissociation phase, D_f and D_{fd}, respectively, is not a typical independent variable, such as analyte concentration, that may be directly manipulated, and one may consider it as a derived variable. It is evaluated from Eqns (6.1)–(6.3). An increase in the fractal dimension or the degree of heterogeneity on the biosensor surface leads, in general, to an increase in the binding and in the dissociation rate coefficient(s).

The binding and dissociation of (1) intracellular glucose in a single cell to the GIPi–Thr sensor using ratiometric FRET measurement is analyzed using fractal analysis. The influence of repeat measurement is presented (Jin et al., 2011), (2) binding of glucose during monitoring through FRET measurement with the GIPi-Thr sensor (Jin et al., 2011), (3) binding and dissociation of glucose during continuous monitoring of glucose by fluorophore-labeled genetically engineered GGBP immobilized on a hydrogel matrix (Weidamaier et al., 2011), (4) binding of different concentrations of glucose in solution in microliter to a boron-doped carbon-coated nickel-modified electrode (Yang et al., 2011b), and (5) binding of glucose and its interferents AP, UA, and AA to GOx immobilized on a biosensor surface (Qiang et al., 2011).

Predictive relations are developed for the binding of glucose to different sensor surfaces. These predictive relations may be used to help manipulate or change the different biosensor parameters such as the binding and the dissociation rate coefficients in desired directions. For example, for the binding and dissociation of intracellular glucose to GIP immobilized on a GIPi-Thr glucose sensor (Jin et al., 2011), the binding rate coefficient, k_1, for a dual-fractal analysis increases as the fractal dimension, D_{f1}, increases on the biosensor surface, and exhibits greater than one-third (equal to 3.191) order of dependence on the fractal dimension, D_{f1}. Similarly, for this case, the binding rate coefficient, k_2, exhibits close to a second (equal to 1.887) order of dependence on D_{f2}. Also, (Jin et al., 2011) noted that the binding rate coefficient, k, for a single-fractal analysis increases with an increase in the glucose concentration in solution in the 12.5−20 mM range. In this case, close to a first (equal to 0.96) order of dependence on the glucose concentration is exhibited. Also, for the binding and dissociation of glucose to a GGBP-based fiber-optic biosensor (Weidamaier et al., 2011) during continuous glucose tracking by a fluorophore-labeled and genetically engineered GGBP, the binding rate coefficient, k, for a single-fractal analysis exhibits slightly higher than one-half (equal to 0.6649) order of dependence on the fractal dimension, D_f, or the degree of heterogeneity that exists on the biosensor surface.

The different examples presented and analyzed in this chapter emphasize that the degree of heterogeneity that exists on the biosensor surface does significantly affect, in general, the rate coefficient and affinity values, and subsequently the kinetics, in general. These are just a few of the representative examples available in the literature. More such studies are required to determine whether the binding and the dissociation rate coefficients, and subsequently the affinity values, are sensitive to their respective fractal dimensions on the biosensor surface for glucose detection.

A better understanding of all possible parameters that influence the kinetics of binding and dissociation of glucose to different biosensor surfaces is critical. More often than not, the influence of diffusion and heterogeneity on the biosensor surface is neglected. As indicated in this chapter, and elsewhere in this book, the degree of heterogeneity significantly influences, in general, the binding as well as the dissociation kinetics occurring on biosensor surfaces. It would behoove the practicing biosensorists to start paying more attention to this aspect of kinetics on biosensor surfaces. One may perhaps argue that the influence of diffusion limitations may be minimized or perhaps even be eliminated if the biosensor is run properly.

Finally, diabetes is reaching epidemic proportions in the world, and though it by itself is not difficult to manage, serious compilations arise due to the "cluster" ailments such as high blood pressure, stroke, heart attack, etc., that are associated with it. Thus, the need to monitor the glucose levels continuously. Due to the invasive nature of the monitoring, there is some hesitancy and noncompliance of patients to monitor the glucose levels. This is where the noninvasive methods of glucose monitoring systems can play a very significant role. There is bound to be, if there is not already, a very significant push and efforts devoted to research in this area.

REFERENCES

Ahmad, F., Christenson, A., Bainbridge, M., Yusof, A.P.M., Abgani, S., 2007. Biosensors and Bioelectronics 22, 1625–1632.

Amiss, T.J., Sherman, D.B., Nycz, C.M., Anadluz, S.A., Pitner, J.B., 2007. Protein Science 16, 2350–2359.

Anonymous, June 2007. Amperometric glucose-biosensor based on multilayer films via layer by layer self-assembly of multi-wall carbon nanotubes, gold particles and glucose oxidase on Pt electrode. Biosensors and Bioelectronics 22 (2), 2854–2860.

Ballerstadt, R., Evans, C., McNichols, R., Gowda, A., 2006. Biosensors and Bioelectronics 22 (2), 275–284.

Ballerstadt, R., Schultz, J.S., 2000. Analytical Chemistry 72, 4185–4192.

Baxter, J., 2012. Nature Photonics 6, 268–269.

Brauker, J., 2009. Diabetes Technology and Therapy 11 (Suppl. 1), S25–S36.

Buckingham, B., Caswell, K., Wilson, D.M., 2007. Current Opinion in Endocrinology Diabetes Obesity 14, 288–295.

Bunn, H.F., Haney, D.N., Kamin, S., Gabbay, K.H., Gallop, P.M., 1976. Journal of Clinical Investigation 57, 1652–1659.

Busk, M., Horeman, M.R., Kristjansen, P.E., van der Kogel, A.J., Bussink, J., Overgaard, J., 2008. International Journal of Cancer 122 (12), 2726–2734.

Caduff, A., Mueller, M., Megej, A., Dewarrat, F., Suri, R.E., Klisic, J., Donath, M., Zakharov, P., Schaub, D., Stahel, W.A., Talary, M.S., May 2011. Characteristics of a multisensory for non-invasive glucose monitoring with external validation and prospective evaluation. Biosensors and Bioelectronics 26 (9), 3794–3800.

Cai, C., Chen, J., 2004. Direct electron transfer of glucose oxidase promoted by carbon nanotubes. Analytical Biochemistry 332, 75–83.

Cash, K.J., Balaconis, M.K., Dubach, J.M., Clark, H.A., November 9, 2010. Fluorescent Nanosensors for Glucose Detection. Annual American Institute of Chemical Engineers, Salt Lake City, Utah paper 262a.

Campbell, R.E., 2009. Fluorescent protein-based biosensors: modulation of energy transfer as a design principle. Analytical chemistry 81 (15), 5972–5979.

Capuccio, F.E., Suri, J.T., Cordes, D.B., Wessling, R.A., Singaram, B., 2004. Journal of Fluorescence 14, 521–533.

Careega, C.L., Falke, J., 1992. Journal of Molecular Biology 226 (4), 1219–1235.

Cella, L.N., Chen, W., Myung, N.V., Mulchandani, A., 2011. Single-walled carbon nanotube-based chemiresistive affinity biosensors for small molecules: ultrasensitive glucose detection. Journal of the American Chemical Society.

Chen, X., Pan, H., Liu, H., Du, M., December 30, 2010. Nonenzymatic glucose sensor based on flower-shaped Au@Pd core-shell nanoparticles-ionic liquids composite film modified glass carbon electrodes. Electrochimica Acta 636–643.

Chen, H., Dong, March 2007. Direct electrochemistry and electrocatalysis of horseradish penoxidase immobilized in sol-gel derived ceramic – carbon nanotube nanocomposite film. Biosensors and Bioelectronics 22 (8), 1811–1815.

Chen, S.P., Yu, X.D., Xu, J.J., Chen, H.Y., August 15 , 2011. Gold nanoparticles-coated magnetic nanospheres as affinity matrix for detection of hemoglobin A1c in blood by microfluidic immunoassay. Biosensors and Bioelectronics 20 (11), 4779–4784.

Chen, X., Parr, H., Liu, H., and Diu, M. Nonenzymatic glucose sensor based on ultrasonic electrodeposition of bimetallic (M = Ru, Pd, and Au) nanoparticles (Journal).

Chen, X.M., Cai, Z.M., Jia, T.E., Liu, Hz, Jiang, Yq, Chen, X., August 15, 2009. A novel non-enzymatic ECL sensor for glucose using palladium nanoparticles supported on functional carbon nanotubes. Biosensors and Bioelectronics 24 (12), 3475–3480.

Claussen, J.C., Franklin, A.D., Haque, A.U., Porterfield, D.M., Fisher, T.S., 2009. ACS Nano 3, 37 e44.

Claussen, J.C., Kim, S.S., Haque, A.U., Artiles, M.S., Porterfield, D.M., Fisher, T.S., 2010. Electrochemical glucose biosensor of platinum nanospheres connected by carbon nanospheres. Journal of Diabetes Science and Technology 4 (2), 312–319.

Claussen, J.C., Widner, M.M., Fisher, T.S., Porterfield, D.M., 2011. Transporting the fabrication and bifunctionalization of gold nano electrodes arrays into versatile electrochemical glucose biosensors. ACS Applied Materials & Interfaces 3 (5), 1765–1770.

Corel Quattro Pro 8.0 Ottawa, Canada, 1997.

Cui, X., Liu, G., Lin, Y., 2005. Amperometric biosensors based on carbon paste electrodes modified with nanostructure mixed-valence manganese oxides and glucose oxidase. Nanomedicine: Nanotechnology, Biology and Medicine 1, 10–15.

Dai, Z.H., Ni, J., Huang, X.H., Lu, G.F., Bao, J.C., 2007. Direct electrochemistry of glucose oxidase immobilized on a hexagonal mesoporous silica-MCM-41 matrix. Bioelectrochemistry 70 (2), 250–256.

De Brock, C., Vertomemmen, J., Manuel-Keeney, B., Gaal, L.V., 2008. Current Diabetes Review 4, 159–168.

Deng, C., Chen, J., Chen, X., Xiao, C., Nie, L., Yao, S., 2008. Biosensors and Bioelectronics 23, 1272–1277.

Ding, Y., Wang, Y., Lei, Y., November 9, 2010a. Non-enzymatic glucose sensing using novel metal oxide-metal composite nanofilms. Annual American Institute of Chemical Engineers Meeting. Salt Lake City, Utah, paper 203g.

Ding, Y., Wang, Y., Su, L., Bellagamba, M., Zhang, H., Lei, Y., October 15, 2010b. Electrospun Co_3O_4 sensitive and selective glucose detection. Biosensors and Bioelectronics 26 (2), 542–548.

Ehlebacher, J., Aziz, M.J.U., Karma, A., Dormitory, N., Sheradzki, K., 2001. Nature 410, 450–453.

Fan, C., Zhuang, Y., Li, G., Zhu, J., Zhu, D., 2000. Direct electrochemistry and enhanced activity for hemoglobin in a sodium montmorillonite film. Electroanalysis 12, 1156–1157.

Fang, H., Kaur, G., Wang, B., 2004. Journal of Fluorescence 14, 481–489.

Fehr, M., Lalonde, S., Ehrhardt, D.W., Frommer, W.B., 2004. Journal of Fluorescence 14, 603–609.

Fehr, J.M., Lalonde, S., Lager, I., Wolff, M.W., Frommer, W.B., 2003. Journal of Biological Chemistry 278 (21), 19127–19133.

Feng, D., Wang, F., Chen, Z., 2009. Sensors and Actuators B: Chemical 138, 539.

Few, J., Hill, H.A., 1987. Analytical Chemistry 59, 933A.

Free, M., Free, H., 1984. Analytical Chemistry 56, 664A.

Fu, C., Yang, W., Chen, X., Evans, D.G., 2009. Electrochemistry Communications 11, 997–1000.

Ginsberg, B.J., Mazze, R., 1994. New Jersey Medicine 91, 221–224.

Giradin, C.M., Huot, C., Gonthier, M., Devlin, E., 2009. Clinical Biochemistry 42, 136–142.

Gong, K., Yan, Y., Zhang, M., Su, L., Xiong, S., Miao, L., 2005. Analytical Science 21, 1383–1394.

Goodall, I., 2005. Chemical and Biochemical Reviews 26, 5–19.

Gooding, J.J., 2005. Electrochimica Acta 50 (15), 3049–3060.

Green, M., Hilditch, P., 1991. Analytical Proceedings 28, 374.

Guan, W.J., Li, Y., Chen, Y.Q., Zhang, X.B., Hu, G.Q., 2005. Glucose biosensor based on multi-wall carbon nanotubes and screen printed carbon electrodes. Biosensors and Bioelectronics 21, 508–512.

Guiseppi-Elie, A., 2011. Bioactive Electroactive Hydrogels, Invited Lecture, Annual American Institute of Chemical Engineers Meeting. October 20, Thursday, Minneapolis, Minnesota.

Gulliam, M.T., Dupraz, P., Thorens, B., 2000. Diabetes 49, 1485–1491.

Harwod, G.W.J., Pouton, C.W., 1996. Amperometric enzyme biosensors for the analysis of drugs and metabolites. Advances in Drug Delivery Reviews 18, 163–191.

Havlin, S., 1989. Molecular diffusion and reaction. In: Avnis, D. (Ed.). The Fractal Approach to Heterogeneous Chemistry: Surfaces, Colloids, Polymers. Wiley, New York, pp. 251–269.

Havlin, S., Ben-Avraham, D., 1987. Diffusion in disordered media. Advances in Physics 36, 695–798.

Heinemann, L., Koschinsky, T., 2002. International Journal of Clinical Practice Supplement 129, 75–79.

Hellman, B., Idahl, L.A., Lenmark, A., Taljedal, I.B., 1974. Proceedings of the National Academy of Sciences of the United States of America 71 (9), 3405–3409.

Hirsch, I.B., 2009. Journal of Clinical Endocrinology and Metabolism 94, 2232–2238.

Hrapovic, S., Liu, Y., Male, K.B., Luong, J.H., 2004. Analytical Chemistry 76 (4), 1083–1088.

Hsieh, H.V., Pfeiffer, Z.A., Hmiss, T.A., Sherman, D.B., Pitner, J.B., 2004. Direct detection of glucose by surface plasmon resonance with hacterial glucose/galactose binging protein. Biosensors and Bioelectronics 19, 653–660.

Huang, Y., Zhang, W., Liao, M.H., Chen, D.H., 2003. Biotechnology Progress 19, 1095–1100.

Huang, Y., Zhang, W., Xiao, H., Li, G., 2005. An electrochemical investigation of glucose oxidase at a CdS nanoparticles modified electrode. Biosensors and Bioelectronics 21, 817–821.

Ikeda, O., Ohtani, M., Yamaguchi, T., Komura, A., 1998. Direct electrochemistry of cytochrome c at glassy electrode covered with a microporous alumina membrane. Electrochimica Acta 43, 83–89.

Jaffe, L.F., Nuccitelli, R., 1974. Journal of Cell Biology 80 (63), 614–628.

Jia, W.Z., Wang, K., Xia, X.H., April 2010. Elimination of electrochemical interferences in glucose biosensors. TrAC Trends in Analytical Chemistry 29 (4), 306–318.

Jin, H.J., Beriet, C., Ferrigno, R., Girault, H.H., 2001. Journal of Electroanalytical Chemistry 502, 138–145.

Jin, S., Veetil, J., Garrett, J.R., Ya, Y., April 15, 2011. Construction of a panel of glucose indicator proteins for continuous glucose monitoring. Biosensors and Bioelectronics 26 (8), 3427–3437.

Josephson, L., Graman, E.V., Men3, E., Lewis, J.M., Bengele, H., 1990. A funtionalized superparamagnetic iron oxide colloid as a receptor directed MR contrast agent, Magnetic Resonance Imaging 8, 637–646.

Kang, K., Wang, J., Wu, H., Aksay, I.A., Lin, Y., 2009. Glucose oxidase-graphene-Chitosan modified electrode for direct electrochemistry and glucose sensing, Bioelectronics 25, 901–905.

Khan, F., Gnudi, L., Pickup, J.C., 2008. Biochemical and Biophysical Research Communications 365 (1), 102–106.

Klonoff, D.C., 2004. Diabetes Care 27, 834–836.

Klonoff, D.C., 2005. Diabetes Care 28, 1231–1239.

Koenig, R.J., Peterson, C.M., Kilo, C., Cerami, A.J., Williamson, R., 1976. Diabetes 25, 230–232.

Kuila, T., Bose, S., Khamra, P., Mishra, A.K., Kim, N.H., Lee, J.H., August 2011. Recent Advances in graphene-based biosensors. Biosensors and Bioelectronics 26 (2), 4637–4648.

Kumar, C.V., Chaudhari, A., 2003. Microporous and Mesoporous Materials 57, 181–190.

Lanevschi, A., Kramer, J.W., 2009. Veterinary Clinical Pathology 25 (1), 10–13.

Lanniello, R.M., Lindsay, T.J., Yacynych, A.M., 1962. Differential pulse voltammetric study of direct electron in glucose oxidase chemically modified graphite electrodes. Analytical Chemistry 54, 1098–1101.

Lee, C.H., Seo, H.I., Lee, Y.C., Cho, B.W., Jeong, H., Sohn, B.K., 2000. All solid type ISFET glucose sensor with fats response and high sensitivity characteristics. Sensors and Actuators B: Chemical 64, 37–43.

Lee, S.Y., May 2010. Glucose biosensors: an overview of use in clinical practice. Biosensors and Bioelectronics 10 (5), 4558–4576.

Lei, C., Lisdat, Wollenberger, U., Scheller, F.W., 1999. Cytochrome c/clay-modified electrode. Electroanalysis 11, 274–276.

Lei, C., Deng, J., 1996. Hydrogen peroxide sensor based on co-immobilized methylene green and horseradish peroxidase in the same montmorillonite-modified bovine serum albumin-glutaraldehyde matrix on glassy carbon electrode surface. Analytical Chemistry 68, 3344–3349.

Li, J., Ng, H.T., Cassell, A., Fan, W., Chen, H., Ye, Q., Koehne, J., Han, J., Meyyappan, M., 2003. Nano Letters 3, 597–602.

Li, F., Wang, Z., Shan, C., Song, J., Han, D., Niu, L., 2009. Biosensors and Bioelectronics 24, 1765–1770.

Li, C.M., Dong, H., Cao, X., Luong, J.H.T., Zhang, X., 2007a. Current Medicinal Chemistry 14, 937–951.

Li, J., Fan, C., Xiao, F., Yan, R., Fan, S., Zhao, F., Zeng, B., 2007b. Electrochimica Acta 52, 8178–8185.

Liao, C.W., Chou, J.C., Sun, T.P., Hsiung, S.K., Hsieh, J.H., May 21, 2007. Preliminary investigations on a glucose biosensor based on potentiometric principle. Sensors and Actuators B: Chemical 123 (2), 720–726.

Lin, Y., Lu, F., Tu, Y., Ren, Z., February 2004. Glucose biosensors based on carbon nanotube nanoelectrode ensembles. Nano Letters 4 (2), 191–195.

Liu, S., Ju, H., 2003. Reagentless glucose biosensor based on direct electron transfer of glucose oxidase immobilized on colloidal gold modified carbon paste electrode. Biosensors and Bioelectronics 19, 177–183.

Lin, S., Lin, Y.S., Chang, J.P., Nov 2003. Ion-enhanced etching of Hfo2 for integration in metal-oxide semiconductor field-effect transistors. Journal of Vacuum Science and technology 2 (6), 2420–2427.

Liu, Y., Wang, M., Zhao, F., Xu, Z., Dong, S., December 2005. The direct electron transfer of glucose oxidase and glucose biosensor based on carbon nanotubes/chitosan matrix. Biosensors and Bioelectronics 21 (5), 984–988.

Lowry, J.P., O'Neill, R.D., Boutelle, M.G., Fillenz, M., 1998. Journal of Neurochemistry 70 (1), 391–396.

Lu, J., Do, I., Drzal, L.T., Wordens, R., Lee, I., 2008a. Nanoparticle decorated exfoliated graphene nanoplatelets based glucose biosensors with large sensitivity and fast response. ACS Nano 2 (9), 1825–1832.

Lu, J., Do, I., Drzal, L.T., Worden, R.M., Lee, I., 2008b. ACS Nano 2, 1825–1832.

Lu, W., Luo, Y., Chang, G., Sun, X., August 15, 2011. Synthesis of functional SIO_2-coated graphene oxide nanosheets decorated with Ag nanoparticles for H_2O_2 glucose detection. Biosensors and Bioelectronics 26 (2), 4791–4797.

Lu, L.H., Sun, G.Y., Zhang, H.J., Wang, H.S., Wang, H.S., Xi, S.H., Hu, J.Q., Tian, Z.Q., Chen, R., 2004. Journal of Material Chemistry 14, 1005.

Lu, S.Y., Li, C.E., Zhang, D.D., Zhang, Y., Mo, Z.H., Cai, Q., Zhu, A.R., 1994. Electron transfer of an electrode of glucose oxidase immobilized in polyaniline. Journal of Electroanalytical Chemistry 364, 1–36.

Marvin, T.S., Hellinga, 2001. Manipulation of ligand binding affinity by exploitation of conformational compling. Nature Structural Biology 8 (9), 795–798.

McLamore, E.S., Shi, J., Jaroch, D., Claussen, J.C., Uhia, A., Jiang, Y., Zhang, W., Donki, S.S., Banks, M.K., Buhman, K.K., Teergarden, D., Ruckus, J.C., Porterfield, D.M., January 15, 2011. A self-referencing platinum nanoparticle decorated enzyme-based microbiosensor for real time measurement of physiological glucose transport. Biosensors and Bioelectronics 2237–2246.

Menon, V.P., Martin, C.R., 1995. Analytical Chemistry 67, 1920–1928.

Miller, D.M., Olson, J.S., Pflugrath, J.W., Quiocho, F.A., 1983. Journal of Biological Chemistry 258 (22), 13655–13672.

Morf, E.W., de Rooij, N.F., 1997. Sensors and Actuators B: Chemical 44, 538–541.

Morris, N., Cardosi, M., Birch, B., Turner, A.P., 1992. Electroanalysis 4, 1.

Morten, B., Horsman, M.R., Kristjansen, P.E.G., van der Kogel, A.J., Bossink, J., Overgaard, J., 2008. International Journal of Cancer 122 (12), 2726–2734.

Moschou, E.A., Sharma, B.V., Deo, S.K., Daunert, S., 2004. Journal of Fluorescence 14, 535–547.

Mousty, C., 2004. Sensors and biosensors based on clay-modified electrode-new trends. Applied Clay Science 27, 159–177.

Musameh, M., Wang, J., Merkoci, A., Lin, Y., et al., 2002. Protein electrochemistry at carbon nanotube electrode Journal of Electroanalysis, 689–694.

Nagata, R., Clark, S.A., Yokoyama, K., Tamiya, E., Karube, I., 1995. Amperometric glucose biosensor manufactured by a printing technique. Analytica Chimica Acta 304, 157–164.

Narsimhan, K., Wimgard Jr., L.B., 1986. Enhanced direct electron transport with glucose oxidase immobilized on (aminophenyl)boronic acid modified glassy carbon electrode. Analytical Chemistry 2984–2987.

Nathan, D.M., Turgeon, H., Regan, S., 2007. Diabetologia 50, 2239–2244.

Ni, J.A., Ju, H.X., Chen, H.Y., Leech, D., 1999. Analytica Chimica Acta 378 (1–3), 151–157.

Pan, D., Chen, J., Yao, S., Nie, L., Xia, J., Tao, W., 2005. Amperometric glucose biosensor based on immobilization of glucose oxidase in electropolymerized *o*-aminophenol film at copper-modified gold electrode. Sensors and Actuators B; Chemical 104, 68–74.

Parente, A.H., Marqueez Jr., E.T., Azevedo, V.M., Diniz, F.B., Hiemel, E., Filko, J.L., December 1994. Glucose biosensor using glucose oxidase immobilized on polyaniline. Analytical Biochemistry and Biotechnology 37 (11), 267–273.

Park, B.W., Ko, K.A., Kim, T.Y., Yoon, D.Y., Kim, D.S., October 11, 2011. Glucose Biosensor Based on Encapsulated Enzyme with Peptide Nanotubes, Annual American Institute of Chemical Engineers Meeting. paper 11g, Monday. Minnesota, Minneapolis.

Park, E.J., Werner, J., Beebe, J., Chan, S., Smith, N.B., 2010. Noninvasive ultrasonic glucose sensing with large pigs (approximately 200 lbs) using a lightweight cymbal transducer array and biosensors. AIP Conference Proceedings 1215, 157–160.

Park, S., Boo, H., Chung, T.D., 2006. Analytica Chimica Acta 556, 46.

Park, S., Lim, J.H., Chung, S.W., Mirkin, C.A., 2004. Science 303, 348.

Patel, H., Li, X., Karan, H.J., 2003. Amperometric glucose sensors based on ferrocene containing polymeric electron transfer systems-a preliminary report. Biosensors and Bioelectronics 18, 1070–1076.

Patolsky, F., Weizmann, Y., Willner, I., 2004. Angewandte Chemie International Edition 43, 2113–2117.

Pickup, J.C., Hussain, F., Evans, N.D., Rolinski, O.J., Birch, D.J.S., June 15, 2005. Fluorescence-based glucose sensors. Biosensors and Bioelectronics 20 (12), 2555–2565.

Piper, H.G., Alexander, J.L., Shukla, A., Pigula, F., Costello, J.M., Laussen, P.C., Jaksic, T., Angus, M.S.D., 2006. Pediatrics 116, 1176–1184.

Poet, P.D.T., Miyamoto, S., Murakami, T., Kimura, J., Karube, I., 1990. Direct electron transfer with glucose oxidase immobilized in n electropolymerized poly(*N*-methylpyrrole) film on a gold microelectrode. Analytica Chimica Acta 235, 255–263.

Psoma, S.D., van der Wal, P.D., Frey, O., de Rooij, N.F., Turner, A.P.F., December 15, 2010. A novel enzyme entrapment in SU-8 microfabricated films for glucose microbiosensors. Biosensors and Bioelectronics 28 (4), 1582–1587.

Pumera, M., Castenada, M.T., Castenada, M.I., Pividori, M.I., Eritja, L., Merkoci, A., Alagret, S., 2005. Langmuir 21, 9625–9629.

Qiang, L., Vaddiraju, S., Patel, D., Papadimitrakopoulos, F., May 15, 2011. Edge-plane microwire electrodes for highly sensitive H_2O_2 and glucose detection. Biosensors and Bioelectronics 26 (9), 3755–3760.

Qiu, B., Lin, Z., Wang, J., Chen, Z., Chen, J., Chen, G., April 15, 2009. An electrochemical luminescent biosensor for glucose based on the electrochemiluminescence of luminol on the nafion/glucose oxidase poly (Nickel (II) tetrasulphothalocyanine)/multi-walled carbon nanotubes modified electrode. Talanta 78 (1), 76–80.

Quinn, C.P., Pathak, C.P., Heller, A., Hubbell, J.A., 1995. Photo-cross linking copolymers of Z-hydroxyethyl methaerylate, poly (ethylene glycol) tetra-acrylate and ethylene dimethacrylate for improving biocompatibility of Biosensors. Biomaterials 16 (5), 389–396.

Ramakrishnan, A., Sadana, A., 2001. A single-fractal analysis of cellular analyte-receptor binding kinetics using biosensors. Biosystems 59, 35–51.

Rodriguez-Gonzalez, B., Pastoria-Santos, I., Liz-Marzan, L.M., 2006. Journal of Physical Chemistry B 110, 11796.

Rollins, D.K., Beverlin, L., Kotz, K., Vyas, N., Andre, D., October 19, 2011. Ultra-sensitive Glucose Detection, Using Copper Nanowires. Annual American Institute of Chemical Engineers Meeting. Minneapolis, Minnesota.

Rossi, L.M., Quach, A.D., Rosenzweig, Z., 2004. Analytical and Bioanalytical Chemistry 380, 606–613.

Sacks, D.B., Bruns, D.B., Goldstein, D.E., MacLaren, N.K., McDonald, J.M., Parrott, M., 2002. Clinical Chemistry 48, 436–472.

Sadana, A., 2005. Fractal Binding and Dissociation Kinetics for Different Biosensor Applications. Elsevier, Amsterdam.

Sadana, A., 2001. A fractal analysis for the evaluation of hybridization kinetics in biosensors. Journal of Colloid and Interface Science 151 (1), 166–177.

Saito, T., Watanabe, M., 1998. Characterization of poly(vinylferrocene-co-2-hydroxymethyl methacrylate) for use as electron mediator in enzymatic glucose sensor. Reactions, Function Polymers 7, 263–269.

Sakamoto, M., Takamura, K., 1982. Bioelectrochemistry and Bioenergetics 9, 571.

Salimi, A., Shanfi, E., Noorbakshsh, A., Soltanian, S., 2006. Electrochemistry Communications 8, 1499–1506.

Sallez, Y., Bianco, P., Lojou, E., 2000. Electrochemical behavior of c-type cytochromes at clay-modified carbon electrodes: a model for re-interaction between proteins and soils. Journal of Electroanalytical Chemistry 493, 37–49.

Savitri, D., Mitra, C.K., 1998. Electrochemistry of reconstituted glucose oxidase on carbon pate electrodes. Bioelectrochemistry and Bioenergetics 47, 67–73.

Saxi, T., Khan, F., Ferla, M., Birch, D., 2011. A fluorescence lifetime-based fiber-optic glucose sensor using glucose/galactose-binding protein. Analyst 136, 968–972.

Schultz, J.S., Mansouri, S., Goldstein, I.J., 1982. Diabetes Care 5 (3), 245–253.

Schultz, J.S., Ye, K., 2003. Analytical Chemistry 75 (14), 3451–3459.

Schwartz, D.E., Gong, P., She, K.L., October 15, 2009. Superior long-term stability of a glucose biosensor based on inserted barrel plating gold electrodes. Biosensors and Bioelectronics i 383–387.

Seo, H.I., Kim, C.S., Sohn, B.K., Yeow, T., Son, M.T., Haskard, M., 1997. ISFET glucose sensor based on a new principle using the electrolysis of hydrogen peroxide. Sensors and Actuators B: Chemical 40, 1–5.

Shangguan, X., Zhang, H., Zheng, J., 2008. Electrochemistry Communications 10, 1140–1143.

Shao, Y., Wang, J., Wu, H., Liu, J., Aksay, I.A., Lin, Y., 2010. Graphene Based electrochemical sensors and Biosensors: A review, electroanalysis 22, 1027–1036.

Singh, M., Kathurojum, P.K., Jampana, N., 2009. Sensors and Actuators B 143, 430–443.

Skeie, S., Kristensen, G.B., Carlsen, S., 2009. Self-monitoring of blood glucose in Type-I diabetes patients with insufficient metabolic control: Focused monitoring of blood glucose intervention can lower glycated hemoglobin A1C. Journal of Diabetes Science and Technology 3, 8–88.

Song, Y.Y., Zhang, D., Gao, W., Xia, X.H., 2005. Chemistry European Journal 11, 2177.

Su, L., Zhang, Y., Hon, C., Huo, D., Lei, Y., Monday October 17, 2011. Ultra-sensitive glucose detection using copper nanowires. Annual American Institute of Chemical Engineers Meeting (Minneapolis, Minnesota).

Sweet, I.R., Li, G., Najafi, H., Berner, D., Matschinsky, F.M., 1996. American Journal of Physiology Endocrinology, and Metabolism 271, E606–E625.

Takanaga, H., Chaudhari, B., Frommer, W.B., 2007. Biochimica Bipohysica Acta.

Toshima, N., Yonezawa, T., 1998. New Journal of Chemistry 22, 1179.

Tsai, T.W., Heckert, G., Nuevesm, L.F., Tan, Y., Kun, D.Y., Harrison, G., Resaica, D.F., Schmidtke, D.W., 2009. Analytical Chemistry 81 (19), 7917–7925.

Tubiana-Rufi, N., Riveline, J.P., Dardari, P., 2007. Diabetes Metabolism 33, 415–420.

Ugo, P., Moretto, L., Vezza, F., 2002. Chemistry and Physical Chemistry 3, 917–925.

Vaddiraju, S., Tomazos, I.I., Burgess, D.J., Jain, F.C., Papadimitrakopoulos, F., 2010. Biosensors and Bioelectronics 25, 1553–1565.

Wang, J., 2005. Electroanalysis 17 (1), 7–14.

Wang, J., Liu, J., Chen, L., Lu, F., 1994. Analytical Chemistry 66, 3600.

Wang, J., Lu, F., Agnes, L., Liu, J., Sakslund, H., Chen, Q., Pedero, M., Chen, L., Himmerich, O., 1995. Analytca Chimica Acta 305, 3–7.

Wang, J., 2008. Electrochemical glucose sensors. Chemical Reviews 108, 814–825.

Wang, J., Li, M., Shi, Z., Li, N., Gu, Z., 2002. Analytical Chemistry 74, 1993–1997.

Wang, J., Musameh, M., Lin, Y., 2003. Journal of the American Chemical Society 125, 2408–2409.

Wang, J., Musamseh, M., 2003. Analytical Chemistry 75, 2075.

Wang, J., Musameh, M., 2003. Analytical Chemistry 75, 2075–2079.

Wang, J., Thomas, D.F., Chen, A., 2008. Analytical Chemistry 80, 997.

Wang, K., Tonney, F., Trewyn, B.T., Lin, V.S.Y., 2007. Mesoporous silica nanoparticles deliver DNA and chemicals into plants. Nature Nanotechnology 2, 295–300.

Wang, S.F., Xie, F., Hu, R.F., April 10, 2007. Carbon-coated nickel magnetic nanoparticles modified electodes as a sensor for determination of acetaminophen. Sensors and Actuators B: Chemical 123 (1), 495–500.

Wang, S.F., Xie, F., Hu, R.F., 2007. Electrochemical study of brucine on an electrode modified with magnetic carbon-coated nickel nanoparticles Analytical and Bioanalytical Chemistry 389, 933–939.

Wang, K., Liu, Q., Guan, Q.M., Wu, J., Li, H.N., Yan, J.J., 2011a. Biosensors and Bioelectronics 26, 2252–2257.

Wang, J., 2008. Electrochemical Glucose Biosensors. Chemical Review 108, 814–825.

Wang, J., 1991. Electroanalysis 3, 255.

Wang, J., Wu, H., 1994. Analytica Chimica Acta 66, 3600.

Wang, K., Yang, H., Zhu, L., Liao, J., Lu, T., Xing, W., Xing, S., Lv, Q., 2009a. Molecular Catalysis B: Enzyme 58, 194–198.

Wang, L., Bai, J., Bo, X., Zhang, X., Guo, L., February 15, 2011b. A novel glucose sensor based on mesoporous carbon-Au nanoparticle nanocomposites. Talanta 8 (5), 1386–1391.

Wang, Y., Zhang, Z., Jain, V., Jinju, Y., Muellere, S., Sokolova, J., Liu, Z., Levon, K., Rigas, B., Rafailovicha, M.H., 8 April 2010. Potentiometric sensors based on surface molecular imprinting: Detection of cancer markers and viruses. sensors and actuators B: Chemical 149 (1), 381–387.

Wang, Z., Yu, C.H., Ba, D.C., Lian, J., 2007. Vacuum 81, 579–582.

Weber, S.G., 1995. Analytical Chemistry 67, 1920–1928.

Weidamaier, K., Lastovich, A., Keith, S., Bruce Pitner, J., Sistare, M., Jacobson, R., Kurisko, D., June 15, 2011. Biosensors and Bioelectronics 26 (10), 4117–4123.

Weiss, J.H., 2000. Ca^{++}-Zn^{++} permeable AMPA or kainate receptors: possible key factors in selective neurodegeneration. Trends in Neuroscience 23, 365–371.

Wilson, G.S., Gifford, R., 2005. Biosensors and Bioelectronics 20, 2388–2403.

Wittstock, G., Strubing, A., Szargan, R., Werner, G., 1998. Journal of Electroanalytical Chemistry 444, 61.

Wu, B.Y., Hou, S.H., Yin, F., Zhao, Z.X., Huang, J.D., Chen, Q., 2007. Biosensors and Bioelectronics 22, 838–844.

Xiao, F., Zhao, F., Mei, D., Mo, Z., Zeng, B., 2009. Biosensors and Bioelectronics 24, 3481.

Xie, F., Gaoz, Shariff M., et al., 2005. A disposable glucose biosensor based on diffusional mediator dispensed in nanoparticlate membrane on screen-printed.

Yang, S., Guo, D., Su, L., Yu, P., Li, D., Ye, J., Mao, L., 2009a. Electrochemistry Communications 11, 1912–1915.

Yang, L., Xiong, H., Zhang, K., Wang, S., Zhang, X., May 15, 2011a. Direct electrochemistry of glucose oxidase and biosensing for glucose based on boron-doped carbon-coated nickel-modified electrode. Biosensors and Bioelectronics 26 (9), 3801–3805.

Yang, H., Yang, X., Pan, M., Chen, Y., Chen, D., 2009b. Functional contrast an kinetic analysis of water-dispersible carbon black nanoparticles and MWNTS in glucose biosensors, Engineering in Medicine and Biology Society, 2009. EMBC. Annual Conference(of IEEE).

Yang, H., Zhu, Y., 2006. Size dependence of SiO_2 particles enhanced glucose biosensor. Talanta 68, 569–574.

Yang, Z., Ren, Y., Zhang, Y., Li, J., Li, H., Hu, X.H.X., Xu, Q., July 2011b. Nano flake-like SnS_2 matrix for glucose biosensing based on direct electrochemistry of glucose oxidase. Biosensors and Bioelectronics 26 (11), 437–4341.

Ye, G., Wang, X., October 15, 2010. Glucose sensing through diffraction grating of hydrogel bearing phenylboronic acid groups. Biosensors and Bioelectronics 26 (2), 772–777.

Ye, K., Schultz, J.S., 2003. Analytical Chemistry 75 (4), 3451–3459.

Yin, L.T., Chou, J.C., Chung, W.Y., Sun, T.P., Hsiung, S.K., 2000. Separate structure extended gate H^+-ion sensitive field effect transistor on a glass substrate. Sensors and Actuators B: Chemical 71, 106–111.

Yin, L.T., Chou, J.C., Chung, W.Y., Sun, T.P., Hsiung, S.K., 2001a. Study of indium tin oxide thin film for separative extended gate ISFET. Materials Chemical Physics 70, 12–16.

Yu, J., Ju, H., 2002. Preparation of porous titania sol-gel matrix for immobilization of horseradish peroxidase by a vapor deposition method. Analytical Chemistry 74, 3579–3583.

Yu, R., Yang, M., Liu, Y., Shen, R., January 2006. Platinum nanoparticle-coped sol-gel carbon nanotubes composite electrochemical sensors and biosensors. Biosensors and Bioelectronics 171, 1125–1131.

Yu, X., Munge, B., Patel, V., Jensen, G., Bhirde, A., Gong, J.D., Kim, S.N., Gillespie, J., Gutkind, J.S., Papadimitrakopoulos, F., Rusling, J.F., 2006. Journal of the American Chemical Society 128, 11199–11205.

Zawalich, W.S., Mitschinsky, F.M., 1977. Endocrinology 100 (1), 1–8.

Zeng, Q., Xing, Y., Gao, J., Wang, Z., Zhang, X., 2010. Langmuir 26, 15022–15026.

Zhang, S., Wang, N., Yu, H., Niu, Y., Sun, C., 2005. Covalent attachment of glucose oxidase to an Au electrode modified with gold nanoparticles for use as glucose biosensor. Bioelectrochemistry 67, 15–22.

Zhang, T., Fang, H.H.P., 2005. Environmental Technology 26 (2), 155–160.

Zhang, C., Li, R., Zhang, D., Shang, Z., Wang, G., Liang, Y., Xing, Y., Pan, Z., Cai, Z., Zhao, X., Liu, C., 2006. Journal of Molecular Structure 785, 1–11.

Zhang, H., Meng, Z., Wang, Q., Zheng, J., November 15, 2011. A novel glucose biosensor based on direct electrochemistry of glucose oxidase incorporated in biomediated gold nanoparticles-carbon nanotubes composite film. Sensors and Actuators B; Chemical 158 (1), 23–27.

Zhang, J., Pearce, M.C., Tang, B.P., Yang, J.P., Septemer 15, 2011. Ultrasensitive electrochemical immunosensor employing glucose oxidase nanoparticles for signal amplification. Biosensors and Bioelectronics 27 (1), 206.

Zhang, X., Chan, K.Y., You, J.K., Lin, Z.G., Tseung, A.C., 1997. Journal of Electroanalytical Chemistry 430, 147.

Zhou, L.Q., Cass, A.E.G., 1991. Biosensors and Bioelectronics 8, 145–150.

Zhu, H., Lu, X., Li, M., Shao, Y., Zhu, Z., 2009. Talanta 79, 1446.

Zuo, S., Tang, Y., Yuan, H., Lan, M., 2008. Sensors and Actuators B 133, 555–560.

FURTHER READING

Gong, K., Du, F., Xia, Z., Durstock, M., Dai, L., 6 February 2009. Nitrogen-doped carbon nanotube arrays with high electrocatalytic Activity for Oxygen Reduction. Science 323 (5915), 760–764.

Chapter 7

A Fractal Analysis of Binding and Dissociation of Glucose to Different Biosensor Surfaces

7.1 INTRODUCTION

In the previous chapter, the detection of glucose by biosensors and the kinetics of binding and dissociation by different biosensors were presented. Since diabetes is such a prevalent disease and has reached almost epidemic proportions in the United States and throughout the world, it is worthwhile to present another chapter on the detection of glucose by different biosensors. These articles have appeared in recent literature. One should remember that the detection of glucose was the first application of biosensors. Since then, of course, there have been tremendous advancements in the detection of glucose by different biosensors.

In this chapter, we present the binding and dissociation kinetics (if applicable) of

1. 100 mM glucose in solution to glucose oxidase immobilized on a microcavity surface plasmon resonance sensor (MSPRS) biosensor (Amarie et al., 2010),
2. different concentrations of glucose in solution to a 25 surface-adsorbed PAH/PAB bilayer biosensor,
3. 5×10^{-3} M glucose to glucose in solution to
   a. polymer (PPD) biosensor,
   b. mineral oil-based biosensor, and
   c. myoglobin mineral oil-based biosensor,
4. glucose at different concentrations in solution to step changes (50, 150, and 300 mg/dl),
5. 10 μM glucose to glucose oxidase—influence of runs in sequence, and
6. glucose in solution to MEBC2/QCM and MEBC1/QCM.

These five additional examples provide further physical insights into the glucose–glucose oxidase reaction when glucose oxidase is immobilized to different surfaces, that is, the reaction is analyzed by different biosensor surfaces.

Biomarkers and Biosensors. http://dx.doi.org/10.1016/B978-0-444-53794-2.00007-0

Some of the recent applications of glucose biosensors that have appeared in the recent literature include

1. Biocompatible graphene oxide-based glucose biosensors (Liu et al., 2010).
2. Carbon nanotube-ionic liquid composite sensors and biosensors (Kachoosangi et al., 2009).
3. Glucose biosensor from covalent immobilization of chitosan-coupled carbon nanotubes on polyaniline-modified gold electrode (Wan et al., 2010).
4. Detection of glucose based on direct electron transfer reaction of glucose oxidase immobilized on highly ordered polyaniline nanotubes (Wang et al., 2009a).
5. Myoglobin-containing carbon-paste enzyme microelectrodes for the biosensing of glucose under oxygen-deficit conditions (Wang et al., 1999).
6. Screen-printing sol-gel enzyme-containing carbon inks (Wang et al., 1999).
7. Exploiting metal-organic coordination polymers (MOCPs) as highly efficient immobilization matrices of enzymes for sensitive electrochemical biosensing (Fu et al., in press).
8. Microscale enzymatic optical biosensors using mass transport limiting nanofilms. 1. Fabrication and characterization using glucose as a model analyte (Stein et al., 2007).
9. Electrochemical biosensor of nanotube-augmented carbon nanotube networks (Claussen et al., 2009).
10. ZnO nanotube arrays as biosensors for glucose (Yang et al., 2009).
11. Nitric oxide-releasing sol-gel particle polyurethane glucose biosensors (Shin et al., 2004).
12. Microfluidic devices integrating microcavity surface-plasmon-resonating sensors: glucose oxidase binding-activity detection (Amarie et al., 2010).
13. High performance, flexible enzymatic glucose biosensor based on ZnO nanowires supported on a gold-coated polyester substrate (Pradhan et al., 2010).
14. Poly(acrylonitrile-co-acrylic acid) for glucose biosensor (Wang et al., 2009a).
15. Carbon nanotube/Teflon composite electrochemical sensors and biosensors (Wang and Musameh, 2003).
16. Microscale enzymatic optical biosensors using mass transport limiting nanofilms. Response modulation by varying analyte transport properties (Stein et al., 2008).
17. Redox mediation and photomechanical oscillations involving photosensitive cyclometalated Ru(II) complexes, glucose oxidase, and peroxidase (Ryabov et al., 2006).
18. Glucose and lactate biosensors for screening electrochemical microscopy imaging of single cells (Giobanu et al., 2008).

19. Coimmobilization of dehydrogenase and their cofactors in electrochemical biosensors (Zhang et al., 2007).
20. Rational design and one-step formation of multifunctional gel transducer for sample fabrication of integrated electrochemical biosensors (Yu et al., 2011).
21. Direct electrochemistry of glucose oxidase biosensing for glucose based on graphene (Shaw et al., 2009).
22. Modeling the response function of dual-enzyme microbiosensors (Masson et al., 2007).
23. Amplified biosensing using the horseradish peroxidase mimicking DNAzyme electrocatalyst (Pelossofor et al., 2010).
24. Nonenzymatic electrochemical detection of glucose based on platinum-single-walled carbon nanotubes hybrid nanostructures (Meng et al., 2009).
25. Electrochemical biosensors (Privett et al., 2008).
26. Fe_3O_4 magnetic nanoparticles as peroxidase mimics and their applications in H_2O_2 and glucose detection.
27. Colorimetric D-fructose biosensor based on direct electron transfer using *D*-fructose dehydrogenase (Tsujimura et al., 2009).
28. Conjugation of glucose oxidase onto Mn-doped ZnS quantum dots for phosphorescent sensing of glucose in biological fluids (Wu et al. (2010)).

7.2 THEORY

7.2.1 Single-Fractal Analysis

7.2.1.1 Binding Rate Coefficient

Havlin (1989) indicates that the diffusion of a particle (analyte [Ag]) from a homogeneous solution to a solid surface (e.g., receptor [Ab]-coated surface) on which it reacts to form a product (analyte—receptor complex, (Ab·Ag)) is given by

$$(\mathrm{Ab}\cdot\mathrm{Ag}) \approx \begin{cases} t^{(3-D_{f,bind})/2} = t^p & t < t_c \\ t^{1/2} & t > t_c \end{cases} \tag{7.1}$$

Here $D_{f,bind}$ or D_f is the fractal dimension of the surface during the binding step and t_c is the crossover value. Havlin (1989) indicates that the crossover value may be determined by $r_c^2 \sim t_c$. Above the characteristic length, r_c, the self-similarity of the surface is lost and the surface may be considered homogeneous. Above time t_c, the surface may be considered homogeneous, since the self-similarity property disappears, and "regular" diffusion is now present. For a homogeneous surface where $D_f = 2$ and when only diffusional limitations are present, $p = ½$ as it should be. Another way of looking at the $p = ½$ case (where $D_{f,bind} = 2$) is that the analyte in solution views the fractal object, in our case, the receptor-coated biosensor surface, from a "large distance." In essence, in the association process, the diffusion of the analyte from the solution to the receptor surface creates a depletion layer of width $(Ðt)^{1/2}$, where Ð is the diffusion constant. This gives rise to the fractal power law, (analyte.receptor) ~ $t^{(3-D)/2}_{f,bind}$. For the present analysis, t_c is arbitrarily chosen and we assume that the value of the t_c is not reached. One may consider the approach as an intermediate "heuristic" approach that may be used in the future to develop an autonomous (and not time-dependent) model for diffusion-controlled kinetics.

7.2.1.2 Dissociation Rate Coefficient

The diffusion of the dissociated particle (receptor [Ab] or analyte [Ag]) from the solid surface (e.g., (analyte [Ag]—receptor [Ab]) complex-coated surface) into solution may be given as a first approximation by

$$(\mathrm{Ab \cdot Ag}) \approx -t^{(3-D_{f,diss})/2} = -t^{p}(t > t_{diss}) \qquad (7.2)$$

Here $D_{f,diss}$ is the fractal dimension of the surface for the dissociation step. This corresponds to the highest concentration of the analyte—receptor complex on the surface.

Henceforth, its concentration only decreases. The dissociation kinetics may be analyzed in a manner "similar" to the binding kinetics.

7.2.2 Dual-Fractal Analysis

7.2.2.1 Binding Rate Coefficient

Sometimes, the binding curve exhibits complexities, and two parameters (k and D_f) are not sufficient to adequately describe the binding kinetics. This is further corroborated by low values of r^2 factor (goodness-of-fit). In that case, one resorts to a dual-fractal analysis (four parameters: k_1, k_2, D_{f1}, and D_{f2}) to adequately describe the binding kinetics. The single-fractal analysis presented above is thus extended to include two fractal dimensions. At present, the time ($t = t_1$) at which the "first" fractal dimension "changes" to the "second" fractal dimension is arbitrary and empirical. For the most part, it is dictated by the data analyzed and experience gained by handling a single-fractal analysis. A smoother curve is obtained in the "transition" region if care is taken to select the correct number of points for the two regions. In this case, the product (antibody—antigen or analyte—receptor complex; Ab.Ag or analyte.receptor) is given by

$$(\mathrm{Ab \cdot Ag}) \approx \begin{cases} t^{(3-D_{f1,bind})/2} = t^{p1} & (t < t_1) \\ t^{(3-D_{f2,bind})/2} = t^{p2} & (t_1 < t < t_2) = t_c \\ t^{1/2} & (t > t_c) \end{cases} \qquad (7.3)$$

In some cases, as mentioned above, a triple-fractal analysis with six parameters (k_1, k_2, k_3, D_{f1}, D_{f2}, and D_{f3}) may be required to adequately model the binding kinetics. This is when the binding curve exhibits convolutions and complexities in its shape due to perhaps the very dilute nature of the analyte (in some of the cases to be presented) or for some other reasons. Also, in some cases, a dual-fractal analysis may be required to describe the dissociation kinetics.

7.3 RESULTS

Amarie et al. (2010) have recently developed a miniature MSPRS integrated with microfluidics. They indicate that their biosensor can detect β-D-glucose at about 10^6 fewer molecules than conventional surface plasmon resonance (SPR) sensors. They indicate that drug discovery requires the interactions of multitudes of small molecules in the libraries against libraries of organisms present in an organism. Research is focused on improving the sensitivity of biosensors (Fan et al., 2008). They indicate that SPR is a proven technology with applications in optoelectronics (Ebbeson et al., 1998), bioanalysis (Karlsson, 2004), and superresolution imaging (Yokota et al., 1999). Bergquist et al. (2007) indicate that the SPR biosensors are known for their ability to monitor the biomolecular interactions in real time.

Figure 7.1 shows the binding of 100 mM glucose in solution to glucose oxidase immobilized on an MSPRS biosensor (Amarie et al., 2010). A dual-fractal analysis is required adequately to describe the binding kinetics. The values of (1) the binding rate coefficient, k, and the fractal dimension, D_f, for a single-fractal analysis and (2) the binding rate coefficients, k_1 and k_2, and the fractal dimensions, D_{f1} and D_{f2}, for a dual-fractal analysis are given in Table 7.1. It is of interest to note that as the fractal dimension increases by a factor of 1.445 from a value of $D_{f1} = 1.9116$ to $D_{f2} = 2.7638$, the binding rate coefficient increases by a factor of 10.68 from a value of $k_1 = 0.9625$ to $k_2 = 10.231$. An increase in the degree of heterogeneity on the surface leads to an increase in the binding rate coefficient.

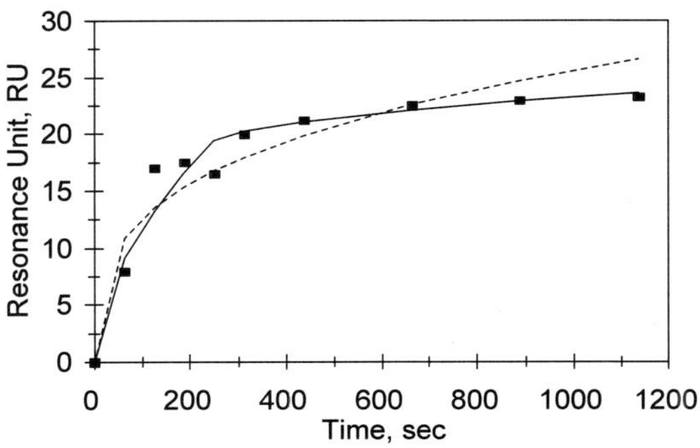

FIGURE 7.1 Binding of 100 mM glucose in solution to glucose oxidase immobilized on a microcavity surface plasmon resonance biosensor. *Amarie et al., 2010.*

TABLE 7.1 Binding rate coefficients for 100 nM β-D-glucose oxidase immobolized onto a MSPRS micro cavity suface plasmon resonance biosensor (Amarie et al., 2010).

Analyte in Solution/ Receptor on Surface	k	k_1	k_2	Df	Df_1	Df_2
Glucose/glucose oxidase	3.0721 ± 0.5776	0.9625 ± 0.2467	10.231 ± 0.157	2.3860 ± 0.1284	1.9116 ± 0.4330	2.7638 ± 0.0290

Biosensors provide the ability to obtain sensitive, real-time, and point-of-care measurements to help manage ailments (Kohli-Seth and Oropello, 2000). Optical biosensors have the capacity to provide selective and sensitive detection of analytes with minimal invasiveness (Wolfbeis, 2005; Pickup et al., 2005; Asian et al., 2005). Specific targets can be minimally invasively interrogated with light to obtain measurements (Stein et al., 2007; McShane, 2006; Ibey et al., 2006; Brown and McShane, 2006; Chinnayelka and McShane, 2005; Russell et al., 1999). Stein et al. (2008) indicate that the intrinsic balance of reaction–diffusion kinetics to realize useful microscale behavior.

It is commonplace to incorporate transport-limiting materials such as polyurethane in the design of enzymatic sensors (Gough et al., 1985). This may be done simply by the deposition of films of polymeric materials to entrap sensor components (Tretnak and Wolfbeis, 1989; Wolfbeis et al., 2000; Rosenzweig and Kopelman, 1996; Choi et al., 2005). Previously, Stein et al. (2008) had employed the transport-limiting properties of polyelectrolyte nanofilms to design glucose-sensitive microspheres toward the goal of minimally invasive diabetic monitoring (Brown et al., 2005). Stein et al. (2008) extend their work, in the present analysis, to modulating the substrate transport into the sensor with variations in surface-adsorbed nanofilm thickness, ionic strength of assembly conditions, and capping layer.

Stein et al. (2008) have analyzed the binding of different concentrations of glucose (in mg/dl) to a 25 surface-adsorbed PAH/PHB bilayer biosensor. Figure 7.2(a) shows the binding of 50 mg/dl of glucose to the 25 surface-adsorbed PAH/PHB bilayer biosensor. A dual-fractal analysis is required to adequately describe the binding kinetics. The values of (1) the binding rate coefficient, k, and the fractal dimension, D_f, for a single-fractal analysis and (2) the binding rate coefficients, k_1 and k_2, and the fractal dimensions, D_{f1} and D_{f2}, are given in Table 7.2. It is of interest to note that as the fractal dimension increases by a factor of 2.485 from a value of $D_{f1} = 1.1491$ to $D_{f2} = 2.5866$, the binding rate coefficient increases by factor of 3.31 from a value of $k_1 = 0.05969$ to $k_2 = 0.1565$. An increase in the fractal dimension leads to an increase in the binding rate coefficient.

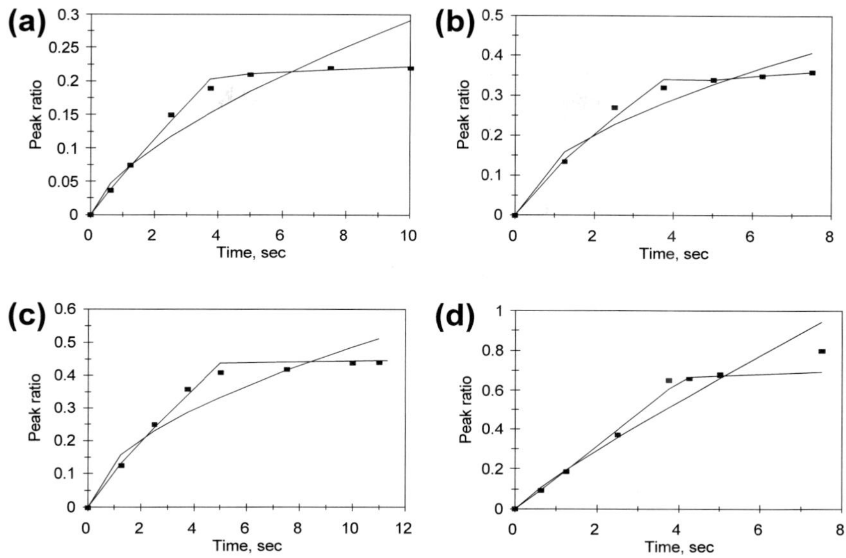

FIGURE 7.2 Binding of different concentrations of glucose in solution to a 25 surface-adsorbed PAH/PHB bilayer biosensor (Stein et al., 2008): (a) 50 mg/dl (b) 75 mg/dl (c) 125 mg/dl (d) 200 mg/dl.

TABLE 7.2 Binding Rate Coefficients and Fractal Dimensions for Different Concentrations of Glucose in Milligrams per deciliter to 25 Surface-Adsorbed PAH/PHB Bilayer Biosensor (Stein et al., 2008)

Analyte in Solution, mg/dl	k	k_1	k_2	D_f	D_{f1}	D_{f2}
50	0.0642 ± 0.01660	0.05969 ± 0.0046	0.1565 ± 0.0030	1.6872 ± 0.1899	1.1491 ± 0.1087	2.5866 ± 0.0637
75	0.1402 ± 0.0225	0.1168 ± 0.0249	0.2709 ± 0.0006	1.9830 ± 0.2008	1.3828 ± 0.3054	2.7290 ± 0.0140
125	0.1403 ± 0.306	0.1060 ± 0.0086	0.3343 ± 0.0031	1.919 ± 0.204	1.2396 ± 0.1436	2.7756 ± 0.0634
200	0.1580 ± 0.0341	0.1507 ± 0.0118	0.6009 ± 0.0082	1.226 ± 0.1845	0.8881 ± 0.1101	2.8598 ± 0.0656

Figure 7.2(b) shows the binding of 75 mg/dl of glucose to the 25 surface-adsorbed PAH/PHB bilayer biosensor. A dual-fractal analysis is, once again, required to adequately describe the binding kinetics. The values of (1) the binding rate coefficient, k, and the fractal dimension, D_f, for a single-fractal analysis and (2) the binding rate coefficients, k_1 and k_2, and the fractal dimensions, D_{f1} and D_{f2}, are given in Table 7.2. It is of interest to note that as the fractal dimension increases by a factor of 1.973 from a value of $D_{f1} = 1.3828$ to $D_{f2} = 2.7290$, the binding rate coefficient increases by factor of 2.319 from a value of $k_1 = 0.1168$ to $k_2 = 0.2709$. An increase in the fractal dimension leads to an increase in the binding rate coefficient.

Figure 7.2(c) shows the binding of 125 mg/dl of glucose to the 25 surface-adsorbed PAH/PHB bilayer biosensor. A dual-fractal analysis is once again required to adequately describe the binding kinetics. The values of (1) the binding rate coefficient, k, and the fractal dimension, D_f, for a single-fractal analysis and (2) the binding rate coefficients, k_1 and k_2, and the fractal dimensions, D_{f1} and D_{f2}, are given in Table 7.2. It is of interest to note that as the fractal dimension increases by a factor of 2.239 from a value of $D_{f1} = 1.2396$ to $D_{f2} = 2.7756$, the binding rate coefficient increases by a factor of 3.15 from a value of $k_1 = 0.1060$ to $k_2 = 0.3343$. An increase in the fractal dimension leads to an increase in the binding rate coefficient.

Figure 7.2(d) shows the binding of 200 mg/dl of glucose to the 25 surface-adsorbed PAH/PHB bilayer biosensor. A dual-fractal analysis is required to adequately describe the binding kinetics. The values of (1) the binding rate coefficient, k, and the fractal dimension, D_f, for a single-fractal analysis and (2) the binding rate coefficients, k_1 and k_2, and the fractal dimensions, D_{f1} and D_{f2}, are given in Table 7.2. It is of interest to note that as the fractal dimension increases by a factor of 3.220 from a value of $D_{f1} = 0.8881$ to $D_{f2} = 2.8598$, the binding rate coefficient increases by a factor of 3.987 from a value of $k_1 = 0.1507$ to $k_2 = 0.6009$. An increase in the fractal dimension leads to an increase in the binding rate coefficient.

Figure 7.3(a) shows the increase in the binding rate coefficient, k_1, with an increase in the glucose concentration in milligrams per deciliter in solution. For the data shown in Figure 7.3(a) and Table 7.2, the binding rate coefficient, k_1, is given by

$$k_1 = (0.00479 \pm 0.001678)[\text{glucose in mg/dl}]^{0.6859 \pm 0.2844} \tag{7.4a}$$

The fit is good. Only four data points are available. The availability of more data points would lead to a more reliable fit. The binding rate coefficient, k_1, exhibits an order of dependence slightly more than one-half (equal to 0.6859) on the glucose concentration in solution.

Figure 7.3(b) shows the increase in the binding rate coefficient, k_2, with an increase in the glucose concentration in milligrams per deciliter in solution.

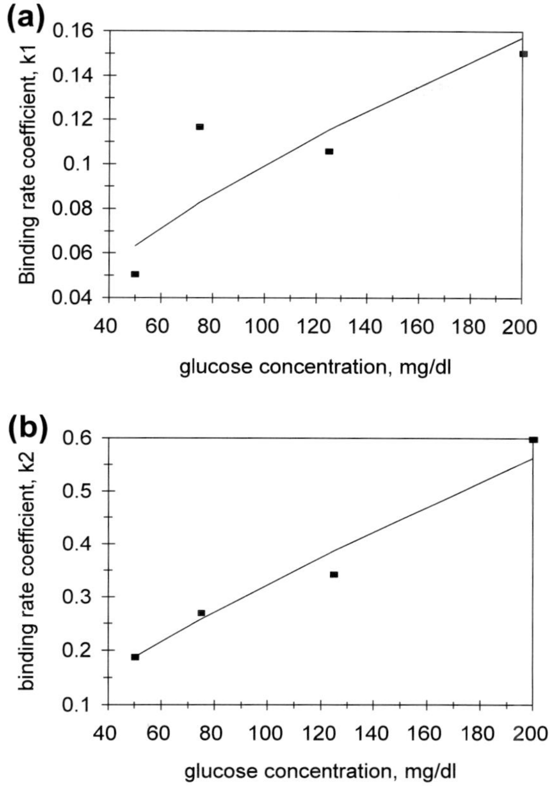

FIGURE 7.3 (a) Increase in the binding rate coefficient, k_1, with an increase in the glucose concentration (in mg/dl) in solution. (b) Increase in the binding rate coefficient, k_2, with an increase in the glucose concentration (in mg/dl) in solution.

For the data shown in Figure 7.3(b) and Table 7.2, the binding rate coefficient, k_2, is given by

$$k_2 = (0.00842 \pm 0.00091)[\text{glucose in mg/dl}]^{0.7927 \pm 0.09872} \tag{7.4b}$$

The fit is very good. Only four data points are available. The availability of more data points would lead to a more reliable fit. The binding rate coefficient, k_2, exhibits an order of dependence between one-half and first (equal to 0.7927) on the glucose concentration in solution.

Wang et al. (1999) have developed a novel myoglobin-containing carbon-paste electrode for the biosensing of glucose under oxygen-deficient conditions. These authors indicate the necessity of tight diabetic control (Henry, 1998). This is achieved by amperometric biosensors for glucose detection (Wilson et al., 1992; Wang et al., 1999). Often, there are severe oxygen limitations that are particularly acute for in vivo applications (Zhang and

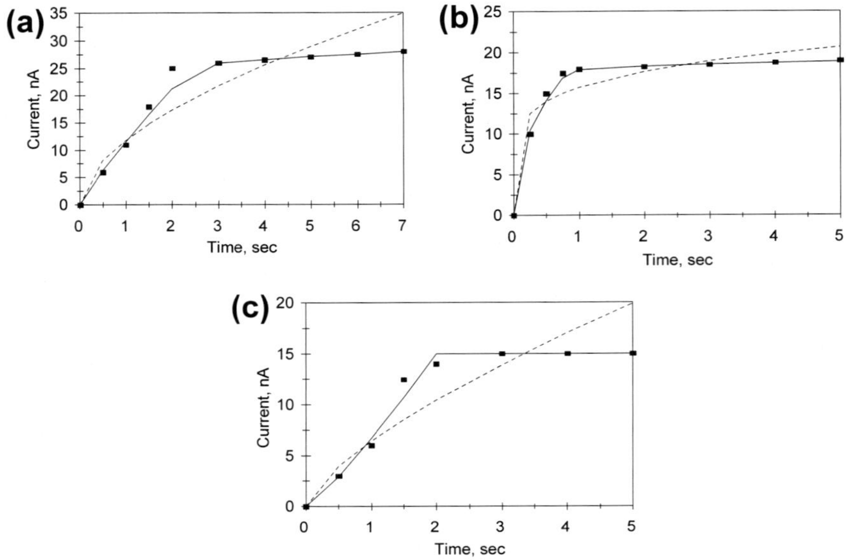

FIGURE 7.4 Binding of 5×10^{-3} M glucose in solution to (a) polymer (PPD) biosensor, (b) mineral oil-based biosensor, and (c) myoglobin-doped mineral oil-based biosensor.

Wilson, 1993). A common approach is to limit the flux of glucose by covering the transducer by semipermeable membrane to ensure the excess oxygen in the enzymatic layer (Fischer et al., 1989). This, however, can compromise the sensitivity. By using oxygen-rich fluorescence carbon pastes, this deficiency may be eliminated (Wang and Lu, 1998). Here, the authors (Wang et al., 1999) describe the use of a myoglobin-modified enzyme microelectrode, which provides a convenient quantitation of glucose under oxygen-deficient conditions.

Figure 7.4(a) shows the binding of 5×10^{-3} M glucose to a polymer (PPD)-based biosensor. A dual-fractal analysis is required to adequately describe the binding kinetics. The values of (1) the binding rate coefficient, k, and the fractal dimension, D_f, for a single-fractal analysis and (2) the binding rate coefficients, k_1 and k_2, and the fractal dimensions for a dual-fractal analysis, D_{f1} and D_{f2}, are given in Table 7.3. It is of interest to note that for a dual-fractal analysis as the fractal dimension increases by a factor of 1.33 from a value of $D_{f1} = 2.116$ to a value of $D_{f2} = 2.8262$, the binding rate coefficient increases by a factor of 2.21 from a value of $k_1 = 11.54$ to $k_2 = 23.553$. An increase in the fractal dimension or the degree of heterogeneity on the surface leads to an increase in the binding rate coefficient.

Figure 7.4(b) shows the binding of 5×10^{-3} M glucose to a mineral oil-based biosensor. A dual-fractal analysis is required to adequately describe the binding kinetics. The values of (1) the binding rate coefficient, k, and

TABLE 7.3 Binding rate coefficients and fractal dimensions for 5.0×10^{-3}M glucose in solution to (a) polymer PPD-based biosensor, (b) mineral oil-based carbon paste biosensor, and (c) myoglobin-doped mineral oil carbon-paste biosensor (Wang et al.,1999)

Analyte in Solution/ Receptor on Surface	k	k_1	k_2	D_f	D_{f1}	D_{f2}
5.0×10^{-3}M glucose/ polymer PPD-based biosensor	11.780 ± 3.044	11.54 ± 2.78	23.553 ± 0.063	1.8864 ± 0.1830	2.116 ± 0.170	2.8262 ± 0.0114
5×10^{-3}M glucose/mineral oil based biosensor	15.751 ± 2.247	19.727 0.833	17.925 ± 0.098	2.6648 ± 0.0946	2.1204 ± 0.1307	2.934 0.00862
5×10^{-3}M glucocse/ myoglobin-doped mineral oil carbon-paste biosensor	6.419 ± 2.314	6.649 ± 1.007	15.0 ± 0	1.5936 0.2886	0.6476 ± 0.2706	3.0– E-14

the fractal dimension, D_f, for a single-fractal analysis and (2) the binding rate coefficients, k_1 and k_2, and the fractal dimensions, D_{f1} and D_{f2}, for a dual-fractal analysis are given in Table 7.3. It is of interest to note that for a dual-fractal analysis as the fractal dimension increases by a factor of 1.38 from a value of $D_{f1} = 2.1204$ to a value of $D_{f2} = 2.934$, the binding rate coefficient increases by a factor of 1.252 from a value of $k_1 = 15.751$ to $k_2 = 19.727$. An increase in the fractal dimension or the degree of heterogeneity on the surface, once again, leads to a decrease in the binding rate coefficient.

Figure 7.4(c) shows the binding of 5×10^{-3} M glucose to a myoglobin-doped mineral oil-based biosensor. A dual-fractal analysis is required to adequately describe the binding kinetics. The values of (1) the binding rate coefficient, k, and the fractal dimension, D_f, for a single-fractal analysis and (2) the binding rate coefficients, k_1 and k_2, and the fractal dimensions, D_{f1} and D_{f2}, for a dual-fractal analysis are given in Table 7.3. It is of interest to note that for a dual-fractal analysis as the fractal dimension increases by a factor of 4.63 from a value of $D_{f1} = 0.6476$ to $D_{f2} = 3.0$, the binding rate coefficient increases by a factor of 2.55 from a value of $k_1 = 6.649$ to $k_2 = 15.0$. An increase in the fractal dimension or the degree of heterogeneity on the surface, once again, leads to an increase in the binding rate coefficient.

Figure 7.5(a) and Table 7.3 show the increase in the binding rate coefficient, k_1, with an increase in the fractal dimension, D_{f1}. For the data shown in Figure 7.5(a) and Table 7.3, the binding rate coefficient, k_1, is given by

$$k_1 = (6.977 \pm 6.028)D_{f1}^{0.692 \pm 0.390} \tag{7.5a}$$

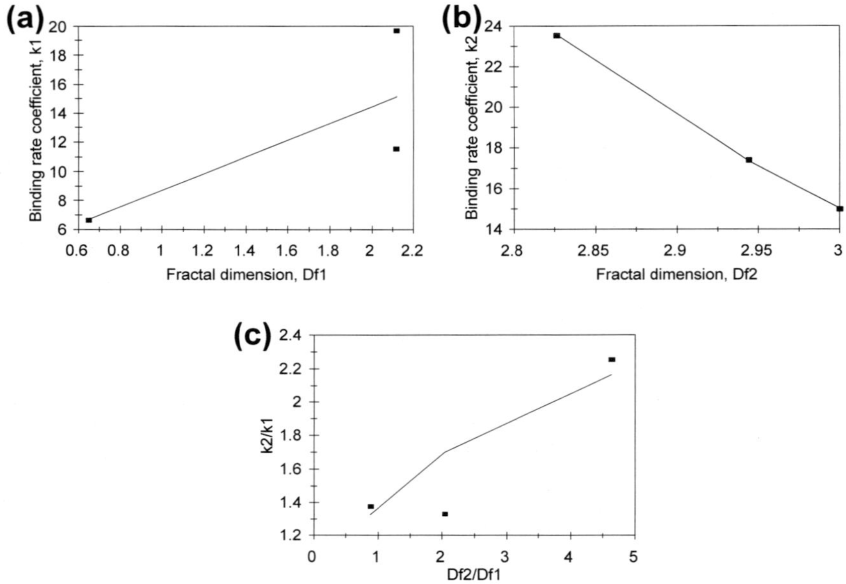

FIGURE 7.5 (a) Increase in the binding rate coefficient, k_1, with an increase in the fractal dimension, D_{f1}. (b) Decrease in the binding rate coefficient, k_2, with an increase in the fractal dimension, D_{f2}. (c) Increase in the binding rate coefficient ratio, k_2/k_1, with an increase in the fractal dimension ratio, D_{f2}/D_{f1}.

The fit is reasonable. Only three data points are available. The availability of more data points would lead to a more reliable fit. The binding rate coefficient, k_1, exhibits an order of dependence between one-half and first (equal to 0.692) on the fractal dimension, D_{f1}.

Figure 7.5(b) and Table 7.3 show the decrease in the binding rate coefficient, k_2, with an increase in the fractal dimension, D_{f2}. For the data shown in Figure 7.5(b) and Table 7.3, the binding rate coefficient, k_2, is given by

$$k_2 = (5.885 \pm 0.035)D_{f2}^{-7.529 \pm 0.137} \quad (7.5b)$$

The fit is very good. Only three data points are available. The availability of more data points would lead to a more reliable fit. The binding rate coefficient, k_2, exhibits a very high negative order (equal to −7.529) of dependence on the fractal dimension, D_{f2}.

Figure 7.5(c) and Table 7.3 show the increase in the binding rate coefficient ratio, k_2/k_1, with an increase in the fractal dimension ratio, D_{f2}/D_{f1}. For the data shown in Figure 7.5(c) and Table 7.3, the binding rate coefficient ratio, k_2/k_1, is given by

$$k_2/k_1 = (1.308 \pm 0.339)(D_{f2}/D_{f1})^{0.293 \pm 0.197} \quad (7.5c)$$

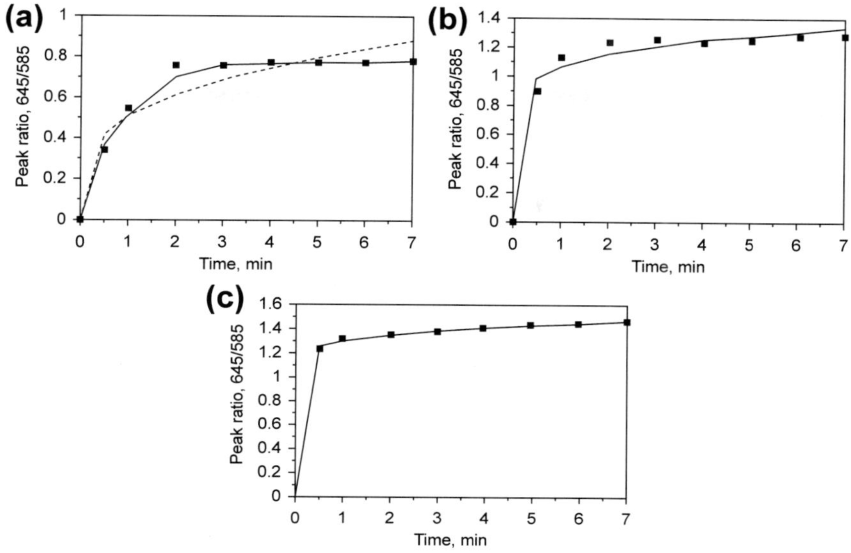

FIGURE 7.6 Binding of glucose at different concentrations in solution to step changes (Stein et al., 2008): (a) 50 mg/dl (b) 150 mg/dl (c) 300 mg/dl.

The fit is reasonable. Only three data points are available. The availability of more data points would lead to a more reliable fit. The binding rate coefficient ratio, k_2/k_1, exhibits an order of dependence of less than one-half (equal to 0.293) on the fractal dimension ratio, D_{f2}/D_{f1}.

Figure 7.6(a) shows the binding of a step change in solution for glucose in solution to the optical enzymatic sensor (Stein et al., 2007). The glucose concentration in solution is 50 mg/dl. A dual-fractal analysis is required to adequately describe the binding kinetics. The values of (1) the binding rate coefficient, k, and the fractal dimension, D_f, for a single-fractal analysis and (2) the binding rate coefficients, k_1 and k_2, and the fractal dimensions, D_{f1} and D_{f2}, are given in Table 7.4. An increase in the fractal dimension or the degree of heterogeneity on the surface leads to an increase in binding rate coefficient.

Figure 7.6(b) shows the binding of a step change in solution for glucose in solution to the optical enzymatic sensor (Stein et al., 2007). The glucose concentration is 150 mg/dl. A single-fractal analysis is required to adequately describe the binding kinetics. The values of the binding rate coefficient, k, and the fractal dimension, D_f, for a single-fractal analysis are given in Table 7.4.

Figure 7.6(c) shows the binding of a step change in solution for glucose in solution to the optical enzymatic sensor (Stein et al., 2007). The glucose concentration is 300 mg/dl. A single-fractal analysis is required to adequately describe the binding kinetics. The values of the binding rate coefficient, k, and the fractal dimension, D_f, for a single-fractal analysis are given in Table 7.4.

TABLE 7.4 Binding Rate Coefficients and Fractal Dimensions for a Step Change in Glucose (in mg/dl) in Solution to a Mass Transport-Limited Optical Enzymatic Sensor (Stein et al., 2007)

Analyte in Solution/ Optical Enzymatic Sensor	k	k_1	k_2	D_f	D_{f1}	D_{f2}
50 mg/dl glucose	0.5094 ± 0.0769	0.5103 ± 0.0623	0.7536 ± 0.0033	2.4562 ± 0.1146	2.1006 ± 0.1686	2.9818 ± 0.0054
150 mg/dl glucose	1.0846 ± 0.0874	NA	NA	2.1806 ± 0.061318	NA	NA
300 mg/dl glucose	1.3191 ± 0.0162	NA	NA	2.9010 ± 0.00912	NA	NA

Optical monitoring techniques are useful as diagnostic techniques. Fluorescence (Pickup et al., 2005), absorbance (Amenrov et al., 2005), and Raman spectroscopy have been researched substantially to monitor analyte levels. "Smart tattoo" sensors are an interesting technique, where micrometer-scale luminescent particles are implanted and noninvasively interrogated with light (Russell et al., 1999; McShane, 2002; McShane et al., 2000; McShane, 2006). Most of these assays are of the competitive type, which rely on fluorescence emission properties to optically transduce glucose levels (Chinnyelka and McShane, 2005; D'Auria et al., 2002, 1999; Barone et al., 2005; Ibey et al., 2005). The resonance energy transfer technique is used, and it relies on the competitive binding of a fluorescence ligand and target analyte to receptor sites on lectins. The ligand is displaced from the receptor in the presence of the analyte. This causes a measurable change in emission spectra (Chinnayelka and McShane, 2006; Ballerstadt et al., 2004; Ballerstadt and Schultz, 2000).

There is a problem with the application of this technique in that one needs to control the reaction–diffusion kinetics of the reactants, which is the key (Zhu et al., 2005). The analyte is measured indirectly, and the local oxygen levels are proportionately reduced and relayed through the fluorescence emission of the oxygen reporter as analyte diffuses into the sensor (Moschou et al., 2004; Xu et al., 2002). It is particularly important to control transport of the low-molecular-weight analytes. This may be accomplished by the deposition of polyelectrolyte nanofilms using layer-by-layer (LbL) self-assembly (Schonhoff, 2003; Liu and Bruening, 2004; Muller and Bruening, 2004). The authors (Stein et al., 2007) focused on developing glucose-sensitive smart tattoos by absorbing critical glucose sensor components, GO_x and PtOEP, into alginate-modified mesoporous silica particles, which were then subsequently coated with fluorescent nanofilms using the LbL technique.

Figure 7.6(a)–(c) shows the sensor response to step changes in glucose concentration. Figure 7.6(a) shows the binding of 50 mg/dl glucose to the sensor. A dual-fractal analysis is required to describe the binding kinetics. The values of (1) the binding rate coefficient, k, and the fractal dimension, D_f, for a single-fractal analysis and (2) the binding rate coefficients, k_1 and k_2, and the fractal dimensions, D_{f1} and D_{f2}, for a dual-fractal analysis are given in Table 7.4. It is of interest to note that as the fractal dimension increases by a factor of 1.419 from $D_{f1} = 2.1006$ to $D_{f2} = 2.9818$, the binding rate coefficient increases by a factor of 1.48 from $k_1 = 0.5103$ to $k_2 = 0.7536$. An increase in the degree of heterogeneity on the surface leads to an increase in the binding rate coefficient.

Figure 7.6(b) shows the sensor response to a step change in the glucose concentration (150 mg/dl) in solution. A single-fractal analysis is adequate to describe the binding kinetics. The values of the binding rate coefficient, k, and the fractal dimension, D_f, are given in Table 7.4.

Figure 7.6(c) shows the sensor response to a step change in the glucose concentration (300 mg/dl) in solution. Once again, a single-fractal analysis is adequate to describe the binding kinetics. The values of the binding rate coefficient, k, and the fractal dimension, D_f, are given in Table 7.4.

One-dimensional (1D) zinc oxide (ZnO) nanostructures find applications in optics, optoelectronics, and sensors due to their semiconducting, piezoelectric, and pyroelectric properties (Tseng et al., 2003; Bao et al., 2006; Li et al., 2004; Wang and Song, 2006; Johnson et al., 2004). Dorfan et al. (2006a,b) have fabricated chemical and biochemical sensors based on ZnO platforms. For enzyme-based nanostructures, these 1D ZnO nanostructures offer biocompatibility, a good surface-to-bulk ratio, relative stability, as well as electrochemical stability (Tian et al., 2002; Sberveglieri et al., 1995; Rodriguez et al., 2000). A major difference between previous work (Zhang et al., 2004; Wang et al., 2006, 2009b; Zhao et al., 2007) and the authors (Yang et al., 2009) work is that in the previous work, the ZnO nanostructures were synthesized first, followed by assembly on the surface, and in the Stein et al. (2007) work, the ZnO nanotubes have been grown directly on a conductive ITO glass to ensure excellent electrical contact between the ZnO nanostructures and the electrode. This technique circumvents the poor contact between the ZnO nanostructure and the electrode. In the present chapter, the authors Yang et al. (2009) report on the sensing of glucose using the ZnO nanotube array-based sensor. The authors emphasize that the large surface-to-bulk ratio of the porous structures of the nanotubes enhances the sensitivity of the analyte.

Figure 7.7(a) shows the binding of 10 μM glucose to GO_x immobilized on a ZnO array. Yang et al. (2009) analyzed the influence of consecutive runs.

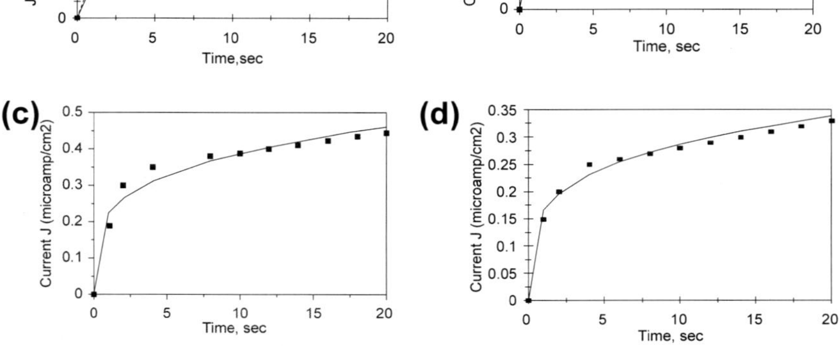

FIGURE 7.7 Binding of 10 μM glucose to glucose oxidase immobilized on a ZnO array. Influence of runs in sequence: (a) Run #1 (b) Run #2 (c) Run #3 (d) Run #4.

This is Run #1. A dual-fractal analysis is required to adequately describe the binding kinetics. The values of (1) the binding rate coefficient, k, and the fractal dimension, D_f, for a single-fractal analysis and (2) the binding rate coefficients, k_1 and k_2, and the fractal dimensions, D_{f1} and D_{f2}, for a dual-fractal analysis are given in Table 7.5. It is of interest to note that as the fractal dimension increases by a factor of 2.063 from $D_{f1} = 1.3374$ to $D_{f2} = 2.7662$, the binding rate coefficient increases by a factor of 3.105 from a value of $k_1 = 0.1064$ to $k_2 = 0.3304$. An increase in the degree of heterogeneity on the surface leads to an increase in the binding rate coefficient.

Figure 7.7(b) shows the binding of 10 μM glucose to GO_x immobilized on a ZnO array. This is Run #2. A dual-fractal analysis is, once again, required to adequately describe the binding kinetics. The values of (1) the binding rate coefficient, k, and the fractal dimension, D_f, for a single-fractal analysis and (2) the binding rate coefficients, k_1 and k_2, and the fractal dimensions, D_{f1} and D_{f2}, for a dual-fractal analysis are given in Table 7.5. It is of interest to note that as the fractal dimension increases by a factor of 2.49 from $D_{f1} = 1.0704$ to $D_{f2} = 2.6724$, the binding rate coefficient increases by a factor of 4.53 from a value of $k_1 = 0.05637$ to $k_2 = 0.2555$. An increase in the degree of heterogeneity on the surface leads, once again, to an increase in the binding rate coefficient.

Figure 7.7(c) shows the binding of 10 μM glucose to GO_x immobilized on a ZnO array. This is Run #3. A single-fractal analysis is adequate to describe the binding kinetics. The values of the binding rate coefficient, k, and the fractal dimension, D_f, for a single-fractal analysis are given in Table 7.5. No explanation is offered at the present time as to why a dual-fractal analysis was required to adequately describe the binding kinetics for Run #s 1 and 2, and a single-fractal analysis was adequate to describe the binding kinetics for Run #3.

Figure 7.7(d) shows the binding of 10 μM glucose to GO_x immobilized on a ZnO array. This is Run #4. A single-fractal analysis is adequate to describe the binding kinetics. The values of the binding rate coefficient, k, and the fractal dimension, D_f, for a single-fractal analysis are given in Table 7.5. No explanation is offered at the present time as to why a dual-fractal analysis was required to adequately describe the binding kinetics for Run #s 1 and 2, and a single-fractal analysis was adequate to describe the binding kinetics for Run #4.

Fu et al. (2011) have recently analyzed MOCPs as efficient immobilization matrices of enzymes for sensitive chemical sensing. According to them, the immobilization of receptors is an important step in developing highly sensitive biosensors. There is a 2D and a 3D protocol to do this. In the 2D protocol, the biorecognition receptors are open to the analyte, but their load is limited and the total activity of the immobilized molecules is limited (Chaki et al., 2002; Jain et al., 2008). In the 3D protocol, entrapment of molecules and simultaneous adsorption through chemical and electrochemical methods can load more biorecognition molecules and more effectively retain their bioactivity

TABLE 7.5 Binding Rate Coefficients and Fractal Dimensions for10 μM of Glucose at 0.02 M pH 7.4 PBS Solution to the GO_x/ZnO Nanotube Array/ITO Electrode. Influence of Consecutive Runs (Yang et al., 2009)

Analyte in Solution/ GO_x/ZnO Array	k	k_1	k_2	D_f	D_{f1}	D_{f2}
10 μM glucose/GO_x/ZnO array; Run #1	0.1547 ± 0.0333	0.1064 ± 0.0196	0.3304 ± 0.0020	1.9830 ± 0.1662	1.3374 ± 0.2448	2.7662 ± 0.0147
10 μM glucose/GO_x/ZnO array; Run #2	0.09975 ± 0.0231	0.05637 ± 0.1114	0.2555 ± 0.0025	1.9348 ± 0.0231	1.0704 ± 0.3458	2.6724 ± 0.0241
10 μM glucose/GO_x/ZnO array; Run #3	0.2243 ± 0.0207	NA	NA	2.5280 ± 0.0582	NA	NA
10 μM glucose/GO_x/ZnO array; Run #4	0.1636 ± 0.0081	NA	NA	2.5256 ± 0.0316	NA	NA

(Cosnier, 1999; Dunn and Zink, 2007). However, there is a limitation with mass transfer in the 3D method when compared with the 2D method. Thus, methods to improve this efficiency in 3D methods are required.

MOCPs with their large surface area have attracted attention for use in chemical sensing (Chen et al., 2010). Electrochemical sensors, especially amperometric sensors, have been considered as efficient sensors for phenol detection due to their simplicity, sensitivity, and ease of miniaturization (Dai et al., 2005; Wang et al., 2008; Lu et al., 2010; Shan et al., 2003; Lakshmi et al., 2009; Zhang and Ji, 2010, Xue and Shen, 2002; Mailley et al., 2003). Even so, newer methods are being sought to improve the biosensing capabilities. Fu et al. (2011) report on using MOCPs as immobilization matrices in aqueous suspensions to develop GO_x-based glucose sensors.

Figure 7.8(a) shows the binding of glucose to MEBC2/QCM. A dual-fractal analysis is required to adequately describe the binding kinetics. The values of (1) the binding rate coefficient, k, and the fractal dimension, D_f, for a single-fractal analysis and (2) the binding rate coefficients, k_1 and k_2, and the

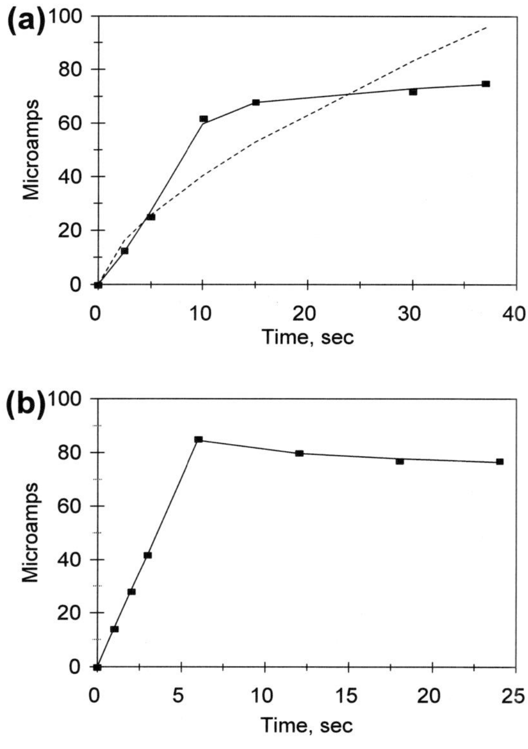

FIGURE 7.8 Binding of glucose in solution to (a) MEBC2/QCM (b) MEBC1/QCM.

TABLE 7.6a Binding and dissociation rate coefficients for glucose to (a) MEBC2/QCM and (b) MEB 1/QCM (Fu et al., 2011)

Analyte in Solution/ Receptor on Surface	k	k_1	k_2	k_d
Glucose/MEBC 2 QCM	8.879 ± 3.305	4.1845 ± 0.3841	51.455 ± 0.731	NA
Glucose/MEBC 1 QCM	13.962 ± 0.062	NA	NA	2.3078 ± 0.3570

TABLE 7.6b Fractal dimensions for the binding and the dissociation phase for glucose (a) MEBC2/QCM and (b) MEBC 1/QCM (Fu et al., 2011)

Analyte in Solution/Receptor on Surface	D_f	D_{f1}	D_{f2}	D_{fd}
Glucose/MEBC 2 QCM	1.6868 ± 0.2718	0.6898 ± 0.1791	2.7960 ± 0.04216	NA
Glucose/MEBC 1 QCM	0.988 ± 0.00674	NA	NA	2.0906 ± 0.3564

fractal dimensions, D_{f1} and D_{f2}, for a dual-fractal analysis are given in Table 7.6(a) and (b). It is of interest to note that as the fractal dimension increases by a factor of 4.053 from a value of $D_{f1} = 0.6898$ to $D_{f2} = 2.7960$, the binding rate coefficient increases by a factor of 12.296 from a value of $k_1 = 4.1845$ to $k_2 = 51.455$. An increase in the fractal dimension on the biosensor surface leads to an increase in the binding rate coefficient.

Figure 7.8(b) shows the binding and dissociation of glucose to MEBC1/QCM. A single-fractal analysis is adequate to describe the binding and the dissociation kinetics. The values of the binding rate coefficient, k, and the fractal dimension, D_f, for a single-fractal analysis are given in Table 7.6.

7.4 CONCLUSIONS

A fractal analysis is, once again, presented for the binding and dissociation of glucose to glucose oxidase immobilized to different biosensor surfaces. The large quantity of papers appearing in the literature necessitated a second chapter. After all, it was the detection of glucose that gave rise to the first application of glucose detection. Glucose detection still remains the biggest market for biosensors, though applications in other areas are fast catching up due to the ease of operation of biosensors.

In this chapter, we present the binding and dissociation kinetics (if applicable) of

1. 100 mM glucose in solution to glucose oxidase immobilized on an MSPRS biosensor (Amarie et al., 2010),
2. different concentrations of glucose in solution to a 25 surface-adsorbed PAH/PAB bilayer biosensor (Stein et al., 2008),
3. 5×10^{-3} M glucose in solution to
   a. polymer (PPD) biosensor,
   b. mineral oil-based biosensor, and
   c. myoglobin mineral oil-based biosensor,
4. glucose at different concentrations in solution to step changes (50, 150, and 300 mg/dl),
5. 10 μM glucose to glucose oxidase—influence of runs in sequence, and
6. glucose in solution to MEBC2/QCM and MEBC1/QCM.

These six additional examples provide further physical insights into the glucose–glucose oxidase reaction, when glucose oxidase is immobilized to different surfaces, that is, the reaction is analyzed by different biosensor surfaces.

Research in the area of glucose detection is bound to be of interest for a while, especially since diabetes is reaching epidemic proportions. Newer techniques are bound to be explored, with the possible emphasis on noninvasive techniques. The invasive nature of present-day techniques limits their compliance, which hinders the daily monitoring of diabetes and could lead to serious consequences. There is, thus, bound to be emphasis on the development of a technique(s) for the noninvasive monitoring of blood sugar levels in diabetes.

REFERENCES

Amarie, D., Alieche, A., Dragnea, B., Glazer, J.A., 2010. Microfluidic devices integrating surface-plasmon resonance sensors: glucose oxidase binding activity detection. Analytical Chemistry 82, 343–352.

Amenrov, A.K., Chen, J., Small, G.W., Arnod, M.A., 2005. Analytical Chemistry 77, 4587–4594.

Asian, K., Lakowicz, J.R., Geddes, C.D., 2005. Analytical Chemistry 77, 2007.

Bao, J.M., Zimmler, M.A., Capasso, F., Wang, X.W., Ren, Z.F., 2006. Nano Letters 6, 1719–1722.

Ballerstadt, R., Polak, A., Beuhler, A., Frye, A., 2004. Biosensors & Bioelectronics 19, 905.

Ballerstadt, R., Schultz, J.S., 2000. Analytical Chemistry 72, 4185.

Barone, F.W., Balk, S., Heller, D.A., Strano, M.S., 2005. Nature Materials 4, 86.

Bergqvist, S., Doyle, M., Glannetti, A.M., Hensley, P., Schwarz, F.P., and Yadav, S.P., 2007. MIRG 2007 Survey on Label-free Biophysical Technologies Used in Characterizing Biomolecular Interactions. The Association of Biomolecular Resource Facilities. http;//www.abrf.org/ (accessed 12.02.08).

Brown, J.Q., McShane, M.J., 2006. Biosensors & Bioelectronics 21, 1760–1769.

Brown, J.Q., Srivastava, K., MShane, M.J., 2005. Biosensors & Bioelectronics 21, 212–216.
Cella, L.C., Chen, W., Myung, N.V., Mulchandani, A.K., 2010. Single-walled carbon nanotube –based affinity biosensor for small molecule ultrasensitive glucose detection. Journal of the American Chemical Society 132 (14), 5024–5026.
Chaki, N.K., Vijayamohan, K., 2002. Biosensors & Bioelectronics 17, 1–12.
Chen, B., Xiang, S.C., Qian, G.D., 2010. Accounts of Chemical Research 43, 1115–1124.
Choi, H.N., Kim, M.A., Lee, W.Y., 2005. Analytica Chimica Acta 537, 179–187.
Chinnayelka, S., McShane, M.J., 2005. Analytical Chemistry 77, 5501–5511.
Chinayleka, S., McShane, M.J., 2006. Diabetes Technology and Therapy 8, 269–278.
Claussen, J.C., Franklin, A.D., Haque, A.U., Porterfield, D.M., Fisher, T.S., 2009. Electrochemical biosensor of nanocube-augmented carbon nanotube networks. ACS Nano 3 (1), 37–44.
Claussen, J.C., Wickner, M., Fisher, T.S., and Porterfield, M., 2009. Transforming the fabricated and bifunctionalization of gold nanoelectrode arrays into versatile electrochemical sensors. ACS Applied Material Interfaces. 3 (5), 1765–1770.
Cosnier, S., 1999. Biosensors & Bioelectronics 14, 443–456.
Dai, Z., Xu, X., Wu, L., Ju, H., 2005. Electroanalysis 17, 1571–1577.
D'Auria, S., DiCesare, N., Stlano, M., Gryczynski, Z., Rossi, M., Lakowicz, J.R., 2002. Analytical Biochemistry 303, 138–144.
D'Auria, S., Herman, P., Rossi, M., Lakowicz, J.R., 1999. Biochemical Biophysical Research Communications 263, 550–553.
Dorfan, A., Kumar, N., Hahm, J., 2006a. Langmuir 22, 4890–4895.
Dorfan, A., Kumar, N., Hahm, J., 2006b. Advances in Materials 18, 2685–2690.
Dunn, B., Zink, J.I., 2007. Accounts of Chemical Research 40, 747–755.
Ebbeson, T.W., Lezec, H.J., Ghaemi, H.F., Thio, P.A., Wolff, P.A., 1998. Nature 391, 667–669.
Fan, X., White, I.M., Shopoa, S.I., Zhu, H., Suter, J.D., Sun, Y., 2008. Analytica Chimica Acta 620, 8–26.
Fischer, V., Hidde, A., Herman, S., von Wiedtke, T., Rebrin, K., Abel, P., 1989. Biomedica Biochimica Acta 48, 965.
Fu, Y., Li, P., Bu, L., Wang, T., Xie, Q., Chen, J., Yao, S., 2011. Exploiting metal-organic coordination polymers as highly efficient immobilization matrices of enzymes for sensitive electrochemical biosensing. Analytical Chemistry.
Giobanu, M., Taylor, D.E., Wilbarn, J.P., Cliffel, D.E., 2008. Glucose and lactate biosensors for scanning electrochemical microscopy imaging of single live cells. Analytical Chemistry 80 (8), 2717–2727.
Gough, D.A., Lucisano, J.Y., Tse, P.H., 1985. Analytical Chemistry 57, 2351–2357.
Havlin, S., 1989. Molecular diffusion and reaction. In: Avnis, D. (Ed.), Polymers. The Fractal Approach to Heterogeneous Chemistry: Surfaces, Colloids. Wiley, New York, pp. 251–269.
Henry, C., 1998. Analytical Chemistry 70, 594A.
Ibey, B.L., Pishko, M., Cote, G.L., 2006. In: Geddes, C.D., Lakowicz, J.R. (Eds.), Topics in Fluorescence Spectroscopy, vol. 11. Springer, New York.
Ibey, B.L., Beier, H.T., Rounds, R.M., Cote, G.L., Yadviili, V.K., Pishko, M.V., 2005. Analytical Chemistry 77, 7039–7046.
Jain, P.K., Huang, X.H., El-Sayed, J.H., El-Sayed, M., 2008. Accounts of Chemical Research 41, 1578–1586.
Johnson, J.C., Knutsen, K.P., Yan, H.Q., Law, M., Zhang, Y.F., Yang, P.D., Saykally, R.J., 2004. Nano Letters 4, 197–204.
Karlsson, R., 2004. Journal of Molecular Recognition 17, 151–161.

Kachoosangi, R.T., Musameh, M.M., Abu-Yousef, I., Yousef, J.M., Kanan, S.M., Xiao, L., Davies, S.G., Russell, A., Compton, R.G., 2009. Analytical Chemistry 81, 435–442.
Kielman, F.J., 1999. Journal of Microstructure 194, 567–570.
Kohli-Seth, R., Oropello, J.M., 2000. Critical Care Clinic 16, 557–578.
Lakshmi, D., Bossi, A., Whitcombe, M.J., Cianella, L., Fowler, S.A., Subrahmanyam, S., Piletska, E.V., Piletsky, S.A., 2009. Analytical Chemistry 81, 3576–3584.
Li, Q.H., Liang, Y.X., Wan, Q., Wan, T.H., 2004. Applied Physics Letters 85, 6389–6391.
Liu, X., Breunling, M.L., 2004. Chemical Materials 16, 351–357.
Liu, Y., Yu, D., Zeng, C., Miao, Z., Dai, L., 2010. Biocompatible graphene oxide-based glucose biosensors. Langmuir 26 (9), 6158–6160.
Lu, L., Zhang, L., Zhang, X., Huan, S., Shen, G., Yu, R., 2010. Analytica Chimica Acta 665, 146–151.
Mailley, P., Cummings, E.A., Mailley, S.C., Eggins, B.R., McAdams, E., Cosnier, S., 2003. Analytical Chemistry 75, 5422–5428.
Masson, J.F., Kranz, C., Mizaikoff, B., 2007. Modelling the response function of dual-enzyme microbiosensors. Analytical Chemistry 79 (22), 8531–8538.
McShane, M.J., 2002. Diabetes Technology and Therapy 4, 533–538.
McShane, M.J., Rastagar, S., Pishko, M., Cote, G.L., 2000. IEEE Transactions in Biomedical Engineering 47, 624–632.
McShane, M.J., 2006. In: Geddes, C.D., Lakowicz, J.R. (Eds.), Topics in Fluorescence Spectroscopy, vol. 11. Springer, New York.
Meng, L., Jim, J., Yang, G., Lu, T., Zhang, H., Cai, C., 2009. Nonenzymatic electrochemical detection of glucose based on platinum-single-walled carbon nanotube hybrid nanostructures. Analytical Chemistry 81 (7), 7271–7280.
Moschou, E.A., Sharma, B.V., Deo, S.K., Daunert, S., 2004. Journal of Fluorescence 14, 535–547.
Muller, M.D., Bruening, M.L., 2004. Langmuir 20, 11545–11551.
Pelossofor, G., Tel-Vared, R., Elberg, J., Willmer, I., 2010. Amplified biosensing using the horseradish peroxidase mimicking DNAzyme electrocatalyst. Analytical Chemistry 82, 4396–4402.
Pickup, J.C., Hussain, F., Evans, N.D., Rolinski, O.J., Birch, D.J., 2005. Biosensors & Bioelectronics 20, 2555.
Pradhan, D., Niroui, F., Leung, K.T., 2010. High-performance, flexible enzymatic glucose biosensors based on ZnO nanowires supported on a gold-coated polyester substrate. ACS Applied Materials and Interfaces 2 (8), 2409–2412.
Privett, B.J., Shim, J.H., Schenfisch, M.H., 2008. Electrochemical biosensors. Analytical Chemistry 80 (12), 4499–4617.
Rodriguez, J.A., Jirsak, T., Dvorak, J., Sambavasian, S., Fischer, D., 2000. Journal of Physical Chemistry B 104, 319–328.
Rosenszweig, Z., Kopelman, R., 1996. Analytical Chemistry 68, 1408–1413.
Russell, R.J., Pishko, M.V., Gefrides, C.C., McShane, M.J., Cote, M., 1999. Analytical Chemistry 71, 3126–3132.
Ryabov, A.D., Kurova, V.S., Ivaniva, E.V., Le Lagadec, R., Alexandrova, L., 2006. Redox mediation and photomechanical oscillations involving photosensitive cyclometalated (RuII) complexes, glucose oxidase, and peroxidase. Analytical Chemistry 77, 1132–1139.
Sberveglieri, G., Gropelli, S., Nelli, F., Tintinnelli, A., Giunta, G., 1995. Sensors & Actuators B 25, 588–590.
Schonoff, M., 2003. Current Opinion in Colloid Interface Science 8, 86–95.

Shaw, C., Yang, H., Song, J., Han, D., Ivaska, A., Niu, L., 2009. Analytical Chemistry 81 (6), 2578–2582.

Shan, D., Cosnier, S., Mousty, C., 2003. Analytical Chemistry 75, 3872–3879.

Shin, J.H., Marxer, S.M., Schoenfisch, M.H., 2004. Nitric oxide-releasing sol-gel particle/polyurethane glucose biosensors. Analytical Chemistry 76, 4543–4549.

Stein, E.W., Grant, P.S., Zhu, H., McShane, M.J., 2007. Microscale enzymatic optical biosensors using mass transport limiting nanofilms. 1.Fabrication and characterization using glucose as a model analyte. Analytical Chemistry 79, 1339–1348.

Stein, E.W., Singh, S., McShane, M.J., 2008. Microscale enzymatic optical biosensors using mass transport limiting nanofilms. 2. Response modulation by varying analyte transport properties. Analytical Chemistry 80, 1408–1417.

Tretnak, W., Wolfbeis, O.S., 1989. Analytica Chimica Acta 221, 195–203.

Tian, Z.R., Voigt, J.A., Liu, J., Mckenzie, B., Medermott, M.J., 2002. Journal of the American Chemical Society 124, 1294–12955.

Tseng, Y.K., Huang, C.J., Cheng, H.M., Lin, I.N., Liu, K.S., Chen, I.C., 2003. Advances in Functional Materials 13, 811–814.

Tsujimura, S., Nishina, A., Kamitaka, Y., Kano, K., 2009. Colometric D-fructose biosensor based on direct electron transfer using D-fructose dehydrogenase. Analytical Chemistry 81 (22), 9383–9387.

Wan, D., Yuan, S., Li, G.L., Neoh, K.G., Kang, E.T., 2010. Glucose biosensor from covalent immobilization of chitosan-coupled carbon nanotubes on polyaniline-modified gold electrode. ACS Applied Materials & Interfaces 2 (11), 3083–3091.

Wan, Q., Cote, G.L., Dixon, J.B., 2005. Journal of Biomedical Optics 10, 024–029.

Wang, J., Chen, L., Jiang, M., Lu, F., 1999. Myoglobin–containing carbon-paste enzyme microelectrodes for the biosensing of glucose under oxygen-deficit conditions. Analytical Chemistry 71, 5009–5011.

Wang, J., Lu, J., 1998. Journal of the American Chemical Society 120, 1048.

Wang, Z., Liu, S., Wu, P., Cai, C., 2009a. Detection of glucose based on direct electron transfer reaction of glucose oxidase immobilized on highly ordered polyaniline nanotubes. Analytical Chemistry 81, 16381645.

Wang, J., Musameh, M., 2003. Carbon nanotube/teflon composite electrochemical sensors and biosensors. Analytical Chemistry 75, 2075–2079.

Wang, J.X., Sun, X.W., Wei, A., Lei, Y., Cai, X.P., Li, C.M., and Dong, Z.L. 2009a. Applied Physics Letters. 88, 233106/1–233106/3.

Wang, S., Tan, Y., Zhao, D., Liu, G., 2008. Biosensors & Bioelectronics 23, 1781–1787.

Wang, Z.G., Wang, Y., Xu, H., Li, G., Xu, Z.K., 2009b. Carbon nanotube-filled nanofibrous membranes electrospun from poly(acrylonitrile-*co*-acrylic acid) for glucose biosensor. Journal of Physical Chemistry C 113, 2955–2960.

Wang, Z.L., Song, J.H., 2006. Science 312, 242–246.

Wilson, G.S., Zhang, Y., Reach, G., Moatti-Sirat, D., Poltout, V., Thevenot, D., Lemonnier, F., Klein, J.C., 1992. Clinical Chemistry 38, 1613.

Wolfbeis, O.S., 2005. Journal of Material Chemistry 15, 2657–2669.

Wolfbeis, O.S., Oehne, I., Papkovskaya, N., Klimani, N., 2000. Biosensors & Bioelectronics 15, 69–76.

Wu, P., He, Y., Wang, H.F., Yan, X.P., 2010. Conjugation of glucose oxidase onto Mn-doped Zn quantum dots for phosphorescent sensing of glucose in biological fluids. Analytical Chemistry 82 (4), 1427–1433.

Xu, H., Aylott, J.W., Kopelman, R., 2002. Analytical Chemistry 127, 1471–1477.

Xue, H., Shen, Z., 2002. Talanta 57, 289–295.

Yang, K., She, G.W., Wang, H., Ou, X.M., Zhang, X.H., Lee, C.S., Lee, S.T., 2009. ZnO nanotube arrays as biosensors for glucose. Journal of Physical Chemistry C 113, 20169–20172.

Yokota, H., Saito, K., Yanagida, T., 1999. Physics Reviews Letters 80, 4606–4609.

Yu, P., Zhou, H., Cheng, H., Qian, Q., Mao, L., 2011. Rational design and one-step formation of multifunctional gel transducer for sample fabrication of integrated electrochemical biosensors. Analytical Chemistry 83 (14), 5715–5720.

Zawadzka, J., Jaroszynski, D.A., Cary, J.J., Wynne, K., 1998. Nuclear Instrumentation Methods Physical Research Section A 445, 324–328.

Zhang, F.F., Wang, X.L., Ali, S.Y., Sun, Z.D., Wan, Q., Zhu, Z.Q., Xian, Y.Z., Jin, L.T., Yamamoto, K., 2004. Analytica Chimica Acta 519, 155–160.

Zhang, M., Mullens, L., Gorski, W., 2007. Coimmobilization of dehydrogenase and their cofactors in electrochemical biosensors. Analytical Chemistry 79 (6), 2446–2450.

Zhang, Y., Ji, C., 2010. Analytical Chemistry 82, 5275–5281.

Zhang, Y., Wilson, G.S., 1993. Analytica Chimica Acta 281, 513.

Zhao, Z.W., Chen, X.J., Tay, B.K., Chen, J.S., Han, Z., Khor, K.A., 2007. Biosensors & Bioelectronics 23, 135–139.

Zhu, H., Srivastava, K., Brown, J.Q., McShane, M.J., 2005. Bioconjugate Chemistry 16, 1451–1458.

Chapter 8

Binding and Dissociation of Biomarkers for Systemic Lupus Erythematosus: A Fractal Analysis of the Kinetics

8.1 INTRODUCTION

Mok (2010) has reviewed the biomarkers for lupus nephritis. He indicates that glomerulonephritis is one of the most serious forms of systemic lupus erythematosus (SLE) (Mok et al., 2005, 2010). It is also a common form of lupus. He indicates that the renal involvement in SLE carries with it significant morbidity and mortality. The 5−10-year renal survival rates of lupus nephritis were 83−92% and 74−84%, respectively. In the 1990s, Mok (2006, 2010) indicates that the prognosis of SLE was low in Hispanics and Africans (Dooley, 1997).

Mok (2010) indicates that in spite of the progress in the overall treatment of SLE, the prognosis of lupus nephritis still leaves much to be desired. He indicates that upto 25% of the patients still develop end-stage renal failure within 10 years (Mok, 2006). Mok (2010) emphasizes the need for improved strategies, combination of effective present-day strategies, or more specific targeting to improve the efficacy of lower toxicities of present-day drugs. Novel biological agents such as rituximab, belimumab, abatacept, and abetimus should be administered to note their effectiveness (Mok, 2006).

Biomarkers for lupus should play an important role in the treatment of SLE. These "newer" biomarkers and even the biomarkers in present-day clinical usage should be sensitive enough and clinically specific to help primarily predict the onset or relapse of renal disease activity in patients with SLE. These biomarkers should go a long way in enhancing the treatment process to help minimize complications arising from this insidious disease.

Lim et al. (1999) indicate that the presence of anti-DNA antibodies in serum is strongly associated with SLE. Thus, these authors indicate that detection of anti-DNA antibodies in serum is expected to be the main approach to SLE diagnosis. Lim et al. (1999) further indicate that different approaches have been developed for the detection, characterization, and making more quantitative antibodies against double-stranded (ds) DNA. However, caution

Biomarkers and Biosensors. http://dx.doi.org/10.1016/B978-0-444-53794-2.00008-2

must be applied here in that antibodies produced against DNA are associated not only with SLE, but with other diseases also (Tomer et al., 1996). Lim et al. (1999) indicate that the Farr assay (Wold et al., 1968) which uses dsDNA as the assay substance, and the indirect immunofluorescence technique (Aarden et al., 1994; Wigand et al., 1997) presently offer the greatest diagnostic specificity for SLE. LeBlanc et al. (1998) indicate that antibodies that bind with DNA with high affinity have been identified as an important component of the autoimmune syndrome of SLE (Arbuckle et al., 2003; Bootsma et al., 1997; Buhl et al., 2007).

Konstantinov et al. (2009) emphasize that the measurement and monitoring of biomarkers are mandatory components in the management of clinical diseases. These authors indicate that different technical approaches and methodologies have been developed to measure antibodies in clinical relevant samples. Privett et al. (2008) indicate that biosensors, particularly electrochemical biosensors, exhibit significant potential for the development of detection devices for relevant analytes.

In this chapter, we present the detection of biomarkers for SLE that have appeared in recent literature. More specifically, we present the fractal analysis for the determination of the kinetics and dissociation (if applicable) of these biomarkers on different biosensor surfaces. As indicated earlier, the fractal analysis provides a convenient means to make more quantitative the heterogeneity that exists on the biosensor surface.

8.2 THEORY

8.2.1 Single-Fractal Analysis

8.2.1.1 Binding Rate Coefficient

Havlin (1989) indicates that the diffusion of a particle (analyte [Ag]) from a homogeneous solution to a solid surface (e.g., receptor [Ab]-coated surface) on which it reacts to form a product (analyte—receptor complex; (Ab·Ag)) is given by:

$$(\mathrm{Ab}\cdot\mathrm{Ag}) \approx \begin{cases} t^{(3-D_{f,\mathrm{bind}})/2} = t^p & t < t_c \\ t^{1/2} & t > t_c \end{cases} \tag{8.1}$$

Here $D_{f,\mathrm{bind}}$ or D_f is the fractal dimension of the surface during the binding step, t_c is the cross-over value. Havlin (1989) indicates that the cross-over value may be determined by $r_c^2 \sim t_c$. Above the characteristic length, r_c, the self-similarity of the surface is lost and the surface may be considered homogeneous. Above time, $t \geq t_c$ to the surface may be considered homogeneous, since the self-similarity property disappears, and "regular" diffusion is now present. For a homogeneous surface where D_f is equal to 2, and when only diffusional limitations are present, $p = 1/2$ as it should be. Another way of looking at the $p = 1/2$ case (where $D_{f,bind}$ is equal to 2) is that the analyte in solution views the fractal object, in our case, the receptor-coated biosensor surface, from a "large distance." In essence, in the association process, the diffusion of the analyte from the solution to the receptor surface creates a depletion layer of width $(Đt)^{1/2}$, where Đ is the diffusion constant. This gives rise to the fractal power law (Analyte.Receptor) $\sim t^{(3-D_{f,bind})/2}$. For the present analysis, t_c is arbitrarily chosen and we assume that the value of the t_c is not reached. One may consider the approach as an intermediate "heuristic" one that may be used in the future to develop an autonomous (and not time-dependent) model for diffusion-controlled kinetics.

8.2.1.2 Dissociation Rate Coefficient

The diffusion of the dissociated particle (receptor [Ab] or analyte [Ag]) from the solid surface (e.g., analyte [Ag]-receptor [Ab] complex coated surface) into solution may be given as a first approximation by:

$$(Ab.Ag) \approx -t^{(3-D_{f,diss})/2} = -t^{p}(t > t_{diss}) \tag{8.2}$$

Here $D_{f,diss}$ is the fractal dimension of the surface for the dissociation step. This corresponds to the highest concentration of the analyte–receptor complex on the surface. Henceforth, its concentration only decreases. The dissociation kinetics may be analyzed in a manner "similar" to the binding kinetics.

8.2.2 Dual-Fractal Analysis

8.2.2.1 Binding Rate Coefficient

Sometimes, the binding curve exhibits complexities and two parameters (k, D_f) are not sufficient to adequately describe the binding kinetics. This is further corroborated by low values of r^2 factor (goodness of fit). In that case, one resorts to a dual-fractal analysis (four parameters: k_1, k_2, D_{f1}, and D_{f2}) to adequately describe the binding kinetics. The single-fractal analysis presented earlier is thus extended to include two fractal dimensions. At present, the time ($t = t_1$) at which the "first" fractal dimension "changes" to the "second" fractal dimension is arbitrary and empirical. For the most part, it is dictated by the data analyzed and experience gained by handling a single-fractal analysis. A smoother curve is obtained in the "transition" region, if care is taken to select the correct number of points for the two regions. In this case, the product (antibody–antigen or analyte–receptor complex, Ab·Ag or analyte.receptor) is given by:

$$(Ab.Ag) \approx \begin{cases} t^{(3-D_{f1,bind})/2} = t^{p1} & (t < t_1) \\ t^{(3-D_{f2,bind})/2} = t^{p2} & (t_1 < t < t_2) = t_c \\ t^{1/2} & (t > t_c) \end{cases} \tag{8.3}$$

In some cases, as mentioned earlier, a triple-fractal analysis with six parameters (k_1, k_2, k_3, D_{f1}, D_{f2}, and D_{f3}) may be required to adequately model the binding kinetics. This is when the binding curve exhibits convolutions and complexities in its shape due to perhaps the very dilute nature of the analyte (in some of the cases to be presented) or for some other reasons. Also, in some cases, a dual-fractal analysis may be required to describe the dissociation kinetics.

8.3 RESULTS

Le et al. (2004) have analyzed lupus-derived autoantibodies from A V gene phage display library specific for the Kringle2 domain of prothrombin. Prothrombin according to these authors is a 579-amino acid plasma glycoprotein and is the precursor of thrombin. Thrombin is the central enzyme in coagulation (Davie, 1995; Degen and Sun, 1994; Mann et al., 1981; Krishnsawamy et al., 1992). Le et al. (2004) indicate that the circulating form of prothrombin contains an amino acid terminal Gla domain, two kringle domains, and a carboxy-terminal catalytic domain. The Gla domain mediates the Ca^{++}-dependent binding to anionic phospholipid vesicles; Krishnaswamy et al. (1992) indicate that thrombin can also cleave peptide bonds, Arg[5]—Ser 152 in intact prothrombin releasing prothrombin I and fragment I, which contain the Gla domain and the first kringle domain.

Le et al. (2004) indicate that antibodies to prothrombin in lupus have been known for almost 50 years, when patients with hypoprothrombinia and circulating anticoagulants were first described (Ley et al., 1951; Bajaj et al., 1994; Edison et al., 1984). Arvieux et al. (1995) indicate that these antibodies have also been found in individuals without lupus. Le et al. (2004) have developed a phage-displayed antibody fragments library from patients with SLE with anti-prothrombin antibodies to analyze structure—function relationships and mechanistic studies. They have isolated two prothrombin-specific ScFvs and localized their epitope and analyzed their effect on prothrombin activation.

Le et al. (2004) analyzed the binding of ScFv in solution for prothrombin activation. Enzyme-linked immunosorbent assays (ELISA) wells were coated with a mixture of prothrombin (10 μg/ml), factor Va (5 nM), phospholipid (10 μM), and ScFvs (10 μg/ml). The reaction was started with 50 pM of factor Va. This was a series of initial runs. Figure 8.1(a) shows the binding and dissociation of 10 μg/ml of ScFv AN78 to the coated ELISA surface. A single-fractal analysis is required to adequately describe the binding and dissociation kinetics. The values of (a) the binding rate coefficient, k, and the fractal dimension, D_f, for a single-fractal analysis, and (b) the dissociation rate coefficient, k_d, and the fractal dimension for dissociation, D_{fd}, are given in Tables 8.1 and 8.2.

Figure 8.1(b) shows the binding of 5 μg/ml of ScFv AN78 to the coated ELISA surface. A dual-fractal analysis is required to adequately describe

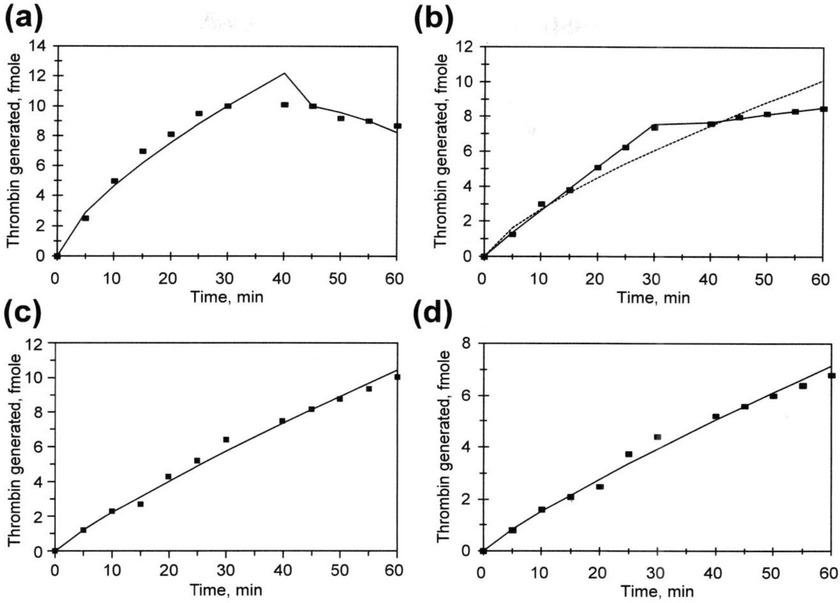

FIGURE 8.1 Binding of different concentrations of ScFv AN78 in solution to prothrombin activation (initial runs) (Le et al., 2004): (a) 10 μg/ml; (b) 5 μg/ml; (c) 1 μg/ml; and (d) 0 μg/ml.

the binding kinetics. The values of (a) the binding rate coefficient, k, and the fractal dimension, D_f, for a single-fractal analysis, and (b) the binding rate coefficients, k_1 and k_2, and the fractal dimensions D_{f1} and D_{f2}, for a dual-fractal analysis are given in Tables 8.1 and 8.2.

Figure 8.1(c) shows the binding of 1 μg/ml of ScFv AN78 to the coated ELISA surface. A dual-fractal analysis is required to adequately describe the binding kinetics. The values of (a) the binding rate coefficient, k, and the fractal dimension, D_f, for a single-fractal analysis, and (b) the binding rate coefficients, k_1 and k_2, and the fractal dimensions, D_{f1} and D_{f2} are given in Tables 8.1 and 8.2.

Figure 8.1(d) shows the binding of 0 μg/ml of ScFv AN78 to the coated ELISA surface. Once again, a single-fractal analysis is required to adequately describe the binding kinetics. The values of the binding rate coefficient, k, and the fractal dimension, D_f, for a single-fractal analysis are given in Tables 8.1 and 8.2.

Le et al. (2004) performed a series of repeat runs with the same experimental variables. Figure 8.2(a) (repeat run) shows the binding and dissociation of 10 μg/ml of ScFv AN78 to the coated ELISA surface. A single-fractal analysis is required to describe the binding and dissociation kinetics. The values of (a) the binding rate coefficient, k, and the fractal dimension, D_f, for a single-fractal analysis, and (b) the dissociation rate coefficient, k_d, and the fractal dimension for dissociation, D_{fd}, are given in Tables 8.1 and 8.2.

TABLE 8.1 Binding and Dissociation Rate Coefficients for ScFv AN78 in Solution to Prothrombin Fragments. ELISA Wells were Coated with a Mixture of Prothrombin (100 μM), Factor Va (5 nM), Phospholipid (10 μM), and ScFv (10 μg/ml). Influence of Different ScFv Concentrations (Le et al., 2004) on Binding and Dissociation

Analyte in Solution (μg/ml)/ Receptor on Surface	k	k_1	k_2	k_d
(a) Initial runs 10	0.944 ± 0.133	NA	NA	0.0062 ± 0.0041
5	0.4954 ± 0.0828	0.2843 ± 0.0231	2.8727 ± 0.0278	NA
1	0.3003 ± 0.0225	NA	NA	NA
0	0.2073 ± 0.0133	0.2843 ± 0.0231	2.873 ± 0.027	NA
(b) Repeat runs 10	1.321 ± 0.130	NA	NA	NA
5	4.7292 ± 0.9359	2.3668 ± 0.4678	7.5323 ± 0.2014	NA
1	2.4795 ± 0.3512	1.4428 ± 0.1685	5.242 ± 0.054	NA
0	0.8060 ± 0.0742	NA	NA	NA

TABLE 8.2 Fractal Dimensions for the Binding and Dissociation Phase for ScFv AN78 in Solution for Prothrombin Activation. ELISA Wells were Coated with a Mixture of Prothrombin (100 μM), Factor Va (5 nM), Phospholipid (10 μM), and ScFv (10 μg/ml). Influence of Different ScFv Concentrations (Le et al., 2004)

Analyte in Solution (ng/ml)/ Receptor on Surface	D_f	D_{f1}	D_{f2}	D_{fd}
(c) Initial runs 10	1.2888 ± 0.2342	NA	NA	0.4224 ± 0.1826
5	1.5274 ± 0.1241	1.0704 ± 0.1056	2.4692 ± 0.0602	NA
1	2.1786 ± 0.1064	NA	NA	NA
0	1.7552 ± 0.07076	1.0702 ± 0.1056	2.4672 ± 0.060	NA
(d) Repeat runs 10	1.612 ± 0.130	NA	NA	NA
5	2.4932 ± 0.1450	1.8994 ± 0.2828	2.7674 ± 6.09565	NA
1	1.2626 ± 0.0591	1.7184 ± 0.1738	2.5922 ± 0.0366	NA
0	1.2682 ± 0.05726	NA	NA	NA

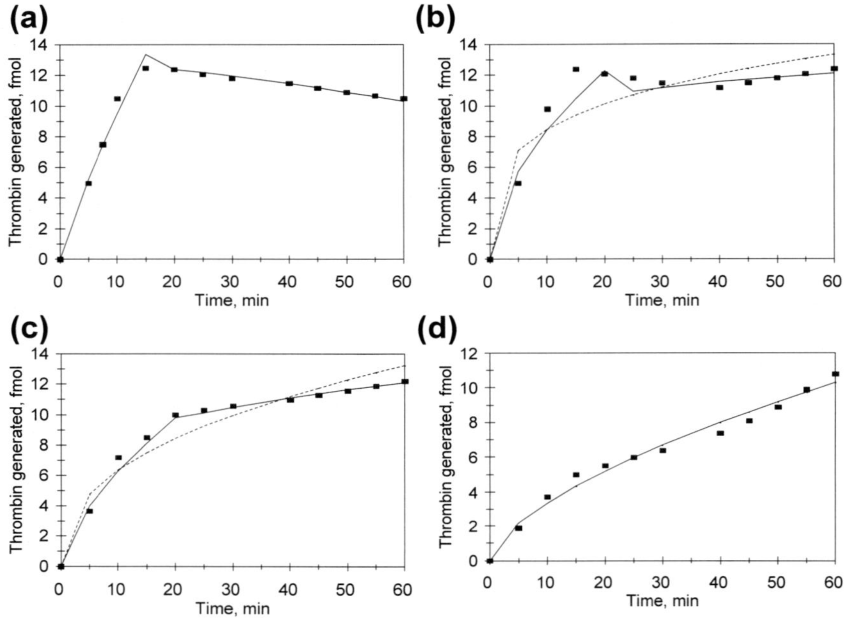

FIGURE 8.2 Binding of different concentrations of ScFv AN78 in solution to prothrombin activation (repeat runs) (Le et al., 2004): (a) 10 μg/ml; (b) 5 μg/ml; (c) 1 μg/ml; and (d) 0 μg/ml.

Figure 8.2(b) (repeat run) shows the binding and dissociation of 5 μg/ml of ScFv AN78 to the coated ELISA surface. A dual-fractal analysis is required to describe the binding kinetics. The values of (a) the binding rate coefficient, k, and the fractal dimension, D_f, for a single-fractal analysis, and (b) the binding rate coefficients, k_1 and k_2, the fractal dimensions for binding, D_{f1} and D_{f2}, for a dual-fractal analysis are given in Tables 8.1 and 8.2.

Figure 8.2(c) (repeat run) shows the binding of 1 μg/ml of ScFv AN78 to the coated ELISA surface. A dual-fractal analysis is required to adequately describe the binding kinetics. The values of the binding rate coefficient, k, and the fractal dimension, D_f, for a single-fractal analysis, and the (b) binding rate coefficients, k_1 and k_2, and the fractal dimensions, D_{f1} and D_{f2}, for a dual-fractal analysis are given in Tables 8.1 and 8.2.

Figure 8.2(d) (repeat run) shows the binding of 0 μg/ml of ScFv AN78 to the coated ELISA surface. Once again, a single-fractal analysis is required to adequately describe the binding kinetics. The values of the binding rate coefficient, k, and the fractal dimension, D_f, for a single-fractal analysis are given in Tables 8.1 and 8.2. Note that in all four cases the binding and dissociation kinetics display similar kinetics, though the values of the rate coefficients obtained for the sequence of initial and repeat runs are different.

Figure 8.3(a) and Table 8.1 show the increase in the binding rate coefficient, k, with an increase in the ScFv AN78 concentration in solution for the

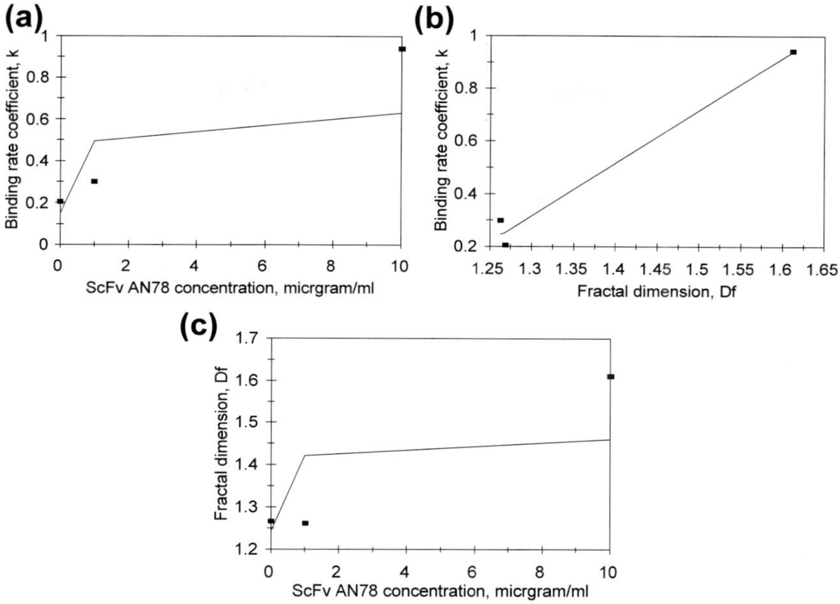

FIGURE 8.3 (a) Increase in the binding rate coefficient, k, with an increase in the ScFv AN78 concentration in solution. (b) Increase in the binding rate coefficient, k, with an increase in the fractal dimension, D_f. (c) Increase in the fractal dimension, D_f, with an increase in the ScFv AN78 concentration in solution.

initial set of runs. For the data shown in Figure 8.3(a), the binding rate coefficient, k is given by:

$$k = (0.4857 \pm 0.4531)[\text{ScFv AN78}]^{0.1057 \pm 0.0754} \tag{8.4a}$$

There is scatter in the data. Only three data points are available. The availability of more data points would lead to a more reliable fit. The binding rate coefficient, k, for a single-fractal analysis exhibits close to a zero (equal to 0.1057) order of dependence on the ScFv AN78 concentration in solution. This indicates that the binding rate coefficient, k, is practically independent of the ScFv concentration in solution in the range studied.

Figure 8.3(b) and Table 8.1 show the increase in the binding rate coefficient, k, with an increase in the fractal dimension, D_f for the initial set of runs. For the data shown in Figure 8.3(b), the binding rate coefficient, k, is given by:

$$k = (0.0689 \pm 0.0222)D_f^{5.474 \pm 1.412} \tag{8.4b}$$

The fit is good. Only three data points are available. The availability of more data points would lead to a more reliable fit. The binding rate coefficient, k, for a single-fractal analysis is very sensitive to the degree of heterogeneity that exists on the surface as noted to be close to five and one-half (equal to

5.474) order of dependence on the fractal dimension or the degree of heterogeneity that exists on the surface.

Figure 8.3(c) and Table 8.1 show the increase in the fractal dimension, D_f with an increase in the ScFv AN78 concentration in solution for the initial set of runs. For the data shown in Figure 8.3(c), the fractal dimension, D_f is given by:

$$D_f = (1.4226 + 0.239)[\text{ScFv AN78}]^{0.0116 \pm 0.0149} \quad (8.4c)$$

There is scatter in the data. Only three data points are available. The availability of more data points would lead to a more reliable fit. The fractal dimension, D_f exhibits close to a zero (equal to 0.0116) order of dependence on the ScFv AN78 concentration in solution. This indicates that the fractal dimension, D_f is practically independent of the ScFv AN78 concentration in solution in the range studied.

Giorgiani et al. (1997) have analyzed the histidine-rich glycoprotein (HRG) binding to human IgG and C1q and noted that it inhibits the formation of insoluble complexes. They indicate that certain autoimmune diseases in which the body "attacks" itself such as SLE and rheumatoid arthritis (RA) are associated with the production of autoantibodies (Mountz et al., 1994). Sittampalam and Wilson (1984) further indicate that these autoantibodies can be cross-linked with their cognate antigens to form insoluble immune complexes. These HCs can precipitate in target tissues and lead to a lack of nutrient transport and consequent tissue injury. Charlesworth et al. (1982) and Giorgiani et al. (1997) suggest that there are two methods by which this may be minimized: inhibit the formation of the HO complexes and promote the solubilization of the IIC complexes. This permits the effective clearance of these circulating complexes.

The complements C3b and C1q can modulate the formation of TIC, in vitro. Upon generation of the complement cascade, the generation of the C3b complement through a complex process inhibits the formation of the IIC and promotes their solubilization (Schifferli, 1987; Taketa et al., 1984), and C1q (a constituent of CI component of the complement) enhances the formation of the EC and leads to the cross-linking of the IgG molecules. Schifferli (1987) also indicated that the interaction of C1q with the Fc region of the IgG molecule immune complexes may also activate the complement cascade by the classical pathway. suggest that C1q inhibits the interaction between IgG and the Fc receptors on neurophils.

Giorgiani et al. (1997) indicate that preliminary data suggest that HRG may associate with the IgG molecules. The precise physiological role of HRG is yet to be determined. These authors have used the ELISA technique and the IAsys resonant mirror biosensor to study the binding of human HRG to human C1q and IgG. They determined the kinetic constants for these interactions. We will reanalyze their data and present an alternate technique (using fractals) to determine the kinetics. Figure 8.4 shows the

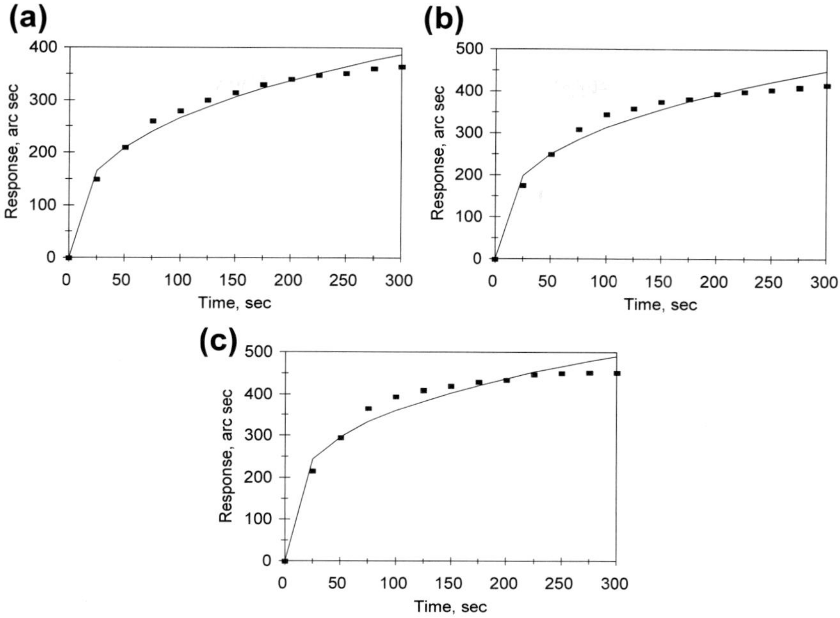

FIGURE 8.4 Influence of Zn^{++} on the binding of HRG in solution to C1q immobilized on an optical IAsys biosensor surface: (a) absence of Zn^{++}; (b) control; and (c) presence of Zn^{++}.

effect of Zn^{++} ions on the binding of HRG to immobilized C1q. HRG binds to divalent metal cations like Zn^{++}. This alters the ability of HRG to interact with heparin and to bind to the T cells and elicit T-cell responses (Kazama and Koide, 1992; Olsen et al., 1996). This was carried out in the presence of 20 μM of added Zn^{++}, with and without EDTA. In the presence of 20 μM Zn^{++} the authors (Giorgiani et al., 1997) noted that the binding of HRG in solution decreased when Zn^{++} was present.

Figure 8.4(a) shows the binding of 600 nM HRG in solution to C1q immobilized on the biosensor surface in the absence of Zn^{++}. A singe-fractal analysis is adequate to describe the binding kinetics. The values of the binding rate coefficient, k, and the fractal dimension, D_f, are given in Table 8.3.

Figure 8.4(b) shows the binding of control in solution to C1q immobilized on the biosensor surface in the absence of Zn^{++}. A singe-fractal analysis is adequate to describe the binding kinetics. The values of the binding rate coefficient, k, and the fractal dimension, D_f, are given in Table 8.3.

Figure 8.4(c) shows the binding of HRG control in solution to C1q immobilized on the biosensor surface in the presence of Zn^{++}. A single-fractal analysis is adequate to describe the binding kinetics. The values of the binding rate coefficient, k, and the fractal dimension, D_f, are given in Table 8.3.

TABLE 8.3 Binding Rate Coefficients and Fractal Dimensions for Native 81 kDa Histidine-Rich Glycoprotein (HRG) in Solution to C1q Immobilized on an IAsys Biosensor Surface (Giorgiani et al., 1997). Influence of Zn^{++}

Analyte in Solution/Receptor on Surface	k	D_f
600 nM HRG + 20 μM Zn^{++} + 1 mM EDTA/C1q immobilized on biosensor surface	54.478 ± 0.029	2.3122 ± 0.0423
Control	69.351 ± 4.82	$2.3458 \pm 0,05748$
600 nM HRG/C1q immobilized on biosensor surface	98.14 ± 7.23	2.4356 ± 0.05674

Figure 8.5 shows an increase in the binding rate coefficient, k, with an increase in the fractal dimension, D_f. For the data shown in Figure 8.5 and in Table 8.3, the binding rate coefficient, k, is given by:

$$k = (0.00653 \pm 0.00040)\, D_f^{10.818 \pm 1.543} \tag{8.5}$$

Only three data points are available. The availability of more data points would lead to a more reliable fit. The binding rate coefficient, k, exhibits close to an eleventh (equal to 10.818) order of dependence on the degree of heterogeneity or the fractal dimension that exists on the biosensor surface. This indicates that the binding rate coefficient, k, is extremely sensitive to the degree of heterogeneity that exists on the biosensor surface.

Figure 8.6(a) shows the binding of 320 nM soluble HRG (in nM) to human complement C1q immobilized on an optical biosensor IAsys surface. A single-fractal analysis is adequate to describe the binding kinetics. The values of the binding rate coefficient, k, and the fractal dimension, D_f, are given in Table 8.4.

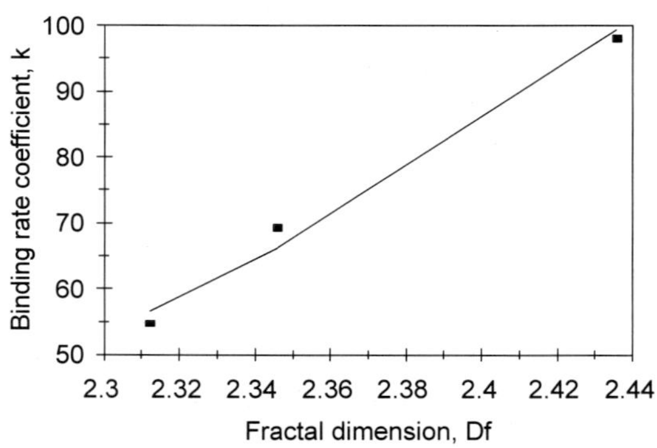

FIGURE 8.5 Increase in the binding rate coefficient, k, with the fractal dimension, D_f.

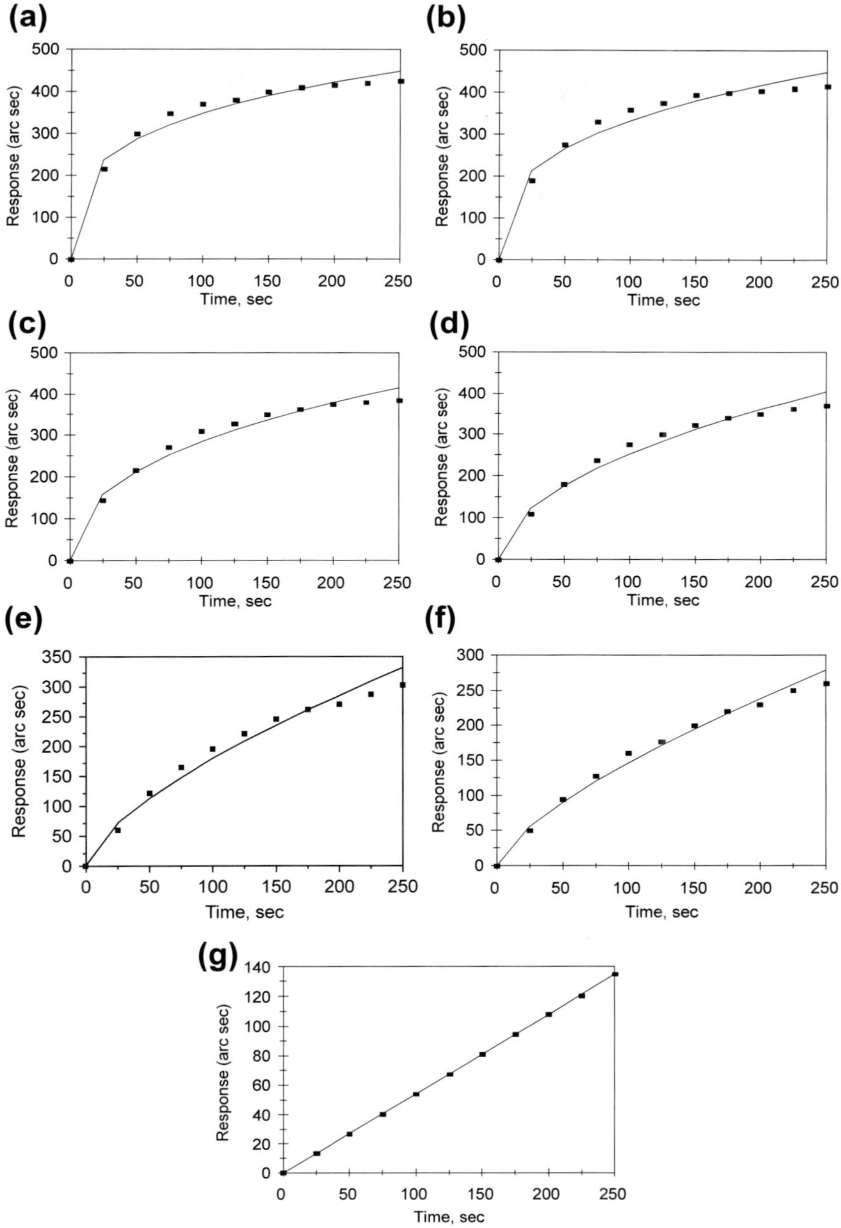

FIGURE 8.6 Binding of different concentrations of soluble histidine glycoprotein (HRG) (in nM) to human complement Clq immobilized on an IAsys optical biosensor surface (Giorgiani et al., 1997): (a) 320; (b) 240; (c) 160; (d) 120; (e) 80; (f) 40; and (g) 20.

TABLE 8.4 Binding Rate Coefficients and Fractal Dimensions for Soluble HRG (Histidine-Rich Glycoprotein) in Solution to Immobilized Clq to an IAsys Resonant Mirror Biosensor Surface (Giorgiani et al., 1997)

Soluble HRG Concentration, nM/Receptor on Surface	k	D_f
20	0.5426 ± 0.0014	1.0024 ± 0.0024
40	5.841 ± 0.392	1.5992 ± 0.05898
80	8.199 ± 0.899	1.6624 ± 0.08758
120	23.379 ± 1.828	1.9668 ± 0.06848
160	41.206 ± 2.762	2.1604 ± 0.0590
240	75.019 ± 5.513	2.3524 ± 0.06448
320	96.799 ± 5.820	2.4426 ± 0.0534

Figure 8.6(b) shows the binding of 240 nM soluble HRG (in nM) to human complement C1q immobilized on an optical biosensor IAsys surface. A single-fractal analysis is adequate to describe the binding kinetics. The values of the binding rate coefficient, k, and the fractal dimension, D_f, are given in Table 8.4.

Figure 8.6(c) shows the binding of 160 nM soluble HRG (in nM) to human complement C1q immobilized on an optical biosensor IAsys surface. A single-fractal analysis is adequate to describe the binding kinetics. The values of the binding rate coefficient, k, and the fractal dimension, D_f, are given in Table 8.4.

Figure 8.6(d) shows the binding of 120 nM soluble HRG (in nM) to human complement C1q immobilized on an optical biosensor IAsys surface. A single-fractal analysis is adequate to describe the binding kinetics. The values of the binding rate coefficient, k, and the fractal dimension, D_f, are given in Table 8.4.

Figure 8.6(e) shows the binding of 80 nM soluble HRG (in nM) to human complement C1q immobilized on an optical biosensor IAsys surface. A single-fractal analysis is adequate to describe the binding kinetics. The values of the binding rate coefficient, k, and the fractal dimension, D_f, are given in Table 8.4.

Figure 8.6(f) shows the binding of 40 nM soluble HRG (in nM) to human complement C1q immobilized on an optical biosensor IAsys surface. A single-fractal analysis is adequate to describe the binding kinetics. The values of the binding rate coefficient, k, and the fractal dimension, D_f, are given in Table 8.4.

Figure 8.6(g) shows the binding of 20 nM soluble HRG (in nM) to human complement C1q immobilized on an optical biosensor IAsys surface. A single-fractal analysis is adequate to describe the binding kinetics. The values of the binding rate coefficient, k, and the fractal dimension, D_f, are given in Table 8.4.

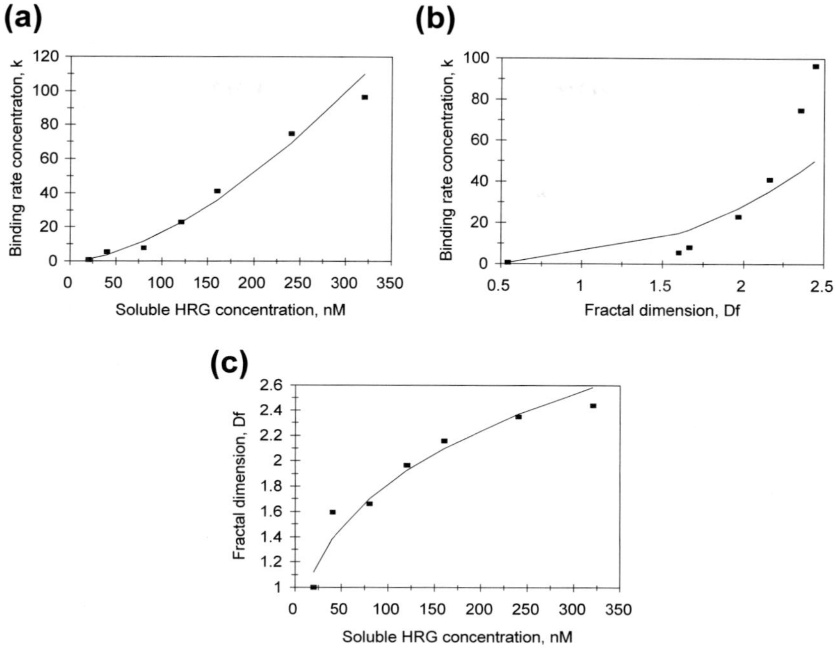

FIGURE 8.7 (a) Increase in the binding rate coefficient, k, with an increase in the soluble HRG (in nM) in solution. (b) Increase in the binding rate coefficient, k, with an increase in the fractal dimension, D_f. (c) Increase in the fractal dimension, D_f, with an increase in the soluble HRG in solution.

Figure 8.7(a) shows the increase in the binding rate coefficient, k, with an increase in the HRG concentration (in nM) in solution. For the data shown in Figure 8.7(a), the binding rate coefficient, k, is given by:

$$k = (0.00955 \pm 0.00310)[HRG]^{1.621 \pm 0.118} \quad (8.6a)$$

The fit is very good. Only seven data points are available. The availability of more data points would lead to a better fit. The binding rate coefficient, k, exhibits an order of dependence between one and one-half and second (equal to 1.621) order of dependence on the HRG concentration in solution. This indicates that the binding rate coefficient, k, is sensitive to the HRG concentration in solution.

Figure 8.7(b) shows the increase in the binding rate coefficient, k, with an increase in the fractal dimension, D_f. For the data shown in Figure 8.7(b), the binding rate coefficient, k, is given by:

$$k = (3.798 \pm 3.659)\ [D_f]^{2.888 \pm 0.529} \quad (8.6b)$$

The fit is good. Only seven data points are available. The availability of more data points would lead to a better fit. The binding rate coefficient, k, exhibits an order of dependence between two and one-half and three (equal to 2.888) on the fractal dimension, D_f. This indicates that the binding rate coefficient, k, is sensitive to the degree of heterogeneity that exists on the IAsys biosensor surface.

Figure 8.7(c) shows the increase in the fractal dimension, D_f, with an increase in the HRG concentration in solution. For the data shown in Figure 8.7(c), the fractal dimension, D_f, is given by:

$$D_f = (0.4556 \pm 0.0422)\,[HRG]^{0.300 \pm 0.0366} \qquad (8.6c)$$

The fit is very good. Only seven data points are available. The availability of more data points would lead to a better fit. The fractal dimension, D_f, exhibits only a mild order of dependence (equal to 0.300) on the HRG concentration in solution.

Figure 8.8(a) shows the binding of 400 nM HRG to IgG immobilized to an IAsys biosensor surface. A dual-fractal analysis is required to adequately describe the binding kinetics. The values of (a) the binding rate coefficient, k, and the fractal dimension, D_f, for a single-fractal analysis, and (b) the binding rate coefficients, k_1 and k_2, and the fractal dimensions, D_{f1} and D_{f2}, for a dual-fractal analysis are given in Table 8.5. It is of note that as the fractal dimension increases by a factor of 1.361 from a value of D_{f1} equal to 2.0244 to D_{f2} equal to 2.7544, the binding rate coefficient increases from k_1 equal to 23.562 to k_2 equal to 127.89, as is generally observed, an increase by a factor of 5.427. An increase in the degree of heterogeneity on the biosensor surface leads to an increase in the binding rate coefficient.

Figure 8.8(b) shows the binding of 300 nM HRG to IgG immobilized to an IAsys biosensor surface. A single-fractal analysis is adequate to describe the binding kinetics. The values of the binding rate coefficient, k, and the fractal dimension, D_f, for a single-fractal analysis are given in Table 8.5.

Figure 8.8(c) shows the binding of 200 nM HRG to IgG immobilized to an IAsys biosensor surface. A single-fractal analysis is adequate to describe the binding kinetics. The values of (a) the binding rate coefficient, k, and the fractal dimension, D_f, for a single-fractal analysis are given in Table 8.5.

Figure 8.8(d) shows the binding of 100 nM HRG to IgG immobilized to an IAsys biosensor surface. A single-fractal analysis is adequate to describe the binding kinetics. The values of (a) the binding rate coefficient, k, and the fractal dimension, D_f, for a single-fractal analysis are given in Table 8.5.

Figure 8.8(e) shows the binding of 50 nM HRG to IgG immobilized to an IAsys biosensor surface. A single-fractal analysis is adequate to describe the binding kinetics. The values of (a) the binding rate coefficient, k, and the fractal dimension, D_f, for a single-fractal analysis are given in Table 8.5.

Figure 8.9(a) shows the increase in the binding rate coefficient, k, for a single-fractal analysis with an increase in the soluble HRG concentration in

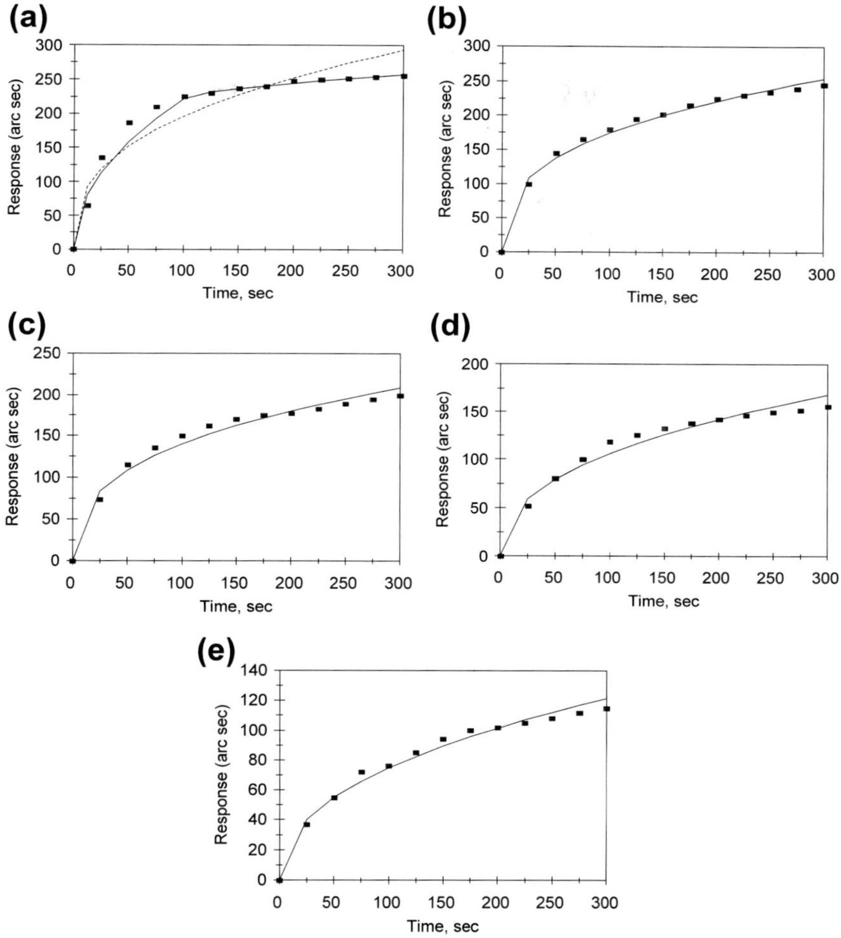

FIGURE 8.8 Binding of different concentrations of HRG in nM in solution to IgG immobilized on an IAsys biosensor surface: (a) 400; (b) 300; (c) 200; (d) 100; and (e) 50.

solution in nM. For the data shown in Figure 8.9(a), the binding rate coefficient, k, is given by:

$$k = (0.5367 \pm 0.0189)\,[\mathrm{HRG}]^{0.734 \pm 0.02543} \tag{8.7a}$$

The fit is good. Only four data points are available. The availability of more data points would lead to a more reliable fit. The binding rate coefficient, k, exhibits an order of dependence between one-half and first (equal to 0.734) order on the HRG concentration in solution. This indicates the binding rate coefficient, k, is quite sensitive to the HRG concentration in solution.

TABLE 8.5 Binding Rate Coefficients and Fractal Dimensions for HRG in Solution to Immobilized IgG (Giorgiani et al., 1997)

Analyte in Solution/ Receptor on Surface	k	k_1	k_2	D_f	D_{f1}	D_{f2}
400 nM HRG/IgG	36.87 ± 3.33	23.562 ± 4.22	127.89 ± 0.87	2.2734 ± 0.09378	2.0244 ± 0.1982	2.7544 ± 0.01688
300 nM HRG/IgG	36.099 ± 1.489	NA	NA	2.3144 ± 0.03224	NA	NA
200 nM HRG/IgG	26.026 ± 1.748	NA	NA	2.2684 ± 0.05186	NA	NA
100 nM HRG/IgG	15.219 ± 1.1481	NA	NA	2.1672 ± 0.0577	NA	NA
50 nM HRG/IgG	9.720 ± 0.532	NA	NA	2.1146 ± 0.0428	NA	NA

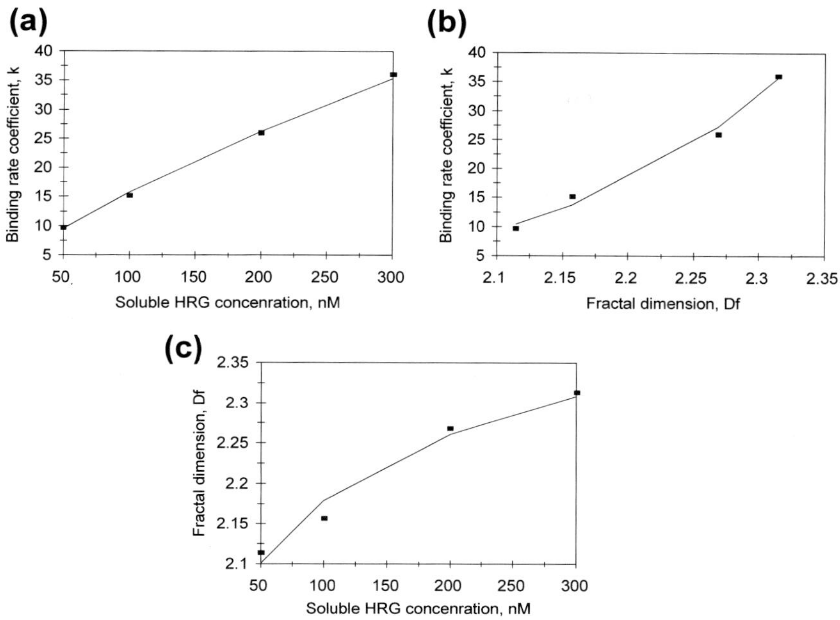

FIGURE 8.9 (a) Increase in the binding rate coefficient, k, with an increase in the HRG concentration in solution in nM. (b) Increase in the binding rate coefficient, k, with an increase in the fractal dimension, D_f. (c) Increase in the fractal dimension, D_f, with an increase in the HRG concentration (in nM) in solution.

Figure 8.9(b) shows the increase in the binding rate coefficient, k, for a single-fractal analysis with an increase in the fractal dimension, D_f. For the data shown in Figure 8.9(b), the binding rate coefficient, k, is given by:

$$k = (0.000389 \pm 0.000039)\, D_f^{13.62 \pm 1.303} \tag{8.7b}$$

The fit is good. Only four data points are available. The availability of more data points would lead to a more reliable fit. The binding rate coefficient, k, exhibits an order of dependence between 13 and one-half and 14 (equal to 13.62) order on the fractal dimension, D_f, that exists on the biosensor surface. This indicates that the binding rate coefficient, k, is very sensitive to the degree of heterogeneity that exists on the IAsys biosensor surface.

Figure 8.9(c) shows the increase in the fractal dimension, D_f, for a single-fractal analysis with an increase in the soluble HRG concentration in solution in nM. For the data shown in Figure 8.9(c), the fractal dimension, D_f, is given by:

$$D_f = (1.710 \pm 0.015)[\mathrm{HRG}]^{0.05267 \pm 0.006575} \tag{8.7c}$$

The fit is good. Only four data points are available. The availability of more data points would lead to a more reliable fit. The fractal dimension, D_f, exhibits an order of dependence of 0.05267 on the HRG concentration in solution. This indicates that the fractal dimension, D_f, is hardly sensitive to the HRG concentration in solution.

8.4 OTHER AUTOIMMUNE REACTIONS

Ortiz et al. (2011) indicate that celiac disease arises in individuals who cannot ingest gluten, and this causes histological changes and leads to maladsorption (Maki and Collin, 1997; Ventura et al., 1999). Apparently, this is one disease that is severely undiagnosed. Ortiz et al. (2011) indicate that only one out of eight celiac cases is diagnosed. Thus, the need for a suitable diagnostic procedure. This is further exacerbated by the fact that there is a variable clinical presentation and symptoms (Fasano and Catassi, 2001). The authors point out that if left untreated, there is significant burden to the patient's family as well as a very significant decrease in the quality of life for the individual and his family. However, Ortiz et al. (2011) do point out that if celiac disease is diagnosed and gluten is removed from the diet, then the prognosis and the long-term benefit for the patient is substantial.

Ortiz et al. (2011) indicate that some diagnostic tests include serum IgA and IgG antibodies against gliadin (Kilander et al., 1983), tissue transglutaminase (Dieterich et al., 1998), reticulin (Eade et al., 1997), and endomysium (Ladinser et al., 1994). This is a component of connective tissue. The above may be detected in patients with celiac disease. Electrochemical biosensors are a popular tool for the diagnosis of celiac disease. These have been used to detect the autoantibodies with regard to different autoimmune diseases which include SLE (Konstantinov et al., 2009), multiple sclerosis (Real-Fernandez et al., 2008), and RA (Drouvalakis et al., 2008). Ortiz et al. (2011) indicate that up until now there is only one published report on the detection of antigliadin autoantibodies (Pereira et al., 2010).

Ortiz et al. (2011) present a novel surface modification strategy based on interfacial self-assembly of bifunctionalized carboxymethylcellulose (CMC). These contain adamantane units and an antigenic fragment onto a cyclodextrin-containing support. This they indicate is capable of detecting antigliadin antibodies in the immunoassay and immunosensor platforms.

The immunoassay platform for the detection of antigliadin antibody is briefly described below (Ortiz et al., 2011). The plate was initially incubated with CDPSH (β-cyclodextrin polymer, molecular weight ~ 18,000 mol/l, degree of substitution: 13 mol SH per mol polymer). This forms the cyclodextrin support layer. After blocking the unreacted maleimide groups with mercaptoethanol, different concentrations of ADA-CMC-GLI and antigliadin antibody are sequentially added to form a matrix of 7 × 7 points. The absorbance signal increases proportionately with both ADA-CMC-GLI and antibody concentrations indicating a good performance of the assay.

Figure 8.10(a) shows the binding of 1 μg/ml of ADA-CMC-GLI to the CDPSH-modified Biacore chip (Ortiz et al., 2011). A dual-fractal analysis is

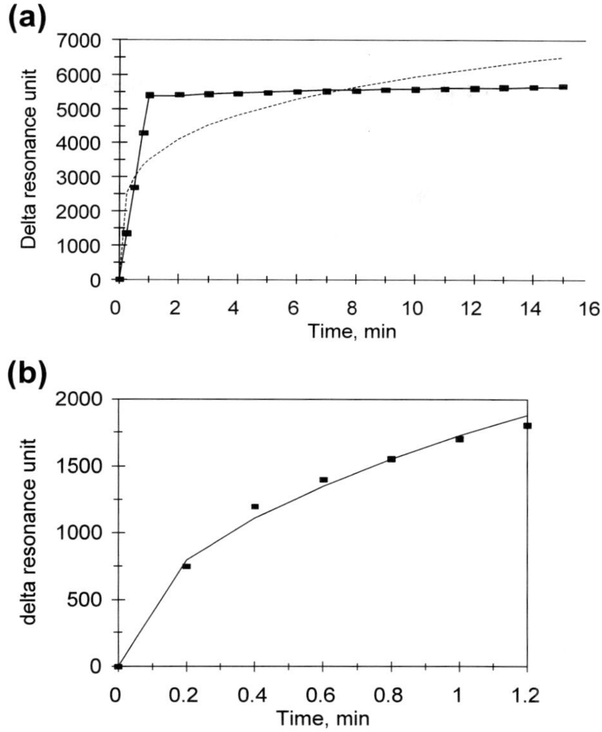

FIGURE 8.10 (a) Binding of 1 μg/ml of ADA-CMC-GLI to CDPSH-modified Biacore chip (Ortiz et al., 2011). (b) Binding of 1 μg/ml of antigliadin to CDPSH-modified Biacore chip.

required to adequately describe the binding kinetics. The values of (a) the binding rate coefficient, k, and the fractal dimension, D_f, for a single-fractal analysis, and (b) binding rate coefficients, k_1 and k_2, and the fractal dimensions, D_{f1} and D_{f2}, for a dual-fractal analysis are given in Table 8.6. It is of interest to note that as the fractal dimension increases by a factor of 2.515 from a value of D_{f1} equal to 1.0030 to D_{f2} equal to 2.5234, the binding rate coefficient decreases slightly from a value of equal 5389.23 to k_2 equal to 5292.01. This goes against the general trend where an increase in the fractal dimension or the degree of heterogeneity on the surface leads to an increase in the binding rate coefficient.

Figure 8.10(b) shows the binding of 1 μg/ml of antigliadin to a CDPSH-modified Biacore chip. A single-fractal analysis is adequate to describe the binding kinetics. The values of the binding rate coefficient, k, and the fractal dimension, D_f, for a single-fractal analysis are given in Table 8.6.

Figure 8.11(a) shows the binding of 1 μg/ml of ADA-GLI to a CDPSH-modified Biacore chip. A single-fractal analysis is adequate to

TABLE 8.6 Binding Rate Coefficients and Fractal Dimensions for (a) ADA-CMC-GLI, (b) Antigliadin, (c) ADAG11, and (d) Antigliadin in Solution to CDPSH-Modified Biacore Chip (Ortiz et al., 2011)

Analyte in Solution/ Receptor on Surface	k	k_1	k_2	D_f	D_{f1}	D_{f2}
1 μg/ml of ADA-CMC-GTLI to CDPSH-modified Biacore chip	3503.99 ± 921.84	5389.23 ± 14.16	5292.01 ± 21.53	2.5422 ± 0.0907	1.0030 ± 0.00494	2.5234 ± 0.003686
1 μg/ml of antigliadin to CDPSH-modified Biacore chip	1721.76 ± 102.45	NA	NA	2.0432 ± 0.07726	NA	NA
1 μg/ml of ADA-GU to CDPSH-modified Biacore chip	1310.43 ± 14.76	NA	NA	2.8968 ± 0.005030	NA	NA
1 μg/ml of antigliadin to CDPSH-modified Biacore chip	1596.84 ± 193.73	NA	NA	2.4904 ± 0.1022	NA	NA

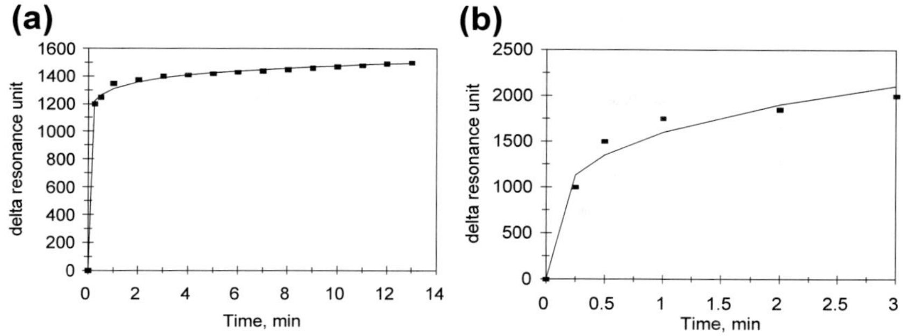

FIGURE 8.11 (a) Binding of 1 μg/ml of ADA-GLI to CDPSH-modified Biacore chip. (b) Binding of antigliadin to CDPSH-modified Biacore chip (Ortiz et al., 2011).

describe the binding kinetics. The values of the binding rate coefficient, k, and the fractal dimension, D_f, for a single-fractal analysis are given in Table 8.6.

Figure 8.11(b) shows the binding of 1 μg/ml of antigliadin to a CDPSH-modified Biacore chip. A single-fractal analysis is adequate to describe the binding kinetics. The values of the binding rate coefficient, k, and the fractal dimension, D_f, for a single-fractal analysis are given in Table 8.6.

Figure 8.12 shows the decrease in the binding rate coefficient, k, for a single-fractal analysis with an increase in the fractal dimension, D_f. For the data shown in Figure 8.12, the binding rate coefficient, k, is given by:

$$k = (3035.54 \pm 102.87) D_f^{-0.762 \pm 0.261} \tag{8.8}$$

The fit is good. Only three data points are available. The availability of more data points would lead to a more reliable fit. The binding rate coefficient, k, exhibits a negative 0.762 order of dependence on the fractal dimension, D_f, or the degree of heterogeneity that exists on the surface. This indicates the

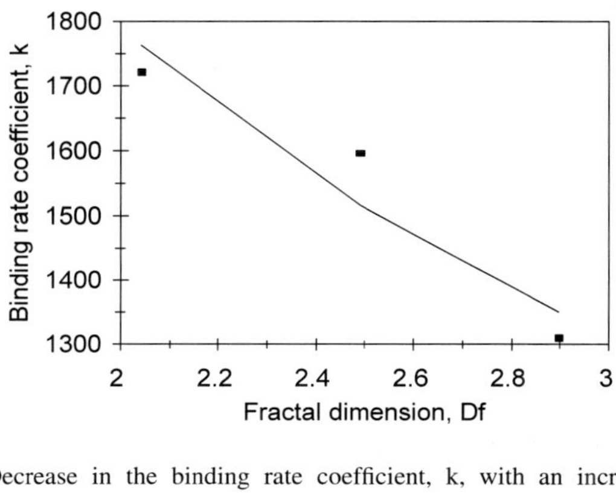

FIGURE 8.12 Decrease in the binding rate coefficient, k, with an increase in the fractal dimension, D_f.

binding rate coefficient, k, is sensitive to the degree of heterogeneity that exists on the surface, and decreases with an increase in the degree of heterogeneity that exists on the surface.

Huang et al. (2008) have analyzed binding of cytokines to photostable single-molecule nanoparticle optical biosensors (SNOBS) for real-time sensing. These authors state "that cytokines are secreted regulatory proteins and play vital roles in controlling cell survival, growth, differentiation, and function by binding with receptors and initiating immune regulation pathways" (Nicola, 1994). Gurevich et al. (2003) indicate that it requires only a few cytokine molecules to induce a significant cellular response. Thus, the need to develop sensitive tools to detect individual cytokine molecules and to characterize their functions in real time. Tagawa (2000) and Haworth et al. (1998) indicate that tumor necrosis factor-α (TNF-α) is a pro-inflammatory cytokine that can mediate a variety of biological effects, for example, immune regulation, antitumor activity, viral replication, and infection resistance. Both these sets of the above-mentioned authors indicate that TNF-α is overproduced in diseases such as cancer, heart disease, diabetes, and autoimmune diseases. However, in spite of this, the role of TNF-α in mediating crucial biological functions is unclear. Thus, the need to develop very sensitive assays to accurately analyze for TNF-α.

Huang et al. (2008) indicate that considerable effort has been spent in developing assays for TNF-α. These include ELISAs (Copeland et al., 2004; Klabusay et al., 2006), radioimmunoassay (de Kossodo et al., 1995), cytotoxicity assay (Watts et al., 2003), flow cytometry (Shin et al., 2003), and RT-PCR (Pico de Coana et al., 2004). Huang et al. (2008) point out that these methods are very time consuming and cannot be used for the detection of TNF-α in real time. Other detection methods have also been developed that include fluorescence quantum dots, protein microarray (Zajac et al., 2007), and electrochemical methods (Wang et al., 2006) for the detection of TNF-α.

Huang et al. (2008) indicate that single-molecule detection (SMD) methods have the advantage that they offer the means to investigate the functions in real time. For example, fluorescence spectroscopy and spectroscopy may be used for the SMD in solution and in living cells (Levisky et al., 2002; Xu et al., 2007; Yu et al., 2006; Zander et al., 2002; Xu et al., 2003; Xu and Yeung 1997). Bohren and Huffman (1983), Kreibig and Vollme (1995), and Mier (1908) indicate that noble metal nanoparticles (Ag, Au) have unique optical properties. These optical properties are dependent on size, shape, surrounding environments, and surface properties (Lee et al., 2007; Nallathamby et al., 2008; Huang et al., 2007). Kyriacou et al. (2004), Xu et al. (2002, 2004) These different authors have demonstrated these optical properties may be used for imaging single living cells in real time.

However, Huang et al. (2008) emphasize the further need to reduce the size of the nanoparticles to prepare smaller single nanoparticle biosensors so that one may detect clusters of multiple individual molecules within an area smaller than, for example, 113 nm^2. Thus, these authors have synthesized smaller Ag nanoparticles, which they claim is a challenging process. They claim that their present study is a major advance in preparing photostable noble metal dots for imaging and sensing proteins of interest.

Figure 8.13(a) shows the binding of TNF-α in solution to SNOBS using a single-monoclonal antibody (mAb) molecule immobilized to the surface. This shows binding to particle I. A single-fractal analysis is required to adequately describe the binding kinetics. The values of the binding rate coefficient, k, and the fractal dimension, D_f, for a single-fractal analysis are given in Tables 8.7 and 8.8.

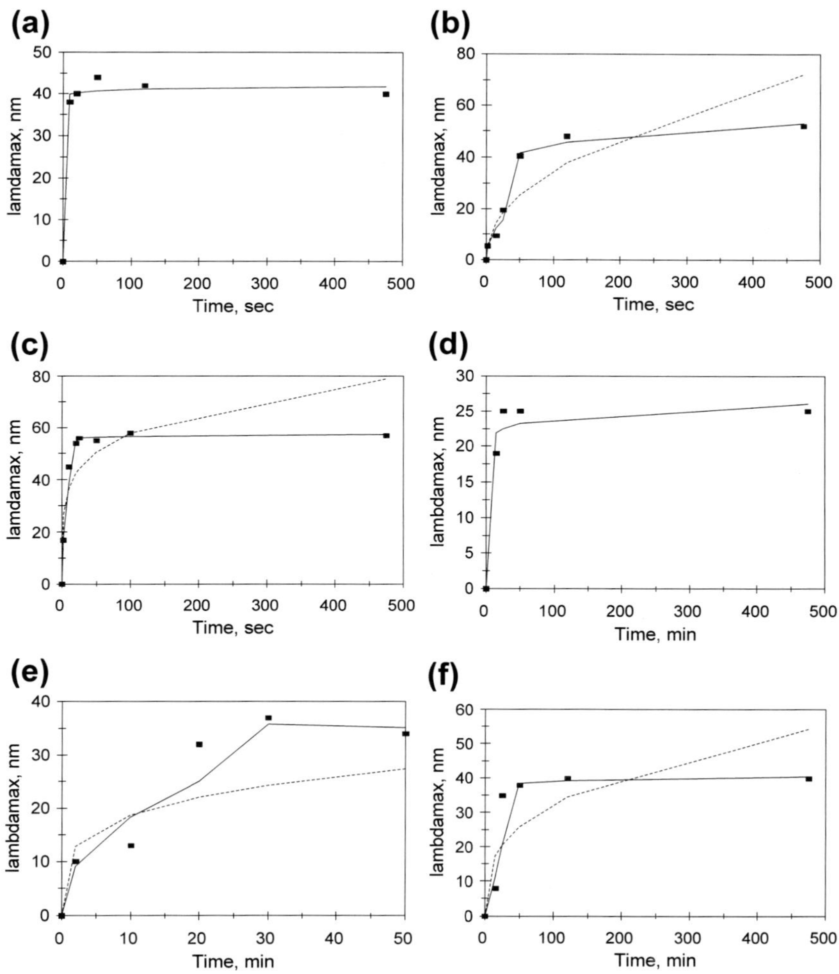

FIGURE 8.13 Binding of TNF-α in solution to different nanoparticles (SNOBS biosensor) Huang et al. (2008): (a) I; (b) II; (c) III; (d) IV; (e) V; and (f) VI.

TABLE 8.7 Binding Rate Coefficients for TNF-α Molecules Using Single-Molecule Nanoparticle Optical Biosensors (SNOBS) to Single-Monoclonal Antibody (mAB) Molecules Binding to Different Single Nanoparticles (Huang et al., 2008)

Particle Number	k	k_1	k_2
I	19.041 ± 2.832	NA	NA
II	10.861 ± 4.841	6.614 ± 3.638	40.087 ± 2.287
III	7.049 ± 5.938	0.3941 ± 0.3704	35.479 ± 0.913
IV	38.759 ± 3.4150	NA	NA
V	4.0778 ± 0.857	3.845 ± 1.650	27.509 ± 1.669
VI	23.273 ± 8.426	12.789 ± 1.683	54.170 ± 1.315

TABLE 8.8 Fractal Dimensions for TNF-α Molecules Using Single-Molecule Nanoparticle Optical Biosensors (SNOBS) to Single-Monoclonal Antibody (mAb) Molecules Binding to Different Single Nanoparticles (Huang et al., 2008)

Particle Number	D_f	D_{f1}	D_{f2}
I	2.8182 ± 0.1050	NA	NA
II	2.5272 ± 0.1714	2.1102 ± 0.5246	$3.0 - 0.5186$
III	2.3376 ± 0.4530	$0.5472 + 1.5510$	2.9584 ± 0.03164
IV	2.9758 ± 0.03986	NA	NA
V	2.004 ± 0.1796	2.1280 ± 0.3670	2.7874 ± 0.07335
VI	2.6264 ± 0.1454	2.0316 ± 0.1236	2.9813 ± 0.0219

Figure 8.13(b) shows the binding of TNF-α in solution to SNOBS using an mAb molecule immobilized to the surface. This shows binding to particle II. In this case, a dual-fractal analysis is required to adequately describe the binding kinetics. The values of (a) the binding rate coefficient, k, and the fractal dimension, D_f, for a single-fractal analysis, (b) and the binding rate coefficients, k_1 and k_2, and the fractal dimensions, D_{f1} and D_{f2}, for a dual-fractal analysis are given in Tables 8.7 and 8.8. It is of interest to note that as the fractal dimension increases by a factor of 1.42 from a value of D_{f1} equal to 2.1102 to D_{f2} equal to 3, the binding rate coefficient, increases by a factor of 6.06 from a value of k_1 equal to 6.614 to a value k_2 equal to 40.087. An increase in the degree of heterogeneity on the nanoparticle surface (SNOBS biosensor) leads to an increase in the binding rate coefficient.

Figure 8.13(c) shows the binding of TNF-α in solution to SNOBS using an mAb molecule immobilized to the surface. This shows binding to particle III. In this case, once again, a dual-fractal analysis is required to adequately describe the binding kinetics. The values of (a) the binding rate coefficient, k, and the fractal dimension, D_f, for a single-fractal analysis, (b) and the binding rate coefficients, k_1 and k_2, and the fractal dimensions, D_{f1} and D_{f2}, for a dual-fractal analysis are given in Tables 8.7 and 8.8. It is of interest to note that as the fractal dimension increases by a factor of 5.41 from a value of D_{f1} equal to 0.5472 to D_{f2} equal to 2.9584, the binding rate coefficient, increases by a factor of 90.05 from a value of k_1 equal to 0.3941 to a value of k_2 equal to 35.479. An increase in the degree of heterogeneity on the nanoparticle surface (SNOBS biosensor) leads to an increase in the binding rate coefficient.

Figure 8.13(d) shows the binding of TNF-α in solution to SNOBS using an mAb molecule immobilized to the surface. This shows binding to particle IV. A single-fractal analysis is required to adequately describe the binding kinetics. The values of the binding rate coefficient, k, and the fractal dimension, D_f, for a single-fractal analysis are given in Tables 8.7 and 8.8.

Figure 8.13(e) shows the binding of TNF-α in solution to SNOBS using an mAB molecule immobilized to the surface. This shows binding to particle V. In this case, once again, a dual-fractal analysis is required to adequately describe the binding kinetics. The values of (a) the binding rate coefficient, k, and the fractal dimension, D_f, for a single-fractal analysis, (b) and the binding rate coefficients, k_1 and k_2, and the fractal dimensions, D_{f1} and D_{f2}, for a dual-fractal analysis are given in Tables 8.7 and 8.8. It is of interest to note that as the fractal dimension increases by a factor of 1.31 from a value of D_{f1} equal to 2.1280 to D_{f2} equal to 2.7874, the binding rate coefficient increases by a factor of 7.15 from a value of k_1 equal to 3.845 to a value of k_2 equal to 27.509. An increase in the degree of heterogeneity on the nanoparticle surface (SNOBS biosensor) once again leads to an increase in the binding rate coefficient.

Figure 8.13(f) shows the binding of TNF-α in solution to SNOBS using an mAb molecule immobilized to the surface. This shows binding to particle VI. In this case, once again, a dual-fractal analysis is required to adequately describe the binding kinetics. The values of (a) the binding rate coefficient, k, and the fractal dimension, D_f, for a single-fractal analysis, (b) and the binding rate coefficients, k_1 and k_2, and the fractal dimensions, D_{f1} and D_{f2}, for a dual-fractal analysis are given in Tables 8.7 and 8.8. It is of interest to note that as the fractal dimension increases by a factor of 1.467 from a value of D_{f1} equal to 2.0316 to equal to 2.9813, the binding rate coefficient increases by a factor of 4.24 from a value of k_1 equal to 12.789 to a value of k_2 equal to 54.170. An increase in the degree of heterogeneity on the nanoparticle surface (SNOBS biosensor) once again leads to an increase in the binding rate coefficient.

It is of interest to note that the binding of TNF-α in solution to the six different nanoparticles is quite different. The binding to particles I and IV exhibit single-fractal analysis, whereas the binding to particles II III, V, and VI exhibit a dual-fractal analysis. For the single-fractal analysis (particles I and IV), there is a difference of the binding rate coefficient, k value by slightly more than a factor of two. For the case when a dual-fractal analysis is required to adequately describe the binding kinetics (particles II, III, V, and VI) the "initial" k_1 values exhibited differ by more than an order of magnitude (0.3941; 12.789), whereas the "saturation" k_2 values differ by about a factor of two. Perhaps, the stochastic nature of the binding process, as mentioned by the original authors (Huang et al., 2008), is partially or completely responsible for this. No other explanation is offered at present to account for this behavior.

Figure 8.14(a) shows the increase in the binding rate coefficient, k_1, for a dual-fractal analysis with an increase in fractal dimension, D_{f1}. For the data shown in Figure 8.14(a), the binding rate coefficient, k_1, is given by:

$$k_1 = (1.44 \pm 1.324)D_{f1}^{2.111 \pm 0.562} \quad (8.9a)$$

There is scatter in the data. Only four data points are available. The availability of more data points would lead to a more reliable fit. The binding rate

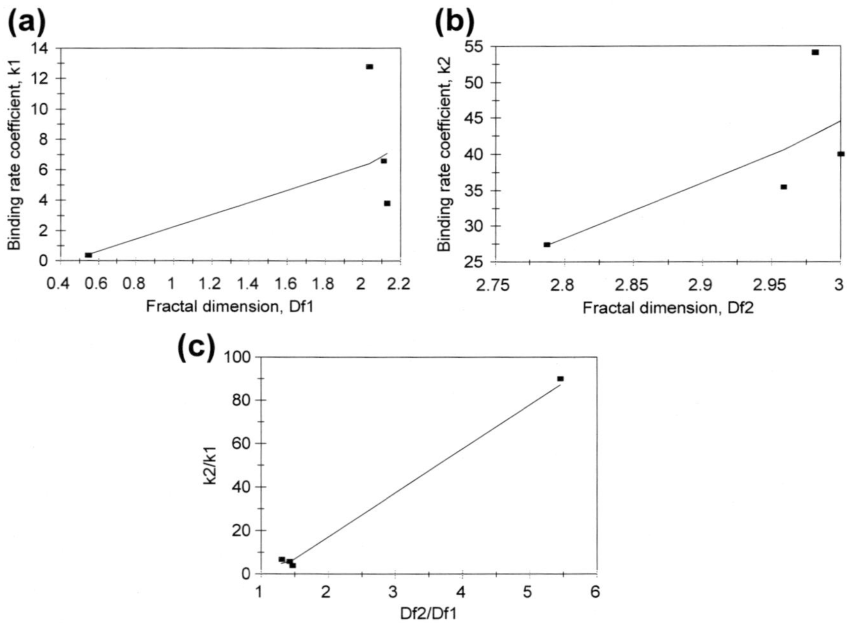

FIGURE 8.14 (a) Increase in the binding rate coefficient, k_1, with an increase in the fractal dimension, D_{f1}. (b) Increase in the binding rate coefficient, k_2, with an increase in the fractal dimension, D_{f2}. (c) Increase in the binding rate coefficient ratio, k_2/k_1, with an increase in the fractal dimension ratio, D_{f2}/D_{f1}.

coefficient, k_1, exhibits slightly more than a second (equal to 2.111) order of dependence on the fractal dimension, D_{f1}, or the degree of heterogeneity that exists on the nanoparticle surface. This indicates that the binding rate coefficient, k_1, is sensitive to the degree of heterogeneity that exists on the SNOBS surface.

Figure 8.14(b) shows the increase in the binding rate coefficient, k_2, for a dual-fractal analysis with an increase in fractal dimension, D_{f2}. For the data shown in Figure 8.14(b), the binding rate coefficient, k_2, is given by:

$$k_2 = (0.0302 \pm 0.0051) D_{f2}^{6.641 \pm 3.519} \tag{8.9b}$$

There is scatter in the data. Only four data points are available. The availability of more data points would lead to a more reliable fit. The binding rate coefficient, k_2, exhibits more than a six and one-half (equal to 6.641) order of dependence on the fractal dimension, D_{f1}, or the degree of heterogeneity that exists on the nanoparticle surface. This indicates that the binding rate coefficient, k_2, is extremely sensitive to the degree of heterogeneity that exists on the SNOBS surface.

Figure 8.14(c) shows the increase in the binding rate coefficient ratio, k_2/k_1, for a dual-fractal analysis with an increase in ratio of fractal dimensions, D_{f2}/D_{fl}. For the data shown in Figure 8.14(c), the ratio of binding rate coefficients, k_2/k_1, is given by:

$$k_2/k_1 = \left(0.2.944 \pm 1.357\right) (D_{f1}/D_{f2})^{1.997 \pm 0.321} \tag{8.9c}$$

The fit is good. Only four data points are available. The availability of more data points would lead to a more reliable fit. The ratio of binding rate coefficients, k_2/k_1, exhibits close to a second (equal to 1.997) order of dependence on the ratio of fractal dimensions, D_{f2}/D_{f1}. This indicates that the binding rate coefficient ratio, k_2/k_1, is sensitive to the degrees of heterogeneity that exists on the SNOBS surface.

LeBlanc et al. (1998) have studied the recognition properties of a sequence-specific DNA-binding antigen. These authors generated a sequence-specific DNA-binding antibody by incorporating a 17-amino acid α-helix from the DNA-binding domain of the transcription factor TFEB into the heavy chain complementary determining region (HCDR3) site of a recombinant human Fab fragment. The authors state that the recombinant DNA-binding antibody, called Fab-E box, binds the TFEB recognition sequence CACGTG with a five—tenfold lower affinity than TFEB. Their results, according to the authors demonstrate the identification of important DNA specificity determinants for high affinity by Fab-box.

LeBlanc et al. (1998) indicate that antibodies that bind with high affinity to DNA are identified as an important component of the autoimmune syndrome of SLE (Deutscher and Keene, 1988; Stollar and Schwartz, 1985). Braun et al. (1986), Braun and Lee (1987), Herron et al. (1991), and Mol et al. (1994) have indicated that autoantibodies that bind dsDNA primarily recognize the helical patterns of the phosphate group with ionic interactions through basic amino acids to their antigen-binding sites. The authors have used phage display to select DNA-binding Fab fragments in vitro from a library which was constructed by a random mutagenesis of the heavy-chain CDR3 (Barbas et al., 1994). The authors designed a new class of DNA-binding antibody by cloning a 17-amino acid DNA recognition (α-helix from a well-characterized eukaryotic transcription factor, TFEB into the HCDCR3 domain of a human fragment. This the authors (LeBlanc et al., 1998) claim recognizes the E-box site CACGTG via interactions in the major groove of the dsDNA (Mclane et al., 1995). This protein the authors call the Fab-E box.

In their study, LeBlanc et al. (1998) have presented evidence that the DNA-binding antibody Fab-E box recognizes DNA with both high affinity and specificity. The authors claim that affinity measurements of the Fab-E box binding to a subset of these variants' binding sites demonstrated that the antibody could specifically recognize DNA sequences with an affinity comparable to naturally occurring DNA-binding sites. The authors determined the binding rate coefficients by measuring the rate of binding of the Fab-E box and the TFEB to the sensor chip surface of a Biacore biosensor at different concentrations.

Figure 8.15(a) shows the binding and dissociation of 400 nM TFEB in solution to E-box DNA immobilized on a surface plasmon resonance (SPR)

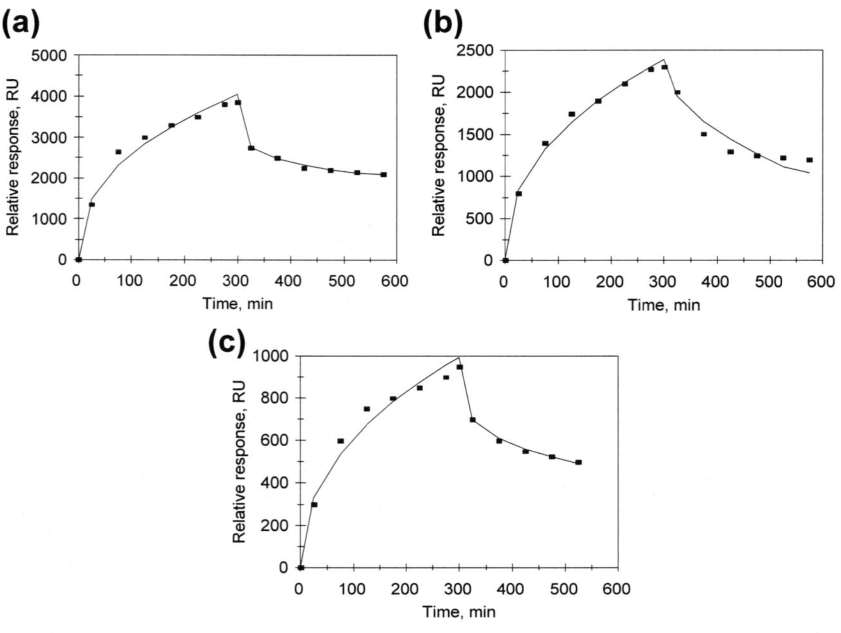

FIGURE 8.15 (a) Binding and dissociation kinetics of TFEB from E-Box DNA using an SPR biosensor (LeBlanc et al., 1998): (a) 400 nM TFEB; (b) 200 nM TFEB; and (c) 100 nM TFEB.

biosensor (LeBlanc et al., 1998). A single-fractal analysis is required to adequately describe the binding and dissociation kinetics. The values of the binding and the dissociation rate coefficients are given in Table 8.9. In this case the affinity, K $(= k/k_d)$ value is 0.707.

Figure 8.15(b) shows the binding and dissociation of 200 nM TFEB in solution to E-box DNA immobilized on an SPR biosensor (LeBlanc et al., 1998). A single-fractal analysis is required to adequately describe the binding and dissociation kinetics. The values of the binding and the dissociation rate coefficients are given in Table 8.9. In this case the affinity, K $(= k/k_d)$ value is 3.68.

Figure 8.15(c) shows the binding and dissociation of 100 nM TFEB in solution to E-box DNA immobilized on an SPR biosensor (LeBlanc et al., 1998). A single-fractal analysis is required to adequately describe the binding and dissociation kinetics. The values of the binding and the dissociation rate coefficients are given in Table 8.9. In this case the affinity, K $(= k/k_d)$ value is 0.740.

Figure 8.16(a) shows the increase in the binding rate coefficient, k, with an increase in the TFEB concentration in solution. For the data shown in Figure 8.16(a), the binding rate coefficient, k, is given by:

$$k = (0.3804 \pm 0.0629)\,[\text{TFEB}]^{\,1.178 \pm 0.1561} \tag{8.10a}$$

The fit is very good. Only three data points are available. The availability of more data points would lead to a better fit. The binding rate coefficient, k, exhibits slightly more than a first (equal to 1.178) order of dependence on the TFEB concentration in solution. This shows that the binding rate coefficient, k, is sensitive to the TFEB concentration in solution.

Figure 8.16(b) shows the increase in the dissociation rate coefficient, k_d, with an increase in the TFEB concentration in solution. For the data shown in Figure 8.16(b), the dissociation rate coefficient, k_d, is given by:

$$k_d = (0.2506 \pm 0.5731)\,[\text{TFEB}]^{\,1.206 \pm 1.1987} \tag{8.10b}$$

The fit is reasonable. Only three data points are available. The availability of more data points would lead to a better fit. The dissociation rate coefficient, k_d, exhibits slightly more than a first (equal to 1.206) order of dependence on the TFEB concentration in solution. This shows that the dissociation rate coefficient, k_d, is sensitive to the TFEB concentration in solution.

Figure 8.16(c) shows the increase in the fractal dimension, D_f, with an increase in the TFEB concentration in solution. For the data shown in Figure 8.16(c), the fractal dimension, D_f, is given by:

$$D_f = (1.871 \pm 0.004)\,[\text{TFEB}]^{\,0.0264 \pm 0.00215} \tag{8.10c}$$

The fit is very good. Only three data points are available. The availability of more data points would lead to a better fit. The fractal dimension, D_f, exhibits a very slight (equal to 0.0264) order of dependence on the TFEB

TABLE 8.9 Binding and Dissociation Rate Coefficients and Fractal Dimensions for TFEB in Solution to E-Box DNA Immobilized on an SPR Biosensor (LeBlanc et al., 1998)

Analyte in Solution/ Receptor on Surface	k	k_d	D_f	D_{fd}
400 nM TFEB/E-box DNA	570.04 ± 15.12	570.04 ± 15.12	2.1906 ± 0.07718	2.5906 ± 0.0270
200 nM TFEB/E-box DNA	215.54 ± 9.68	58.573 ± 8.950	2.1564 ± 0.0400	1.8900 ± 0.01706
100 nM TFEB/E-box DNA	79.258 ± 9.388	107.04 ± 2.63	2.1118 ± 0.08252	2.4626 ± 0.02794

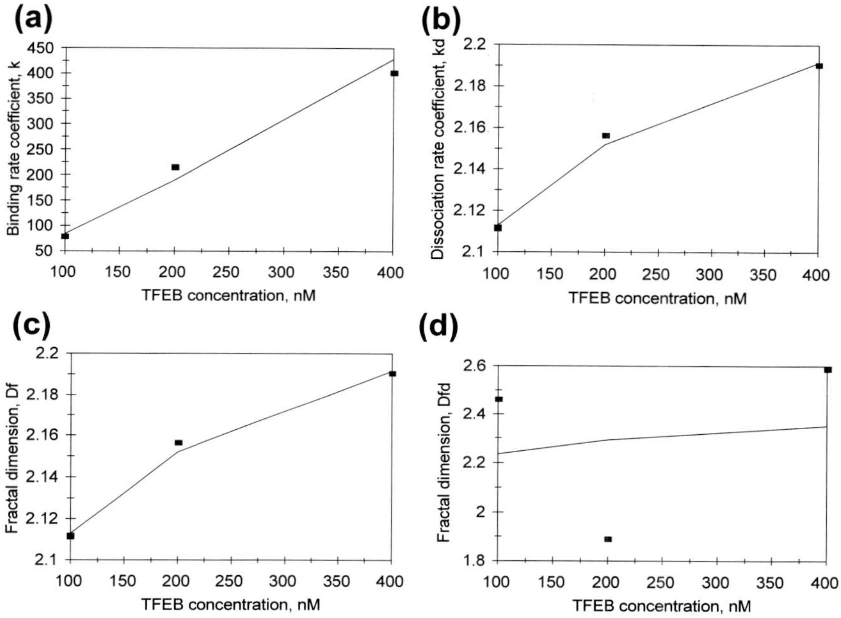

FIGURE 8.16 (a) Increase in the binding rate coefficient, k, with an increase in the TFEB concentration in solution. (b) Increase in the dissociation rate coefficient, k_d, with an increase in TFEB concentration in solution. (c) Increase in the fractal dimension, D_f, with an increase in the TFEB concentration in solution. (d) Increase in the fractal dimension for dissociation, D_{fd}, with an increase in the TFEB concentration in solution. (e) Increase in the binding rate coefficient, k, with an increase in the fractal dimension, D_f. (f) Increase in the dissociation rate coefficient, k_d, with an increase in the fractal dimension for dissociation, D_{fd}. (g) Increase in the affinity, K (= k/k_d), with an increase in the fractal dimension ratio, D_f/D_{fd}.

concentration in solution. This shows that the fractal dimension is only mildly sensitive to the TFEB concentration in solution.

Figure 8.16(d) shows the increase in the fractal dimension for dissociation, D_{fd}, with an increase in the TFEB concentration in solution. For the data shown in Figure 8.16(d), the fractal dimension for dissociation, D_{fd}, is given by:

$$D_{fd} = (1.889 \pm 0.505)\,[\text{TFEB}]^{0.0365+0.241} \qquad (8.10\text{d})$$

There is scatter in the data. The fit is very good. Only three data points are available. The availability of more data points would help define a better fit and direction. The fractal dimension for dissociation, D_{fd}, exhibits a very slight (equal to 0.0365) order of dependence on the TFEB concentration in solution. This shows that the fractal dimension for dissociation, D_{fd}, is only mildly sensitive to the TFEB concentration in solution.

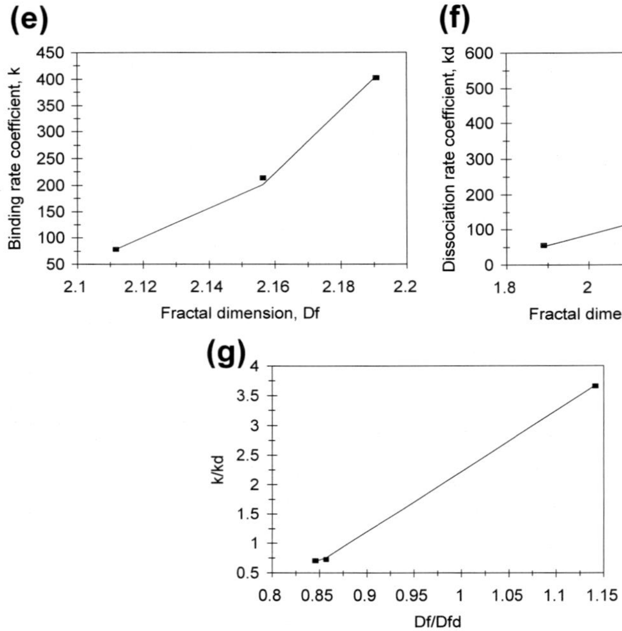

FIGURE 8.16 cont'd

Figure 8.16(e) shows the increase in the binding rate coefficient, k, with an increase in the fractal dimension, D_f. For the data shown in Figure 8.16(e), the binding rate coefficient, k, is given by:

$$k = (2.7E - 13 + 2.9E - 13) D_f^{44.569 \pm 2.280} \tag{8.10e}$$

The fit is very good. Only three data points are available. The availability of more data points would lead to a better fit. The binding rate coefficient, k, exhibits an extreme (equal to 44.569) order of dependence on the fractal dimension, D_f, or the degree of heterogeneity that exists on the surface. This shows that the binding rate coefficient, k, is extremely sensitive to the state of disorder that exists on the surface.

Figure 8.16(f) shows the increase in the dissociation rate coefficient, k_d, with an increase in the fractal dimension, D_{fd}. For the data shown in Figure 8.16(f), the dissociation rate coefficient, k_d, is given by:

$$k_d = (1.476 + 2.512) D_{fd}^{5.590 \pm 4.148} \tag{8.10f}$$

There is scatter in the data. Only three data points are available. The availability of more data points would lead to a better fit. The dissociation rate coefficient, k_d, exhibits a very high (equal to 5.590) order of dependence on the fractal dimension for dissociation, D_{fd}, or the degree of heterogeneity that exists on the surface. This shows that the dissociation rate coefficient, k_d, is very sensitive to the state of disorder that exists on the surface in the dissociation phase.

Figure 8.16(g) shows the increase in the affinity, K (= k/k_d) with an increase in the fractal dimension ratio, D_f/D_{fd}. For the data shown in Figure 8.16(g), the affinity, K is given by:

$$K(=k/k_d) = (1.765 \pm 0.137)\,(D_f/D_{fd})^{5.541 \pm 0.0845} \quad (8.10g)$$

The fit is very good. Only three data points are available. The availability of more data points would lead to a better fit. The affinity, K exhibits a very high (equal to 5.541) order of dependence on the ratio of fractal dimensions, D_f/D_{fd}, or the ratio of the degree of heterogeneity that exists on the surface in the binding and in the dissociation phase. This shows that the affinity, K, is very sensitive to the state of disorder that exists on the surface.

Figure 8.17(a) shows the binding and dissociation of 800 nM Fab-E box in solution to E-box DNA immobilized on an SPR biosensor surface (LeBlanc et al., 1998). The binding and dissociation kinetics may be described by a single-fractal analysis. The values of (a) the binding rate coefficient, k, and the fractal dimension, D_f, for a single-fractal analysis, and (b) the dissociation rate

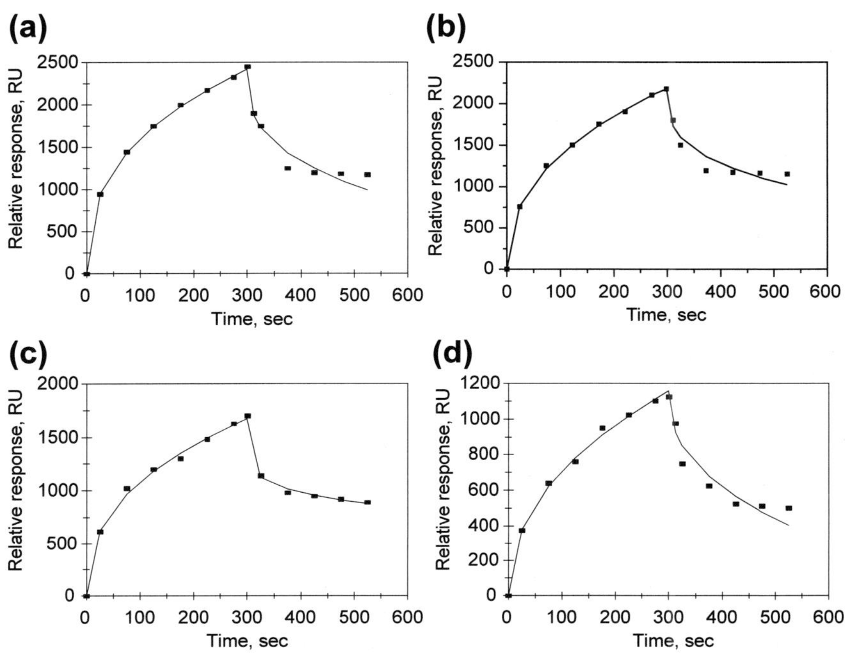

FIGURE 8.17 Binding and dissociation of Fab-E box in solution (in nM) to E-box DNA immobilized on an SPR biosensor surface (LeBlanc et al., 1998): (a) 800 nm; (b) 600 nM; (c) 400 nM; and (d) 200 nM.

TABLE 8.10 Binding Rate Coefficients and Fractal Dimensions for Fab-E Box in Solution to E-Box DNA Immobilized on an SPR Biosensor Surface (LeBlanc et al., 1998)

Analyte in Solution	k	k_d	D_f	D_{fd}
800 nM Fab-E Box/E-box DNA	283.41 ± 2.55	256.61 ± 0.41	2.2462 ± 0.0072	2.3590 ± 0.08486
600 nM Fab-E Box/E-box DNA	195.05 ± 3.40	203.54 ± 38.90	2.1536 ± 0.0160	2.3588 ± 0.1364
400 nM Fab-E Box/E-box DNA	176.53 ± 6.06	337.49 ± 11.18	2.2104 ± 0.03124	2.6708 ± 0.0374
200 nM Fab-E Box/E-box DNA	90.559 ± 2.832	66.643 ± 17.629	2.102 ± 0.0286	2.1688 ± 0.1862

coefficient, k_d, and the fractal dimension for dissociation, D_{fd}, are given in Table 8.10.

Figure 8.17(b) shows the binding and dissociation of 600 nM Fab-E box in solution to E-box DNA immobilized on an SPR biosensor surface (LeBlanc et al, 1998). The binding and dissociation kinetics may be described by a single-fractal analysis. The values of (a) the binding rate coefficient, k, and the fractal dimension, D_f, for a single-fractal analysis, and (b) the dissociation rate coefficient, k_d, and the fractal dimension for dissociation, D_{fd}, are given in Table 8.10.

Figure 8.17(c) shows the binding and dissociation of 400 nM Fab-E box in solution to E-box DNA immobilized on an SPR biosensor surface (LeBlanc et al., 1998). The binding and dissociation kinetics may be described by a single-fractal analysis. The values of (a) the binding rate coefficient, k, and the fractal dimension, D_f, for a single-fractal analysis, and (b) the dissociation rate coefficient, k_d, and the fractal dimension for dissociation, D_{fd}, are given in Table 8.10.

Figure 8.17(d) shows the binding and dissociation of 200 nM Fab-E box in solution to E-box DNA immobilized on an SPR biosensor surface (LeBlanc et al., 1998). The binding and dissociation kinetics may be described by a single-fractal analysis. The values of (a) the binding rate coefficient, k, and the fractal dimension, D_f, for a single-fractal analysis, and (b) the dissociation rate coefficient, k_d, and the fractal dimension for dissociation, D_{fd}, are given in Table 8.10.

Figure 8.18(a) shows the increase in the binding rate coefficient, k, for a single-fractal analysis with an increase in the Fab-E box concentration in solution in nM. For the data shown in Figure 8.18(a), the binding rate coefficient, k, is given by:

$$k = (0.771 \pm 0.416)\,[\mathrm{Fab-E\ box}]^{0.853\pm0.414} \qquad (8.11a)$$

The fit is good. Only four data points are available. The availability of more data points would lead to a more reliable fit. The binding rate coefficient, k, exhibits an order of dependence between one-half and first (equal to 0.853) on the Fab-E box concentration in solution.

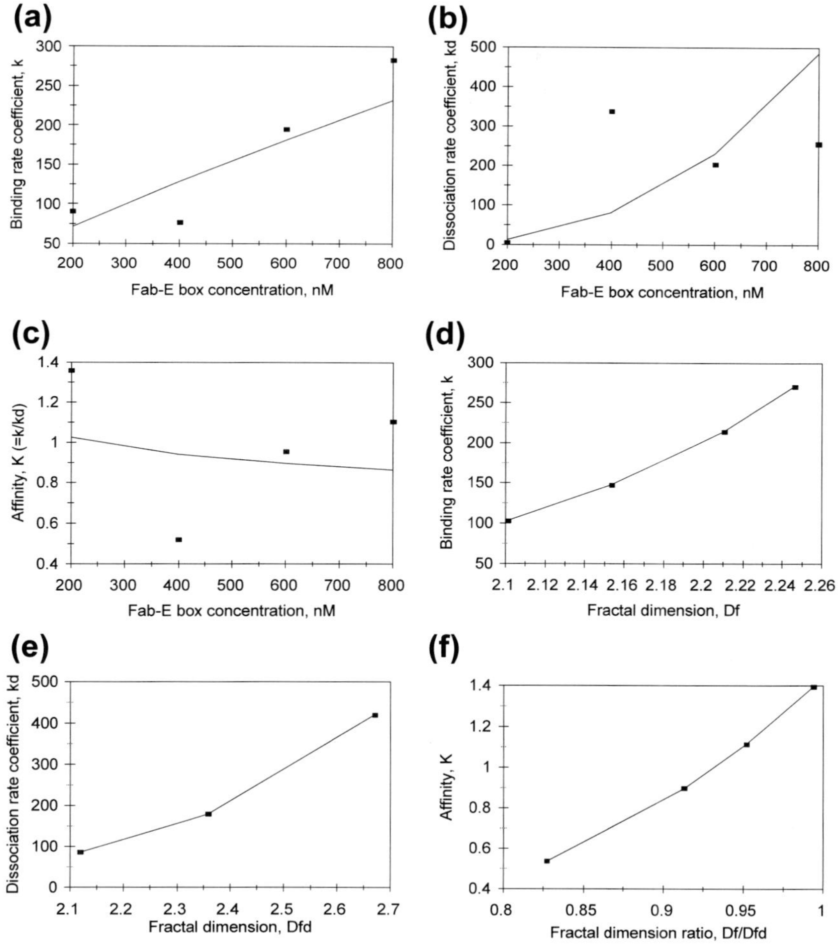

FIGURE 8.18 (a) Increase in the binding rate coefficient, k, with an increase in the Fab-E box concentration in solution in nM. (b) Increase in the dissociation rate coefficient, k_d, with an increase in the Fab-E box concentration in solution in nM. (c) Decrease in the affinity, K (= k/k_d), with an increase in the Fab-E box concentration in solution. (d) Increase in the binding rate coefficient, k, with an increase in the fractal dimension, D_f. (e) Increase in the dissociation rate coefficient, k_d, with an increase in the fractal dimension, D_{fd}. (f) Increase in the affinity, K (= k/k_d), with an increase in the fractal dimension ratio, D_f/D_{fd}. (g) Increase in the fractal dimension, D_f, with an increase in the Fab-E box concentration, nM. (h) Increase in the fractal dimension for dissociation, D_{fd}, with an increase in the Fab-E box concentration, nM. (i) Decrease in the fractal dimension ratio, D_f/D_{fd}, with an increase in the Fab-E box concentration, nM.

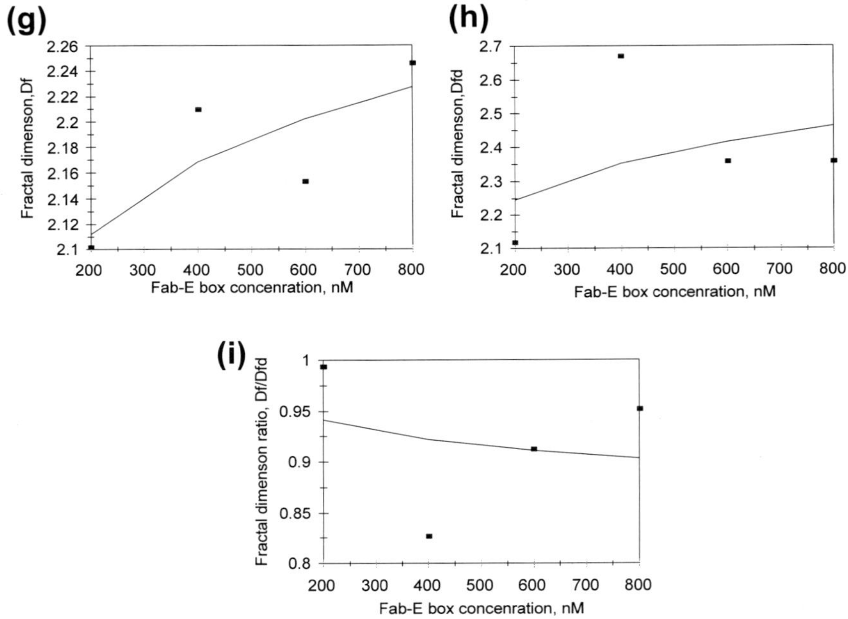

FIGURE 8.18 cont'd

Figure 8.18(b) shows the increase in the dissociation rate coefficient, k_d, for a single-fractal analysis with an increase in the Fab-E box concentration in solution in nM. For the data shown in Figure 8.18(b), the dissociation rate coefficient, k_d, is given by:

$$k_d = (1.5E - 05 + 3.5E - 05)\ [\text{Fab-E box}]^{2.587 \pm 1.168} \quad (8.11b)$$

The fit is reasonable. Only four data points are available. The availability of more data points would lead to a more reliable fit. The dissociation rate coefficient, k_d, exhibits slightly more than two and one-half (equal to 2.587) order on the Fab-E box concentration in solution. This indicates that the dissociation rate coefficient, k_d, is sensitive to the Fab-E box concentration in solution.

Figure 8.18(e) shows the increase in the affinity K (= k/k_d) for a single-fractal analysis with an increase in the Fab-E box concentration in solution in nM. For the data shown in Figure 8.18(c), the affinity, K is given by:

$$K(= k/k_d) = (1.961 \pm 1.254)\ [\text{Fab-E box}]^{-0.1222 \pm 0.475} \quad (8.11c)$$

There is scatter in the data. Only four data points are available. The availability of more data points would lead to a more reliable fit. The affinity, K decreases with the Fab-Box concentration in solution, and exhibits a very mild (equal to −0.1222) negative order of dependence on the Fab-E box concentration in solution.

Figure 8.18(d) shows the increase in the binding rate coefficient, k, for a single-fractal analysis with an increase in the fractal dimension, D_f. For the data shown in Figure 8.18(d), the binding rate coefficient, k, is given by:

$$k = (0.002218 \pm 0.000661)\, D_f^{14.47 \pm 5.132} \quad (8.11d)$$

The fit is very good. Only four data points are available. The availability of more data points would lead to a more reliable fit. The binding rate coefficient, k, exhibits a very strong dependence on the degree of heterogeneity or the fractal dimension that exists on the surface, as it exhibits close to 14 and one-half (equal to 14.47) order of dependence.

Figure 8.18(e) shows the increase in the dissociation rate coefficient, k_d, for a single-fractal analysis with an increase in the fractal dimension, D_{fd}. For the data shown in Figure 8.18(e), the dissociation rate coefficient, k_d, is given by:

$$k_d = (0.5006 \pm 0.2164)\, D_{fd}^{6.856 \pm 2.186} \quad (8.11e)$$

The fit is very good. Only four data points are available. The availability of more data points would lead to a more reliable fit. The dissociation rate coefficient, k_d, exhibits a very strong dependence on the degree of heterogeneity or the fractal dimension that exists on the surface, as it exhibits close to seventh (equal to 6.856) order of dependence.

Figure 8.18(f) shows the increase in the affinity K (= k/k_d) for a single-fractal analysis with an increase in the fractal dimension ratio, D_f/D_{fd}. For the data shown in Figure 8.18(f), the affinity, K is given by:

$$K = (k/k_d) = (1.440 \pm 0.079)(D_f/D_{fd})^{5.187 \pm 0.394} \quad (8.11f)$$

The fit is good. Only four data points are available. The availability of more data points would lead to a more reliable fit. The affinity, K exhibits slightly more than a fifth (equal to 5.187) order of dependence on the fractal dimension ratio, D_f/D_{fd}. This shows the affinity, K is very sensitive to the ratio of the fractal dimensions.

Figure 8.18(g) shows the increase in the fractal dimension, D_f, for a single-fractal analysis with an increase in the Fab-E box concentration in solution in nM. For the data shown in Figure 8.18(g), the fractal dimension, D_f, is given by:

$$D_f = (1.724 \pm 0.038)\, [\text{Fab-E box}]^{0.038 \pm 0.021} \quad (8.11g)$$

There is scatter in the data. Only four data points are available. The availability of more data points would lead to a more reliable fit. The fractal dimension, D_f, exhibits a very mild order (equal to 0.038) of dependence on the Fab-E box concentration in solution.

Figure 8.18(h) shows the increase in the fractal dimension for dissociation, D_{fd}, for a single-fractal analysis with an increase in the Fab-E box concentration in solution in nM. For the data shown in Figure 8.18(h), the fractal dimension, D_{fd}, is given by:

$$D_{fd} = (1.574 \pm 0.200)\ [\text{Fab-E box}]^{0.067 \pm 0.100} \qquad (8.11h)$$

There is scatter in the data. Only four data points are available. The availability of more data points would lead to a more reliable fit. The fractal dimension for dissociation, D_{fd}, exhibits a very mild order (equal to 0.067) of dependence on the Fab-E box concentration in solution.

Figure 8.18(i) shows the decrease in the ratio, D_f/D_{fd}, for a single-fractal analysis with an increase in the Fab-E box concentration in solution in nM. For the data shown in Figure 8.18(i), the fractal dimension ratio, D_f/D_{fd}, is given by:

$$D_f/D_{fd} = (1.105 \pm 0.108)\ [\text{Fab-E box}]^{-0.0302 \pm 0.0.090} \qquad (8.11i)$$

There is scatter in the data. Only four data points are available. The availability of more data points would lead to a more reliable fit. The affinity, K decreases with the Fab-Box concentration in solution, and exhibits a very mild (equal to −0.0302) negative order of dependence on the Fab-E box concentration in solution.

White et al. (2011) have recently developed a wash-free electrochemical platform for the quantitative multiplexed detection of antibodies. This they claim significantly improves the sensitivity and specificity to help diagnose diseases like SLE. This disease requires four or more biomarkers to be above a certain level for the medical personnel to say with any degree of confidence that a person is suffering from SLE. These authors claim that the detection of multiple specific antibodies plays a significant role in the diagnosis, detection, and monitoring of rheumatic and other autoimmune diseases (Baud et al., 2010; Burbelo et al., 2010; Chan et al., 2008; Gill et al., 2003; Murray et al., 2008; Korponay-Szabo et al., 2007; Parekh et al., 2002; Pfafflin and Schleicher, 2009; Posthuma-Trumpie et al., 2009; Brennan et al., 2010; Tang et al., 2003; Rouquette et al., 2003; Bissonnette and Bergeron, 2010). Chan et al. (2008) point out that current methods for the detection of antibodies such as ELISA are slow and cumbersome, and often require hours to get any sort of useful medial information.

To enable point-of-care antibody detection in recent years, the dip-stick assay method has been developed (Bissonnette and Bergeron, 2010; Yaeger et al., 2008). However, Postuma-Trumple et al. (2009) provide only qualitative and semiquantitative information regarding the targeted antibodies in serum. Thus, White et al. (2011) indicate the need for improved methods to monitor autoimmune status and disease progression (Osterland, 1994; Nakamura et al., 2005) and to discriminate between active and prior infections (Yaeger et al., 2008; Goudsmit et al., 1987).

White et al. (2011) state that to overcome these drawbacks, a number of reagent- and wash-free sensors have been developed and to meet these challenges these authors state the standard techniques such as SPR, QCM (quartz crystal microbalance), field-effect transistor, and microcantilever - and electrochemical impedance spectroscopy-based biosensors are not very efficient for antibody detection due to nonspecific adsorption. White et al. (2011) have developed a quantitative, electrochemical platform for the rapid, wash-free detection of multiple specific antibodies in submillimeter clinical samples.

A FLAG-tag of FLAG octapeptide is a polypeptide protein tag that can be added to a protein using recombinant technology. A FLAG-tag can be used in many different assays that require recognition by an antibody. The peptide sequence of the FLAG-tag is N-DYKDDDDK-C. The anti-FLAG antibody recognizes the FLAG epitope located on the FLAG-tagged fusion proteins at the N-terminus or C-terminus I in different types of assays.

Figure 8.19 shows the binding of the anti-DNA sensor to the anti-FLAG antibody in solution. A single-fractal analysis is adequate to describe the binding kinetics. The values of the binding rate coefficient, k, and the fractal dimension, D_f, are given in Table 8.11.

White et al. (2011) state that the ability of their sensor is particularly useful for the detection of multiple markers in a single sample. This would be of significant advantage for use in the diagnosis of diseases such as SLE, wherein no single antibody test achieves clinical specificity (Gill et al., 2003). At least, four or more biomarkers have to be in a certain range when the medical personnel can say with any degree of certainty that a person has SLE. Furthermore, the authors (White et al., 2011) conclude by indicating that their E-DNA platform exhibits potential for the detection of antibodies at the point of care.

Lim et al. (1999) have recently developed an automated system to detect anti-dsDNA antibody in SLE serum by flow immunoassay. These authors indicate that autoimmune diseases are characterized by the biosynthesis of

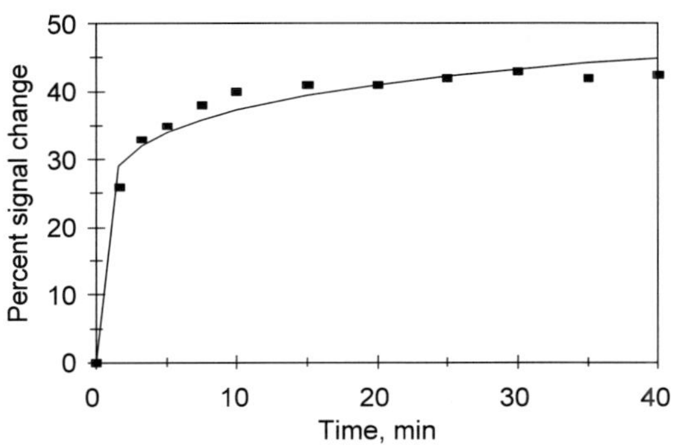

FIGURE 8.19 Binding of anti-DNA sensor to anti-FLAG in solution (White et al., 2011).

TABLE 8.11 Binding Rate Coefficient, k, and the Fractal Dimension, D_f, for the Anti-DNA Sensor Binding to Anti-FLAG Antibody in Solution (White et al., 2011)

Analyte in Solution/Receptor on Surface	k	D_f
Anti-FLAG antibody/anti-DNA sensor	27.333 ± 1.634	2.7302 ± 0.03484

antibodies against one or more components of the patient. In simple language, the antibodies "attack" the patient itself. It is systemic because the lupus affects several organs (Tan et al., 1982; Ravirajan et al., 1996; Koutouzov et al., 2004). Typically, according to Lim et al. (1999) the antibodies generated attack the self-DNA and other nuclear antigens (Termaat et al., 1993; Sanchez-Guerrero et al., 1996). These nuclear antigens may be histones (Chabre et al., 1995; Mitzushiam et al., 1996; Spronk et al., 1995) and soluble nuclear proteins (Gulalis et al., 1994; Williams et al., 1997). Lim et al. (1999) indicate that SLE is characterized by anti-DNA antibodies and an increase in the natural polyreactive antibodies in serum (Adyel et al., 1996), besides an increase in the serum and kidney eluates (Amoura et al., 1994). The diagnosis of SLE mainly relies on the detection of anti-DNA antibodies.

Lim et al. (1999) point out that different techniques have been developed to detect and characterize these antibodies against dsDNA, for example, anti-dsDNA antibodies (Ter Borg et al., 1990; Budhai et al., 1996; Froelich et al., 1990). Tomer et al. (1996) caution, however, that antibodies produced against DNA are not only associated with SLE but other diseases too. Lim et al. (1999) indicate that several methods have been proposed using liquid chromatography to detect anti-dsDNA antibodies (Kubota et al., 1985; Lafer et al., 1983; Miller and Herman, 1996). Lim et al. (1999) point out that these are difficult to automate. Thus, the need to develop a simple and reliable method for a specific assay for anti-dsDNA antibodies.

Lim et al. (1999) report on the diagnostic performance of their immunoassay from patients suffering from SLE. Their novel approach does not require prior immobilization of antigen or antibody to a solid phase. Figure 8.20 shows the anti-dsDNA antibody levels determined by their flow immunoassay method in relation to disease activity in a patient with SLE. The patient was being treated with prednisolone (a steroid) and exhibited a biphasic activity. On increasing the dosage of the drug, the serum revealed a decrease in the antibody levels. The patient recovered gradually. However, later the patient revealed (about 6 months later) an increase in the levels of anti-dsDNA levels in the serum. Thus, the biphasic nature was exhibited. Table 8.12 shows the binding and the dissociation rate coefficients along with the corresponding fractal dimensions exhibited during these phases.

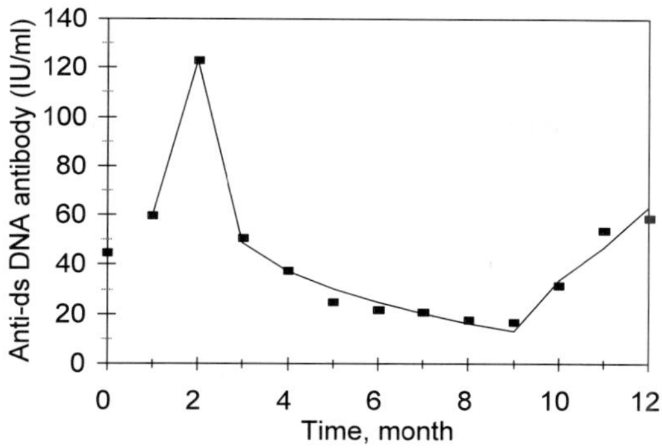

FIGURE 8.20 Binding of anti-dsDNA levels determined by flow immunoassay in relation to disease activity in patients with SLE (Lim et al., 1999).

TABLE 8.12 Binding and Dissociation Rate Coefficients and the Corresponding Fractal Dimensions during the Biphasic Nature of Anti-dsDNA Exhibited by a Patient with SLE (Lim et al., 1999)

Phase	k	D_f	k_d	D_{fd}
First phase	60 ± 0	0.9287 ± 0	74.144 ± 2.716	2.5966 ± 0.8429
Second phase	0.01392 ± 0.0252	$0 + 2.574$	NA	NA

It may be noted that the binding rate coefficient for the second phase is considerably less than that of the first phase. In other words, the anti-dsDNA levels shoot up much faster in the first phase, whereas there is a much slower rise of anti-dsDNA levels in the serum in the second. No more data is available. Thus, any significant trends and conclusions are not possible here.

Recently, Ali et al. (2005) have analyzed hydrophobic transmembrane peptide-linked conjugations to enhance membrane binding and functional activity in T cells. These authors state that T-cell receptor (TCR) serves as a critical function in the immune system. This is also one of most complex receptor structures in biology. By blocking the TCR and thereby inhibiting the T cells provides for a novel method of treating autoimmune diseases such as RA, psoriasis, and SLE at an early stage in the progression of the disease. According to Ali et al. (2005) this provides for a whole new avenue to treat these types of diseases. Manolios et al. (1990, 1991) indicate that the hydrophobic transmembrane peptide which acts as a core peptide (CP) contains two positively charged amino acids, which are critical for receptor assembly. CP significantly reduced the induction of T-cell mediated inflammation in animal models with adjuvant-induced arthritis, allergic encephalomyelitis, and delayed contact hypersensitivity (Manolios et al., 1997, 2002). Gollner et al. (2000) have noted that CP has been tested in humans too, and it was seen that CP was effective in the treatment of skin conditions. Mahnke et al. (2003) have shown that genetically engineered cells secrete this peptide, and subsequently reduce inflammation. However, the therapeutic effect is antigen specific.

Ali et al. (2005) C-terminally attached CP to the amino group of the Tris molecule, and the hydroxyl group of the Tris was attached to long-chain fatty acyl molecules in order to produce lipopeptides. This increases the bioavailability of CP by conjugating lipids to CP (Whittaker et al., 1993). Tris technology has also been used by Wells et al. (2000) to prepare fatty acid conjugates of the analgesic morphine and nonsteroidal anti-inflammatory drugs like indomethacin and the antivirial drug, AZT (3′-azido-3′-deoxythymidine). The authors (Ali et al., 2005) studied the effect of lipopeptides on the modulation of T-cell function in vitro by measuring the 1L-2 production. These authors noted that the biological activity of the lipopeptides increased when it was compared to the parent peptide. These authors reported on the biological properties in their work.

Figure 8.21(a) shows the binding of XP1 to DMPC.DMPG (50:50) liposomes immobilized on an LI sensor chip (Ali et al., 2005). A dual-fractal analysis is required to adequately describe the binding kinetics. The values of (a) the binding rate coefficient, k, and the fractal dimension, D_f, for a single-fractal analysis, and (b) the binding rate coefficients, k_1 and k_2, and the fractal dimensions, D_{f1} and D_{f2}, for a dual-fractal analysis are given in Tables 8.13 and 8.14. It is of interest to note that as the fractal dimension increases by a factor of 1.405 from a value of D_{f1} equal to 1.9542 to D_{f2} equal to 2.7464, the binding rate coefficient increases by a factor of 8.94 from a value of k_1 equal to 275.12 to k_2 equal to 2460.36. An increase in the degree of heterogeneity on the biosensor (LI sensor chip) surface leads to an increase in the binding rate coefficient.

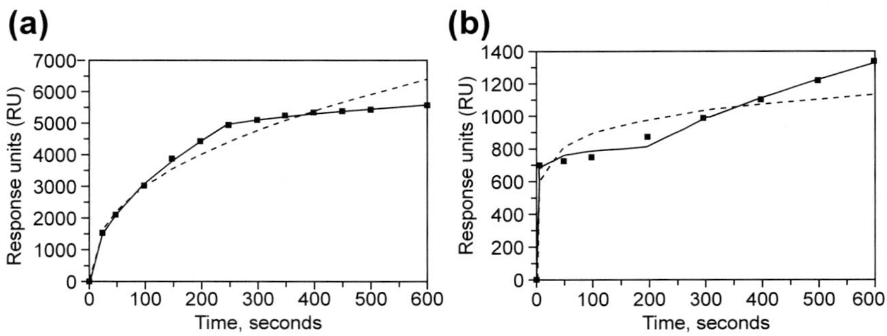

FIGURE 8.21 Binding of (a) XP1 to DMPC.DMPG (50:50) liposome immobilized on an L1 sensor chip and (b) XP2 to DMPC.DMPG (50:50) liposome immobilized on an L1 sensor chip (Ali et al., 2005).

TABLE 8.13 Binding Rate Coefficients for XP1 and XP2 to DMFC.DMPG (50:50) Liposomes Immobilized on an LI Sensor Chip (Ali et al., 2005)

Analyte in Solution/Receptor on Surface	k	k_1	k_2
XP1/DMPC.DMPG (50:50) Liposome on an LI sensor chip	414.36 ± 38.41	275.12 ± 5.17	2460.36 ± 19.14
XP2/DMPC.DMPG (50:50) Liposome on an LI sensor chip	486.09 ± 74.51	631.05	81.76 ± 1.03

TABLE 8.14 Fractal Dimensions for XP1 and XP2 to DMPC.DMPG (50:50) Liposomes Immobilized on an LI Sensor Chip (Ali et al., 2005)

Analyte in Solution/ Receptor on Surface	D_f	D_{f1}	D_{f2}
XP1/DMPC.DMPG (50:50) Liposome on an LI sensor chip	2.1464 ± 0.05468	1.9542 ± 0.0221	2.7404 ± 0.0210
XP2/DMPC.DMPG (50:50) Liposome on an LI sensor chip	1.7326 ± 0.1418	2.1282 ± 0.04850	2.9045 ± 00210

Figure 8.21(b) shows the binding of XP2 to DMPC.DMPG liposomes (70:30) immobilized on an LI sensor chip (Ali et al., 2005). A dual-fractal analysis is, once again, required to adequately describe the binding kinetics. The values of (a) the binding rate coefficient, k, and the fractal dimension, D_f, for a single-fractal analysis, and (b) the binding rate coefficients, k_1 and k_2, and the fractal dimensions, D_{f1} and D_{f2}, for a dual-fractal analysis are given in Tables 8.13 and 8.14. It is of interest to note that as the fractal dimension decreases by a factor of 0.732 from a value of D_{f1} equal to 2.9045 to D_{f2} equal to 2.1282, the binding rate coefficient increases by a factor of 0.129 from a value of 631.05 to a value of k_2 equal to 81.72. This time a decrease in the degree of heterogeneity on the LI sensor chip surface leads to a decrease in the binding rate coefficient.

Figure 8.22(a) shows the binding of LP3 (70:30) liposome immobilized on an LI sensor chip (Ali et al., 2005). A dual-fractal analysis is, once again, required to adequately describe the binding kinetics. The values of (a) the binding rate coefficient, k, and the fractal dimension, D_f, for a single-fractal analysis, and (b) the binding rate coefficients, k_1 and k_2, and the fractal dimensions, D_{f1} and D_{f2}, for a dual-fractal analysis are given in Table 8.15 and 8.16. It is of interest to note that as the fractal dimension decreases by a factor of 0.757 from a value of D_{fl} equal to 2.8338 to D_{f2} equal to 2.1450, the

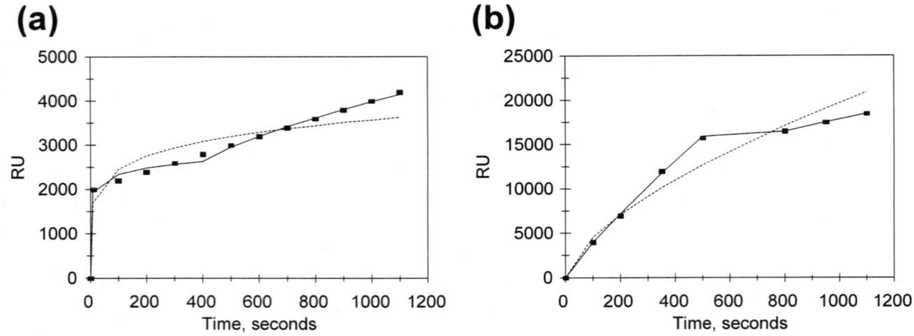

FIGURE 8.22 Binding of (a) XP1 to DMPC.DMPG (70:30) liposome immobilized on an L1 sensor chip and (b) XP2 to DMPC:DMPG (70:30) liposome immobilized on an L1 sensor chip (Ali et al., 2005).

TABLE 8.15 Binding Rate Coefficients for XP1 and XP2 to DMPC.DMPG (50:50) Liposomes Immobilized on an LI Sensor Chip (Ali et al., 2005)

Analyte in Solution/Receptor on Surface	k	k_1	k_2
XP1/DMPC.DMPG (70:30) Liposome on an LI sensor chip	1167.15 ± 141.94	1602.07 ± 96.61	208.28 ± 2.80
XP2/DMPC.DMPG (70:30) Liposome on an LI sensor chip	247.00 ± 40.93	74.25 ± 2.23	1497.97 ± 3.56

TABLE 8.16 Fractal Dimensions for XP1 and XP2 to DMPC.DMPG (70:30) Liposomes Immobilized on an LI Sensor Chip (Ali et al., 2005)

Analyte in Solution/ Receptor on Surface	D_f	D_{f1}	D_{f2}
XP1/DMPC.DMPG (70:30) Liposome on an LI sensor chip	2.6762 ± 0.5170	2.8338 ± 0.03956	2.1450 ± 0.02492
XP2/DMPC.DMPG (70:30) Liposome on an LI sensor chip	1.7326 ± 0.1418	1.2732 ± 0.04850	2.2824 ± 0.0210

binding rate coefficient decreases by a factor of 0.13 from a value of 1602.07 to a value of k_2 equal to 208.28. This time a decrease in the degree of heterogeneity on the LI sensor chip surface leads to a decrease in the binding rate coefficient.

Figure 8.22(b) shows the binding of LP1 (50:50) liposome immobilized on an LI sensor chip (Ali et al., 2005). A dual-fractal analysis is, once again, required to adequately describe the binding kinetics. The values of (a) the binding rate coefficient, k, and the fractal dimension, D_f, for a single-fractal analysis, and (b) the binding rate coefficients, k_1 and k_2, and the fractal dimensions, D_{f1} and D_{f2}, for a dual-fractal analysis are given in Tables 8.15 and 8.16. It is of interest to note that as the fractal dimension increases by a factor of 1.79 from a value of D_{f1} equal to 1.2732 to D_{f2} equal to 2.2826, the binding rate coefficient increases by a factor of 20.17 from a value of 74.25 to a value of k_2 equal to 1497.97. This time an increase in the degree of heterogeneity on the LI sensor chip surface leads to an increase in the binding rate coefficient.

Figure 8.23(a) shows the increase in the binding rate coefficient, k_1, with an increase in the fractal dimension, D_{f1}. For the data shown in Figure 8.23(a), the binding rate coefficient, k_1, is given by:

$$k_1 = (34.04 \pm 22.38)D_{f1}^{3.194 \pm 0.753} \quad (8.12a)$$

There is some scatter in the data. Only four data points are available. The availability of more data would lead to a more reliable fit. The binding rate coefficient, k_1, is very sensitive to the degree of heterogeneity, D_{f1}, on the surface as it exhibits slightly more than a third (equal to 3.194) order of dependence on the degree of heterogeneity that exists on the LI sensor chip.

Figure 8.23(b) shows the increase in the binding rate coefficient, k_2, with an increase in the fractal dimension, D_{f2}. For the data shown in Figure 8.23(b), the binding rate coefficient, k_2, is given by:

$$k_2 = (0.064 \pm 0.050)D_{f2}^{10.05 \pm 2.171} \quad (8.12b)$$

The fit is good. Only four data points are available. The availability of more data would lead to a more reliable fit. The binding rate coefficient, k_2, is

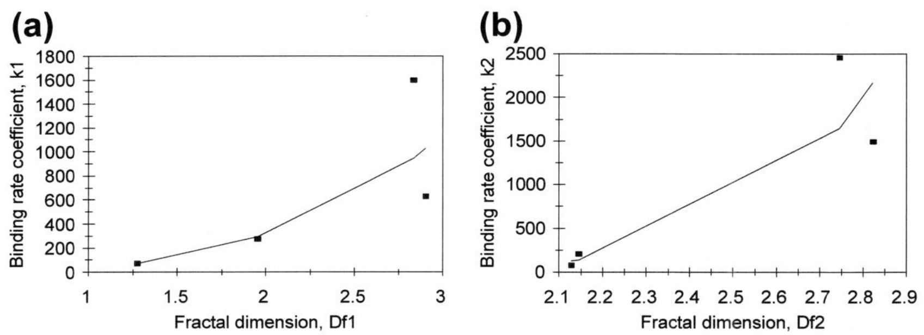

FIGURE 8.23 (a) Increase in the binding rate coefficient, k_1, with an increase in the fractal dimension, D_{f1}. (b) Increase in the binding rate coefficient, k_2, with an increase in the fractal dimension, D_{f2}.

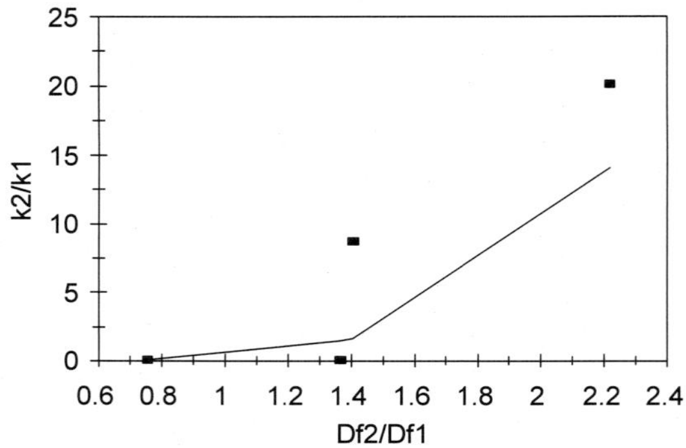

FIGURE 8.24 Increase in the binding rate coefficient ratio, k_2/k_1, with an increase in the fractal dimension ratio, D_{f2}/D_{f1}.

extremely sensitive to the degree of heterogeneity on the surface as it exhibits slightly more than a tenth (equal to 10.05) order of dependence on the degree of heterogeneity that exists on the LI sensor chip.

Figure 8.24 shows the increase in the ratio of the binding rate coefficients, k_2/k_1, with an increase in the ratio of the fractal dimensions, D_{f2}/D_{f1}. For the data shown in Figure 8.24, the ratio of the binding rate coefficients, k_2/k_1, is given by:

$$k_2/k_1 = (0.331 + 2.726)D_{f2}/D_{f1}^{4.713 \pm 2.762} \tag{8.12c}$$

There is scatter in the data. This is indicated in the estimate of constant. Only the positive is given since the ratio of the binding rate coefficients, k_2/k_1 cannot have a negative value. Only four data points are available. The availability of more data would lead to a more reliable fit. The ratio of the binding rate coefficients, k_2/k_1, is very sensitive to the ratio of fractal dimensions, D_{f2}/D_{f1}, or the degree of heterogeneity on the biosensor surface since it exhibits an order of dependence between four arid one-half and five (equal to 4.713) on D_{f2}/D_{f1}.

Kakuyama et al. (2005) have analyzed the binding of CLAC to aggregated Aβ and Aβ fragments and noted their influence on the attenuation of fibril elongation. They state that Alzheimer's disease (AD) is the most prevalent cause of dementia in elderly people. They indicate that one of the neuropathology features of AD is the occurrence of senile plaques (SPs). These are mainly composed of fibrils formed from the 40–42 residue amyloid peptide (Aβ). They state that Aβ is derived from the β-amyloid precursor protein (APP). This is a transmembrane protein (Hardy and Selkoe, 2002). Sisodia and St. George-Hyslop (2002) indicate that the final step in the production of Aβ is executed by the γ-secretase complex, which is critically dependent on presenilin 1 or 2. AD progresses by the polymerization of Aβ onto oligomers and the larger species is supported by the identification of familial AD mutations into APP and the presenilin genes (Selkoe, 2002; St. George-Hyslop, 2002). Kakuyama et al. (2005) indicate that most of the mutations increase the production of the more fibrillogenic Aβ42 species. This Hardy and Selkoe, (2002) indicate is the initial Aβ species to be deposited on the SPs. Gotz et al. (2004) suggest that data toxicity studies in cell culture also support the involvement of Aβ in AD.

Kakuyama et al. (2005) suggest that it is not clear whether the mechanism of Aβ monomers assemble into oligomers, larger aggregates, and eventually into SPs. According to these authors, perhaps these 'diffuse plaques' represent an early stage of Aβ deposition. Other proteins have also been identified with SPs: glucosamine proteoglycans (Castillo et al., 1997; Snow et al., 1988), complement factor C1q (Webster et al., 1995), apolipoprotein (apoE) (Namba et al., 1991; Gomez-Isla et al., 1997; Raber et al., 2004), and laminin (Palu and Liesi, 2002). Kakuyama et al. (2005) suggest that the proteins mentioned earlier have been known to modulate Aβ aggregation. Also, these authors suggest that others not characterized as yet as proteins may also affect the aggregation process.

One such plaque-associated protein is collagenous Alzheimer's amyloid plaque component (CLAC) (Hashimoto et al., 2002). It was identified during screening by monoclonal antibodies raised against crude amyloid preparations isolated from the human AD brain. Kakuyama et al. (2005) indicate that AMY (also a plaque associated protein) (Schmidt et al., 1999; Lemere et al., 1999) was identical to CLAC (Soderberg et al., 2003). Kakuyama et al. (2005) indicate that CLAC is expressed in neurons and is a collagenous transmembrane group of proteins which consists of eight members (Franzke et al., 2005).

Wu et al. (2006) have recently analyzed the real-time kinetic analysis of the interaction between interleukin (IL)-1-α and soluble interleukin-1 receptor using a resonant mirror biosensor. Stylianou and Saklatvala (1998) indicate that IL-1 is a primary regulator of inflammatory and immune response. It is an important mediator of pathogenesis of many inflammatory and immunologically mediated disease. The family consists of IL-1α and IL-1β-1 receptor antagonist. These forms can bind to the same receptor, IL-1 (Dinarello, 2004). Arend (2003) indicates that there are two types of receptors that can be released from the cell surface in soluble form to become soluble IL-1 receptor (sILR) 1 and sIL-1R. However, this degrades quickly in vivo, with several hours of half-life (Arner et al., 1995).

Wu et al. (2006) indicate that the binding analysis may be performed on an optical biosensor conveniently and rapidly. This permits a real-time kinetic analysis with a lot of time saved and the use of a small amount of reagents (Leopoldo et al., 1997; Dmitriev et al., 2003). These authors indicate that resonant mirror devices are surface-sensitive analytical devices which are capable of online monitoring (Hirmo et al., 1999; Wilson, 2002; Chang and Chu, 2005). Wu et al. (2006) further indicate that the resonant optical biosensor provides quantitative and qualitative kinetic information for biomolecular interaction analysis and has been applied to different types of studies (Buckle et al., 1993; Chen et al., 2004; Huang et al., 2005).

Figure 8.25 shows the binding of different concentrations of IL-1α in solution to immobilized sIL-1R-1 (Wu et al., 2006). Figure 8.25(a) shows the binding of 2400 nM in solution to immobilized sIL-1R. A single-fractal analysis is adequate to describe the binding kinetics. The values of the binding rate coefficient, k, and the fractal dimension, D_f, are given in Table 8.17.

Figure 8.25(b) shows the binding of 1200 nM in solution to immobilized sIL-1R. Once again, a single-fractal analysis is adequate to describe the binding kinetics. The values of the binding rate coefficient, k, and the fractal dimension, D_f, are given in Table 8.17.

Figure 8.25(c) shows the binding of 600 nM in solution to immobilized sIL-1R. Once again, a single-fractal analysis is adequate to describe the binding kinetics. The values of the binding rate coefficient, k, and the fractal dimension, D_f, are given in Table 8.17.

Figure 8.25(d) shows the binding of 300 nM in solution to immobilized sIL-1R. Once again, a single-fractal analysis is adequate to describe the binding kinetics. The values of the binding rate coefficient, k, and the fractal dimension, D_f, are given in Table 8.17.

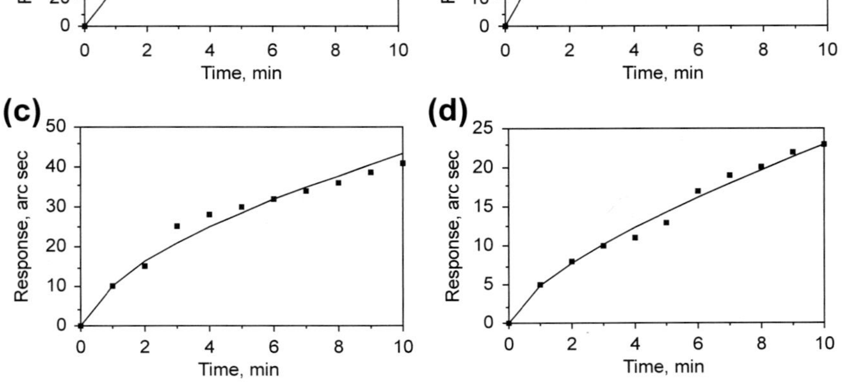

FIGURE 8.25 Binding of different concentrations of IL-1α in solution to immobilized sIL-1R-I (Wu et al., 2006): (a) 2400 nM; (b) 1200 nM; (c) 600 nM; and (d) 300 nM.

TABLE 8.17 Binding Rate Coefficients and Fractal Dimensions for Different Concentrations of IL-Iα in Solution to Immobilized SIL-IR-I Immobilized on a Biosensor (Wu et al., 2006)

Analyte in Solution/ Receptor on Surface	k	D_f
2400 nM IL-Iα/sIL-IR-I	29.898 ± 1.334	1.7918 ± 0.0408
1200 nM IL-Iα/sIL-IR-I	20.332 ± 1.117	1.9960 ± 0.0486
600 nM IL-Iα/sIL-IR-I	12.212 ± 1.645	1.8207 ± 0.0516
300 nM IL-Iα/sIL-IR-I	4.829 ± 0.320	1.6486 ± 0.0572

Figure 8.26(a) shows the increase in the binding rate coefficient, k, with an increase in the IL-1α concentration in solution in nM. For the data shown in Figure 8.26(a), the binding rate coefficient, k, is given by:

$$k = (0.038 \pm 0.0068)\,[IL - 1\alpha]^{0.865 \pm 0.104} \tag{8.13a}$$

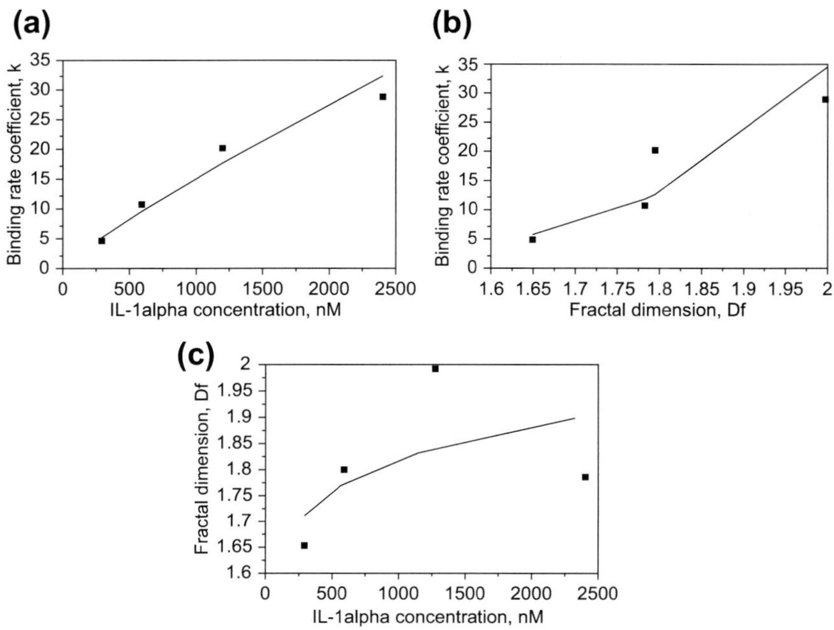

FIGURE 8.26 (a) Increase in the binding rate coefficient, k, with an increase in the IL-1α concentration in solution. (b) Increase in the binding rate coefficient, k, with an increase in the fractal dimension, D_f. (c) Increase in the fractal dimension, D_f, with an increase in the IL-1α concentration in solution.

The fit is very good. Only four data points are available. The availability of more data points would lead to a more reliable fit. The binding rate coefficient, k, exhibits an order of dependence between half and first (equal to 0.865) order on the 1L-1α concentration in solution.

Figure 8.26(b) shows the increase in the binding rate coefficient, k, with an increase in fractal dimension, D_f. For the data shown in Figure 8.26(b), the binding rate coefficient, k, is given by:

$$k = \left(0.0592 \pm 0.0297\right)\ D_f^{9.19 \pm 2.78} \tag{8.13b}$$

The fit is good. Only four data points are available. The availability of more data points would lead to a more reliable fit. The binding rate coefficient, k, exhibits an order of dependence between nine and nine and one-half (equal to 9.19) order on the fractal dimension or the degree of heterogeneity that exists on the biosensor surface.

Figure 8.26(c) shows the increase in the fractal dimension, D_f, with an increase in the IL-1α concentration in solution in nM. For the data shown in Figure 8.26(c), the fractal dimension, D_f, is given by:

$$D_f = (1.294 \pm 0.0.108)\ [\text{IL-1}\alpha]^{0.049 \pm 0.051} \tag{8.13c}$$

There is scatter in the data. Only four data points are available. The availability of more data points would lead to a more reliable fit. The fractal dimension, D_f, exhibits almost zero (0.049) order of dependence on the IL-1α concentration in solution.

Affinity interactions are important in biochemical processes involved in nature (Skladal, 2003; Halamek et al., 2005). ELISA yields useful information regarding binding assays but usually takes a long time to reach equilibrium. ELISA takes a long incubation time, typically several hours to reach equilibrium and cannot measure in real time (Leopoldo et al., 1997; Liu et al., 2004). The advantage of the optical biosensor is that it does not have to reach equilibrium, performs the analysis rapidly, and permits real-time monitoring. There is also no radiolabeling or biochemical tagging and there is flexibility in the design. Besides, it provides for rapid qualitative and quantitative information about the kinetic analysis involved in biomolecular interactions (Myszka et al., 1997; Tschmeiak et al., 2005). The optical biosensor has been widely used in chemical sensing and medicinal chemistry (Arakawa et al., 2005; Fisher and Tjarnhage, 2000; Green et al., 2000). Wu et al. (2006) state that the optical biosensor technology has led to the development of the resonant mirror technology and the SPR techniques (Hirmo et al., 1999; Wilson, 2002; Minunni et al., 2005).

Wu et al. (2006) state that interleukin-1 (IL-1) is a primary regulator of inflammatory and immune response and is an important mediator in the pathogenesis of many inflammatory and immunologically mediated disease. The soluble 1L-1 receptor (sIL-1R) can block IL-1 activity (Fanslow et al.,

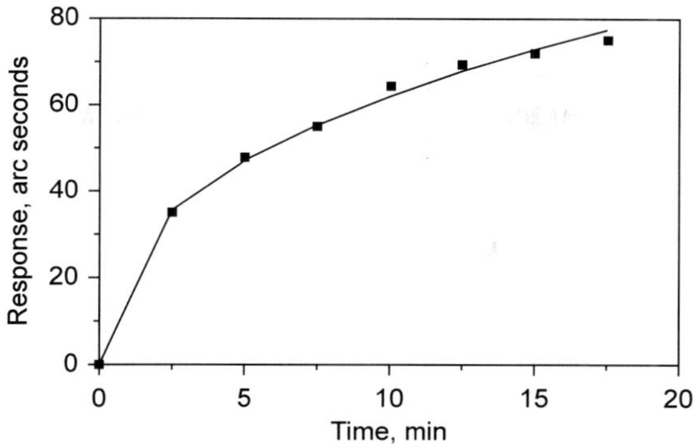

FIGURE 8.27 Binding of IL-1α to immobilized in solution sIL-1RI (Wu et al., 2006).

1990; Klasing and Peng, 2001) and is thought to be an effective regulator of IL-1 signaling activity (Sporri et al., 2001). Wu et al. (2006) analyzed the kinetics of sIL-1RI binding to immobilized IL-1α in order to get better physical insights into the sIL-1RI and IL-α interactions using a resonant mirror biosensor.

Figure 8.27 shows the binding of IL-1α in solution to sIL-1RI immobilized on a biosensor surface. A single-fractal analysis is adequate to describe the binding kinetics.

Figure 8.28(a) shows the binding of 200 nM sIL-1RI in solution to IL-1α immobilized on a biosensor surface. This is cycle two of a sequence of four cycles. A single-fractal analysis is adequate to describe the binding kinetics.

Figure 8.28(b) shows the binding of 400 nM sIL-1RI in solution to IL-lα immobilized on a biosensor surface. This is cycle three of a sequence of four cycles. A single-fractal analysis is adequate to describe the binding kinetics.

Figure 8.28(c) shows the binding of 800 nM sIL-1RI in solution to IL-1α immobilized on a biosensor surface. This is cycle four of a sequence of four cycles. A single-fractal analysis is adequate to describe the binding kinetics.

Figure 8.29(a) shows the increase in the binding rate coefficient, k, with an increase in the sIL-1RI concentration in nM. For the data shown in Figure 8.29(a), the binding rate coefficient, k, is given by:

$$k = (0.426 \pm 0.141)\,[\text{sIL-lRI}]^{0.705 \pm 0.291} \tag{8.14a}$$

The fit is good. Only three data points are available. The availability of more data points would lead to a more reliable fit. The binding rate coefficient, k, exhibits an order of dependence between one-half and first (equal to 0.705) order on the sIL-1RI concentration in solution. This shows that the binding rate coefficient is sensitive to the sIL-1RI concentration in solution.

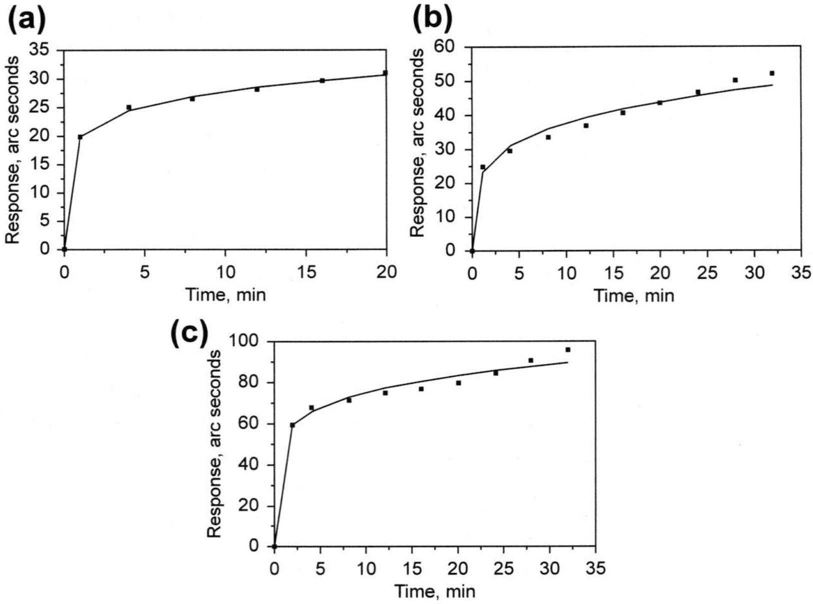

FIGURE 8.28 Binding of different concentrations of sIL-1RI in solution to IL-1α immobilized to a biosensor surface in consecutive cycles (Wu et al., 2006): (a) cycle 200 nM; (b) cycle 400 nM; and (c) cycle 800 nM.

Figure 8.29(b) shows the increase in the binding rate coefficient, k, for a single-fractal analysis with an increase in the fractal dimension, D_f. In Figure 8.29(b), the binding rate coefficient, k, is given by:

$$k = (0.01668 \pm 0.008009)D_f^{8.289 \pm 6.641} \tag{8.14b}$$

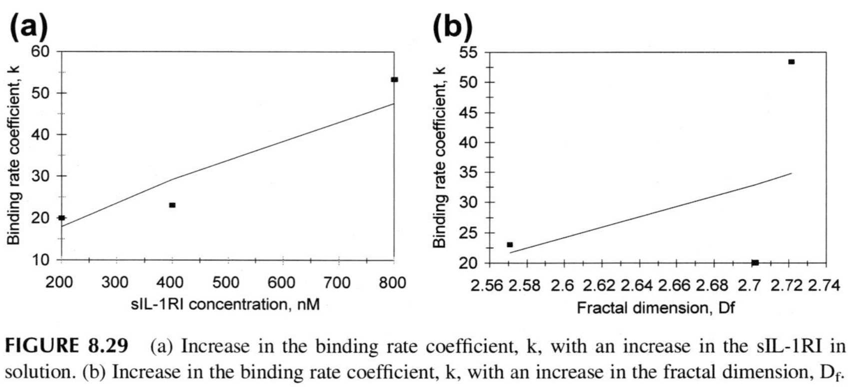

FIGURE 8.29 (a) Increase in the binding rate coefficient, k, with an increase in the sIL-1RI in solution. (b) Increase in the binding rate coefficient, k, with an increase in the fractal dimension, D_f.

There is scatter in the data. Only three data points are available. The availability of more data points would lead to a more reliable fit. The binding rate coefficient, k, exhibits an order of dependence between eight and eight and one-half (equal to 8.289) order on the fractal dimension. This shows that the binding rate coefficient, k, is extremely sensitive to the degree of heterogeneity that exists on the biosensor surface.

Figure 8.30(a) shows the increase in the binding rate coefficient, k, with an increase in the ScFv concentration in nM in solution. Figures for binding and dissociation not shown. For the data shown in Figure 8.30(a), the binding rate coefficient, k, is given by:

$$k = (0.177 \pm 0.039)[\text{ScFv}]^{0.719 \pm 0.119} \tag{8.14c}$$

The fit is good. Six data points are available. The availability of more data points would lead to a better fit. The binding rate coefficient, k, exhibits an order of dependence between one-half and first (equal to 0.719) order on the ScFv concentration in solution. The binding rate coefficient, k, is quite sensitive to the ScFv concentration in solution.

Figure 8.30(b) shows the increase in the dissociation rate coefficient, k_d, with an increase in the ScFv concentration in nM in solution. For the data shown in Figure 8.30(b), the dissociation rate coefficient, k_d, is given by:

$$k_d = (0.167 \pm 0.099)[\text{ScFv}]^{0.616 \pm 0.283} \tag{8.14d}$$

The fit is good. Five points are available. The availability of more data points would lead to a better fit. The dissociation rate coefficient, k_d, exhibits an order of dependence between one-half and first (equal to 0.616) order on the ScFv concentration in solution. The binding rate coefficient, k, is quite sensitive to the ScFv concentration in solution.

Figure 8.3l(a) shows the binding of the complement C_4 in an immune sandwich assay for human C_4. The antibody was immobilized on the biosensor

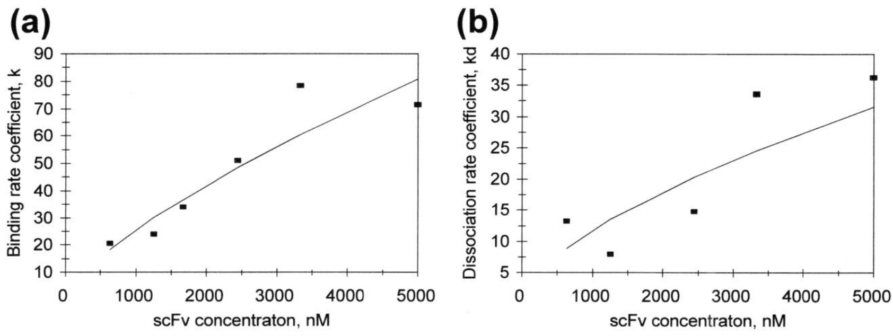

FIGURE 8.30 (a) Increase in the binding rate coefficient, k, with an increase in the ScFv concentration in solution in nM. (b) Increase in the dissociation rate coefficient, k_d, with an increase in the ScFv concentration in solution.

surface prepared with the amine-coupling method. The solution contained 80 μg/ml of human C_4. In this case, we have signal enhancement. An increase in the reaction time beyond 40 min resulted in a virtually constant resonance wavelength. This suggested to the authors (Liu et al., 2004) that the reaction had been completed. A single-fractal analysis is adequate to describe the binding kinetics. The values of the binding rate coefficient, k, and the fractal dimension, D_f, for a single-fractal analysis are given in Table 8.18.

Figure 8.31(b) shows the binding of the complement C_4 in an immune sandwich assay for human C_4. A secondary antibody was injected by the authors (Liu et al., 2004) (1:10 v/v) for 1 h. The authors note that at 0.2 μg/ml human C_4 cannot be determined by the SPR method, but the sandwich method does reveal a 0.47 nm shift of the resonance wavelength. A single-fractal analysis is adequate to describe the binding kinetics. The values of the binding rate coefficient, k, and the fractal dimension, D_f, for a single-fractal analysis are given in Table 8.18. It is of interest to note that as one compares the signal enhancement with the case of no signal enhancement, the

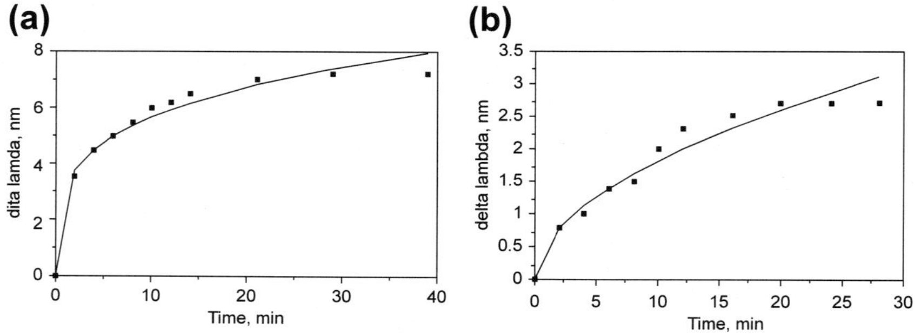

FIGURE 8.31 (a) Binding of C_4 in solution to antibody immobilized on a sandwich immunoassay (Liu et al., 2004). (b) Binding of C_4 in solution to antibody immobilized to a sandwich immunoassay (signal enhancement) (Liu et al., 2004).

TABLE 8.18 Binding Rate Coefficients and Fractal Dimensions for Complement C_4 in an Immune Sandwich Assay to Human C_4 Immobilized by an Amine-Coupling Method (Liu et al., 2004)

Analyte in Solution/Receptor on Surface	k	D_f
C_4/human C_4 (signal enhancement)	3.185 ± 0.189	2.4996 ± 0.422
C_4/human C_4	0.5496 ± 0.0575	1.9638 ± 0.0794

fractal dimension increases by a factor of 1.273 from a value of 1.9638 to 2.4996 and the binding rate coefficient, k, increases by a factor 5.79 from a value of k equal to 0.5496 to a value of 3.185. For the signal enhancement case, some more antibody was immobilized on the biosensor surface.

Buhl et al. (2009) have analyzed an optical biosensor for anti-dsDNA monoclonal antibodies as new standards for laboratory tests. Buhl et al. (2009) indicate that SLE is a systemic autoimmune disorder that has the potential to cause damage to the major organs in the body. Kotzm (1996) and D'Cruz et al. (2007) indicate that SLE can be lethal when developing lupus nephritis. As SLE is systemic, the clinical manifestations are different. Buhl et al. (2009) emphasize that the early diagnosis of SLE is important. The detection of antinuclear antibodies and findings of antibodies to native DNA in abnormal titer is contained in the American College of Rheumatology classification criteria (Tan et al., 1982; Hochberg, 1997). Arbuckle et al. (2003) suggest that these are present often before the onset of symptoms. Therefore, Buhl et al. (2009) indicate their importance in diagnostic tests. Anti-dsDNA autoantibodies are due to their high specificity for SLE. Isenberg (2004) indicate the importance of these in guiding the diagnosis in the right direction. Riboldi et al. (2005) and Bootsma et al. (1997) indicate that they are useful in predicting the flares of SLE.

Reveille (2004) indicate that ssDNA are less disease specific than dsDNA. Simon et al. (2001) indicate that mAb standards, in this context, are useful. Buhl et al. (2007) have recently introduced a biosensor device for the detection of anti-dsDNA in serum based on SPR technology. This they indicate has superior discriminating power between sera from SLE patients and controls. Jonsson and Malmqvist (1992) and Van Regenmortel et al. (1968) indicate that the change in mass concentration at the interface resulting from specific binding of the autoantibodies at the surface of the SPR is detected as changes in the refractive index. Buhl et al. (2009) applied the SPR biosensor technology for characterizing the putative anti-dsDNA mAb standards. The authors also obtained kinetic and equilibrium standards.

Figure 8.32(a) shows the binding of 14.4 nM anti-dsDNA mAb 32 in solution to dsDNA antigen immobilized on a sensor chip (Buhl et al., 2009). A dual-fractal analysis is required to adequately describe the binding kinetics. The values of (a) the binding rate coefficient, k, and the fractal dimension, D_f, for a single-fractal analysis, and (b) binding rate coefficients, k_1 and k_2, and the fractal dimensions, D_{f1} and D_{f2}, for a dual-fractal analysis are given in Tables 8.19 and 8.20.

Figure 8.32(b) shows the binding of 7.2 nM anti-dsDNA mAb 32 in solution to dsDNA antigen immobilized on a sensor chip. A dual-fractal analysis is required to adequately describe the binding kinetics. The values of (a) the binding rate coefficient, k, and the fractal dimension, D_f, for a single-fractal analysis, and (b) binding rate coefficients, k_1 and k_2, and the fractal

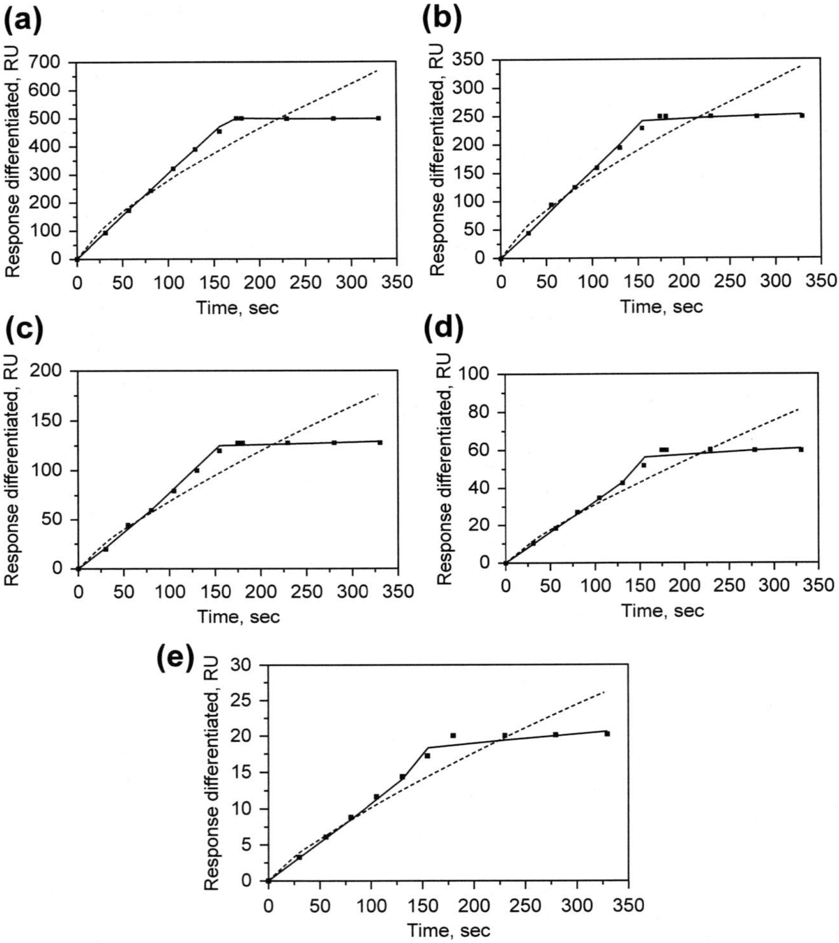

FIGURE 8.32 Binding of anti-dsDNA mAb 32 in solution to immobilized dsDNA antigen on a sensor chip (Buhl et al., 2009): (a) 14.4 nM; (b) 7.2 nM; (c) 3.6 nM; (d) 1.8 nM; and (e). 0.9 nM.

dimensions, D_{f1} and D_{f2}, for a dual-fractal analysis are given in Tables 8.19 and 8.20.

Figure 8.32(c) shows the binding of 3.6 nM anti-dsDNA mAb 32 in solution to dsDNA antigen immobilized on a sensor chip. A dual-fractal analysis is required to adequately describe the binding kinetics. The values of (a) the binding rate coefficient, k, and the fractal dimension, D_f, for a single-fractal analysis, and (b) binding rate coefficients, k_1 and k_2, and the fractal

TABLE 8.19 Binding Rate Coefficients for Different Concentrations of Anti-dsDNA mAb32 in Solution to dsDNA Antigen Immobilized on a Sensor Chip (Buhl et al., 2009)

Analyte in Solution/ Receptor on Surface	k	k_1	k_2
14.4 nM anti-dsDNA 32/dsDNA	9.784 ± 1.850	3.598 ± 0.057	500 ± 0
7.2 nM anti-dsDNA 32/dsDNA	4.894 ± 0.976	1.689 ± 0.118	174.95 ± 5.63
3.6 nM anti-dsDNA 32/dsDNA	1.815 ± 0.406	0.5456 ± 0.0406	97.099 ± 1.408
1.8 nM anti-dsDNA 32/dsDNA	0.7746 ± 0.1483	0.3308 ± 0.0044	32.515 ± 1.716

TABLE 8.20 Fractal Dimensions for the Binding of Different Concentrations of Anti-dsDNA mAb 32 in Solution to dsDNA Antigen Immobilized on a Sensor Chip (Buhl et al., 2009)

Analyte in Solution/ Receptor on Surface	D_f	D_{f1}	D_{f2}
14.4 nM anti-dsDNA 32/dsDNA	1.5416 ± 0.1522	1.070 ± 0.0267	3.0 ± 0
7.2 nM anti-dsDNA 32/dsDNA	1.5402 ± 0.1596	1.0372 ± 0.1160	2.8726 ± 0.956
3.6 nM anti-dsDNA 32/dsDNA	1.4184 ± 0.1776	0.850 ± 0.1210	2.9014 ± 0.074
1.8 nM anti-dsDNA 32/dsDNA	1.3938 ± 0.1538	0.9978 ± 0.02256	2.7812 ± 0.1640

dimensions, D_{f1} and D_{f2}, for a dual-fractal analysis are given in Tables 8.19 and 8.20.

Figure 8.32(d) shows the binding of 1.8 nM anti-dsDNA mAb 32 in solution to dsDNA antigen immobilized on a sensor chip. A dual-fractal analysis is required to adequately describe the binding kinetics. The values of (a) the binding rate coefficient, k, and the fractal dimension, D_f, for a single-fractal analysis, and (b) binding rate coefficients, k_1 and k_2, and the fractal dimensions, D_{f1} and D_{f2}, for a dual-fractal analysis are given in Tables 8.19 and 8.20.

Figure 8.32(e) shows the binding of 0.9 nM anti-dsDNA mAb 32 in solution to dsDNA antigen immobilized on a sensor chip. A dual-fractal analysis is required to adequately describe the binding kinetics.

Figure 8.33(a) shows the increase in the binding rate coefficient, k_1, with an increase in the dsDNA concentration in solution for a dual-fractal analysis. For the data shown in Figure 8.33(a), the binding rate coefficient, k_1, is given by:

$$k_1 = (0.146 \pm 0.032)\,[\text{ds DNA}]^{1.206 \pm 0.1303} \qquad (8.15a)$$

The fit is very good. Only four data points are available. The availability of more data points would lead to a more reliable fit. The binding rate coefficient, k_1, exhibits an order of dependence between first and one and one-half (equal to

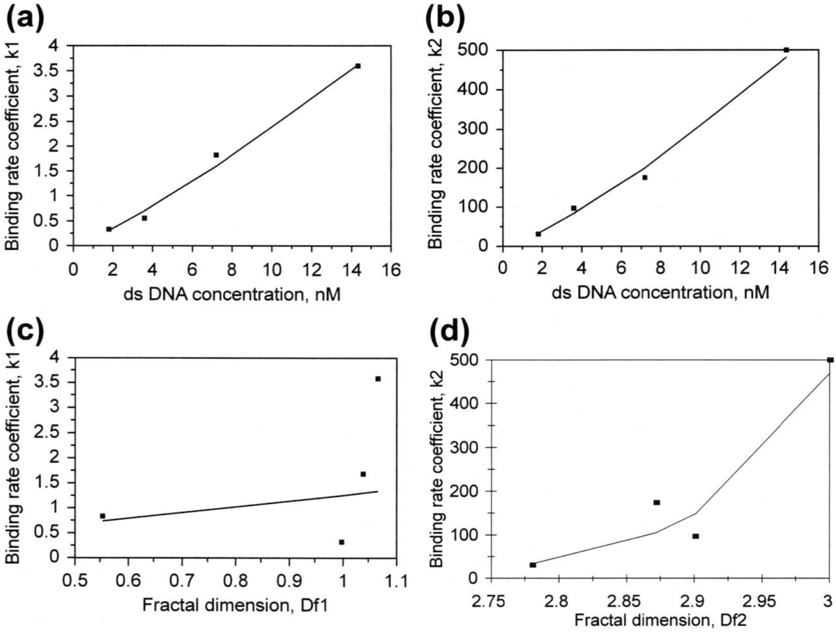

FIGURE 8.33 (a) Increase in the binding rate coefficient, k_1, with an increase in the dsDNA concentration (in nM) in solution. (b) Increase in the binding rate coefficient, k_2, with an increase in the dsDNA concentration (in nM) in solution. (c) Increase in the binding rate coefficient, k_1, with an increase in the fractal dimension, D_{f1}. (d) Increase in the binding rate coefficient, k_2, with an increase in the fractal dimension, D_{f2}. (e) Increase in the ratio of the binding rate coefficients, k_2/k_1, with an increase in the fractal dimension ratio, D_{f2}/D_{f1}. (f) Increase in the fractal dimension, D_{f1}, with an increase in the dsDNA concentration (in nM). (g) Increase in the fractal dimension, D_{f2}, with an increase in the dsDNA concentration in solution (in nM). (h) Decrease in the ratio of the fractal dimensions, D_{f2}/D_{f1}, with an increase in the dsDNA concentration in solution. (i) Increase in D_{f2} with an increase in ds-DNA concentration in solution.

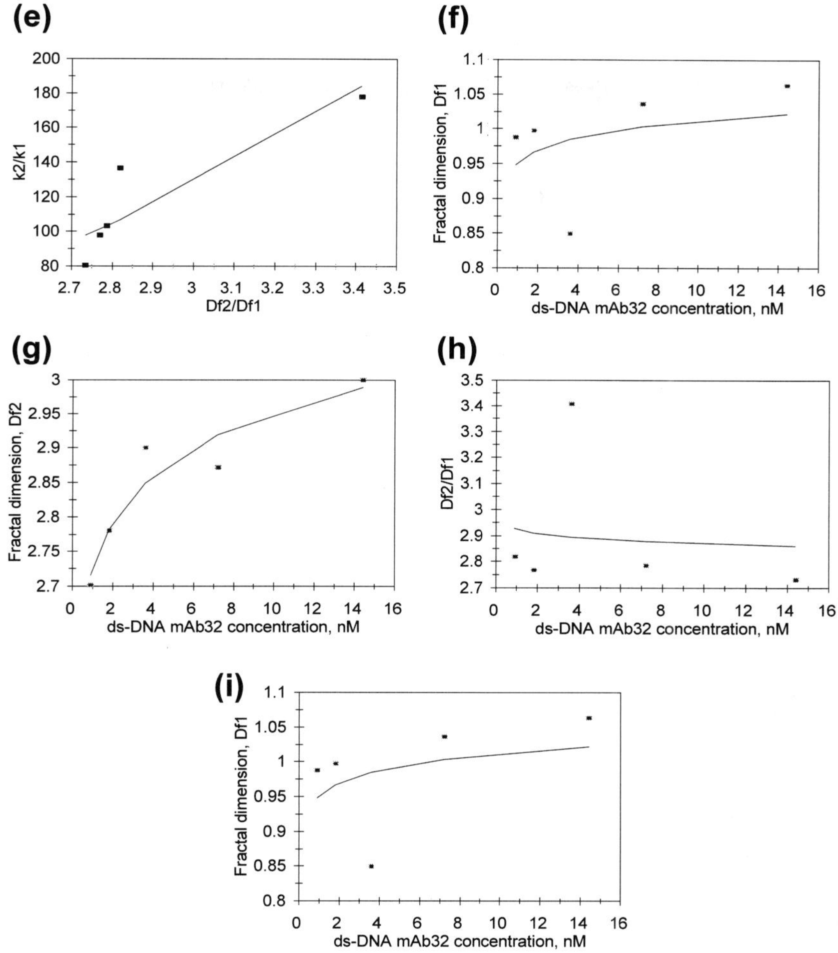

FIGURE 8.33 cont'd

1.206) on the dsDNA concentration in solution. This indicates that the binding rate coefficient, k_1, is sensitive to the dsDNA concentration in solution.

Figure 8.33(b) shows the increase in the binding rate coefficient, k_2, with an increase in the dsDNA concentration in solution for a dual-fractal analysis. For the data shown in Figure 8.33(b), the binding rate coefficient, k_2, is given by:

$$k_2 = (16.37 \pm 2.719)\,[\text{ds DNA}]^{1.268 \pm 0.0991} \tag{8.15b}$$

The fit is very good. Only four data points are available. The availability of more data points would lead to a more reliable fit. The binding rate coefficient, k_2, exhibits an order of dependence between first and one and one-half (equal to 1.268) on the dsDNA concentration in solution. This indicates that the binding rate coefficient, k_2, is sensitive to the dsDNA concentration in solution.

Figure 8.33(c) shows the increase in the binding rate coefficient, k_1, with an increase in the fractal dimension, D_{f1}. For the data shown in Figure 8.33(c), the binding rate coefficient, k_1, is given by:

$$k_1 = (1.277 + 2.943)D_{f1}^{1.268 \pm 0.0991} \quad (8.15c)$$

The fit is very good. Only four data points are available. The availability of more data points would lead to a more reliable fit. The binding rate coefficient, k_2, exhibits an order of dependence between first and one and one-half (equal to 1.268) on the fractal dimension, D_{f1}, that exists on the surface. This indicates that the binding rate coefficient, k_1, is sensitive to the fractal dimension that exists on the surface.

Figure 8.33(d) shows the increase in the binding rate coefficient, k_2, with an increase in the fractal dimension, D_{f2}. For the data shown in Figure 8.33(d), the binding rate coefficient, k_2, is given by:

$$k_2 = (1.8E - 14 + 1.5E - 14)\ D_{f2}^{34.41 \pm 8.664} \quad (8.15d)$$

The fit is very good. Only four data points are available. The availability of more data points would lead to a more reliable fit. The binding rate coefficient, k_2, exhibits an order of dependence close to 34 and one-half (equal to 34.41) on the fractal dimension, D_{f2}, that exists on the surface. This indicates that the binding rate coefficient, k_2, is extremely sensitive to the degree of heterogeneity that exists on the surface.

Figure 8.33(e) shows the increase in the binding rate coefficient ratio, k_2/k_1, with an increase in the fractal dimension ratio, D_{f2}/D_{f1}. For the data shown in Figure 8.33(e), the binding rate coefficient ratio, k_2/k_1, is given by:

$$k_2/k_1 = (5.579 \pm 1.116)(D_{f2}/D_{f1})^{2.84 \pm 0.981} \quad (8.15e)$$

The fit is good. Only five data points are available. The availability of more data points would lead to a more reliable fit. The binding rate coefficient ratio, k_2/k_1, exhibits an order of dependence between two and a half and third (equal 2.84) order on the ratio of fractal dimension ratio, D_{f2}/D_{f1}. This indicates that the binding rate coefficient ratio, k_2/k_1, is very sensitive to the degree of heterogeneity that exists on the surface.

Figure 8.33(f) shows the increase in the fractal dimension, D_{f2}, with an increase in the dsDNA mAB 32 concentration in nM in solution. For the data shown in Figure 8.33(f), the fractal dimension, D_{f1}, is given by:

$$D_{f1} = (2.725 \pm 0.040)[\text{ds-DNA mAb 32}]^{0.0347 \pm 0.0065} \quad (8.15f)$$

There is scatter in the data. Only five data points are available. The availability of more data points would lead to a more reliable fit. The fractal dimension, D_{f2}, exhibits close to a zero order of dependence (equal to 0.0347) on the dsDNA mAb 32b concentration in solution. This indicates that the fractal dimension, D_{f1}, is hardly dependent on the dsDNA mAb 32b concentration in solution.

Figure 8.33(g) shows the increase in the fractal dimension ratio, D_{f2}/D_{f1}, with an increase in the dsDNA mAB 32 concentration in nM in solution. For the data shown in Figure 8.33(g), the fractal dimension ratio, D_{f2}/D_{f1}, is given by:

$$D_{f2}/D_{f1} = (0.951 \pm 0.095)[\text{ds-DNA mAb } 32]^{0.0268 \pm 0.043} \qquad (8.15g)$$

There is scatter in the data. Only five data points are available. The availability of more data points would lead to a more reliable fit. The fractal dimension, D_{f2}, exhibits close to a zero order of dependence (equal to 0.0268) on the dsDNA mAb 32b concentration in solution. This indicates that the fractal dimension, D_{f1}, is hardly dependent on the dsDNA mAb 32b concentration in solution.

Figure 8.33(h) shows the increase in the fractal dimension ratio, D_{f2}/D_{f1}, with an increase in the dsDNA mAB 32 concentration in nM in solution. For the data shown in Figure 8.33(g), the fractal dimension ratio, D_{f2}/D_{f1}, is given by:

$$D_{f2}/D_{f1} = (2.725 \pm 0.040)[\text{ds-DNA mAb } 32]^{0.0347 \pm 0.0065} \qquad (8.15h)$$

There is scatter in the data. Only five data points are available. The availability of more data points would lead to a more reliable fit. The fractal dimension, D_{f2}, exhibits to a zero order of dependence (equal to 0.0347) on the dsDNA mAb 32b concentration in solution. This indicates that the fractal dimension, D_{f1}, is hardly dependent on the dsDNA mAb 32b concentration in solution.

Figure 8.34(a) shows the binding of 2.5 nM dsDNA mAb 333 in solution to a receptor immobilized on an SPR biosensor surface (Buhl et al., 2009). A single-fractal analysis is adequate to describe the binding and the dissociation kinetics. The values of (a) the binding rate coefficient, k, and the fractal dimension, D_f, and (b) the dissociation rate coefficient, k_d, and the fractal dimension for dissociation, D_{fd}, are given in Tables 8.19 and 8.20. In this case the affinity, K (= k/k_d) value is 0.631.

Figure 8.34(b) shows the binding of 1.25 dsDNA mAb 333 in solution to a receptor immobilized on an SPR biosensor surface. A single-fractal analysis is adequate to describe the binding and the dissociation kinetics. The values of (a) the binding rate coefficient, k, and the fractal dimension, D_f, and (b) the dissociation rate coefficient, k_d, and the fractal dimension for dissociation, D_{fd}, are given in Table 8.21. In this case the affinity, K (= k/k_d) value is 1.49.

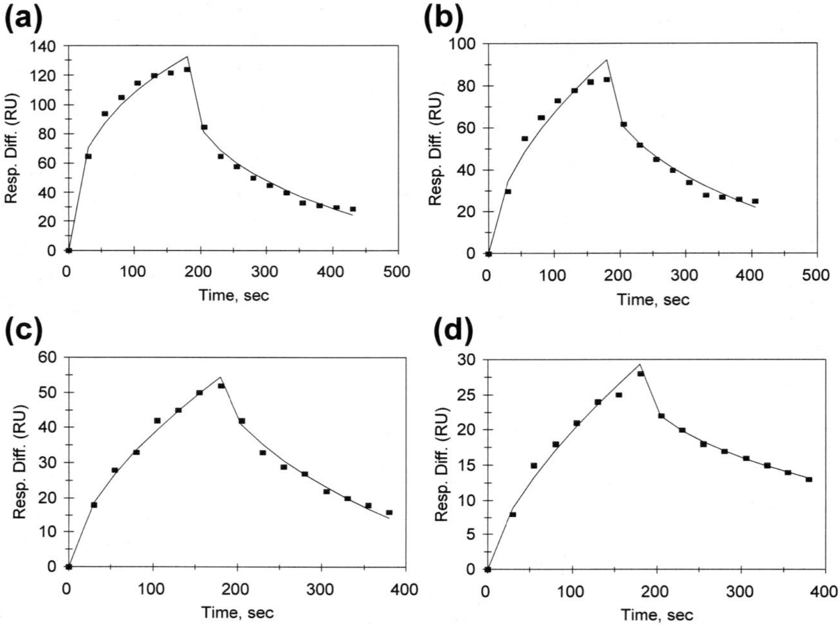

FIGURE 8.34 Binding of different concentrations of dsDNA mAb 33 in solution (in nM) to receptor immobilized on an SPR surface (Buhl et al., 2009): (a) 2.5; (b) 1.25; (c) 0.62; and (d) 0.31.

TABLE 8.21 Binding and Dissociation Rate Coefficients for dsDNA mAB 33 Concentration in Solution to Receptor Immobilized on Biosensor Surface (Buhl et al., 2009)

Analyte in Solution/ Receptor on Surface	k	k_d	D_f	D_{fd}
2.5 nM dsDNA mAb 33	21.433 ± 1.495	12.988 ± 0.582	2.2980 ± 0.0870	2.2612 ± 0.04634
1.25 nM dsDNA mAb 33	5.376 ± 0.638	5.0489 ± 0.2827	1.9046 ± 0.1446	2.0760 ± 0.0494
0.62 nM dsDNA mAb 33	2.4434 ± 0.1088	1.6387 ± 0.1309	2.8076 ± 0.0564	1.8222 ± 0.08632
0.31 nM dsDNA mAb 33	0.929 ± 0.081	1.472 ± 0.023	1.6710 ± 0.0762	2.1206 ± 0.0166

Figure 8.34(c) shows the binding of 0.62 dsDNA mAb 333 in solution to a receptor immobilized on an SPR biosensor surface. A single-fractal analysis is adequate to describe the binding and the dissociation kinetics. The values of (a) the binding rate coefficient, k, and the fractal dimension, D_f, and (b) the dissociation rate coefficient, k_d, and the fractal dimension for dissociation, D_{fd}, are given in Table 8.21, In this case the affinity, K (= k/k_d) value is 1.065.

Figure 8.34(d) shows the binding of 0.31 dsDNA mAb 333 in solution to a receptor immobilized on an SPR biosensor surface. A single-fractal analysis is adequate to describe the binding and the dissociation kinetics. The values of (a) the binding rate coefficient, k, and the fractal dimension, D_f, and (b) the dissociation rate coefficient, k_d, and the fractal dimension for dissociation, D_{fd}, are given in Table 8.21. In this case the affinity, K (= k/k_d) value is 1.65.

Figure 8.35(a) shows the increase in the binding rate coefficient, k, with an increase in the dsDNA mAb 33 concentration in solution. For the data shown in Figure 8.35(a), the binding rate coefficient, k, is given by:

$$k = (4.847 \pm 1.035)[\text{ds-DNA mAb 33}]^{1.465 \pm 0.1243} \quad (8.16a)$$

The fit is very good. Only four data points are available. The availability of more data points would lead to a more reliable fit. The binding rate coefficient exhibits close to a one and one-half (equal to 1.465) order of dependence on the (dsDNA) concentration in solution. This indicates that the binding rate coefficient, k, is sensitive to the dsDNA concentration in solution.

Figure 8.35(b) shows the increase in the binding rate coefficient, k, with an increase in the fractal dimension, D_f, or the degree of heterogeneity that exists on the biosensor surface. For the data shown in Figure 8.35(b), the binding rate coefficient, k, is given by:

$$k = (0.007914 \pm 0.0025) D_f^{9.654 \pm 1.143} \quad (8.16b)$$

The fit is very good. Only four data points are available. The availability of more data points would lead to a more reliable fit The binding rate coefficient exhibits an order of dependence between nine and one-half and 10 (equal to 9.654) on the fractal dimension or the degree of heterogeneity that exists on the biosensor surface. This indicates that the binding rate coefficient, k, is extremely sensitive to the degree of heterogeneity that exists on the biosensor surface.

Figure 8.35(c) shows the increase in the dissociation rate coefficient, k_d, with an increase in the dsDNA mAb 33 concentration in solution. For the data shown in Figure 8.35(c), the dissociation rate coefficient, k_d, is given by:

$$k_d = (4.105 \pm 1.635)[\text{ds-DNS mAb33}]^{1.096 \pm 0.2152} \quad (8.16c)$$

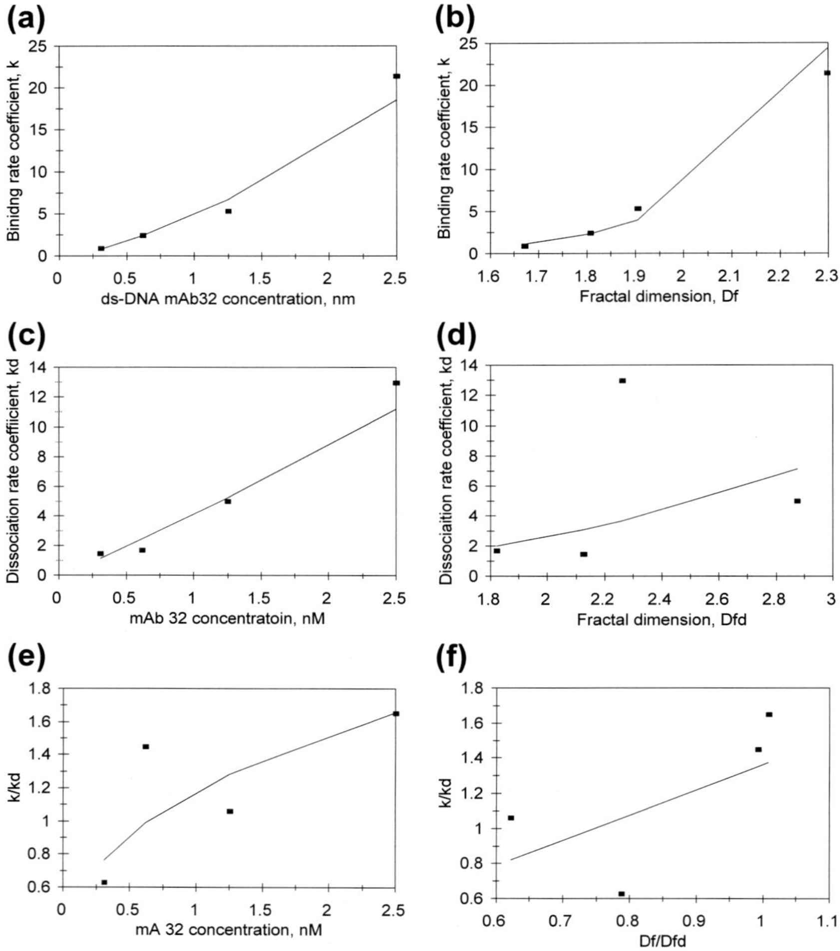

FIGURE 8.35 (a) Increase in the binding rate coefficient, k, with an increase in the dsDNA mAb 33 concentration in nM in solution. (b) Increase in the binding rate coefficient, k, with an increase in the fractal dimension, D_f. (c) Increase in the dissociation rate coefficient, k_d, with an increase in the dsDNA mAb 33 concentration in nM in solution. (d) Increase in the dissociation rate coefficient, k_d, with an increase in the fractal dimension for dissociation, D_{fd}. (e) Increase in the affinity, K (= k/k_d) with an increase in the dsDNA mAb 33 concentration in nM in solution. (f) Increase in the affinity, K (= k/k_d) with an increase in the fractal dimension ratio, D_f/D_{fd}.

The fit is very good. Only four data points are available. The availability of more data points would lead to a more reliable fit. The dissociation rate coefficient exhibits close to a first-order (equal to 1.096) dependence on the [dsDNA] concentration in solution. This indicates that the dissociation rate coefficient, k_d, is sensitive to the dsDNA mAb 33 concentration in solution.

Figure 8.35(d) shows the increase in the dissociation rate coefficient, k_d, with an increase in the fractal dimension for dissociation, D_{fd}, or the degree of heterogeneity that exists on the biosensor surface in the dissociation phase. For the data shown in Figure 8.35(d), the dissociation rate coefficient, k_d, is given by:

$$k_d = (0.3788 + 0.7291)D_{fd}^{2.780+3.267} \quad (8.16d)$$

The fit is poor. Only four data points are available. The availability of more data points would lead to a more reliable fit. The poor fit is also demonstrated by the error in the estimate of the rate coefficient and in the order of the dependence exhibited.

Figure 8.35(e) shows the increase in the affinity, $K (= k/k_d)$ with an increase in the ds mAb 33 concentration in solution. For the data shown in Figure 8.35(e), the affinity, $K (= k/k_d)$ is given by:

$$K(= k/k_d) = (1.1811 \pm 0.462)[\text{ds mAb } 33]^{0.3698\pm0.212} \quad (8.16e)$$

The fit is reasonable. Only four data points are available. The availability of more data points would lead to a more reliable fit. The affinity, $K (= k/k_d)$ exhibits a 0.369 order of dependence on the ds mAb 33 concentration in solution. This indicates that the affinity, K is only mildly dependent on the ds mAb 33 concentration in solution.

Figure 8.35(f) shows the increase in the affinity, $K (= k/k_d)$ with an increase in the ratio of fractal dimensions, D_f/D_{fd}. For the data shown in Figure 8.35(f), the affinity, $K (= k/k_d)$ is given by:

$$K(k/k_d) = (1.364 \pm 0.735)\ (D_f/D_{fd})^{1.074+1.094} \quad (8.16f)$$

There is scatter in the data. Only four data points are available. The availability of more data points would lead to a more reliable fit. The affinity, $K (= k/k_d)$ exhibits close to a first (equal to 1.074) order of dependence on the ratio of fractal dimensions, D_f/D_{f1}. This indicates that the affinity, $K (= k/k_d)$ is sensitive to the degree of heterogeneities in the binding and in the dissociation phase.

Figure 8.36(a) shows the binding and dissociation of 16.4 nM anti-ds HYB331-01 in solution to ds HYB331-01 immobilized on a biosensor chip SPR surface (Buhl et al., 2009). The binding kinetics is described by a dual-fractal analysis. The dissociation kinetics is described by a single-fractal analysis. The values of (a) the binding rate coefficient, k, and the fractal dimension, D_f, for a single-fractal analysis, (b) the binding rate coefficients, k_1 and k_2, and the fractal dimensions, D_{f1} and D_{f2}, for a dual-fractal analysis, and (c) the dissociation rate coefficient, k_d, and the fractal dimension for dissociation, D_{fd}, for a single-fractal analysis are given in Tables 8.22 and 8.23. In this case the affinity, $K (= k/k_d)$ is equal to 641.8.

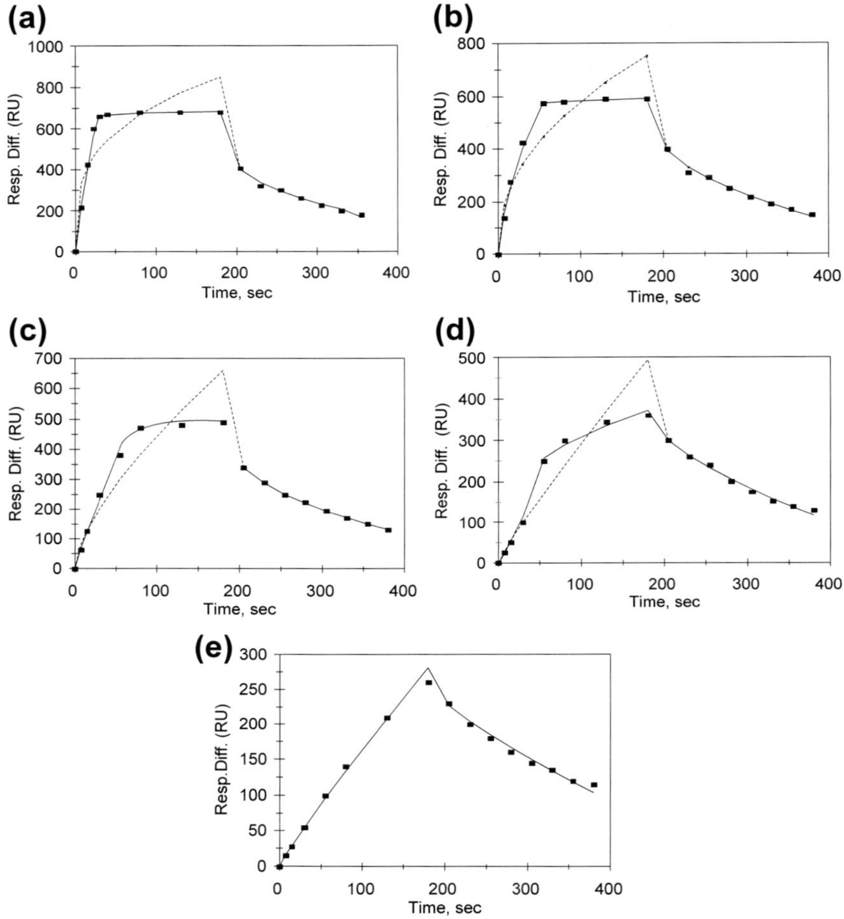

FIGURE 8.36 Binding and dissociation of different concentrations of anti-ds HYB331-01 to ds HYB331-01 immobilized on a biosensor chip SPR surface (Buhl et al., 2009): (a) 16.4 nM; (b) 8.2 nM; (c) 4.1 nM; (d) 2.05 nM; and (e) 1.02 nM.

Figure 8.36(b) shows the binding and dissociation of 8.2 nM anti-ds HYB331-01 in solution to ds HYB331-01 immobilized on a biosensor chip SPR surface (Buhl et al., 2009). The binding kinetics is described by a dual-fractal analysis. The dissociation kinetics is described by a single-fractal analysis. The values of (a) the binding rate coefficient, k, and the fractal dimension, D_f, for a single-fractal analysis, (b) the binding rate coefficients, k_1 and k_2, and the fractal dimensions, D_{f1} and D_{f2}, for a dual-fractal analysis, and (c) the dissociation rate coefficient, k_d, and the fractal dimension for dissociation, D_{fn}, for a single-fractal analysis are given in Tables 8.22 and 8.23. In this case the affinity, K $(= k/k_d)$ is equal to 1.37.

TABLE 8.22 Binding and Dissociation Rate Coefficients for Different Concentrations (in nM) of Anti-Double Stranded HYB331-01 to ds HYB331-10 Immobilized on a Biosensor Chip Surface (Buhl et al., 2009)

Analyte in Solution/ Receptor on Surface	k	k_1	k_2	k_d
16.4 nM anti-double stranded HYB331-01/ds HYB331-01	185.14 ± 58.13	41.61 ± 3.89	629.95 ± 4.38	0.2885 ± 0.01647
8.2 nM anti-double stranded HYB331-01/ds HYB331-01	77.33 ± 10.54	38.894 ± 2.43	523.09 ± 2.12	56.485 ± 2.0988
4.1 nM anti-double stranded HYB331-01/ds HYB331-01	21.044 ± 5.833	10.33 ± 0.79	186.21 ± 14.51	38.603 ± 2.267
2.05 nM anti-double stranded HYB331-01/ds HYB331-01	4.899 ± 3.396	2.545 ± 0.367	75.532 ± 3.421	7.239 ± 0.332
1.02 nM anti-double stranded HYB331-01/ds HYB331-01	2.411 ± 0.123	NA	NA	3.026 ± 0.138

TABLE 8.23 Fractal Dimensions for the Binding and the Dissociation Phase for Different Concentrations (in nM) of Anti-Double Stranded HYB331-01 to ds HYB331-10 Immobilized on a Biosensor Chip Surface (Buhl et al., 2009)

Analyte in Solution/ Receptor on Surface	D_f	D_{f1}	D_{f2}	D_{fd}
16.4 nM anti-double stranded HYB331-01/ds HYB331-01	2.413 ± 0.1894	1.3274 ± 0.1717	2.9686 ± 0.00914	2.430 ± 0.03294
8.2 nM anti-double stranded HYB331-01/ds HYB331-01	2.0232 ± 0.1664	1.5934 ± 0.1519	2.9524 ± 0.00888	2.2194 ± 0.0392
4.1 nM anti-double stranded HYB331-01/ds HYB331-01	1.6734 ± 0.1730	1.1714 ± 0.09850	2.6148 ± 0.1658	2.1576 ± 0.00742
2.05 nM anti-double stranded HYB331-01/ds HYB331-01	1.2232 ± 0.1772	0.7646 ± 0.1086	2.3868 ± 0.09778	1.6734 ± 0.0482
1.02 nM anti-double stranded HYB331-01/ds HYB331-01	1.1680 ± 0.03514	NA	NA	1.5104 ± 0.0812

Figure 8.36(c) shows the binding and dissociation of 4.1 nM anti-ds HYB331-01 in solution to ds HYB331-01 immobilized on a biosensor chip SPR surface (Buhl et al., 2009). The binding kinetics is described by a dual-fractal analysis. The dissociation kinetics is described by a single-fractal

analysis. The values of (a) the binding rate coefficient, k, and the fractal dimension, D_f, for a single-fractal analysis, (b) the binding rate coefficients, k_1 and k_2, and the fractal dimensions, D_{f1} and D_{f2}, for a dual-fractal analysis, and (c) the dissociation rate coefficient, k_d, and the fractal dimension for dissociation, D_{fd}, for a single-fractal analysis are given in Tables 8.22 and 8.23. In this case the affinity, K (= k/k_d) is equal to 0.545.

Figure 8.36(d) shows the binding and dissociation of 2.05 nM anti-ds HYB331-01 in solution to ds HYB331-01 immobilized on a biosensor chip SPR surface (Buhl et al., 2009). The binding kinetics is described by a dual-fractal analysis. The dissociation kinetics is described by a single-fractal analysis. The values of (a) the binding rate coefficient, k, and the fractal dimension, D_f, for a single-fractal analysis, (b) the binding rate coefficients, k_1 and k_2, and the fractal dimensions, D_{f1} and D_{f2}, for a dual-fractal analysis, and (c) the dissociation rate coefficient, k_d, and the fractal dimension for dissociation, D_{fd}, for a single-fractal analysis are given in Tables 8.22 and 8.23. In this case the affinity, K (= k/k_d) is equal to 0.677.

Figure 8.36(e) shows the binding and dissociation of 1.02 nM anti-ds HYB331-01 in solution to ds HYB331-01 immobilized on a biosensor chip SPR surface (Buhl et al., 2009). Both the binding and the dissociation kinetics is described by a single-fractal analysis. The values of (a) the binding rate coefficient, k, and the fractal dimension, D_f, for a single-fractal analysis, and (b) the dissociation rate coefficient, k_d, and the fractal dimension for dissociation, D_{fd}, for a single-fractal analysis are given in Tables 8.22 and 8.23. In this case the affinity, K (= k/k_d) is equal to 0.797.

Figure 8.37(a) shows the increase in the binding rate coefficient, k_1, with an increase in the anti-ds HYB331-01 concentration in solution. For the data shown in Figure 8.37(a), the binding rate coefficient, k_1, is given by:

$$k_1 = (1.215 \pm 0.807)\,[\text{anti-ds HYB331} - 01]^{1.409 \pm 0.3309} \tag{8.17a}$$

The fit is reasonable. Only four data points are available. The availability of more data points would lead to a more reliable fit. The binding rate coefficient, k_1, exhibits close to a one and one-half (equal to 1.409) order of dependence on the anti-ds HYB331-01 concentration in solution. This indicates that the binding rate coefficient, k_1, is sensitive to the anti-ds HYB331-01 concentration in solution. Figure 8.37(b) shows the increase in the binding rate coefficient, k_2, with an increase in the anti-ds HYB331-01 concentration in solution. For the data shown in Figure 8.37(b), the binding rate coefficient, k_2, given by:

$$k_2 = (38.695 \pm 4.788)\,[\text{anti-ds HYB331} - 01]^{1.085 \pm 0.195} \tag{8.17b}$$

The fit is good. Only four data points are available. The availability of more data points would lead to a more reliable fit. The binding rate coefficient, k_2, exhibits close to a first- (equal to 1.085) order of dependence on the anti-ds HYB331-01 concentration in solution. This indicates that the binding rate coefficient, k_1, is sensitive to the anti-ds HYB331-01 concentration in solution.

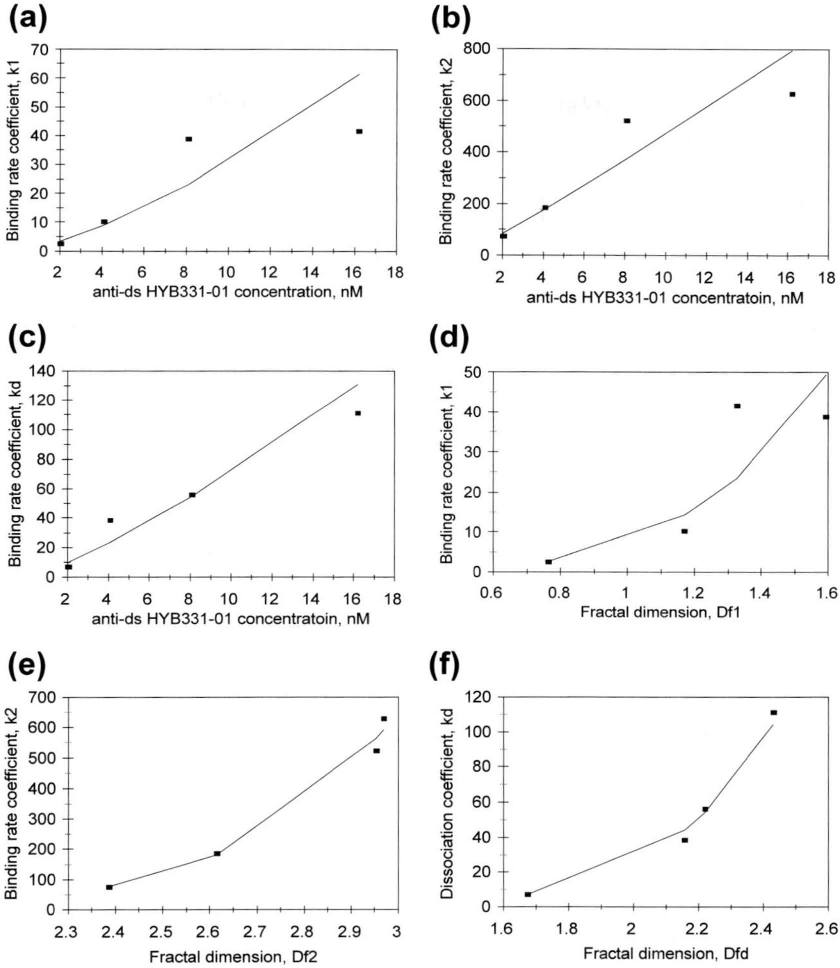

FIGURE 8.37 (a) Increase in the binding rate coefficient, k_1, with an increase in the anti-ds HYB331 = 01 concentration in nM in solution. (b) Increase in the binding rate coefficient, k_2, with an increase in the anti-ds HYB331-01 concentration in 1 nM in solution. (c) Increase in the dissociation rate coefficient, k_d, with an increase in the anti-ds HYB331-01 concentration in solution in 1 nM. (d) Increase in the binding rate coefficient, k_1, with an increase in the fractal dimension, D_{f1}. (e) Increase in the binding rate coefficient, k_2, with an increase in the fractal dimension, D_{f2}. (f) Increase in the dissociation rate coefficient, k_d, with an increase in the fractal dimension, D_{fd}. (g) Increase in the fractal dimension, D_{f1}, with an increase in the anti-ds HYB331-01 concentration in solution in nM. (h) Increase in the fractal dimension, D_{f2}, with an increase in the anti-ds HYB3331-01 concentration in solution in nM. (i) Increase in the fractal dimension for dissociation, D_{fd}, with an increase in the anti-ds HYB331-01 concentration in solution in nM. (j) Increase in K_1 with an increase in HYB331-01 concentration in solution (k) Increase in D_{f2}/D_{fd} with an increase in HYB331-01 concentration in solution.

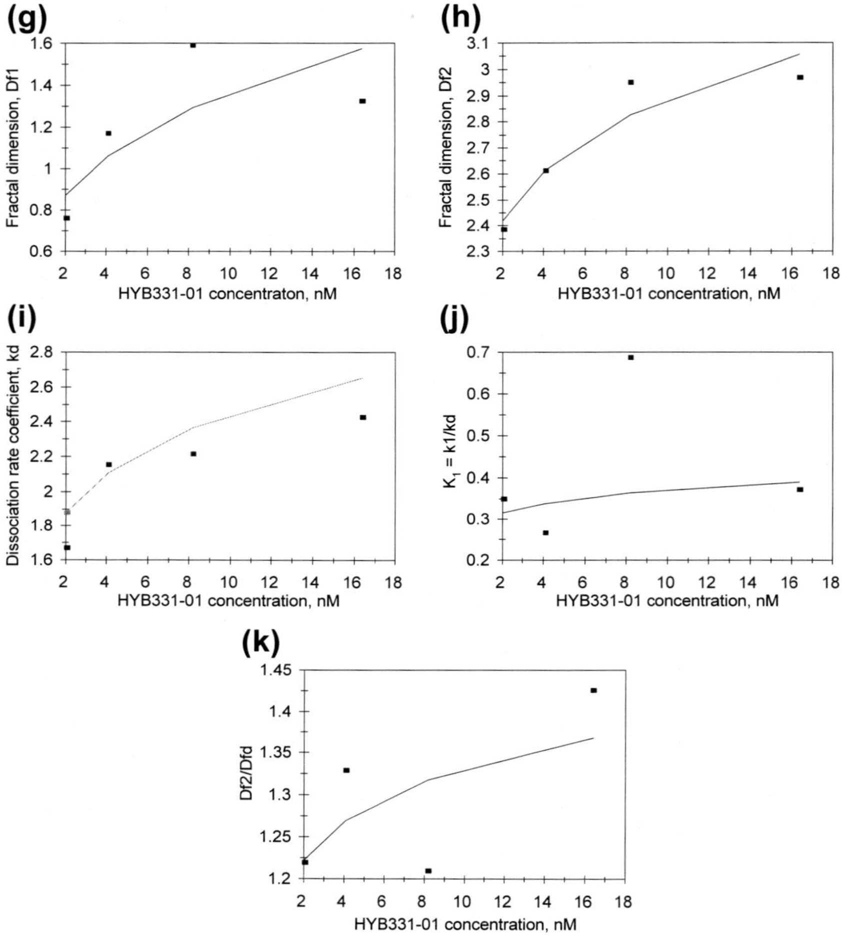

FIGURE 8.37 cont'd

Figure 8.37(c) shows the increase in the dissociation rate coefficient, k_d, with an increase in the anti-ds HYB331-01 concentration in solution. For the data shown in Figure 8.37(c), the dissociation rate coefficient, k_d, is given by:

$$k_d = (4.094 \pm 2.197)\,[\text{anti-dsHYB331} - 01]^{1.247 \pm 0.279} \tag{8.17c}$$

The fit is good. Only four data points are available. The availability of more data points would lead to a more reliable fit. The dissociation rate coefficient, k_d, exhibits an order of dependence between first- and one-half (equal to 1.247) order of dependence on the anti-ds HYB331-01 concentration in solution. This indicates that the dissociation rate coefficient, k_d, is sensitive to the anti-ds HYB331-01 concentration in solution.

Figure 8.37(d) shows the increase in the binding rate coefficient, k_1, with an increase in the fractal dimension, D_{f1}, or the degree of heterogeneity on the biosensor surface. For the data shown in Figure 8.37(d), the binding rate coefficient, k_1, is given by:

$$k_1 = (7.545 \pm 4.79)\, D_{f1}^{4.029 \pm 0.907} \tag{8.17d}$$

The fit is good. Only four data points are available. The availability of more data points would lead to a more reliable fit. The binding rate coefficient, k_1, exhibits slightly higher than a fourth (equal to 4.029) order of dependence on the fractal dimension, D_{f1}, or the degree of heterogeneity that exists on the SPR biosensor surface. This indicates that the binding rate coefficient, k_1, is very sensitive to the degree of heterogeneity that exists on the biosensor surface.

Figure 8.37(e) shows the increase in the binding rate coefficient, k_2, with an increase in the fractal dimension, D_{f2}, or the degree of heterogeneity on the biosensor surface. For the data shown in Figure 8.37(e), the binding rate coefficient, k_2, is given by:

$$k_2 = (0.022 \pm 0.0016)\, D_{f2}^{9.375 \pm 0.396} \tag{8.17e}$$

The fit is very good. Only four data points are available. The availability of more data points would lead to a more reliable fit. The binding rate coefficient, k_2, exhibits higher than a ninth (equal to 9.375) order of dependence on the fractal dimension, D_{f2}, or the degree of heterogeneity that exits on the SPR biosensor surface. This indicates that the binding rate coefficient, k_2, is extremely sensitive to the degree of heterogeneity that exists on the biosensor surface.

Figure 8.37(f) shows the increase in the dissociation rate coefficient, k_d, with an increase in the fractal dimension in the dissociation phase, D_{fd}, or the degree of heterogeneity on the biosensor surface. For the data shown in Figure 8.37(f), the dissociation rate coefficient, k_d, is given by:

$$k_d = (0.1673 \pm 0.0.198) D_{fd}^{7.235 \pm 0.4038} \tag{8.17f}$$

The fit is very good. Only four data points are available. The availability of more data points would lead to a more reliable fit. The dissociation rate coefficient, k_d, exhibits higher than a seventh (equal to 7.235) order of dependence on the fractal dimension in the dissociation phase, D_{fd}, or the degree of heterogeneity that exits on the SPR biosensor surface. This indicates that the dissociation rate coefficient, k_d, is extremely sensitive to the degree of heterogeneity that exists on the biosensor surface.

Figure 8.37(g) shows the increase in the fractal dimension, D_{f1}, with an increase in the HYB 331-01 concentration in nM in solution. For the data shown in Figure 8.37(g), the fractal dimension, D_{f1}, is given by:

$$D_{f1} = (0.7132 \pm 0.1784)\, [\text{HYB } 331 - 01]^{0.283 \pm 0.144} \tag{8.17g}$$

The fit is reasonable. Only four data points are available. The availability of more data points would lead to a more reliable fit. The fractal dimension, D_{f1}, exhibits a mild dependence (equal to 0.283) on the HYB 331-01 concentration in nM in solution.

Figure 8.37(h) shows the increase in the fractal dimension, D_{f2}, with an increase in the HYB 331-01 concentration in nM in solution. For the data shown in Figure 8.37(h), the fractal dimension, D_{f2}, is given by:

$$D_{f2} = (2.234 \pm 0.0.007)[\text{HYB}\ \ 331 - 01]^{0.1193 \pm 0.02459} \tag{8.17h}$$

The fit is reasonable. Only four data points are available. The availability of more data points would lead to a more reliable fit. The fractal dimension, D_{f2}, exhibits a mild dependence (equal to 0.1193) on the HYB 331-01 concentration in nM in solution.

Figure 8.37(i) shows the increase in the fractal dimension, with an increase in the HYB 331-01 concentration in nM in solution. Figure 8.37(j) shows the increase in the affinity, K_1 (= k/k_d) with an increase in the HYB 331-01 concentration in nM in solution. For the data shown in Figure 8.37(j), the affinity, K (= k/k_d) is given by:

$$K_1(= k/k_d) = (9.421 \pm 4.806)[\text{HYB } 331 - 01]^{-0.159 \pm 0.2659} \tag{8.17j}$$

There is scatter in the data. Only four data points are available. The availability of more data points would lead to a more reliable fit. The affinity, K (= k/k_d) exhibits a mild order of dependence (equal to 0.159) on the HYB 331-01 concentration in nM in solution.

Figure 8.37(k) shows the increase in the fractal dimension ratio (D_{f2}/D_{fd}) with an increase in the HYB 331-01 concentration in nM in solution. For the data shown in Figure 8.37(k), the fractal dimension ratio, D_{f2}/D_{fd}, is given by:

$$D_{f2}/D_{fd} = (11766 \pm 0.092)\ [\text{HYB } 331 - 0]^{0.0538 \pm 0.0482} \tag{8.17k}$$

There is scatter in the data. Only four data points are available. The availability of more data points would lead to a more reliable fit. The fractal dimension ratio, D_f/D_{fd}, exhibits a very mild order of dependence (0.05386) on the HYB 331-01 concentration in solution.

Konstantinov et al. (2009) have recently developed a biosensor for the rapid detection of antichromatin antibodies in human serum using a portable electrochemical biosensor. They indicate that monitoring of biomarkers is a mandatory component in the management of clinical diseases. These authors emphasize that though solid-phase immunoassays are extensively used, they are time consuming and involve nonportable equipment that requires technical expertise in the reporting and operation aspects. Thus, they indicate there is an urgent need to develop a device that is inexpensive and easy to operate.

Privett et al. (2008) indicate that biosensors demonstrate the potential to fill that need. Moreover, Konstantinov et al. (2009) indicate that they are ideally suitable to measure and detect clinical analytes. Besides being highly sensitive and selective they do exhibit a rapid response time (Wilkins and Sitdikov, 2006). However, Wang et al. (2008) indicate that there are various issues with regard to the use of biosensors that need to be addressed. Some of the limitations pointed out by these authors include portability, robustness for real-life samples, and operational stability of the biological receptor, poor reproducibility between sensors, and reduced specificity in complex matrices. Andreescu and Sadik (2004) emphasize that many a time a problem arises when the biosensor is used outside the laboratory.

Davidson and Diamond (2001) indicate that autoimmune and rheumatic diseases are a vexing problem and impose serious financial crisis on families which have a family member afflicted by these diseases. Lernmark (2001) indicate that these types of diseases are associated with antibody biomarkers which define the disease. These authors indicate that they are also used to monitor the efficacy of the therapy. Konstantinov et al. (2009) emphasize that the detection of these biomarkers may be particularly amenable to biosensor detection. These authors indicate that antichromatin (-nuclesome) autoantibodies are a sensitive indicator of SLE and related diseases (Rahman and Isenberg, 2008; Rubin and Fritzler, 2007). They have devised a rapid electrochemical-based biosensor for the detection and measurement of antichromatin antibodies, and validated their results to standard and well-accepted assay procedures.

Konstantinov et al. (2009) analyzed the response of antichromatin antibody from the sera of five patients with SLE. These were applied to the antichromatin biosensor and the signal output was measured over time. Their electrochemical biosensor required less than 20 min of processing time and used a hand-held apparatus with a disposable electrode. Chromatin was purified from calf thymus and stripped of histone HI (Lutter, 1978). The chromatin protein was bound to the PVDF membrane, and the membrane was placed over a plastic strip on which electrode sensors were screen printed.

Figure 8.38(a) shows the binding of antichromatin antibody from the sera of patient # 1 with SLE. The binding kinetics is adequately described by a dual-fractal analysis. The values of (a) the binding rate coefficient, k, and the fractal dimension, D_f, for a single-fractal analysis, and (b) the binding rate coefficients, k_1 and k_2, and the fractal dimensions, D_{f1}, and D_{f2} are given in Tables 8.24 and 8.25.

Figure 8.38(b) shows the binding of antichromatin antibody from the sera of patient # 2 with SLE. Here too, the binding kinetics is adequately described by a dual-fractal analysis. The values of (a) the binding rate coefficient, k, and the fractal dimension, D_f, for a single-fractal analysis, and (b) the binding rate coefficients, k_1 and k_2, and the fractal dimensions, D_{f1} and D_{f2}, are given in Tables 8.24 and 8.25.

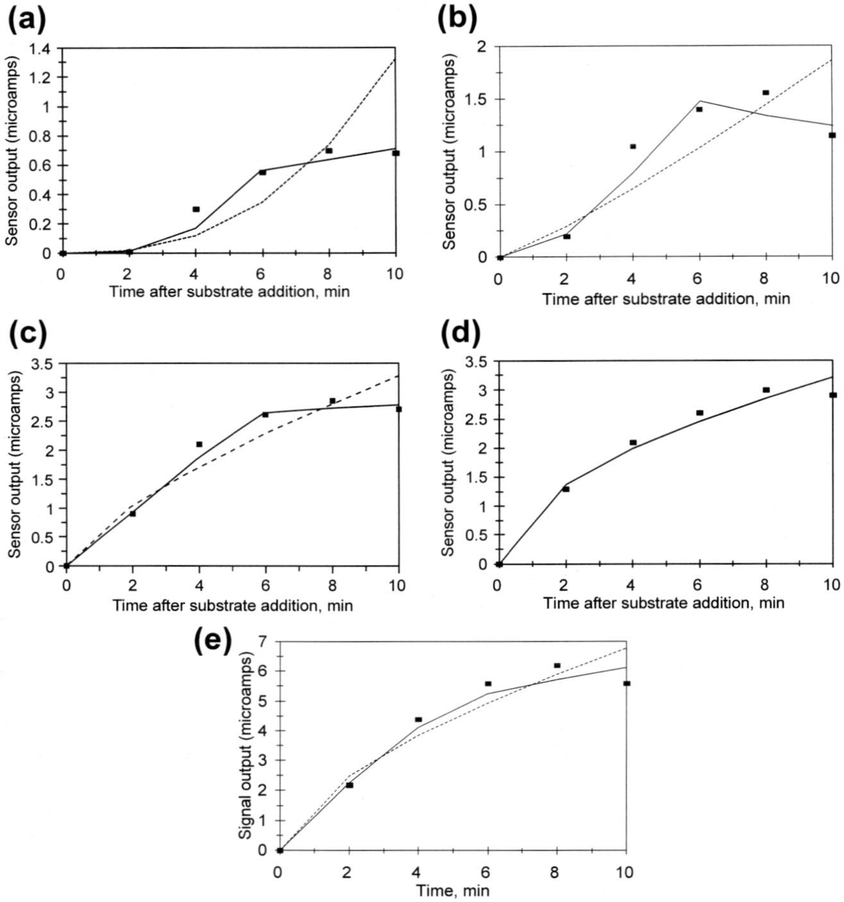

FIGURE 8.38 Binding of antichromatin antibody from the sera of five patients with SLE to a chromatin-coated membrane on an electrochemical biosensor (Konstantinov et al., 2009): (a) patient #1; (b) patient #2; (c) patient #3; (d) patient #4; and (e) patient #5.

Figure 8.38(c) shows the binding of antichromatin antibody from the sera of patient # 3 with SLE. Once again, the binding kinetics is adequately described by a dual-fractal analysis. The values of (a) the binding rate coefficient, k, and the fractal dimension, D_f, for a single-fractal analysis, and (b) the binding rate coefficients, k_1 and k_2, and the fractal dimensions, D_{f1} and D_{f2}, are given in Tables 8.24 and 8.25.

Figure 8.38(d) shows the binding of antichromatin antibody from the sera of patient # 4 with SLE. In this case, the binding kinetics is adequately described by a single-fractal analysis. The values of the binding rate coefficient, k, and the fractal dimension, D_f, for a single-fractal analysis are given in

TABLE 8.24 Binding Rate Coefficients for Chromatin Response Using an Anti-Chromatin Electrochemical Biosensor from the Sera of Five Patients with SLE (Konstantinov et al., 2009)

Chromatin in Solution from Sera of Patients with SLE	k	k_1	k_2
Patient #1	0.003148 ± 0.003894	0.000896 ± 0.00097	0.2611 ± 0.0272
Patient #2	0.1321 ± 0.824	0.0620 ± 0.0258	2.7886 ± 0.5270
Patient #3	0.6476 ± 0.1431	0.4718 ± 0.0730	2.2739 ± 1.1373
Patient #4	0.9544 ± 0.0869	NA	NA
Patient #5	1.6229 ± 0.3058	1.2371 ± 0.1080	3.068 ± 0.336

TABLE 8.25 Fractal Dimensions During the Binding of Chromatin Response Using an Anti-Chromatin Electrochemical Biosensor from the Sera of Five Patients with SLE (Konstantinov et al., 2009)

Chromatin in Solution from Sera of Patients with SLE	D_f	D_{f1}	D_{f2}
Patient #1	0 + 1.2670	0 + 1.7938	2.1294 ± 0.5470
Patient #2	0.7040 + 0.7626	0 + 0.8848	0
Patient #3	1.5934 ± 0.3142	1.0136 ± 0.3656	2.8291 ± 0.3172
Patient #4	1.9444 ± 0.1369	NA	NA
Patient #5	1.7596 ± 0.2716	1.7330 ± 0.2030	2.4004 ± 0.3024

Tables 8.24 and 8.25. No explanation is offered at present as to why the binding kinetics of antichromatin antibody from the sera of patient number 4 may be adequately described by a single-fractal analysis, and the binding kinetics of the antichromatin antibody from the sera of patients 1, 2, and 3 require a dual-fractal analysis.

Figure 8.38(e) shows the binding of antichromatin antibody from the sera of patient # 5 with SLE. Once again, the binding kinetics is adequately described by a dual-fractal analysis. The values of (a) the binding rate

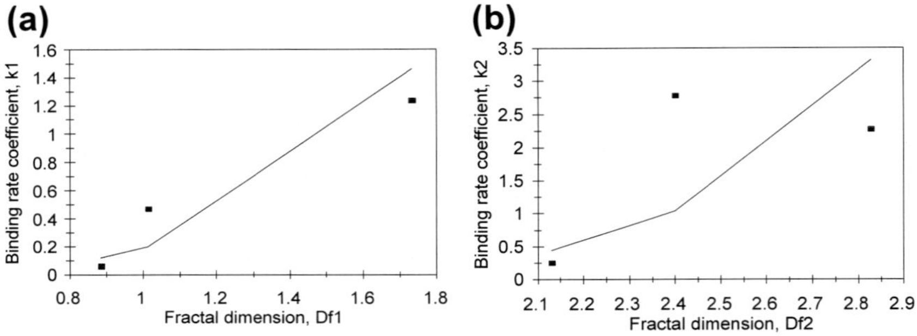

FIGURE 8.39 (a) Increase in the binding rate coefficient, k_1, with an increase in the fractal dimension, D_{f1}. (b) Increase in the binding rate coefficient, k_2, with an increase in the fractal dimension, D_{f2}.

coefficient, k, and the fractal dimension, D_f, for a single-fractal analysis, and (b) the binding rate coefficients, k_1 and k_d, and the fractal dimensions, D_{f1} and D_{f2}, are given in Tables 8.24 and 8.25.

Figure 8.39(a) shows the increase in the binding rate coefficient, k_1, with an increase in the fractal dimension, D_{f1}, for a dual-fractal analysis. For the data shown in Figure 8.39(a), the binding rate coefficient, k_1, is given by:

$$k_1 = \left(0.191 + 0.3846\right)D_{f1}^{3.70\pm2.187} \tag{8.18a}$$

The fit is reasonable. Only three data points are available. The availability of more data points would lead to a more reliable fit. The binding rate coefficient, k_1, exhibits an order of dependence between three and one-half and fourth (equal to 3.7) order on the fractal dimension, D_{f1}, or the degree of heterogeneity that exists on the surface. This indicates that the binding rate coefficient, k_1, is very sensitive to the degree of heterogeneity that exists on the electrochemical biosensor surface.

Figure 8.39(b) shows the increase in the binding rate coefficient, k_2, with an increase in the fractal dimension, D_{f2} for a dual-fractal analysis. For the data shown in Figure 8.39(b), the binding rate coefficient, k_2, is given by:

$$k_2 = (0.0021 + 0.00490)\ D_{f2}^{7.08\pm5.87} \tag{8.18b}$$

There is scatter in the data. This is reflected in the error in the estimate of the binding rate coefficient. Only three data points are available. The availability of more data points would lead to a more reliable fit. The binding rate coefficient, k_2, exhibits an order of dependence slightly more than seventh (equal to 7.08) order on the fractal dimension, D_{f2}, or the degree of heterogeneity that exists on the surface. This indicates that the binding rate coefficient, k_2, is extremely sensitive to the degree of heterogeneity that exists on the electrochemical biosensor surface.

Thaler et al. (2009) have recently reviewed the detection of serum antibodies by biosensors. They analyzed biomarkers for antiphospholipid syndrome and SLE. These authors state that autoimmune diseases are rare human diseases that affect 1.5% of the Caucasian population. They are systemic and include RA, antiphospholipid syndrome, and SLE (Tozzoli, 2008). These diseases, the authors claim, are characterized by the presence of circulating antibodies which are bound to specific antigens in the human body (Atassi and Casali, 2008). Thaler et al. (2009) indicate that some of the autoantibodies are closely linked to characterized antigenic compounds (Eaton et al., 2007; Cooper and Strohla, 2003), for example, cyclic citrullinated peptide with RA. Thaler et al. (2009) indicated that during the natural course of these diseases, organ destructions are mediated by T-cell cytotoxicity (Watts, 2002).

Thaler et al. (2009) emphasize that the measurement of these autoantibodies in affected patients is an important guide to the diagnosis and disease progression. The tests are not straight-forward and frequently lead to confusion. The authors point out that many serum autoantibodies may be detected a long time before the clinical onset of the disease. Scofield (2004) emphasizes that a follow-up of antibody levels is a useful means of disease activity.

Thaler et al. (2009) emphasize the need to reliably differentiate binding characteristics. Leon et al. (1977), Cui and Zhao (2005), and Saraux et al. (1997) have indicated the associations between the affinites/avidites of autoantibodies and their pathogenic potentials for different autoimmune disorders. Thaler et al. (2009) emphasize the need for improved analytical approaches to improve the diagnoses of autoimmune disorders.

They emphasize that the routine determination of autoantibodies is performed by immunofluorescence microscopy and/or immunoassays (von Landenberg et al., 2001). These methods are aided in their application due to the availability of highly purified or recombinant autoantigens used in these methods (Tozzoli, 2007). Optical immunosensors are particularly useful in determining immunoglobulin subtypes which contribute to clinical signs of SLE and Antiphospholipid Antibody Syndrome (APS) (De Laat et al., 2006; Nossent et al., 1989). McDonnell (2001) emphasize the importance of analyzing the association and dissociation phases, which provide insights into the binding characteristics of the respective antibodies. Finally, Sedel (2005) emphasize that biosensors are well suited for multiplexed analysis.

Thaler et al. (2009) indicate that applications of immunosensors for the detection of autoimmune diseases are available, for example, identification of relevant subtypes of autoantigens in RA (Hueber et al., 2005; Lyons et al., 2005) and the detection of autoantibody clusters that predict lupus nephritis disease activity using glomerular arrays (Li et al., 2005).

Thaler et al. (2009) have obtained binding curves by injecting 1:100-diluted sera from patients with antiphospholipid syndrome and systemic lupus erythemoatosus. Figure 8.40(a) shows the binding and dissociation of 1:100-diluted sera from patient #1 with APS onto a surface coated with both

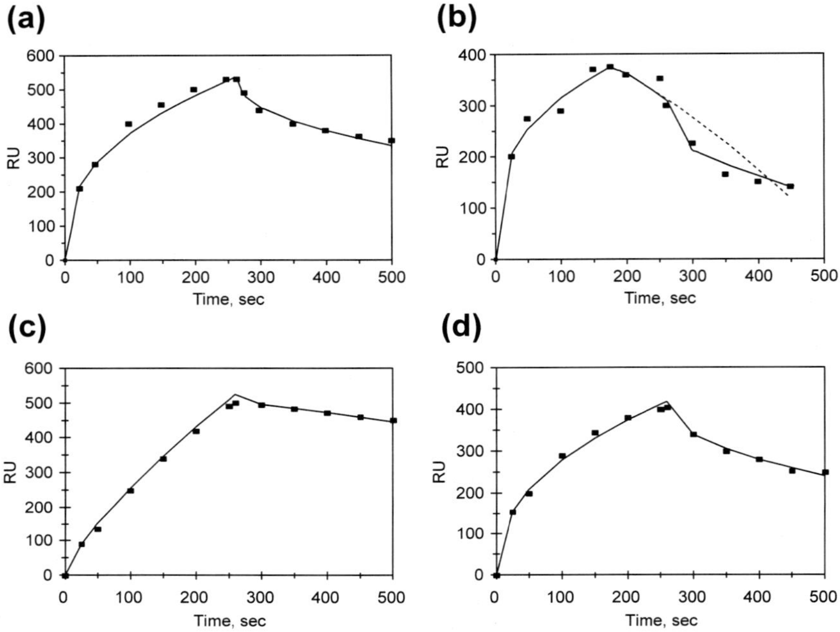

FIGURE 8.40 Binding and dissociation curves for 1:100 diluted sera from patients with anti-phospholipid syndrome onto a biosensor surface coated with both 745 RU pelCL (1,12,2′-tetra-9-oxopelargonoyl cardiolipin) and 2052 RU β2-GPl (β-glycoprotein 1) (Thaler et al., 2009): (a) patient #1; (b) patient #2; (c) patient #3; and (d) patient #4.

745 RU of pelCL (1,12,2′-tetra-9-oxopelargonoyl cardiolipin) and 2052 RU β2-GP1 (β2 glycoprotein 1). A single-fractal analysis is adequate to describe the binding and the dissociation kinetics. The values of (a) the binding rate coefficient, k, and the fractal dimension, D_f, for a single-fractal analysis, and (b) the dissociation rate coefficient, k_d, and the fractal dimension for dissociation, D_{fd}, for a single-fractal analysis are given in Tables 8.26 and 8.27. The affinity, K (= k/k_d) is equal to 4.27.

Figure 8.40(b) shows the binding and dissociation of 1:100-diluted sera from patient #2 with APS onto a surface coated with both 745 RU of pelCL (1,12,2′-tetra-9-oxopelargonoyl cardiolipin) and 2052 RU β2-GP1 (β2 glycoprotein 1). A single-fractal analysis is adequate to describe the binding kinetics. The dissociation kinetics is adequately described by a dual-fractal analysis. The values of (a) the binding rate coefficient, k, and the fractal dimension, D_f, for a single-fractal analysis, (b) the dissociation rate coefficient, k_d, and the fractal dimension for dissociation, D_{fd}, for a single-fractal analysis, and (c) the dissociation rate coefficients, k_1 and k_2, and the fractal dimensions for dissociation, D_{f1} and D_{f2}, are given in Tables 8.26 and 8.27. In this case, the affinity, K_1 (= k/k_{d1}) and K_2 (= k/k_{d2}) are 642.17 and 4.415, respectively.

TABLE 8.26 Binding and Dissociation Rate Coefficients from the Sera of Patients with APS and SLE to 745 RU of PelCL and β2-GP1 Immobilized to a SPR Biosensor Surface (Thaler et al., 2009)

Patient #	k	k_d	k_{d1}	k_{d2}
1	64.066 ± 4.095	14.988 ± 1.531	NA	NA
2	76.226 ± 5.568	0.2516 ± 0.1363	NA	NA
3	7.6126 ± 0.319	0.04598 ± 0.00464	NA	NA
4	39.654 ± 1.481	10.541 ± 5.568	NA	NA

TABLE 8.27 Fractal Dimensions for the Binding and Dissociation Phase from the Sera of Patients with APS and SLE to 745 RU of PelCL and β2-GP1 Immobilized to a SPR Biosensor Surface (Thaler et al., 2009)

Patient #	D_f	D_{fd}	D_{fd1}	D_{fd2}
1	2.2380 ± 0.05322	2.622 ± 0.07286	NA	NA
2	2.3832 ± 0.08692	0.5316 ± 0.388	NA	NA
3	1.4910 ± 0.03792	0.04598 ± 0.00464	NA	NA
4	2.1530 ± 0.0338	10.541 ± 5.568	NA	NA

Figure 8.40(c) shows the binding and dissociation of 1:100-diluted sera from patient #3 with APS onto a surface coated with both 745 RU of pelCL (1,12,2′-tetra-9-oxopelargonoyl cardiolipin) and 2052 RU β2-GP1 (β2 glycoprotein 1). A single-fractal analysis is adequate to describe the binding and the dissociation kinetics. The values of (a) the binding rate coefficient, k, and the fractal dimension, D_f, for a single-fractal analysis, and (b) the dissociation rate coefficient, k_d, and the fractal dimension for dissociation, D_{fd}, for a single-fractal analysis are given in Tables 8.26 and 8.27. The affinity, K (= k/k_d) is equal to 172.76.

Figure 8.40(d) shows the binding and dissociation of 1:100-diluted sera from patient #4 with APS onto a surface coated with both 745 RU of pelCL (1,12,2′-tetra-9-oxopelargonoyl cardiolipin and 2052 RU β2-GPl (β2 glycoprotein 1)). A single-fractal analysis is adequate to describe the binding and the dissociation kinetics. The values of (a) the binding rate coefficient, k, and the fractal dimension,

D_f, for a single-fractal analysis, and (b) the dissociation rate coefficient, k_d, and the fractal dimension for dissociation, D_{fd} for a single-fractal analysis are given in Tables 8.26 and 8.27. The affinity, K (= k/k_d) is equal to 3.762.

Figure 8.41(a) shows the increase in the binding rate coefficient, k, with an increase in the fractal dimension, D_f for a single-fractal analysis. For the data presented in Figure 8.41(a), the binding rate coefficient, k, is given by:

$$k = \left(1.125 \pm 0.179\right) D_f^{4.848 \pm 0.401} \tag{8.19a}$$

The fit is good. Only four data points are available. The availability of more data points would lead to a more reliable fit. The binding rate coefficient, k, is very sensitive to the fractal dimension, D_f, or the degree of heterogeneity on the SPR surface as it exhibits an order of dependence between four and one-half and five (equal to 4.848) on the degree of heterogeneity on the SPR biosensor surface.

Figure 8.41(b) shows the increase in the dissociation rate coefficient, k_d, with an increase in the fractal dimension, D_{fd}, for a single-fractal analysis. For the data presented in Figure 8.40(b), the dissociation rate coefficient, k_d is given by:

$$k_d = (1.017 \pm 0.186) D_{fd}^{3.554 \pm 1.203} \tag{8.19b}$$

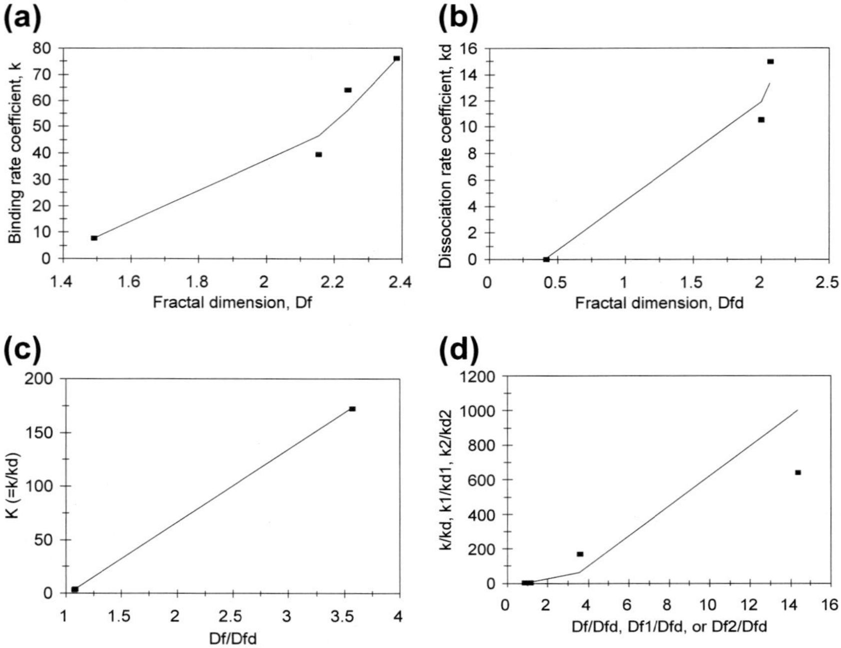

FIGURE 8.41 (a) Increase in the binding rate coefficient, k, with an increase in the fractal dimension, D_f. (b) Increase in the dissociation rate coefficient, k_d, with an increase in fractal dimension, D_{fd}. (c) Increase in affinity, K with an increase in fractal dimension ratio, D_f/D_{fd} (d) Increase in affinities, K, K_1, K_2 with increase in different fractal dimension ratios.

The fit is good. Only three data points are available. The availability of more data points would lead to a more reliable fit. The dissociation rate coefficient, k_d, is very sensitive to the fractal dimension, for dissociation, D_{fd}, or the degree of heterogeneity on the SPR surface as it exhibits an order of dependence slightly greater than three and one-half order (equal to 3.554) on the degree of heterogeneity on the SPR biosensor surface.

Figure 8.41(c) shows the increase in the affinity, K (= k/k_d) with an increase in the fractal dimension ratio, D_f/D_{fd}. For the data shown in Figure 8.41(c), the affinity, K (= k/k_d) is given by:

$$K = (3.131 \pm 0.247)\left(D_f/D_{fd}\right)^{3.154 \pm 0.0781} \tag{8.19c}$$

The fit is good. Only three data points are available. The availability of more data points would lead to a more reliable fit. The affinity, K exhibits slightly greater than the third (equal to 3.154) order of dependence on the ratio of fractal dimensions, D_f/D_{fd}. This indicates that the affinity, K is very strongly dependent on the ratio of fractal dimensions, D_f/D_{fd}, that exists on the biosensor surface.

Figure 8.41(d) shows the increase in the different ratios of the binding and dissociation rate coefficients with an increase in the different fractal dimension ratios. For the data shown in Figure 8.41(d), the different affinities, are given by:

$$\text{Different affinties, } K_i = (4.835 + 5.181)\,(\text{different fractal dimension ratios})^{2.00 \pm 0.308} \tag{8.19d}$$

The fit is good. Only four data points are available. The availability of more data points would lead to a more reliable fit. The different affinities, K, K_1, K_2 exhibit a second order on the different ratio of fractal dimensions that exist on the biosensor surface.

Figure 8.42 shows the binding and the dissociation curves for 1:100 diluted sera from patients with antiphospholipid syndrome onto a biosensor coated with 1161 RU pelCL (Thaler et al., 2009). Figure 8.42(a) shows the data from the sera from patient #1. A single-fractal analysis is adequate to describe the binding and the dissociation kinetics. The values of the (a) binding rate coefficient, k, and the fractal dimension, D_f, for a single-fractal analysis, and (b) the dissociation rate coefficient, k_d, and the fractal dimension for dissociation, D_{f1}, are given in Table 8.28. In this case the affinity, K (= k/k_d) value is 6.60.

Figure 8.42(b) shows the binding and the dissociation curves for 1:100 diluted sera from patient #2 with antiphospholipid syndrome onto a biosensor coated with 1161 RU pelCL (Thaler et al., 2009). A single-fractal analysis is adequate to describe the binding and the dissociation kinetics. The values of the (a) binding rate coefficient, k, and the fractal dimension, D_f, for a single-fractal analysis, and (b) the dissociation rate coefficient, k_d, and the fractal dimension for dissociation, D_{fd}, are given in Table 8.28. In this case the affinity, K (= k/k_d) value is 4.0.

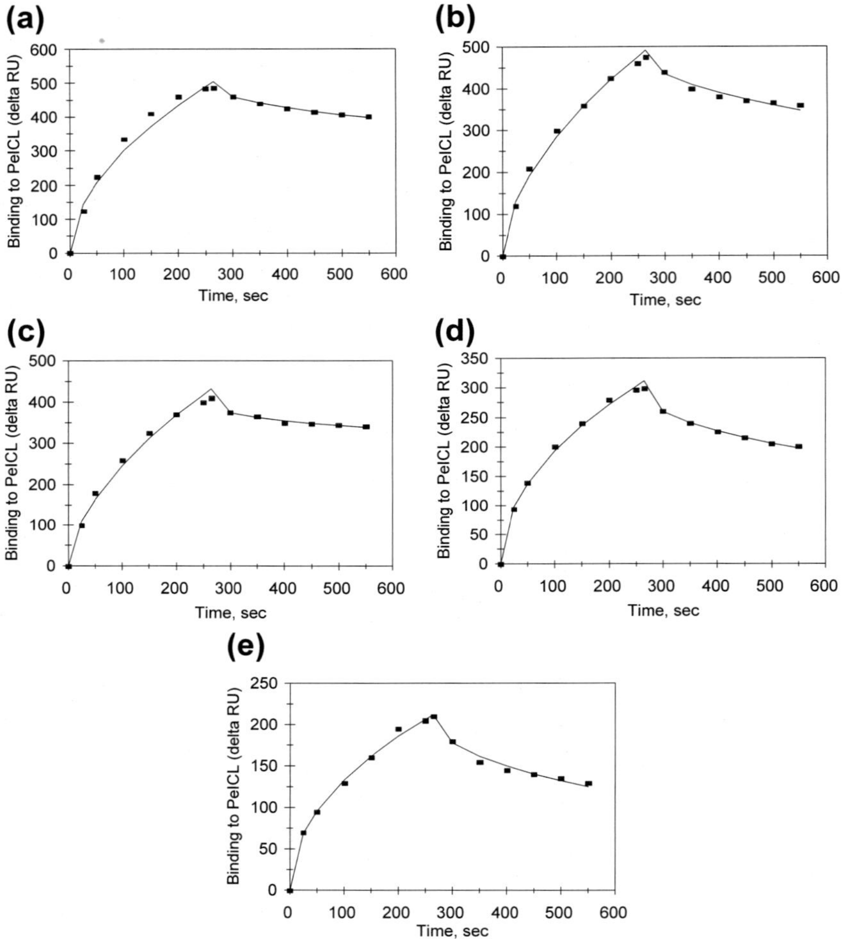

FIGURE 8.42 Binding and dissociation curves for 1:100 diluted sera from patients with antiphospholipid syndrome onto a biosensor surface coated with 1161 RU pelCL (1,1′2,2′-tetra-9-oxopelargonoyl cardiopolin): (a) patient #1; (b) patient #2; (c) patient #3; (d) patient #4; and (e) patient #5.

Figure 8.42(c) shows the binding and the dissociation curves for 1:100 diluted sera from patient #3 with antiphospholipid syndrome onto a biosensor coated with 1161 RU pelCL (Thaler et al., 2009). A single-fractal analysis is adequate to describe the binding and the dissociation kinetics. The values of the (a) binding rate coefficient, k, and the fractal dimension, D_f, for a single-fractal analysis, and (b) the dissociation rate coefficient, k_d, and the fractal dimension for dissociation, D_{fd}, are given in Table 8.28. In this case the affinity, K (= k/k_d) value is 1.623.

TABLE 8.28a Binding and Dissociation Rate Coefficients from the Sera of Patients with APS and SLE to 745 RU of PelCL and β2-GP1 Immobilized to a SPR Biosensor Surface (Thaler et al., 2009)

Patient #	k	k_d	k_{d1}	k_{d2}
1	64.066 ± 4.095	14.988 ± 1.531	NA	NA
2	76.226 ± 5.568	0.2516 ± 0.1363	NA	NA
3	7.6126 ± 0.319	0.04598 ± 0.00464	NA	NA
4	39.654 ± 1.481	10.541 ± 5.568	NA	NA

TABLE 8.28b Fractal Dimensions for the Binding and Dissociation Phase from the Sera of Patients with APS and SLE to 745 RU of PelCL and β2-GP1 Immobilized to a SPR Biosensor Surface (Thaler et al., 2009)

Patient #	D_f	D_{fd}	D_{fd1}	D_{fd2}
1	2.2380 ± 0.05322	2.622 ± 0.07286	NA	NA
2	2.3832 ± 0.08692	0.5316 ± 0.388	NA	NA
3	1.4910 ± 0.03792	0.04598 ± 0.00464	NA	NA
4	2.1530 ± 0.0338	10.541 ± 5.568	NA	NA

Figure 8.42(d) shows the binding and the dissociation curves for 1:100 diluted sera from patient #4 with antiphospholipid syndrome onto a biosensor coated with 1161 RU pelCL (Thaler et al., 2009). A single-fractal analysis is adequate to describe the binding and the dissociation kinetics. The values of the (a) binding rate coefficient, k, and the fractal dimension, D_f, for a single-fractal analysis, and (b) the dissociation rate coefficient, k_d, and the fractal dimension for dissociation, D_{fd}, are given in Table 8.28. In this case the affinity, K (= k/k_d) value is 2.53.

The binding and the dissociation curves for 1:100 diluted sera from patient #5 with antiphospholipid syndrome onto a biosensor coated with 1161 RU pelCL (Thaler et al., 2009). A single-fractal analysis is adequate to describe the binding and the dissociation kinetics. The values of the (a) binding rate coefficient, k, and the fractal dimension, D_f, for a single-fractal analysis, and (b) the dissociation rate coefficient, k_d, and the fractal dimension for dissociation, D_{fd}, are given in Table 8.28. In this case the affinity, K (= k/k_d) value is 2.53.

Figure 8.42(f) shows the binding and the dissociation curves for 1:100 diluted sera from patient #6 with antiphospholipid syndrome onto a biosensor coated with 1161 RU pelCL was also analyzed (Thaler et al., 2009). Figure not shown. A single-fractal analysis is adequate to describe the binding and the dissociation kinetics. The values of the (a) binding rate coefficient, k, and the fractal dimension, D_f, for a single-fractal analysis, and (b) the dissociation rate coefficient, k_d, and the fractal dimension for dissociation, D_{fd}, are given in Table 8.28. In this case the affinity, K (= k/k_d) value is 2.53.

Figure 8.43(a) shows the decrease in the binding rate coefficient, k, with an increase in the fractal dimension, D_f, for a single-fractal analysis. For the data presented in Figure 8.43(a), the binding rate coefficient, k, is given by:

$$k = \left(12349.7 + 18651.6\right)D_f^{-10.38 \pm 9.5261} \tag{8.20a}$$

There is scatter in the data. Only five data points are presented. The availability of more data points would lead to a more reliable fit. The binding rate coefficient, k, is very sensitive to the fractal dimension, D_f, or the degree of heterogeneity on the SPR surface as it exhibits a negative order of dependence between 10 and then and one-half (equal to −10.38) on the degree of heterogeneity on the SPR biosensor surface.

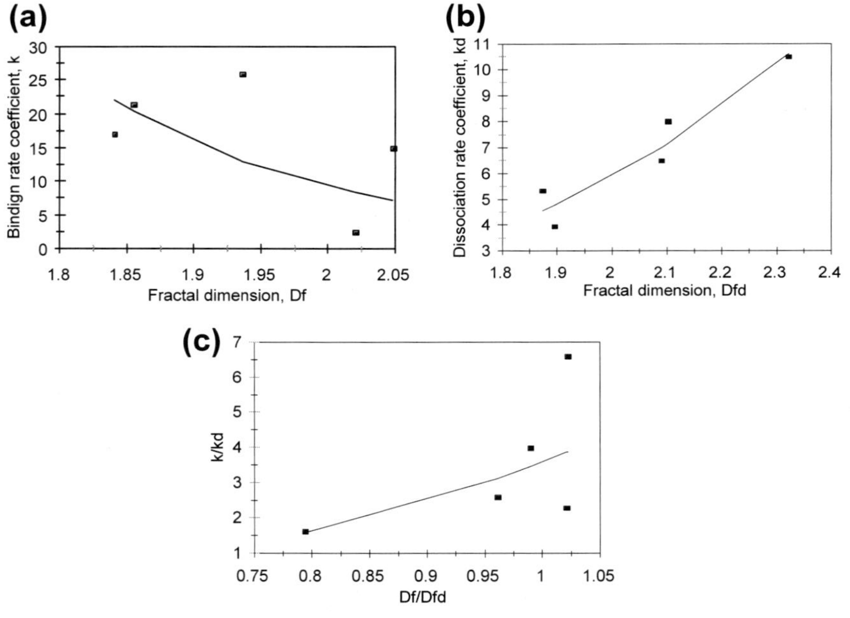

FIGURE 8.43 (a) Decrease in the binding rate coefficient, k, with an increase in the fractal dimension, D_f. (b) Increase in the dissociation rate coefficient, k_d, with an increase in fractal dimension, D_{fd} (c) Increase in the affinity, K (= k/k_d) with an increase in the fractal dimension ratio, D_f/D_{fd}.

Figure 8.43(b) shows the increase in the dissociation rate coefficient, k_d, with an increase in the fractal dimension, D_{fd}, for a single-fractal analysis. For the data presented in Figure 8.43(b), the dissociation rate coefficient, k_d, is given by:

$$k_d = \left(0.3779 \pm 0.0670\right) D_{fd}^{3.96 \pm 0.931} \qquad (8.20b)$$

The fit is good. Only five data points are available. The availability of more data points would lead to a more reliable fit. The dissociation rate coefficient, k_d, is very sensitive to the fractal dimension for dissociation, D_{fd}, or the degree of heterogeneity on the SPR surface as it exhibits an order of dependence close to fourth (equal to 3.96) on the degree of heterogeneity on the SPR biosensor surface.

Figure 8.43(c) shows the increase in the affinity $K (= k/k_d)$ with an increase in the fractal dimension ratio, D_f/D_{fd}. For the data presented in Figure 8.43(c), the affinity K is given by:

$$K = (3.5835 \pm 2.041)\,(D_f/D_{fd})^{3.558 \pm 2.138} \qquad (8.20c)$$

There is scatter in the data. Only five data points are presented. The availability of more data points would lead to a more reliable fit. The affinity, K is sensitive to the ratio of fractal dimensions, D_f/D_{fd}, or the degree of heterogeneity on the SPR surface as it exhibits slightly greater than three and one-half order (equal to 3.558) of dependence on the degree of heterogeneity on the SPR biosensor surface.

Figure 8.44(a) shows the binding and dissociation of 1:100 diluted sera from patients with APS (Thaler et al., 2009). Figure 8.44(a) shows the binding and dissociation of the sera from patient #1. A single-fractal analysis is required to adequately describe the binding and dissociation kinetics. The values of the binding rate coefficient, k, and the fractal dimension, D_f, for a single-fractal analysis, and (b) the dissociation rate coefficient, k_d, and the fractal dimension for dissociation, D_{fd}, are given in Table 8.29 a and b. In this case, the affinity $K (= k/k_d)$ is equal to 0.115.

Figure 8.44(b) shows the binding and dissociation of the sera of patient #2. A single-fractal analysis is required to adequately describe the binding and dissociation kinetics. The values of the binding rate coefficient, k, and the fractal dimension, D_f, for a single-fractal analysis, and (b) the dissociation rate coefficient, k_d, and the fractal dimension for dissociation, D_{fd}, are given in Table 8.29 a and b. In this case, the affinity $K (= k/k_d)$ is equal to 1.009.

Figure 8.44(c) shows the binding and dissociation of the sera from patient #3. A single-fractal analysis is, once again, required to adequately describe the binding and dissociation kinetics. The values of the binding rate coefficient, k, and the fractal dimension, D_f, for a single-fractal analysis, and (b) the dissociation rate coefficient, k_d, and the fractal dimension for dissociation, D_{fd}, are given in Table 8.29 a and b. In this case, the affinity $K (= k/k_d)$ is equal to 1.67.

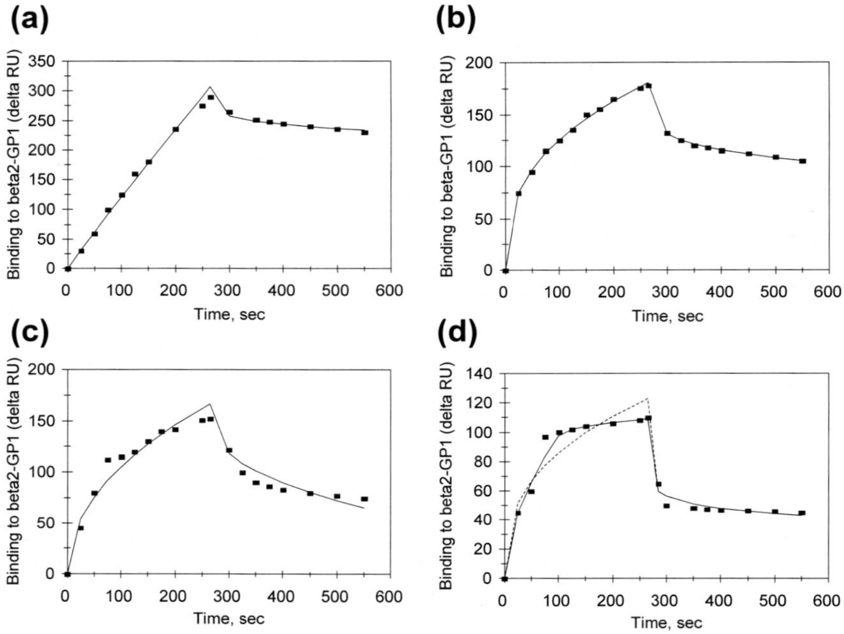

FIGURE 8.44 Binding and dissociation curves for 1:100 diluted sera from patients with anti-phospholipid syndrome onto a biosensor surface coated with 1161RU β2-GP19 (β-glycoprotein-1): (a) patient #1; (b) patient #2; (c) patient #3; and (d) patient #4.

Figure 8.44(d) shows the binding and dissociation of the sera of patient #4. A single-fractal analysis is required to adequately describe the binding and dissociation kinetics. The values of the binding rate coefficient, k, and the fractal dimension, D_f, for a single-fractal analysis, the binding rate coefficients, k_1 and k_2 and the fractal dimensions, D_{f1} and D_{f2}, for a dual-fractal analysis and (b) the dissociation rate coefficient, k_d, and the fractal dimension for dissociation, D_{fd}, are given in Table 8.29 a and b. In this case, the affinity K (= k/k_d) is equal to 3.76.

TABLE 8.29a Binding and Dissociation Rate Coefficients from the Sera of Patients with APS and SLE to 1161 RU of β2-GP1 Immobilized to a SPR Biosensor Surface (Thaler et al., 2009)

Patient #	k	k_1	k_2	k_d
1	1.4575 ± 0.0900	NA	NA	12.708 ± 0.804
2	22.505 ± 0.373	NA	NA	22.299 ± 0.360
3	11.716 ± 1.413	NA	NA	7.0307 ± 0.8154
4	15.89 ± 2.51	7.339 ± 0.908	22.652	65.93 ± 0.49

TABLE 8.29b Fractal Dimensions for the Binding and Dissociation Phase from the Sera of Patients with APS and SLE to 1161 RU of β2-GP1 Immobilized to a SPR Biosensor Surface (Thaler et al., 2009)

Patient #	D_f	D_{f1}	D_{f2}	D_{fd}
1	1.0822 ± 0.0542	NA	NA	2.4722 ± 0.07052
2	2.254 ± 0.01458	NA	NA	2.5806 ± 0.0170
3	2.0462 ± 0.1010	NA	NA	2.1146 ± 0.1170
4	2.2266 ± 0.0639	1.8758 ± 0.1834	2.487	2.7192 ± 0.0334

Figure 8.45(a) shows the increase in the binding rate coefficient, k, for a single-fractal analysis with an increase in the fractal dimension, D_f. For the data shown in Figure 8.45(a), the binding rate coefficient, k, is given by:

$$k = (1.1769 \pm 0.212) D_f^{2.98 \pm 0.387} \tag{8.21a}$$

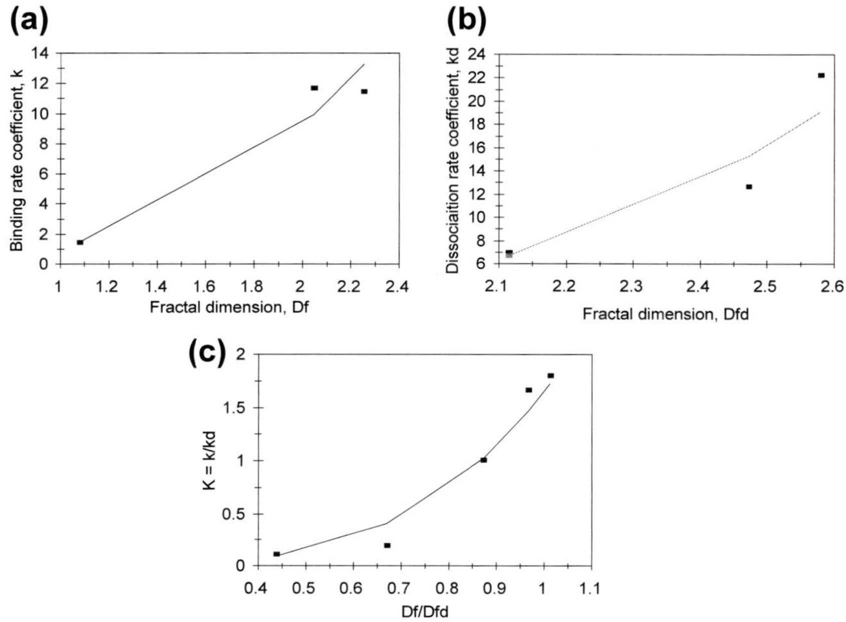

FIGURE 8.45 (a) Increase in the binding rate coefficient, k, with an increase in the fractal dimension, D_f. (b) Increase in the dissociation rate coefficient, k_d, with an increase in fractal dimension, D_{fd}. (c) Increase in the affinity K (= k/k_d) with an increase in the fractal dimension ratio, D_f/D_{fd}.

The fit is good. Only three points are available. The availability of more data points would lead to a more reliable fit. The binding rate coefficient, k, exhibits close to a third (equal to 2.98) order of dependence on the fractal dimension, D_f, or the degree of heterogeneity on the biosensor surface. This indicates that the binding rate coefficient, k, is sensitive to the degree of heterogeneity on the biosensor surface.

Figure 8.45(b) shows the increase in the dissociation rate coefficient, k_d, for a single-fractal analysis with an increase in the fractal dimension, D_{fd}. For the data shown in Figure 8.45(b), the dissociation rate coefficient, k_d, is given by:

$$k_d = (0.1323 \pm 0.0365) D_{fd}^{5.25 \pm 1.642} \tag{8.21b}$$

The fit is good. Only three points are available. The availability of more data points would lead to a more reliable fit. The dissociation rate coefficient, k_d, exhibits close to a five and one-half order (equal to 5.25) order of dependence on the fractal dimension, D_{fd}, or the degree of heterogeneity on the biosensor surface. This indicates that the dissociation rate coefficient, k_d, is very sensitive to the degree of heterogeneity on the biosensor surface.

Figure 8.45(c) shows the increase in the affinity ($K = k/k_d$) with an increase in the fractal dimension ratio, D_f/D_{f1}. For the data shown in Figure 8.45(c), the affinity, K, is given by:

$$K = (1.550 \pm 0.831)(D_f/D_{fd})^{3.51 \pm 0.0619} \tag{8.21c}$$

The fit is very good. Only five points are available. The availability of more data points would lead to a more reliable fit. The affinity, K exhibits close to a three and one-half (equal to 3.51) order of dependence on the ratio of fractal dimension, D_f/D_{fd}, or the ratio of degrees of heterogeneity on the biosensor surface. This indicates that the affinity, K is very sensitive to the degree of heterogeneities on the biosensor surface. It is of interest to note that in this case the data for both the single- and the dual-fractal analysis are plotted together.

Thaler et al. (2009) indicate that immunoglobulin isotypes IgG, IgM, and IgA are reported to be present in parallel in sera from SLE patients (Forger et al., 2004). Thaler thought it worthwhile to clarify the role of each of them and their distribution. The authors removed the isotypes. They compared the depleted sera with no antibodies (positive control), IgG and IgM (IgA response), IgA and IgM (IgG response), IgG and IgA (IgM response), or all antibodies (negative control). Figure 8.46(a) shows the binding and dissociation of isotypes in serum a-Q (positive control). A single-fractal analysis is required to adequately describe the binding and dissociation kinetics. The values of the (a) binding rate coefficient, k, and the fractal dimension, D_f, for a single-fractal analysis, and (b) the dissociation rate coefficient, k_d, and the fractal dimension for dissociation, D_{fd}, are given in Table 8.30. The affinity K value in this case K ($= k/k_d$) is 2.94.

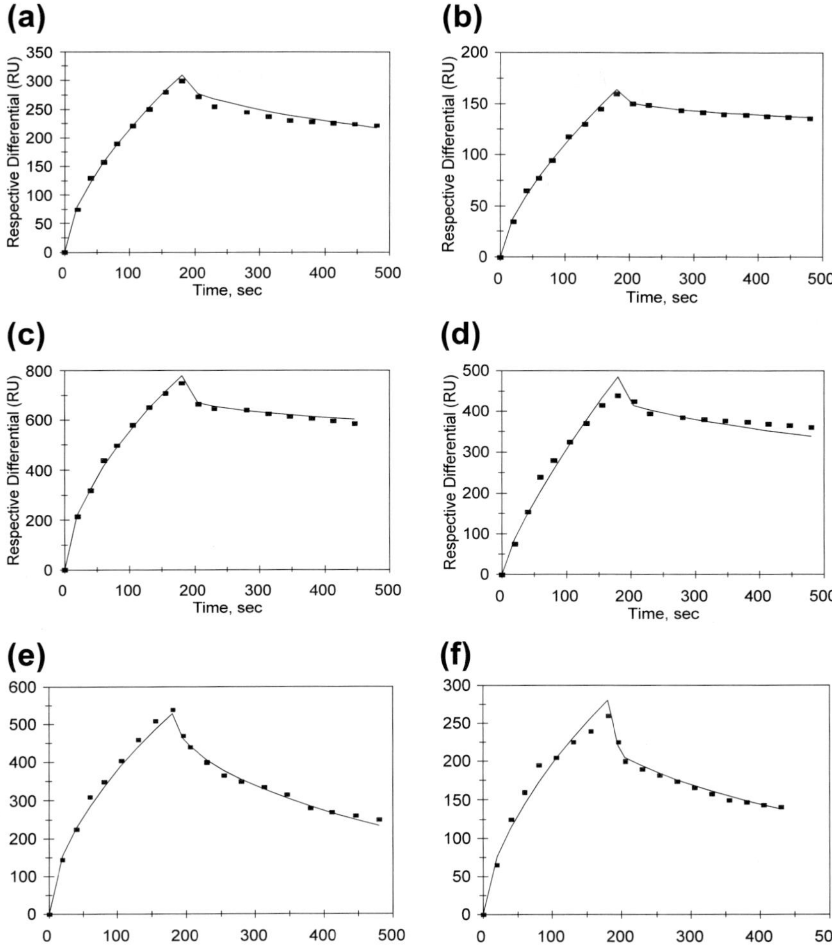

FIGURE 8.46 Binding and dissociation of 1:100 dilution of isotypes in serum a-Q (Thaler et al., 2009): (a) positive control; (b) minus IgG/IgM. Binding and dissociation of 1:100 dilution of isotypes in serum c-l: (c) positive control; (d) minus IgA/IgM. Binding and dissociation of 1:100 dilution of isotypes in serum k-1: (e) positive control; (f) minus IgG/IgA. Binding and dissociation of 1:100 dilution of isotypes in serum n: (g) positive control; (h) minus IgG/IgA/IgM.

Figure 8.46(b) shows the binding and dissociation of isotypes in serum a-Q (minus IgG/IgM). A single-fractal analysis, once again, adequately describes the binding and dissociation kinetics. The values of the (a) binding rate coefficient, k, and the fractal dimension, D_f, for a single-fractal analysis, and the (b) the dissociation rate coefficient, k_d, and the fractal dimension for dissociation, D_{fd}, are given in Table 8.30. The affinity, K value in this case is $K (= k/k_d)$ is 1.51.

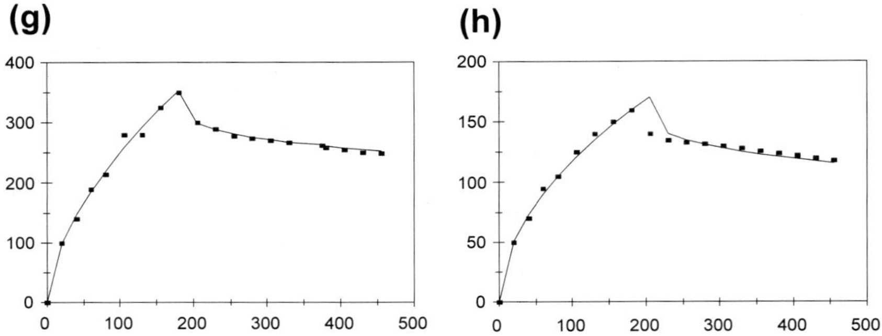

FIGURE 8.46 cont'd

Figure 8.46(c) shows the binding and dissociation of isotypes in serum c-1 (positive control). A single-fractal analysis is required to adequately describe the binding and dissociation kinetics. The values of the (a) binding rate coefficient, k, and the fractal dimension, D_f, for a single-fractal analysis, and the (b) the dissociation rate coefficient, k_d, and the fractal dimension for dissociation, D_{fd}, are given in Table 8.30. The affinity, K value in this case is K (= k/k_d) is 1.15.

Figure 8.46(d) shows the binding and dissociation of isotypes in serum k-1 (minus IgA/IgM). A single-fractal analysis is required to adequately describe the binding and dissociation kinetics. The values of the (a) binding rate coefficient, k, and the fractal dimension, D_f, for a single-fractal analysis, and the (b) the dissociation rate coefficient, k_d, and the fractal dimension for dissociation, D_{fd}, are given in Table 8.30. The affinity, K value in this case is K (= k/k_d) is 2.38.

Figure 8.46(e) shows the binding and dissociation of isotypes in serum cl (positive control). A single-fractal analysis is required to adequately describe the binding and dissociation kinetics. The values of the (a) binding rate coefficient, k, and the fractal dimension, D_f, for a single-fractal analysis, and the (b) the dissociation rate coefficient, k_d, and the fractal dimension for dissociation, D_{fd}, are given in Table 8.30. The affinity, K value in this case is K (= k/k_d) is 1.32.

Figure 8.46(f) shows the binding and dissociation of isotypes in serum cl (minus IgA/IgM). A single-fractal analysis is, once again, required to adequately describe the binding and dissociation kinetics. The values of the (a) binding rate coefficient, k, and the fractal dimension, D_f, for a single-fractal analysis, and the (b) the dissociation rate coefficient, k_d, and the fractal dimension for dissociation, D_{fd}, are given in Table 8.30. The affinity, K value in this case is K (= k/k_d) is 1.44.

Figure 8.46(g) shows the binding and dissociation of isotypes in serum n (positive control). A single-fractal analysis is required to adequately describe the binding and dissociation kinetics. The values of the (a) binding rate coefficient, k, and the fractal dimension, D_f, for a single-fractal analysis, and the (b) the dissociation rate coefficient, k_d, and the fractal dimension for dissociation, D_{fd}, are given in Table 8.30. In this case the affinity value K (= k/k_d) is 0.799.

Figure 8.46 (h) shows the binding and dissociation of isotypes in serum n (minus IgG/IgG/IgM). A single-fractal analysis is required to adequately describe the binding and dissociation kinetics. The values of the (a) binding rate coefficient, k, and the fractal dimension, D_f, for a single-fractal analysis, and the (b) the dissociation rate coefficient, k_d, and the fractal dimension for dissociation, D_{f1}, are given in Table 8.30. In this case the affinity value K (= k/k_d) is 1.69.

Figure 8.47(a) shows the increase in the binding rate coefficient, k, with an increase in the fractal dimension, D_f. For the data shown in Figure 8.47(a), the binding rate coefficient, k, (positive controls) is given by:

$$k = (5.639 \pm 2.824)\ D_f^{2.536 \pm 1.514} \tag{8.22a}$$

There is scatter in the data. More data points are required to obtain a better fit. The availability of more data points restricts one to this conclusion. The binding rate coefficient, k (for positive controls), exhibits slightly more than two and one-half (equal to 2.536) order of dependence on the fractal dimension, D_f, or the degree heterogeneity that exists on the surface. This indicates

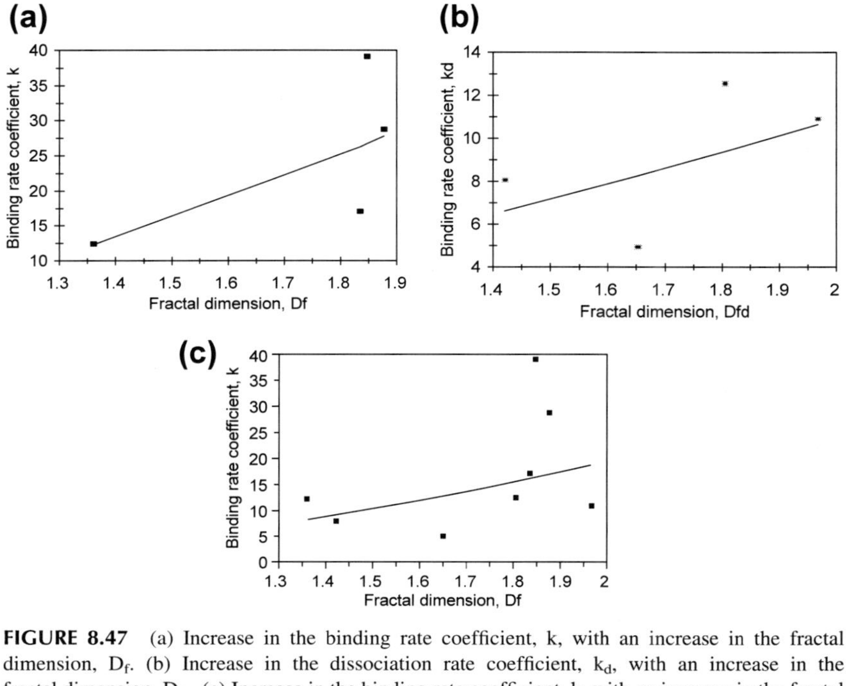

FIGURE 8.47 (a) Increase in the binding rate coefficient, k, with an increase in the fractal dimension, D_f. (b) Increase in the dissociation rate coefficient, k_d, with an increase in the fractal dimension, D_{fd}. (c) Increase in the binding rate coefficient, k, with an increase in the fractal dimension, D_f (all the points taken together).

that the binding rate coefficient, k, is quite sensitive to the degree of heterogeneity that exists on the biosensor surface.

Figure 8.47(b) shows the decrease in the dissociation rate coefficient, k_d, with an increase in the fractal dimension, D_{fd} (positive control for the four cases). For the data presented in Figure 8.47(b), the dissociation rate coefficient, k_d, is given by:

$$k_d = (3.971 \pm 2.199) D_{fd}^{1.459 \pm 1.831} \tag{8.22b}$$

There is scatter in the data. Only four data points are available. The availability of more data points would lead to a more reliable fit. The scatter in the data is reflected in both the coefficients (3.971 ± 2.199 and 1.459 ± 1.831). The dissociation rate coefficient, k_d, is sensitive to the degree of heterogeneity that exists on the biosensor surface since it exhibits close to a one and one-half (equal to 1.459) order of dependence on the fractal dimension, D_{fd}, or the degree of heterogeneity that exists on the biosensor surface in the dissociation phase.

Figure 8.47(c) shows the increase in the binding rate coefficient, k, with an increase in the fractal dimension, D_f (both sets of data taken together, positive control and minus BOT). For the data shown in Figure 8.47(c), the binding rate coefficient, k, is given by:

$$k = (5.639 \pm 2.824)\ D_f^{2.536 \pm 1.514} \tag{8.22c}$$

There is scatter in the data. More data points are required to obtain a better fit. The availability of more data points restricts one to this conclusion. The binding rate coefficient, k (for positive controls), exhibits slightly more than two and one-half (equal to 2.536) order of dependence on the fractal dimension, D_f, or the degree of heterogeneity that exists on the surface. This indicates that the binding rate coefficient, k, is quite sensitive to the degree of heterogeneity that exists on the biosensor surface.

Figure 8.48(a) shows the increase in the dissociation rate coefficient, k_d, with an increase in the fractal dimension, D_{fd}. For the data shown in Figure 8.48(a), the dissociation rate coefficient, k_d (positive controls), is given by:

$$k_d = (0.1323 + 0.1437)\ D_{fd}^{5.988 \pm 3.399} \tag{8.23a}$$

There is scatter in the data. More data points are required to obtain a better fit. The availability of more data points restricts one to this conclusion. The dissociation rate coefficient, k_d (for positive controls), exhibits close to a sixth (equal to 5.988) order of dependence on the fractal dimension, for dissociation, D_{fd}, or the degree of heterogeneity that exists on the surface. This indicates that the dissociation rate coefficient, k_d, is quite sensitive to the degree of heterogeneity that exists on the biosensor surface.

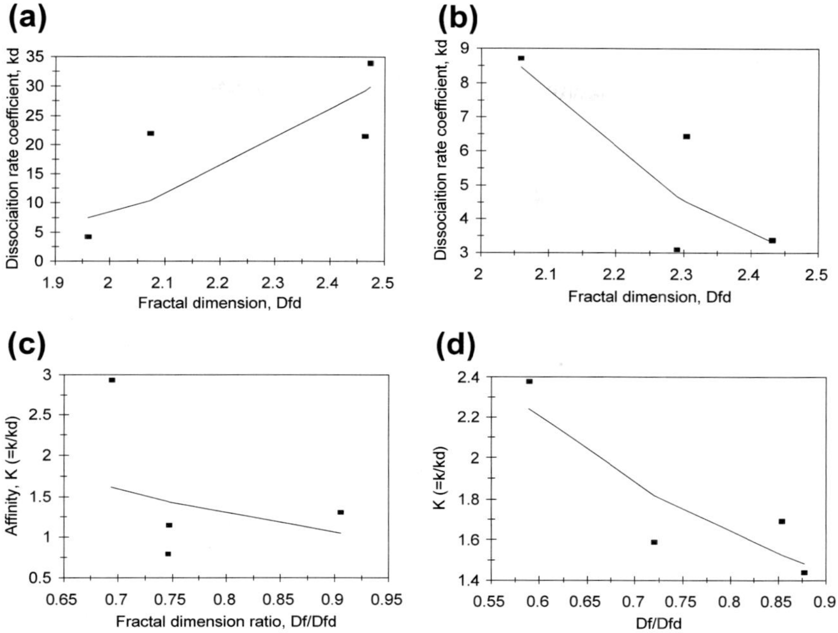

FIGURE 8.48 (a) Increase in the dissociation rate coefficient, k_d, with an increase in the fractal dimension for dissociation, D_{fd} (positive control only). (b) Decrease in the dissociation rate coefficient, k_d, with an increase in the fractal dimension for dissociation, D_{fd} (minus immunoglobulin isotypes). (c) Decrease in the affinity K (= k/k_d) with an increase in the fractal dimension ratio, D_f/D_{fd} (positive control). (d) Decrease in the affinity K (= k/k_d) with an increase in the fractal dimension ratio, D_f/D_{fd} (minus immunoglobulin isotypes).

Figure 8.48(b) shows the increase in the dissociation rate coefficient, k_d, with an increase in the fractal dimension, D_{fd}. For the data shown in Figure 8.48(b), the dissociation rate coefficient, k_d, (minus immunoglobulin isotype) is given by:

$$k_d = (3.971 \pm 2.200)\, D_{fd}^{-1.459\pm+1.83} \tag{8.23b}$$

There is scatter in the data. More data points are required to obtain a better fit. The availability of more data points restricts one to this conclusion. The dissociation rate coefficient, k_d (minus immunoglobulin isotype) exhibits close to a one and one-half (equal to 1.459) order of dependence on the fractal dimension, D_{fd}, or the degree of heterogeneity that exists on the surface in the dissociation phase. This indicates that the dissociation rate coefficient, k_d, is quite sensitive to the degree of heterogeneity that exists on the biosensor surface in the dissociation phase.

Figure 8.48(c) shows the decrease in the affinity K (= k/k_d) with an increase in the fractal dimension ratio, D_f/D_{fd} (minus immunoglobulin isotype). For the data shown in Figure 8.48(c), the affinity, IC (positive controls) is given by:

$$K(k/k_d) = (0.891 \pm 0.788)\ D_f/D_{fd}^{-1.644 \pm 1.588} \quad (8.23c)$$

There is scatter in the data. Only four data points are available. More data points are required to obtain a better fit. The availability of more data points would lead to a more reliable fit. The affinity, K (minus immunoglobulin isotype), exhibits slightly more than a negative one and one-half (equal to −1.643) order of dependence on the fractal dimension ratio, D_f/D_{fd}, or the ratio of the degrees of heterogeneity that exists on the surface in the binding and in the dissociation phases. This indicates that the affinity, K, is quite sensitive to the ratio of the degree of heterogeneity that exists on the biosensor surface in the binding and in the dissociation phase. Figure 8.48(d) shows the decrease in the affinity K (= k/k_d) with an increase in the fractal dimension ratio, D_f/D_{fd} (minus immunoglobulin isotypes). For the data shown in Figure 8.48(d), the affinity, K (minus immunoglobulin isotypes), is given by:

$$K\ (k/k_d) = (0.891 \pm 0.788)(D_f/D_{fd})^{-1.644 \pm 1.588} \quad (8.23d)$$

There is scatter in the data. More data points are required to obtain a better fit. The availability of more data points would lead to a more reliable fit. The affinity, K (minus immunoglobulin isotypes), exhibits slightly more than negative one and one-half (equal to −1.644) order of dependence on the ratio of fractal dimensions, D_f/D_{fd}, or the ratio of the degrees of heterogeneity that exists on the surface in the binding and the dissociation phases. This indicates that the affinity, K, is quite sensitive to the degrees of heterogeneity that exist on the biosensor surface in the binding and the dissociation phases.

8.5 CONCLUSIONS

A fractal analysis is presented for the binding of SLE biomarkers to biosensor surfaces. A single-fractal analysis is initially used to model the binding and dissociation (if applicable) kinetics. If a single-fractal analysis does not provide an adequate fit, then a dual-fractal analysis is used to model the kinetics. This is done using Corel Quattro 8.0 (1997). The fractal dimension provides a quantitative measure of the degree of heterogeneity on the biosensor surface.

The binding and dissociation kinetics of some of the biomarkers analyzed include (a) ScFv in solution to prothrombin activation (Le et al., 2004), (b) influence of Zn^{++} binding on HRG to complement C1q immobilized on an optical IAsys biosensor surface (Giorgiani et al., 1997), (c) binding of 1 μg/ml of ADA-CMC-GL1 to CDPSH-modified Biacore chip (Ortiz et al., 2011), (d) binding of TNF-(in solution to different nanoparticles (SNOBS biosensor)

(Huang et al., 2008), (e) binding and dissociation of transcription factor, TFEB from E-Box DNA using an SPR biosensor (LeBlanc et al., 1998), (f) binding and dissociation of Fab-E box in solution (in mM) to E-box DNA immobilized on an SPR biosensor surface (LeBlanc et al., 1998), binding of anti-FLAG (FLAG tag is N-DYKDDK-C, peptide sequence) in solution to an anti-DNA sensor (White et al., 2011), (h) binding of XP1 and XP2 to DMPC.DMPG (50:50) liposome immobilized on an LI sensor chip, (i) binding of different concentrations of IL (interleukin)-1α in solution to immobilized sTL-IR-l (soluble IL-1 receptor (sILR) 1 on a biosensor surface, (j) binding of C4 (complement) in solution to antibody immobilized on a sandwich immunoassay (Liu et al., 2004), (k) binding of anti-dsDNA-rn Ab32 in solution to immobilized dsDNA antigen on a sensor chip (Buhl et al., 2009), (1) binding of different concentrations of dsDNA mAb 33 concentration in solution (1 nM) to receptor immobilized on an SPR surface (Buhl et al., 2009), (m) binding of antichromatin antibody from the sera of five patients with SLE to a chromatin-coated membrane on an electrochemical biosensor (Konstantinov et al., 2009), (n) binding and dissociation of 1:100 diluted sera from patients with antiphospholipid syndrome onto a biosensor surface coated with pelCLl, 1′2,2′-tetra-9-oxopelargonoyl cardiolipin and β2-GP1 (glycoprotein 1) (Thaler et al., 2009), (o) binding and dissociation of 1:100 diluted sera from patients with antiphospholipid syndrome onto a biosensor surface coated with β-glycoprotein-1) (Thaler et al., 2009), and (p) binding and dissociation of 1:100 dilution of isotypes I serum a-Q (Thaler et al., 2009).

An increase in the fractal dimension value or the degree of heterogeneity on the surface leads, in general, to an increase in the binding rate coefficient. For example, (a) for the binding of ScFv AN78 for prothrombin activation coated on an ELISA surface the binding rate coefficient, k, increases as the fractal dimension, D_f, increases for a single-fractal analysis (Le et al., 2004), (b) the binding rate coefficient, k, of HRG in solution to C1q immobilized on the biosensor surface in the absence and in the presence of Zn^{++} increases as the fractal dimension, D_f increases (Giorgiani et al., 1997), (c) decrease in the binding rate coefficient, k, with an increase in the fractal dimension for ADA-CMC-GLI in solution to the CDPSH-modified Biacore chip (Ortiz et al., 2011), (d) increase in the binding rate coefficient, k_1 and k_2, with an increase in the fractal dimension, D_{f1} and D_{f2}, for a dual-fractal analysis for TNF-α in solution to a single-molecule nanoparticle biosensor (SNOBS) using an mAb molecule immobilized to the surface (Huang et al., 2008), and (e) increase in the binding rate coefficient, k_1, with an increase in the fractal dimension, D_{f1}, for a dual-fractal analysis for the binding of XP1 and XP2 to DMPC.DMPG (70:30) liposomes immobilized on an LI sensor chip (Ali et al., 2005).

The fractal dimension, as indicated earlier, is not a classical independent variable such as analyte, antibody, or other biological molecule concentration in solution. Nevertheless, the expressions obtained for the binding (and the dissociation) rate coefficients for a single- and a dual-fractal analysis as a function of the fractal dimension indicate a high degree of sensitivity of these rate coefficients on their respective fractal dimensions on the SPR sensor chip or other biosensor surface. Since lupus is a particularly insidious disease, it behooves one to use all possible means available to gather information as possible from the kinetics occurring on the biosensor surface (as well as other sources). Molecular interactions play a myriad of important roles in drug discovery and development. Lupus is particularly difficult to treat since the etiology of lupus is not well known. This plays particular emphasis on the different avenues that should be explored in that they may provide suitable physical insights into the biomarkers for lupus.

Predictive relations need to developed, and they are developed and presented. These relationships for the binding and dissociation rate coefficients are presented as a function of analyte concentration and the nontraditional variable such as the fractal dimension, D_f, or the degree of heterogeneity on the biosensor surface. These predictive relations should prove invaluable in helping to understand the physical insights involved in these types of injections (namely, those between the different biomarkers for lupus and the corresponding receptors involved in the different biosensor surfaces).

The earlier timewise one may detect and diagnose the probable onset of diseases, the earlier one can begin the medical protocols necessary to alleviate or correct the onset of especially debilitating and intractable diseases such as lupus. Particularly exacerbating is the general cure for the treatment of lupus which includes steroids (prenosil, for example). To minimize the deleterious effects of steroid treatment, one starts with a relative high dose of steroids (based generally on the patient's weight), especially when we have a 'flare up' of lupus, and then one begins to minimize the administration of the steroids when the medical personnel feel that lupus is "under control." For the patients afflicted with lupus, even when the flare up of lupus has been controlled, the patient has to continuously have his steroid treatment. Needless to say, one has to periodically monitor the progress of the disease. All of this points to the fact that one should aim to diagnose the biomarkers for lupus, as quickly as is possible. The development of more sensitive biosensors would significantly assist in this endeavor. Needless to say that the biosensor should be able to detect the lupus biomarkers in a reasonable amount of tine, say 15−30 min to permit the medical personnel to take the necessary corrective action.

REFERENCES

Aarden, E.M., Berger, E.H., Nyweide, P., 1994. Function of osteocytes in bone. Journal of Cell Biochemistry 55, 287−999 [Pub Med].

Ali, M., Anion, M., Bender, V., Manolios, N., 2005. Hydorphobic transmembrane-peptide lipid conjugations enhance membrane binding and functional activity in T-cells. Bioconjugate Chemistry 16, 1556−1563.

Amoura, Z., Chabre, H., Koutouzov, S., Lotton, C., Cabrespines, A., Bach, J.F., Jacob, L., 1994. Arthritis and Rheumatology 37, 1684–1688.

Andreescu, S., Sadik, A.O., 2004. Pure and Applied Chemistry 76, 861–878.

Arakawa, T., Kawbayashi, A., Saga, T., 2005. 'ensors & Actuators B 108, 899.

Arbuckle, M.R., McClain, M.T., Rubertione, M.V., Scofield, R.H., Dennis, G.J., James, J.A., Harley, J.B., 2003. New England Journal of Medicine 349, 1526–1533.

Arend, W.P., 2003. Cytokine growth factor. Cytokine 13, 323–333.

Arner, E.C., Harris, R.R., Dimeo, T.M., Collins, R.C., Galbraith, W., 1995. Journal of Rheumatology 22, 1338–1346.

Arvieux, J., Darnigge, L., Caron, O., Reber, G., Bensa, J.C., Colomb, M.G., 1995. Development of an ELISA for autoantibodies to prothrombin showing their prevalence in patients with lupus anticoagulants. Thrombosis Hemostasis 74, 1120–1125.

Atassi, M.Z., Casali, P., 2008. Autoimmunity 41, 123–132.

Adyel, F.Z., Hentati, B., Boulila, A., Hachicha, J., Ternynck, T., Avrameas, S., Ayadi, H., 1996. Clinical Laboratory Analysis 10, 451–457.

Bajaj, S.P., Rapaport, S.I., Fierer, D.S., Herbst, K.D., Schwartz, D.B., 1994. A mechanism for the hypothrombinemia of the acquired hyperthrombinemia-lupus anticoagulant syndrome. Journal of the American Chemical Society 116, 2161–2162.

Barbas, S.M., Ghazel, P, Barbas III, C.F., Burton, D.R., 1994. Journal of the American Chemical Society 116, 2161–2162.

Baud, D., Regan, L., Greub, G., 2010. European Journal of Clinical Microbiology and Infection 29, 669–675.

Bissonnette, L., Bergeron, M.G., 2010. Clinical Microbiology Infection 16, 1044–1053.

Bohren, C.F., Huffman, D.R., 1983. Adsorption and Scattering of Light by Small Particles. Wiley, New york.

Bootsma, H., Spronk, P.E., Ten Borg, E.J., Hummel, E.J., de Boer, G., Linburg, P.C., Kallenberg, C.G.M., 1997. The predictive value of fluctuations in IgM and IgG class anti-ds DNA antibodies for replaces in systemic lupus erythematosus. A prospective long term observation. Journal of Rheumatic Diseases 56, 661–666.

Braun, R.P., Woodsworth, M.L., Lee, J.S., 1986. Molecular Immunology 23, 685–691.

Braun, R.P., Lee, J.S., 1987. Journal of Immunology 139, 161.

Brennan, D.J., O'Connor, D.P., Rexhepaj, E., Ponten, F., Gallagher, W.M., 2010. National Reviews on Cancer 10, 605–617.

Buckle, P.E., Davies, R.J., Kinning, T., Yeung, D., Edwards, P.R., Pollard-Knight, D., 1993. Biosensors & Bioelectronics 8, 355–363.

Buhl, A., Metzger, J.H., Heegaard, N.H., von Landenberg, P., Fleck, M., Luppa, P.B., 2007. Clinical Chemistry 53, 334–341.

Buhl, A., Page, S., Heegaard, N.H.H., von Landenberg, P., Luppa, P.A., 2009. Optical biosensor-based characterization of anti-double-stranded DNA monoclonal antibodies as possible new standards for laboratory tests. Biosensors & Bioelectronics 25, 198–203.

Burbelo, P.D., Issa, A.T., Ching, K.H., Cohen, J.I., Iadarola, M.J., Marques, A., 2010. Clinical Vaccine Immunology 17, 904–909.

Budhai, L., Oh, K., Davidson, A., 1996. Journal of Clinical Investigation 98, 1585–1593.

Castillo, G.M., Ngo, C., Cummings, J., Wight, T.N., Snow, A.D., 1997. Perlecan binds to the β-amyloid proteins (Aβ) of Alzheimer's disease, accelerates Aβ fibril formation, and maintains Aβ fibril stability. Journal of Neurochemistry 69, 2452–2465.

Chabre, H., Amoura, Z., Pitte, J.C., Codeau, P., Bach, J.F., Koutouzov, S., 1995. Arthritis and Rheumatology 38, 1485–1491.

Chang, Y.P., Chu, Y.H., 2005. Analytical Biochemistry 340, 74–79.
Chan, C.P., Cheung, Y., Renneberg, R., Seydack, M., 2008. Advances in Biochemical Engineering and Biotechnology 109, 123–154.
Charlesworth, J.A., Endre, Z.H., Pusslell, B.A., Yasmeen, D., Peake, P.W., 1982. Infectious Diseases 143, 505–513.
Chen, Q., Li, J., Yin, H.J., Huang, J.D., Wu, B.Y., Jiang, Y.R., Chen, K.J., 2004. Chinese Journal of Integral Methods 24, 746–748.
Cooper, G.S., Strohla, B.C., 2003. Autoimmunity Reviews 2, 119–125.
Copeland, S., Siddiqui, J., Remick, D., 2004. Journal of Immunological Methods 284, 99–106.
Corel Quattro Pro 8.0, 1997. Corel Corporation, Ottawa, Canada.
Cui, Z., Zhao, M.H., 2005. Clinical Immunology 116, 77–82.
Davidson, A., Diamond, B., 2001. New England Journal of Medicine 345 (5), 340–350.
Davie, E.W., 1995. Biochemical and molecular aspects of the coagulation cascade. Thrombosis Homeostasis 34, 1–6.
D'Cruz, D.P., Kharmashta, M.A., Hughes, G.R., 2007. Lancet 369, 587–596.
Degen, S.J., Sun, W.Y., 1994. The biology of prothrombin. CRC Review Eulmyotic Expression 8, 203–224.
De Laat, B., Derksen, R.H.W.M., de Groot, P.G., 2006. Journal of Thrombosis and Haemostasis 4, 1619–1621.
Deutscher, S.L., Keene, J.D., 1988. Proceedings of the National Academy of Sciences USA 85, 3299–3303.
Dmitriev, D.A., Massino, O.L., Segal, J., 2003. Journal of Immunological Methods 280, 183–202.
Dinarello, C.A., 2004. Current Opinion in Pharmacology 4, 378–385.
Dieterich, W., Laag, E., Schopper, H., 1998. Gastroenterology 115, 1317.
Dooley, M.A., Hoga, S., Jennette, Falk, R., 1997. Cyclophosphamide therapy for lupus nephritis; poor renal survival in black Americans. Kidney International 51 (2), 1188–1195.
Drouvalakis, K.A., Bangsaruntip, S., Hueber, W., Kozar, L.G., Utz, P.J., Dai, H.J., 2008. Biosensors & Bioelectronics 23, 1413.
Eade, O.E., Lloyd, R.S., Lang, C., Wright, R., 1997. Gut 18, 99.
Eaton, W.W., Rose, N.R., Kalaydjian, A., Pedersen, M.G., Mortensen, P.B., 2007. Journal of Autoimmunity 29, 1–9.
Edison, J.R., Vogt, J.M., Hasgawa, D.K., 1984. Abnormal prothrombin crossed immunoelectrophoresis in patients with lupus inhibitors. Blood 64, 807–816.
Fanslow, W.C., Sims, J.E., Sassenfeld, H., Morrissey, P.J., Gillis, S., Dower, S.K., Widmer, M.B., 1990. Science 248, 739.
Fasano, A., Catassi, C., 2001. Gastroenterology 120, 636.
Fisher, M., Tjarnhage, T., 2000. Biosensors & Bioelectronics 15, 463.
Forger, F., Matthias, T., Uppermann, M., Becker, H., Helmke, K., Clinical Significance of anti-ds DNA antibody isotypes: Igt/IgM ratio of anti-ds DNA antibodies as a prognostic marker for Lupus nephritis. 13, 364–410.
Franzke, C.W., Bruckner, P., Bruckner-tuderman, I., 2005. Collagenous transmembrane proteins: recent insights into biology and pathology. Journal of Biological Chemistry 280, 4005–4008.
Froelich, C.J., Wallman, J., Skosey, J.L., Terodorescu, M., 1990. Journal of Rheumatology 17, 192–200.
St. George Hyslop, April 2002. γ-secretase, notch, Aβ and Alzheimer's disease: where do presenihis fit in? Nature Reviews Neuroscience 3, 281–290.
Gill, J.M., Quisel, A.M., Rocca, P.V., Walters, D.T., 2003. American Family Physician 68, 2179–2186.

Giorgiani, N.N., Parish, C.R., Easterbrook Smith, S.B., Altin, J.G., 1997. Histidine-rich glycoprotein binds to human Ig and Clq and inhibits the formation of insoluble immune complexes. Biochemistry 36, 6653–6662.

Gollner, G.P., Muller, G., Alt, R., Knop, J., Enk, A.H., 2000. Therapeutic application of T cell receptor mimic peptides of cDNA in the treatment of T cell-mediated skin diseases. Gene Therapy 7, 1000–1004.

Gomez-Isla, T., Hollister, R., West, H., Mui, S., Growdon, J.H., Petersen, R.C., Parisi, J.E., Hyman, B.T., 1997. Neuronal loss correlates with but exceeds neurofibrillary tangles in Alzheimer's disease. Annals of Neurology 41, 17–24.

Gotz, J., Streffer, J.R., David, D., Schild, A., Hoernle, L.F., Pennanen, L., Chen, F., July 2004. Transgenic animal modes of Alzheimer's disease and related disorders: histopathology, behavior and therapy. Molecular Psychiatry 9 (7), 664–683.

Goudsmit, J., Lange, J.M.A., Paul, D.A., Dawson, G.J., Mar, N., 1987. Infection 155, 558–560.

Green, R.J., Frazier, R.A., Shakessheff, K.M., Davies, M.C., Roberts, C.J., Tendler, S.J.B., 2000. Biomaterials 21, 1823.

Gulalis, A., Patrinou-Georgoula, M., Tsifetaki, N., Aldints, V., Sekeris, C.E., Moutsopoulos, H.M., 1994. Clinical Experiments in Immunology 95, 385–389.

Gurevich, K.G., Agutter, P.S., Wheatley, D.N., 2003. Cell Signal 15, 447–453.

Hardy, J., Selkoe, D.J., 2002. The amyloid hypothesis of Alzheimer's disease: progress and problems on the road to therapeutics. Science 297–353.

Hashimoto, T., Wakabayashi, T., Watanabe, A., Kowa, H., Hosoda, R., Nakamura, A., Kanazawa, I., Arai, T., Takio, K., Mann, D.M., Iwatsubo, T., 2002. CLAC: a novel Alzheimer amyloid plaque component derived from a transmembrane precursor, CLAC-P/collagen type XXV. EMBO Journal 21, 1524–1534.

Havlin, S., 1989. Molecular diffusion and reaction. In: Avnis, D. (Ed.), The Fractal Approach to Heterogeneous Chemistry: Surfaces, Colloids, Polymers. Wiley, New York, pp. 251–269.

Haworth, C., Maini, R.N., Feldman, M., 1998. In: Thomson, A. (Ed.), The Cytokine Handbook. Academic Press, New York, pp. 777–801.

Herron, J.N., He, X.H., Ballard, D.W., Blier, P.R., Pace, P.E., Bothwell, A.L.M., Voss Jr., E.W., Edundson, A.B., 1991. Proteins; Structure, Function, Genetics 11, 159–175.

Hirmo, S., Artursson, E., Puu, G., Wadstrom, T., Nilsson, B., 1999. Journal of Microbiological Methods 37, 177–182.

Halamek, J., Makower, A., Knosche, K., Skaldal, P., Scheller, F.W., 2005. Talanta 65, 337.

Hochberg, M.C., 1997. Arthritis and Rheumatism 40, 1725.

Huang, I.D., Zhao, S., Li, J., Wu, B.Y., Yin, H.J., Chen, K.J., 2005. Chinese Journal of Integral Methods 11, 37–40.

Huang, T., Nallathamby, P.D., Gillet, Xu, X.H.N., 2007. Analytical Chemistry 79, 7708–7718.

Huang, T., Nallathamby, P.D., Xu, X.H.N., 2008. Photostable single-molecule nanoparticle optical biosensors for real-time sensing of single cytokine molecules and their binding response. Journal of the American Chemical Society 130, 17095–17105.

Hueber, W., Kidd, B.A., Tomooka, B.H., Lee, B.J., Bruce, B., Fries, J.F., Sonderstrup, G., Monach, P., Drijfhout, J.W., van Venrooij, W.J., Utz, P.J., Genovese, M.C., Robinson, W.H., 2005. Arthritis and Rheumatology 52, 2645–2655.

Isenberg, D., 2004. Lupus 13, 881–885.

Jonsson, U., Malmqvist, M., 1992. Real Time Biospecific Interaction Analysis-The Integration of Surface Plasmon Resonance Detection, General Biospecific Inter-Intergration of Surface Plasmon Resonance Detection, General Biospecific Interface Chemistry, Microfluidics into One Analytical System. Elsevier Science & Technology, The Netherlands.

Kakuyama, H., Soderberg, L., Horigome, K., et al., 2005. CLAC binds to aggregated Abeta and Abeta fragments and attenuates fibril elongation. Biochemistry 44, 15602–15609.

Kazama, Y., Koide, T., 1992. Thrombosis Homeostasis 67, 50–55.

Kilander, A.F., Dotevall, G., Fallstrom, S.P., Gillberg, R.E., Nilsson, L.A., Tarkowski, A., 1983. Scandinavian Journal of Gastroenterology 18, 377.

Klabusay, M., Khutova, V., Coupek, P., Nenicckova, M., Tesarova, E., 2006. Mediators: Inflammation 1–7.

Klasing, K.C., Peng, R.K., 2001. Development and Comparative Biology 25, 345.

Konstantinov, K.N., Sitdikov, R.A., Lopez, G.P., Atanasov, P., Robin, R.I., 2009. Rapid detection of anti-chromatin autoantibodies in human serum using a portable electrochemical biosensor. Biosensors & Bioelectronics 24, 1949.

Kotzm, B.L., 1996. Cell 85, 303–306.

Korponay-Szabo, I.R., Szabados, K., Pusztai, J., Uhrin, K., Ludmany, E., Nemes, E., Kauinen, K., Kapitany, A., Koskinen, L., Spikan, S., Imre, A., Makis, L., 2007. British Medical Journal 335, 1244–1247.

Koutouzov, S., Jenonimo, A., Campos, H., Amoura, Z., 2004. Nucleosomes in the pathogenesis of systemic lupus erythematosus. Rheumatology Diseases Clinical North America 30, 529–558.

de Kossodo, S., Houba, V., Grau, G.E., 1995. Journal of Immunological Methods 182, 107–114.

Kreibig, U., Vollme, M., 1995. Optical Properties of Clusters. Springer, Berlin.

Krishnaswamy, S., Field, K.A., Edgington, T.S., Morrssay, J.H., Mann, K.G., 1992. Role of the membrane surface in the activation of human coagulation factor X. Journal of Biological Chemistry 267, 26110–26120.

Kubota, T., Akatsuka, T., Kanai, Y., 1985. Clinical Experiments in Immunology 62, 321–328.

Kyriacou, S.V., Brownlow, W.J., Xu, X.-H.N., 2004. Biochemistry 43, 140–147.

Ladinser, B., Rossipal, B., Pittsschicler, K., 1994. Gut 35, 776.

Lafer, E.M., Valle, R.P.C., Mslier, A., Nordheim, A., Schur, P.H., Rich, A., Stollar, B.D., 1983. Journal of Clinical Investigation 71, 314–319.

Le, A., Dasgupta, S., Plaque, S., Paul, S., Thiagrajan, P., 2004. Lupus-derived anti-prothrombin autoantibodies from a V gene phage display library are specific for the krongle 2 domain of prothrombin. Biochemistry 4047–4054.

LeBlanc, I.F., McLane, K.H., Parren, P.W.H.I., Burton, D.R., Ghazal, P., 1998. Recognition properties of a sequence-specific DNA binding antibody. Biochemistry 37, 6015–6022.

Lee, K.J., Nallathamby, P.D., Browning, L.M., Osgood, C.J., Xu, X.H.N., 2007. ACS Nano 1, 133–143.

Lemere, C.A., Grefell, T.J., Selkoe, D.J., 1999. The AMY antigen co-occurs with Aβ and follows its deposition in the amyloid plaques of Alzheimer's disease and Down's syndrome. American Journal of Pathology 155, 29–37.

Leon, S.A., Green, A., Ehrlich, G.E., Poland, M., Shapiro, B., 1977. Arthritis and Rheumatology 20, 23–29.

Leopoldo, L.R., Stefania, M., Alessandra, V., Stefania, R., Paolo, R., Roberto, P.R., 1997. Biosensors & Bioelectronics 12, 765–778.

Lernmark, A., 2001. Clinical Investigations 108 (8), 1091–1096.

Levisky, J.M., Shenoy, S.M., Pezo, R.C., Singer, R.H., 2002. Science 297, 836–840.

Ley, A.B., Reader, G.G., Sorenson, C.W., Overman, R.S., 1951. Idiopathic hypothrombinemia associated with hemorrhagic diathesis and the effect of vitamin K. Blood 6, 740–746.

Li, Q.Z., Xie, C., Wu, T., Mackay, M., Aranow, C., Putterman, C., Mohan, C., 2005. Journal of Clinical Investigation 115, 3428–3439.

Lim, T.-k., Komodo, Y., Nakamura, N., Matsunaga, T., 1999. Automated detection of anti-double-stranded antibody in systemic Lupus erythematosus serum by flow immunoassay. Analytical Chemistry 71, 1298–1302.

Liu, X., Song, D.Q., Zhang, Q.L., Ttan, Y., Zhang, H.Q., 2004. Talanta 62, 773.

Lutter, L.C., 1978. Journal of Molecular Biology 124, 391–420.

Lyons, R., Narian, S., Nichols, C., Satoh, M., Reeves, W.H., 2005. Annals of the New York Academy of Sciences 1050, 217–228.

Maki, M., Collin, P., 1997. Lancet 34, 1755.

Mahnke, K., Qian, Y., Knop, J., Enk, A.H., 2003. Dendritic cells, engineered to secrete a T cell-receptor mimic peptide, induce antigen-mediated specific immunosuppression in vivo. Nature Biotechnology 21, 903–908.

Mann, K.G., Elicon, J., Butkowski, R.J., Downing, M., Nesham, M.E., 1981. Prothrombin. Methods in Enzymology 80, 28–302.

Manolios, N., Bonifacino, J.S., Klausner, R.D., 1991. Pairwise, cooperative and inhibitory interactions describe the assembly and probable structure of the T cell antigen receptor. EMBO Journal 10, 1643–1651.

Manolios, N., Bonifacino, J.S., Klausner, R.D., 1990. Transmembrane helical interactions and the assembly of the T cell receptor complex. Science 249, 274–277.

Manolios, N., Collier, S., Taylor, J., Pollard, J., Harrison, L., Bender, V., 1997. T-cell receptor antigen receptor transmembrane peptides modulate T-cell function and T cell-mediated disease. Nature Medicine 3, 84–88.

Manolios, N., Huynh, N., Collier, S., 2002. Peptides in the treatment of inflammatory skin disease. Australian Journal of Dermatology 43, 226–227.

McDonnell, J.M., 2001. Current Opinion in Chemical Biology 5, 572–577.

Mclane K.E., Burton D.R., and Ghazal P., Proceedings of the National Academy of Sciences; 92,5214–5218 (1995).

Mier, G., 1908. Annals of Physics 25, 377–445.

Miller, K.J., Herman, A.C., 1996. Analytical Chemistry 68, 3077–3082.

Minunni, M., Tombelli, M., Mascini, A., Bilia, A., Bergonzi, M.C., Voncieri, F.F., 2005. Talania 65, 578.

Mitzushiam, N., Kubota, T., Nanki, T., Kolke, R., Kohsaka, H., Miyasaka, N., 1996. Rinsho Biyuri 44, 585–589.

Mok, C.C., 2010. Biomarkers for lupus nephritis: a critical appraisal. Journal of Biomedicine and Biotechnology. Article ID 638413, 11 pages.

Mok, C.C., Tang, S.S.K., To, C.H., Petri, M., 2005. Incidence and risk factors of thromboembolism in systemic lupus erythematosus: a comparison of three ethnic groups. Arthritis and Rheumatism 52 (9), 2774–2782.

Mok, C.C., 2006. Therapeutic options for resistant lupus nephritis. Seminars in Arthritis and Rheumatism 36 (2), 71–81.

Mol, C.D., Muir, A.K.S., Lee, J.S., Anderson, W.F., 1994. Journal of Biological Chemistry 269, 3605–3610.

Mok, C.C., 2005. Prognostic factors in lupus nephritis. Lupus 1, 39–44.

Mountz, J.D., Wu, J., Cheng, J., Zhou, T., 1994. Arthritis and Rheumatology 37, 1415–1420.

Murray, C.K., Gasser, R.A., Magill, A.J., Miler, R.S., 2008. Clinical Microbiology Reviews 21, 97–110.

Myszka, D.G., Morton, T.A., Doyle, M.L., Chaiken, I.M., 1997. Biophysical Chemistry 64, 127.

Nakamura, M., Shiizu-Yoshida, Y., Takii, Y., Komori, A., Yokoyama, T., Ueki, T., Daikoku, M., Yano, K., Matsumoto, T., Migita, K., Yatushashi, H., Ito, M., Masaki, N., Adachi, H.,

Watanabe, Y., Nakamura, Y., Saoshiro, T., Sodeyama, T., Koga, M., Shimoda, S., Ishibashi, H., 2005. Journal of Hepatology 42, 386–392.
Nallathamby, P.D., Lee, K.J., Xu, X.H.N., 2008. ACS Nano 2, 1371–1380.
Namba, Y., Tomonaga, M., Kawasaki, H., Otomo, E., Ikeda, K., 1991. Apolipoprotien E immunoreactivity in cerebral amyloid deposits and neurofibrillary tangles in Alzheimer's disease and kuru plaque amyloid in Creutzfeldt-Jakob disease. Brain Research 541, 163–166.
Nicola, N.A., 1994. Guidebook to Cytokines and Their Receptors. A Samlook and Tooze Publication at Oxford University Press, Oxford.
Nossent, J.C., Huysen, Y., Smeenk, R.J., Swaak, A.J., 1989. Annals of Rheumatic Diseases 48, 677–682.
Olsen, H.M., Parish, C.R., Altin, J.G., 1996. Immunology 88, 198–206.
Ortiz, M., Fragoso, A., O'Sullivan, C.K., 2011. Detection of Antigladin Autoantibodies in Celiac Patient Samples using a cyclodextrin-based Supramolecular Based Biosensor. Analytical Chemistry 83, 2931–2938.
Osterland, C.K., 1994. Clinical Chemistry 40, 2146–2153.
Palu, E., Liesi, P., 2002. Differential distribution of laminins in Alzheimer's disease and normal human brain tissue. Journal of Neuroscience Research 69, 243–256.
Parekh, B.S., Kennedy, M.S., Dobbs, T., Pau, C.P., Byers, R., Green, T., Hu, D.J., Vanichseni, S., Young, N.L., Choopanya, K., Mastro, T.D., McDougal, J.S., 2002. AIDS Research Human Retroviruses 18, 295–307.
Pereira, S.V., Raba, J., Messina, G.A., 2010. Analytical and Bioanalytic Chemistry 396, 2921.
Pfafflin, A., Schleicher, E., 2009. Analytical Bioanalytical Chemistry 393, 1473–1480.
Pico de Coana, Y., Barrero, C., Cajiao, I., Mosquera, C., Patarroya, M.E., Patarroyo, M.A., 2004. Cytokine 27, 129–133.
Posthuma-Trumple, G.A., Korf, J., van Amerongen, A., 2009. Analytical Bioanalytical Chemistry 393, 569–582.
Privett, B.J., Shin, J.H., Schoenfish, M.H., 2008. Analytical Chemistry 80 (12), 4499–4517.
Raber, J., Huang, Y., Ashford, J.W., 2004. ApoE genotype account for the vast majority of AD risk and AD pathology. Neurobiology of Aging 25, 641–650.
Rahman, A., Isenberg, D.A., 2008. New England Journal of Medicine 358 (9), 929–939.
Ravirajan, C.T., Saraf, C.E., Anilkumar, T.V., Golding, M.C., Alison, M.R., Isenberg, D.A., 1996. Clinical Experiments in Immunology 26, 472–486.
Real-Fernandez, F., Colson, A., Bayardon, J., Nuti, F., Peroni, E., Meunier-Prest, R., Lolli, F., Chelli, M., Darcel, C., Juge, S., Papini, A.M., 2008. Biopolymers 90, 488.
Reveille, J.D., 2004. Lupus 13, 290–297.
Riboldi, P., Gerosa, M., Moroni, G., Radice, A., Allegri, F., Simico, A., Tincani, A., Meroni, P.L., 2005. Autoimmunity 38, 39–45.
Rouquette, A.M., Desgruelles, C., Laroche, P., 2003. American Journal of Clinical Pathology 120, 676–681.
Rubin, R.L., Fritzler, M.J., 2007. In: Wallace, D.J., Hahn, B.H. (Eds.), Dubois' Lupus Erythemtaosus, seventh ed. Lippincott Williams and Wilkins, Philadelphia ,PA, pp. 464–486.
Sanchez-Guerrero, J., Lewe, K.A., Fossel, A.H., Schur, P.H., 1996. Arthritis and Rheumatology 39, 1055–1161.
Saraux, A., Bendaoud, B., Dueymes, M., Le Goff, P., Youinou, P., 1997. Annals of Rheumatic Diseases 56, 126–129.
Schifferli, J.A., 1987. Immunology Letters 14, 225–228.
Scofield, R., 2004. Lancet 363, 1544–1546.

Schmidt, M.L., Saido, T.C., Lee, V.M., Trojanowski, J.Q., 1999. Spatial relationship of AMY protein deposits and different species of Aβ peptides in amyloid plaques of Alzheimer's disease brain. Neuropathology and Experimental Neurology 58, 1227–1233.

Selkoe, D.J., 25 October 2002. Alzheimer's disease is a synaptic failure. Science 298 (5594), 789–791.

Shin, A.H., Bravo, N.C., Maecker, H.T., Smith, J.W., 2003. Journal of Immunological Methods 282, 169–174.

Simon, T., Kneusel, R., Haubruck, H., Liedvogel, B., Diaect, A.G., 2001. Scandinavian Journal of Clinical Laboratory Investigation Supplement 235, 91–97.

Sittampalam, G., Wilson, G.S., 1984. Analytical Chemistry 56, 2170–2175.

Skladal, P., 2003. Brazil Chemical Society 14, 491.

Snow, A.D., Mar, H., Noclin, D., Kimata, K., Kato, M., Suzuki, S., Hassel, J., Wight, T.N., 1988. The presence of heparin sulfate proteoglycans in the neuritic plaques and congophilic angiopathy in Alzheimer's disease. American Journal of Pathology 133, 456–463.

Soderberg, L., Zhukareva, V., Bogdanovic, N., Hashimoto, T., Winbad, B., Iwatsubo, T., Lee, V.M., Trojanowski, J.Q., Naslund, J., 2003. Molecular identification of AMY, an Alzheimer's disease amyloid-associated protein. Journal of Neuropathology and Experimental Neurology 62, 1108–1117.

Sporri, B., Bickel, M., Dobbelaere, D., Machado, J., Lottaz, J.D., 2001. Cytokine Growth Factor Reviews 12, 27.

Spronk, P.E., Limburg, P.C., Kallenberg, C.G., 1995. Lupus 4, 86–94.

Stollar, B.D., Schwartz, R.S., 1985. Journal of Clinical Investigation 75, 321.

Stylianou, J., Saklatvala, 1998. International Journal of Cell Biology 30, 1075–1079.

Tagawa, M., 2000. Current Pharmacy Design 6, 681–689.

Taketa, Y., Tamura, N., Fujita, T., 1984. Journal of Immunology 132, 2531–2537.

Tan, E.M., Cohen, A.S., Fries, J.S., Masi, A.T., McShane, D.J., Rothfield, N.F., Schaller, J.G., Talal, N., Winchetser, R.J., 1982. Arthritis and Rheumatology 24, 1271–1277.

Tang, R., Yeh, C., Wang, J.Y., Changchien, C., Chen, J.S., Hsieh, L., 2003. Annals of Surgery and Oncology 16, 2516–2523.

Ter Borg, E.J., Horst, G., Himmel, E.J., Limburg, P.C., Kalienberg, C.G.M., 1990. Arthritis and Rheumatology 33, 634–643.

Termaat, R.M., Asseman, K.J.M., van Son, J.P.H.F., Dijkman, H.B.P.M., Koene, R.A.P., Berden, JHM, K.M., 1993. Laboratory Investigation 68, 164–173.

Thaler, M., Buhl, A., Welter, H., Schrelegg, A., Kehrel, M., Alber, B., Metzger, J., Luppa, P.B., 2009. Biosensor analyses of serum autoantibodies: application to antiphospholipid syndrome and systemic lupus erythematosus. Analytical and Bioanalytical Chemistry 393, 1417–1429.

Tomer, Y., Gilburd, B., Sack, J., Davies, T.F., Meshorer, A., Burek, C.L., Rose, N.R., Shoenfeld, Y., 1996. Clinical Immunology and Immunopathology 78, 180–187.

Tozzoli, R., 2007. Autoimmunity Reviews 6, 334–340.

Tozzoli, R., 2008. Clinical Chemistry and Laboratory Medicine 46, 577–587.

Tschmeiak, J., Proll, G., Gauglitz, G., 2005. Talanta 65, 313.

Van Regenmortel, M.H., Altschuh, D., Chatelleir, J., Christensen, L., Rauffer-Bruyere, N., Richahlet-Secordel, P., Witz, J., Zeder-Lutz, G., 1968. Journal of Molecular Recognition 11, 163–167.

Ventura, A., Magazzu, G., Greco, L., 1999. Gastroenterology 117, 297.

Von Landenberg, P., Scholmerich, J., Lackner, K.J., 2001. Medical Klinic 96, 331–342.

Wang, J., Liu, G., Engelhard, M.H., Lin, Y., 2006. Analytical Chemistry 78, 6974–6979.

Wang, Y., Xu, H., Zhang, J., Guang, L., 2008. Sensors 8, 2043–2081.

Watts, A.D., Onier-Cherix, N., Hunt, N.H., Chaudhari, R., 2003. Journal of Immunological Methods 282, 169–174.

Watts, R.A., 2002. Medicine 30, 2–6.

Webster, S., Glabe, C., Rogers, J., 1995. Multivalent binding of complement protein C1Q to the amyloid β-peptide (Aβ) promotes the nucleation phase of Aβ aggregation. Biochemical Biophysical Research Communications 217, 869–875.

Wells, X., Bender, V., Francis, C., He-Williams, H.M., Manthey, M.K., Moghaddam, M.J., Reilly, W.G., Whittaker, R.G., 2000. Tris lipidation: a chemically flexible technology for modifying the delivery of drugs and genes. Clinical Experiments in Pharmacology and Physiology 27, 563–567.

White, R.J., Kallewaard, H.M., Hsieh, W., Patterson, A.S., Kasahagen, J.B., Cash, K.J., Uzawa, T., Soh, H.T., Plaxco, K.W., 2011. Wash-free electrochemical platform for the quantitative multiplexed detection of specific antibodies. Analytical Chemistry 84, 1098–1103.

Whittaker, R.G., Hayes, P.J., Bender, V., 1993. A gentle method of linking tris to amino acids and peptides. Peptide Research 6, 125–128.

Wigand, R., Gottschalk, R., Falkenbach, A., Matthias, T., Kaltwasser, J.P., Hoelzer, D., 1997. Z Rheumatology 56, 53–62.

Wilkins, E., Sitdikov, E., 2006. In: Knopf, C.-K., Bassi, A.S. (Eds.), Smart Biosensor Technology. CRC Press, Florida, USA.

Williams, Jr., R.C., Malone, C.C., Cimbalnik, K., Presley, M.A., Roux, K.,H., Strelets, L., Silvestris, F., 1997. Arthritis and Rheumatology 40, 109–123.

Wilson, W.D., 2002. Science 295, 2103–2105.

Wold, R.T., Young, F.E., Tan, E.M., Fair, R.S., 1968. Science 161, 806–809.

Wu, B.Y., Li, J., Huang, J.D., Wang, Y.Y., Yin, H.J., Chen, K.J., Chen, Q., 2006. Real time kinetic analysis of the interaction between interleukin-1α and soluble interleukin-1 receptor 1 using a resonant mirror biosensor. Analytica Chimica Acta 557, 106–110.

Xu, X.H.N., Song, Y., Nallathamby, P.D., 2007. In: Xu, X.H.N. (Ed.), New Frontiers in Ultrasensitive Bioanalysis: Advanced Analytical Chemistry Applications in Nanobiotechnolgy, Single Molecule Detection, and Single Cell Analysis. Wiley, New York, pp. 41–65.

Xu, X.H.N., Brownlow, W.J., Huang, S., Chen, J., 2003. Biochemical and Biophysical Research Communications 305, 79–86.

Xu, X.H.N., Brownlow, W.J., Kyriacon, S.V., Wan, Q., Viola, J.J., 2004. Biochemistry 43, 10400–10413.

Xu, X.H.N., Chen, J., Jeffers, R.B., Kyriacon, S.V., 2002. Nano Letters 2, 175–182.

Xu, X.-H.N., Yeung, E.S., 1997. Science 275, 1106–1109.

Yaeger, P., Domingo, G.J., Gerdes, J., 2008. Annual Reviews in Biomedical Engineering 10, 107–144.

Yu, J., Xiao, J., Ren, Y., Lao, K., Xie, X.S., 2006. Science 311, 1600–1603.

Zajac, A., Song, D., Qian, W., Zhukov, T., 2007. Colloids, Surfaces B: Biointerfaces 58, 309–314.

Zander, C., Enderlein, J., Keller, R.A., 2002. Single Molecule Detection in Solution: Methods and Applications. Wiley-VCH, Berlin.

Chapter 9

Binding and Dissociation of Biomarkers for Alzheimer's Disease on Biosensor Surfaces: A Fractal Analysis

9.1 INTRODUCTION

Dementia is a loss of brain function that may occur with certain diseases. Alzheimer's disease (AD) is one form of dementia. AD worsens with time, and it affects memory, thinking, and behavior (PubMed Health, 2011). These authors indicate that as one grows older the risk of developing AD goes up. These authors emphasize that AD is not a part of the normal aging process. The risk of AD increases if a close blood relative (such as a brother, sister, or parent) has it. These authors further indicate that AD may be characterized as early onset and late onset. *Early onset* of AD occurs before 60 years of age, and *late onset* of AD develops in people of 60 years of age and older. AD may be briefly described in undergoing three stages: (1) early symptoms (misplacing items, getting lost in familiar routes, etc.), (2) "intermediate stage" symptoms (forgetting details about current events, difficulty in reading and writing, etc.), and (3) severe AD symptoms (can no longer understand language, recognize family members, or perform daily living activities).

The Alzheimer's Association (2011) indicates that "Alzheimer's is a brain disease that causes problems with memory, thinking, and behavior. These authors indicate that Alzheimer's is the most common form of dementia and accounts for 50–70% of dementia cases. These authors emphasize that the greatest known risk factor is increasing age. However, younger onset AD often appears when someone is in their 40s or 50s (5% of the population with AD). The authors further add that those with AD live an average of 8 years after their symptoms become noticeable. This survival period can range from 3 to 20 years depending on the health and other conditions.

The Alzheimer's Association (2011) emphasizes that presently there is no cure for AD, but treatment for symptoms are available, and intensive research continues. The present day AD treatments can, however, slow down the worsening of AD symptoms, and help improve the quality of life for those with

Biomarkers and Biosensors. http://dx.doi.org/10.1016/B978-0-444-53794-2.00009-4

AD and their caregivers (either family or outside medically trained personnel). Research is ongoing worldwide to help find better ways to treat the disease, delay its onset, and prevent it from developing.

The Alzheimer's Research Forum (2011) indicates that an estimated 10% of Americans over 65 years of age and half of those over 85 years of age have AD. This forum indicates that presently 4 million Americans currently suffer from Alzheimer's and the number may increase to 10–15 million in the next 20–30 years. According to these authors this disease is the third most expensive disease to treat in the United States. The estimated cost to the society is around $100 billion per year.

Salamon (2011) indicates that MRI scans assist in the prediction of Alzheimer's risk. The author indicates that MRI brain scans of individuals afflicted with mild cognitive impairment (MCI) may predict the onset of AD. The author cautions that MCI does not always progress to AD. Apparently, there is a thinning in the brain's cerebral cortex. This is the outermost layer of the cerebral hemispheres, which plays a key role in memory, thought, language, and attention. Apparently, the disease has a long asymptomatic period of the order of decades. In any case, biomarkers should assist in the early detection of AD, which would permit patients to have access to clinical trials and families to prepare for caregiving.

Forshee (2011) in an article entitled "Study Aims to Identify Biomarkers of Alzheimer's for Early Detection" indicates that according to Dr Mike Huntington of the Huntington Medical Research Institute in Pasadena, California, if every human lived till 126 years old, they would all develop AD at some point. Huntington and his colleagues are looking for AD biomarkers before symptoms occur. It typically takes 20 years to develop Alzheimer's before there are signs of memory loss. Huntington indicates that tracking chemical changes in the body is a potential means to permit the early detection of the onset of AD.

Research and Markets (2011) in a recent review of AD biomarkers found in the cerebrospinal fluid (CSF) and blood identified 58 candidates in AD biomarkers—40 in CSF, and 18 in blood. The authors indicate that 30% of these were linked with early detection of AD. The report emphasizes that the following two points were paid careful attention (1) to identification of biomarkers for the early detection of AD, and (2) on differentiating the different stages or types of neurodegenerative disease, which may be incorrectly related to AD.

The National Institute of Aging in the United Sates (US NIH, 2011) indicates that experts suggest that as many as 5.1 million Americans may have AD (though estimates vary). The disease is named after Dr Alois Alzheimer, who, in 1906, noticed that there were changes in the brain tissue of a woman who had been affected and died of a mental illness. Dr Alzheimer noted the presence of clumps (now called amyloid plaques) and tangled bundles (neurofibrillary tangles). These are the two main characteristics of AD.

Another distinct feature is the loss of connections between nerve cells (neurons) in the brain. These authors indicate that though the initial start of the AD process is not known precisely as yet; however, estimates are that it takes 10–20 years for AD to develop before symptoms and problems are noticed.

A possible avenue that is gaining considerable momentum research wise is the early diagnosis of AD. One way of doing this is the identification and validation of biomarkers for AD. The early detection or diagnosis of AD using biomarkers will go a long way in helping the management of AD and its prognosis. A better understanding of the kinetics of the binding of different biomarkers to different biosensor surfaces will considerably help better understand the disease and its progression, and simultaneously lead to the development of better biosensors for the early detection of AD. Needless to say, the earlier one is able to detect AD (using AD biomarkers or otherwise) the better will be the prognosis and management of AD.

Steenhuysen (2011) very recently indicates that an Alzheimer's drug hits two forms of the toxic protein. Apparently, the experimental drug attacks the free-floating bits of a protein called β-amyloid. This amyloid is considered by some researchers to be the main culprit for AD. This author emphasizes that AD is a brain-wasting disease that affects 26 million people worldwide, and costs $600 billion to treat per year. Furthermore, current drugs only treat symptoms. The ideal drug should, however, engage both the soluble and the insoluble forms of β-amyloid.

Doong et al. (2010) have recently developed a solgel-derived optical array biosensor for the simultaneous determination of biomarkers for AD. The biomarkers detected were β-amyloid, acetylcholine, and glutamate. Doong et al. (2010) indicate that AD is characterized by selective neuronal death and by the abnormal formation of neurotic and core plaques in the cerebral cortex (Schoonenboom et al., 2004; Bennow et al., 2010). Cannon et al. (2004) indicate that the diagnosis of AD is confirmed by the existence of β-amyloid peptide growth and aggregation in the diseased tissue. Doong et al. (2010) indicate that the β-amyloid peptide, which is the most proteinaceous component of the extracellular amyloid deposits, is apparently one of the key components that is responsible for the pathogenesis of AD (Brzyska et al., 2001). Bush (2003) further indicates that the accumulation of β-amyloid induces the accumulation of hydrogen peroxide and the over production of free radicals in neurons.

In this chapter, we use fractal analysis to analyze the binding and dissociation (if applicable) kinetics of different AD biomarkers to different biosensor surfaces. Note that, and as indicted in the earlier chapters in this book, the fractal analysis method presented is not the only means by which to analyze the external diffusion-limited binding and dissociation (if applicable) kinetics on biosensor surfaces. However, this technique does provide the values of the binding rate coefficient, as well as the degree of heterogeneity present on the biosensor surface (by the fractal dimension). The binding (and the dissociation, if applicable) rate coefficient is, in general, quite sensitive to the degree of that exists on the biosensor surface.

9.2 THEORY

9.2.1 Single-Fractal Analysis

9.2.1.1 Binding Rate Coefficient

Havlin (1989) indicates that the diffusion of a particle (analyte [Ag]) from a homogeneous solution to a solid surface (e.g., receptor [Ab]-coated surface) on which it reacts to form a product (analyte–receptor complex, (Ab·Ag)) is given by:

$$(\mathrm{Ab}\cdot\mathrm{Ag}) \approx \begin{cases} t^{(3-D_{f,bind})/2} = t^p & t < t_c \\ t^{1/2} & t > t_c \end{cases} \tag{9.1}$$

Here $D_{f,bind}$ or D_f is the fractal dimension of the surface during the binding step and t_c is the crossover value. Havlin (1989) indicates that the crossover value may be determined by $r_c^2 \sim t_c$. Above the characteristic length, r_c, the self-similarity of the surface is lost and the surface may be considered homogeneous. Above time, t_c of the surface may be considered homogeneous, since the self-similarity property disappears, and "regular" diffusion is now present. For a homogeneous surface where D_f is equal to 2, and when only diffusional limitations are present, $p = ½$ as it should be. Another way of looking at the $p = ½$ case (where $D_{f,bind}$ is equal to two) is that the analyte in solution views the fractal object, in our case, the receptor-coated biosensor surface, from a "large distance." In essence, in the association process, the diffusion of the analyte from the solution to the receptor surface creates a depletion layer of width $(Đt)^{½}$, where Đ is the diffusion constant. This gives rise to the fractal power law, (analyte·receptor) $\sim t^{(3-D_{f,bind})/2}$. For the present analysis, t_c is arbitrarily chosen and we assume that the value of the t_c is not reached. One may consider the approach as an intermediate "heuristic" approach that may be used in the future to develop an autonomous (and not time-dependent) model for diffusion-controlled kinetics.

9.2.1.2 Dissociation Rate Coefficient

The diffusion of the dissociated particle (receptor [Ab] or analyte [Ag]) from the solid surface (e.g., analyte [Ag]–receptor [Ab] complex-coated surface) into solution may be given, as a first approximation by:

$$(\mathrm{Ab}\cdot\mathrm{Ag}) \approx -t^{(3-D_{f,diss})/2} = -t^p \quad (t > t_{diss}) \tag{9.2}$$

Here $D_{f,diss}$ is the fractal dimension of the surface for the dissociation step. This corresponds to the highest concentration of the analyte–receptor complex on the surface. Henceforth, its concentration only decreases. The dissociation kinetics may be analyzed in a manner "similar" to the binding kinetics.

9.2.2 Dual-Fractal Analysis

9.2.2.1 Binding Rate Coefficient

Sometimes, the binding curve exhibits complexities and two parameters (k and D_f) are not sufficient to adequately describe the binding kinetics. This is further corroborated by low values of r^2 factor (goodness-of-fit). In that case, one resorts to a dual-fractal analysis (four parameters; k_1, k_2, D_{f1}, and D_{f2}) to adequately describe the binding kinetics. The single-fractal analysis presented above is thus extended to include two fractal dimensions. At present, the time ($t = t_1$) at which the "first" fractal dimension "changes" to the "second" fractal dimension is arbitrary and empirical. For the most part, it is dictated by the data analyzed and experience gained by handling a single-fractal analysis. A smoother curve is obtained in the "transition" region, if care is taken to select the correct number of points for the two regions. In this case, the product (antibody—antigen or analyte—receptor complex, Ab·Ag or analyte.receptor) is given by:

$$(Ab\cdot Ag) \approx \begin{cases} t^{(3-D_{f1,bind})/2} = t^{p1} & (t < t_1) \\ t^{(3-D_{f2,bind})/2} = t^{p2} & (t_1 < t < t_2) = t_c \\ t^{1/2} & (t > t_c) \end{cases} \tag{9.3}$$

In some cases, as mentioned above, a triple-fractal analysis with six parameters (k_1, k_2, k_3, D_{f1}, D_{f2}, and D_{f3}) may be required to adequately model the binding kinetics. This is when the binding curve exhibits convolutions and complexities in its shape due to perhaps to the very dilute nature of the analyte (in some of the cases to be presented) or for some other reasons. Also, in some cases, a dual-fractal analysis may be required to describe the dissociation kinetics.

9.3 RESULTS

A fractal analysis is used to model the binding and the dissociation (if applicable) kinetics of (1) different concentrations of APP770 (amyloid precursor protein) in complete medium (CM) to the DE antibodies immobilized on a gold surface using total internal reflection ellipsometry (TIRE) (Mustafa et al., 2010), (2) binding of different concentrations of prion protein (PrP) variant (ARQ) (in nM) in solution to asolectin liposome immobilized on the surface of a commercial L1 biosensor chip (Steunou et al., 2010), (3) binding of different concentrations of O_3 (in nM) in solution to immobilized asolectin liposomes on a surface plasmon resonance (SPR) biosensor chip (Steunou et al., 2010), (4) binding of 10 nM ARQ-Cu(II) to chip-immobilized PC liposomes with or without lipid rafts (Steunou et al., 2010), (5) binding of 100 nM ARQ in solution to chip-immobilized PC liposomes with or without lipid rafts (Stenou et al., 2010), (6) binding of different concentrations in micrometer of Aβ in solution to sonicated β-amyloid fibrils (fAβ) immobilized on a flow cell at a density of 2000 RU (Hasegawa et al., 2002), (7) binding of 10 μM of Aβ in solution to 1000 RU of fAβ immobilized on a flow cell (repetition runs) (Hasegawa et al., 2010), and (8) binding of acetylcholine and acetylcholinesterase (AChE) inhibitor to AChE and choline oxidase immobilized on a working electrode surface of a three-electrode surface (Lenigk et al., 2000).

Mustafa et al. (2010) have recently developed a highly sensitive method using spectroscopic ellipsometry in total internal reflection mode (TIRE) to detect β-amyloid peptide (Aβ1-16) using monoclonal antibodies (mAbs) raised against Aβ1-6. These DE antibodies were immobilized on the surface of gold. These authors indicate that a major neuropathological hallmark of AD is the deposition of β-amyloid (Aβ). These deposits occur frequently in the walls of cerebral and meningeal blood vessels. Clark et al. (1985) and Glenner and Wong (1984) indicate that these deposits consist of an insoluble deposit of a 1−42 amino acid peptide known as β-amyloid or Aβ. Kang et al. (1987) indicate that Aβ is derived from a larger transmembrane-spanning protein known as amyloid precursor protein (APP). The role of APP has not been clearly identified, however.

Ibach et al. (2006) indicate that an ELISA sandwich immunoassay is used in current clinical practice of Aβ diagnostics. This is based on the detection of Aβ1−12. This procedure, according to Mustafa et al. (2010), is quite reliable though it is expensive and laborious. Thus, it is not suitable for the early diagnosis of Aβ. Mustafa et al. (2010) emphasize the need for the development of a procedure for the early diagnostics of AD.

Westpal and Bornmann (2002) and Arwin et al. (2004) indicate that the TIRE method has been used to detect different analytes. This TIRE method is closely related to the SPR method but exhibits greater sensitivity than SPR. The TIRE method has been used by Nabok and coworkers (Nablok et al., 2007a,b; Nabok and Tsargorodskaya, 2008) for the detection of low molecular weight toxins. Mustafa et al. (2010) have used quartz crystal microbalance (QCM) microfluidic measurements to monitor in situ the reaction between APP770 and DE antibodies in CM. CM is a complex mixture of different salts, amino acids, and proteins (other than APP).

Figure 9.1(a) shows the binding of APP in a 1:2 dilution in CM to the antibodies immobilized on a gold surface using TIRE (Mustafa et al., 2010). A dual-fractal analysis is required to adequately describe the binding kinetics. The values of (1) the binding rate coefficient, k, and the fractal dimension, D_f, for a single-fractal analysis, and (2) the binding rate coefficients, k_1 and k_2, and the fractal dimensions, D_{f1} and D_{f2}, for a dual-fractal analysis are given in Table 9.1. It is of interest to note that as the

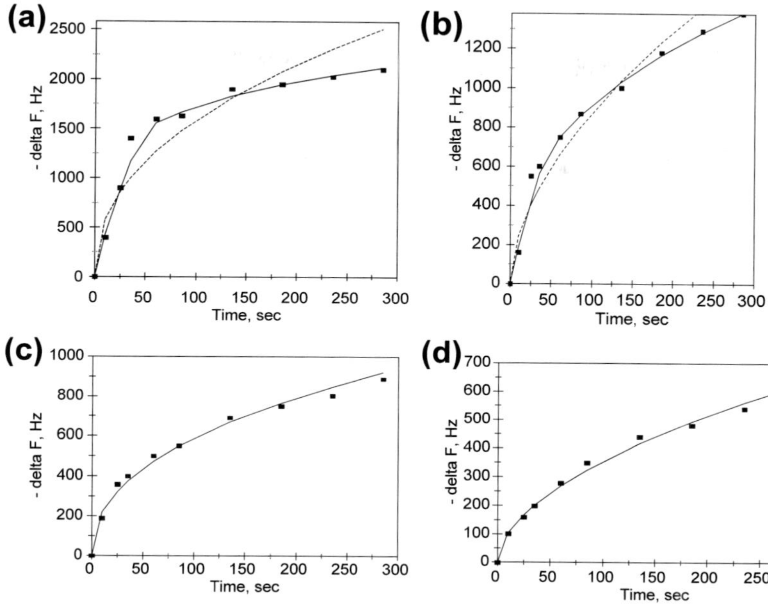

FIGURE 9.1 Binding of different concentrations of APP770 (amyloid precursor protein) in CM (complete medium) to the DE antibodies immobilized on a gold surface using total internal reflection ellipsometry (TIRE) (Mustafa et al., 2010): (a) APP 1:2; (b) 1:4; (c) 1:8; (d) 1:16.

fractal dimension increases by a factor of 1.524 from a value of D_{f1} equal to 1.7738 to D_{f2} equal to 2.7038, the binding rate coefficient increases by a factor of 10.62 from a value of k_1 equal to 65.240 to k_2 equal to 691.93. Note that for a dual-fractal analysis an increase in the fractal dimension or the degree of heterogeneity on the sensing surface leads to an increase (in the same direction) in the binding rate coefficient.

Figure 9.1(b) shows the binding of APP in a 1:4 dilution in CM to the antibodies immobilized on a gold surface using TIRE (Mustafa et al., 2010). A dual-fractal analysis is, once again, required to adequately describe the binding kinetics. The values of (1) the binding rate coefficient, k, and the fractal dimension, D_f, for a single-fractal analysis, and (2) the binding rate coefficients, k_1 and k_2, and the fractal dimensions, D_{f1} and D_{f2}, for a dual-fractal analysis are given in Table 9.1. It is of interest to note that as the fractal dimension increases by a factor of 1.7854 from a value of D_{f1} equal to 1.2396 to D_{f2} equal to 2.2132, the binding rate coefficient increases by a factor of 6.08 from a value of k_1 equal to 24.584 to k_2 equal to 149.573. Note, once again for a dual-fractal analysis that an increase in the fractal dimension or the degree of heterogeneity on the sensing surface leads to an increase (in the same direction) in the binding rate coefficient.

TABLE 9.1 Binding Rate Coefficients and Fractal Dimensions for Different Concentrations (Dilutions) of APP_{770} in Solution (Complete Medium, CM) to DE Antibodies Immobilized on a Biosensor Surface (Mustafa et al., 2010)

APP_{770} Dilution	k	k_1	k_2	D_f	D_{f1}	D_{f2}
1:2	212.28 ± 54.77	65.140 ± 11.277	691.93 ± 18.86	2.1260 ± 0.1440	1.7738 ± 0.2450	2.7038 ± 0.0563
1:4	67.634 ± 17.286	24.584 ± 7.380	149.573 ± 3.062	1.8874 ± 0.1428	1.2396 ± 0.4030	2.2132 ± 0.0424
1:8	81.534 ± 6.712	NA	NA	2.1414 ± 0.0496	NA	NA
1:16	29.344 ± 1.477	NA	NA	1.8186 ± 0.0378	NA	NA

Figure 9.1(c) shows the binding of APP in a 1:8 dilution in CM to the antibodies immobilized on a gold surface using TIRE (Mustafa et al., 2010). A single-fractal analysis is adequate to describe the binding kinetics. The values of the binding rate coefficient, k, and the fractal dimension, D_f, for a single-fractal analysis are given in Table 9.1.

Figure 9.1(d) shows the binding of APP in a 1:16 dilution in CM to the antibodies immobilized on a gold surface using TIRE (Mustafa et al., 2010). Once again, a single-fractal analysis is adequate to describe the binding kinetics. The values of the binding rate coefficient, k, and the fractal dimension, D_f, for a single-fractal analysis are given in Table 9.1.

Steunou et al. (2010) have recently analyzed the charge effect of Cu(II) ions on the binding and prion oligomerization on an SPR biosensor. These authors indicate that PrPs are directly involved in a transmissible spongiform encephalopathies that affect a number of mammalian species (Caughey and Baron, 2006; Caughey et al., 2009). Stenou et al. (2010) emphasize that the disease leads to a progressive dementia and probably involves the conversion of endogeneous cellular prion protein, PrPC into the infectious pathological scrapie form, PrPSc. Fischer et al. (1996) and Pruisner et al. (1983) have indicated that the PrPSc form has a high content of β-sheet structure with a less α-helix structure (when compared with the PrPC form). The high β-sheet structure facilitates the polymerization into amyloid fibrils. Noinvliie et al. (2008) emphasize that the PrPC conversion to PrPSc is enhanced by prion interaction with different molecules such as nucleic acids, lipids, copper ions, etc.

Steunou et al. (2010) indicate that recombinant PrP binds to small unilamellar liposomes immobilized on sensor chips. These authors have characterized the binding of monomeric and oligomeric forms of PrP variant, ARQ, either in full-length or in N-terminal truncated forms (that is delta ARQ) with lipids of different composition. The authors noticed that upon membrane-binding ARQ can switch from normal to β-sheet-rich structure.

Figure 9.2(a) shows the binding of 100 mM ARQ in solution at a pH of 7 to asolectin liposomes immobilized on an SPR biosensor chip surface (Steunou et al., 2010). A dual-fractal analysis is required to adequately describe the binding kinetics. The values of (1) the binding rate coefficient, k, and the fractal dimension, D_f, for a single-fractal analysis, and (2) the binding rate coefficients, k_1 and k_2, and the fractal dimensions, D_{f1} and D_{f2}, for a dual-fractal analysis are given in Table 9.2. It is of interest to note that as the fractal dimension increases by a factor of 1.15 from a value of D_{f1} equal to 2.3090 to D_{f2} equal to 2.7578, the binding rate coefficient increase by a factor of 2.92 from a value of k_1 equal to 1219.81 to k_2 equal to 3556.61. Note that an increase in the fractal dimension or the degree of heterogeneity on the SPR biosensor surface leads to an increase in the binding rate coefficient.

Figure 9.2(b) shows the binding of 20 mM ARQ in solution at a pH of 7 to asolectin liposomes immobilized on an SPR biosensor chip surface

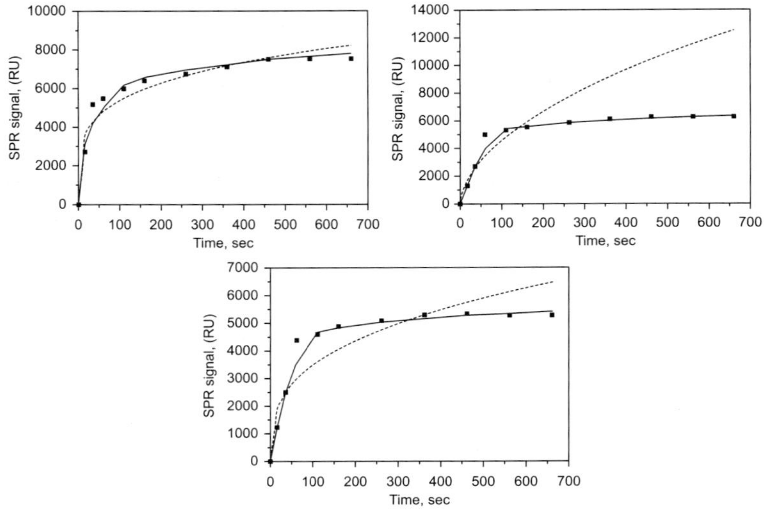

FIGURE 9.2 Binding of different concentrations of PrP (prion protein) variant, ARQ (in nM) in solution to asolectin liposome antibodies immobilized to an SPR L1 biosensor chip (Steunou et al., 2010): (a) 200; (b) 20; (c) 10.

(Steunou et al., 2010). A dual-fractal analysis is required to adequately describe the binding kinetics. The values of (1) the binding rate coefficient, k, and the fractal dimension, D_f, for a single-fractal analysis, and (2) the binding rate coefficients, k_1 and k_2, and the fractal dimensions, D_{f1} and D_{f2}, for a dual-fractal analysis are given in Table 9.2. It is of interest to note that as the fractal dimension increases by a factor of 1.99 from a value of D_{f1} equal to 1.4170 to D_{f2} equal to 2.8202, the binding rate coefficient increase by a factor of 22.86 from a value of k_1 equal to 153.89 to k_2 equal to 3517.94. Note that an increase in the fractal dimension or the degree of heterogeneity on the SPR biosensor surface leads to an increase in the binding rate coefficient.

Figure 9.2(c) shows the binding of 10 mM ARQ in solution at a pH of 7 to asolectin liposomes immobilized on an SPR biosensor chip surface (Steunou et al., 2010). A dual-fractal analysis is required to adequately describe the binding kinetics. The values of (1) the binding rate coefficient, k, and the fractal dimension, D_f, for a single-fractal analysis, and (2) the binding rate coefficients, k_1 and k_2, and the fractal dimensions, D_{f1} and D_{f2}, for a dual-fractal analysis are given in Table 9.2. It is of interest to note that as the fractal dimension increases by a factor of 1.85 from a value of D_{f1} equal to 1.533 to D_{f2} equal to 2.8411, the binding rate coefficient increases by a factor of 18.6 from a value of k_1 equal to 174.27 to k_2 equal to 3242.04. Note that an increase in the fractal dimension or the degree of heterogeneity on the SPR biosensor surface leads to an increase in the binding rate coefficient.

TABLE 9.2 Binding Rate Coefficients and Fractal Dimensions for Different Concentrations of PrP (Prion Protein) Variant, ARQ in Solution to an Surface Plasmon Resonance Biosensor Chip (Steunou et al., 2010)

ARQ Concentration, mM	k	k_1	k_2	D_f	D_{f1}	D_{f2}
100	1943.71 ± 303.23	1219.81 ± 246.81	3556.61 ± 932.53	2.554 ± 0.07742	2.3090 ± 0.2056	2.7578 ± 0.05206
20	390.76 + 462.08	153.89 ± 39.61	3517.94 ± 75.06	1.9324 ± 0.4168	1.4170 ± 0.3330	2.8202 ± 0.03608
10	778.24 ± 226.99	174.26 ± 42.71	3242.04 ± 69.59	2.3466 ± 0.1367	1.5332 ± 0.3186	2.8411 ± 0.02614

Table 9.2 and Figure 9.3(a) show the increase in the binding rate coefficient, k_1, with an increase in the ARQ concentration in solution (in nM). For the data shown in Figure 9.3(a), the binding rate coefficient, k_1, is given by:

$$k_1 = (15.20 \pm 11.55)\,[ARQ]^{0.923 \pm 0.338} \tag{9.3a}$$

The fit is good. Only three data points are available. The availability of more data points would lead to a more reliable fit. The binding rate coefficient, k_1, exhibits close to a first order (equal to 0.923) of dependence on the ARQ concentration in solution.

Table 9.2 and Figure 9.3(b) show the increase in the binding rate coefficient, k_2, with an increase in the ARQ concentration in solution (in nM). For the data shown in Figure 9.3(b), the binding rate coefficient, k_2, is given by:

$$k_2 = (3067.12 \pm 134.87)\,[ARQ]^{0.0343 \pm 0.0256} \tag{9.3b}$$

The fit is reasonable. Only three data points are available. The availability of more data points would lead to a more reliable fit. The binding rate coefficient, k_2, exhibits close to a zero (equal to 0.0343) order of dependence on the ARQ concentration in solution.

Table 9.2 and Figure 9.3(c) show the increase in the binding rate coefficient, k_1, with an increase in the fractal dimension, D_{f1}. For the data shown in Figure 9.3(c), the binding rate coefficient, k_1, is given by:

$$k_1 = \left(29.91 \pm 5.17\right) D_{f1}^{4.41 \pm 0.43} \tag{9.3c}$$

The fit is very good. Only three data points are available. The availability of more data points would lead to a more reliable fit. The binding rate coefficient, k_1, exhibits close to a four and one-half (equal to 4.41) order of dependence on the fractal dimension, D_{f1}.

Table 9.2 and Figure 9.3(d) show the decrease in the binding rate coefficient, k_2, with an increase in the fractal dimension, D_{f2}. For the data shown in Figure 9.3(d), the binding rate coefficient, k_2, is given by:

$$k_2 = \left(45378.18 \pm 2137.01\right) D_{f2}^{-2.501 \pm 2.099} \tag{9.3d}$$

The fit is reasonable. Only three data points are available. The availability of more data points would lead to a more reliable fit. The binding rate coefficient, k_2, exhibits close to a negative two and one-half order of dependence on the fractal dimension, D_{f2}.

Table 9.2 and Figure 9.3(e) show the increase in the fractal dimension, D_{f1}, with an increase in the ARQ concentration in nanomoles in solution. For the data shown in Figure 9.3(e), the fractal dimension, D_{f1}, is given by:

$$D_{f1} = (0.8847 \pm 0.1543)\,[ARQ]^{0.199 \pm 0.0962} \tag{9.3e}$$

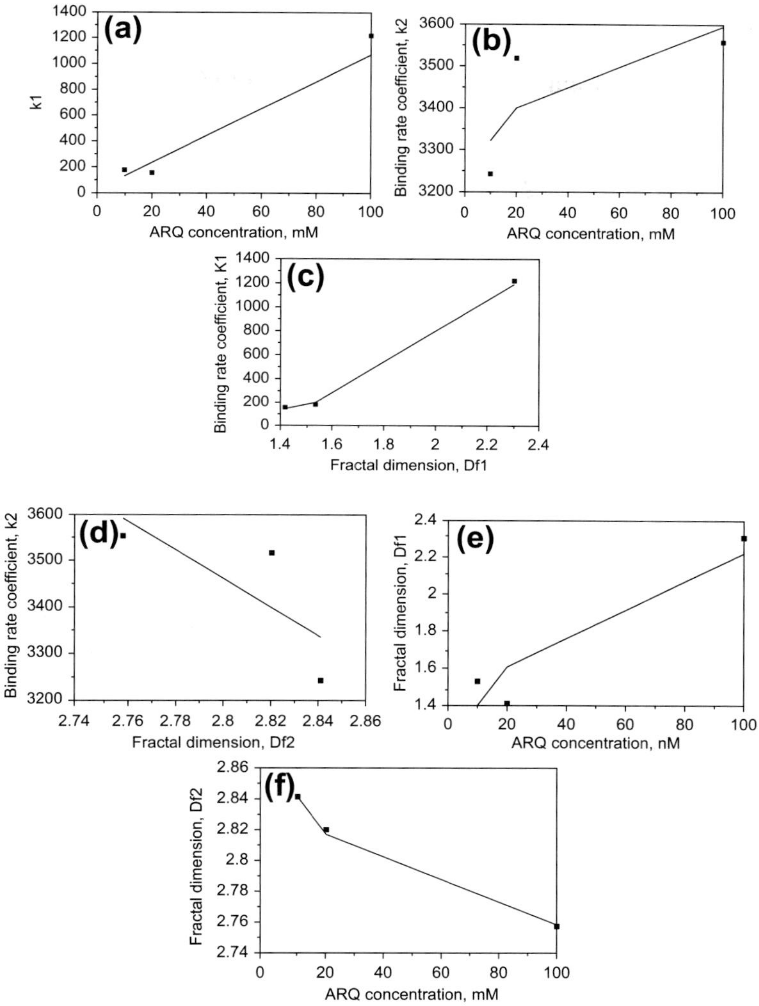

FIGURE 9.3 (a) Increase in the binding rate coefficient, k_1, with an increase in the ARQ concentration (in mM) in solution. (b) Increase in the binding rate coefficient, k_2, with an increase in the ARQ concentration (in mM) in solution. (c) Increase in the binding rate coefficient, k_1, with an increase in the fractal dimension, D_{f1}. (d) Decrease in the binding rate coefficient, k_2, with an increase in the fractal dimension, D_{f2}. (e) Increase in the fractal dimension, D_{f1}, with an increase in the ARQ concentration (in mM) in solution. (f) Decrease in the fractal dimension, D_{f2}, with an increase in the ARQ concentration (in mM) in solution. (g) Decrease in the ratio of the binding rate coefficients, k_2/k_1, with an increase in the ARQ concentration (in mM) in solution. (h) Decrease in the ratio of the fractal dimensions, D_{f2}/D_{f1}, with an increase in the ARQ concentration (in mM) in solution. (i) Increase in binding rate coefficient ratio, k_2/k_1, with an increase in the ratio of the fractal dimensions, D_{f2}/D_{f1}.

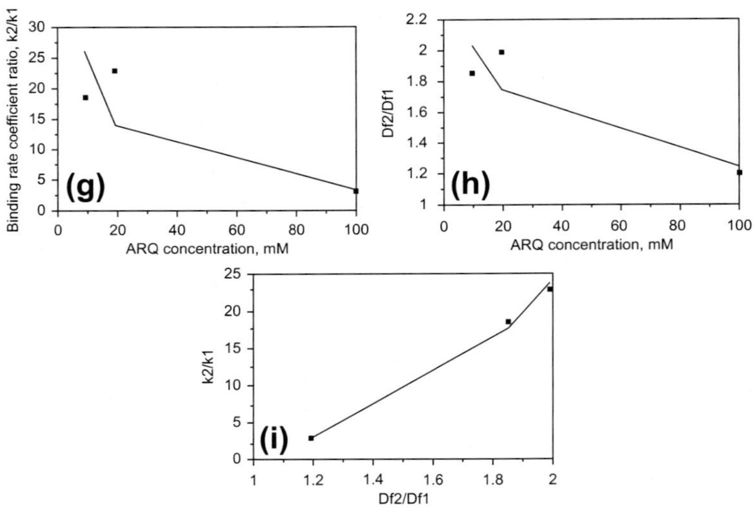

FIGURE 9.3 cont'd

The fit is good. Only three data points are available. The availability of more data points would lead to a more reliable fit. The fractal dimension, D_{f1}, exhibits close to a zero (equal to 0.199) order of dependence on the ARQ concentration in solution.

Table 9.2 and Figure 9.3(f) show the decrease in the fractal dimension, D_{f2}, with an increase in the ARQ concentration in nanomoles in solution. For the data shown in Figure 9.3f, the fractal dimension, D_{f2}, is given by:

$$D_{f2} = (2.930 \pm 0.003)\,[\mathrm{ARQ}]^{-0.031\,+\,0.00075} \tag{9.3f}$$

The fit is very good. Only three data points are available. The availability of more data points would lead to a more reliable fit. The fractal dimension, D_{f2}, exhibits close to a zero (equal to −0.031) order of dependence on the ARQ concentration in solution.

Table 9.2 and Figure 9.3(g) show the decrease in the binding rate coefficient ratio, k_2/k_1, with an increase in the ARQ concentration in nanomoles in solution. For the data shown in Figure 9.3(g), the binding rate coefficient ratio, k_2/k_1, is given by:

$$k_2/k_1 = (202.21 \pm 169{,}37)\,[\mathrm{ARQ}]^{-0.8893 \pm 0.364} \tag{9.3g}$$

The fit is good. Only three data points are available. The availability of more data points would lead to a more reliable fit. The binding rate coefficient ratio, k_2/k_1, exhibits close to a negative first (equal to −0.8893) on the ARQ concentration in solution.

Table 9.2 and Figure 9.3(h) show the decrease in the fractal dimension ratio, D_{f2}/D_{f1}, with an increase in the ARQ concentration in nanomoles (nM) in solution. For the data shown in Figure 9.3(h), the fractal dimension ratio, D_{f2}/D_{f1}, is given by:

$$D_{f2}/D_{f1} = (3.313 \pm 0.582)\,[ARQ]^{-0.213 \pm 0.0969} \tag{9.3h}$$

The fit is good. Only three data points are available. The availability of more data points would lead to a more reliable fit. The fractal dimension ratio, D_{f2}/D_{f1}, exhibits close to a negative zero (equal to −0.213) on the ARQ concentration in solution.

Steunou et al. (2010) indicate that O_3 binds in a quite stable and quasi-irreversible manner to immobilized liposomes. They point out that oligomeric forms have 12 times higher molecular weights than their monomeric counterparts. They emphasize that conformational structures of PrP in 12 mer oligomers prevent some domain interactions with membranes. They further emphasize that the binding of O_3 is facilitated compared to binding of delta O_3 to liposomes.

Figure 9.4(a) shows the binding of 100 mM O_3 in solution to the asolectin liposome immobilized on the SPR sensor chip (Steunou et al., 2010). A dual-fractal analysis is required to adequately describe the binding kinetics. The values of (1) the binding rate coefficient, k, and the fractal dimension, D_f, for a single-fractal analysis, and (2) the binding rate coefficients, k_1 and k_2, and the fractal dimensions, D_{f1} and D_{f2}, for a dual-fractal analysis are given in Table 9.3. It is of interest to note that as the fractal dimension increases by a

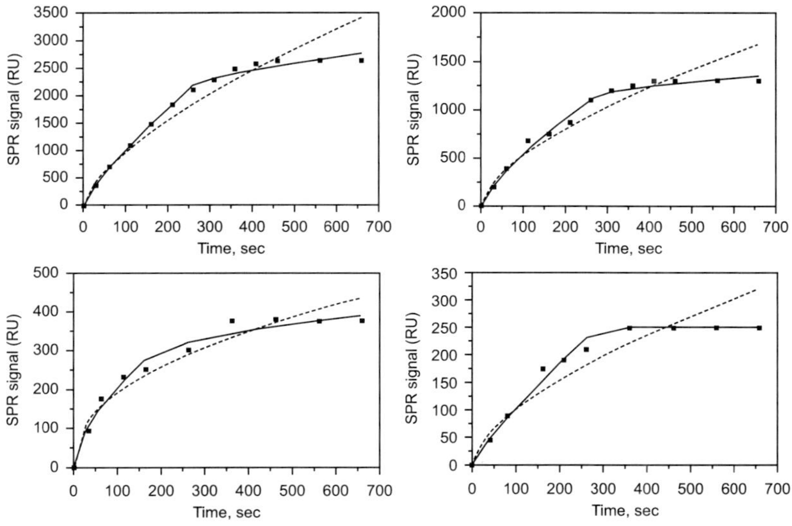

FIGURE 9.4 Binding of different concentrations of O_3 (in nM) in solution to a asolectin liposomes immobilized on the surface of an SPR L1 biosensor ship (Stenou et al., 2010): (a) 100 nM; (b) 50 nM; (c) 20 nM; (d) 10 nM.

TABLE 9.3 Binding Rate Coefficients and Fractal Dimensions for PrP 12 mers O_3 in Solution to Liposomes Immobilized on an SPR Biosensor Chip Surface (Steunou et al., 2010)

O_3 Concentration, nM	k	k_1	k_2	D_f	D_{f1}	D_{f2}
100	49.48 ± 7.437	35.558 ± 1.177	605.35 ± 27.26	1.6946 ± 0.0902	1.3976 ± 0.0504	2.5304 ± 0.1102
50	32.64 ± 5.41	17.07 ± 1.93	463.75 ± 17.50	1.7870 ± 0.09876	1.4964 ± 0.1171	2.6712 ± 0.0926
20	25.70 ± 3.87	12.387 ± 1.936	97.19 ± 7.38	2.1262 ± 0.09312	1.7774 ± 0.2288	2.5696 ± 0.1991
10	6.264 ± 1.393	2.1413 ± 1.393	250 ± 0	1.7876 ± 0.1528	1.3156 ± 0.1355	3.0 ± 0

SPR, surface plasmon resonance; PrP, prion protein.

factor of 1.81 from a value of D_{f1} equal to 1.3976 to D_{f2} equal to 2.5304, the binding rate coefficient increases by a factor of 17 from a value of k_1 equal to 35.558 to k_2 equal to 605.35. Note that an increase in the fractal dimension or the degree of heterogeneity on the SPR sensor chip surface leads to an increase in the binding rate coefficient.

Figure 9.4(b) shows the binding of 50 mM O_3 in solution to the asolectin liposome immobilized on the SPR sensor chip (Steunou et al., 2010). A dual-fractal analysis is, once again, required to adequately describe the binding kinetics. The values of (1) the binding rate coefficient, k, and the fractal dimension, D_f, for a single-fractal analysis, and (2) the binding rate coefficients, k_1 and k_2, and the fractal dimensions, D_{f1} and D_{f2}, for a dual-fractal analysis are given in Table 9.3. It is of interest to note that as the fractal dimension increases by a factor of 1.785 from a value of D_{f1} equal to 1.4964 to D_{f2} equal to 2.6712, the binding rate coefficient increases by a factor of 27.17 from a value of k_1 equal to 17.07 to k_2 equal to 463.75. Note that an increase in the fractal dimension or the degree of heterogeneity on the SPR sensor chip surface, once again, leads to an increase in the binding rate coefficient.

Figure 9.4(c) shows the binding of 20 mM O_3 in solution to the asolectin liposome immobilized on the SPR sensor chip (Steunou et al., 2010). A dual-fractal analysis is, once again, required to adequately describe the binding kinetics. The values of (1) the binding rate coefficient, k, and the fractal dimension, D_f, for a single-fractal analysis, and (2) the binding rate coefficients, k_1 and k_2, and the fractal dimensions, D_{f1} and D_{f2}, for a dual-fractal analysis are given in Table 9.3. It is of interest to note that as the fractal dimension increases by a factor of 1.445 from a value of D_{f1} equal to 1.7774 to D_{f2} equal to 2.5696, the binding rate coefficient increases by a factor of 7.846 from a value of k_1 equal to 12.387 to k_2 equal to 97.19. Note that an increase in the fractal dimension or the degree of heterogeneity on the SPR sensor chip surface, once again, leads to an increase in the binding rate coefficient.

Figure 9.4(d) shows the binding of 10 mM O_3 in solution to the asolectin liposome immobilized on the SPR sensor chip (Steunou et al., 2010). A dual-fractal analysis is, once again, required to adequately describe the binding kinetics. The values of (1) the binding rate coefficient, k, and the fractal dimension, D_f, for a single-fractal analysis, and (2) the binding rate coefficients, k_1 and k_2, and the fractal dimensions, D_{f1} and D_{f2}, for a dual-fractal analysis are given in Table 9.3. It is of interest to note that as the fractal dimension increases by a factor of 2.28 from a value of D_{f1} equal to 1.3156 to D_{f2} equal to 3.0, the binding rate coefficient increases by a factor of 116.75 from a value of k_1 equal to 2.1413 to k_2 equal to 250. Note that an increase in the fractal dimension or the degree of heterogeneity on the SPR sensor chip surface, once again, leads to an increase in the binding rate coefficient.

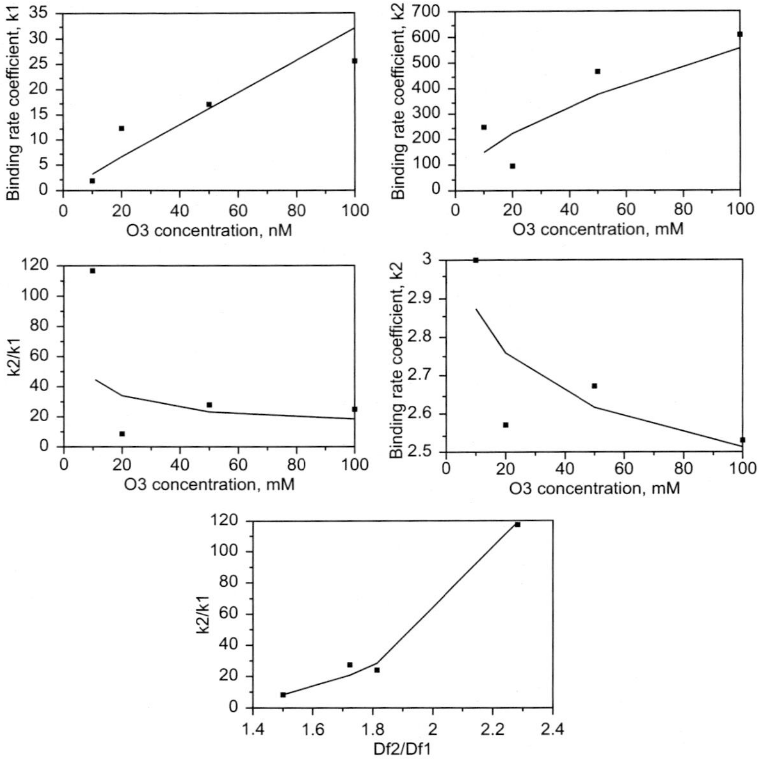

FIGURE 9.5 (a) Increase in the binding rate coefficient, k_1, with an increase in the O_3 concentration (in mM) in solution. (b) Increase in the binding rate coefficient, k_2, with an increase in the O_3 concentration (in mM) in solution. (c) Decrease in the binding rate coefficient ratio, k_2/k_1, with an increase in the O_3 concentration. (d) Decrease in the binding rate coefficient, k_2, with an increase in the O_3 concentration (in mM) in solution. (e) Increase in the binding rate coefficient ratio, k_2/k_1, with an increase in the fractal dimension ratio, D_{f2}/D_{f1}.

Figure 9.5(a) and Table 9.3 show for a dual-fractal analysis the increase in the binding rate coefficient, k_1, with an increase in the O_3 concentration in solution. For the data shown in Figure 9.5(a), the binding rate coefficient, k_1, is given by:

$$k_1 = (0.3545 \pm 0.2719)\,[O_3]^{0.9775 \pm 0.3248} \qquad (9.4a)$$

The fit is good. Only four data points are available. The availability of more data points would lead to a more reliable fit. The binding rate coefficient, k_1, exhibits close to a first (equal to 0.9775) order of dependence on the O_3 concentration in solution.

Figure 9.5(b) and Table 9.3 show for a dual-fractal analysis the increase in the binding rate coefficient, k_2, with an increase in the O_3 concentration in solution. For the data shown in Figure 9.5(b), the binding rate coefficient, k_2, is given by:

$$k_2 = (40.871 + 42.005)\ [O_3]^{0.5647 \pm 0.4034} \tag{9.4b}$$

The fit is reasonable. Only four data points are available. The availability of more data points would lead to a more reliable fit. The binding rate coefficient, k_2, exhibits slightly higher than one-half (equal to 0.5647) order of dependence on the O_3 concentration in solution.

Figure 9.5(c) and Table 9.3 show for a dual-fractal analysis the decrease in the binding rate coefficient ratio, k_2/k_1, with an increase in the O_3 concentration in solution. For the data shown in Figure 9.5(c), the binding rate coefficient ratio, k_2/k_1, is given by:

$$k_2/k_1 = (115.32 + 290.33)\ [O_3]^{-0.4130-0.718} \tag{9.4c}$$

The fit is poor. This is reflected in the error in the estimated binding rate coefficient ratio as well as in error in the exponent. For the binding rate coefficient ratio, only the positive error is given since the binding rate coefficient ratio cannot have a negative value. Similarly, for the exponent only the negative error is given, since the slope is negative in Figure 9.5(c). Only four data points are available. The availability of more data points would lead to a more reliable fit. The binding rate coefficient ratio, k_2/k_1, exhibits a negative (equal to −0.414−0.718) order of dependence on the O_3 concentration in solution.

Figure 9.5(d) and Table 9.3 show for a dual-fractal analysis the decrease in the fractal dimension, D_{f2}, for a dual-fractal analysis with an increase in the O_3 concentration in solution. For the data shown in Figure 9.5(d), the fractal dimension, D_{f2}, is given by:

$$D_{f2} = (3.283 + 0.206)\ [O_3]^{-0.058-0.0348} \tag{9.4d}$$

The fit is reasonable. Only four data points are available. The availability of more data points would lead to a more reliable fit. The fractal dimension, D_{f2}, exhibits close to a negative zero (equal to −0.058) order of dependence on the O_3 concentration in solution.

Figure 9.5(e) and Table 9.3 show for a dual-fractal analysis the increase in the binding rate coefficient ratio, k_2/k_1, with an increase with the fractal dimension ratio, D_{f2}/D_{f1}. For the data shown in Figure 9.5(e), the binding rate coefficient ratio, k_2/k_1, is given by:

$$k_2/k_1 = (0.6747 \pm 0.1902)\ [D_{f2}/D_{f1}]^{6.2778 \pm 0.8256} \tag{9.4e}$$

The fit is very good. Only four data points are available. The availability of more data points would lead to a more reliable fit. The binding rate coefficient ratio, k_2/k_1, exhibits a very high, and greater than sixth (equal to 6.2778) order of dependence on the fractal dimension ratio, D_{f2}/D_{f1}.

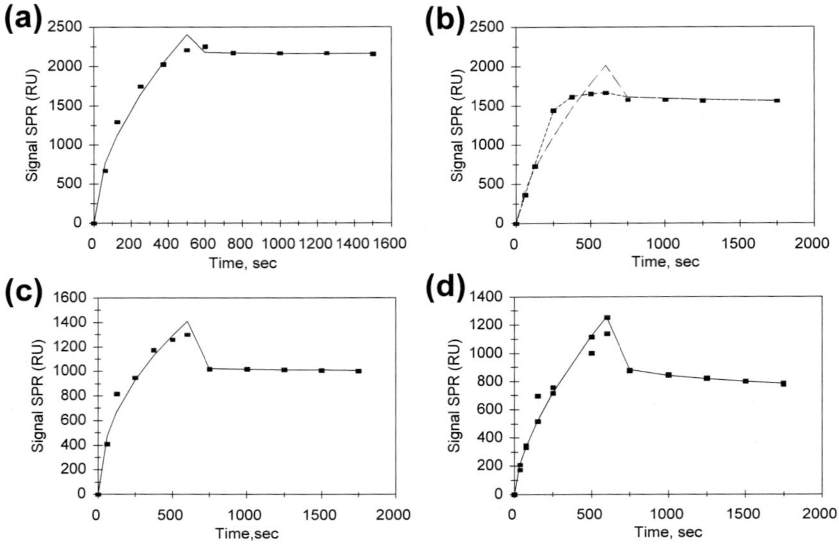

FIGURE 9.6 Binding of 100 nM ARQ-Cu(II) to chip-immobilized PC liposomes with or without lipid rafts (Steunou et al., 2010): (a) PC/cholesterol, 50/50; PC/Chol; (b) PC binding of 100 nM ARQ to chip-immobilized PC liposomes with or without lipid rafts (Steunou et al., 2010); (c) PC/cholesterol, 50/50; (d) PC.

Figure 9.6(a) shows the binding of 100 nM ARQ-Cu(II) in solution to chip-immobilized PC liposomes with or without lipid rafts (PC/cholesterol, 50/50, PC/Chol) (Steunou et al., 2010). A single-fractal analysis is required to adequately describe the binding and the dissociation kinetics. The values of (1) the binding rate coefficient, k, and the fractal dimension, D_f, for a single-fractal analysis, and (2) the dissociation rate coefficient, k_d, and the fractal dimension, D_{fd}, for a single-fractal analysis are given in Table 9.4(a) and (b). In this case, the affinity, K ($=k/k_d$), value is equal to 1.96.

Figure 9.6(b) shows the binding of 100 nM ARQ-Cu(II) in solution to chip-immobilized PC liposomes with or without lipid rafts (PC) (Steunou et al., 2010). A dual-fractal analysis is required to adequately describe the binding kinetics. The dissociation kinetics is adequately described by a single-fractal analysis. The values of (1) the binding rate coefficient, k, and the fractal dimension, D_f, for a single-fractal analysis, (2), the binding rate coefficients, k_1 and k_2, and the fractal dimensions, D_{f1} and D_{f2}, for a dual-fractal analysis, and (3) the dissociation rate coefficient, k_d, and the fractal dimension, D_{fd}, for a single-fractal analysis are given in Table 9.4(a) and (b).In this case, note that for the binding phase, an increase in the fractal dimension by a factor of 2.754 from D_{f1} equal to 1.0410 to D_{f2} equal to 2.8670, leads to an increase in the

TABLE 9.4a Binding and Dissociation Rate Coefficients for ARQ-Cu(II) and ARQ in Solution to Chip-Immobilized PC Liposomes with or without Lipid Rafts (PC/Cholesterol) (Steunou et al., 2010)

Analyte	k	k_1	k_2	k_d
100 nM ARQ-Cu(II) (PC/cholesterol 50:50); PC/chol	76.737 ± 10.546	NA	NA	39.142 ± 2.978
100 nM ARQ-Cu(II) (PC/cholesterol 50:50); PC	25.902 ± 5.470	64.78 ± 0.039	1094.01 ± 5.14	9.250 ± 1.075
100 nM ARQ (PC/cholesterol 50:50); PC/chol	66.996 ± 9.851	NA	NA	245.37 ± 3.24
100 nM ARQ (PC/cholesterol 50:50); PC	21.289 ± 4.627	NA	NA	120.34 ± 4.08

TABLE 9.4b Fractal Dimensions for the Binding and the Dissociation Phase for ARQ-Cu(II) and ARQ in Solution to Chip-Immobilized PC Liposomes with or without Lipid Rafts (PC/Cholesterol) (Steunou et al., 2010)

Analyte	D_f	D_{f1}	D_{f2}	D_{fd}
100 nM ARQ-Cu(II) (PC/cholesterol 50:50); PC/chol	1.9912 ± 0.1519	NA	NA	2.753 ± 0.9478
100 nM ARQ-Cu(II) (PC/cholesterol 50:50); PC	1.638 ± 0.194	1.0410 ± 0.0060	2.8670 ± 0.02796	2.2948 ± 0.1364
100 nM ARQ (PC/cholesterol 50:50); PC/chol	2.0470 ± 0.1400	NA	NA	2.9512 ± 0.008146
100 nM ARQ (PC/cholesterol 50:50); PC	1.7252 ± 0.1618	NA	NA	2.6948 ± 0.04342

binding rate coefficient by a factor of 16.89 from a value of k_1 equal to 64.78 to k_2 equal to 1094.0. An increase in the fractal dimension leads to an increase in the binding rate coefficient. In this case, the affinity values are given by K_1 ($=k_1/k_d$) is equal to 7.0, and the K_2 ($=k_2/k_d$) is equal to 118.3.

Figure 9.6(c) shows the binding of 100 nM ARQ-Cu(II) in solution to chip-immobilized PC liposomes with or without lipid rafts (PC/cholesterol, 50/50) (Steunou et al., 2010). A single-fractal analysis is required to adequately describe the binding and the dissociation kinetics. The values of (1) the binding rate coefficient, k, and the fractal dimension, D_f, for a single-fractal analysis, and (2) the dissociation rate coefficient, k_d, and the fractal dimension, D_{fd}, for a single-fractal analysis are given in Table 9.4(a) and (b). In this case, the affinity, K ($=k/k_d$), value is equal to 0.273.

Figure 9.6(d) shows the binding of 100 nM ARQ-Cu(II) in solution to chip-immobilized PC liposomes with or without lipid rafts (PC) (Steunou et al., 2010). A single-fractal analysis is, once again, required to adequately describe the binding and the dissociation kinetics. The values of (1) the binding rate coefficient, k, and the fractal dimension, D_f, for a single-fractal analysis, and (2) the dissociation rate coefficient, k_d, and the fractal dimension, D_{fd}, for a single-fractal analysis are given in Table 9.4(a) and (b). In this case, the affinity, K ($=k/k_d$), value is equal to 0.177.

Figure 9.7(a) and Table 9.4(a) and (b) show for a dual-fractal analysis the increase in the binding rate coefficient, k_1, with an increase in the fractal dimension, D_{f1}. For the data shown in Figure 9.7(a), the binding rate coefficient, k_1, is given by:

$$k_1 = (0.6083 \pm 0.4393)\, D_{f1}^{6.895 \pm 4.490} \tag{9.5a}$$

The fit is reasonable. Only three data points are available. The availability of more data points would lead to a more reliable fit. The binding rate coefficient, k_1, is very sensitive to the degree of heterogeneity that exists on the biosensor surface since it exhibits close to a seventh (equal to 6.895) order of dependence on the fractal dimension, D_{f1}.

Figure 9.7(b) and Table 9.4(a) and (b) show for a dual-fractal analysis the increase in the binding rate coefficient, k_2, with an increase in the fractal

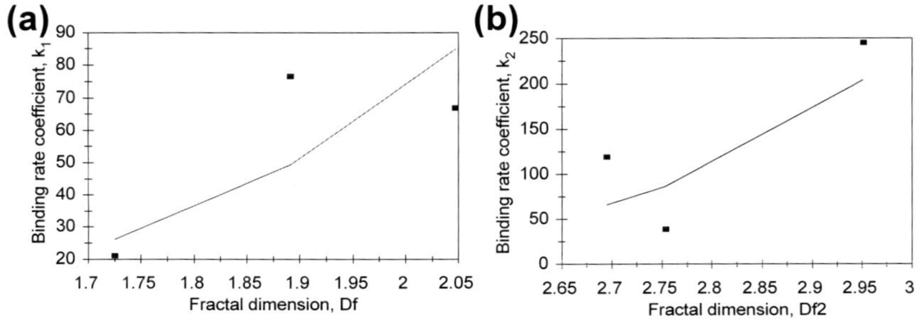

FIGURE 9.7 (a) Increase in the binding rate coefficient, k_1, with an increase in the fractal dimension, D_{f1}. (b) Increase in the binding rate coefficient, k_2, with an increase in the fractal dimension, D_{f2}.

dimension, D_{f2}. For the data shown in Figure 9.7(b), the binding rate coefficient, k_2, is given by:

$$k_2 = (0.000298 + 0.000518)\, D_{f2}^{12.415+15.00} \quad (9.5b)$$

The fit is reasonable. Only three data points are available. The availability of more data points would lead to a more reliable fit. The binding rate coefficient, k_2, is extremely sensitive to the degree of heterogeneity that exists on the biosensor surface since it exhibits close to a twelve and one-half (equal to 12.415) order of dependence on the fractal dimension, D_{f2}.

Hasegawa et al. (2002) indicate that the deposition of fAβ in the brain is suggested to play a central role in the pathogenesis of AD. This is an early event in the amyloid cascade (Selkoe, 2001; Hardy and Higgins, 1992; Sommer, 2002). There is an extensive ongoing research to help reduce fAβ deposition in the brain (Ono et al., 2002; Findeis, 2000; Naiki et al., 1997, 1998; Janus et al., 2000). Kelly (1998) indicates that amyloid fibril formation is considered as a protein misfolding and aggregation event. A first-order kinetics model indicates that fAβ extension proceeds via the consecutive association of Aβ into the ends of existing fibrils (Naiki and Nakakuki, 1996; Lomakin et al., 1996; Esler et al., 1996, 1997).

Esler et al. (2000) have indicated that fAβ extension may be explained by a "clock-dock" mechanism that involves reversible and irreversible association steps of Aβ onto the ends of fAβ. Myszka et al. (1999) indicate that the SPR biosensor is useful for the analysis of the extension and dissociation reaction of fAβ.

Figure 9.8(a) shows the binding of 50 μl of 10 μM Aβ in solution to sonicated fAβ immobilized on a flow cell at a density of 2000 RU (Hasegawa et al., 2002). A dual-fractal analysis is required to adequately describe the binding kinetics. The values of the binding rate coefficient, k, and the fractal dimension, D_f, for a single-fractal analysis are given in Table 9.5. The affinity, K $(=k/k_d)$, is 8.46.

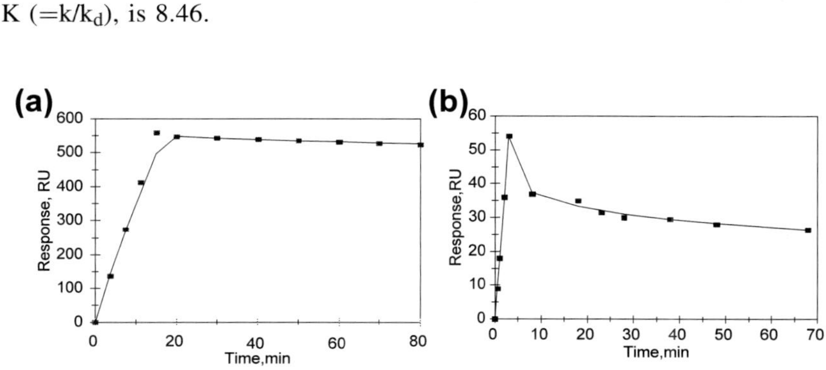

FIGURE 9.8 Binding of different concentrations in micromoles of Aβ in solution to sonicated fAβ immobilized on a flow cell at a density of 2000 RU (Hasegawa et al., 2002): (a) 10; (b) 2.

TABLE 9.5 Binding and Dissociation Rate Coefficients and Fractal Dimensions for the Binding and Dissociation Phase for Two Different Concentrations of Aβ (in μM) in Solution to Sonicated fAβ Immobilized on a Flow Cell at a Density of 2000 RU (Hasegawa et al., 2002)

fAβ Concentration, μM	k	k_d	D_f	D_{fd}
10	46.328 ± 6.132	5.477 ± 0.492	1.250 ± 0.189	2.1414 ± 0.0786
2	18 ± 0	12.078 ± 0.573	2.0 ± 0	2.6012 ± 0.04506

Figure 9.8(b) shows the binding of 30 μl of 2 μM Aβ in solution to sonicated fAβ immobilized on a flow cell at a density of 2000 RU (Hasegawa et al., 2002). A single-fractal analysis is required to adequately describe the binding and dissociation kinetics. The values of (1) the binding rate coefficient, k, and the fractal dimension, D_f, for a single-fractal analysis, and (2) the dissociation rate coefficient, k_d, are given in Table 9.5. In this case, the affinity, K ($=k/k_d$), value is 1.49.

Figure 9.9(a) shows the binding of 50 μl (of 30 μM) Aβ in solution to 1000 RU of fAβ immobilized on a flow cell (Hasegawa et al., 2002). A single-fractal analysis is adequate to describe the binding kinetics. The values of (1) the binding rate coefficient, k, and the fractal dimension, D_f, for a single-fractal analysis are given in Table 9.6.

Figure 9.9(b) shows the binding of 50 μl (of 20 μM) Aβ in solution to 1000 RU of fAβ immobilized on a flow cell (Hasegawa et al., 2002). A single-fractal analysis is adequate to describe the binding kinetics. The values of (1) the binding rate coefficient, k, and the fractal dimension, D_f, for a single-fractal analysis are given in Table 9.6.

Figure 9.9(c) shows the binding of 50 μl (of 10 μM) Aβ in solution to 1000 RU of fAβ immobilized on a flow cell (Hasegawa et al., 2002). A single-fractal analysis is adequate to describe the binding kinetics. The values of (1) the binding rate coefficient, k, and the fractal dimension, D_f, for a single-fractal analysis are given in Table 9.6.

Figure 9.9(d) shows the binding of 50 μl (of 6.5 μM) Aβ in solution to 1000 RU of fAβ immobilized on a flow cell (Hasegawa et al., 2002). A single-fractal analysis is adequate to describe the binding kinetics. The values of (1) the binding rate coefficient, k, and the fractal dimension, D_f, for a single-fractal analysis are given in Table 9.6.

Figure 9.9(e) shows the binding of 50 μl (of 3 μM) Aβ in solution to 1000 RU of fAβ immobilized on a flow cell (Hasegawa et al., 2002).

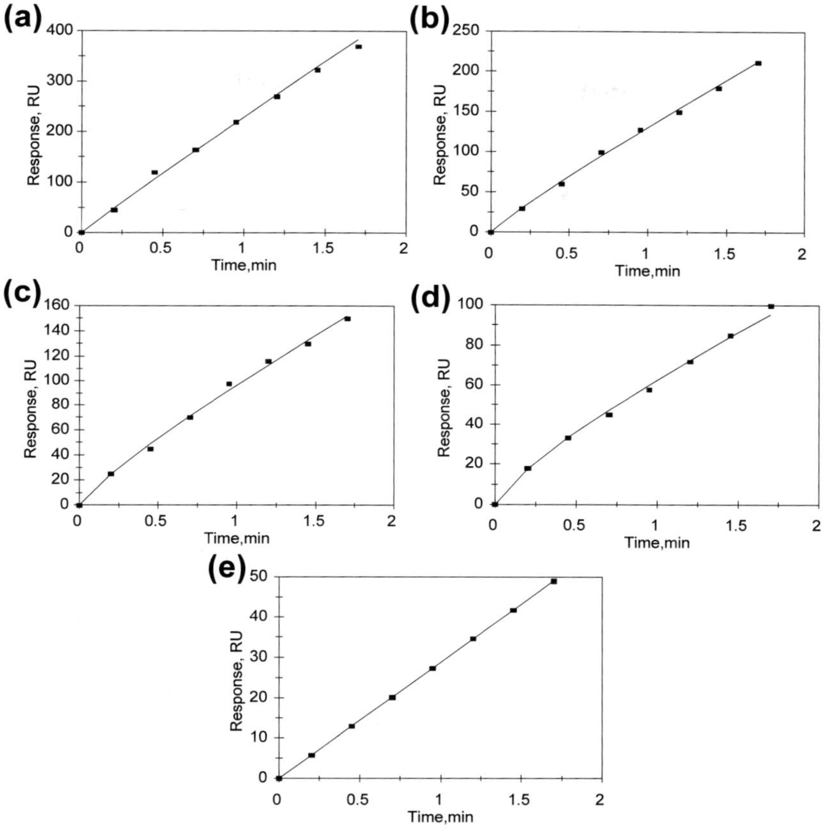

FIGURE 9.9 Binding of different concentrations in micromoles of Aβ in solution to 1000 RU of fAβ immobilized on a flow cell (Hasegawa et al., 2002): (a) 30; (b) 20; (c) 10; (d) 6.5; (e) 3.

TABLE 9.6 Binding Rate Coefficients and Fractal Dimensions for Different Concentrations of Aβ in Solution to fAβ Immobilized on a Flow Cell (Hasegawa et al., 2002)

Analyte in Solution, μM	k	D_f
30	229.47 ± 15.97	1.174 ± 0.07312
20	130.43 ± 5.20	1.172 ± 0.0424
10	96.375 ± 4.860	1.284 ± 0.05349
6.5	62.562 ± 2.360	1.4122 ± 0.0402
3	28.818 ± 0.100	0.992 ± 0.00377

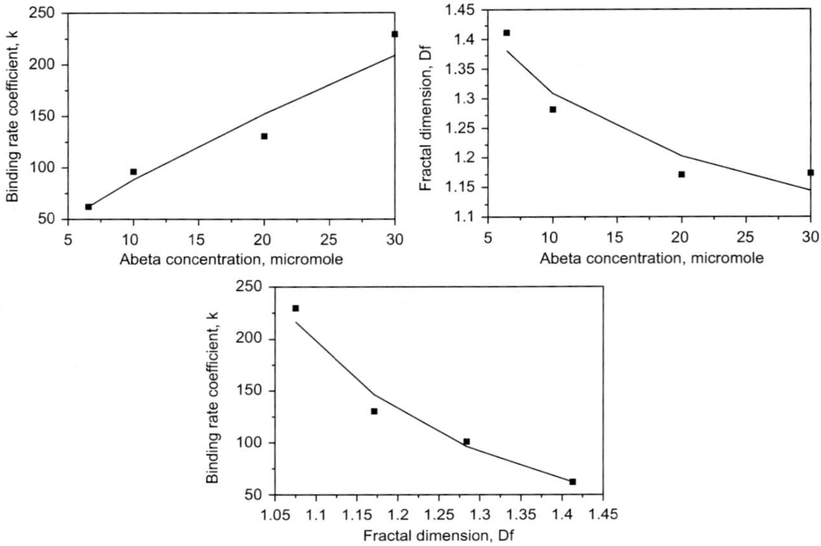

FIGURE 9.10 (a) Increase in the binding rate coefficient, k, with an increase in the Aβ concentration in solution. (b) Decrease in the fractal dimension, D_f, with an increase in the Aβ concentration in solution. (c) Decrease in the binding rate coefficient, k, with an increase in the fractal dimension, D_f.

A single-fractal analysis is adequate to describe the binding kinetics. The values of (1) the binding rate coefficient, k, and the fractal dimension, D_f, for a single-fractal analysis are given in Table 9.6.

Figure 9.10(a) and Table 9.6 show the increase in the binding rate coefficient, k, with an increase in the Aβ concentration in micromole in solution. For the data shown in Figure 9.10(a), the binding rate coefficient, k, is given by:

$$k = (14.804 \pm 2.244)\,[A\beta]^{0.7786 \pm 0.1189} \quad (9.6a)$$

The fit is very good. Only four data points are available. The availability of more data points would lead to a more reliable fit. The binding rate coefficient, k, exhibits an order of dependence between one-half and first (equal to 0.7786) order on the Aβ concentration in solution in micromole.

Figure 9.10(b) and Table 9.6 show the decrease in the fractal dimension, D_f, with an increase in the Aβ concentration in micromole in solution. For the data shown in Figure 9.10(b), the fractal dimension, D_f, is given by:

$$D_f = (1.7379 \pm 0.0596)\,[A\beta]^{-0.123 \pm 0.0283} \quad (9.6b)$$

The fit is good. Only four data points are available. The availability of more data points would lead to a more reliable fit. The fractal dimension, D_f, exhibits close to a negative zero (equal to −0.123) order of dependence on the Aβ concentration in solution in micromole.

Figure 9.10(c) and Table 9.6 show the decrease in the binding rate coefficient, k, with an increase in the fractal dimension, D_f. For the data shown in Figure 9.10(c), the binding rate coefficient, k, is given by:

$$k = (299.41 \pm 29.72)\,[D_f]^{-4.549 \pm 0.464} \quad (9.6c)$$

The fit is very good. Only four data points are available. The availability of more data points would lead to a more reliable fit. The binding rate coefficient, k, exhibits close to a negative four and one-half (equal to −4.549) order of dependence on the fractal dimension, D_f, or the degree of heterogeneity that exists on the biosensor surface.

Hasegawa et al. (2002) have analyzed the reproducibility of the extension reaction. These authors injected a 10 μM Aβ solution (50 μl) into a flow cell on which 1000 RU of Aβ was immobilized. Ten repeat runs were performed in sequence. As each run in the sequence was performed, the response decreased. The binding kinetics of a few of these runs is presented below.

Figure 9.11(a) shows the binding of 10 μM Aβ in solution (50 μl) to 1000 RU of Aβ immobilized on a flow cell surface. This was run #1 (start). A single-fractal analysis is adequate to describe the binding kinetics. The values of the binding rate coefficient, k, and the fractal dimension, D_f, for a single-fractal analysis are given in Table 9.7.

Figure 9.11(b) shows the binding of 10 μM Aβ in solution (50 μl) to 1000 RU of Aβ immobilized on a flow cell surface. This was repeat run #2 (in sequence). A single-fractal analysis is, once again, adequate to describe the binding kinetics. The values of the binding rate coefficient, k, and the fractal dimension, D_f, for a single-fractal analysis are given in Table 9.7.

Figure 9.11(c) shows the binding of 10 μM Aβ in solution (50 μl) to 1000 RU of Aβ immobilized on a flow cell surface. This was repeat run #3 (in sequence). A single-fractal analysis is, once again, adequate to describe the binding kinetics. The values of the binding rate coefficient, k, and the fractal dimension, D_f, for a single-fractal analysis are given in Table 9.7.

Figure 9.11(d) shows the binding of 10 μM Aβ in solution (50 μl) to 1000 RU of Aβ immobilized on a flow cell surface. This was repeat run #5 (in sequence). A single-fractal analysis is, once again, adequate to describe the binding kinetics. The values of the binding rate coefficient, k, and the fractal dimension, D_f, for a single-fractal analysis are given in Table 9.7.

Figure 9.11(e) shows the binding of 10 μM Aβ in solution (50 μl) to 1000 RU of Aβ immobilized on a flow cell surface. This was repeat run #10 (in sequence). A single-fractal analysis is, once again, adequate to describe the binding kinetics. The values of the binding rate coefficient, k, and the fractal dimension, D_f, for a single-fractal analysis are given in Table 9.7.

Lenigk et al. (2000) have developed an enzyme biosensor for studying the therapeutics of AD. These authors indicate that acetylcholinesterase (AChE)

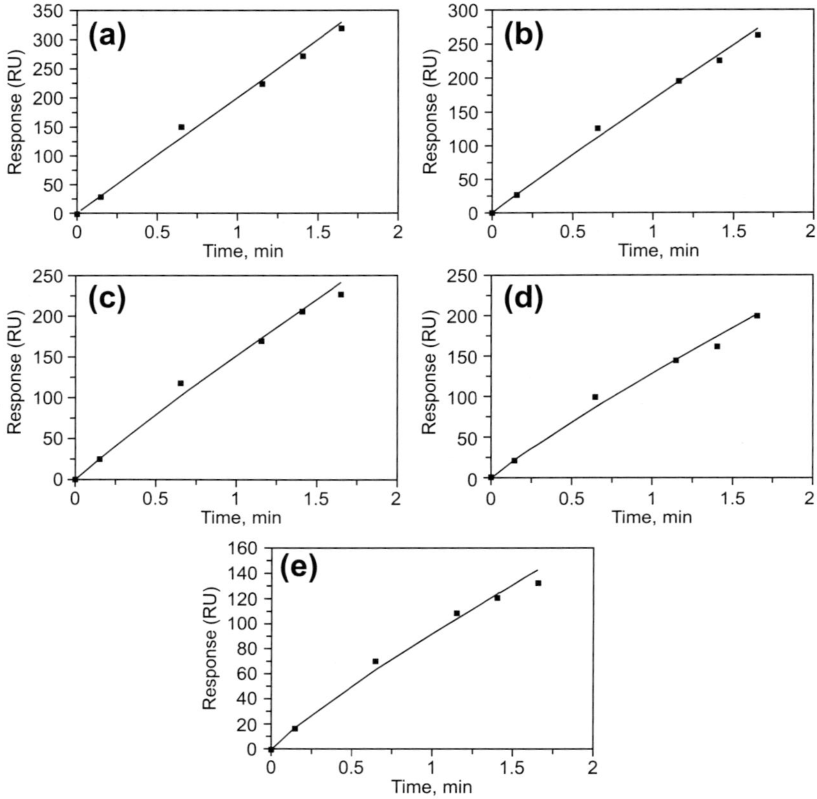

FIGURE 9.11 Binding of 10 μM Aβ in solution to 1000 RU of fAβ immobilized on a flow cell (repetition runs) (Hasegawa et al., 2002): (a) run #1; (b) run #2; (c) run #3; (d) run #5; (e) run #10.

TABLE 9.7 Influence of Repetition Runs on the Binding Rate Coefficient, k and Fractal Dimension D_f for the fAβ Immobilized on a Flow Cell (Hasegawa et al., 2002)

Run Number	k	D_f
1	201.42 ± 12.39	1.1386 ± 0.0849
2	169.4 ± 12.47	1.1054 ± 0.1286
3	152.24 ± 15.39	1.1506 ± 0.09876
5	128.55 ± 12.82	1.8104 ± 0.0974
10	91.93 ± 7.72	1.2866 ± 0.0818

inhibitors are regarded as very promising remedies against AD. Giacobini (1996), Mimori et al. (1997), Robbins et al. (1997), and Valenti et al. (1997) indicate that these AChE inhibitors prolong the effects of endogeneously released acetylcholine. This alleviates the defects in the central cholinergic system. The inhibition reaction of acetylcholinesterase has been used for pesticide detection (Kumaran and Tran-Minh, 1992; Marty et al., 1992; Pachetti et al., 1997; Simonian et al., 1997). Lenigk et al. (2000) indicate that they have circumvented the problems of the currently used tests by making use of an electrochemical detection method that uses acetylcholine (AChCl) as substrate for the enzyme-modified electrode, on which two enzymes are coimmobilized.

Lenigk et al. (2000) indicate that AChE is sensitive for the inhibitor. The second enzyme, choline oxidase, oxidizes the choline (Ch) generated by the first enzyme that leads to the production of hydrogen peroxide. This hydrogen peroxide is oxidized at the working electrode. Therefore, Lenigk et al. (2000) indicate that the current generated is inversely proportional to the amount of inhibitor present in the solution. These authors indicate that the following reactions are taking place at the enzyme-coated working electrode:

1. Hydrolysis of acetylcholine:

$$\text{Acetylcholine} + H_2O \xleftrightarrow{\text{Acetylcolinesterase}} \text{Acetate} + \text{Choline}$$

2. Reaction of Choline:

$$\text{Choline} + 2O_2 + H_2O \xleftrightarrow{\text{Cholinesterase}} \text{Betaine} + 2H_2O$$

Figure 9.12 shows the binding of acetylcholine and AChE inhibitor and AChE and choline oxidase immobilized on a working electrode surface of a three-electrode system using gel entrapment (Lenigk et al., 2002). A single-fractal analysis is adequate to describe the binding kinetics. The binding rate coefficient, k, is equal to 1.3477 ± 0.3378, and the fractal dimension, D_f, is equal to 2.1294 ± 0.1344.

Figure 9.13 shows the binding (digestion of substrate) of 10 μM Aβ(10−20) to 0.025 μM trypsin (protease). Peptide was monitored by recording peptide translocation through an ion channel in the absence and in the presence of an enzyme (protease) (Zhao et al., 2008). A dual-fractal analysis is required to adequately describe the binding kinetics. The values of (1) the binding rate coefficient, k, and the fractal dimension, D_f, for a single-fractal analysis, and (2) the binding rate coefficients, k_1 and k_2, and the fractal dimensions, D_{f1} and D_{f2}, for a dual-fractal analysis are given in Table 9.8(a) and (b). It is of interest to note that as the fractal dimension increases by a factor 1.76 from a value of D_{f1} equal to 1.4586 to a value of D_{f2} equal to 2.5586 the binding rate coefficient increases by a factor of 3.42 from a value of

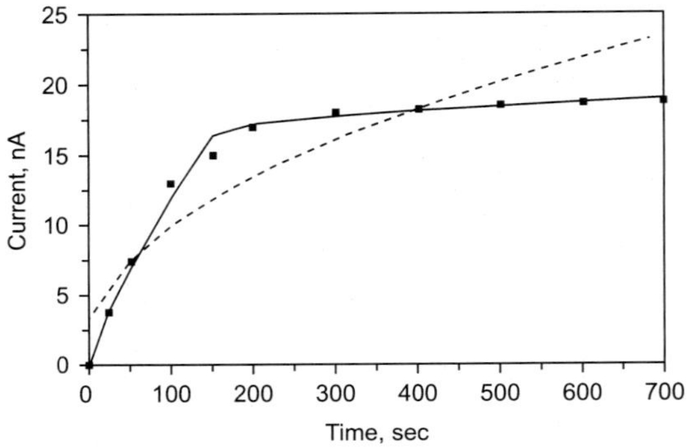

FIGURE 9.12 Binding of acetylcholine and AChE inhibitor to acetylcholinesterase (AChE) and choline oxidase immobilized on a working electrode surface of a three-electrode system using gel entrapment (Lenigk et al., 2000).

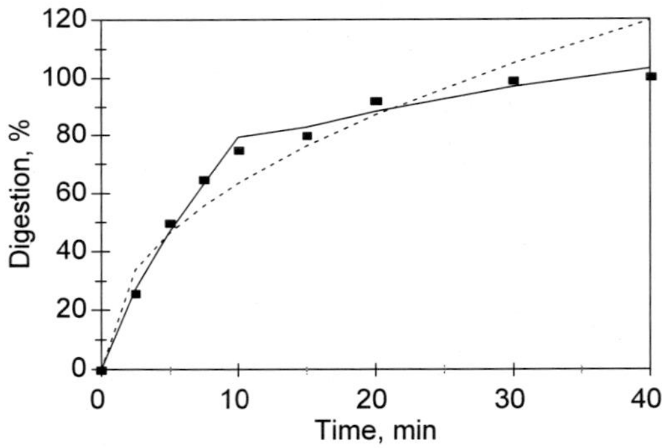

FIGURE 9.13 Binding (digestion of substrate) of 10 μM Aβ(10–20) in solution to 0.025 μM trypsin (cleavage) peptide cleavage was monitored by recording peptide translocation through an ion channel in the absence and in the presence of an enzyme (Zhao et al., 2008).

TABLE 9.8a Binding Rate Coefficients for the Aβ(10–20)/Trypsin Interaction (Zhao et al., 2008)

Aβ(10–20)/Nanopore Probe	k	k_1	k_2
10 μM Aβ(10–20)/(M113F) protein for peptide cleavage	22.136 ± 4.0	13.345 ± 1.033	45.610 ± 1.2180

TABLE 9.8b Fractal Dimensions for the Aβ(10–20)/Trypsin Interaction (Zhao et al., 2008)

Aβ(10–20)/Nanopore Probe	D_f	D_{f1}	D_{f2}
10 μM Aβ(10–20)/(M113F) protein for peptide cleavage	2.0846 ± 0.1348	1.4518 ± 0.1432	2.5586 ± 1.2442

k_1 equal to 13.345 to k_2 equal to 45.610. An increase in the degree of heterogeneity on the biosensor surface leads to an increase in the binding rate coefficient.

Xia et al. (2009) have analyzed the regenerable and simultaneous surface plasmon detection of Aβ(1–4) and Aβ(1–42) peptide in CSF along with signal amplification by streptavidin conjugated to N-terminus specific antibody. These authors indicate that AD is a progressive neurodegenerative disease wherein neuritic (senile) plaques in the brain are involved (Hardy and Selkoe, 2002; Masters et al., 1985). These neurotic peptides include amyloid Aβ peptides containing 39–43 amino acid residues (Shoji et al., 1992). Thinakran and Kao (2008) Golde et al. (2005) indicate that these segments are cleaved from APP. Xia et al. (2010) indicate that Aβ peptides are important biomarkers and drug targets for AD and therapy. Indicate that methods have been developed to detect Aβ monomers and its aggregates. However, Golde et al. (2005) emphasize that these methods are applicable only to postmortem analysis. Golde et al. (2005) emphasize that for the early diagnosis of Aβ sensitive and accurate methods are required to detect Aβ and other biomarkers in body fluids.

ELISA has been shown to be an effective method to detect Aβ species in body fluids and cell media with high sensitivity and versatility. However, it requires 1–2 days and the use of a relatively expensive enzyme-linked antibody. SPR has recently been shown to be an effective technique for the detection of Aβ (Lee et al., 2008). Homola (2006) has employed the SPR technique to detect 17β-hydroxysteroid dehydrogenase type 10 (17-HSD10) enzyme and an analogue in artificial CSF solutions. Gravina et al. (1995) indicate that the interaction between 17β-HSD10 and Aβ has been suggested as a cause for mitochondrial dysfunction in AD.

Xia et al. (2010) have used a dual-channel SPR for determining the ratio of Aβ(1–40) to Aβ(1–42) concentration by immobilizing antibody that recognized the respective peptides. The size of Aβ(1–40) and Aβ(1–42) is relatively small, 4330 and 4514 Da, respectively. They have analyzed multiple sets of CSF samples from AD patients and healthy donors to obtain Aβ(1–40)/Aβ(1–42) concentration ratios.

Figure 9.14(a) shows the binding of Aβ(1–40) for the CSF from a healthy donor with signal amplification by streptavidin conjugated to an N-terminus

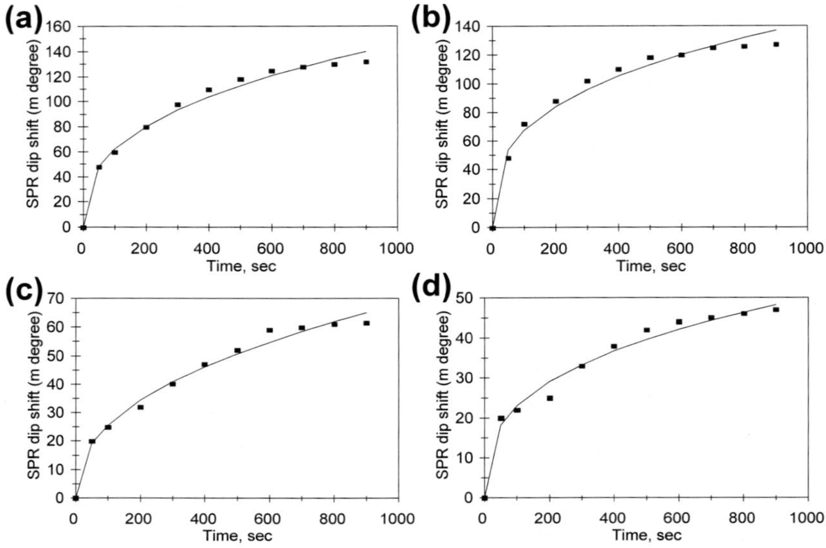

FIGURE 9.14 SPR-binding curves for the simultaneous detection of Aβ(1−40) and Aβ(1−42) concentration in CSF during with signal amplification by streptavidin conjugated to an N-terminus specific antibody (Xia et al., 2010): (a) healthy donor, Aβ(1−40); (b) healthy donor, Aβ(1−42); (c) patient, Aβ(1−40); (d) patient, Aβ(1−42).

antibody immobilized on a biosensor surface. A single-fractal analysis is required to adequately describe the binding kinetics. The values of the binding rate coefficient, k, and the fractal dimension, D_f, are given in Table 9.9.

Figure 9.14(b) shows the binding of Aβ(1−40) for the CSF from a patient with signal amplification by streptavidin conjugated to an N-terminus antibody

TABLE 9.9 Binding Rate Coefficients and Fractal Dimensions for Aβ(1−40) and Aβ(1−42) from a Patient and a Healthy Donor on Binding to Streptavidin Conjugated to an N-terminus Antibody (Xia et al., 2010)

Analyte in Solution/Receptor on Surface	k	D_f
Aβ(1−40) from CSF (healthy donor)/streptavidin conjugated to an N-terminus antibody	11.314 ± 0.494	2.2582 ± 0.02928
Aβ(1−40) from CSF (patient)/streptavidin conjugated to an N-terminus antibody	15.115 ± 1.023	2.3508 ± 0.04596
Aβ(1−42) from CSF (healthy donor)/streptavidin conjugated to an N-terminus antibody	3.665 ± 0.152	2.1548 ± 0.03392
Aβ (1−42) form CSF (patient)/streptavidin conjugated to an N-terminus antibody	4.915 ± 0.366	2.3286 ± 0.0452

CSF, cerebrospinal fluid.

immobilized on a biosensor surface. A single-fractal analysis is required to adequately describe the binding kinetics. The values of the binding rate coefficient, k, and the fractal dimension, D_f, are given in Table 9.9.

Figure 9.14(c) shows the binding of Aβ(1−42) for the CSF from a healthy donor with signal amplification by streptavidin conjugated to an N-terminus antibody immobilized on a biosensor surface. A single-fractal analysis is required to adequately describe the binding kinetics. The values of the binding rate coefficient, k, and the fractal dimension, D_f, are given in Table 9.9.

Figure 9.14(d) shows the binding of Aβ(1−42) for the CSF from a patient with signal amplification by streptavidin conjugated to an N-terminus antibody immobilized on a biosensor surface. A single-fractal analysis is required to adequately describe the binding kinetics. The values of the binding rate coefficient, k, and the fractal dimension, D_f, are given in Table 9.9.

Figure 9.15 shows the increase in the binding rate coefficient, k, with an increase in the fractal dimension, D_f, when the data for Aβ(1−40) and Aβ(1−42) for healthy donors and patients are plotted together. For the data plotted in Figure 9.15, the binding rate coefficient, k, is given by:

$$k = \left(0.0017 \pm 0.00158\right) D_f^{10.23 \pm 9.606} \quad (9.9)$$

The fit is quite good. Only four data points are available. There is scatter in the data. The availability of more data points would lead to a more reliable fit. The binding rate coefficient, k, is very sensitive to the fractal dimension or the degree of heterogeneity that exists on the SPR biosensor surface as is noted by the greater than 10th (equal to 10.23) order of dependence exhibited.

Figure 9.16(a) shows the binding of 30 nM of Aβ(1−16) + 1 nM of Aβ(1−40) on injection to a precoated channel with Aβ(1−40) antibody (Xia et al., 2010). A dual-fractal analysis is required to adequately describe the binding kinetics. The values of (1) the binding rate coefficient, k, and the

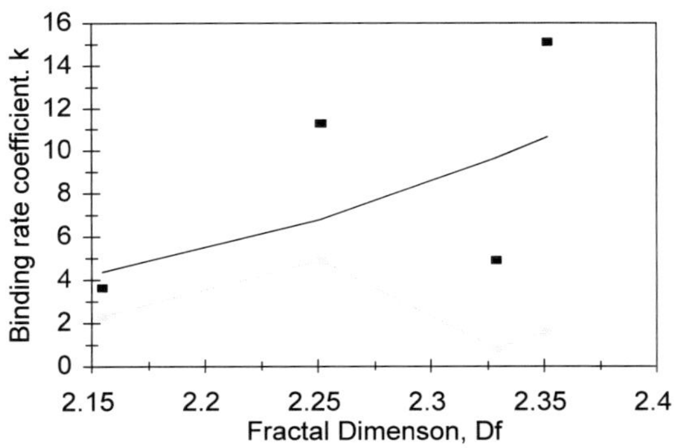

FIGURE 9.15 Increase in the binding rate coefficient, k, with an increase in the fractal dimension, D_f.

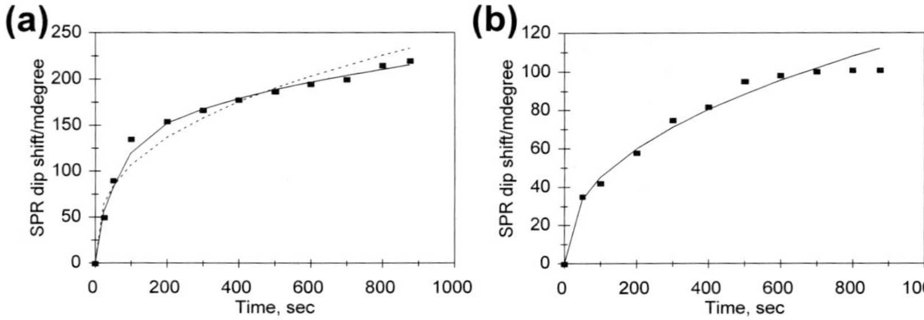

FIGURE 9.16 (a) Binding of 30 nM of Aβ(1−16) + 1 nM of Aβ(1−40) on injection to precoated channel with Aβ(1−40) capture antibody (SPR sensorgram) (Xia et al., 2010). (b) Binding of 30 nM of Aβ(1−16) on injection into precoated channel with Aβ(1−40) capture antibody + 30 nM of detection conjugate (SPR sensorgram) (Xia et al., 2010).

fractal dimension, D_f, for a single-fractal analysis, and (2) the binding rate coefficients, k_1 and k_2, and the fractal dimensions, D_{f1} and D_{f2}, for a dual-fractal analysis are given in Table 9.10(a) and (b). It is of interest to note that as the fractal dimension increases by a factor of 1.33 from a value of D_{f1} equal to 1.9038 to D_{f2} equal to 2.5318, the binding rate coefficient increases by a factor of 4.63 from a value k_1 equal to 9.541 to k_2 equal to 44.115.

TABLE 9.10a Binding Rate Coefficients for the Detection of Different Forms of Aβ in Solution to Appropriate Antibodies Immobilized on an Antibody-Coated Channel (Xia et al., 2010)

Analyte in Solution/Receptor on Surface	k	k_1	k_2
30 nM of Aβ(1−16) + 1 nM Aβ/precoated channel with Aβ(1−40) capture antibody	20.0431 ± 2.912	9.541 ± 1.651	44.115 ± 0.785
30 nM of Aβ(1−16)/precoated channel with Aβ capture antibody + 30 nM of detection conjugate	6.384 ± 0.470	NA	NA
30 nM detection conjugate/ Aβ(1−40) capture antibody (before exposure to 0.2 nM Aβ(1−40)	14.053 ± 0.344	NA	NA
30 nM detection conjugate/ Aβ(1−40) capture antibody (after exposure to 0.2 nM Aβ(1−40)	14.267 ± 0.581	NA	NA

TABLE 9.10b Fractal Dimensions for the Detection of Different Forms of Aβ in Solution to Appropriate Antibodies Immobilized on an Antibody-Coated Channel (Xia et al., 2010)

Analyte in Solution/ Receptor on Surface	D_f	D_{f1}	D_{f2}
30 nM of Aβ(1−16) + 1 nM Aβ/precoated channel with Aβ(1−40) capture antibody	2.2816 ± 0.07066	1.9038 ± 0.2058	2.5318 ± 0.02608
30 nM of Aβ(1−16)/precoated channel with Aβ capture antibody + 30 nM of detection conjugate	2.1542 ± 0.1039	NA	NA
30 nM detection conjugate/ Aβ(1−40) capture antibody (before exposure to 0.2 nM Aβ(1−40)	2.4372 ± 0.0166	NA	NA
30 nM detection conjugate/ Aβ(1−40) capture antibody (after exposure to 0.2 nM Aβ(1−40)	2.4410 ± 0.02742	NA	NA

Figure 9.16(b) shows the binding of 30 nM of Aβ(1−16) + 1 nM of Aβ(1−16) on injection to a precoated channel with Aβ(1−40) capture antibody + 30 nM of detection conjugate (Xia et al., 2010). A single-fractal analysis is adequate to describe the binding kinetics. The values of (1) the binding rate coefficient, k, and the fractal dimension, D_f, for a single-fractal analysis are given in Table 9.10(a) and (b).

Figure 9.17(a) shows the binding of 30 nM of detection conjugate to Aβ(1−40) capture antibody before exposure to 0.2 nM Aβ(1−40) (Xia et al.,

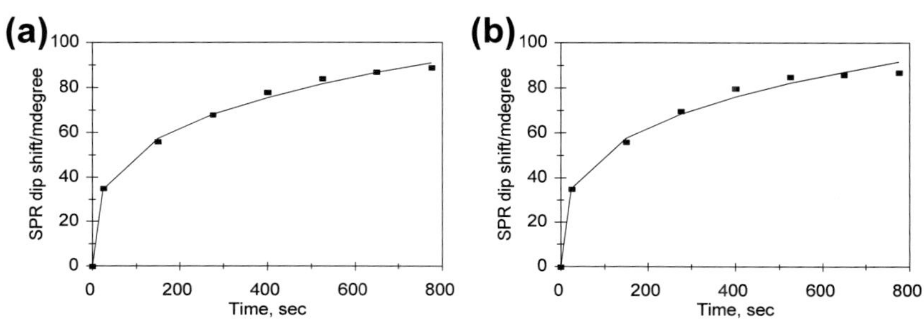

FIGURE 9.17 (a) Binding of 30 nM detection conjugate to Aβ(1−40) capture antibody before exposure to 0.2 nM Aβ(1−40) (SPR sensorgram) (Xia et al., 2010). (b) Binding of 30 nM detection conjugate on injection to Aβ(1−40) antibody that had been previously exposed to 0.2 nM Aβ(1−42) (Xia et al., 2010).

2010). A single-fractal analysis is adequate to describe the binding kinetics. The values of (1) the binding rate coefficient, k, and the fractal dimension, D_f, for a single-fractal analysis are given in Table 9.10(a) and (b).

Figure 9.17(b) shows the binding of 30 nM of detection conjugate on injection to Aβ(1−42) capture antibody that had been previously exposed to 0.2 nM Aβ(1−42) (Xia et al., 2010). A single-fractal analysis is adequate to describe the binding kinetics. The values of (1) the binding rate coefficient, k, and the fractal dimension, D_f, for a single-fractal analysis are given in Table 9.10(a) and (b).

Figure 9.18 shows the increase in the binding rate coefficient, k, with an increase in the fractal dimension, D_f. For the data shown in Figure 9.16 and 9.17 the binding rate coefficient, k, is given by:

$$k = \left(0465 \pm 0.001\right) D_f^{6.413 \pm 0.0357} \tag{9.10a}$$

The binding rate coefficient, k, exhibits a very close fit. More data points would lead to a more reliable fit. The binding rate coefficient, k, exhibits a very strong dependence on the fractal dimension or the degree of heterogeneity that exists on the biosensor surface as is noted by the close to the sixth and one-half order (equal to 6.413) of dependence exhibited.

Xia et al. (2010) analyzed the influence of repeated runs on the binding of repeated runs on the binding of 0.5 nM Aβ(1−40) followed by 30 nM detection conjugate with an SPR channel covered with Aβ(1−40) capture antibody followed by regeneration of surface by NaOH. Figure 9.19(a) shows the first run in a sequence of three repeated runs. A first-order fractal analysis is required to describe the binding kinetics. The values of the binding rate coefficient, k, and the fractal dimension, D_f, are given in Table 9.11.

Figure 9.19(b) shows the second run in a sequence of three repeated runs. A first-order fractal analysis is required to describe the binding kinetics. The values of the binding rate coefficient, k, and the fractal dimension, D_f, are given in Table 9.11.

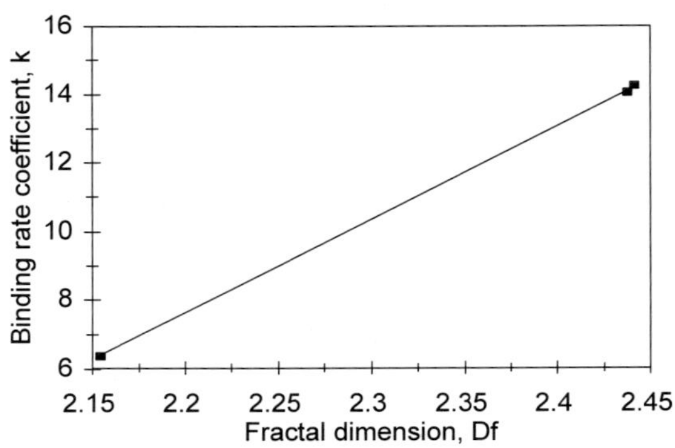

FIGURE 9.18 Increase in the binding rate coefficient, k, with an increase in the fractal dimension, Df.

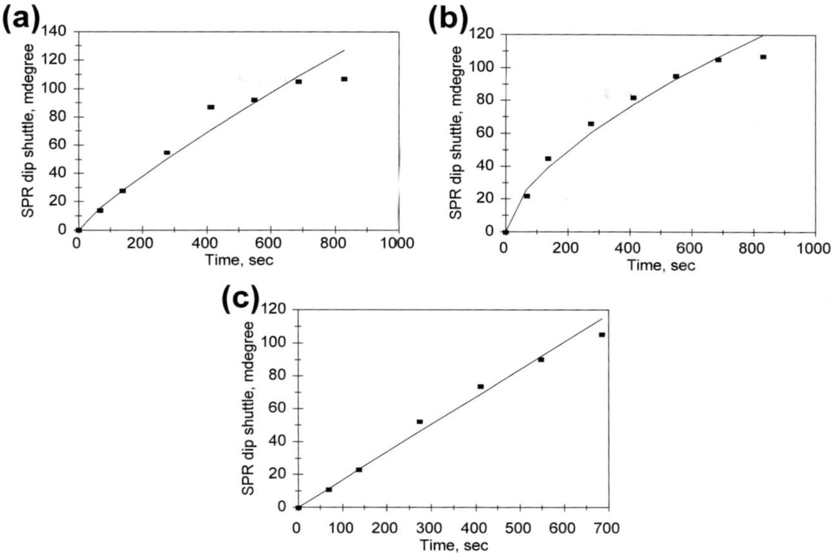

FIGURE 9.19 Binding of 0.5 nM Aβ(1−40) followed by 30 nM detection conjugate into an SPR channel covered with Aβ(1−40) capture antibody followed by regeneration of surface by NaOH (Xia et al., 2010). Influence of repeated runs.

TABLE 9.11 Binding of 0.5 nM Aβ(1−40) Followed by 30 nM Detection Conjugate with an SPR Channel Covered with Aβ(1−40) Captured Antibody Followed by Regeneration of Surface by a NaOH (Xia et al., 2010). Influence of Repeated Runs

Run#	k	D_f
1	0.4356 ± 0.0642	1.3084 ± 0.1232
2	1.859 ± 0.233	1.7586 ± 0.1014
3	0.1734 ± 0.056	1.0042 ± 0.08716

SPR, surface plasmon resonance.

Figure 9.19(c) shows the third run in a sequence of three repeated runs. A first-order fractal analysis is required to describe the binding kinetics. The values of the binding rate coefficient, k, and the fractal dimension, D_f, are given in Table 9.11.

Figure 9.20 shows the increase in the binding rate coefficient, k, with an increase in the fractal dimension, D_f. For the data shown in Figure 9.20 the binding rate coefficient, k, is given by:

$$k = (0.1555 \pm 0.0268)\, D_f^{4.313 \pm 0.4047} \quad (9.10b)$$

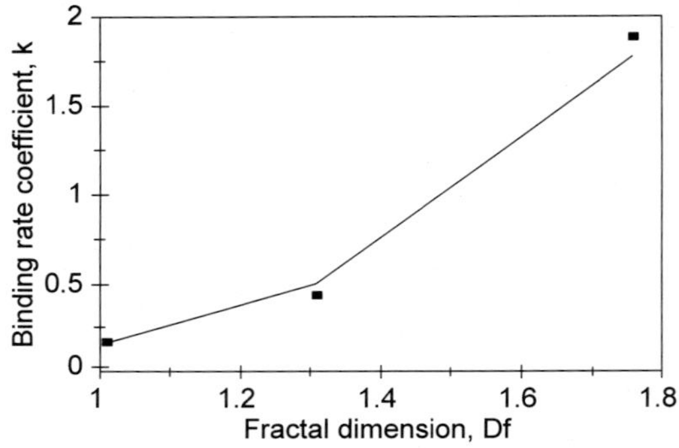

FIGURE 9.20 Increase in the binding rate coefficient, k, with an increase in the fractal dimension, D_f.

The fit is very good. Only three data points are available. The availability of more data points would lead to a more reliable fit. The binding rate coefficient, k, is very sensitive to the degree of heterogeneity that exists on the biosensor surface as is noted by the order of dependence between four and four and one-half (equal to 4.313) exhibited.

Figure 9.21(a) shows the binding and dissociation of Ch (internal standard) to the PB (Prussian blue/AChE-CHO (acetylcholine−choline oxidase) electrode (Mukherjee and Kirchoff, 2009). A first-order fractal analysis is adequate to describe the binding kinetics. The values of (1) the binding rate coefficient, k, and the fractal dimension, D_f, and (2) the dissociation rate coefficient, k_d, and the fractal dimension for dissociation, D_{fd}, are given in Table 9.12(a) and (b). The affinity, K ($=k/k_d$), value is 0.851.

Figure 9.21(b) shows the binding and dissociation of butyrylcholine (BuCh) (internal standard) to the PB (Prussian blue/AChE-CHO (acetylcholine-choline oxidase) electrode. A first-order fractal analysis is adequate to describe the binding kinetics. The values of (1) the binding rate coefficient, k, and the fractal

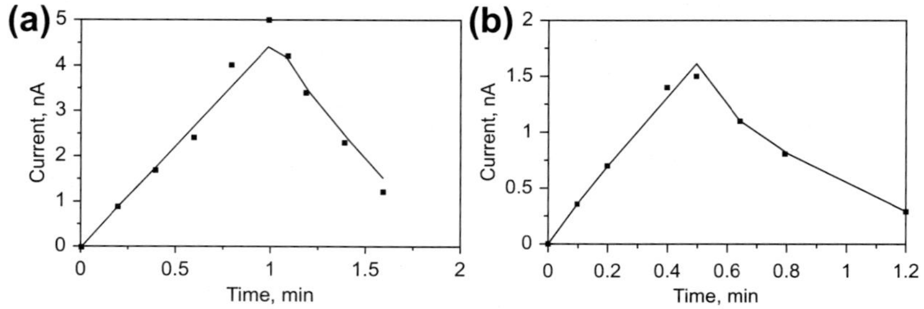

FIGURE 9.21 Binding and dissociation of (a) butyrylcholine (BuCh) (internal standard) and (b) choline (Ch) to PB (Prussian blue)/AChE-CHO (acetylcholine−choline esterase oxidase) electrode (Mukherjee and Kirchoff, 2009).

TABLE 9.12a Binding and Dissociation Rate Coefficients and Affinity Values for Butyrylcholine (BuCh) and Choline (Ch) to PB/AChE-CHO (Prussian Blue/Acetylcholinesterase–Choline Oxidase) at 0.10 V (Mukherjee and Kirchof, 2009)

Analyte in Solution/Receptor on Surface	k	k_d	K (=k/k_d)
BuCh/PB/AChE-CHO	4.411 ± 0.592	5.185 ± 0.610	0.851
Ch/PB/AChE-CHO	3.071 ± 0.234	1.516 ± 0.087	2.025

TABLE 9.12b Fractal Dimensions for the Binding and the Dissociation Phase and the Ratio of the Fractal Dimensions for the Binding and the Dissociation Phase for Butyrylcholine (BuCh) and Choline (Ch) to PB/AChE-CHO (Prussian Blue/Acetylcholinesterase–Choline Oxidase) at 0.10 V (Mukherjee and Kirchof, 2009)

Analyte in Solution/Receptor on Surface	D_f	D_{fd}	D_f/D_{fd}
BuCh/PB/AChE-CHO	0.994 ± 0.1748	1.454 ± 0.1226	0.684
Ch/PB/AChE-CHO	1.1334 ± 0.1164	1.5580 ± 0.0974	0.717

dimension, D_f, and (2) the dissociation rate coefficient, k_d, and the fractal dimension for dissociation, D_{fd}, are given in Table 9.12(a) and (b). The affinity, K (=k/k_d), value is 2.015.

Figure 9.22(a) shows the binding and dissociation of Ch to an HRP-Os/AChE-CHO (horseradish peroxidase cross-linked in an osmium-based redox polymer hydrogel-modified Au (silver)) electrode at −0.1 V (Mukherjee and Kirchoff, 2009). A single-fractal analysis is adequate to describe the binding and the dissociation kinetics. The values of (1) the binding rate coefficient, k, and the fractal dimension, D_f, for a single-fractal analysis is given in Table 9.13(a) and (b). The affinity, K, value is 1.013.

Figure 9.22(b) shows the binding and dissociation of BuCh to an HRP-Os/AChE-CHO (horseradish peroxidase cross-linked in an osmium-based redox polymer hydrogel-modified Au (silver) electrode at −0.1 V (Mukherjee and Kirchoff, 2009). A single-fractal analysis is, once again, adequate to describe the binding and the dissociation kinetics. The values of (1) the binding rate coefficient, k, and the fractal dimension, D_f, for a single-fractal analysis is given in Table 9.13(a) and (b). The affinity, K, value is 2.025.

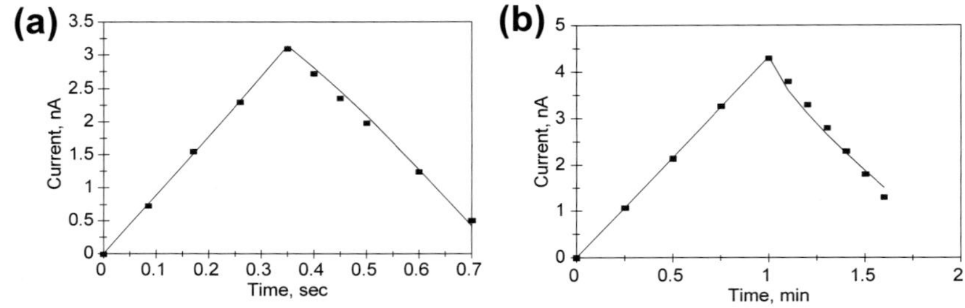

FIGURE 9.22 Binding and dissociation of (a) choline (Ch) and (b) butyrylcholine (BuCh) to an HRP-Os/AChE-CHO (horseradish peroxidase cross-linked in an osmium-based redox polymer hydrogel-modified Au (silver) electrode at −0.1 V (Mukherjee and Kirchoff, 2009).

TABLE 9.13a Binding and Dissociation Rate Coeffients and Affinity Values for Butyrycholine (BuCh) and Choline (ch) to HRP-Os/AChE-CHO (Horseradish Peroxidase Cross-linked in an Osmium Based Redox Polymer Hydrogel Modified Au Electrode at −0.10 V (Mukherjee and Kirchof, 2009)

Analyte in Solution/Receptor on Surface	k	k_d	K (=k/k_d)
Ch/HRP-Os/AChE-CHO	9.129±0.249	9.0076±0.6178	1.013
BuCh/HRP-Os/AChE-CHO	4.325±0.034	4.148±0.851	2.025

TABLE 9.13b Fractal Dimensions for the Binding and the Dissociation Phase and the Ratio of the Fractal Dimensions for the Binding and the Dissociation Phase for Butyrylcholine (BuCh) and Choline (Ch) to HRP-Os/AChE-CHO (Horseradish Peroxidase Cross-linked in an Osmium Base Redox Polymer Hydogel Modified Au Electrode at −0.10 V (Mukherjee and Kirchof, 2009)

Analyte in Solution/Receptor on Surface	D_f	D_{fd}	D_f/D_{fd}
Ch/HRP-Os/AChE-ChO	2.036±0.0496	2.3074±0.0946	0.882
BuCh/HRP-Os/AChE-CHO	0.9922±0.01474	1.4488±0.1776	0.685

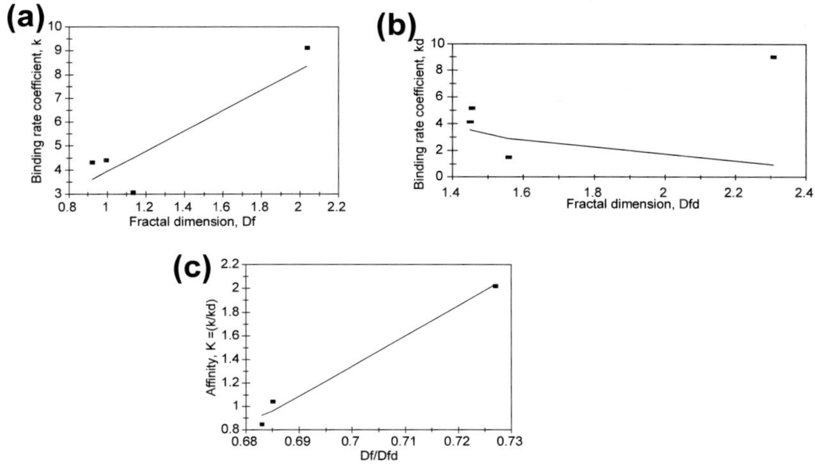

FIGURE 9.23 (a) Increase in the binding rate coefficient, k, with an increase in the fractal dimension, D_f. (b) Decrease in the dissociation rate coefficient, k_d, with an increase in the fractal dimension, D_{fd}. (c) Increase in the affinity, K ($=k/k_d$), with an increase in the fractal dimension ratio, D_f/D_{fd}.

Figure 9.23(a) shows the increase in the binding rate coefficient, k, with an increase in the fractal dimension, D_f. Both sets of data are plotted together (Tables 9.12(a,b) and 9.13(a,b)). For the data plotted in Figure 9.23(a) the binding rate coefficient, k, is given by:

$$k = (3.94 \pm 1.47)\, D_f^{1.054 \pm 0.507} \tag{9.10c}$$

The fit is quite good. Only four data points are available (as indicated above from two sets of data). The availability of more data points would lead to a more reliable fit. The binding rate coefficient, k, exhibits a slightly more than first-order dependence (equal to 1.054) on the fractal dimension or the degree of heterogeneity that exists on the biosensor surface.

Figure 9.23(b) shows the decrease in the dissociation rate coefficient, k_d, with an increase in the fractal dimension for dissociation, D_{fd}. Both sets of data are plotted together (Tables 9.12(a,b) and 9.13(a,b)). For the data plotted in Figure 9.23(b) the dissociation rate coefficient, k_d, is given by:

$$k_d = (10.55 \pm 5.36)\, D_{fd}^{-2.934 \pm 1.412} \tag{9.10d}$$

Figure 9.23(c) shows the increase in the affinity, K ($=k/k_d$) with an increase in the fractal dimension ratio, D_f/D_{fd}. Both sets of data are plotted together (Tables 9.12(a,b) and 9.13(a,b)). For the data plotted in Figure 9.23(c) the affinity K is given by:

$$K(=k/k_d) = (113.20 \pm 14.08) D_f/D_{fd}{}^{12.61 \pm 2.35} \tag{9.10e}$$

The fit is quite very good. Only three data points are available (as indicated above from two sets of data). The availability of more data points would lead

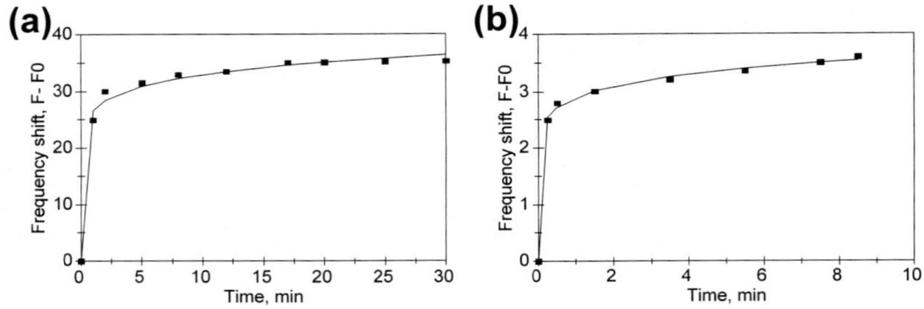

FIGURE 9.24 (a) QCM measurement of P13K-SH3 amyloid growth. Covalent attachment of the fibrils to the biosensor surface after flushing the cell with 40 mM NaCl (Buell et al., 2009). (b) Fibrils attached via a modification scheme.

to a more reliable fit. The affinity, K is very sensitive to the ratio of fractal dimensions that are present on the biosensor surface as noted by the greater than twelve (equal to 12.61) order of dependence exhibited.

Figure 9.24(a) shows the binding of fibrils to the biosensor surface after flushing the cell with 40 mM NaCl. This is measured by a QCM biosensor (Buell et al., 2010). A single-fractal analysis is required to adequately describe the binding and dissociation kinetics. The values of the (a) the binding rate coefficient, k and the fractal dimension, D_f for a single-fractal analysis, and (b) the dissociation rate coefficient, k_d and the fractal dimension for dissociation, D_{fd} are given in Table 9.14.

Figure 9.24(b) shows the binding of fibrils to the biosensor surface after flushing the cell with 40 mM NaCl. This is measured by a QCM biosensor (Buell et al., 2010). In this case there were fibrils attached by a modification scheme. A single-fractal analysis is required to adequately describe the binding and dissociation kinetics. The values of the (a) the binding rate coefficient, k and the fractal dimension, D_f for a single-fractal analysis, and (b) the dissociation rate coefficient, k_d and the fractal dimension for dissociation, D_{fd} are given in Table 9.14.

Figure 9.25(a) shows the frequency shift (F-F0) due to fibril elongation in the absence of NaCl after being flushed repeatedly with solution of moderate

TABLE 9.14 Binding Rate Coefficients and Fractal Dimensions for QCM Measurement of P13K-3H3 Amyloid Growth and Fibrils Covalently Attached via a Modification Scheme; Fibril Elongation in the Presence of 20 mM NaCl (Buell et al., 2009)

	k	D_f
QCM measurement of P13K-3H3 amyloid growth	26.559 ± 0.996	2.8169 ± 0.02206
Fibril elongation in the presence of 20 nM NaCl	2.896 ± 0.059	2.6270 ± 0.01203

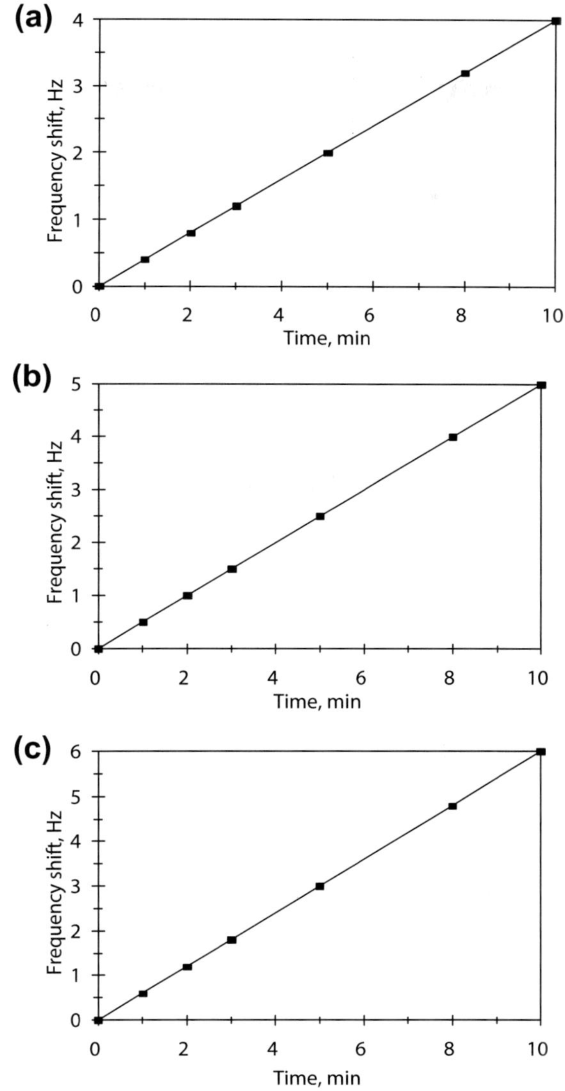

FIGURE 9.25 Frequency shift (F−F0) due to fibril elongation in the absence of NaCl after being flushed repeatedly with solutions of moderate ionic strength. Influence of repeated runs (Buell et al., 2009). (a) Run #1. (b) Run #2. (c) Run #3.

ionic strength. The influence of repeated runs was analyzed. This is run #1. A single-fractal analysis is required to adequately describe the binding kinetics. The values of the binding rate coefficient, k and the fractal dimension, Df are given in Table 9.15.

Figure 9.25(b) shows the frequency shift (F-F0) due to fibril elongation in the absence of NaCl after being flushed repeatedly with solution of moderate

TABLE 9.15 Binding Rate Coefficient and Fractal Dimension for Fibril Elongation Rate (Buell et al., 2009)

Ionic Strength, mM	k	D_f
5	0.4	1.0
10	0.5	1.0
20	0.6	1.0

Influence of ionic concentration in mM.

ionic strength. The influence of repeated runs was analyzed. This is run #2. A single-fractal analysis is required to adequately describe the binding kinetics. The values of the binding rate coefficient, k and the fractal dimension, D_f are given in Table 9.15.

Figure 9.25(c) shows the frequency shift (F-F0) due to fibril elongation in the absence of NaCl after being flushed repeatedly with solution of moderate ionic strength. The influence of repeated runs was analyzed. This is run #3. A single-fractal analysis is required to adequately describe the binding kinetics. The values of the binding rate coefficient, k and the fractal dimension, D_f are given in the Table 9.15.

Figure 9.26 shows the increase in the binding rate coefficient, k with an increase in the ionic strength (NaCl, 5, 10, 20 mM). For the data shown in Figure 9.26 the binding rate coefficient, k is given by:

$$k = (0.251 \pm 0.004)\,[\mathrm{NaCl}]^{0.292 \pm 0.017} \tag{9.10e}$$

The fit is quite good. Only three data points are available (as indicated above from two sets of data). The availability of more data points would lead to a more reliable fit. The dissociation rate coefficient, k_d, exhibits a slightly less negative third-order dependence (equal to −2.934) on the fractal dimension or the degree of heterogeneity that exists on the biosensor surface.

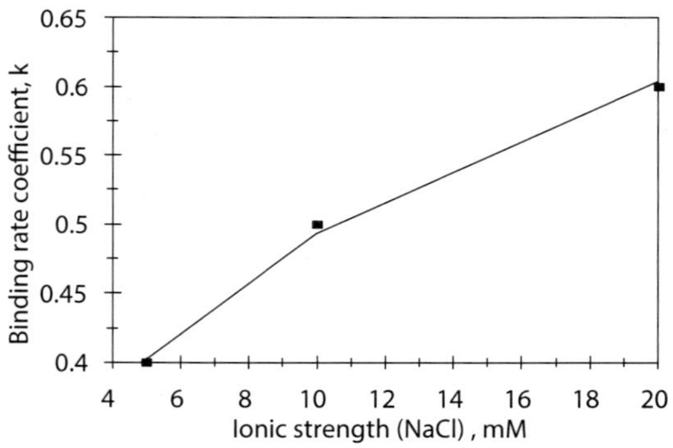

FIGURE 9.26 Increase in the binding rate coefficient, k, with an increase in the ionic strength (NaCl, 5, 10, 20 mM).

Liu et al. (2004) have analyzed single chain variable fragments (scFv) against Aβ, and they state that these scFv can inhibit aggregation and prevent Aβ-induced neurotoxicity. These authors indicate that AD is a neurodegenerative disease, which is characterized by neuronal loss and the presence of amyloid plaques and neurofibrillary tangles. Aβ is a primary component of these amyloid plaques and contain 40–42 amino acids. These amino acids can assemble in different morphologies all of which seem to be neurotoxic (Klein et al., 2001; Koo et al., 1999; Selkoe, 1999; Caughey and Lansbury, 2003). The role that these different morphologies play in the development of AD is not known (Walsh et al., 1999; Lashuel et al., 2002; Nilsberth et al., 2001; Kayed et al., 2003).

Liu et al. (2004) indicate that the potential therapeutic approaches to AD include targeting the enzymes that process APP to Aβ peptides (Esler and Wolfe, 2001; Dewachter and Van Leuven, 2002), increasing the clearance of Aβ peptides from the brain (Lemere et al., 2003: Bard et al., 2000), inhibiting Aβ aggregation (Levine, 2002; Du et al., 2003; Schenk, 2002), and reducing inflammation associated with Aβ deposition (Jantzen et al., 2002). Liu et al. (2004) indicate that antibodies can alter Aβ-induced aggregation through several different mechanisms including blocking of nucleation sites, disrupting existing fibrils or inhibiting fibrillogenesis (Solomon et al., 1997). Also, passive immunization with antibodies against Aβ reduced amyloid deposition (Bard et al., 2003). Indicate that immunotherapy exhibits great potential for treating or preventing AD.

Figure 9.27(a) shows the binding and the dissociation kinetics of 4988 nM scFv (H1v2) in solution to Aβ peptide immobilized on a SPR using a BIAcore X biosensor. A single-fractal analysis is adequate to describe the binding and the dissociation kinetics. The values of (1) the binding rate coefficient, k, and the fractal dimension, D_f, for a single-fractal analysis, and (2) the dissociation rate coefficient, k_d, and the fractal dimension for dissociation, D_{fd}, are given in Table 9.16.

In this case, the affinity, K ($=k/k_d$), is equal to 1.97.

Figure 9.27(b) shows the binding and the dissociation kinetics of 3325 nM scFv (H1v2) in solution to Aβ peptide immobilized on a SPR biosensor. A single-fractal analysis is adequate to describe the binding and the dissociation kinetics. The values of (1) the binding rate coefficient, k, and the fractal dimension, D_f, for a single-fractal analysis, and (2) the dissociation rate coefficient, k_d, and the fractal dimension for dissociation, D_{fd}, are given in Table 9.16.

In this case, the affinity, K ($=k/k_d$), is equal to 2.33.

Figure 9.27(c) shows the binding and the dissociation kinetics of 2494 nM scFv (H1v2) in solution to Aβ peptide immobilized on a SPR biosensor. A single-fractal analysis is adequate to describe the binding and the dissociation kinetics. The values of (1) the binding rate coefficient, k and the fractal dimension, D_f, for a single-fractal analysis, and (2) the dissociation rate coefficient, k_d, and the fractal dimension for dissociation, D_{fd}, are given in Table 9.16.

In this case, the affinity, K ($=k/k_d$), is equal to 3.44.

Figure 9.27(d) shows the binding and the dissociation kinetics of 1663 nM scFv (H1v2) in solution to Aβ peptide immobilized on a SPR biosensor.

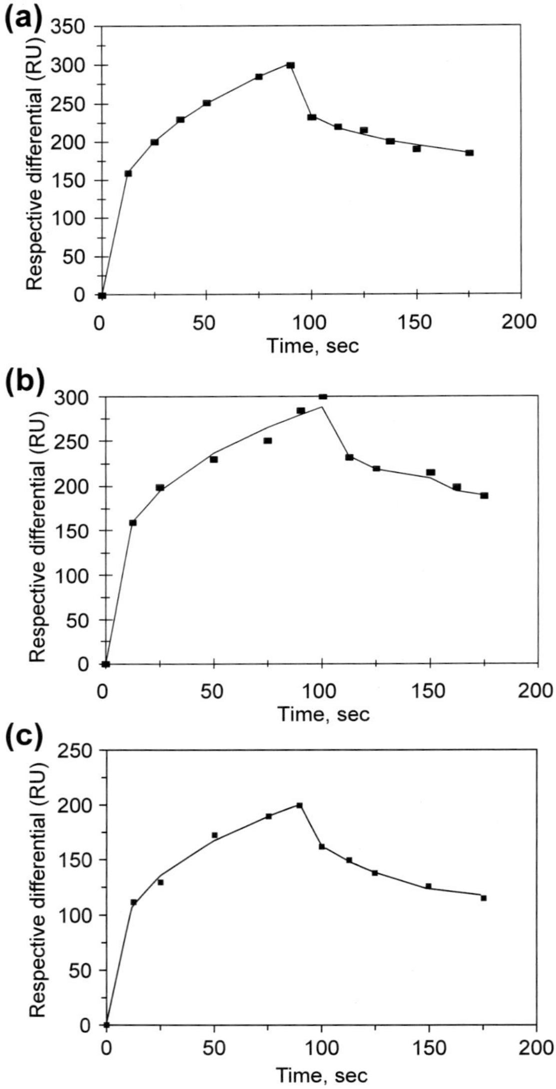

FIGURE 9.27 Binding kinetics of different concentrations (in nM) of scFv to Aβ peptide immobilized on a biosensor surface: (a) 4988; (b) 3325; (c) 2494; (d) 1663; (e) 1248; (f) 632.

A single-fractal analysis is adequate to describe the binding and the dissociation kinetics. The values of (1) the binding rate coefficient, k, and the fractal dimension, D_f, for a single-fractal analysis, and (2) the dissociation rate coefficient, k_d, and the fractal dimension for dissociation, D_{fd}, are given in Table 9.16. In this case the affinity, K $(=k/k_d)$, is equal to 236.7. In this case, the affinity, K, value is very high, since the dissociation rate coefficient, k_d, is very high.

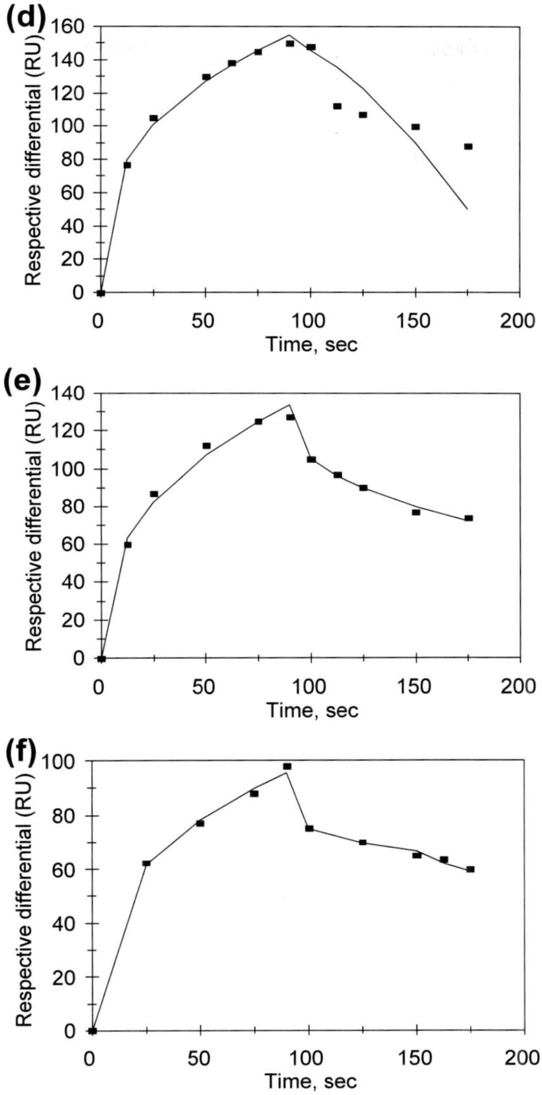

FIGURE 9.27 cont'd

Figure 9.27(e) shows the binding and the dissociation kinetics of 1248 nM scFv (H1v2) in solution to Aβ peptide immobilized on a SPR biosensor. A single-fractal analysis is adequate to describe the binding and the dissociation kinetics. The values of (1) the binding rate coefficient, k, and the fractal dimension, D_f, for a single-fractal analysis, and (2) the dissociation rate coefficient, k_d and the fractal dimension for dissociation, D_{fd}, are given in Table 9.16.

TABLE 9.16 Binding and Dissociation Rate Coefficients and Fractal Dimensions for scFv in Solution to Aβ40 Peptide on a CM5 Sensor Chip of a Surface Plasmon Resonance Biosensor Using BIAcore X Biosensor (Liu et al., 2004)

Analyte in Solution/Receptor on Surface	k	k_d	D_f	D_{fd}	$K = k/k_d$
4988 nM scFv/Aβ40 on CM sensor chip	71.52 ± 0.41	36.35 ± 1.73	2.360 ± 0.0072	2.4820 ± 0.05424	1.97
3325 nM scFv/Aβ40 on CM sensor chip	78.54 ± 3.30	33.66 ± 1.29	2.4440 ± 0.0448	2.479 ± 0.0412	2.33
2494 nM scFv/Aβ40 on CM sensor chip	51.13 ± 1.66	14.885 ± 0.520	2.3926 ± 0.0390	2.202 ± 0.0430	3.44
1663 nM scFv/Aβ40 on CM sensor chip	34.09 ± 1.18	0.143 + 0.183	2.328 ± 0.0402	0.054 + 0.972	236.7
1248 nM scFv/Aβ40 on CM sensor chip	24.10 ± 1.18	8.042 ± 0.355	2.2374 ± 0.0742	2.1364 ± 0.514	3.0
632 nM scFv/Aβ40 on CM sensor chip	20.778 ± 0.663	13.283 ± 0.454	2.514 ± 0.0398	2.3222 ± 0.0542	1.564

scFv, single chain variable fragments.

In this case, the affinity, K $(=k/k_d)$, is equal to 3.0.

Figure 9.28(a) shows the increase in the binding rate coefficient, k, with an increase in the scFv (H1v2) concentration in nanomoles in solution. For the data shown in Figure 9.28(a), the binding rate coefficient, k, is given by:

$$k = (0.1805 \pm 0.0376)\,(\text{scFv})^{0.717 \pm 0.114} \tag{9.13a}$$

The fit is good. Only six data points are available. The availability of more data points would lead to a more reliable fit. The binding rate coefficient, k, for a single-fractal analysis exhibits an order of dependence between on-half and first order (equal to 0.717) on the scFv concentration in solution.

Figure 9.28(b) shows the increase in the binding rate coefficient, k, with an increase in the fractal dimension, D_f. For the data shown in Figure 9.28(b), the binding rate coefficient, k, is given by:

$$k = (8.8E - 0.5 \pm 3.9E - 0.5)\,(D_f)^{15.31 \pm 5.56} \tag{9.13b}$$

The fit is good. Only six data points are available. The availability of more data points would lead to a more reliable fit. The binding rate coefficient, k, for a single-fractal analysis exhibits an order of dependence slightly higher than 15th order (equal to 15.31) on the fractal dimension that exists on the biosensor surface. This indicates that the binding rate coefficient, k, is very sensitive to the degree of heterogeneity that exists on the biosensor surface.

Figure 9.28(c) shows the increase in the dissociation rate coefficient, k_d, with an increase in the scFv (H1v2) concentration in nanomoles in solution. For the data shown in Figure 9.28(c) the dissociation rate coefficient, k_d, is given by:

$$k_d = \left(0.171 \pm 0.0103\right)(\text{scFv})^{0.612 \pm 0.284} \tag{9.13c}$$

There is some scatter in the data. Only five data points are available. The availability of more data points would lead to a more reliable fit. The dissociation binding rate coefficient, k_d, for a single-fractal analysis exhibits slightly more than a half-order dependence (equal to 0.612) on the scFv concentration in solution.

Figure 9.28(d) shows the increase in the dissociation rate coefficient, k_d, with an increase in the fractal dimension, D_{fd}. For the data shown in Figure 9.28(d), the dissociation rate coefficient, k_d, is given by:

$$k_d = \left(0.1117 \pm 0.0074\right)(D_{fd})^{5.93 \pm 3.42} \tag{9.13d}$$

There is some scatter in the data. Only five data points are available. The availability of more data points would lead to a more reliable fit. The dissociation binding rate coefficient, k_d, for a single-fractal analysis exhibits close to a sixth order of dependence (equal to 5.93) on the fractal dimension that exists on he surface. This indicates that the dissociation rate coefficient, k_d, is very sensitive to the degree of heterogeneity that exists on the biosensor surface.

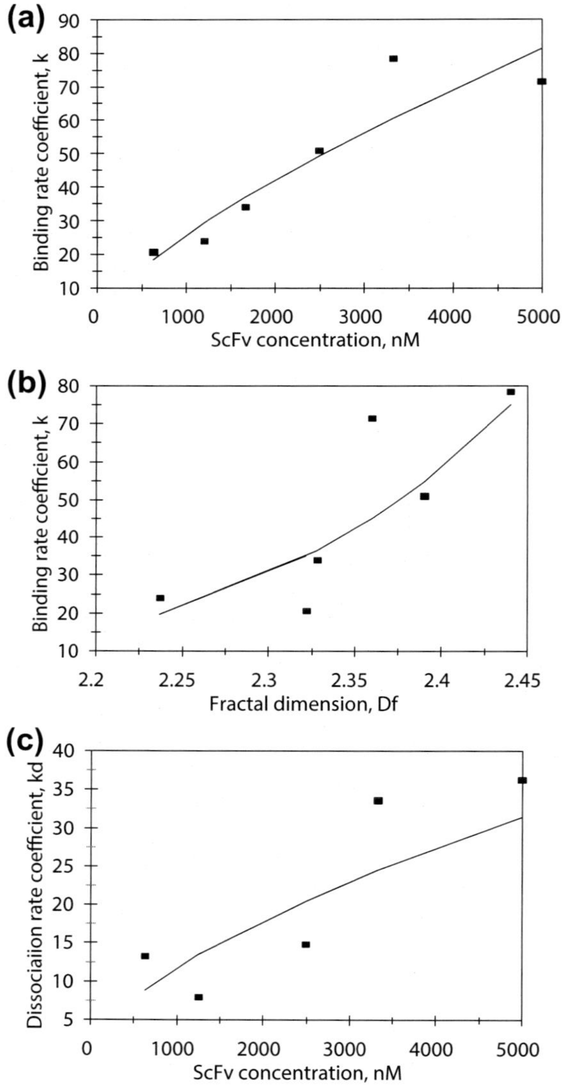

FIGURE 9.28 (a) Increase in the binding rate coefficient, k, with an increase in the scFv concentration in nanometer in solution. (b) Increase in the binding rate coefficient, k, with an increase in the fractal dimension, D_f. (c) Increase in the dissociation rate coefficient, k_d, with an increase in the scFv concentration in nanomter in solution. (d) Increase in the dissociation rate coefficient, k_d, with an increase in the fractal dimension, D_{fd}. (e) Increase in the fractal dimension, D_f with an increase in the scFv concentration in solution.

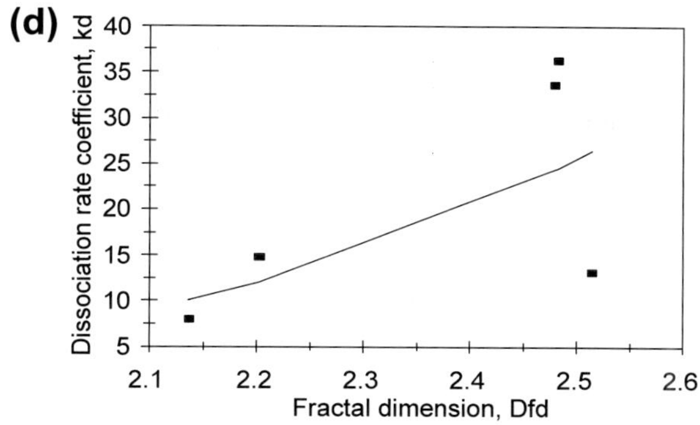

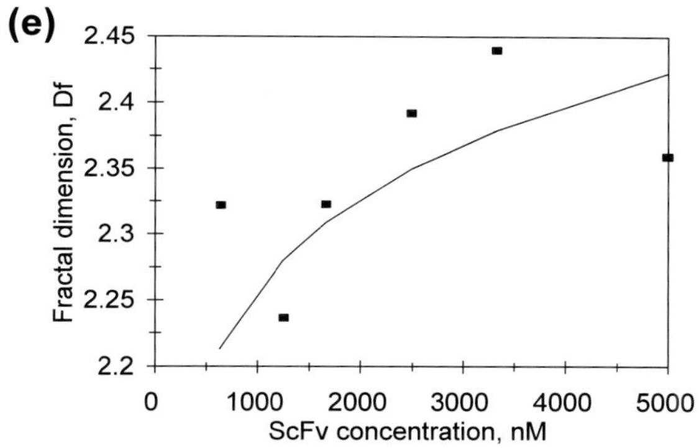

FIGURE 9.28 cont'd

Figure 9.28(e) shows the increase in the fractal dimension, D_f, with an increase in the scFv (H1v2) concentration in nanomoles in solution. For the data shown in Figure 9.28(e) the fractal dimension, D_f, is given by:

$$D_f = \left(1.672 \pm 0.044\right) (\text{scFv})^{0.0435 \pm 0.023} \tag{9.13e}$$

There is some scatter in the data. Only five data points are available. The availability of more data points would lead to a more reliable fit. The fractal dimension, D_f, for a single-fractal analysis exhibits only a very mild dependence (equal to 0.0435) on the scFv concentration in solution.

Figure 9.29(a) shows the binding and dissociation kinetics for 17,241 scFv (C1) nanomoles concentration in solution to Aβ peptide immobilized to a SPR biosensor surface. A single-fractal analysis is adequate to describe the binding and the dissociation kinetics.

Figure 9.29(b) shows the binding and dissociation kinetics for 12,931 scFv (C1) nanomoles concentration in solution to Aβ peptide immobilized to a SPR biosensor surface. A single-fractal analysis is, once again, adequate to describe the binding and the dissociation kinetics. In this case, the affinity, K ($=k/k_d$), is equal to 3.21.

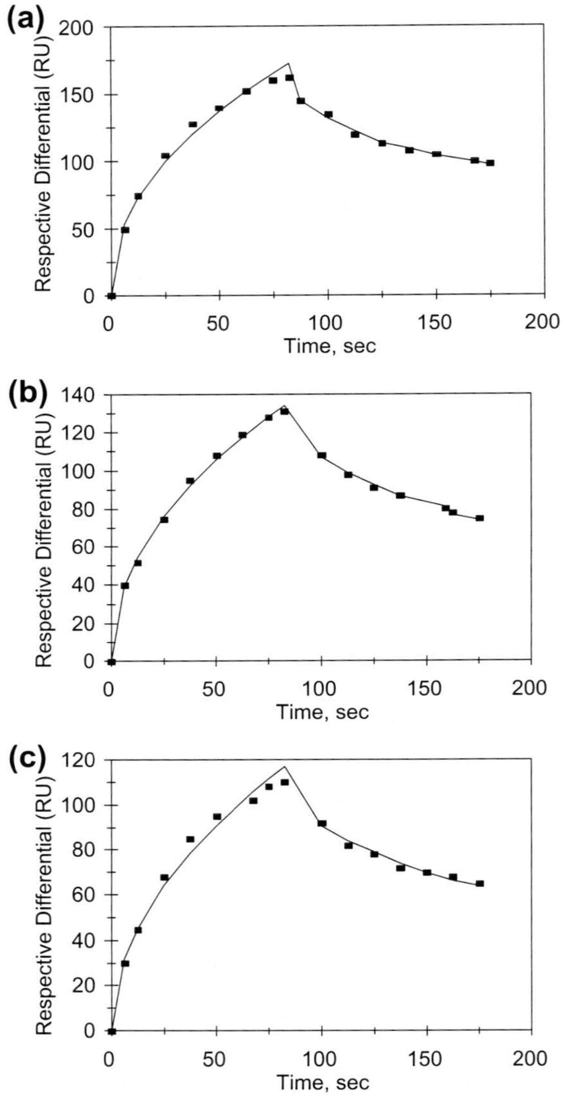

FIGURE 9.29 Binding and dissociation kinetics of scFv (C1) concentration (in nM) in solution to Aβ peptide immobilized on an SPR biosensor surface (Ramakrishnan et al., 2009): (a) 17,241; (b) 12,931; (c) 10,191; (d) 6465; (e) 5170; (f) 3977.

Figure 9.29(c) shows the binding and dissociation kinetics for 10,191 scFv (C1) nanomoles concentration in solution to Aβ peptide immobilized to a SPR biosensor surface. A single-fractal analysis is, once again, adequate to describe the binding and the dissociation kinetics. In this case the affinity, K ($=k/k_d$), is equal to 3.05.

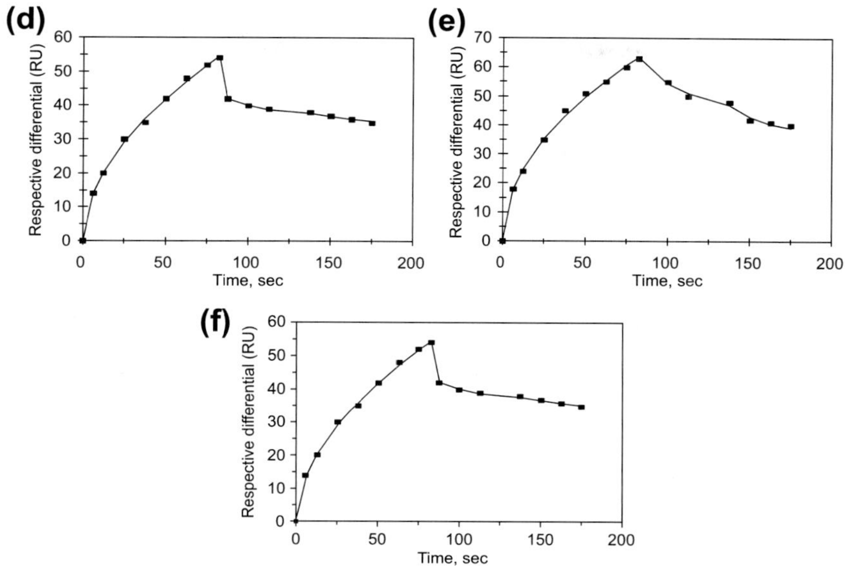

FIGURE 9.29 cont'd

Figure 9.29(d) shows the binding and dissociation kinetics for 6465 scFv (C1) nanomoles concentration in solution to Aβ peptide immobilized to a SPR biosensor surface. A single-fractal analysis is, once again, adequate to describe the binding and the dissociation kinetics. In this case, the affinity, K ($=k/k_d$), is equal to 3.21.

Figure 9.29(e) shows the binding and dissociation kinetics for 5170 scFv (C1) nanomoles concentration in solution to Aβ peptide immobilized to a SPR biosensor surface. A single-fractal analysis is, once again, adequate to describe the binding and the dissociation kinetics. In this case, the affinity, K ($=k/k_d$), is equal to 5.174.

Figure 9.29(f) shows the binding and dissociation kinetics for 3977 scFv (C1) nanomoles concentration in solution to Aβ peptide immobilized to a SPR biosensor surface. A single-fractal analysis is, once again, adequate to describe the binding and the dissociation kinetics. In this case, the affinity, K ($=k/k_d$), is equal to 0.968.

Figure 9.30(a) shows the increase in the binding rate coefficient, k, with an increase in the scFv concentration in nanomoles in solution. For the data given in Figure 9.30(a), the binding rate coefficient, k, is given by:

$$k = (0.1773 \pm 0.0385)\,[\text{scFv}]^{0.7197 \pm 0.1192} \tag{9.14a}$$

The fit is good. Only six data points are available. The availability of more data points would lead to a more reliable fit. The binding rate coefficient, k,

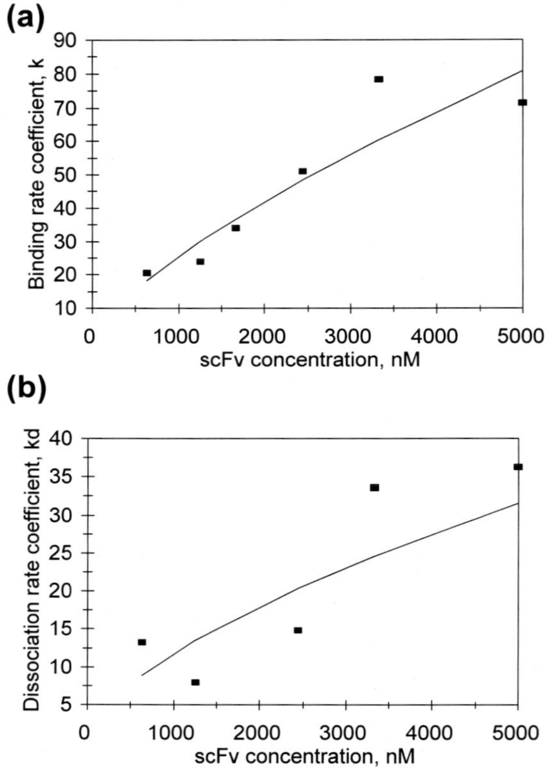

FIGURE 9.30 (a) Increase in the binding rate coefficient, k, with an increase in the scFv concentration in solution. (b) Increase in the dissociation rate coefficient, k_d, with an increase in the scFv concentration in solution. (c) Increase in the affinity, K (=k/k_d), with an increase in the scFv concentration in solution. (d) Increase in the dissociation rate coefficient, k_d, with an increase in the fractal dimension for dissociation, D_{fd}. (e) Increase in the affinity, K (=k/k_d), with an increase in the fractal dimension ratio, D_f/D_{fd}.

exhibits an order of dependence between one-half and first (equal to 0.7197) order on the scFv concentration in solution.

Figure 9.30(b) shows the increase in the dissociation rate coefficient, k_d, with an increase in the scFv concentration in nanomoles in solution. For the data given in Figure 9.30(b), the dissociation rate coefficient, k_d, is given by:

$$k_d = \left(0.1678 \pm 0.0991\right) [\text{scFv}]^{0.6158 \pm 0.283} \tag{9.14b}$$

The fit is good. Only five data points are available. The availability of more data points would lead to a more reliable fit. The dissociation rate coefficient, k_d, exhibits an order of dependence between one-half and first (equal to 0.6158) order on the scFv concentration in solution.

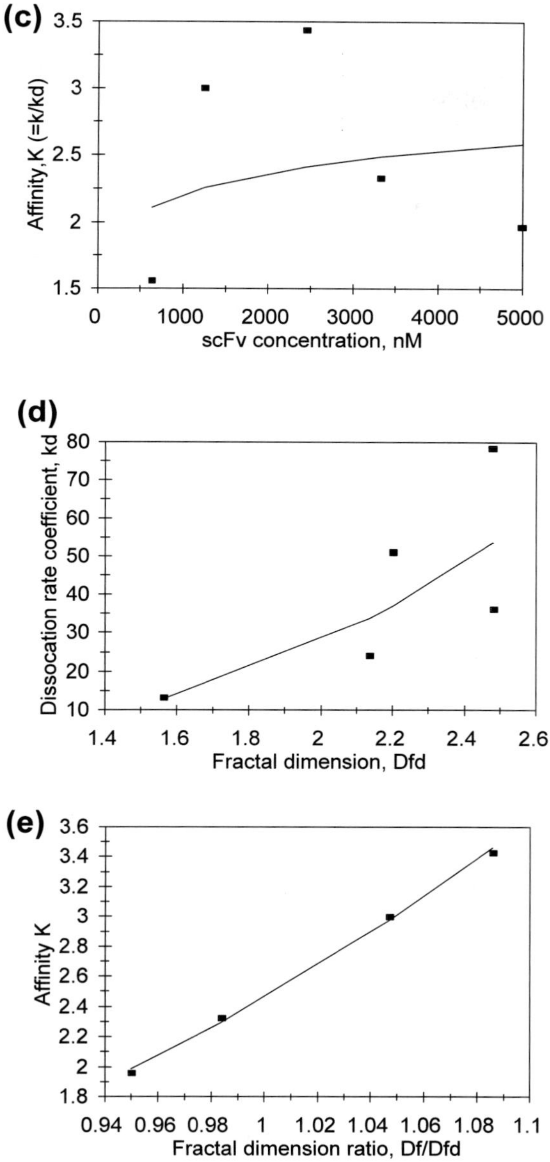

FIGURE 9.30 cont'd

Figure 9.30(c) shows the increase in the affinity, K (=k/k_d), with an increase in the scFv concentration in nanomoles in solution. For the data given in Figure 9.30(c), the affinity, K (=k/k_d), is given by:

$$K(=k/k_d) = \left(1.117 \pm 0.474\right) [\mathrm{scFv}]^{0.0986 \pm 0.215} \qquad (9.14c)$$

There is considerable scatter in the data. Only five data points are available. The availability of more data points would lead to a more reliable fit. The affinity, K, exhibits close to a zeroth (equal to 0.0986) order of dependence on the scFv concentration in solution.

Figure 9.30(d) shows the increase in the dissociation rate coefficient, k_d, with an increase in the fractal dimension for dissociation, D_{fd}. For the data given in Figure 9.30(d), the dissociation rate coefficient, k_d, is given by:

$$k_d = (3.259 \pm 0.678)\, D_{fd}^{3.082 \pm 1.099} \qquad (9.14d)$$

The fit is good. Only five data points are available. The availability of more data points would lead to a more reliable fit. The dissociation rate coefficient, k_d, exhibits close to a third (equal to 3.082) order of dependence on the fractal dimension in the dissociation phase, D_{fd}. This indicates that the dissociation rate coefficient, k_d, is sensitive to the degree of heterogeneity presents on the biosensor surface.

Figure 9.30(e) shows the increase in the affinity, K (=k/k_d), with an increase in the fractal dimension ratio, D_f/D_{fd}. For the data given in Figure 9.30(e), the affinity, K (=k/k_d), is given by:

$$K\,(=k/k_d) = \left(2.457 \pm 0.040\right) [D_f/D_{fd}]^{4.163 \pm 0.154} \qquad (9.14e)$$

The fit is very good. Only four data points are available. The availability of more data points would lead to a more reliable fit. The affinity, K, exhibits a dependence between four and four and one-half (equal to 4.163) order of dependence on the ratio of fractal dimensions present on the biosensor surface.

Figure 9.31(a) shows the binding of 67 nM monoclonal anti-Aβ antibody to immobilized Aβ peptide on an SPR biosensor surface. This recognizes the middle part of the hydrophobic amino acids (Ramakrishnan et al., 2009). A dual-fractal analysis is required to adequately describe the binding kinetics. The values of (1) the binding rate coefficient, k, and the fractal dimension, D_f, for a single-fractal analysis, and (2) the binding rate coefficients, k_1 and k_2, and the fractal dimensions, D_{f1} and D_{f2}, are given in Table 9.17(a) and (b).

Figure 9.31(b) shows the binding of 33 nM monoclonal anti-Aβ antibody to immobilized Aβ peptide on an SPR biosensor surface. This recognizes the middle part of the hydrophobic amino acids. A single-fractal analysis is adequate to describe the binding kinetics. The values of (1) the binding rate coefficient, k, and the fractal dimension, D_f, for a single-fractal analysis are given in Table 9.17(a) and (b).

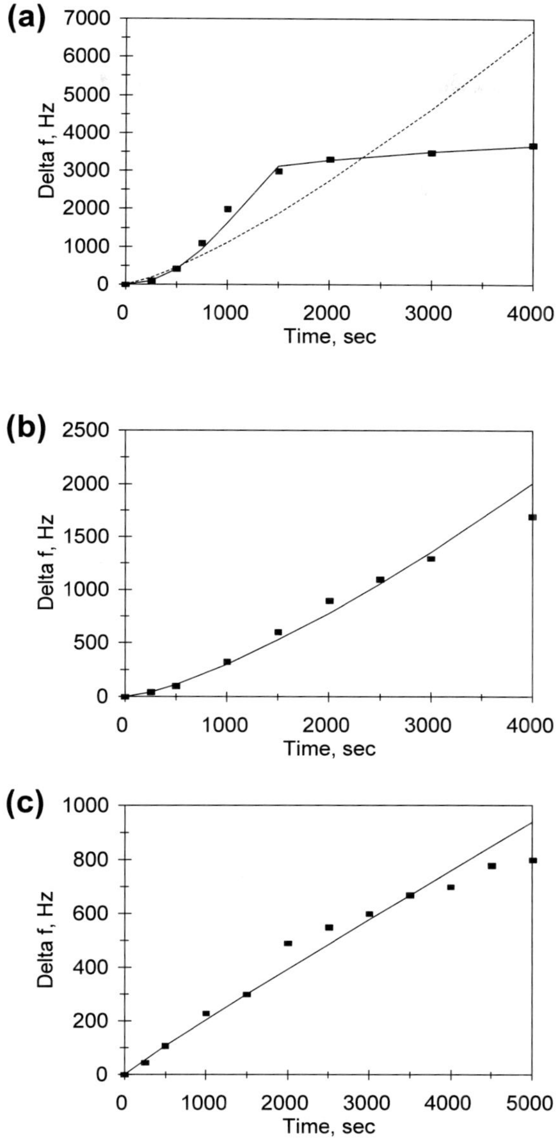

FIGURE 9.31 Binding kinetics for different concentrations of monoclonal anti-Aβ antibody in solution to Aβ peptide immobilized on an SPR biosensor surface (Ramakrishnan et al., 2009): (a) 67 nM (b) 33 nM (c) 6.7 nM (d) 0.67 nM (e) 0.067 nM (f) 0.0067 nM.

Figure 9.31(c) shows the binding of 6.7 nM monoclonal anti-Aβ antibody to immobilized Aβ peptide on an SPR biosensor surface. This recognizes the middle part of the hydrophobic amino acids. A single-fractal analysis is, once

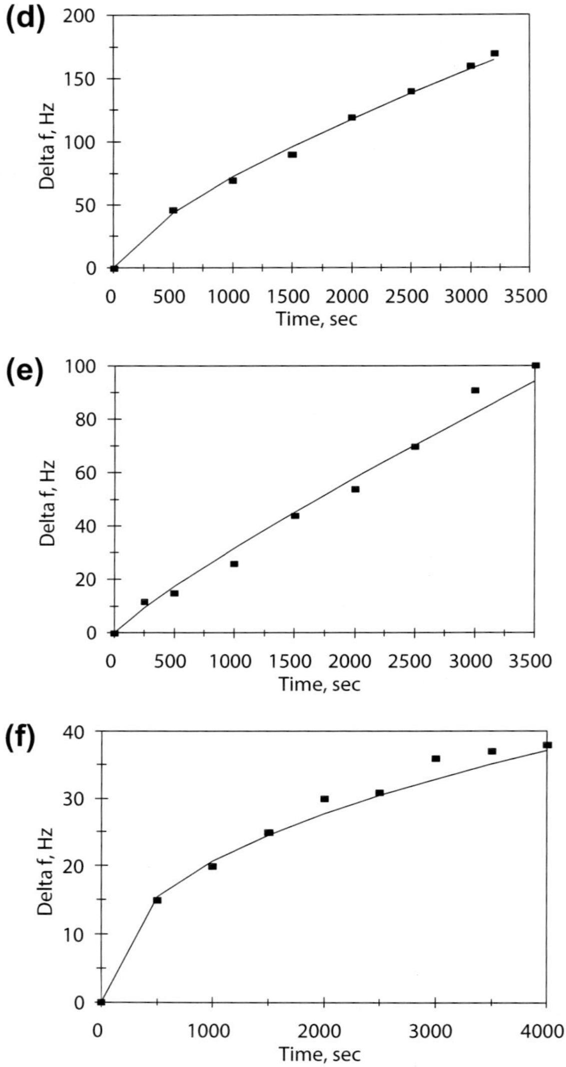

FIGURE 9.31 cont'd

again, adequate to describe the binding kinetics. The values of (1) the binding rate coefficient, k, and the fractal dimension, D_f, for a single-fractal analysis are given in Table 9.17a and b.

Figure 9.31(d) shows the binding of 0.67 nM monoclonal anti-Aβ antibody to immobilized Aβ peptide on an SPR biosensor surface. This recognizes the middle part of the hydrophobic amino acids. A single-fractal analysis is, once again, adequate to describe the binding kinetics. The values of (1) the binding

TABLE 9.17a **Binding Rate Coefficients and Fractal Dimensions for Monoclonal Anti-Aβ Antibody to Recognize the Middle Part of the Hydrophobic Amino Acids by Immobilized Aβ Peptide (Ramakrishnan et al., 2009)**

Analyte in Solution/Receptor on Surface	k	k_1	k_2
67 nM monoclonal anti-Aβ antibody/immobilized peptide	0.1498 ± 0.0987	0.002078 ± 0.000428	948.43 ± 6.25
33 nM monoclonal anti-Aβ antibody/immobilized peptide	0.02373 ± 0.00327	NA	NA
6.7 nM monoclonal anti-Aβ antibody/immobilized peptide	0.2938 ± 0.0373	NA	NA
0.67 nM monoclonal anti-Aβ antibody/immobilized peptide	0.5220 ± 0.0223	NA	NA
0.067 nM monoclonal anti-Aβ antibody	0.07959 ± 0.01109	NA	NA
0.0067 nM monoclonal anti-Aβ antibody	1.129 ± 0.060	NA	NA

TABLE 9.17b Fractal Dimensions for Monoclonal Anti-Aβ Antibody to Recognize the Middle Part of the Hydrophobic Amino Acids by Immobilized Aβ Peptide (Ramakrishnan et al., 2009)

Analyte in Solution/Receptor on Surface	D_f	D_{f1}	D_{f2}
67 nM monoclonal anti-Aβ antibody/immobilized peptide	0.4154 ± 0.4108	0.00 + 0.2356	2.6370 ± 0.2668
33 nM monoclonal anti-Aβ antibody/immobilized peptide	0.2582 ± 0.1041	NA	NA
6.7 nM monoclonal anti-Aβ antibody/immobilized peptide	1.1036 ± 0.08496	NA	NA
0.67 nM monoclonal anti-Aβ antibody/immobilized peptide	1.5328 ± 0.05090	NA	NA
0.067 nM monoclonal anti-Aβ antibody	1.2658 ± 0.0913	NA	NA
0.0067 nM monoclonal anti-Aβ antibody	2.1546 ± 0.1305	NA	NA

rate coefficient, k, and the fractal dimension, D_f, for a single-fractal analysis are given in Table 9.17(a) and (b).

Figure 9.31(e) shows the binding of 0.067 nM monoclonal anti-Aβ antibody to immobilized Aβ peptide on an SPR biosensor surface. This recognizes the middle part of the hydrophobic amino acids. A single-fractal analysis is, once again, adequate to describe the binding kinetics. The values of (1) the binding rate coefficient, k, and the fractal dimension, D_f, for a single-fractal analysis are given in Table 9.17(a) and (b).

Figure 9.31(f) shows the binding of 0.0067 nM monoclonal anti-Aβ antibody to immobilized Aβ peptide on an SPR biosensor surface. This recognizes the middle part of the hydrophobic amino acids. A single-fractal analysis is, once again, adequate to describe the binding kinetics. The values of (1) the binding rate coefficient, k, and the fractal dimension, D_f, for a single-fractal analysis are given in Table 9.17(a) and (b).

Note that in this case, except for the highest concentration in solution (67 nM monoclonal anti-Aβ antibody, the binding kinetics for all the monoclonal anti-Aβ antibody concentration in solution is described by a single-fractal analysis. It seems like at the highest monoclonal anti-Aβ concentration the active sites on the SPR surface are saturated. This is exhibited by the "saturation curve" characteristics exhibited by the curve. In other words, the immobilized Aβ peptide is completely saturated by the monoclonal anti-Aβ antibody in solution.

Figure 9.32(a) shows the decrease in the binding rate coefficient, k, with an increase in the anti-Aβ concentration in solution. For the data shown in Figure 9.32(a) the binding rate coefficient, k, is given by:

$$k = \left(0.170 + 0.519\right)\left[\text{anti-A}\beta\right]^{-0.282 \pm 0.204}$$

There is scatter in the data. Only five data points are available. The availability of more data points would lead to a more reliable fit. The binding rate coefficient, k, exhibits a mild negative order (equal to −0.282) of dependence on the Aβ concentration in solution.

Figure 9.32(b) shows the decrease in the fractal dimension, D_f, with an increase in the anti-Aβ concentration in solution. For the data shown in Figure 9.32(b) the fractal dimension, D_f, is given by:

$$D_f = (0.938 + 0.692)\left[\text{anti-A}\beta\right]^{-0.194 \pm 0.0801}$$

There is scatter in the data. Only five data points are available. The availability of more data points would lead to a more reliable fit. The fractal dimension, D_f, exhibits a mild negative order (equal to −0.194) of dependence on the Aβ concentration in solution.

Ramakrishnan et al. (2009) have analyzed the binding kinetics of AD peptide-capturing and plaque-binding mAbs. They state that AD is an incurable neurodegenerative disease that primarily affects old people. The disease

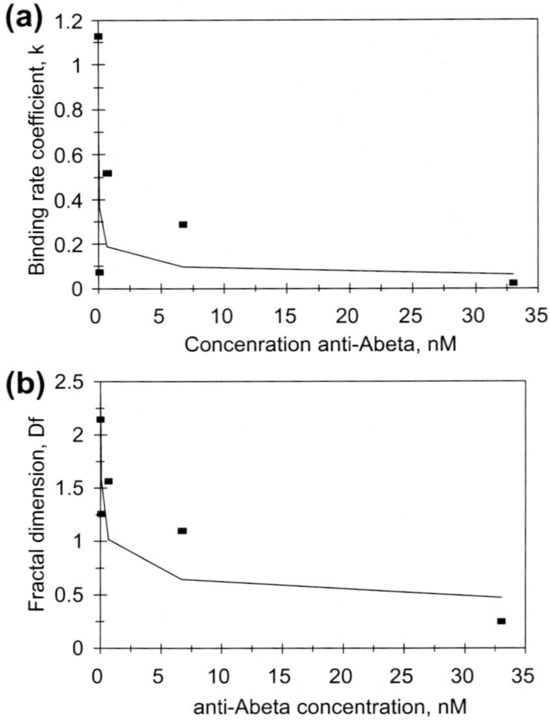

FIGURE 9.32 (a) Decrease in the binding rate coefficient, k, with an increase in the monoclonal anti-Aβ antibody concentration in solution. (b) Decrease in the fractal dimension, D_f, with an increase in the monoclonal anti-Aβ antibody concentration in solution.

pathology is not known, but it is commonly believed that toxic Aβ (amyloid peptides) accumulation primarily in the brain leads to cerebrovascular deposits (Selkoe, 2001). Aβ is randomly oriented in the monomeric form, but in fibrillar form it exhibits different morphological structures (Paravastu et al., 2008; Petkova et al., 2005; Tyco, 2006). Ramakrishnan et al. (2009) indicate that though mechanism of neuronal dysfunction is not clear it is a particular form of the amyloid assembly causes a loss of memory in the AD patients (Shankar et al., 2008; Lesne et al., 2006).

There have been efforts to minimize the effects of these amyloid deposits. DeMattos et al. (2001) demonstrated that a high-affinity Aβ peptide-capturing antibody (m266) targets the central domain of Aβ but lacks binding to plaques, and thereby reduces the Aβ deposits. However, Seubert et al. (2008) reported that the peptide-capturing antibody (m266) is not effective in clearing plaques as it increases the level of cerebral amyloidosis. Ramkrishnan et al. (2009) indicate that mAbs are used to quantify Aβ peptide in tissues, plasma, and CSF (Abdullah et al., 2007; Ertekin-Taner et al., 2008; Ringman et al., 2008) and also to develop novel diagnostic imaging methods (Agayre et al., 2008). Ramakrishnan et al. (2009) have characterized the binding of mAbs to monomeric and fibrillar human Aβ40 using an SPR biosensor. These authors indicate these mAbs have fundamental differences in their interactions with the different structures of monomeric and fibrillar structures.

Ramakrishnan et al. (2009) have analyzed the binding and dissociation kinetics for different antibodies in solution to amyloid peptide immobilized on an SPR biosensor surface (Ramakrishanan et al., 2009). Figure 9.33(a) shows the binding and dissociation kinetics of antibody 11A50 in solution to amyloid peptide immobilized on an SPR biosensor surface. A single-fractal analysis is adequate to describe the binding and dissociation kinetics. The values of (1) the binding rate coefficient, k, and the fractal dimension, D_f, and (2) the dissociation rate coefficient, k_d, and the fractal dimension for dissociation, D_{fd}, for a single-fractal analysis are given in Table 9.18(a) and (b).

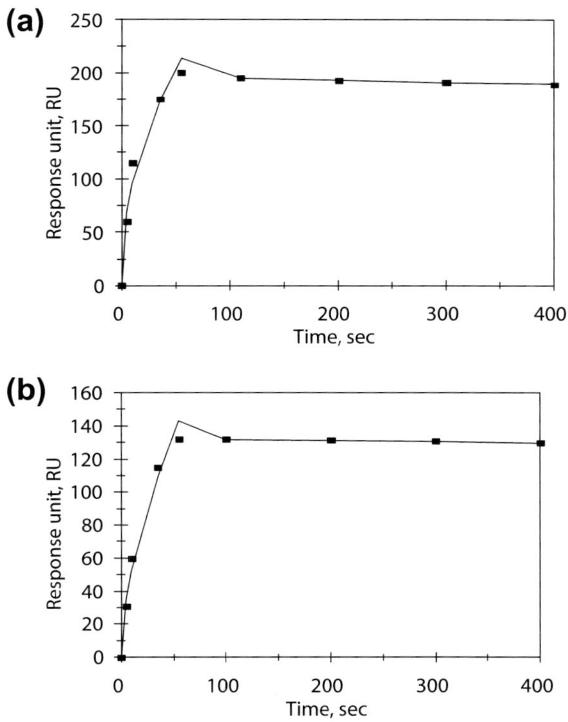

FIGURE 9.33 Binding and dissociation of different antibodies in solution to amyloid β peptide immobilized on an SPR biosensor surface (Ramakrishnan et al., 2009): (a) 11A50; (b) 6E10; (c) IgG4.1; (d) F(ab′)$_2$4.1.

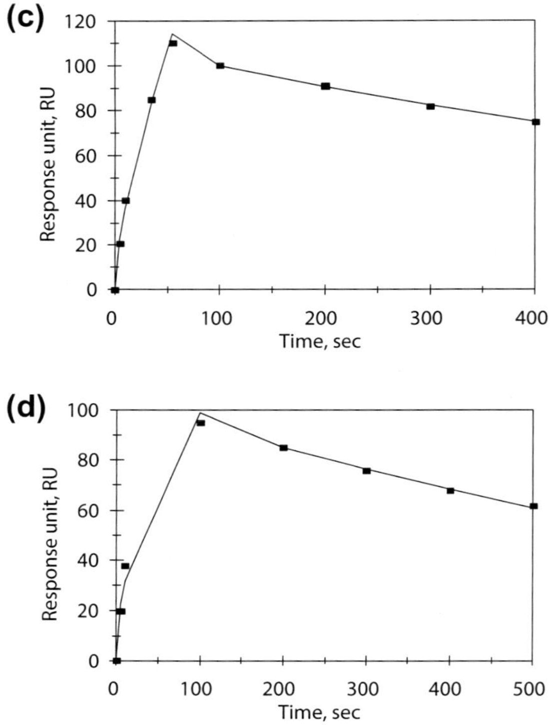

FIGURE 9.33 cont'd

Figure 9.33(b) shows the binding and dissociation kinetics of antibody 6E10 in solution to amyloid peptide immobilized on an SPR biosensor surface. A single-fractal analysis is adequate to describe the binding and dissociation kinetics. The values of (1) the binding rate coefficient, k, and the fractal dimension, D_f, and (2) the dissociation rate coefficient, k_d, and the fractal dimension for dissociation, D_{fd}, for a single-fractal analysis are given in Table 9.18(a) and (b).

Figure 9.33(c) shows the binding and dissociation kinetics of antibody lgG4.1 in solution to amyloid peptide immobilized on an SPR biosensor surface. A single-fractal analysis is adequate to describe the binding and dissociation kinetics. The values of (1) the binding rate coefficient, k, and the fractal dimension, D_f, and (2) the dissociation rate coefficient, k_d, and the fractal dimension for dissociation, D_{fd}, for a single-fractal an analysis are given in Table 9.18(a) and (b).

Figure 9.33(d) shows the binding and dissociation kinetics of antibody $F(ab')_2 4.1$ in solution to amyloid peptide immobilized on an SPR biosensor surface. A single-fractal analysis is adequate to describe the binding and dissociation kinetics. The values of (1) the binding rate coefficient, k, and the

TABLE 9.18a Binding and Dissociation Rate Coefficients and Affinity Value for Different Antibodies to Immobilized and Sonicated Aβ40 Fibrils. Freshly Prepared Monomeric Aβ40 was Immobilized at a Ligand Density of Approximately 60 RU (Ramakrishnan et al., 2009)

Analyte in Solution/Receptor on Surface	k	k_d	K (=k/k_d)
11A50/Aβ fibrils	32.101 ± 5.957	0.8982 ± 0.1514	35.72
6E10/Aβ fibrils	13.330 ± 2.0242	0.000362 + 1.177	13.99
IgG4.1/Aβ fibrils	7.6306 ± 0.7467	0.6372 ± 0.3779	11.97
F(ab')$_2$4.1/Aβ fibrils	10.376 ± 2.545	0.1847 + 0.8476	56.17

TABLE 9.18b Fractal Dimensions for the Binding and the Dissociation Phase, and Ratio of the Fractal Dimensions for These Phases for Different Antibodies to Immobilized and Sonicated Aβ40 Fibrils (Ramakrishnan et al., 2009)

Analyte in Solution/ Receptor on Surface	D_f	D_{fd}	D_f/D_{fd}
11A50/Aβ fibrils	0 + 0.005624	2.1758 ± 0.06988	~0
6E10/Aβ fibrils	1.8156 ± 0.1487	2.980 ± 0.234	0.609
IgG4.1/Aβ fibrils	1.6496 ± 0.0974	1.7078 ± 0.0216	0.9619
F(ab')$_2$4.1/Aβ fibrils	2.0212 ± 0.1878	1.2596 ± 0.0612	1.598

Freshly prepared monomeric Aβ40 was immobilized at a ligand density of approximately 60 RU.

fractal dimension, D_f, and (2) the dissociation rate coefficient, k_d, and the fractal dimension for dissociation, D_{fd}, for a single-fractal an analysis are given in Table 9.18(a) and (b).

Figure 9.34(a) and Table 9.18(a) and (b) show the increase in the binding rate coefficient, k, with an increase in the fractal dimension, D_f. For the data shown in Figure 9.34(a), the binding rate coefficient, k, is given by:

$$k = (0.875 \pm 0.611) D_f^{4.357 \pm 3.0121} \tag{9.14f}$$

The fit is quite good. Only four data points are available. The availability of more data points would lead to a more reliable fit. The binding rate coefficient, k, is very sensitive to the degree of heterogeneity or the fractal dimension on

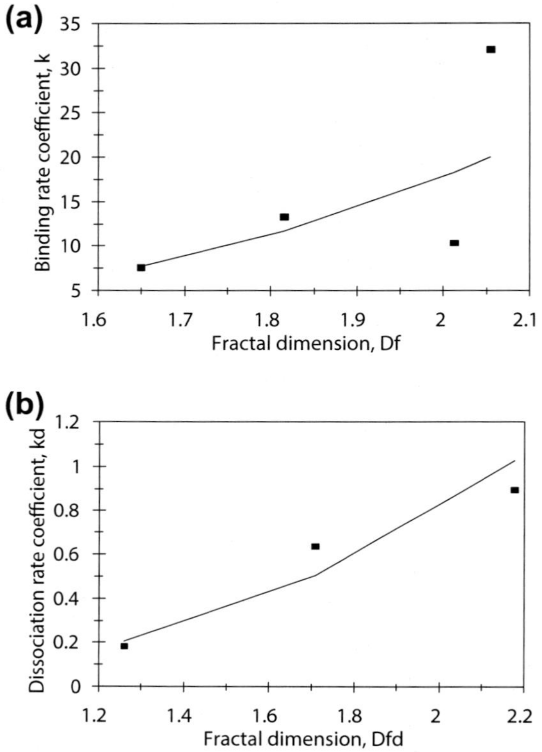

FIGURE 9.34 (a) Increase in the binding rate coefficient, k, with an increase in the fractal dimension, D_f. (b) Increase in the dissociation rate coefficient, k_d, with an increase in the fractal dimension for dissociation, D_{fd}.

the biosensor surface. The binding rate coefficient, k, exhibits an order of dependence on the fractal dimension between four and four and a half (equal to 4.357) on the fractal dimension on the biosensor surface.

Figure 9.34(b) and Table 9.18(a) and (b) show the increase in the dissociation rate coefficient, k_d, with an increase in the fractal dimension, D_{fd}. For the data shown in Figure 9.34(b), the dissociation rate coefficient, k_d, is given by:

$$k_d = (0.104 \pm 0.0352) D_{fd}^{2.943 \pm 0.752} \quad (9.14g)$$

The fit is good. Only three data points are available. The availability of more data points would lead to a more reliable fit. The dissociation rate coefficient, k_d, is very sensitive to the degree of heterogeneity or the fractal dimension on the biosensor surface. The dissociation rate coefficient, k_d, exhibits close to a third order of dependence on the fractal dimension (in the dissociation phase (equal to 2.943) in the dissociation phase on the biosensor surface.

Figure 9.35 shows the binding and dissociation of antibody IgG4.1 in solution to immobilized fibrillar Aβ on an SPR biosensor surface (Ramakrishanan et al., 2009). Figure 9.35(a) shows the binding of 200 nM IgG4.1 in solution to immobilized fibrillar Aβ on an SPR biosensor surface. The binding rate coefficient, k, and the fractal dimension, D_f, for a single-fractal analysis, and the dissociation rate coefficient, k_d, and the fractal dimension for dissociation, D_{fd}, are given in Table 9.19(a) and (b).

Figure 9.35(b) shows the binding of 166 nM IgG4.1 in solution to immobilized fibrillar Aβ on an SPR biosensor surface. The binding rate coefficient, k, and the fractal dimension, D_f, for a single-fractal analysis, and the dissociation rate coefficient, k_d, and the fractal dimension for dissociation, D_{fd}, are given in Table 9.19(a) and (b).

Figure 9.35(c) shows the binding of 100 nM IgG4.1 in solution to immobilized fibrillar Aβ on an SPR biosensor surface. The binding rate coefficient, k, and the fractal dimension, D_f, for a single-fractal analysis, and the

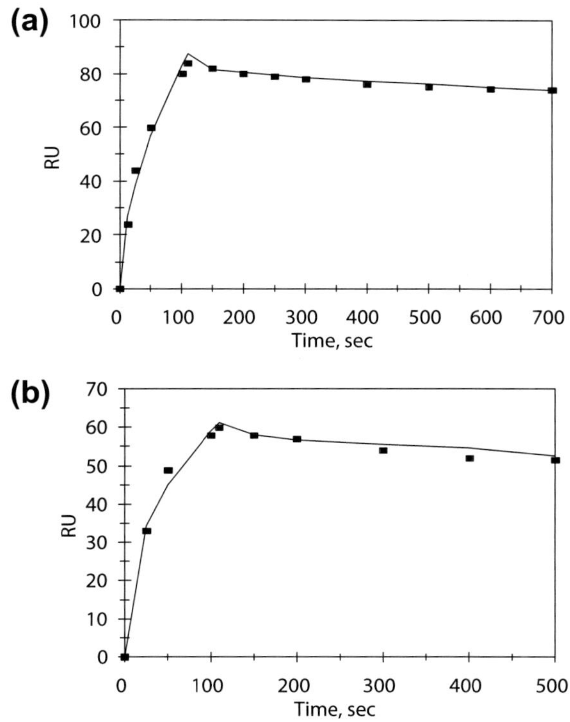

FIGURE 9.35 Binding and dissociation of different concentrations of IgG4.1 in solution to immobilized fibrillar Aβ40 on an SPR biosensor surface (Ramakrishan et al., 2009): (a) 200 nM; (b) 166 nM; (c) 100 nM; (d) 33 nM.

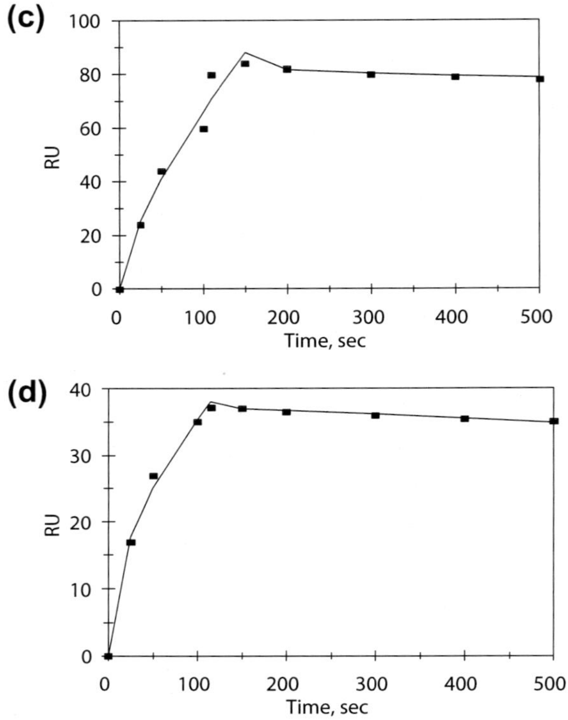

FIGURE 9.35 cont'd

TABLE 9.19a Binding and Dissociation Rate Coefficients of Different Concentrations of IgG4.1 in Solution to Immobilized Fibrillar Aβ on an SPR Biosensor Surface (Ramakrishnan et al., 2009)

Concentration of IgG4.1 in Solution in nM/Receptor on Surface	k	k_d	$K = k/k_d$
200	6.681 ± 0.726	0.3248 ± 0.0362	20.56
166	9.846 ± 0.7018	0.1495 ± 0.01681	65.84
100	2.615 ± 0.297	0.3248 ± 0.0362	8.05
33	3.5335 ± 0.2309	0.001674 ± 0.00052	597.43

SPR, surface plasmon resonance.

TABLE 9.19b Binding and Dissociation Rate Coefficients of Different Concentrations of IgG4.1 in Solution to Immobilized Fibrillar Aβ on an SPR Biosensor Surface (Ramakrishnan et al., 2009)

Concentration of IgG4.1 in Solution in nM/Receptor on surface	D_f	D_{fd}	D_f/D_{fd}
200	1.9060 ± 0.1017	1.9276 ± 0.0861	0.991
166	2.22218 ± 0.1153	1.6280 ± 0.1106	1.365
100	1.5956 ± 0.1490	1.9276 ± 0.08694	0.828
33	1.9990 ± 0.1040	0.5778 ± 0.2456	8.13

SPR, surface plasmon resonance.

dissociation rate coefficient, k_d, and the fractal dimension for dissociation, D_{fd}, are given in Table 9.19(a) and (b).

Figure 9.35(d) shows the binding of 33 nM IgG4.1 in solution to immobilized fibrillar Aβ on an SPR biosensor surface. The binding rate coefficient, k, and the fractal dimension, D_f, for a single-fractal analysis, and the dissociation rate coefficient, k_d, and the fractal dimension for dissociation, D_{fd}, are given in Table 9.19(a) and (b).

Figure 9.36(a) shows the binding and dissociation of 200 nM IgG4.1 in solution to fibrillar Aβ on the SPR biosensor surface (Ramakrishnan et al., 2009). A single-fractal analysis is adequate to describe the binding and dissociation kinetics. The values of the binding rate coefficient, k, and the fractal dimension, D_f, for a single-fractal analysis, and the dissociation rate coefficient, k_d, and the fractal dimension for dissociation, D_{fd}, for a single-fractal analysis are give in Table 9.20(a) and (b).

Figure 9.36(b) shows the binding and dissociation of 200 nM $F(ab')_2 4.1$ in solution to fibrillar Aβ on the SPR biosensor surface (Ramakrishnan et al., 2009). A single-fractal analysis is adequate to describe the binding and dissociation kinetics. The values of the binding rate coefficient, k, and the fractal dimension, D_f, for a single-fractal analysis, and the dissociation rate coefficient, k_d, and the fractal dimension for dissociation, D_{fd}, for a single-fractal analysis are give in Table 9.20(a) and (b).

Figure 9.37(a) and Table 9.19(a) and (b) show the increase in the binding rate coefficient, k, with an increase in the IgG4.1 concentration in solution (in nM). For the data shown in Figure 9.37(a), the binding rate coefficient, k, is given by:

$$k = (0.551 \pm 0.420)\,[\text{IgG4.1}]^{0.478 \pm 0.404} \qquad (9.14h)$$

The fit is reasonable. There is scatter in the data. Only four data points are available. The availability of more data points would lead to a more reliable fit.

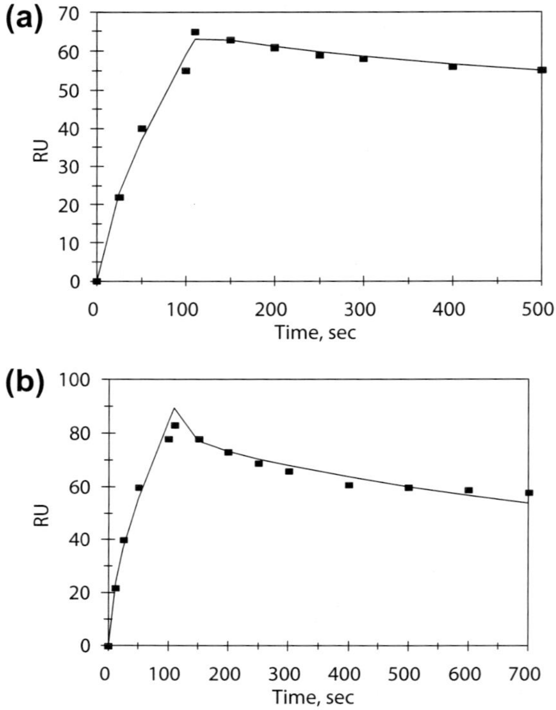

FIGURE 9.36 (a) Binding and dissociation of IgG44.1 and (b) F(ab')$_2$4.1 in solution to immobilized Fibrillar Aβ immobilized on an SPR Biosensor Surface (Ramakrishnan et al., 2009).

This is also reflected in the estimated error of the binding rate coefficient, k. The binding rate coefficient, k, exhibits less than one-half (equal to 0.478) order of dependence on the IgG4.1 concentration in solution (in nM). This indicates that the binding rate coefficient, k, is only mildly sensitive to the IgG4.1 in solution.

TABLE 9.20a Binding and Dissociation of IgG4.1 and F(ab′)$_2$4.1 in Solution to Immobilized Fibrillar Aβ Immobilized on an SPR Biosensor Surface (Ramakrishnan et al., 2009)

Analyte in Solution/Receptor on Surface	k	k_d	K (=k/k_d)
200 nM IgG4.1/fibrillar Aβ on SPR biosensor surface	5.186 ± 0.012	0.6674 ± 0.0977	7.77
200 nM F(ab′)$_2$4.1/fibrillar Aβ on SPR biosensor surface	6.724 ± 0.923	0.2557 ± 0.0248	2.630

SPR, surface plasmon resonance.

TABLE 9.20b Fractal Dimensions for the Binding and Dissociation of IgG4.1 and F(ab′)$_2$4.1 in Solution to Immobilized Fibrillar Aβ Immobilized on an SPR Biosensor Surface (Ramakrishnan et al., 2009)

Analyte in Solution/Receptor on Surface	D_f	D_{fd}	D_f/D_{fd}
200 nM IgG4.1/fibrillar Aβ on SPR biosensor surface	1.9060 ± 0.10134	1.9276 ± 0.0869	0.984
200 nM F(ab′)$_2$4.1/fibrillar Aβ on SPR biosensor surface	1.8994 ± 0.1388	1.8158 ± 0.07624	1.046

SPR, surface plasmon resonance.

Figure 9.37(b) and Table 9.19(a) and (b) show the increase in the dissociation rate coefficient, k_d with an increase in the IgG4.1 concentration in solution (in nM). For the data shown in Figure 9.37(b), the dissociation rate coefficient, k_d, is given by:

$$k_d = \left(7.3E - 08 \pm 1.4E - 08\right) [IgG4.1]^{2.867 \pm 0.08122} \tag{9.15}$$

The fit is very good. Only three data points are available. The availability of more data points would lead to a more reliable fit. The dissociation rate coefficient, k_d, exhibits an order of dependence on the IgG4.1 concentration in solution (in nM) between two and one-half and three (equal to 2.867). This indicates that the dissociation rate coefficient, k_d, is sensitive to the IgG4.1 in solution.

Figure 9.37(c) and Table 9.19(a) and (b) show the increase in the fractal dimension, D_{fd}, with an increase in the IgG4.1 concentration in solution (in nM). For the data shown in Figure 9.37(c), the fractal dimension for dissociation, D_{fd}, is given by:

$$D_{fd} = \left(5.5E - 0.5 \pm 0.8E - 0.5\right) [IgG4.1]^{1.99 \pm 0.0103} \tag{9.16}$$

The fit is good. Only three data points are available. The availability of more data points would lead to a more reliable fit. The fractal dimension for dissociation, D_{fd}, exhibits very close to a second order of dependence (equal to 1.99) on the IgG4.1 concentration in solution (in nM). This indicates that the D_{fd} is sensitive to the IgG4.1 in solution.

Figure 9.37(d) and Table 9.19(a) and (b) show the increase in the binding rate coefficient, k, with an increase in the fractal dimension, D_f. For the data shown in Figure 9.37(d), the binding rate coefficient, k, is given by:

$$k = \left(0.018 \pm 0.0005\right) D_f^{4.359 \pm 0.0305} \tag{9.17}$$

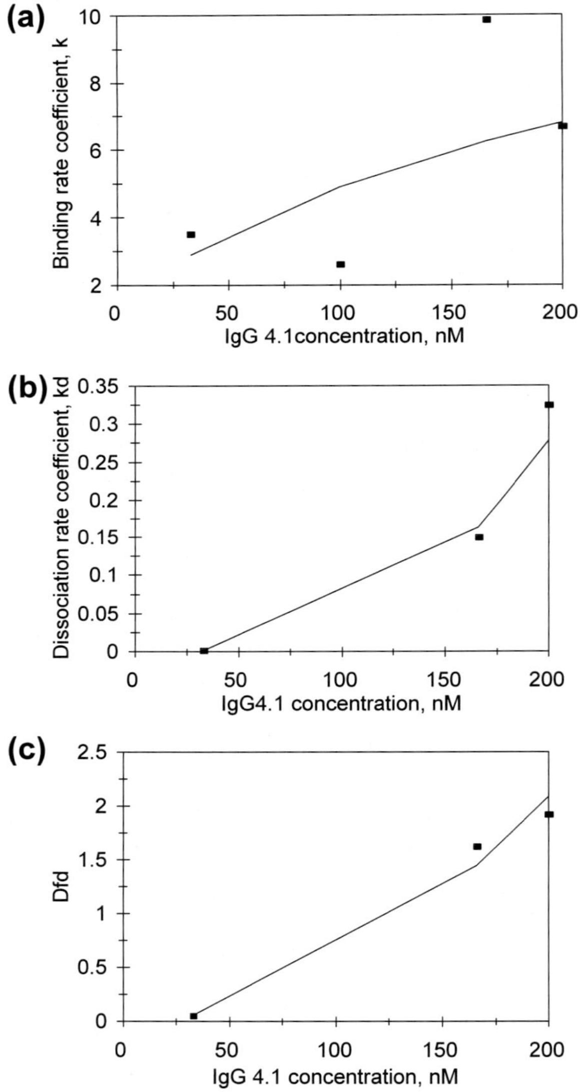

FIGURE 9.37 (a) Increase in the binding rate coefficient, k, with an increase in the IgG4.1 concentration in solution. (b) Increase in the dissociation rate coefficient, k_d, with an increase in the IgG4.1 concentration in solution. (c) Increase in the fractal dimension, D_{fd}, with an increase in the IgG4.1 concentration in solution. (d) Increase in the binding rate coefficient, k, with an increase in the fractal dimension, D_f. (e) Increase in the dissociation rate coefficient, k_d, with an increase in the fractal dimension for dissociation, D_{fd}. (f) Increase in the affinity, K, with an increase in the fractal dimension ratio, D_f/D_{fd}. (g) Decrease in the ratio, D_f/D_{fd} with an increase in the IgG4.1 concentration in solution.

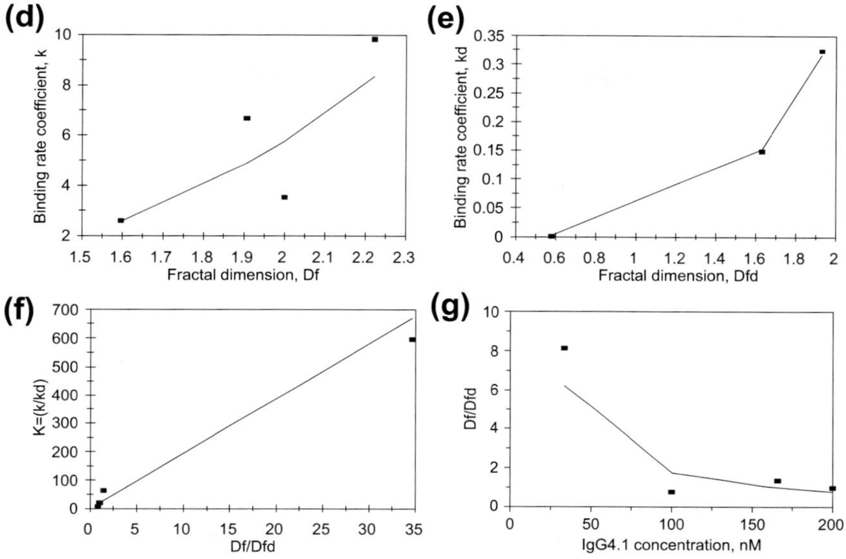

FIGURE 9.37 cont'd

The fit is very good. Only three data points are available. The availability of more data points would lead to a more reliable fit. The binding rate coefficient, k, is very sensitive to the fractal dimension, D_f, or the degree of heterogeneity that exists on the biosensor surface, since it exhibits an order of dependence between four and four and one-half (equal to 4.359) order of dependence on the fractal dimension or the degree of heterogeneity that exists on the surface.

Figure 9.37(e) and Table 9.19(a) and (b) show the increase in the dissociation rate coefficient, k_d, with an increase in the fractal dimension for dissociation, D_{fd}. For the data shown in Figure 9.37(d), the dissociation rate coefficient, k_d, is given by:

$$k_d = \left(0.495 \pm 0.264\right) D_{fd}^{3.541 \pm 1.785} \tag{9.18}$$

The fit is good. Only four data points are available. The availability of more data points would lead to a more reliable fit. The dissociation rate coefficient, k_d, is very sensitive to the fractal dimension, D_f, or the degree of heterogeneity that exists on the biosensor surface, since it exhibits an order of dependence close to three and one-half (equal to 3.541) order of dependence on the fractal dimension or the degree of heterogeneity that exists on the surface.

Figure 9.37(f) and Table 9.19(a) and (b) show the increase in the affinity, K ($=k/k_d$), with an increase in the fractal dimension ratio (D_f/D_{fd}). For the data shown in Figure 9.37(f), the affinity, K, is given by:

$$K = \left(20.56 \pm 15.18\right)\left(D_f/D_{fd}\right)^{0.983 \pm 0.265} \tag{9.19a}$$

The fit is good. Only four three data points are available. The availability of more data points would lead to a more reliable fit. The affinity, K, exhibits close to a first (equal to 0.983) order of dependence on the ratio of the fractal dimensions, D_f/D_{fd}. This indicates that the affinity, K, is sensitive to the ratio of fractal dimensions that exists on the biosensor surface.

Figure 9.37(g) and Table 9.19(a) and (b) show the increase in the ratio, D_f/D_{fd} with an increase in the IgG4.1 concentration in nM in solution. For the data shown in Figure 9.37(g), the ratio is given by:

$$D_f/D_{fd} = (327.35 \pm 286.52)\,[\mathrm{IgG4.1}]^{-1.31 \pm 0.4456} \tag{9.19b}$$

The fit is good. Only four three data points are available. The availability of more data points would lead to a more reliable fit. The affinity, K exhibits close to a negative one and one-half (−1.31) order of dependence on the ratio of the fractal dimensions, D_f/D_{fd}. This indicates that the affinity, K is sensitive to the ratio of fractal dimensions that exists on the biosensor surface.

Figure 9.38(a) shows the binding and dissociation of 200 nM IgG4.1 in solution to fibrillar Aβ immobilized on an SPR biosensor surface (Ramakrishnan et al., 2009). A single-fractal analysis is adequate to describe the binding and dissociation kinetics. The value of (1) the binding rate coefficient, k, and the fractal dimension, D_f, for a single-fractal analysis, and (2) the dissociation rate coefficient, k_d, and the fractal dimension for dissociation, D_{fd}, for a single-fractal analysis are given in Table 9.20(a) and (b). In this case, the affinity, K (=k/k_d), is equal to 0.988.

Figure 9.38(b) shows the binding and dissociation of 200 nM $F(ab')_2$4.1 in solution to fibrillar Aβ immobilized on an SPR biosensor surface (Ramakrishnan et al., 2009). A single-fractal analysis is adequate to describe the binding and dissociation kinetics. The value of (1) the binding rate coefficient, k, and the fractal dimension, D_f, for a single-fractal analysis, and (2) the dissociation rate coefficient, k_d, and the fractal dimension for dissociation, D_{fd}, for a single-fractal analysis are given in Table 9.20(a) and (b). In this case, the affinity, K (=k/k_d), is equal to 1.046.

Ramakrishnan et al. (2009) analyzed the binding and dissociation of Aβ monomers to immobilized antibodies. Figure 9.39(a) shows the binding and dissociation of 200 nM IgG 6E10 monoclonal antibody to a monomeric Aβ40 immobilized on an SPR biosensor surface. A single-fractal analysis is required to adequately describe the binding and the dissociation kinetics. The values of (1) the binding rate coefficient, k, and the fractal dimension, D_f, for a single-fractal analysis, and (2) the dissociation rate coefficient, k_d, and the fractal dimension, D_{fd}, for as ingle-fractal analysis are given in Table 9.21. The affinity, K (=k/k_d), is equal to 9.01.

Ramakrishnan et al. (2009) analyzed the binding and dissociation of Aβ monomers to immobilized antibodies. Figure 9.39(b) shows the binding and dissociation of 66 nM IgG 6E10 monoclonal antibody to a monomeric Aβ40 immobilized on an SPR biosensor surface. A single-fractal analysis is, once

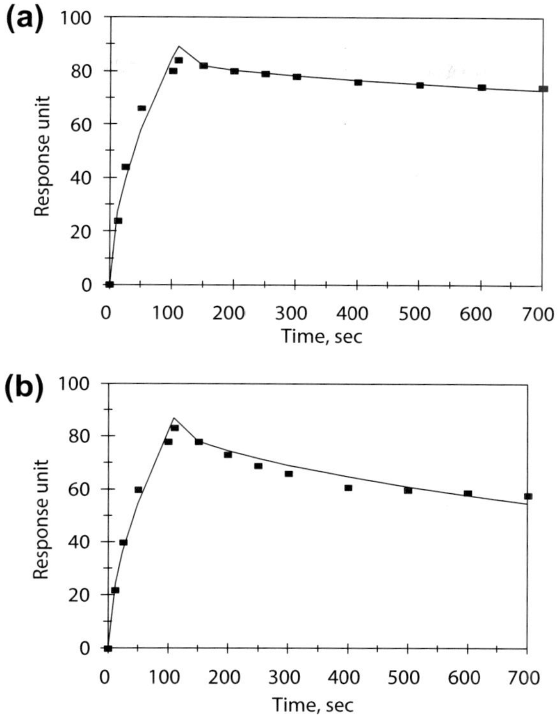

FIGURE 9.38 (a) Binding and dissociation of 200 nM IgG4.1 in solution to immobilized Aβ on an SPR biosensor surface. (b) Binding and dissociation of 200 nM $F(ab')_2$4.1 in solution to immobilized Aβ on an SPR biosensor surface.

again, required to adequately describe the binding and the dissociation kinetics. The values of (1) the binding rate coefficient, k, and the fractal dimension, D_f, for a single-fractal analysis, and (2) the dissociation rate coefficient, k_d, and the fractal dimension, D_{fd}, for a single-fractal analysis are given in Table 9.21. The affinity, K ($=k/k_d$), is equal to 3.27.

Ramakrishnan et al. (2009) analyzed the binding and dissociation of Aβ monomers to immobilized antibodies. Figure 9.39(c) shows the binding and dissociation of 100 nM IGg 6E10 monoclonal antibody to a monomeric Aβ40 immobilized on an SPR biosensor surface. A single-fractal analysis is, once again, required to adequately describe the binding and the dissociation kinetics. The values of (1) the binding rate coefficient, k, and the fractal dimension, D_f, for a single-fractal analysis, and (2) the dissociation rate coefficient, k_d, and the fractal dimension, D_{fd}, for as single-fractal analysis are given in Table 9.21. The affinity, K ($=k/k_d$), is equal to 3.27.

Ramakrishnan et al. (2009) analyzed the binding and dissociation of Aβ monomers to immobilized antibodies. Figure 9.39(d) shows the binding and

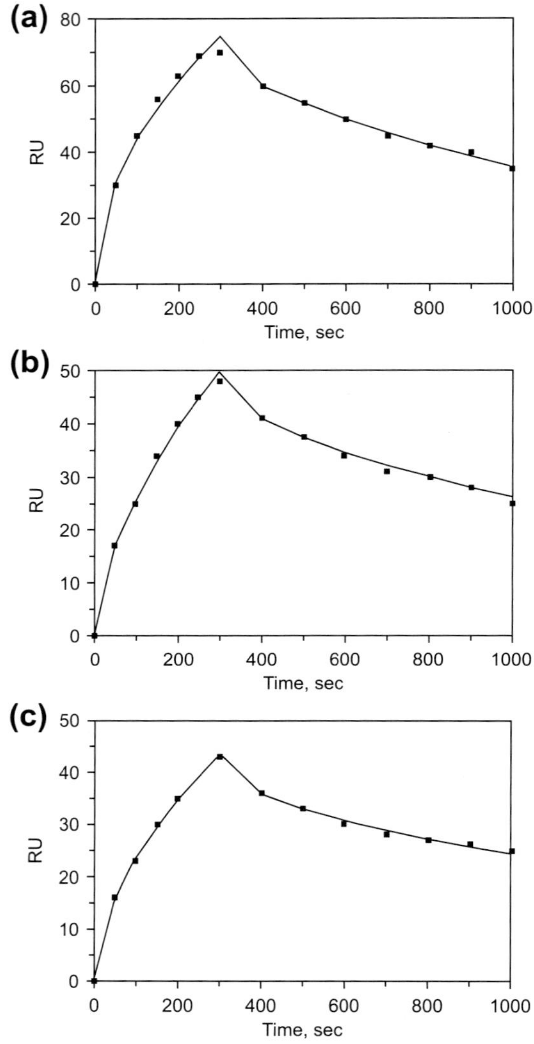

FIGURE 9.39 Binding and dissociation kinetics for different concentrations of Antibody 6E10 in nanomoles to receptor immobilized on SPR surface (Ramakrishnan et al., 2009): (a) 200; (b) 166; (c) 100; (d) 66; (e) 33.

dissociation of 66 nM IgG 6E10 monoclonal antibody to a monomeric Aβ40 immobilized on an SPR biosensor surface. A single-fractal analysis is, once again, required to adequately describe the binding and the dissociation kinetics. The values of (1) the binding rate coefficient, k, and the fractal dimension, D_f, for a single-fractal analysis, and (2) the dissociation rate

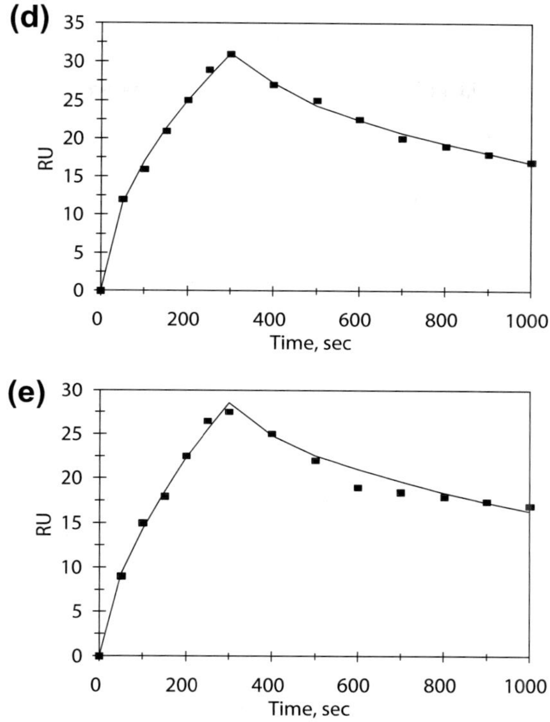

FIGURE 9.39 Cont'd

TABLE 9.21 Binding and Dissociation Rate Coefficients and Fractal Dimensions for the Binding and the Dissociation Phase for Antibody 6E10 in Solution in Nanomoles to Receptor Immobilized on an SPR Surface (Ramakrishnan et al., 2009)

Analyte in Solution	k	k_d	D_f	D_{fd}
200	4.819 ± 0.820	0.0763 ± 0.0031	2.1790 ± 0.1859	1.5912 ± 0.1858
166	6.839 ± 1.740	0.1087 ± 0.0034	2.312 ± 0.207	1.7374 ± 0.0408
100	7.079 ± 1.638	0.1 ± 0	2.3482 ± 0.1898	$1.0 \pm$ 1.38E-15
66	1.699 ± 0.331	0.005 ± 0	1.7036 ± 0.2366	$1.0 \pm$ E-14
33	2.326 ± 0.376	0.00398 ± 0.00018	1.9308 ± 0.1932	2.02 ± 0.0293

SPR, surface plasmon resonance.

coefficient, k_d, and the fractal dimension, D_{fd}, for as single-fractal analysis are given in Table 9.21. The affinity, K ($=k/k_d$), is equal to 2.40.

Ramakrishnan et al. (2009) analyzed the binding and dissociation of Aβ monomers to immobilized antibodies. Figure 9.39(e) shows the binding and dissociation of 33 nM IgG 6E10 monoclonal antibody to a monomeric Aβ40 immobilized on an SPR biosensor surface. A single-fractal analysis is, once again, required to adequately describe the binding and the dissociation kinetics. The values of (1) the binding rate coefficient, k, and the fractal dimension, D_f, for a single-fractal analysis, and (2) the dissociation rate coefficient, k_d, and the fractal dimension, D_{fd}, for as single-fractal analysis are given in Table 9.21. The affinity, K ($=k/k_d$), is equal to 2.40.

Figure 9.40(a) and Table 9.21 show the increase in the binding rate coefficient, k, with an increase in the Aβ concentration in solution in nanomoles. For the data shown in Figure 9.40(a), the binding rate coefficient, k, is given by:

$$k = \left(0.03582 \pm 0.00812\right) [A\beta]^{0.875 \pm 0.1429} \tag{9.20}$$

The fit is very good. Only five data points are available. The availability of more data points would lead to a more reliable fit. The binding rate coefficient,

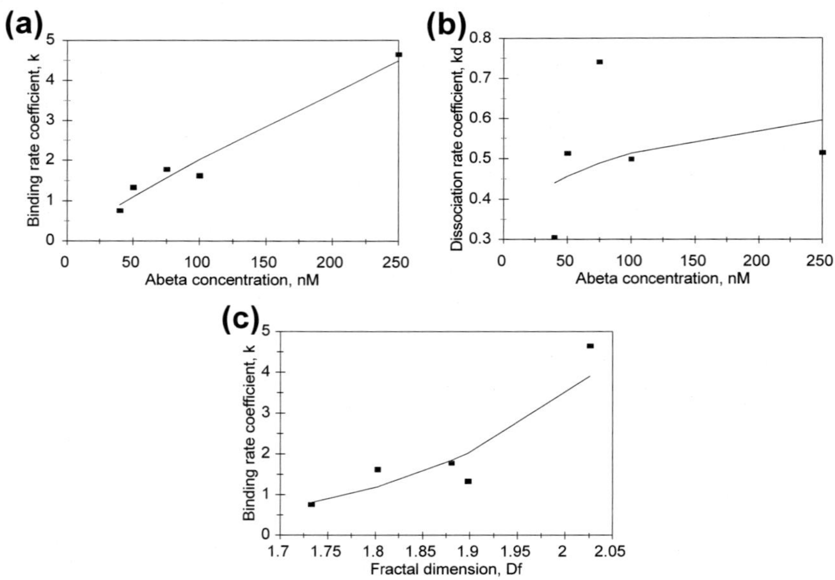

FIGURE 9.40 (a) Increase in the binding rate coefficient, k, with an increase in the Aβ concentration in solution. (b) Increase in the dissociation rate coefficient, k_d, with an increase in the Aβ concentration in solution (c) increase in the binding rate coefficient, k with an increase in the fractal dimension, D_f. (d) Increase in the fractal dimension, D_f with an increase in Aβ concentration in solution (e) Increase in D_{fd} with increase in Aβ concentration in solution (f) Increase in affinity, K with an increase in D_f/D_{fd}.

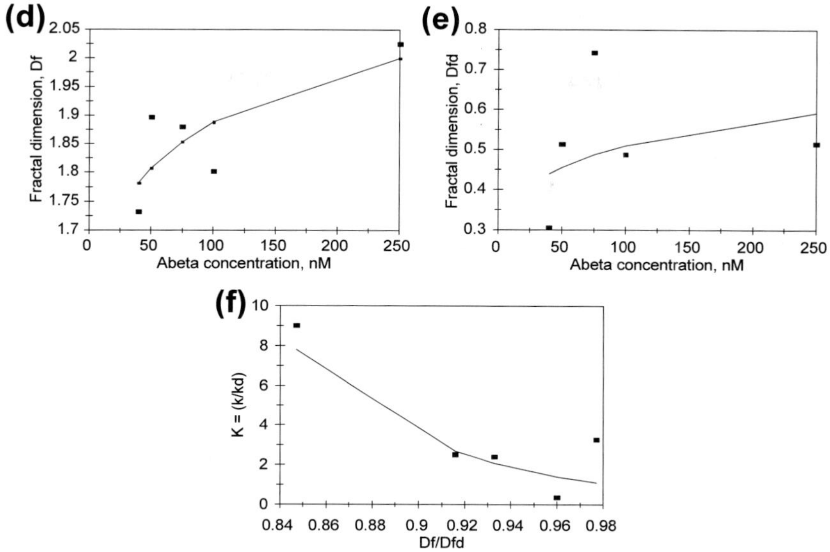

FIGURE 9.40 cont'd

k, exhibits an order of dependence between one-half and first (equal to 0.875) order on the Aβ concentration in solution. This indicates that the binding rate coefficient, k, is sensitive to the Aβ concentration in solution.

Figure 9.40(b) and Table 9.21 show the increase in the dissociation rate coefficient, k_d, with an increase in the Aβ concentration in solution in nanometer. For the data shown in Figure 9.40(b), the dissociation rate coefficient, k_d, is given by:

$$k_d = (0.2397 \pm 0.0980)\,[A\beta]^{0.165+0.235} \qquad (9.21)$$

There is scatter in the data. Only five data points are available. The availability of more data points would lead to a more reliable fit. The dissociation rate coefficient, k_d, exhibits close to zero (equal to 0.165) order of dependence on the Aβ concentration in solution. This indicates that the dissociation rate coefficient, k_d, is not sensitive to the Aβ concentration in solution.

Figure 9.40(c) and Table 9.21 show the increase in the binding rate coefficient, k, with an increase in the fractal dimension, D_f. For the data shown in Figure 9.40(c), the binding rate coefficient, k, is given by:

$$k = (0.003205 \pm 0.001182)\,[D_f]^{10.06 \pm 2.67} \qquad (9.22)$$

The fit is good. Only five data points are available. The availability of more data points would lead to a more reliable fit. The binding rate rate coefficient, k, exhibits greater than a 10th (equal to 10.06) order of dependence on the fractal dimension, D_f, or the degree of heterogeneity that exists on the surface.

Figure 9.40(d) and Table 9.21 show the increase in the fractal dimension, D_f, with an increase in the Aβ concentration in solution in nanomoles. For the data shown in Figure 9.40(d), the fractal dimension, D_f, is given by:

$$D_f = (1.413 \pm 0.053)\ [A\beta]^{0.063 \pm 0.034} \qquad (9.23)$$

There is scatter in the data. Only five data points are available. The availability of more data points would lead to a more reliable fit. The fractal dimension, D_f, close to zeroth (equal to 0.063) order of dependence on the Aβ concentration in solution. This indicates that fractal dimension, D_f, is insensitive to the Aβ concentration in solution.

Figure 9.40(e) and Table 9.21 show the increase in the fractal dimension for dissociation, D_{fd}, with an increase in the Aβ concentration in solution in nanometer. For the data shown in Figure 9.40(e), the fractal dimension for dissociation, D_{fd}, is given by:

$$D_{fd} = (0.241 \pm 0.0970)\ [A\beta]^{0.0162 + 0.236} \qquad (9.24)$$

There is scatter in the data. Only five data points are available. The availability of more data points would lead to a more reliable fit. The fractal dimension for dissociation, D_{fd}, exhibits close to zeroth (equal to 0.0162) order of dependence on the Aβ concentration in solution. This indicates that fractal dimension, D_f, is insensitive to the Aβ concentration in solution. Note that both fractal dimensions, D_f and D_{fd}, are insensitive to the Aβ concentration in solution.

Figure 9.40(f) and Table 9.21 show the increase in the affinity, K (k/k_d), with an increase in the ratio of fractal dimension ratio, D_f/D_{fd}. For the data shown in Figure 9.40(f) the affinity, K, is given by:

$$K(=k/k_d) = (0.807 \pm 1.355)\ [D_f/D_{fd}]^{-13.67 \pm 8.88} \qquad (9.25)$$

The fit is good. Only five data points are available. The availability of more data points would lead to a more reliable fit. The affinity, K, exhibits less than a negative 13th (equal to −13.67) order of dependence on the ratio of fractal dimension ratio, (D_f/D_{fd}).

Figure 9.41(a) shows the binding of 200 nM antibody 11,450-B1 in solution to immobilized on an SPR biosensor surface (Ramakrishnan et al., 2009). A dual-fractal analysis is required to adequately describe the binding kinetics. A single-fractal analysis is adequate to describe the dissociation kinetics.

Figure 9.41(b) shows the binding of 100 nM antibody 11,450-B1 in solution to immobilized on an SPR biosensor surface (Ramakrishnan et al., 2009). A dual-fractal analysis is required to adequately describe the binding kinetics. A single-fractal analysis is adequate to describe the dissociation kinetics.

Figure 9.41(c) shows the binding of 75 nM antibody 11,450-B1 in solution to immobilized on an SPR biosensor surface (Ramakrishnan et al., 2009). A dual-fractal analysis is required to adequately describe the binding kinetics. A single fractal analysis is adequate to describe the dissociation kinetics.

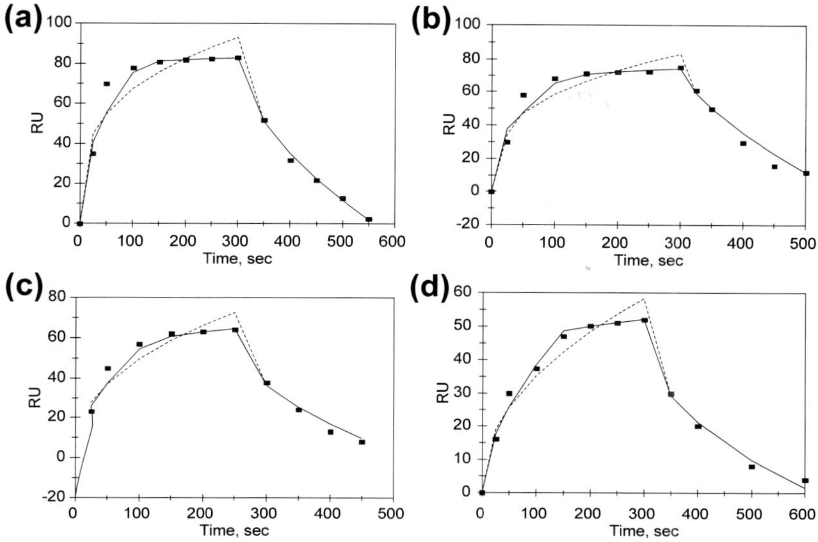

FIGURE 9.41 Binding of different concentrations (in nM) of antibody 11,450-B1 in solution to immobilized on an SPR biosensor surface (Ramakrishnan et al., 2009): (a) 200; (b) 100; (c) 75; (d) 50.

Figure 9.41(d) shows the binding of 50 nM antibody 11,450-B1 in solution to immobilized on an SPR biosensor surface (Ramakrishnan et al., 2009). A dual-fractal analysis is required to adequately describe the binding kinetics. A single-fractal analysis is adequate to describe the dissociation kinetics.

Figure 9.42(a) shows the increase in the binding rate coefficient, k_1, with an increase in the antibody 11,450-B1 concentration in nanomoles in solution. For the data shown in Figure 9.42(a), the binding rate coefficient, k_1, is given by:

$$k_1 = (0.1757 \pm 0.0565)\ [\text{antibody } 11{,}450\text{-B1}]^{0.7947 \pm 0.235} \tag{9.26}$$

The fit is good. Only four data points are available. The availability of more data points would lead to a more reliable fit. The binding rate coefficient, k_1, exhibits an order of dependence between one-half and first (equal to 0.7497) order on the antibody 11,450-B1 concentration in nanometer in solution. This indicates that the binding rate coefficient, k_1, is somewhat sensitive to the antibody 11,450-B1 in solution.

Figure 9.42(b) shows the increase in the binding rate coefficient, k_2, with an increase in the antibody 11,450-B1 concentration in nanomoles in solution. For the data shown in Figure 9.42(b), the binding rate coefficient, k_2, is given by:

$$k_2 = (0.3367 + 0.470)\ [\text{antibody } 11{,}450\text{-B1}]^{0.554 \pm 0.110} \tag{9.27}$$

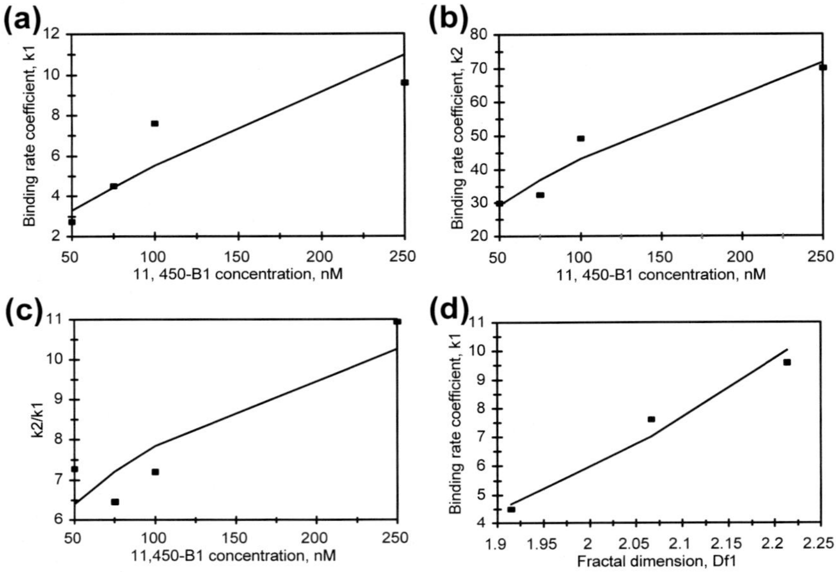

FIGURE 9.42 (a) Increase in the binding rate coefficient, k_1, with an increase in the antibody 11,450-B1 concentration in solution in nanomoles. (b) Increase in the binding rate coefficient, k_2, with an increase in the antibody 11,450-B1 concentration in solution in nanomoles. (c) Increase in the binding rate coefficient ratio, k_2/k_1, with an increase in the antibody 11,450-B1 concentration in solution in nanomoles. (d) Increase in the binding rate coefficient, k_1, with an increase in the fractal dimension, D_{f1}. (e) Increase in the binding rate coefficient, k_2, with an increase in the fractal dimension, D_{f2}. (f) Increase in the dissociation rate coefficient, k_d, with an increase in the fractal dimension, D_{fd}. (g) Increase in the affinity, K_1 ($=k_1/k_d$), with an increase in the antibody 11,450-B1 concentration in solution in nanomoles. (h) Increase in the fractal dimension ratio, (D_{f2}/D_{f1}), with an increase in the antibody 11,450-B1 concentration in solution in nanomoles. (i) Increase in the affinity, K_2 ($=k_2/k_d$), with an increase in the antibody 11,450-B1 concentration in solution in nanomoles. (j) Increase in the fractal dimension, D_{f1}, with an increase in the antibody 11,450-B1 concentration in solution in nanomoles. (k) Increase in the fractal dimension, D_{f2}, with an increase in the antibody 11,450-B1 concentration in solution in nanomoles.

The fit is good. Only four data points are available. The availability of more data points would lead to a more reliable fit. The binding rate coefficient, k_1, exhibits close to one-half order of dependence (equal to 0.554) on the antibody 11,450-B1 concentration in nanomoles in solution. This indicates that the binding rate coefficient, k_1, is mildly sensitive to the antibody 11,450-B1 in solution.

Figure 9.42(c) shows the increase in the ratio of the binding rate coefficients, k_2/k_1, with an increase in the antibody 11,450-B1 concentration in nanomoles in solution. For the data shown in Figure 9.42(c), the ratio of the binding rate coefficients, k_2/k_1, is given by:

$$k_2/k_1 = (0.2036 + 0.308)\,[\text{antibody } 11{,}450\text{-B1}]^{0.293 \pm 0.119} \quad (9.28)$$

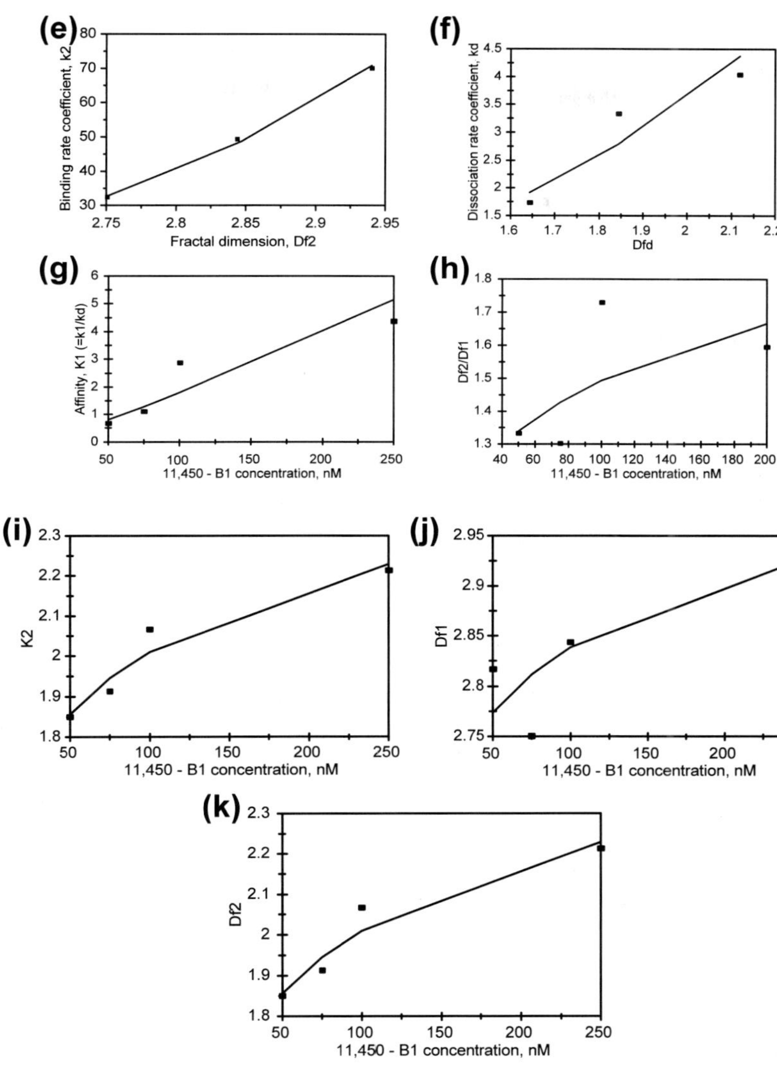

FIGURE 9.42 cont'd

The fit is good. Only four data points are available. The availability of more data points would lead to a more reliable fit. The ratio of the binding rate coefficients, k_2/k_1, exhibits an order of dependence between zero and one-half (equal to 0.293) order on the antibody 11,450-B1 concentration in nanomoles in solution. This indicates that the binding rate coefficient, k_1, is mildly sensitive to the antibody 11,450-B1 in solution.

Figure 9.42(d) shows the increase in the binding rate coefficient, k_1, with an increase in the fractal dimension, D_{f1}. For the data shown in Figure 9.42(d), the binding rate coefficient, k_1, is given by:

$$k_1 = (0.1559 \pm 0.01711)\, D_{f1}^{5.241 \pm 1.0156} \tag{9.29}$$

The fit is very good. Only three data points are available. The availability of more data points would lead to a more reliable fit. The binding rate coefficient, k_1, is very sensitive to the fractal dimension, D_{f1}, or the degree of heterogeneity that exists on the biosensor surface as noted by the order of dependence between five and five and one-half (equal to 5.241) exhibited.

Figure 9.42(e) shows the increase in the ratio of the binding rate coefficient, k_2, with an increase in the fractal dimension, D_{f2}. For the data shown in Figure 9.42(e), the ratio of the binding rate coefficient, k_2, is given by:

$$k_2 = (0.000285 \pm 0.000007)\, D_{f2}^{11.52 \pm 1.0571} \tag{9.30}$$

The fit is very good. Only three data points are available. The availability of more data points would lead to a more reliable fit. The binding rate coefficient, k_2, is extremely sensitive to the fractal dimension, D_{f2}, or the degree of heterogeneity that exists on the biosensor surface as noted by the close to eleven and one-half order of dependence (equal to 11.52) exhibited.

Figure 9.42(f) shows the increase in the ratio of the dissociation rate coefficient, k_d, with an increase in the fractal dimension for dissociation, D_{fd}. For the data shown in Figure 9.42(f), the ratio of the dissociation rate coefficient, k_d, is given by:

$$k_d = (0.3815 \pm 0.0931)\, D_{fd}^{3.25 \pm 1.22} \tag{9.31}$$

The fit is good. Only three data points are available. The availability of more data points would lead to a more reliable fit. The dissociation rate coefficient, k_d, is very sensitive to the fractal dimension for dissociation, D_{fd}, or the degree of heterogeneity that exists on the biosensor surface as noted by the order of dependence between three and three and one-half (equal to 3.25) exhibited.

Figure 9.42(g) shows the increase in the ratio of the affinity, K_1 ($=k_1/k_d$), with an increase in the antibody 11,450-B1 concentration in nanomoles in solution. For the data shown in Figure 9.42(g), the affinity, K, is given by:

$$K_1(=k_1/k_d) = (0.008812 \pm 0.000415)\,[\text{antibody 11,450-B1}]^{1.155 \pm 0.326} \tag{9.32}$$

The fit is good. Only four data points are available. The availability of more data points would lead to a more reliable fit. The affinity, K, exhibits close to a first order of dependence (equal to 1.155) on the antibody 11,450-B1 concentration in nanomoles in solution. This indicates that the affinity, K_1, is sensitive to the antibody 11,450-B1 in solution.

Figure 9.42(h) shows the increase in the ratio of the fractal dimension ratio, D_{f2}/D_{f1}, with an increase in the antibody 11,450-B1 concentration in nanomoles in solution. For the data shown in Figure 9.42(h) the ratio of the binding rate coefficients, D_{f2}/D_{f1}, is given by:

$$D_{f2}/D_{f1} = (0.7253 \pm 0.973)\,[\text{antibody } 11{,}450\text{-B1}]^{\,0.157 \pm 0.124} \tag{9.33}$$

The fit is good. Only four data points are available. The availability of more data points would lead to a more reliable fit. The ratio of the fractal dimensions, D_{f2}/D_{f1}, exhibits an order of dependence between zero and one-half (equal to 0.157) order on the antibody 11,450-B1 concentration in nanomoles in solution. This indicates that the binding rate coefficient, k_1, is only mildly sensitive to the antibody 11,450-B1 in solution.

Figure 9.42(i) shows the increase in the affinity, K_2 ($=k_2/k_d$), with an increase in the antibody 11,450-B1 concentration in nanomoles in solution. For the data shown in Figure 9.42(i), the affinity is given by:

$$K_2 = (0.3547 \pm 0.2713)\,[\text{antibody } 11{,}450\text{-B1}]^{\,0.804 \pm 0.562} \tag{9.34}$$

There is scatter in the data. Only four data points are available. The availability of more data points would lead to a more reliable fit. The affinity, K_2, exhibits an order of dependence between one-half and first (equal to 0.804) order on the antibody 11,450-B1 concentration in nanometer in solution. This indicates that the binding rate coefficient, k_1, is sensitive to the antibody 11,450-B1 in solution.

Figure 9.42(j) shows the increase in the fractal dimension, D_{f1}, with an increase in the antibody 11,450-B1 concentration in nanometer in solution. For the data shown in Figure 9.42(j), the fractal dimension, D_{f1}, is given by:

$$D_{f1} = (1.191 \pm 0.028)\,[\text{antibody } 11{,}450\text{-B1}]^{\,0.1136 + 0.1968} \tag{9.35}$$

The fit is very good. Only four data points are available. The availability of more data points would lead to a more reliable fit. The fractal dimension, D_{f1}, exhibits close to a zero order of dependence (equal to 0.1136) on the antibody 11,450-B1 concentration in nanomoles in solution. This indicates that the binding fractal dimension exhibits a very slight dependence on the antibody 11,450-B1 concentration in solution in the range analyzed.

Figure 9.42(k) shows the increase in the fractal dimension, D_{f2}, with an increase in the antibody 11,450-B1 concentration in nanometer in solution. For the data shown in Figure 9.42(k), the fractal dimension, D_{f2}, is given by:

$$D_{f2} = (0.2.436 \pm 0.048)[\text{antibody } 11{,}450\text{-B1}]^{\,0.033 \pm 0.116} \tag{9.36}$$

The fit is good. Only four data points are available. The availability of more data points would lead to a more reliable fit. The fractal dimension, D_{f1}, exhibits close to a zero order (equal to 0.033) of dependence on the antibody

11,450-B1 concentration in nanometer in solution. This indicates that the binding rate coefficient, k_1, is not sensitive to the antibody 11,450-B1 in solution.

Figure 9.43(a) shows the binding and dissociation of 200 nM antibody 6E10 in solution to receptor immobilized on an SPR biosensor surface. A single-fractal analysis is required to adequately describe the binding and dissociation kinetics. The values of (1) the binding rate coefficient, k, and the fractal dimension, D_f, for a single-fractal analysis, and (2) the dissociation rate coefficient, k_d, and the fractal dimension for dissociation, D_{fd}, are given in Table 9.21. In this case the affinity, K ($=k/k_d$), value is 63.29.

Figure 9.43(b) shows the binding and dissociation of 166 nM antibody 6E10 in solution to receptor immobilized on an SPR biosensor surface. Once again, a single-fractal analysis is required to adequately describe the binding

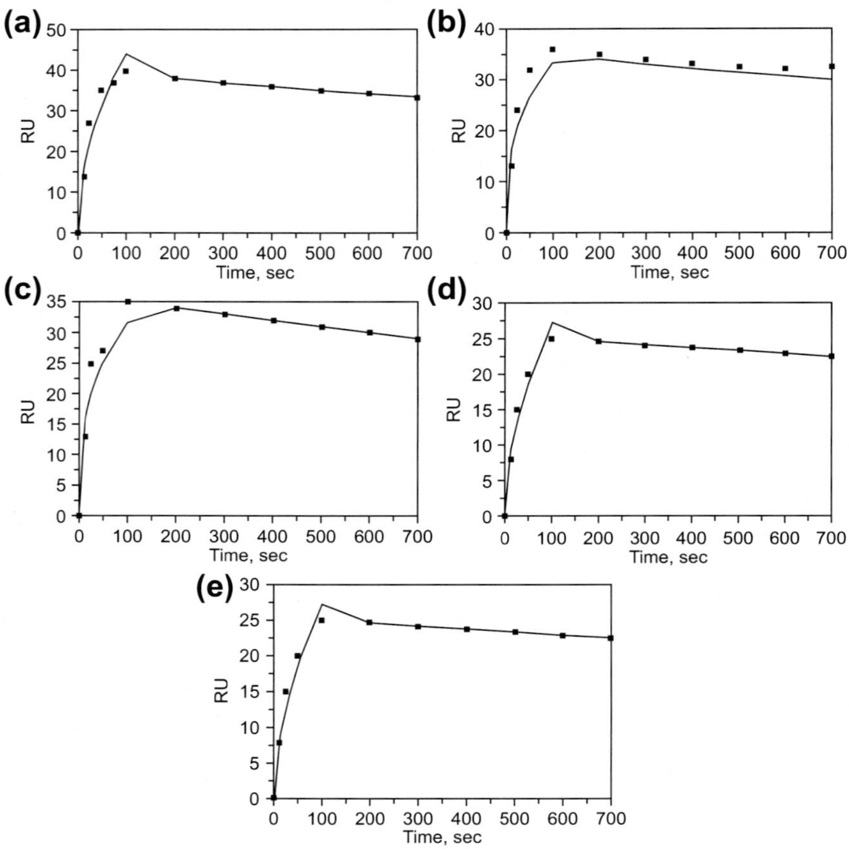

FIGURE 9.43 Binding of different concentrations of antibody 6E10 to receptor immobilized on an SPR biosensor surface: (a) 200 nM; (b) 166 nM; (c) 100 nM; (d) 66 nM; (e) 33 nM.

and dissociation kinetics. The values of (1) the binding rate coefficient, k, and the fractal dimension, D_f, for a single-fractal analysis, and (2) the dissociation rate coefficient, k_d, and the fractal dimension for dissociation, D_{fd}, are given in Table 9.21. In this case, the affinity, K (=k/k_d), value is 63.32. This value is very close to the previous value.

Figure 9.43(c) shows the binding and dissociation of 100 nM antibody 6E10 in solution to receptor immobilized on an SPR biosensor surface. Once again, a single-fractal analysis is required to adequately describe the binding and dissociation kinetics. The values of (1) the binding rate coefficient, k, and the fractal dimension, D_f, for a single-fractal analysis, and (2) the dissociation rate coefficient, k_d, and the fractal dimension for dissociation, D_{fd}, are given in Table 9.21. In this case, the affinity, K (=k/k_d), value is 70.79.

Figure 9.43(d) shows the binding and dissociation of 66 nM antibody 6E10 in solution to receptor immobilized on an SPR biosensor surface. Once again, a single-fractal analysis is required to adequately describe the binding and dissociation kinetics. The values of (1) the binding rate coefficient, k, and the fractal dimension, D_f, for a single-fractal analysis, and (2) the dissociation rate coefficient, k_d, and the fractal dimension for dissociation, D_{fd}, are given in Table 9.21. In this case the affinity, K (=k/k_d), value is 339.8.

Figure 9.43(e) shows the binding and dissociation of 33 nM antibody 6E10 in solution to receptor immobilized on an SPR biosensor surface. Once again, a single-fractal analysis is required to adequately describe the binding and dissociation kinetics. The values of (1) the binding rate coefficient, k, and the fractal dimension, D_f, for a single-fractal analysis, and (2) the dissociation rate coefficient, k_d, and the fractal dimension for dissociation, D_{fd}, are given in Table 9.21. In this case, the affinity, K (=k/k_d), value is 583.4.

Figure 9.44(a) and Table 9.21 shows the increase in the binding rate coefficient, k, with an increase in the antibody 6E10 concentration in solution. For the data shown in Figure 9.44(a), the binding rate coefficient, k, is given by:

$$k = (0.2163 \pm 0.1477)[6E10]^{8.6378+0.3373} \tag{9.38}$$

There is scatter in the data. Only five data points are available. The availability of more data points would lead to a more reliable fit. The binding rate coefficient, k, exhibits an order of dependence between one-half and first (equal to 0.6378) order on the antibody 6E10 concentration in solution. This indicates that the binding rate coefficient, k, is somewhat sensitive to the antibody 6E10 concentration in solution.

Figure 9.44(b) and Table 9.21 shows the increase in the dissociation rate coefficient, k_d, with an increase in the antibody 6E10 concentration in solution. For the data shown in Figure 9.44(b), the dissociation rate coefficient, k_d, is given by:

$$k_d = \left(0.52E - 6 + 2.7E - 6\right)[6E10]^{1.946 \pm 0.498} \tag{9.39}$$

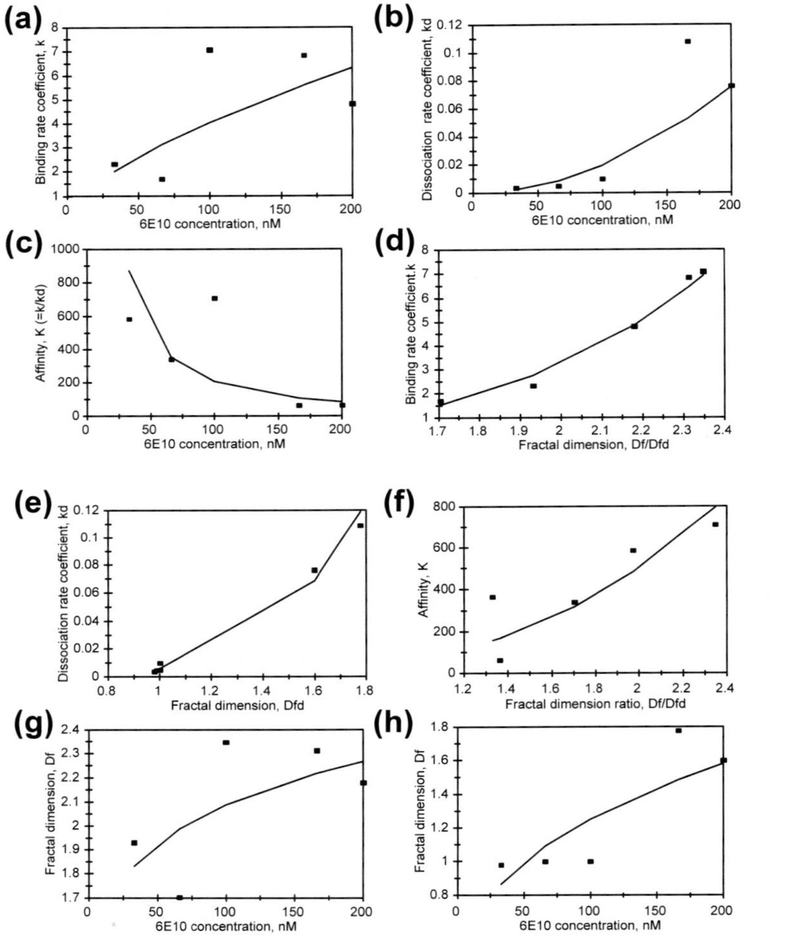

FIGURE 9.44 (a) Increase in the binding rate coefficient, k, with an increase in the antibody 6E10 concentration in nanomoles. (b) Increase in the dissociation rate coefficient, k_d, with an increase in the antibody 6E10 concentration in nanomoles. (c) Decrease in the affinity, K, with an increase in the antibody 6E10 concentration in nanomoles. (d) Increase in the binding rate coefficient, k, with an increase in the fractal dimension, D_f. (e) Increase in the dissociation rate coefficient, k_d, with an increase in the fractal dimension for dissociation, D_{fd}. (f) Increase in the affinity, K, with an increase in the fractal dimension ratio, D_f/D_{fd}. (g) Increase in the fractal dimension, D_f, with an increase in the antibody 6E10 concentration in solution. (h) Increase in the fractal dimension for dissociation, D_{fd}, with an increase in antibody 6E10 concentration in solution in nanometer.

The fit is good. Only five data points are available. The availability of more data points would lead to a more reliable fit. The dissociation rate coefficient, k_d, exhibits close to a second order of dependence (equal to 1.946) on the antibody 6E10 concentration in solution. This indicates that the dissociation rate coefficient, k_d, is sensitive to the antibody 6E10 concentration in solution.

Figure 9.44(c) and Table 9.21 show the decrease in the affinity, K (=k/k_d), with an increase in the antibody 6E10 concentration in solution. For the data shown in Figure 9.44(c), the affinity, K, is given by:

$$K = (82419.06 \pm 5663.24)[6E10]^{-1.302 \pm 0.5669} \tag{9.40}$$

The fit is good. Only five data points are available. The availability of more data points would lead to a more reliable fit. The affinity, K, decreases with an increase in the antibody 6E10 concentration in solution. The affinity, k, exhibits an order of dependence between minus first and minus one and one-half (equal to −1.302) order of dependence on the 6E10 concentration in solution. This indicates that the affinity, K, is sensitive to the antibody 6E10 concentration in solution.

Figure 9.44(d) and Table 9.21 show the increase in the binding rate coefficient, k, with an increase in the fractal dimension, D_f. For the data shown in Figure 9.44(d), the binding rate coefficient, k, is given by:

$$k = (0.1245 \pm 0.0161)\, D_f^{4.709 \pm 0.4478} \tag{9.41}$$

The fit is very good. Only five data points are available. The availability of more data points would lead to a more reliable fit. The binding rate coefficient, k, exhibits an order of dependence between four and one-half and fifth (equal to 4.709) order of dependence on the fractal dimension, D_f, or the degree of heterogeneity that exits on the surface. This indicates that the binding rate coefficient, k, is very sensitive to the heterogeneity that exists on the biosensor surface.

Figure 9.44(e) and Table 9.21 show the increase in the dissociation, k_d, with an increase in the fractal dimension for dissociation, D_{fd}. For the data shown in Figure 9.44(e), the dissociation rate coefficient, k_d, is given by:

$$k_d = (0.00603 \pm 0.00272)[D_{fd}]^{5.14 \pm 0.635} \tag{9.42}$$

The fit is good. Only five data points are available. The availability of more data points would lead to a more reliable fit. The dissociation rate coefficient, k_d, exhibits an order of dependence between five and five and one-half (equal to 5.14) order on the fractal dimension in the dissociation phase, D_{fd}. This indicates the dissociation rate coefficient, k_d, is very sensitive to the degree of heterogeneity on the SPR biosensor surface in the dissociation phase, D_{fd}.

Figure 9.44(f) and Table 9.21 show the increase in the affinity, K (=k/k_d), with an increase in the fractal dimension ratio, D_f/D_{fd}. For the data shown in Figure 9.44(f), the affinity, K, is given by:

$$K = (69.73 + 79.19)[D_f/D_{fd}]^{2.85 \pm 1.564} \tag{9.43}$$

The fit is good. Only five data points are available. The availability of more data points would lead to a more reliable fit. The affinity, K, increases with an increase in the fractal dimension ratio, D_f/D_{fd}. The affinity, K, exhibits an order of dependence between two and one-half and third (equal to 2.85) order of dependence on the ratio of fractal dimensions. This indicates that the affinity, K, is sensitive to the ratio of fractal dimension ratio.

Figure 9.44(g) and Table 9.22 shows the increase in the fractal dimension, D_f, with an increase in the antibody 6E10 concentration in solution. For the data shown in Figure 9.44(g), the fractal dimension, D_f, is given by:

$$D_f = (1.214 \pm 0.155)[6E10]^{0.1186 \pm 0.0828} \quad (9.44)$$

There is scatter in the data. Only five data points are available. The availability of more data points would lead to a more reliable fit. The fractal dimension, D_f, exhibits close to a zero (equal to 0.1186) order of dependence on the antibody 6E10 concentration ni solution. This indicates that the fractal dimension, D_f, is hardly dependent on the 6E10 concentration in solution.

Figure 9.44(h) and Table 9.21 show the increase in the fractal dimension for dissociation, D_{fd}, with an increase in the antibody 6E10 concentration in solution. For the data shown in Figure 9.44(c), the fractal dimension for dissociation, D_{fd}, is given by:

$$D_{fd} = (0.2695 \pm 0.0557)[6E10]^{0.3336 \pm 0.12935669} \quad (9.45)$$

The fit is reasonable. Only five data points are available. The availability of more data points would lead to a more reliable fit. The fractal dimension for dissociation, D_{fd}, exhibits an order of dependence between zero and one-half (equal to 0.3336) order on the fractal dimension for dissociation. This indicates that the fractal dimension for dissociation is only mildly dependent on the antibody 6E10 concentration in solution.

Kakuyama et al. (2005) have analyzed the binding and dissociation kinetics for different CLAC concentration in solution to aggregated Aβ using SPR spectroscopy. Figure 9.45(a) shows the binding and dissociation of 3.12 nM CLAC in solution to aggregated Aβ. A single-fractal analysis is required to adequately describe the binding and dissociation kinetics. The values of (1) the binding rate coefficient, k, and the fractal dimension, D_f, and (2) the dissociation rate coefficient, k_d, and the fractal dimension for dissociation, D_{fd}, are given in Table 9.22(a) and (b).

Figure 9.45(b) shows the binding and dissociation of 6.25 nM CLAC in solution to aggregated Aβ. A single-fractal analysis is required to adequately describe the binding and dissociation kinetics. The values of (1) the binding rate coefficient, k, and the fractal dimension, D_f, and (2) the dissociation rate coefficient, k, and the fractal dimension for dissociation, D_{fd}, are given in Table 9.22(a) and (b).

TABLE 9.22a Binding and Dissociation Rate Coefficients for Different Concentrations of CLAC to Aggregated Aβ Using SPR Spectroscopy (Kakuyama et al., 2005)

Analyte in Solution/Receptor on Surface	k	k_1	k_2	k_d	$K = k/k_d$
3.12 nM CLAC/aggregated Aβ	7.953 ± 0.700	–	–	0.1302 ± 0.0056	61.08
6.25 nM CLAC/aggregated Aβ	6.0962 ± 0.5496	–	–	0.04310 ± 0.00972	141.4
12.5 nM CLAC/aggregated Aβ	9.9402 ± 2.2975	4.5864 ± 0.3657	32.746 ± 0.635	0.01723 ± 0.00234	266.18; 1990.5
25 nM CLAC/aggregated Aβ	7.7999 ± 1.2553	–	–	0.3682 ± 0.0779	21.18

SPR, surface plasmon resonance.

TABLE 9.22b Fractal Dimensions for the Binding and Dissociation Phases for Different Concentrations of CLAC to Aggregated Aβ Using SPR Spectroscopy (Kakuyama et al., 2005)

Analyte in Solution/Receptor on Surface	D_f	D_{f1}	D_{f2}	D_{fd}	D_f/D_{fd}
3.12 nM CLAC/aggregated Aβ	2.1734 ± 0.1318	–	–	1.8406 ± 0.1494	1.527
6.25 nM CLAC/aggregated Aβ	2.0468 ± 0.1420	–	–	1.3272 ± 0.2062	1.542
12.5 nM CLAC/aggregated Aβ	2.3462 ± 0.1414	1.7184 ± 0.1520	2.8788 ± 0.1211	0.9132 ± 0.2450	1.881; 3.152
25 nM CLAC/aggregated Aβ	2.6667 ± 0.0769	–	–	1.8404 ± 0.1595	21.18

SPR, surface plasmon resonance.

Figure 9.45c shows the binding and dissociation of 12.5 nM CLAC in solution to aggregated Aβ, A single-fractal analysis is required to adequately describe the binding and dissociation kinetics. The values of (1) the binding rate coefficient, k, and the fractal dimension, D_f, and (2) the dissociation rate coefficient, k_d, and the fractal dimension for dissociation, D_{fd}, are given in Table 9.22(a) and (b).

Figure 9.45(d) shows the binding and dissociation of 25 nM CLAC in solution to aggregated Aβ. A single-fractal analysis is required to adequately describe the binding and dissociation kinetics. The values of (1) the binding rate coefficient, k, and the fractal dimension, D_f, and (2) dissociation rate coefficient, k_d, and the fractal dimension for dissociation, D_{fd}, are given in Table 9.22(a) and (b).

Figure 9.46(a) shows the increase in the dissociation rate coefficient, k_d, with an increase in the CLAC concentration in solution. For the data shown in Figure 9.46(a), the dissociation rate coefficient, k_d, is given by:

$$k_d = (0.033 \pm 0.74)\,[\mathrm{CLAC}]^{0.648 \pm 0.777} \tag{9.46a}$$

The fit is quite good. Only three data points are available. The availability of more data points would lead to a more reliable fit. The dissociation rate coefficient, k_d, exhibits an order of dependence between one-half and first (equal to 0.648) on the CLAC concentration in solution.

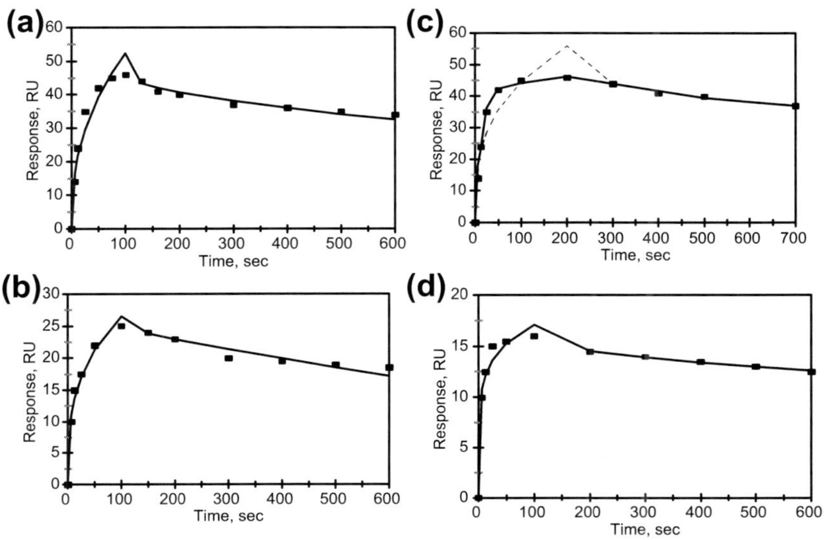

FIGURE 9.45 Binding and dissociation kinetics for different concentrations of CLAC in solution to aggregated Aβ using SPR spectroscopy (Kakuyama et al., 2005): (a) 3.12 nM (b) 6.25 nM (c) 12.5 nM (d) 25 nM.

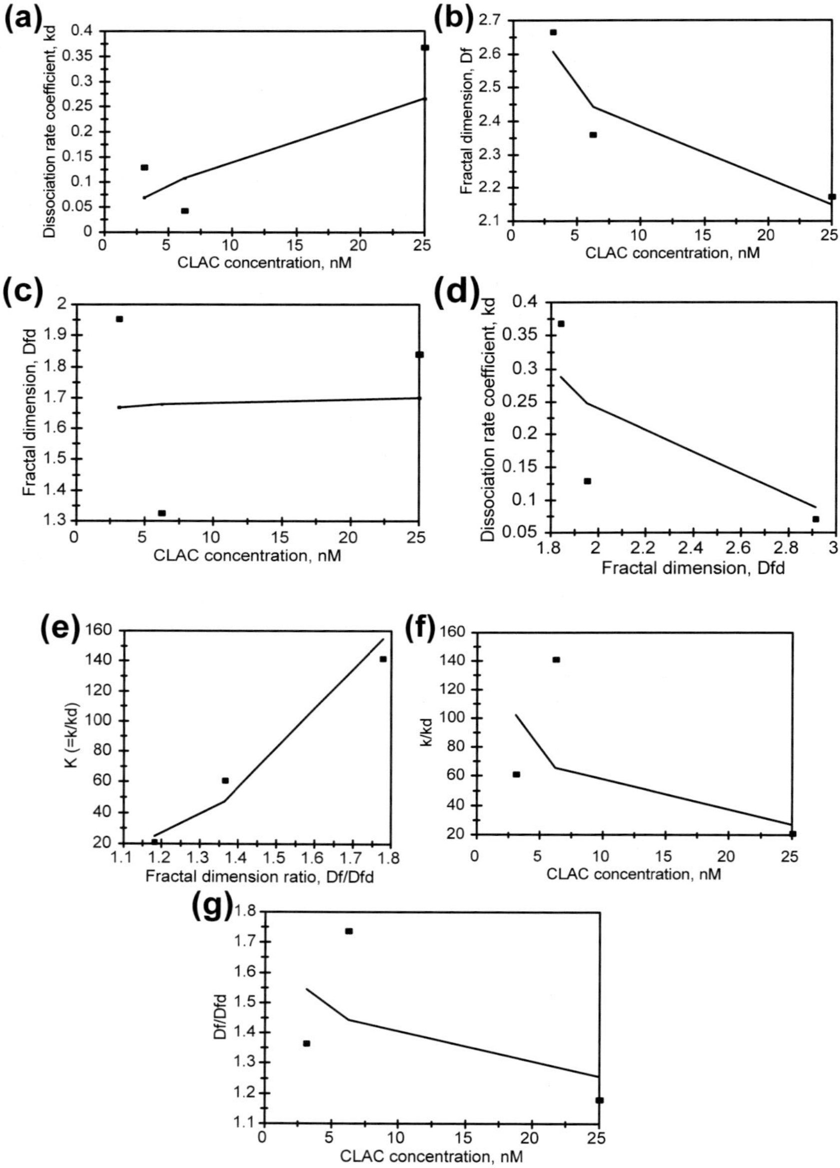

FIGURE 9.46 (a) Increase in the dissociation rate coefficient, k_d with an increase in the CLAC concentration in solution, nM. (b) Decrease in the fractal dimension, D_f with an increase in the CLAC concentration in solution. (c) Effect of CLAC concentration in solution on the fractal dimension for dissociation, $D_{fd.}$ (d) Decrease in the dissociation rate coefficient, k_d with an increase in the fractal dimension, $D_{fd.}$ (e) Increase in the affinity, K (=k/k_d) with an increase in the fractal dimension ratio, $D_f/D_{fd.}$ (f) Decrease in the affinity, K with an increase in the CLAC concentration, nM. (g) Decrease in the fractal dimension ratio, D_f/D_{fd} with an increase in the CLAC concentration in solution.

Figure 9.46(b) shows the decrease in the fractal dimension, D_f, with an increase in the CLAC concentration in solution. For the data shown in Figure 9.46(b), the fractal dimension, D_f, is given by:

$$D_f = (2.08 \pm 0.127)\ [\mathrm{CLAC}]^{-0.0928} \tag{9.46b}$$

The fit is quite good. Only three data points are available. The availability of more data points would lead to a more reliable fit. The fractal dimension, D_f, exhibits close to a zero order of dependence on the CLAC concentration in solution.

Figure 9.46(c) shows the effect of CLAC concentration in solution on the fractal dimension for dissociation, D_{fd}. For the data shown in Figure 9.46(c), the fractal dimension for dissociation, D_{fd}, is given by:

$$D_{fd} = (1.652 \pm 0.546)\ [\mathrm{CLAC}]^{0.0091+0.196} \tag{9.46c}$$

The fit is not good. Only three data points are available. The availability of more data points would lead to a more reliable fit. The fractal dimension for dissociation, D_{fd}, hardly shows a dependence on the CLAC concentration in solution.

Figure 9.46(d) shows the decrease in the dissociation rate coefficient, k_d, with an increase in the fractal dimension, D_{fd}. For the data shown in Figure 9.46(d), the dissociation rate coefficient, k_d, is given by:

$$k_d = \left(1.374 \pm 1.184\right) D_{fd}^{-2.852 \pm 1.758} \tag{9.46d}$$

The fit is quite good. Only three data points are available. The availability of more data points would lead to a more reliable fit. The dissociation rate coefficient, k_d, exhibits close to a negative third order of dependence (equal to −2.825) order of dependence on the fractal dimension for dissociation, D_{fd}, or the degree of heterogeneity present in the dissociation phase on the biosensor surface.

Figure 9.46(e) shows the increase in the affinity, K, with an increase in the fractal dimension ratio, D_f/D_{fd}. For the data shown in Figure 9.46(e), the affinity, K (=k/k_d), is given by:

$$K(=k/k_d) = (11.852 \pm 4.346)(D_f/D_{fd})^{4.466 \pm 1.064} \tag{9.46e}$$

The fit is very good. Only three points are available. The availability of more data points would lead to a more reliable fit. The affinity, K, is very sensitive to the fractal dimension ratio, D_f/D_{fd}, as it exhibits an order of dependence close to four and one-half order (equal to 4.466) on the fractal dimension ratio.

Figure 9.46(f) shows the decrease in the affinity, K, with an increase in the CLAC concentration in solution. For the data shown in Figure 9.46(f), the affinity, K (=k/k_d), is given by:

$$K(=k/k_d) = (209.67 \pm 235.12)\ [\mathrm{CLAC}]^{-0.633+0.637} \tag{9.46f}$$

There is scatter in the data. Only three points are available. The availability of more data points would lead to a more reliable fit. The affinity, K, exhibits an order of dependence between negative one-half and first (equal to −0.633) on the CLAC concentration in solution.

Figure 9.46(g) shows the decrease in the fractal dimension ratio, (D_f/D_{fd}), with an increase in the CLAC concentration in solution. For the data shown in Figure 9.46(g), the fractal dimension ratio, D_f/D_{fd}, is given by:

$$D_f/D_{fd} = (1.731 \pm 0.455)\,[\mathrm{CLAC}]^{-0.0998 \pm 0.155} \tag{9.46g}$$

There is scatter in the data. Only three points are available. The availability of more data points would lead to a more reliable fit. The fractal dimension ratio, D_f/D_{fd}, exhibits an order of dependence less than negative 0.1 (equal to −0.0998) on the CLAC concentration in solution.

Figure 9.47 shows the elongation of Aβ fibrils in the presence of CLAC (Kakuyama et al., 2005). Figure 9.47(a) shows the binding of 50 nM CLAC which was injected and the elongation rate was calculated (cycle 4) over aggregated Aβ(l−40) immobilized over a CM3 chip (1580 RU), A dual-fractal analysis is required to adequately describe the binding kinetics. The values of (1) binding rate coefficient, k, and the fractal dimension, D_f, for a single-fractal analysis, (2) the binding rate coefficients, k_1 and k_2, and the fractal dimensions, D_{f1} and D_{f2}, for a dual-fractal analysis are given in Table 9.23(a) and (b).

Figure 9.47(b) shows the binding of 100 nM CLAC which was injected and the elongation rate was calculated (cycle 4) over aggregated Aβ(l−40) Immobilized over a CM3 chip (1580 RU). A single-fractal analysis is adequate to describe the binding kinetics. The values of the binding rate coefficient, k, and the fractal dimension, D_f, for a single-fractal analysis are given in Table 9.23(a) and (b).

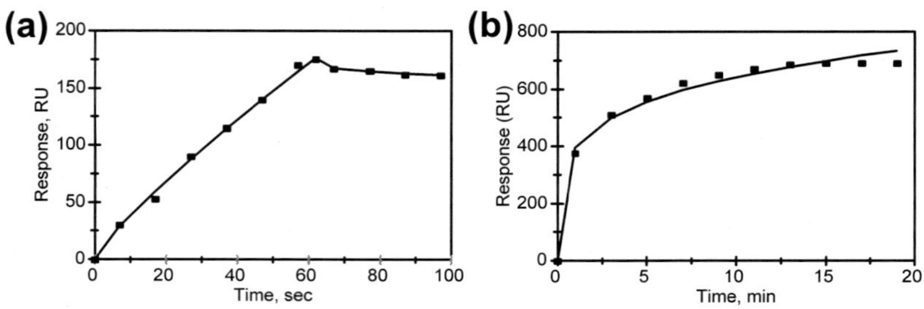

FIGURE 9.47 (a) Elongation of Aβ 1−40 fibrils in the presence of CLAC. Aggregated Aβ1−40 immobilized on a CM3 chip (1580 RU). 50 nM CLAC was injected and the elongation rate was calculated (Cycle 4) (Kakuyama et al., 2005). (b) Elongation of Aβ 1−40 fibrils in the presence of CLAC. Aggregated Aβ1−40 immobilized on a CM3 chip (1580 RU). 100 nM CLAC was injected and the elongation rate was calculated (Cycle 2) (Kakuyama et al., 2005).

TABLE 9.23a Binding Rate Coefficients for CLAC to Aggregated Aβ(1−40) Immobilized on a CM3 Chip and the Elongation Rate was Calculated (Kakuyama et al., 2005)

Analyte in Solution	k	k_1	k_2
50 nM CLAC/cycle 4/aggregated Aβ(l−40) on a CM3 chip	200.59 ± 17.24	181.89 ± 4.07	394.95 ± 0.84
100 nM CLAC/cycle 2/aggregated Aβ(l−40) on a CM3 chip	395.79 ± 16.39	NA	NA

TABLE 9.23b Fractal Dimensions for the Binding of CLAC to Aggregated Aβ(1−40) Immobilized on a CM3 Chip and the Elongation Rate was Calculated (Kakuyama et al., 2005)

Analyte in Solution	D_f	D_{f1}	D_{f2}
50 nM CLAC/cycle 4/aggregated Aβ(l−40) on a CM3 chip	2.3008 ± 0.05996	2.0962 ± 0.0300	2.8131 ± 0.07528
100 nM CLAC/cycle 2/aggregated Aβ(l−40) on a CM3 chip	2.5812 ± 0.02952	NA	NA

Kakuyama et al. (2005) have characterized Aβ fibril elongation. Figure 9.48(a) shows the binding and dissociation of 30 μM freshly dissolved Aβ(l−40) on immobilized Aβ(1−40) (2120 RU) on an SPR biosensor surface. A single-fractal analysis is adequate to describe the binding and dissociation kinetics. The values of (1) the binding rate coefficient, k, and the fractal dimension, D_f, for a single-fractal analysis, and (2) the dissociation rate coefficient, k_d, and the fractal dimension for dissociation, D_{fd}, for a single-fractal analysis are given in Table 9.24.

Figure 9.48(b) shows the binding and dissociation of 15 μM freshly dissolved Aβ(l-40) on immobilized Aβ(l-40) (2120 RU) on an SPR biosensor surface. A single-fractal analysis is adequate to describe the binding and dissociation kinetics. The values of (1) the binding rate coefficient, k, and the fractal dimension, D_f, for a single-fractal analysis, and (2) the dissociation rate coefficient, k_d, and the fractal dimension for dissociation, D_{fd}, for a single-fractal analysis are given in Table 9.24.

Figure 9.48(c) shows the binding and dissociation of 7.5 μM freshly dissolved Aβ(1−40) on immobilized Aβ(l−40) (2120 RU) on an SPR biosensor

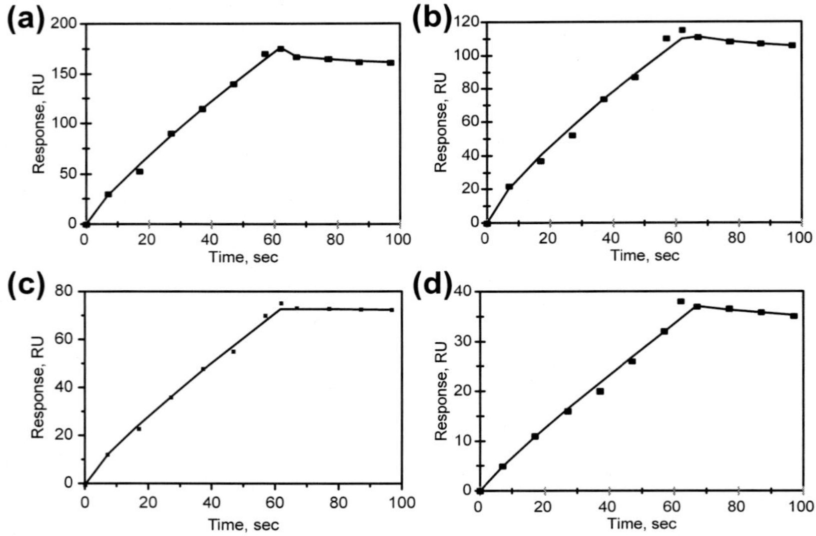

FIGURE 9.48 Characterization of Aβ fibril elongation. Binding and dissociation of different concentrations of freshly dissolved Aβ (1−40) on immobilized. Aβ1−40(2120) RU: (a) 30 μm; (b) 15 μm; (c) 7.5 μm; (d) 3.75 μm.

surface. A single-fractal analysis is adequate to describe the binding and dissociation kinetics. The values of (1) the binding rate coefficient, k, and the fractal dimension, D_f, for a single-fractal analysis, and (2) the dissociation rate coefficient, k_d, and the fractal dimension for dissociation, D_{fd}, for a single-fractal analysis are given in Table 9.24.

Figure 9.48(d) shows the binding and dissociation of 3.75 μM freshly dissolved Aβ(1−40) on immobilized Aβ(l−40) (2120 RU) on an SPR biosensor surface. A single-fractal analysis is adequate to describe the binding and dissociation kinetics. The values of (1) the binding rate coefficient, k, and the fractal dimension, D_f, for a single-fractal analysis, and (2) the dissociation rate coefficient, k_d, and the fractal dimension for dissociation, D_{fd}, for a single-fractal analysis are given in Table 9.24.

Figure 9.49(a) and Table 9.24 show the increase in the binding rate coefficient, k, with an increase in the Aβ concentration in solution. For the data shown in Figure 9.49(a), the binding rate coefficient, k, is given by:

$$k = (0.3206 \pm 0.1508)\,[A\beta]^{0.895 \pm 0.184} \qquad (9.47a)$$

The fit is good. Only four data points are available. The availability of more data points would lead to a more reliable fit. The binding rate coefficient, k, exhibits an order of dependence between one-half and first (equal to 0.895) on the Aβ concentration in solution.

TABLE 9.24 Characterization of Aβ Fibril Elongation. Binding and Dissociation of Different Concentrations of Freshly Dissolved Aβ(1–40) to Immobilized Aggregated Aβ(1–40) (2120 RU) (Kakuyama et al., 2005)

Analyte in Solution/Receptor on Surface	k	k_d	D_f	D_{fd}
30 μM Aβ(1–40)/aggregated Aβ(1–40)	5.548 ± 0.351	4.847 ± 0.312	1.3242 ± 0.0645	2.4100 ± 0.08468
15 μM Aβ(1–40)/aggregated Aβ(1–40)	4.4455 ± 0.3687	2.100 ± 0.1288	1.4456 ± 0.08350	2.4090 ± 0.08468
7.5 μM Aβ(1–40)/aggregated Aβ(1–40)	2.369 ± 0.125	1.5346 ± 0.0669	1.3420 ± 0.05372	2.6974 ± 0.09796
3.75 μM Aβ(1–40)/aggregated Aβ(1–40)	0.8564 ± 0.0540	0.3899 ± 0.0525	1.2096 ± 0.06418	1.9074 ± 0.1712

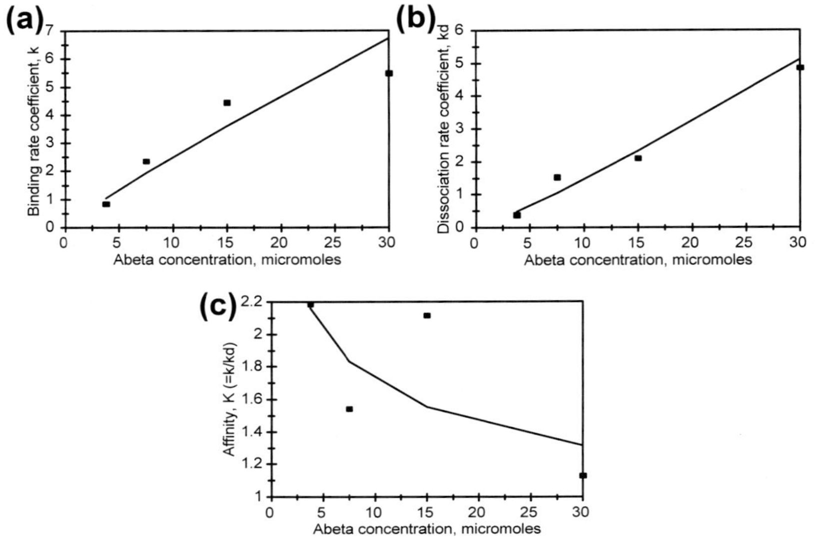

FIGURE 9.49 (a) Increase in the binding rate coefficient, k with an increase in the concentration of Aβ in nM. (b) Increase in the dissociation rate coefficient, k_d with an increase in the concentration of Aβ in nM. (c) Decrease in the affinity, K with an increase in the concentration of Aβ in solution.

Figure 9.49(b) and Table 9.24 show the increase in the dissociation rate coefficient, k_d, with an increase in the Aβ concentration in solution. For the data shown in Figure 9.49(b), the dissociation rate coefficient, k_d, is given by:

$$k_d = (0.107 \pm 0.039)\,[A\beta]^{1.136 \pm 0.202} \tag{9.47b}$$

The fit is very good. Only four data points are available. The availability of more data points would lead to a more reliable fit. The dissociation rate coefficient, k_d, exhibits an order of dependence slightly more than first (equal to 1.136) order on the Aβ concentration in solution.

Figure 9.49(c) and Table 9.24 show the decrease in the affinity, K (=k/k_d), with an increase in the Aβ concentration in solution. For the data shown in Figure 9.49(c), the affinity, K, is given by:

$$K(=k/k_d) = (2.967 \pm 0.927)\,[A\beta]^{-0.239 \pm 0.175} \tag{9.47c}$$

The fit is reasonable. Only four data points are available. The availability of more data points would lead to a more reliable fit. The affinity, K, exhibits an order of negative dependence between zero and one-half (−0.239) order on the Aβ concentration in solution.

Figure 9.50(a) and Table 9.24 show the increase in the fractal dimension, D_f, as a function of Aβ concentration in solution in micromole. For the data

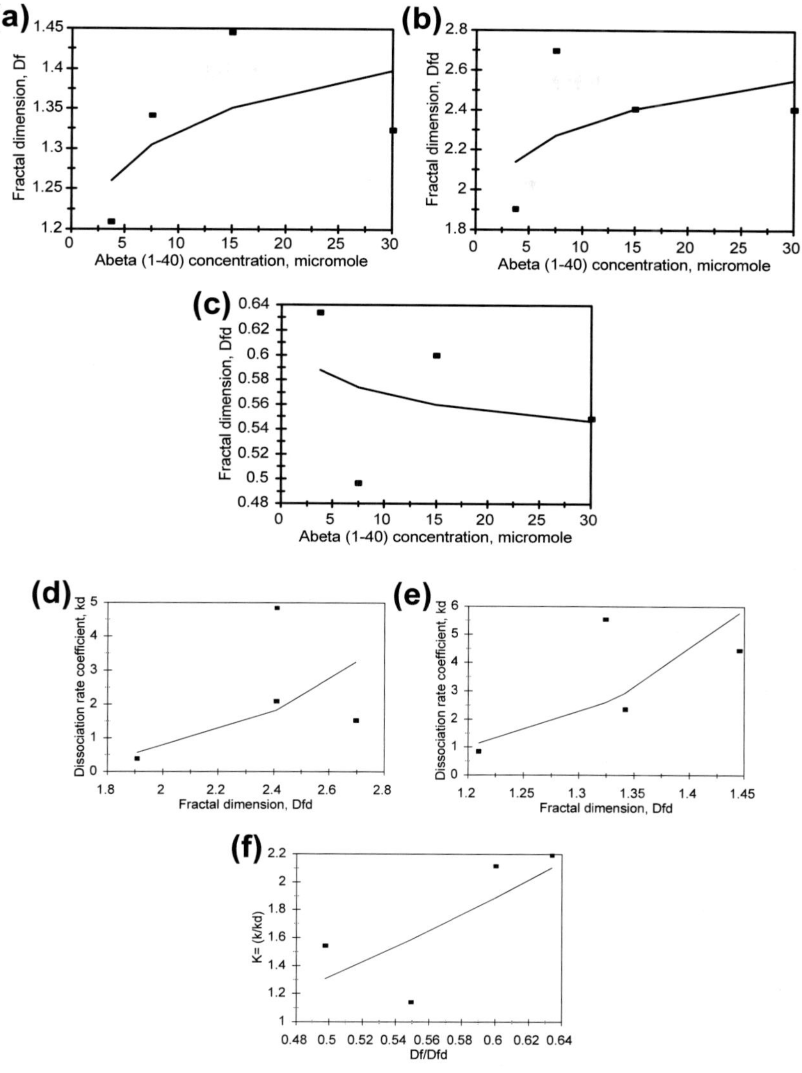

FIGURE 9.50 (a) Increase in the fractal dimension, D_f with an increase in the Aβ(1−40) concentration in solution. (b) Increase in the fractal dimension for dissociation, D_{fd} with an increase In the Aβ(1−40) concentration in solution. (c) Decrease in the fractal dimension for ratio, D_f/D_{fd} with an increase in the Aβ(1−40) concentration in solution. (d) Increase in the dissociation rate coefficient, k with an increase in the fractal dimension D_f. (e) Increase in the affinity, K with an increase in the fractal dimension ratio, D_{fd}.(f) Increase in affinity, K with an increase in the dimension ratio, D_f/D_{fd}.

shown in Figure 9.50(a) and in Table 9.24, the fractal dimension, D_f, is given by:

$$D_f = (1.180 \pm 0.167)\,[A\beta]^{0.0499 \pm 0.0457} \qquad (9.47d)$$

There is scatter in the data. Only four data points are available. The availability of more data points would lead to a more reliable fit. The fractal dimension, D_f, exhibits close to a zero (equal to 0.0499) order of dependence on the Aβ concentration in solution.

Figure 9.50(b) and Table 9.24 show the increase in the fractal dimension for dissociation, D_{fd}, as a function of Aβ concentration in solution in micromole. For the data shown in Figure 9.50(b) and in Table 9.24, the fractal dimension, D_{fd}, is given by:

$$D_{fd} = (1.913 \pm 0.314)\,[A\beta]^{0.08491 + 0.098} \qquad (9.47e)$$

There is scatter in the data. Only four data points are available. The availability of more data points would lead to a more reliable fit. The fractal dimension for dissociation, D_{fd}, exhibits close to a zero (equal to 0.08491) order of dependence on the Aβ concentration in solution.

Figure 9.50(c) and Table 9.24 shows the decrease in the ratio of fractal dimensions, D_f/D_{fd}, as a function of Aβ concentration in solution in micromole. For the data shown in Figure 9.50(c) and in Table 9.24, the ratio of the fractal dimensions, D_f/D_{fd}, is given by:

$$D_f/D_{fd} = (0.617 \pm 0.082)\,[A\beta]^{-0.035 + 0.080} \qquad (9.47f)$$

There is scatter in the data. Only four data points are available. The availability of more data points would lead to a more reliable fit. The fractal dimension ratio, D_f/D_{fd}, exhibits close to a zero (equal to −0.035) order of dependence on the Aβ concentration in solution.

Figure 9.50(d) and Table 9.24 show the increase in the binding rate coefficient, k, with an increase in the fractal dimension, D_f. For the data shown in Figure 9.50(d) and in Table 9.24, the binding rate coefficient, k, is given by:

$$k = (0.201 \pm 0.173)\,D_f^{9.10 \pm 4.90} \qquad (9.47g)$$

There is scatter in the data. Only four data points are available. The availability of more data points would lead to a more reliable fit. The binding rate coefficient, k, is very sensitive to the fractal dimension, D_f, or the degree of heterogeneity on the biosensor surface as noted by slightly greater than the ninth (equal to 9.10) order of dependence on the fractal dimension.

Figure 9.50(e) and Table 9.24 show the increase in the dissociation rate coefficient, k_d, with an increase in the fractal dimension, D_{fd}. For the data shown in Figure 9.50(e) and in Table 9.24, the dissociation rate coefficient, k_d, is given by:

$$k_d = (0.0201 \pm 0.0314)(D_{fd})^{5.074\pm3.606} \quad (9.47h)$$

There is scatter in the data. Only four data points are available. The availability of more data points would lead to a more reliable fit The dissociation rate coefficient, k_d, is very sensitive to the fractal dimension, D_f, or the degree of heterogeneity on the biosensor surface as noted by slightly greater than the fifth (equal to 5.074) order of dependence on the fractal dimension in the dissociation phase, D_{fd}.

Figure 9.50(f) and Table 9.24 show the increase in the affinity, $K = (k/k_d)$, with an increase in the fractal dimension ratio, D_f/D_{fd}. For the data shown in Figure 9.50(f) and in Table 9.22(a) and (b), the affinity, K, is given by:

$$K(=k/k_d) = (5.132 \pm 1.615)(D_f/D_{fd})^{1.9587\pm1.488} \quad (9.47i)$$

There is scatter in the data. Only four data points are available. The availability of more data points would lead to a more reliable fit. The affinity, K, is sensitive to the fractal dimension ratio, D_f/D_{fd}, or the ratio of the degree of heterogeneities that exists on the biosensor surface in the binding and in the dissociation phases as it exhibits close to a second (equal to 1.9587) order of dependence on the ratio, D_f/D_{fd}.

9.4 CONCLUSIONS

A fractal analysis is presented for the binding and dissociation (if applicable) of AD biomarkers on biosensor surfaces. The AD biomarkers analyzed include (1) the binding of different concentrations of APP770 in CM to the DE antibodies immobilized on a solid surface using TIRE (Mustafa et al., 2010), (2) binding of different concentrations of PrP variant, ARQ (in nM) in solution to asolectin liposome antibodies immobilized on an SPR L1 biosensor chip (Steunou et al., 2010), (3) binding of different concentrations of O_3 (in nM) in solution to asolectin liposomes immobilized on the surface of an SPR L1 biosensor chip (Steunou et al., 2010), (4) binding of 100 nM ARQ-Cu (II) to chip-immobilized PC liposomes with or without lipid rafts (Steunou et al., 2010), (5) binding of different concentrations of Aβ in micrometer in solution to sonicated fAβ immobilized on a flow cell at a density of 2000 RU (Hasegawa et al., 2002), (6) binding of different concentrations of Aβ (in μM) in solution to 1000 RU of fAβ immobilized on a flow cell (repetition runs) (Hasegawa et al., 2002), (7) and binding of acetylcholine and AChE inhibitor to AChE and choline oxidase immobilized on a working electrode using gel entrapment (Lenigk et al., 2000). Both a single- and a dual-fractal analysis are used to model the binding and dissociation (if applicable) kinetics.

The dual-fractal analysis is used only if the single-fractal analysis did not provide an adequate fit that is the regression coefficient is less than 0.95.

The fractal dimension is not a classical independent variable such as analyte (antigen, antibody, or other biological molecule) concentration in solution. Nevertheless, the expressions obtained for the binding (and the dissociation) rate coefficients for a single- and a dual-fractal analysis as a function of the fractal dimension indicate a high sensitivity of these rate coefficients on their respective fractal dimensions on the SPR sensor chip surface. It can be seen that the data analysis in itself does not provide any evidence for surface roughness or heterogeneity, and the existence of surface roughness or heterogeneity assumed may not be correct. However, considering the complexity involved on the SPR chip surface, this is not an unreasonable assumption. Furthermore, there is deviation in the data that may be minimized by providing a correction for the depletion of the analyte.

Predictive relations are presented for (1) the binding rate coefficients, k_1 and k_2, with an increase in the ARQ concentration in solution (in nM) (Steunou et al., 2010), (2) the binding rate coefficients, k_1 and k_2, with an increase in the fractal dimensions, D_{f1} and D_{f2}, respectively (Steunou et al., 2010), (3) the fractal dimensions, D_{f1} and D_{f2}, as a function of ARQ concentration in nanomoles in solution (Steunou et al., 2010), (4) the binding rate coefficient ratio, k_2/k_1, as a function of ARQ concentration in nanomoles in solution (Steunou et al., 2010), (5) the binding rate coefficients, k_1 and k_2, as a function of O_3 concentration in solution (Steunou et al., 2010), (6) the binding rate coefficients, k_1 and k_2, as a function of the fractal dimensions, D_{f1} and D_{f2}, respectively, for the binding of 100 nM ARQ-Cu (II) to chip-immobilized PC liposomes with or without lipid rafts (Steunou et al., 2010), (7) the binding rate coefficient, k, as a function of fAβ concentration in solution (Hasegawa et al., 2002), and (8) the binding rate coefficient, k, as a function of the fractal dimension, D_f, for the binding of 3 μM Aβ in solution to 1000 RU of fAβ immobilized on a flow cell (Hasegawa et al., 2002).

It is suggested that the fractal surface (roughness) leads to turbulence, which enhances mixing, decreases diffusional limitations, and leads to an increase in the binding rate coefficient (Martin et al., 1991). For this to occur, the characteristic length of the turbulent boundary layer may have to extend a few monolayers above the sensor chip surface to affect bulk diffusion to and from the surface. However, given the extremely laminar flow regimes in most biosensors this may not actually take place. The sensor chip surface is characterized by grooves and ridges and this surface morphology may lead to eddy diffusion. This eddy diffusion can then help to enhance the mixing and extend the characteristic length of the boundary layer to affect the bulk diffusion to and from the surface.

AD is a very serious pathological disease. Any insights that may be gained into the further progression of AD will lead to significant management, control of, and alleviation of this very serious debilitating disease. Different approaches may be used, and are being aggressively researched and applied to better understand this disease. An attempt is made in this chapter to better understand the binding of AD biomarkers to different biosensor surfaces is a step in this direction. A better understanding of the binding and the dissociation (if applicable) kinetics will lead to better insights into the progression of AD, and hopefully to the alleviation of AD's debilitating effects; and eventually in the not too distant future, to a better management and possible cure of AD itself. Any sort of progress, no matter how small, should prove extremely beneficial.

REFERENCES

Abdullah, L., Paris, D., Luis, C., Quadrus, A., OParish, J., Valdes, L., Keegan, A.P., Mathurn, V., Crawford, F., Mullan, M., 2007. The influence of diagnosing intra- and inter-fusion variability in serum and plasma Aβ levels. Neuroscience Letters 428, 53–58.

Agayre, E.K., Curran, G.L., Ramakrishnan, M., Yu, C.C., Poduslo, J.F., Kandimaller, K.K., 2008. Development of a smart nano-vehicle to target cerebrovascular amyloid deposits and brain parenchymal and cerebral amyloid angiopathy. Pharmacology Research 25, 2674–2684.

Alzheimer's Association, January 2011. http://www.alz.org/alzheimers_disease_what_is_alzheimers_.asp, downloaded March 11, 2011.

Alzheimer's Research Forum, 2011. About Alzheimers. http://www.alzforum.org/dis/abo/default.asp, downloaded April 6, 2011.

Arwin, H., Pokskinski, M., Johansen, K., 2004. Applied Optics 43, 3028–3036.

Bard, F., Cannon, C., Barbour, R., Burke, R.L., Gammes, D., Gradeja, H., Guido, T., Hu, K., Huang, J., Johnson–Wood, K., Khan, K., Kholodenko, D., Lee, M., Lieberburg, I., Motter, R., Nguyen, M., Soriano, F., Vasquez, N., Weiss, Welch, B., Seubert, P., Schlenk, P., Yednock, T., 2000. Peripherally administered antibodies against amyloid beta-peptide enter the central nervous system and reduce pathology in a mouse model of Alzheimer's disease. Nature Medicine 6, 916–919.

Bard, F., Barbour, R., Cannon, C., Carreto, C., Fox, M., Games, D., Guido, T., Hoenow, K., Hu, K., Johnson-Wood, K., Khan, K., Kholodenko, D., Lee, C., Lee, M., Motter, R., Nguyen, M., Reed, A., Kholodenko, D., Lee, C., Lee, M., Motter, R., Nguyen, M., Reed, A., Schenk, D., Tang, P., Vasquez, N., Seubert, P., Yednock, T., 2003. Epitope and isotype specificities of antibodies to beta-amyloid peptide for protection against Alzheimer's disease–like neuropathology. Proceedings of the National Academy of Sciences USA 100, 2023–2028.

Bennow, K., Hampel, H., Weiner, M., Zetterberg, H., March 2010. Cerebrospinal fluid and plasma biomarkers in Alzheimer disease. Nature reviews Neurology 6, 131–144.

Brzyska, M., Bacia, A., Elbaumi, D., 2001. European Journal of Biochemistry 268 (12), 3443–3454.

Buell, A.K., Tartaglia, G.G., Birkett, N.R., Waudby, C.A., Vendruscolo, M., Salvatella, X., Welland, M.E., Dobson, C.M., Knowles, T.P.J., 2009. FEBS Letters 583, 2587–2592.

Buell, A.K., White, D.A., Dobson, C.M., 2010. Surface attachment of protein fibrils via covalent modification strategies. Journal of Physical Chemistry B 114, 10925–10938.

Bush, A.I., 2003. Trends in Neuroscience 26 (4), 207–214.

Cannon, M.J., Williams, A.D., Wetzel, R., Myszka, D.G., 2004. Analytical Biochemistry 328 (1), 67–75.

Caughey, B., Baron, G.S., 2006. Nature 443 (7113), 803–810.

Caughey, B., Baron, G.S., Chesebro, B., Jeffrey, M., 2009. Annual Reviews of Biochemistry 78, 177–204.

Caughey, B., Lansbury Jr., P.T., 2003. Protofibrils, pores, fibrils, and neurodegeneration: separating the responsible protein aggregates fro the innocent bystanders. Annual Reviews in Neuroscience.

Clark, M.J., Gagnon, J., Williams, A.F., Barclay, A.N., 1985. EMBO 4, 113–118.

Dewachter, I., Van Leuven, F., 2002. Secretases as targets for the treatment of Alzheimer's disease: the prospects. Lancet Neurology 1, 409–416.

DeMattos, R.B., Bales, K.R., Cummins, D.J., Didart, J.C., Paul, S.M., Holtzman, D.M., 2001. Peripheral anti-Aβ antibody alters CNS and plasma Aβ burden in a mouse model of Alzheimer's disease. Proceedings of the National Academy of Sciences USA 98, 3850–3855.

Doong, R.A., Lee, P.S., Anitha, K., July 15, 2010. Simultaneous determination for Alzheimer's disease using sol-gel-derived optical array biosensor. Biosensors & Bioelectronics 25 (11), 2464–2469.

Du, Y., Wei, X., Dodel, R., Sommer, N., Hampel, H., Gao, F., Ma, Z., Zhao, L., Oertel, W.H., Farlow, M., 2003. Human anti-β-amyloid antibodies block β-amyloid fibril formation and prevent β-amyloid-induced neurotoxicity. Brain 126, 1935–1939.

Ertekin-Taner, Y.N., Younkin, L.H., Yager, D.M., Parfitt, F., Baker, M.C., Asthana, S., Hutton, M.L., Younkin, S.D., Graff-Radford, N.R., 2008. Plasma amyloid β-protein is elevated in late-onset Alzheimer's families. Neurology 70, 596–606.

Esler, W.P., Stimson, E.R., Ghilardi, J.R., Vinters, H.V., Lee, J.P., Maniyh, Maggio, J.E., 1996. Biochemistry 35, 749–757.

Esler, W.P., Stimson, E.R., Ghilardi, J.R., Felix, A.M., Lu, Y.A., Vinters, H.V., Mantyh, P.W., Maggio, J.E., 1997. Nature Biotechnology 15, 258–263.

Esler, W.P., Stimson, E.R., Jennings, J.M., Vinters, H.V., Ghiraldi, J.R., Lee, J.P., Mantyh, P.W., Maggio, J.E., 2000. Biochemistry 39, 6288–6295.

Esler, W.P., Wolfe, M.S., 2001. A portrait of Alzheimer secretases – new features and familiar faces. Science 293, 1449–1454.

Findeis, M.A., 2000. Biochimica Biophysica Acta 1502, 76–84.

Fischer, M., Ruliche, T., Raeber, A., Sailer, A., Moser, M., Oesch, B., Brandner, S., Aguzzi, A., Weissman, C., 1996. EMBO Journal 15 (6), 1255–1264.

Forshee, S., 2011. Study Aims to Identify Biomarkers of Alzheimer's for Early Detection. DailyBreeze.com. https://www.linkedin.com/20/690/655, downloaded April 6, 2011.

Giacobini, E., 1996. Cholinesterase inhibitors do more than inhibit cholinesterase. Alzheimer's Disease. From Molecular Biology to Therapy. Birkhauser, Boston, 187–204.

Glenner, G.G., Wong, C.W., 1984. Biochemical and Biophysical Research Communications 120, 885–890.

Golde, T.E., January 2005. The Abeta hypothesis leading us to rationally-designed therapeutic strategies for the treatment or prevention of Alzheimer's disease. Brain Pathology 15 (10), 85–87.

Gravina, S.A., Ho, L., Eckman, C.R., Long, K.E., Younkin, S.G., 1995. Amyloid beta protein (Abeta) in Alzheimer's brain: biochemical and immunocytochemical analysis with antibodies specific for forms ending at Abeta 40 or Abeta 42 (43). Journal of Biological Chemistry 270, 7013–7016 (1995).

Haes, A.J., Chang, L., Klein, W.L., et al., 2005. Detection of a biomarker for Alzheimer's disease from synthetic and clinical samples using a nanoscale optical biosensor. Journal of the American Chemical Society 127 (7), 2264–2271.

Havlin, S., 1989. Molecular diffusion and reaction. In The Fractal Approach to Heterogeneous Chemistry: Surfaces, Colloids, Polymers. In: Avnin, D. (Ed.). Wiley, New York, pp. 251–269.

Hardy, J.A., Higgins, G.A., 1992. Science 256, 184–185.

Hardy, J., Selkoe, D.J., 2002. The amyloid hypothesis of Alzheimer's hypothesis of Alzheimer's disease: progress and problems on the road to therapeutics. Science 297, 353–356.

Hasegawa, K., Ono, K., Yamada, M., Naiki, H., 2002. Kinetic modeling and determination of reaction constants of Alzheimer's β-amyloid fibril extension and dissociation using surface plasmon resonance. Biochemistry 41, 13489–13498.

Hasegawa, K., Monnaga, A., Ngmura, R., Ookoshi, T., Ozawa, D., Goto, Y., Yamada, M., Naiki, H., April 2010. Critical role of interfaces and agitation on the nucleation of Aβ amyloid fibrils at low concentrations of Aβ monomers. Biochimica Biophysica Acta (BBA) – Proteins and Proteomics 1804 (4), 986–995.

Homola, J., Pillarik, M., 2006. Surface plasmon resonance (SPR) sensors, Springer Series, Chemical Sensors and Biosensors, 4, 45-67, 8 July 2006. Springer-Verlag, Berlin, Heidelberg.

Ibach, B., Binder, H., Dragon, M., Poijansky, H.E., et al., 2006. Neurobiological Aging 27, 1202–1211.

Janus, C., Pearson, J., Mclaurin, J., Matthews, P.M., Jiang, Y., Schmidt, S.D., Azkhar Chishti, M., Horne, P., Heslin, D., French, J., Mount, H.T.J., Nixon, R.A., Merchen, Bergron, C., Eraser, P.E., St George-Hyslop, P., Westaway, D., 2000. Nature 408, 979–982.

Jantzen, P.T., Connor, K.E., DiCarlo, G., Wenk, G.L., Wallace, J.L., Rojiani, A.M., Coppola, D., Morgan, D., Gordon, N., 2002. Microbial activation and β-amyloid deposit reduction caused by nitric oxide-releasing nonsteroidal anti-inflammatory drug in amyloid precursor protein plus presenilin-1 transgenic mice. Journal of Neuroscience 22, 2246–2254.

Kakuyama, H., Soderberg, L., Horigome, K., Winblad, B., Dahlqvist, C., Naslund, J., Tjernberg, L.O., 2005. CLAC binds to aggregated Aβ and Aβ fragments, and attenuates fibril elongation. Biochemistry 44, 15602–15609.

Kang, J., Lemeiree, H.G., Utterbeck, A., Salbaum, J.M., Masters, C.L., et al., 1987. Nature 325, 733–736.

Kayed, R., Head, E., Thompson, J.L., McIntire, T.M., Milton, S.C., Cotman, C.W., Glabe, C.G., 2003. Common structure of soluble amyloid oligomers implies common mechanism of pathogenesis. Science 300, 486–489.

Kelly, J.W., 1998. Current Opinions in Structural Biology 8, 101–1065.

Klein, W.L., Krafft, G.A., Finch, C.E., 2001. Targeting small Aβ oligomers: on the solution to an Alzheimer's disease conundrum? Trends in Neuroscience 24, 219–224.

Koo, F.H., Lansbury Jr., P.T., Kelly, J.W., 1999. Amyloid diseases: abnormal protein aggregation in neurodegeneration. Proceedings National Academy of Sciences USA 96, 9989–9990.

Kotarek, J.A., Johnson, K.C., Moss, M.A., 2008. Analytical Biochemistry 378, 15–24.

Lashuel, H.A., Hartley, D., Petre, B.M., Walz, T., Lansbury Jr., P.T., 2002. Neurodegenerative disease: amyloid pores from pathogenic mutations. Nature 418, 291.

Lee, H., Sun, E., Han, D., Weissloder, R., 2008. Nature Medicine 14, 869–874.

Lemere, C.A., Spooner, E.T., LaFrancois, J., Malester, B., Mori, C., Leverone, J.F., Matsuoka, Y., Taylor, J.W., DeMattos, R.B., Holtzman, D.M., Clements, J.D., Selkoe, D.J., Duff, K.E., 2003. Evidence for peripheral clearance of cerebral Aβ protein following chronic, active Aβ immunization in PSAPP mice. Neurobiological Disorders 14, 10–18.

Lenigk, R., Lam, E., Lai, A., Wang, H., Han, Y., Carlier, P., 2000. Enzyme biosensor for studying therapeutics of Alzheimer's disease. Biosensors & Bioelectronics 15, 541–547.

Lesne, S., Koh, M.T., Kotilinek, L., Kayed, R., Giabe, C.G., Yang, A., Gallagher, M., Ashe, K.H., 2006. A specific amyloid-β protein assembly in the brain impairs memory. Nature 440, 352–357.

LeVine, H., 2002. The challenge of inhibiting Aβ polymerization. Current Medical Chemistry 9, 1121–1133.

Liu, R., McAllister, C., Lyubchenko, Y., Sierks, M.R., July 16, 2004. Proteolytic antibody light chains alter β-amyloid aggregation and prevent cytotoxicity. Biochemistry 43 (31), 9066–9075.

Lomakin, A., Chang, D.S., Benedek, G.B., Kirschner, D.A., Teplow, D.B., 1996. Proceedings of the National Academy of Science, USA 93, 1125–1129.

Martin, S.T., Granshoff, V.E., Frye, G.C., 1991. Effect of surface roughness on the response of thickness-shear mode resonators in liquids. Analytical Chemistry 65, 2910–2922.

Marty, J.L., Sode, Karube, I., 1992. Biosensor for detection of organophosphate and carbamate pesticides. Electroanalysis 4, 240–252.

Masters, C.L., Multhaup, G., Simms, G., Pottgiesser, J., Martins, R.N., Beyreuther, K., 1985. Neuronal origin of a cerebral amyloid: neurofibrillary tangles of Alzheimer's disease contain the same protein as the amyloid of plaque cores and blood vessels. The EMBO Journal 4 (11), 2757–2763.

Mimori, Y., Nakamura, S., Yukawa, M., 1997. Abnormalities of acetylcholinesterase in Alzheimer's disease with special reference to effect of acetylcholinesterase inhibitor. Behavorial Brian Research 83, 25–30.

Mukherjee, J., Kirchoff, J., 2009. Electrocatalytic microelectro detectors fro choline and acetylcholine following separation by electrophoresis. Analytical Chemistry 6996–7002.

Mustafa, M.K., Nabok, A., Parkinson, D., Tothill, I.E., Salam, F., Tsargorodskaya, A., 2010. Detection of β-amyloid peptide (1-16) and amyloid precursor protein (APP770) using spectroscopic ellipsometry and QCM techniques: a step forward towards Alzheimer's disease diagnostics. Biosensors & Bioelectronics 26, 1332–1336.

Myszka, D.G., Wood, S.J., Biere, A.L., 1999. Methods in Enzymology 309, 386–402.

Nabok, A., Tsargorodskaya, A., Holloway, A., Starodub, N.F., Gojster, O., 2007a. Biosensors & Bioelectronics 22, 885–890.

Nabok, A., Tsargorodskaya, A., Holloway, A., Starodub, N.F., Demchenko, 2007b. Langmuir 23, 8485–8490.

Nabok, A., Tsargorodskaya, A., 2008. Thin Solid Films 516, 8993–9001.

Naiki, H., Gejyo, F., Narakuki, K., 1997. Biochemistry 36, 6243–6250.

Naiki, H., Hasegawa, K., Yamaguchi, I., Nakamura, H., Gejyo, F., Nakakuki, K., 1998. Biochemistry 37, 17882–17889.

Naiki, H., Nakakuki, K., 1996. Laboratory Investigation 74, 374–383.

The National Institute of Aging, US NIH, 2011, http://www.nia.nih.gov/Alzheimer's/Publicatios/adfact.htm.

Nilsberth, C., Westlind-Danielsson, A., Eckman, C.B., Condron, M.M., Axelman, K., Forsell, C., Stenh, C., Luthman, J., Teplow, D.B., Younkin, S.G., Naslund, J., Lannfelt, L., 2001. The "Arctic" APP mutation (693G) causes Alzheimer's disease by enhanced Aβ protofibril formation. Nature Neuroscience 4, 887–893.

Noinvliie, S., Chich, J.F., Rezai, H., 2008. Veternarian Research 39 (4), 48.

Okuno, H., Mori, K., Okada, T., Yokoyama, Y., Suzuki, H., 2007. Chemistry and Biology of Drug Description 69, 356–361.

Ono, K., Hasegawa, K., Yoshiike, Y., Takashima, A., Yamada, M., Naiki, H., 2002. Journal of Neurochemistry 81, 434–440.

Pachetti, I., Cagini, A., Del Carlo, M., Coppi, C., Mascim, M., Turner, A.P.F., 1997. Determination of acetylcholinesterase pesticides in real samples using a disposable biosensor. Analytica Chimica Acta 337, 315–321.

Paravastu, A.K., Leapman, R.D., Yau, W.M., Tycko, R., 2008. Molecular structural basis for polymorphism in Alzheimer's β-amyloid fibrils. Proceedings National Academy of Sciences USA 105, 18349–18354.

Petkova, A.T., Leapman, R.D., Guo, Z., Yau, W.M., Mattson, M.P., Tyco, R., 2005. Self-propagating, molecular-level polymorphism in Alzheiemr's β-amyloid fibrils. Science 307, 262–265.

Pruisner, S.B., McKinley, M.P., Bowman, K.A., Bolton, D.C., Bendheim, P.E., Groth, D.F., Glenner, G.C., 1983. Cell 35 (2 Pt 1), 349–358.

Ramakrishnan, M., Kandimalla, K.K., Wengenack, T.M., Howll, K.G., Poduslo, J.F., 2009. Surface plasmon resonance binding kinetics of Alzheimer's disease amyloid β peptide capturing and plaque-binding monoclonal antibodies. Biochemistry 48, 10405–10415.

PubMed Health, July 2011. Journal of Medical Library Association 99 (3), 265–266. http://dx.doi.org/10.3163/1536-5050.99.3.018.

Research and Markets, 2011. http://www.researchandmarkets.com/research1e52657/2011_technology/.

Ringman, J.M., Younkin, S.G., Pratco, D., Seltzer, W., Cole, G.M., Sokolow, S., Rosario, E.R., Gylys, K.H., Varpetian, A., Medina, L.D., Cummins, J.L., 2008. Biochemical markers in persons with preclinical familial Alzheimer's disease. Neurology 71, 85–92.

Robbins, T.W., McAlonan, G., Muir, U., Everitt, B.J., 1997. Cognitive enhancers in theory and practice: studies of the cholinergic hypothesis of cognitive deficits in Alzheimer's disease. Behaviorial Brain Research 83, 15–23.

Salamon, J., 2011. MRI Scans Assist in the Prediction of Alzheimer's Risk. HealthDay downloaded April 6, 2010.

Schenk, D., 2002. Amyloid-beta immunotherapy for Alzheimer's disease: the end of the beginning. Nature Reviews in Neuroscience 3, 824–828.

Schoonenboom, N.S.M., Pijnenburg, Y.A.L., Muler, C., Rosso, S.M., Van Elk, G.J., Van Kamp, G.C., Van Swelten, J.C., Scheltens, P., 2004. Neurology 62 (9), 1580–1584.

Selkoe, D.J., 2001. Alzheimer's disease: genes, proteins, and therapy. Physiological Reviews 81, 741–766.

Selkoe, D.J., 1999. Translating cell biology into therapeutic advances in Alzheimer's disease. Nature 399, A23–A31.

Seubert, P., Barbour, R., Khan, K., Motter, R., Tang, P., Kholodenko, D., King, K., Schenk, D., Johnson-Wood, K., Schroeter, S., Gill, D., Jacobsen, J.S., Pangalos, M., Basi, G., Games, D., 2008. Antibody capture of soluble Aβ does not reduce cortical Aβ amyloidosis in the PDAPP mouse. Neurodegenrative Diseases 5, 65–71.

Shankar, G.M., Li, S., Mehta, T.H., Garcia-Munoz, A., Shepardin, N.E., Smith, I., Brett, F.M., Farell, M.A., Rowan, M.J., Lemere, C.A., Regan, C.M., Walsh, D.M., Sabintini, B.I., Selko, D.J., 2008. Amyloid β-protein dimmers isolated directly from Alzheimer's brains impair synergistic plasticity and memory. Nature Medicine 14, 837–842.

Shoji, M., Golde, T.E., Ghiso, J., Cheung, T.T., Estus, S., Shaffer, L.M., Cai, X.D., McKay, D.M., Tintner, R., Frangione, B., et al., 2 October 1992. Production of the Alzheimer amyloid beta protein by normal proteolytic processing Science 258 (5079), 126–129.

Simonian, A.L., Rainina, E.I., Wild, J.R., 1997. A new approach for discriminative detection of organophosphate neurotoxins in the presence of other cholinesterase inhibitors. Analytical Letters 30, 2453–2468.

Solomon, B., Koppel, R., Frankel, D., Hanan-Ahron, E., 1997. Disaggregation of Alzheimer β-amyloid by site-directed mAb. Proceedings National Academy of Sciences USA 94, 4109–4112.

Sommer, B., 2002. Current Opinions in Pharmacology 2, 87–92.

Steenhuysen, 2011. http://www.vancouversun.com/health/Alzheimer + dry + hits + forms + toxic + protein +, downloaded March 11, 2011.

Steunou, S., Chich, J.-F., Rezai, H., Vidic, J., 2010. Biosensing of lipid-prion interactions: insights on charge effects, Cu(II)-ions binding and prion oligomerization. Biosensors & Bioelectronics 26, 1399−1406.

Thinakaran, G., Kao, E.H., 31 October 2008. Amyloid precursor protein trafficking, processing, and function. Journal of Biological Chemistry 283 (44).

Tyco, R., 2006. Molecular structure of amyloid fibrils. Insights from solid-state NMR. Quarterly Reviews in Biophysics 39, 1−55.

Valenti, P., Rampa, A., Bisi, A., Andrsano, V., Cavrini, V., Fin, L., Buriani, A., Giusti, P., 1997. Acetylcholinesterase inhibition by tacrine analogues. Bioorganic Medical Chemistry Letters 7, 2599−2602.

Walsh, D.M., Hartley, D.M., Kusumoto, Y., Fezoui, Y., Condron, M.M., Lomakin, A., Benedek, G.B., Selkoe, D.J., Teplow, D.B., 1999. Amyloid beta-protein fibrillogenesis. Structure and biological activity of protofibrillar intermediates. Journal of Biological Chemistry 274, 25945−25952.

Westpal, P., Bornmann, A., 2002. Sensors & Actuators B: Chemical 84 (2), 278−282.

Xia, W., Yang, T., Smith, I.M., Shen, Y., Walsh, D.M., Selkoe, D.J., 2009. A specific ELISA for measuring amyloid β-protein oligomers in human plasma and the brains of Alzheimer patients. Archives Neurology 66 (2), 190−199.

Xia, W., Yang, T., Shankar, G., Smith, I.M., Shen, Y., Walsh, D.M., Selkoe, D.J., 2010. A specific enzyme-linked immunosorbent assay for measuring beta-amyloid protein oligomers in human plasma and brain tissue of patients with Alzheimer disease. Archives of Neurology 66, 190−199.

Zhao, H., Werringloer, D., Marambaud, P., 2008. Bio Med Central Research Notes 1, 38.

Chapter 10

Detection of Biomarkers for Different Diseases on Biosensor Surfaces

10.1 INTRODUCTION

Wild et al. have recently analyzed the prevalence of diabetes, and made projections for the year 2030. The biggest commercial market for biosensors is for diabetes mellitus. These authors analyzed the data from a few countries based on age and sex. They considered urban and rural populations separately for developing countries. Their analysis indicates that it is higher in men than women, though there are more women with diabetes than men. Their results also indicate that there is an increase in diabetes with individuals over 65 years of age and also that the diabetic epidemic will continue to increase even if the levels of obesity remains the same. King et al. (1998) as early as in the year 1998 indicated the global burden of diabetes. They had indicated the prevalence, numerical estimates, and projections for the year 2010. Amos et al. (1997) have also indicated the rising global burden of diabetes and its complications. They provide the estimates and projections for the year 2010.

Malic et al. (2011) indicate that genomics and proteonomics have revealed a number of disease biomarkers that may be used for diagnostic and prognostic indicators of infectious diseases. The current trend for biosensors to be rapid, label-free and measurements in real time has led to the sensitive detection of biomarkers. The SPRi (surface plasmon resonance imaging) technique is ideally suited for the detection of biomarkers. However, it could be improved with regard to an increase in sensitivity. Malic et al. (2011) have developed a digital microfluidic platform with an integrated nanostructured biosensor interface that permits the sensitive and rapid detection of DNA hybridization. The authors indicate that the novel microfluidic and nanotechnology has led to an increase in SPRi detection capabilities. This is especially useful for the detection of biomarkers in low concentrations. The authors emphasize that their proposed platform is very versatile and allows for the manipulation of multiple samples which is very useful for diagnostic applications.

Biomarkers and Biosensors. http://dx.doi.org/10.1016/B978-0-444-53794-2.00010-0

10.2 THEORY

Havlin (1987) has reviewed and analyzed the diffusion of reactants toward fractal surfaces. The details of the theory and the equations involved for the binding and the dissociation phases for analyte–receptor binding are available (Sadana, 2001). The details are not repeated here, except that just the equations are given to permit an easier reading. These equations have been applied to other biosensor systems (Sadana, 2001; Ramakrishnan and Sadana, 2001; Sadana, 2005). For most applications, a single- or a dual-fractal analysis is often adequate to describe the binding and the dissociation kinetics. Peculiarities in the values of the binding and the dissociation rate coefficients, as well as in the values of the fractal dimensions with regard to the dilute analyte systems being analyzed will be carefully noted, if applicable.

10.2.1 Single-Fractal Analysis

10.2.1.1 Binding Rate Coefficient

Havlin (1989) indicates that the diffusion of a particle (analyte [Ag]) from a homogeneous solution to a solid surface (e.g., receptor [Ab]-coated surface) on which it reacts to form a product (analyte–receptor complex (Ab.Ag)) is given by:

$$(\mathrm{Ab\cdot Ag}) \approx \begin{cases} t^{(3-D_{f,bind})/2} = t^{p} & t < t_c \\ t^{1/2} & t > t_c \end{cases} \tag{10.1}$$

Here $D_{f,bind}$ or D_f is the fractal dimension of the surface during the binding step. t_c is the cross-over value. Havlin (1989) indicates that the cross-over value may be determined by $r_c^2 \sim t_c$. Above the characteristic length, r_c, the self-similarity of the surface is lost and the surface may be considered homogeneous. Above time, t_c of the surface may be considered homogeneous, since the self-similarity property disappears, and "regular" diffusion is now present. For a homogeneous surface where D_f is equal to 2, and when only diffusional limitations are present, $p = ½$ as it should be. Another way of looking at the $p = ½$ case (where $D_{f,bind}$ is equal to 2) is that the analyte in solution views the fractal object, in our case, the receptor-coated biosensor surface, from a "large distance." In essence, in the association process, the diffusion of the analyte from the solution to the receptor surface creates a depletion layer of width $(Đ_t)^{½}$ where Đ is the diffusion constant. This gives rise to the fractal power law, (Analyte.Receptor) $\sim t^{(3-D_{f,diss})/2}$. For the present analysis, t_c is arbitrarily chosen and we assume that the value of t_c is not reached. One may consider the approach as an intermediate "heuristic" approach that may be used in the future to develop an autonomous (and not time-dependent) model for diffusion-controlled kinetics.

10.2.1.2 Dissociation Rate Coefficient

The diffusion of the dissociated particle (analyte [Ag]) from the solid surface (e.g., analyte [Ag]—receptor [Ab] complex coated surface) into solution may be given, as a first approximation by:

$$(\mathrm{Ab \cdot Ag}) \approx -t^{(3-D_{f,diss})/2} = -t_p \quad (t > t_{diss}) \tag{10.2}$$

Here $D_{f,diss}$ is the fractal dimension of the surface for the dissociation step. This corresponds to the highest concentration of the analyte—receptor complex on the surface. Henceforth, its concentration only decreases. The dissociation kinetics may be analyzed in a manner “similar” to the binding kinetics.

10.2.2 Dual-Fractal Analysis

10.2.2.1 Binding Rate Coefficient

Sometimes, the binding curve exhibits complexities and two parameters (k, D_f) are not sufficient to adequately describe the binding kinetics. This is further corroborated by low values of r^2 factor (goodness-of-fit). In that case, one resorts to a dual-fractal analysis (four parameters; k_1, k_2, D_{f1}, and D_{f2}) to adequately describe the binding kinetics. The single-fractal analysis presented above is thus extended to include two fractal dimensions. At present, the time ($t = t_1$) at which the “first” fractal dimension “changes” to the “second” fractal dimension is arbitrary and empirical. For the most part, it is dictated by the data analyzed and experience gained by handling a single-fractal analysis. A smoother curve is obtained in the “transition” region, if care is taken to select the correct number of points for the two regions. In this case, the product (antibody—antigen or analyte—receptor complex, Ab.Ag or analyte.receptor) is given by:

$$(\mathrm{Ab \cdot Ag}) \approx \begin{cases} t^{(3-D_{f1,bind})/2} = t^{p1} & (t < t_1) \\ t^{(3-D_{f2,bind})/2} = t^{p2} & (t_1 < t < t_2) = t_c \\ t^{1/2} & (t > t_c) \end{cases} \tag{10.3}$$

In some cases, as mentioned above, a triple-fractal analysis with six parameters (k_1, k_2, k_3, D_{f1}, D_{f2}, and D_{f3}) may be required to adequately model the binding kinetics. This is when the binding curve exhibits convolutions and complexities in its shape perhaps due to the very dilute nature of the analyte (in some of the cases to be presented) or for some other reasons. Also, in some cases, a dual-fractal analysis may be required to describe the dissociation kinetics.

10.3 RESULTS

Noah et al. (2011) have developed a suite of biosensors comprising of electrochemical, fluorescence, and SPR biosensors for detecting inducible nitric oxide synthase. Nitric oxide (NO) is a highly reactive free radical species, which is an important intra- and intercellular signaling molecule that is used in the regulation of a variety of physiological and pathophysiological mechanisms such as cardiovascular, nervous, and immunological systems (Aklan, 2004; Cho et al., 2004). Noah et al. (2011) state that the overproduction of NO leads to it being a major cytotoxic mediator in pathological processes. It plays a major role in inflammatory processes. Tannous et al. (2001) indicate that excess NO is known to interact with oxygen radicals and to form reactive peroxynitrite. This induces inflammatory cellular cytokines. These lead to cell death via apoptosis and necrosis.

Noah et al. (2011) state that inducible (iNOS) is expressed in microphages and other tissues in response to infection of inflammation (Coleman, 2001). Franco et al. (1994) indicate that it is expressed after an exposure to diverse stimuli such as inflammatory cytokines and lipopolysaccharides (LPSs). Nathan and Xie (1994) indicate that once iNOS is expressed it generates NO in the blood. This leads to pathological effects (Franco et al., 1994).

Previous approaches have attempted to detect these pathological processes. For example, refer Njagi et al. (2010). Noah et al. (2011) proposed to monitor iNOS as a viable surrogate for the short-lived species. The goal was to develop and analyze the three biosensors to detect iNOS using electrochemical, fluorescence, and SPR principles. They developed a suite of advanced biosensors to detect iNOS.

Figure 10.1(a) shows the binding of 540 ng/ml of iNOS in solution to an SPR biosensor surface. A single-fractal analysis is adequate to describe the

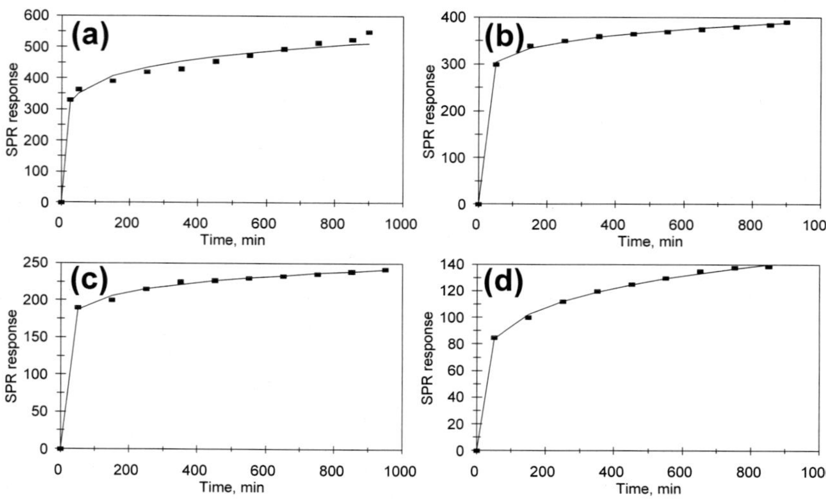

FIGURE 10.1 Binding of different concentrations of iNOS in solution to a surface plasmon resonance (SPR) biosensor surface (Noah et al., 2011): (a) 540 ng/ml (b) 270 ng/ml (c) 135 ng/ml (d) 67.5 ng/ml.

TABLE 10.1 Binding Rate Coefficients and Fractal Dimensions for Different Concentrations of NOS to iNOS Antibody Immobilized on a Surface Plasmon Resonance (SPR) Biosensor Surface (Noah et al., 2011)

Analyte in Solution/Receptor on Surface	k	D_f
540 ng/ml NOS/iNOS antibody immobilized on SPR biosensor surface	210.55 ± 8.82	2.7364 ± 0.0268
270 ng/ml NOS/iNOS antibody immobilized on SPR biosensor surface	218.68 ± 2.11	2.8316 ± 0.0074
135 ng/ml NOS/iNOS antibody immobilized on SPR biosensor surface	134.10 ± 1.74	2.8269 ± 0.00936
67.5 ng/ml NOS/iNOS antibody immobilized on SPR biosensor surface	41.257 ± 0.497	2.6370 ± 0.00930

binding kinetics. The values of the binding rate coefficient, k and the fractal dimension, D_f for a single-fractal analysis are given in Table 10.1.

Figure 10.1(b) shows the binding of 270 ng/ml of iNOS in solution to the SPR biosensor surface. A single-fractal analysis is, once again, adequate to describe the binding kinetics. The values of the binding rate coefficient, k and the fractal dimension, D_f for a single-fractal analysis are given in Table 10.1.

Figure 10.1(c) shows the binding of 135 ng/ml of iNOS in solution to the SPR biosensor surface. A single-fractal analysis is, once again, adequate to describe the binding kinetics. The values of the binding rate coefficient, k and the fractal dimension, D_f for a single-fractal analysis are given in Table 10.1.

Figure 10.1(d) shows the binding of 67.5 ng/ml of iNOS in solution to the SPR biosensor surface. A single-fractal analysis is, once again, adequate to describe the binding kinetics. The values of the binding rate coefficient, k and the fractal dimension, D_f for a single-fractal analysis are given in Table 10.1.

Figure 10.2(a) shows the increase in the binding rate coefficient, k for a single-fractal analysis with an increase in the COX-2 concentration in solution. The binding rate coefficient, k is given by:

$$k = (12.803 + 16.416)[\text{iNOS}]^{0.4358+0.5324} \quad (10.4)$$

The fit is reasonable. Only four data points are available. The availability of more data points would lead to a more reliable fit. The binding rate coefficient,

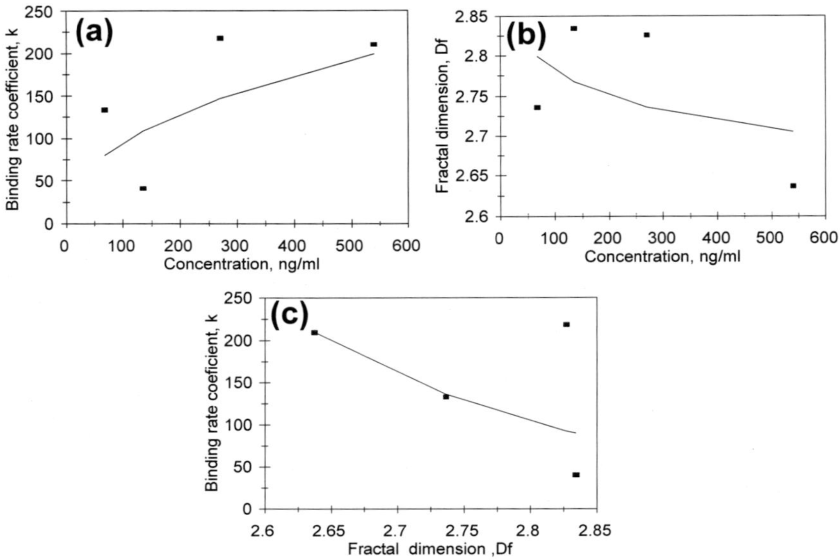

FIGURE 10.2 (a) Increase in the binding rate coefficient, k with an iNOS concentration in solution (b) Decrease in the fractal dimension, D_f with an increase in the iNOS concentration in solution (c) Decrease in the binding rate coefficient, k with an increase in the fractal dimension, D_f.

k exhibits less than one-half (equal to 0.4354) order of dependence on the iNOS concentration in solution. This indicates that the binding rate coefficient, k is sensitive to the iNOS concentration in solution.

Figure 10.2(b) shows the decrease in the fractal dimension, D_f for a single-fractal analysis with an increase in the iNOS concentration in solution. The fractal dimension, D_f is given by:

$$D_f = (3.00 + 0.12)[COX - 2]^{-0.1644 \pm 0.024} \tag{10.5}$$

The fit is reasonable. Only four data points are available. The availability of more data points would lead to a more reliable fit. The fractal dimension, D_f exhibits close to a zero-order (equal to −0.01644) order of dependence on the iNOS concentration in solution. This indicates that the fractal dimension, D_f is only mildly sensitive to the iNOS concentration in solution.

Figure 10.2(c) shows the decrease in the binding rate coefficient, k with an increase in the fractal dimension, D_f for a single-fractal analysis. The binding rate coefficient, k is given by:

$$k = (1.9E + 07 + 2.5E + 07 + 0.12)[D_f]^{-11.768 + 14.01} \tag{10.6}$$

The fit is reasonable. Only four data points are available. The availability of more data points would lead to a more reliable fit. The binding rate coefficient, k exhibits a negative (equal to −11.768) order of dependence on the fractal dimension, D_f or the degree of heterogeneity on the biosensor surface. This indicates that the binding rate coefficient, k is very sensitive to the degree of heterogeneity on the biosensor surface.

Ladd et al. (2009) have developed a direct detection method for carcinoembryonic antigen autoantibodies in clinical human serum using an SPR biosensor. As mentioned earlier throughout the book the SPR technique is a quantitative method which allows for real-time, label-free detection of interactions occurring at interfaces. The authors indicate that the SPR method has been applied to medical diagnostics (Cherif et al., 2006; Thaler et al., 2005; Englebienne et al., 2003). Ladd et al. (2009) emphasize that few studies on SPR in clinically relevant matrices have been previously reported.

It is rather well known that the earlier one is able to detect cancer, the better the prognosis is (Draghici et al., 2005; Jacobs and Menn, 2004). Simply speaking, on cancer progress, the treatment methods change and the effectiveness decreases. Cancers present aberrant protein, or an increasing number of proteins. Cancer progression and type can lead to varying in vivo concentrations (Wang et al., 2005). Thus, it is essential to detect cancer at an early stage. One technique relies on the detection of immune responses to antigen (Kinazawa et al., 1982; Sahin et al., 1995; Zhong et al., 2003; Lechpamme et al., 2004; Gadducci et al., 1999).

emphasize that the detection of serum antibody response to tumor antigens may be a more reliable method for the detection of the antigen itself. Scanian et al. (1989) indicate that serum antibodies are more stable than the antigens particularly during the early stages of the disease (Disis et al., 1997, 1994).

Carcinoembryonic antigen (CEA) is a tumor-associated antigen and has been identified as a biomarker for colorectal, gastric, and pancreatic cancer (Aquino et al., 2004). Besides, CEA is a protein that is expressed in normal cells and overexpressed in adenocarcinoma. Thus, elevated levels of CEA can be used as a biomarker for cancer, and is a useful biomarker for early cancer detection, as well as diagnosis of the disease progression. Ladd et al. (2009) have developed an assay for differentiating anti-CEA levels in clinical and colorectal serum samples using an SPR sensor.

The kinetic analysis for the binding of anti-CEA autoantibody in a sandwich assay in solution to CEA immobilized on an SPR biosensor surface. A single-fractal analysis is adequate to describe the binding and the dissociation kinetics. The values of the binding rate coefficient, k and the dissociation rate coefficient k_d, and the fractal dimension, D_f for binding and the fractal dimension for dissociation, D_{fd} for a first-order analysis are given in Tables 10.2 and 10.3.

Figure 10.3(a) shows the binding of anti-CEA in serum to a functionalized CEA surface. A dual-fractal analysis is required to adequately describe the binding kinetics. The values of (a) the binding rate coefficient, k and the fractal

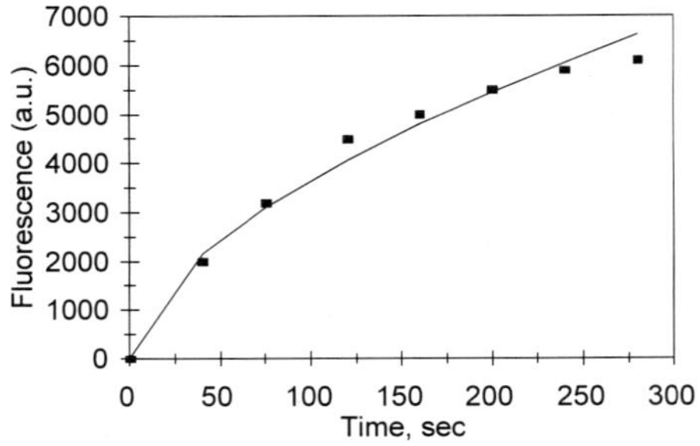

FIGURE 10.3 Binding of anticarcinoembryonic antigen (anti-CEA) in a sandwich assay in solution to CEA immobilized on a surface plasmon resonance (SPR) biosensor surface.

TABLE 10.2 Binding Rate Coefficient, k and the Dissociation Rate Coefficient, k_d for Anticarcinoembryonic Antigen (Anti-CEA) in Solution to CEA Autoantibody Immobilized on the Surface Plasmon Resonance (SPR) Biosensor Surface

Analyte in Solution/Receptor on SPR Biosensor Surface	k	k_d
Anti-CEA/CEA on SPR surface	2.197 ± 0.180	1.3496 ± 0.0672

TABLE 10.3 Fractal Dimensions for the Binding and the Dissociation Phase for Anticarcinoembryonic Antigen (Anti-CEA) in Solution to CEA Autoantibody Immobilized on the Surface Plasmon Resonance (SPR) Biosensor Surface

Analyte in Solution/Receptor on SPR Biosensor Surface	D_f	D_{fd}
Anti-CEA/CEA on SPR surface	1.3946 ± 0.0672	0 + 0.6582

dimension, D_f for a single-fractal analysis, and for a dual-fractal analysis are given in Tables 10.4 and 10.5.

Figure 10.4(a) shows the binding of 74.6 ng/ml of COX-2 in solution to goat anti-COX-2 antibody immobilized on an SPR biosensor surface. A single-fractal analysis is adequate to describe the binding kinetics. The values of the

TABLE 10.4 Binding Rate Coefficients for Anti-CEA Autoantibodies in Serum to a CEA-Functionalized Surface Plasmon Resonance (SPR) Biosensor Surface (Ladd et al., 2009)

Analyte in Serum/ Antibody on Surface	k	k_1	k_2
16 ng/ml anti-CEA autoantibodies/CEA on SPR surface	2.745 ± 0.702	1.7886 ± 0.2756	8.819 ± 0.123

TABLE 10.5 Fractal Dimensions for Binding for Anti-CEA Autoantibodies in Serum to a CEA-Functionalized Surface Plasmon Resonance (SPR) Biosensor Surface (Ladd et al., 2009)

Analyte in Serum/ Antibody on Surface	D_f	D_{f1}	D_{f2}
16 ng/ml anti-CEA autoantibodies/CEA on SPR surface	2.0616 ± 0.1298	1.5720 ± 0.01872	2.7832 ± 0.03644

binding rate coefficient, k and the fractal dimension, D_f for a single-fractal analysis are given in Table 10.4.

Figure 10.4(b) shows the binding of 7.46 ng/ml of COX-2 in solution to goat anti-COX-2 antibody immobilized on an SPR biosensor surface. A single-fractal analysis is, once again, adequate to describe the binding kinetics. The values of the binding rate coefficient, k and the fractal dimension, D_f for a single-fractal analysis are given in Table 10.4.

Figure 10.4(c) shows the binding of 0.746 ng/ml of COX-2 in solution to goat anti-COX-2 antibody immobilized on an SPR biosensor surface. A single-fractal analysis is, once again, adequate to describe the binding kinetics. The values of the binding rate coefficient, k and the fractal dimension, D_f for a single-fractal analysis are given in Table 10.4.

Figure 10.4(d) shows the binding of 0.0746 ng/ml of COX-2 in solution to goat anti-COX-2 antibody immobilized on an SPR biosensor surface. A single-fractal analysis is, once again, adequate to describe the binding kinetics. The values of the binding rate coefficient, k and the fractal dimension, D_f for a single-fractal analysis are given in Table 10.4.

Figure 10.4(e) shows the binding of 0.00746 ng/ml of COX-2 in solution to goat anti-COX-2 antibody immobilized on an SPR biosensor surface.

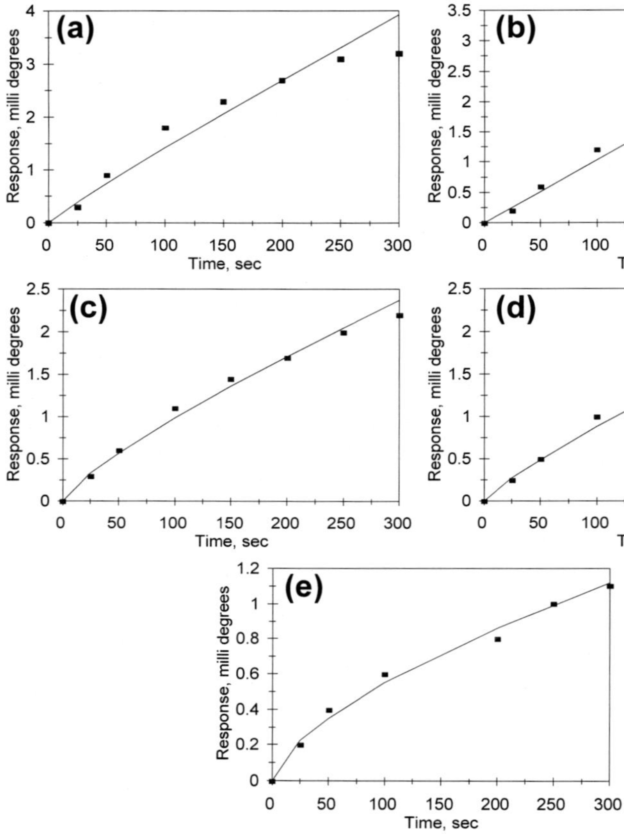

FIGURE 10.4 Binding of different concentrations of COX-2 in solution to goat anti-COX-2 antibody immobilized on a surface plasmon resonance (SPR) biosensor surface: (a) 74.6 ng/ml (b) 7.46 ng/ml (c) 0.746 ng/ml (d) 0.0746 ng/ml (e) 0.00746 ng/ml.

A single-fractal analysis is, once again, adequate to describe the binding kinetics. The values of the binding rate coefficient, k and the fractal dimension, D_f for a single-fractal analysis are given in Table 10.4.

Figure 10.5(a) shows the increase in the binding rate coefficient, k with an increase in the fractal dimension, D_f. For the data shown in Figure 10.5(a) the binding rate coefficient, k is given by:

$$k = (0.01199 \pm 0.002\,84) D_f^{1.808 \pm 0.486}$$

The fit is good. Only five data points are available. The availability of more data points would lead to a more reliable fit. The binding rate coefficient exhibits an order of dependence between one and one-half and second (equal to 1.808) order on the fractal dimension or the degree of heterogeneity on the biosensor surface.

TABLE 10.6 Binding Rate Coefficients and Fractal Dimensions for COX-2 in Solution to Anti-COX Antibody Immobilized on a Surface Plasmon resonance (SPR) Biosensor Surface (Noah et al., 2011)

Analyte in Solution/Receptor on Surface	k	D_f
74.6 ng/ml COX-2/goat anti-COX-2 immobilized on SPR biosensor surface	0.0198 ± 0.0047	1.1446 ± 0.1896
7.46 ng/ml COX-2/goat anti-COX-2 immobilized on SPR biosensor surface	0.0092 ± 0.008	0.9528 ± 0.1528
0.746 ng/ml COX-2/goat anti-COX-2 immobilized on SPR biosensor surface	0.254 ± 0.00421	1.4092 ± 0.0718
0.0746 ng/ml COX-2/goat anti-COX-2 immobilized on SPR biosensor surface	0.0179 ± 0.0018	1.3092 ± 0.0870
0.00746 ng/ml COX-2/goat anti-COX-2 immobilized on SPR biosensor surface	0.0274 ± 0.0037	2.3528 ± 0.0527

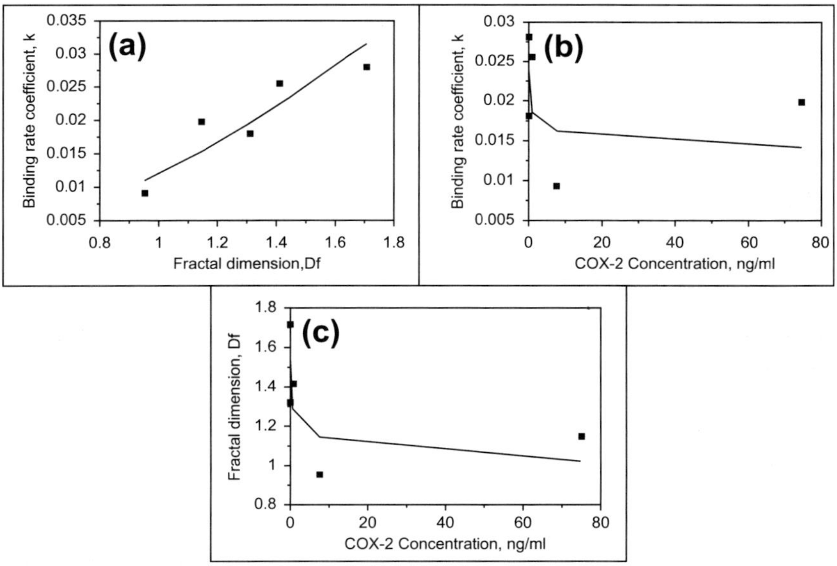

FIGURE 10.5 (a) Increase in the binding rate coefficient, k with an increase in the fractal dimension, D_f (b) Increase in the binding rate coefficient, k with an increase in the COX-2 concentration in solution (c) Decrease in the fractal dimension, D_f with an increase in the COX-2 concentration in solution.

Figure 10.5(b) shows the increase in the binding rate coefficient, k with an increase in the COX-2 concentration in solution. For the data shown in Figure 10.5(b) the binding rate coefficient, k is given by:

$$k = (0.01842 \pm 0.01017)[COX - 2]^{-0.0587 - 0.060}$$

The fit is poor. There is scatter in the data. Only five data points are available. The availability of more data points would lead to a more reliable fit. The binding rate coefficient exhibits almost a negative zero (equal to −0.0587) order of dependence on the COX-2 concentration in solution. This indicates that the binding rate coefficient, k is almost insensitive to the COX-2 concentration in solution.

Figure 10.5(c) shows the decrease in the fractal dimension, D_f with an increase in the COX-2 concentration in solution. For the data shown in Figure 10.5(c) the fractal dimension is given by:

$$D_f = 1.2619 \pm 0.205[COX - 2]^{-0.0483 \pm 0.0206}$$

The fit is poor. There is scatter in the data. Only five data points are available. The availability of more data points would lead to a more reliable fit. The fractal dimension, D_f exhibits close to a negative zero (equal to −0.04834) order of dependence on the COX-2 concentration in solution. This indicates that the fractal dimension, D_f is almost insensitive to the COX-2 concentration in solution.

Wolf et al. (2004) have recently analyzed the simultaneous detection of C-reactive protein and other cardiac markers in human plasma using micromosaic immunoassays. They indicate a particular challenge in the medical area is differentiating between diseases that exhibit similar symptoms. For example, myocardial infarction (MI) and pulmonary embolism both lead to chest pain in patients. Biomarkers specific for these diseases should aid medical personnel to distinguish between these diseases (McCord et al., 2001). Wolf et al. (2004) have provided a proof-of-concept device to rapidly screen for MI biomarkers in small volumes of samples using a simple method based on microfluidic networks. The authors indicate that microfluidic networks in which the liquid flow occurs due to capillary forces do have a self-regulating capability and are autonomous (Juncker et al., 2002).

Wolf et al. (2004) indicate that micromosiac immunoassays (μMIAs) are based on patterning lines of capture antibodies. These lines are exposed orthogonally to solutions containing the analytes (Bernard et al., 2001). At the intersection of these regions the captured analytes are detected with fluorescently labeled detection antibodies. The authors indicate that a micromosaic of fluorescent zones reveals the binding events. This combinatorial method of screening has been applied previously to detect bacterial antigens, and epitope antibody binding, or DNA.

Wolf et al. (2004) indicate that C - Reactive Protein (CRP) is a widely used biomarker for inflammation (Gabay and Kushner, 1999). CRP is a good

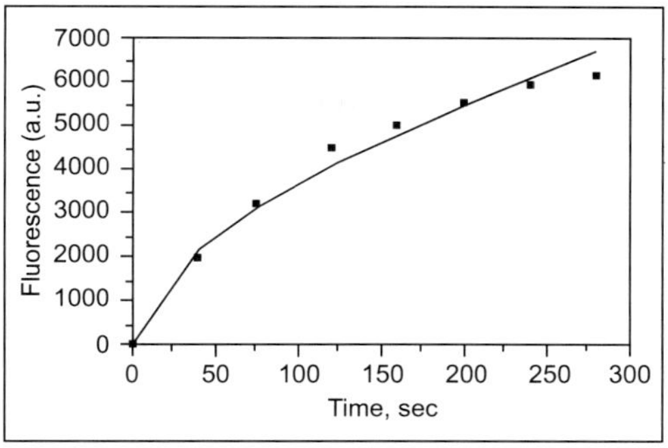

FIGURE 10.6 Binding of CRP in solution in a sandwich assay while emitting a fluorescent signal.

prognostic indicator of CHD (coronary heart disease) (Rifai and Ridker, 2001; Ridker et al., 2002). Wolf et al. (2004) indicate that CRP is a 144-kDa pentameric protein which is synthesized in the liver upon stimulation by IL-6 and IL-1β (Weinhold and Ruther, 1997). Wolf et al. (2004) have characterized the best conditions for performing the μMIA to detect CRP. They also determined the dynamic range of the assay as well as the quality of the micromosaic when CRP is detected.

Figure 10.6 shows the binding of CRP in solution in sandwich-type assay while emitting a fluorescent signal (Wolf et al., 2004). A single-fractal analysis is required to describe the binding kinetics. The values of the binding rate coefficient, k and the fractal dimension, D_f for a single-fractal analysis are given in Table 10.7.

Interleukin-6 (IL-6) is a multifunctional cytokine which acts, for example, on inflammation, growth regulation, and differentiation of various cell types (Chou et al., 2010). These authors indicate that this cytokine may be produced by lymphocytes, fibroblasts, endothelial cells, and keratinocytes in response to injury, infection, burns, and trauma. It is also produced in response to a number of stimulants including tumor necrosis factor alpha (TNF-α) and platelet-derived growth factor (PDGF) (Kishimoto, 1989; Tosato and Jones, 1999; Hellgott et al., 1987). Chou et al. (2010) indicate that elevated levels have been noted in most types of cancer (Lukaszewicz et al., 2007).

TABLE 10.7 Binding Rate Coefficient, k and Fractal Dimension, D_f for CRP in Solution to Anti-CRP in a Sandwich-Type Immunoassay (Wolf et al., 2004)

Analyte in Solution/Receptor on SPR Biosensor Surface	k	D_f
0.1 μg/ml CRP in PBS/anti-CRP on biosensor surface	258.93 ± 19.52	1.850 ± 0.0856

Chou et al. (2010) indicate that SPR biosensors have been used to detect analytes, and kinetics of binding (association), and dissociation (Homola, 2003; Green et al., 2000). Homola et al. (1999) indicate that SPR biosensors are highly sensitive, provide rapid detection without labeling, and are very accurate, precise, and robust. However, when the SPR biosensor is used in serum, blood plasma, and saliva they suffer from the effects of nonspecific binding which leads to a decrease in the analysis of analytes at low concentrations. Battaglia et al. (2005) emphasize that this is a major drawback in the application of SPR biosensors. Chou et al. (2010) have analyzed IL-6 secreted by human fibroblast MRC5-CVI cells induced by LPS using the SPR technique. They have used a sandwich-type immunoassay which was adapted in this study since the secondary antibody, which had to be bound to the primary antibody on the SPR sensor chip.

Andersson et al. (2010) have analyzed cardiac biomarkers for toxicity testing using biosensors and cardiomyocytes derived from human-derived embryonic cells (Andersson et al., 2008). There is need for cardiomyocytes for toxicity testing but where the excess of primary cells from the human heart is limited. Steel et al. (2009) indicate that human embryonic stem cells (hESC) differentiated into cardiomyocyte-like cells can replace the animal-derived cell lines currently used in toxicity testing.

hESC-derived cardiomyocytes have been used to assess drug-derived perturbations of cells' electrophysiological properties (Salin et al., 2013; Caspi et al., 2009). Andersson et al. (2010) indicate that these studies provide support to validate assays for these hESC-derived cardiomyocytes for safety assessments in pharmacology.

Dolci et al. (2008).indicate that established biomarkers of cardiac damage may permit large-scale cardiotoxicity screening. However, Andersson et al. (2010) indicate that they are issues with regard to predictivity, standardization, and validation which have still to be addressed carefully.

Troponin T (cTNT) and fatty acid binding protein 3 (FABP3 or FABP) are critical biomarkers for assaying adverse cardiac toxic responses of drugs. These biomarkers are clinical indicators of MI, but are also considered sensitive translational safety biomarkers for drug-reduced toxicity (Nakata et al., 2003; Key et al., 1999a,b) indicate that FABP3 levels are high during the early stages of MI.

Andersson et al. (2010) indicate that SPR biosensors have not been used to detect hESC-derived cells for developing in vitro toxicity assays. Andersson et al. (2010) have used an SPR biosensor to monitor cardiomyocytes. They have taken advantage of the high specificity achieved on the sensor chip which is coupled with ligands to permit the parallel monitoring of biomarkers on a miniaturized format.

Figure 10.7(a) shows the binding of FABP3 in solution to cTNT 7G7 antibody immobilized on an SPR biosensor surface. A dual-fractal analysis is required to adequately model the binding kinetics. The values of the binding rate coefficients, k_1 and k_2 and the fractal dimensions, D_{f1} and D_{f2} are given in Table 10.7. It is of interest to note that as the fractal dimension increases by a

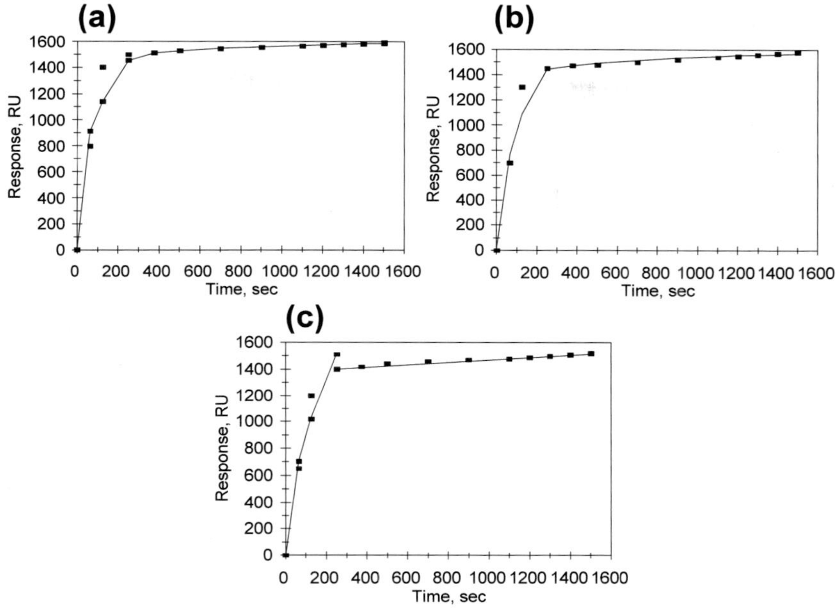

FIGURE 10.7 (a) Binding of fatty acid binding protein (FABP) in solution to cTNT 7G7 antibody immobilized on a surface plasmon resonance (SPR) biosensor surface (b) Binding of cTNT 7G7 in solution to cTNT 7G7 antibody immobilized on an SPR biosensor surface (c) Binding of cTNT lF11 in solution to cTNT 7G7 antibody immobilized on an SPR biosensor surface.

factor of 1.27 from a value of D_{f1} equal to 2.31 to a value of D_{f2} equal to 2.93, the binding rate coefficient increases by a factor of 5.71 from a value of k_1 equal to 216.72 to k_2 equal to 1238.37. An increase in the fractal dimension or the degree of heterogeneity on the SPR biosensor surface leads to an increase in the binding rate coefficient.

Figure 10.7(b) shows the binding of cTNT 7G7 in solution to cTNT 7G7 antibody immobilized on an SPR biosensor surface. A dual-fractal analysis is, once again, required to adequately model the binding kinetics. The values of the binding rate coefficients, k_1 and k_2 and the fractal dimensions, D_{f1} and D_{f2} are given in Table 10.7. It is of interest to note that as the fractal dimension increases by a factor of 1.53 from a value of D_{f1} equal to 1.9266 to a value of D_{f2} equal to 2.93, the binding rate coefficient increases by a factor of 13.56 from a value of k_1 equal to 81.57 to k_2 equal to 1106.15. An increase in the fractal dimension or the degree of heterogeneity on the SPR biosensor surface leads to an increase in the binding rate coefficient.

Figure 10.7(c) shows the binding of cTNT lF11 in solution to cTNT 7G7 antibody immobilized on an SPR biosensor surface. A dual-fractal analysis is, once again, required to adequately model the binding kinetics. The values of the binding rate coefficients, k_1 and k_2 and the fractal dimensions, D_{f1} and D_{f2} are given in Table 10.6. It is of interest to note that as the fractal dimension increases by a factor of 1.56 from a value of D_{f1} equal to 1.8678 to a value of

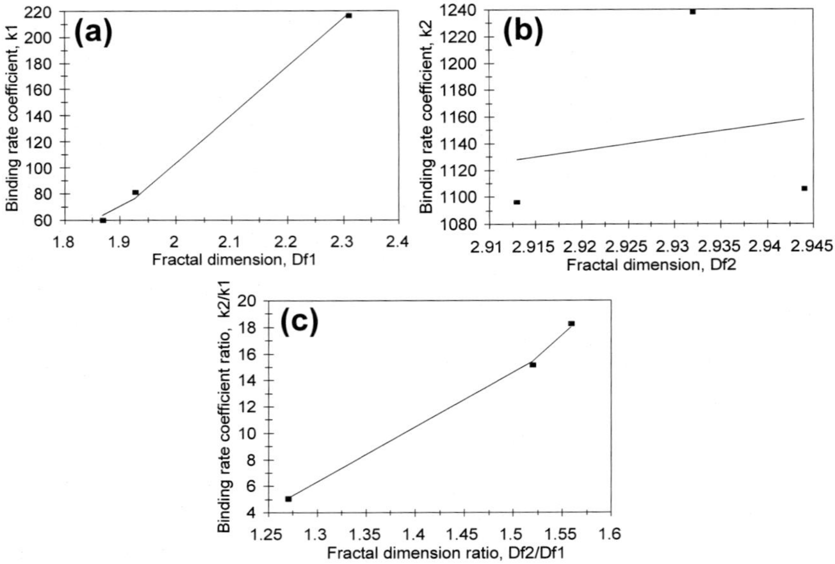

FIGURE 10.8 (a) Increase in the binding rate coefficient, k_1 with an increase in the fractal dimension, D_{f1} (b) Increase in the binding rate coefficient, k_2 with an increase in the fractal dimension, D_{f2} (c) Increase in the binding rate coefficient ratio, k_2/k_1 with an increase in the fractal dimension ratio, D_{f2}/D_{f1}.

D_{f2} equal to 2.913, the binding rate coefficient increases by a factor of 18.15 from a value of k_1 equal to 60.44 to k_2 equal to 1096.85. An increase in the fractal dimension or the degree of heterogeneity on the SPR biosensor surface leads to an increase in the binding rate coefficient.

Figure 10.8(a) shows an increase in the binding rate coefficient, k_1 with an increase in the fractal dimension, D_{f1}. For the data shown in Figure 10.8(a) the binding rate coefficient, k_1 is given by:

$$k_1 = (1.691 \pm 0.160)D_{f1}^{5.806 \pm 0.555}$$

The fit is very good. Only three data points are available. The availability of more data points would lead to a more reliable fit. The binding rate coefficient, k_1 exhibits an order of dependence between five and one-half and sixth (equal to 5.806) order of dependence on the fractal dimension, D_{f1} or the degree of heterogeneity on the surface. This indicates that the binding rate coefficient, k_1 is very sensitive to the degree of heterogeneity on the biosensor surface.

Figure 10.8(b) shows an increase in the binding rate coefficient, k_2 with an increase in the fractal dimension, D_{f2}. For the data shown in Figure 10.8(b) the binding rate coefficient, k_2 is given by:

$$k_2 = (80.79 \pm 7.98)D_{f2}^{2.467+1246}$$

The fit is poor. Only three data points are available. There is scatter in the data. The availability of more data points would lead to a more reliable fit.

The binding rate coefficient, k_2 exhibits an order of dependence close to two and one-half (equal to 2.467) order of dependence on the fractal dimension, D_{f2} or the degree of heterogeneity on the surface. This indicates that the binding rate coefficient, k_2 is very sensitive to the degree of heterogeneity on the biosensor surface.

Figure 10.8(c) shows an increase in the binding rate coefficient ratio, k_2/k_1 with an increase in the fractal dimension ratio, D_{f2}/D_{f1}. For the data shown in Figure 10.8(a) the binding rate coefficient ratio, k_2/k_1 is given by:

$$k_2/k_1 = (1.166 \pm 0.024)D_{f2}^{6.165 \pm 0.085}$$

The fit is very good. Only three data points are available. The availability of more data points would lead to a more reliable fit. The binding rate coefficient ratio, k_2/k_1 exhibits an order of dependence between six and six and one-half (equal to 6.165) order of dependence on the ratio of fractal dimensions, D_{f2}/D_{f1} or the ratio of the degrees of heterogeneity on the surface. This indicates that the binding rate coefficient ratio, k_2/k_1 is very sensitive to the ratio of degree of heterogeneities on the biosensor surface.

Chou et al. (2010) have analyzed the binding kinetics of interleukin-6 in cell culture medium using an SPR biosensor. They state that IL-6 is multifunctional cytokine and is widely used as a biomarker for inflammatory and autoimmune diseases, coronary artery diseases, neurological diseases, and gestational problems. These authors emphasize that it is important to quantify the IL-6 levels for disease diagnosis and subsequent therapy.

Chou et al. (2010) indicate that interleukin-6 is a multifunctional cytokine which acts on inflammation, acute phase reaction, growth regulation, and differentiation of various cell types (Deckert and Legay, 2000). Chou et al. (2010) indicate that this cytokine is produced by lymphocytes, fibroblasts, endothelial cells, and keratinocytes in response to injury, infection, burns, trauma, and a variety of stimulants such as LPS, IL-1, TNF-α, and PDGF (Kishimoto, 1989; Tosata and Jones, 1999; Hellgott). Chou et al. (2010) indicate that the aberrant expression of IL-6 leads to different autoimmune and inflammatory diseases (Hong et al., 2007). Lukaszewicz et al. (2007) indicate that elevated levels of IL-6 have been observed in most types of cancer.

Figure 10.9 shows the binding of 12 ng/ml IL-6 antibody directly immobilized to an SPR biosensor surface (Chou et al., 2010). A single-fractal analysis is adequate to describe the binding kinetics. The values of the binding coefficient, k and the fractal dimension, Df for a single-fractal analysis are given in Table 10.8.

Figure 10.10 shows the binding of IL-6 during the enhancement assay. A single-fractal analysis is, once again, required to describe the binding kinetics. The values of the binding rate coefficient, k and the fractal dimension, D_f are given in Table 10.9

Figure 10.11 shows the binding and dissociation kinetics for IL-6 detection in solution to indirect antibody immobilization. A single-fractal is adequate to

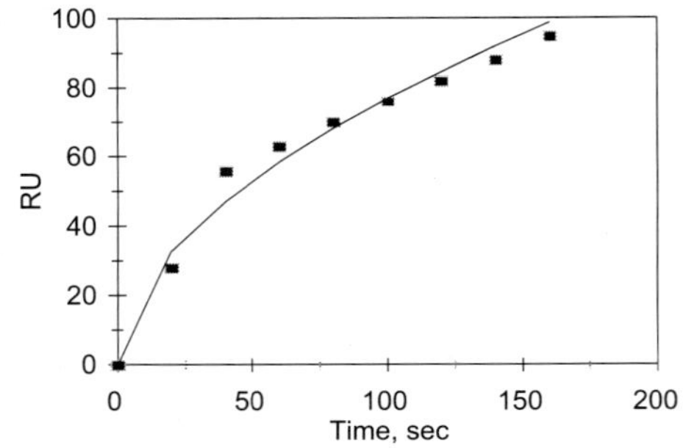

FIGURE 10.9 Binding of IL-6 to antibody directly immobilized to a surface plasmon resonance (SPR) biosensor surface (Chou et al. (2010)).

TABLE 10.8 Binding Rate Coefficient, k and the Fractal Dimension, D_f for IL-6 Binding in Solution to Antibody Directly Immobilized to a Surface Plasmon Resonance (SPR) Biosensor Surface (Chou et al., 2010)

Analyte in Solution/Receptor on Surface	k	D_f
12.5 ng/ml IL-6	6.571 ± 0.686	1.934 ± 0.05516

describe the binding and the dissociation kinetics. The values of the binding and dissociation rate coefficients and the corresponding fractal dimensions are given in Table 10.10.

Figure 10.10 shows the binding of 10 μg/ml anti-IL-6 in solution during the enhancement assay to IL-6 immobilized on an SPR biosensor surface (Chou et al., 2010). The binding rate coefficient, k and the fractal dimension, D_f are given in Table 10.9.

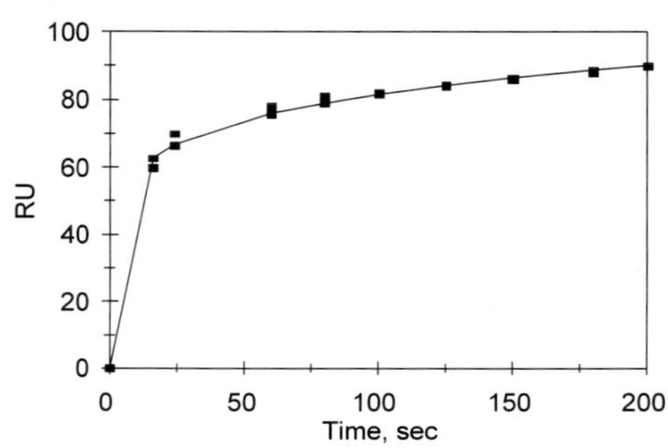

FIGURE 10.10 Binding of IL-6 in solution to a surface plasmon resonance (SPR) biosensor during the enhancement assay (Chou et al., 2010).

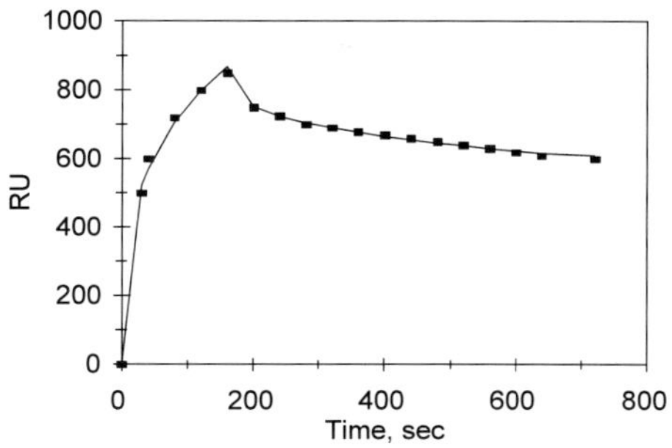

FIGURE 10.11 Binding and dissociation of IL-6 in solution to indirect antibody immobilization on a surface plasmon resonance (SPR) biosensor surface (Chou et al., 2010).

TABLE 10.9 Binding of Anti-IL-6 in Solution to IL-6 During an Enhancement Assay to a Surface Plasmon Resonance (SPR) Biosensor Surface (Chou et al., 2010)

Analyte in Solution/Receptor on Surface	k	D_f
10 μg/ml anti-IL-6/IL-6 immobilized on an SPR biosensor surface	42.153 ± 1.198	2.7130 ± 0 0.02238

Figure 10.11 shows the binding and dissociation of 2nd anti-IL-6 to the indirect IL-6 immobilized on an SPR biosensor surface (Chou et al., 2010). A single-fractal analysis is adequate to describe the binding and the dissociation kinetics. The values of (a) the binding rate coefficient, k and the fractal dimension, D_f for a single-fractal analysis, and (b) the dissociation rate coefficient, k_d and the fractal dimension for dissociation, Dfd are given in Table 10.10. In this case, the affinity, K (=k/k_d) value is equal to 6.86.

Figure 10.12 shows the binding of 16 ng/ml anti-CEA autoantibodies in serum to a CEA functionalized SPR biosensor surface (Ladd et al., 2009). A dual-fractal analysis is required to describe the binding kinetics. The values of (a) the binding rate coefficient, k and the fractal dimension, D_f for a single-fractal analysis, and (b) the binding rate coefficients, k_1 and k_2, and the fractal dimensions, D_{f1} and D_{f2} are given in Tables 10.4 and 10.5.

Figure 10.13(a) shows the effect of incubation time on GC/ILs/MCNTs/anti-PSS and Prostate Specific Antigen (PSA) anti-PSA labeled horse radish peroxidase (HRP) (Salimi et al., 2013). A dual-fractal analysis is required to describe the binding kinetics. The values of (a) the binding rate coefficient, k and the fractal dimension, D_f for a single-fractal analysis, and (b) the binding rate coefficients, k_1 and k_2 and the fractal dimensions, D_{f1} and D_{f2} are given in Tables 10.11 and 10.12.

TABLE 10.10 Binding and Dissociation Rate Coefficients and Fractal Dimensions for the Binding and the Dissociation Phase for 2nd Anti-IL-6 to the Indirect IL-5 Immobilized on a Surface Plasmon Resonance (SPR) Biosensor Surface (Chou et al., 2010)

Analyte in Solution/ Antibody on Surface	k	k_d	D_f	D_{fd}
2nd anti-IL-6/IL-6	187.94 ± 9.23	27.38 ± 0.60	2.9364 ± 0.0602	2.304 ± 0.0162

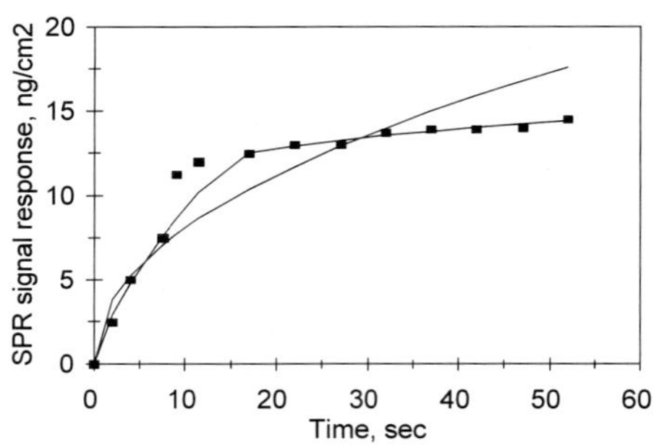

FIGURE 10.12 Binding of 16 ng/ml anticarcinoembryonic antigen (anti-CEA) autoantibodies in serum to a CEA functionalized surface (Ladd et al., 2009).

Salimi et al. (2013) developed a highly sensitive immunosensing biosensor for detecting prostate-specific antigen based on an ionic liquid-carbon nanotubes modified electrode. PSA antibody (anti-PSA) was immobilized onto a robust nanocomposite imide. Thionine was used as the redox systems for the electrochemical probe. The PSA antibody (anti-PSA) was immobilized or entrapped in the nanocomposite and used in a sandwich-type complex immunoassay. Anti-PSA labeled HRP was used as the secondary antibody and H_2O_2 was used as the substrate.

Wu et al. (2010) indicate that tumor markers play a significant role in screening, diagnosing, and evaluating the prognosis of diseases. For prostate cancer prostate-specific antigen has been identified as a reliable marker for the early diagnosis of cancer (Panini et al., 2008). Electrochemical biosensors have been used as biosensors due to their ease of use in antibody−antigen reaction, low cost, use of simple instrumentation, ease of miniaturization, high sensitivity, fast response, and low detection limit (Wang et al., 2006a; Okuno et al., 2007; Mantzila et al., 2008; Zhou et al., 2009; Du et al., 2010;

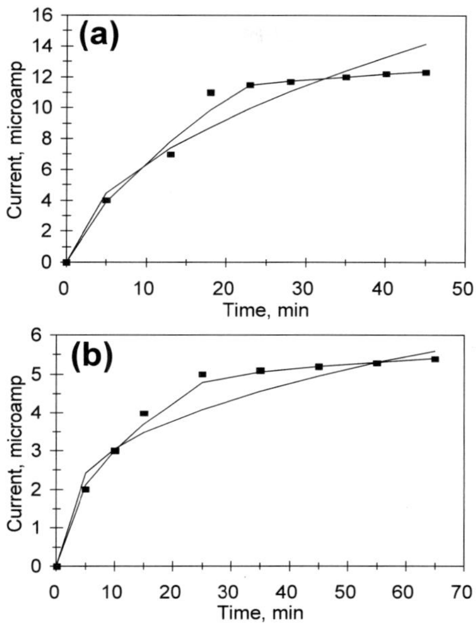

FIGURE 10.13 Binding of (a) GC/ILs/MWCNTs/anti-PSA and PSA/PSA and anti-PSA with anti-PSA/labeled HRP, and (b) PSA with anti-labeled HRP on immunosensor surface (Salimi et al., 2013).

TABLE 10.11 Binding Rate Coefficients for 16 ng/ml Anti-CEA Autoantibodies in Serum to CEA Immobilized on a Surface Plasmon Resonance (SPR) Biosensor Surface (Salimi et al., 2013)

Analyte in Serum/ Antibody on Surface	k	k_1	k_2
16 ng/ml anti-CEA autoantibodies/CEA on SPR surface	1.924 ± 0.2202	1.222 ± 0.144	8.170 ± 0.019

Huang et al., 2010; Tang and Ren, 2010). Hayes et al. (2007) emphasize that they provide a sensitive and selective tool for detecting immuno reagents by detecting changes in either potential, current, capacitance, conductance, or impedance.

Nanoparticle-based amplification processes have enhanced the sensitivity of electrochemical processes on using electrochemical immunoassays. They have been used for the detection of biomarkers (Daniel and Anstruc, 2004; Tang and Ren, 2010), The following have been used: carbon nanotubes (CNTs; Chen et al., 2009), gold nanoparticles (Zhou et al., 2009), silica nanoparticles (Wang et al., 2008), Au$-Fe_3O_4$ (Wei et al., 2010), silica nanoparticles

TABLE 10.12 Fractal Dimensions for Binding of 16 ng/ml anti-CEA Autoantibodies in Serum to CEA Immobilized on a Surface Plasmon Resonance (SPR) Biosensor Surface (Salimi et al., 2013)

Analyte in Serum/ Antibody on Surface	D_f	D_{f1}	D_{f2}
16 ng/ml anti-CEA autoantibodies/CEA on SPR surface	1.9314 ± 0.1478	1.1548 ± 0.1914	2.7840 ± 0.01250

(Wang et al., 2010), carbon nanoparticles (Ho et al., 2009), carbon nanospheres (Du et al., 2010), and nanowires (NWs).

In the present study, the nanocomposite containing multiwalled carbon nanotubes (MWCNTs) was used for entrapment of a model cancer biomarker and demonstrated the amplification using sandwich detection. The biosensor was used to determine PSA in human serum samples and prostate cancer cells.

Figure 10.13(a) shows the effect of incubation time on the binding of GC/ILs/MCNTs/anti-PSS and PSA/anti-PSA labeled HRP (Salimi et al., 2013). A dual-fractal analysis is required to describe the binding kinetics. The values of (1) the binding rate coefficient, k and the fractal dimension, D_f for a single-fractal analysis, and (2) the binding rate coefficients, k_1 and k_2 and the fractal dimensions D_{f1} and D_{f2} for a dual-fractal analysis are given in Tables 10.11 and 10.12. An increase in the fractal dimension by a factor of 1.79 from a value of D_{f1} equal to 1.155 to D_{f2} equal to 2.784 leads to an increase in the binding rate coefficient by a factor of 6.86 from a value k_1 equal to 1.222 to a value of k_2 equal to 8.170. An increase in the degree of heterogeneity on the biosensor surface leads to an increase in the binding rate coefficient.

Figure 10.13(b) shows the effect of incubation time on the binding of PSA with anti-labeled HRP immobilized on a biosensor surface (Salimi et al., 2013). A dual-fractal analysis is required to describe the binding kinetics. The values of (1) the binding rate coefficient, k and the fractal dimension, D_f for a single-fractal analysis, and (2) the binding rate coefficients, k_1 and k_2 and the fractal dimensions D_{f1} and D_{f2} for a dual-fractal analysis are given in Tables 10.11 and 10.12.

An increase in the fractal dimension by a factor of 1.4 from a value of D_{f1} equal to 1.9916 to D_{f2} equal to 2.7878 leads to an increase in the binding rate coefficient by a factor of 3.72 from a value k_1 equal to 0.9347 to a value of k_2 equal to 3.4794. An increase in the degree of heterogeneity on the biosensor surface once again, leads to an increase in the binding rate coefficient.

have analyzed inflammatory biomarkers using rectangular polycrystalline silicon NWs made by etching. They presented poly-silicon NWs and demonstrated their functionality by measuring inflammatory biomarkers. They indicate that their fabrication process is based on an advanced spacer etch process and also that in the last decade silicon (Si) NWs have been used as biosensors. They have been used to detect proteins, DNA, and viruses (Ragmir et al., 2010; Squires et al., 2008). There are shortcomings with regard to NWs used as biosensors, specifically in determining the binding kinetics. Commercially available NWs are produced by a top-down fabrication process, and Gao et al. (2007) indicate that they are produced by advanced lithographic techniques such as deep-UV, electron beam (Park et al., 2010), and nano-printing lithography (Melosh et al., 2003). An alternate approach is to use spacer NWs patterned by a conventional etch (Chang et al., 2008). Used fabricated silicon NWs with a spacer etch process to analyze the binding kinetics of inflammatory biomarkers.

Figure 10.14(a) shows the binding kinetics of IL-8 to an NW biosensor functionalized for these inflammatory biomarkers. A dual-fractal analysis is required to adequately describe the binding kinetics. The values of (1) the binding rate coefficient, k and the fractal dimension, D_f and the (2) binding rate coefficients k_1 and k_2 and the fractal dimensions, D_{f1} and D_{f2} for a dual-fractal analysis are given in Tables 10.13 and 10.14 It is of interest to note that

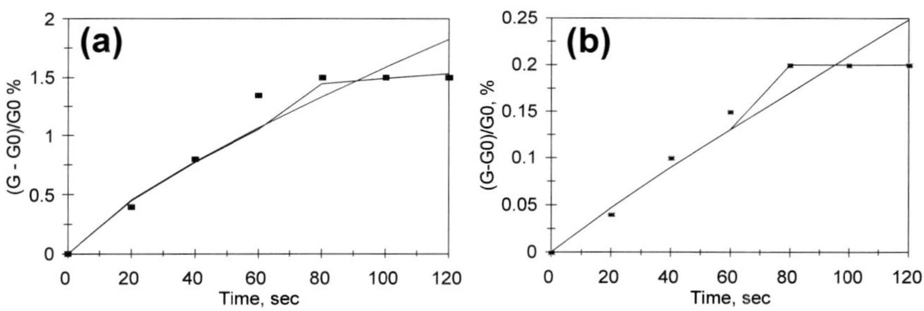

FIGURE 10.14 Binding of (a) IL-8 and (b) TNF-α in solution to surface functionalized nanowires (NWs).

TABLE 10.13 Binding of IL-8 and TNF-α in Solution to Nanowires (NWs) Functionalized for These Specific Inflammatory Biomarkers

Analyte in Solution/ Receptor on Surface	k	k_1	k_2
IL-8/functionalized NWs	0.04456 ± 0.00874	0.01475 ± 0.01566	0.7587 ± 0.0277
TNF-α/functionalized NWs	0.002878 ± 0.000574	0.001068 ± 0.000074	2.0 ± 0

TABLE 10.14 Fractal Dimensions for the Binding of IL-8 and TNF-α in Solution to Nanowires (NWs) Functionalized for These Specific Inflammatory Biomarkers

Analyte in Solution/ Receptor on Surface	D_f	D_{f1}	D_{f2}
IL-8/functionalized NWs	1.5490 ± 0.2418	0.8088 ± 0.1527	2.7068 ± 0.1383
TNF-α/functionalized NWs	1.172 ± 0.2456	0.5682 ± 0.1693	3.0 ± 1.5 E−14

as the fractal dimension increases by a factor 1.71 from a value of D_{f1} equal to 1.6556 the binding rate coefficient increases by a factor of 103 from a value of k_1 equal to 0.002009 to a value of k_2, equal to 0.02066. An increase in the degree of heterogeneity on the functionalized NW surface leads to an increase in the binding rate coefficient Figure 10.14b shows the binding of TNF-α to the surface functionalized nanowires (NWs) (Lombardini et al., 2001), A dual-fractal analysis is required to adequately describes the binding kinetics. The values of the binding rate coefficients k, k_1 and k_2 and the fractal dimensions, D_f, D_{f1} and D_{f2} are given in Tables 10.15 and 10.16.).

Figure 10.16(a) shows the binding of 5 pg/ml cTnI in solution to a functionalized NW biosensor surface. A single-fractal analysis is required to adequately describe the binding kinetics. The values of the binding rate coefficient, k and the fractal dimension, D_f for a single-fractal analysis are given in Tables 10.15 and 10.16.

Figure 10.16(b) shows the binding of 500 pg/ml cTnI in solution to a functionalized NW biosensor surface. A dual-fractal analysis is required to adequately describe the binding kinetics. The values of (1) the binding rate

TABLE 10.15 Binding Rate Coefficients for Different Concentrations of cTnI Protein to Nanowire Biosensor Surface

cTnI Concentration/ Nanowire Surface	k	k_1	k_2
5 pg/ml cTnI protein/nanowire surface	0.0121 ± 0.0006	NA	NA
500 pg/ml cTnI protein/ nanowire surface	0.0300 ± 0.0076	0.00759 ± 0.00171	0.07770 ± 0.00002

TABLE 10.16 Fractal Dimensions for the Binding of Different Concentrations of cTnI Protein to Nanowire Biosensor Surface

cTnI Concentration/ Nanowire Surface	D_f	D_{f1}	D_{f2}
5 pg/ml cTnI protein/ nanowire surface	2.4140 ± 0.0444	NA	NA
500 pg/ml cTnI protein/ nanowire surface	2.5588 ± 0.2232	1.6526 ± 0.5208	2.95466 ± 0.000684

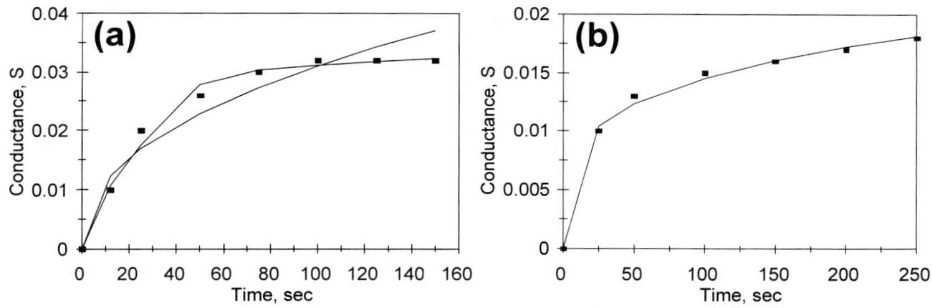

FIGURE 10.15 Binding of different concentrations of BNP to a PANI nanowire biosensor (a) 50 fg/ml (b) 1 pg/ml.

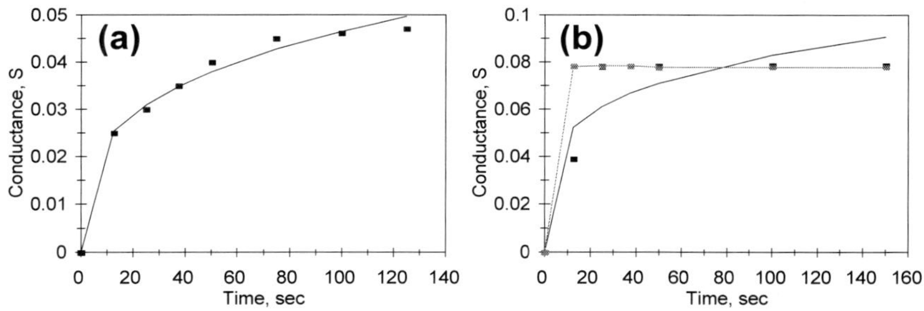

FIGURE 10.16 Binding of different concentrations of cTnI proteins in solution to a PANI nanowire biosensor (a) 5 pg/ml (b) 500 pg/ml.

coefficient, k and the fractal dimension, D_f for a single-fractal analysis, and (2) the binding rate coefficients, k_1 and k_2 and the fractal dimensions, D_{f1} and D_{f2} are given in Tables 10.15 and 10.16. It is of interest to note that as the fractal dimension increases by a factor of 1.79 from a value of D_{f1} equal to 1.6526 to a value of D_{f2} equal to 2.95466 the binding rate coefficient increases by a factor of 10.2 from a value of k_1 equal to 0.00759 to a value of k_2 equal to 0.07770. An increase in the fractal dimension or the degree of heterogeneity on the biosensor surface leads to an increase in the binding rate coefficient.

Figure 10.17(a) shows the binding of cTnI with BSA in solution to a NW biosensor surface. The values of (a) the binding rate coefficient, k and the fractal dimension, D_f, and (b) the binding rate coefficients, k_1 and k_2 and the fractal dimensions, D_{f1} and D_{f2} are given in Tables 10.17 and 10.18. An increase in the fractal dimension leads to an increase in the binding rate coefficient.

Figure 10.17b shows the binding of CK-MB with BSA in solution to an NW biosensor surface. The values of the binding rate coefficient, k and the fractal dimension, D_f for a single-fractal analysis are given in Tables 10.17 and 10.18.

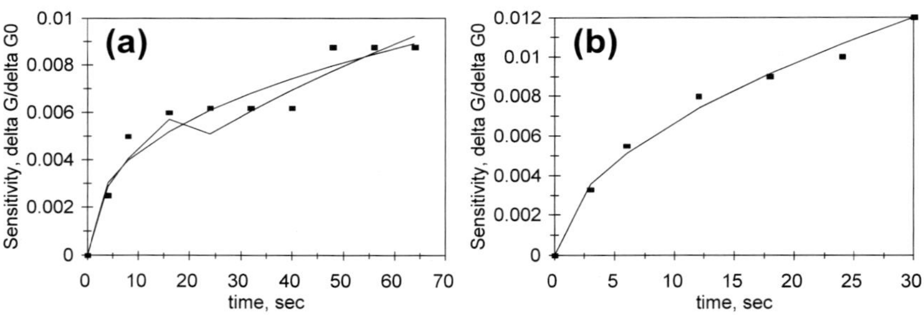

FIGURE 10.17 Binding of (a) cTnI and (b) CK-MB with BSA to a nanowire biosensor surface.

TABLE 10.17 Binding Rate Coefficients for cTnI and CK-MB in Solution to a Nanowire Biosensor Surface

Analyte in Solution/ Receptor on Surface	k	k_1	k_2
cTnI + BSA/ functionalized sensor surface	0.000294 ± 0.000294	0.001447 ± 0.000321	0.000737 ± 0.0008
CK-MB + BSA/ Functionalized sensor surface	0.001984 ± 0.000149	NA	NA

TABLE 10.18 Fractal Dimensions for Binding of cTnI and CK-MB in Solution to a Nanowire Biosensor Surface

Analyte in Solution/ Receptor on Surface	D_f	D_{f1}	D_{f2}
cTnI + BSA/functionalized sensor surface	2.2250 ± 0.1144	2.2350 ± 0.1144	1.784 ± 0.374
CK-MB + BSA/functionalized sensor surface	1.9420 ± 0.07368	NA	NA

10.4 CONCLUSIONS

A fractal analysis is presented for the binding and dissociation of different biomarkers to different biosensor surfaces, in this chapter, but we do not characterize the biomarkers like the previous chapters which dealt with specific diseases. This is sort of a "cleanup" chapter wherein we present the binding and dissociation of biomarkers to different biosensor surfaces that have appeared recently in the open literature. Otherwise, the analysis presented is similar to that which is presented in the previous chapters. Initially, a single-fractal analysis is used to model the binding and dissociation (if applicable) kinetics. Only if the single-fractal analysis does not provide an adequate fit, the dual-fractal analysis is used to model the binding kinetics.

Some of the biomarkers kinetics for the different diseases analyzed include: (1) the binding of different concentrations (in ng/ml) of iNOS in solution to an SPR biosensor surface (Noah et al., 2011), (2) detection of anti-CEA autoantibodies in clinical human serum using CEA immobilized to an SPR biosensor (Ladd et al., 2009), (3) the binding of different concentrations (in ng/ml) of COX-2 in solution to goat anti-COX-2 antibody immobilized in a sandwich-type immunoassay (Wolf et al., 2004), (4) binding of FABP, TNT 7G7, and cTNT lF11 in solution to cTNT 7G7 antibody immobilized on an SPR biosensor surface (Andersson et al., 2010), (5) binding of IL-6 antibody directly immobilized on an SPR biosensor surface (Chou et al., 2010), (6) binding of IL-6 during the enhancement assay, (7) binding and dissociation for IL-6 in solution to indirect antibody immobilization, (8) effect of incubation time on GC/ILs/MCNts/anti-PSS and PSA/anti-PSA labeled HRP (Salami et al., 2013).

The kinetics of other biomarkers include: (1) IL-8 and TNF-α in solution to surface functionalized NWs, (2) binding of GSH to QC/Au/MES/AuNP in consecutive cycles (3) binding (resonant frequency transient) of 0.95 nM AuNP@GSH in solution to a QC/Au/AHT/Ab, BSA on a gold quartz piezoimmunosensor, (4) binding and dissociation of clusterin (CLU) to anti-clusterin immobilized on an SPR biosensor surface, (5) binding and dissociation of thrombin in solution to Aptl and Apt2 immobilized on an SPR biosensor surface.

The fractal dimension is not a classical independent variable (as indicated in the previous chapters) such as analyte, antigen, antibody, or any other biological molecule concentration in solution. The expressions obtained for the binding and dissociation rate coefficients, and affinity (wherever applicable) for a single- and dual-fractal analysis as a function of the fractal dimension indicate generally a high degree of sensitivity of these biosensor parameters on their respective fractal dimensions or the ratio of the fractal dimensions on the SPR or other biosensor chip surface. It may be noted, as pointed out throughout the chapters in the book that the data analysis in itself does not provide evidence for the surface heterogeneity and the existence of surface heterogeneity assumed may be incorrect. It would perhaps be more appropriate to provide for a qualitative and quantitative measure of heterogeneity that exists on the biosensor surface by some surface characterization methods. However, in lieu of that we presume that a degree of heterogeneity exists on the surface (not an unreasonable assumption) even though it has not been characterized independently.

Predictive relations are presented for (1) k as function of iNOS concentration in solution (Noah et al., 2011), (2) k as a function of the degree of heterogeneity as on the biosensor Surface (Noah et al., 2011). It may be noted that the binding rate coefficient, k is very sensitive to the degree of heterogeneity on the surface, and exhibits an order of dependence between negative 11 and one-half and 12 (-11.68) order of dependence on the degree of heterogeneity that exists on the biosensor surface, (3) the binding rate coefficient, k exhibits an order of dependence between one and half and two (equal to 1.808) on the degree of heterogeneity or the fractal dimension, D_f for COX-2 (4) the binding rate coefficient, k as a function of the COX-2 concentration, (5) the fractal dimension, D_f as a function of the COX-2 concentration in solution (6) the binding rate coefficient, k_1 as a function of the fractal dimension, D_{f1} for a dual-fractal analysis for the binding of cTNT lF11 in solution to CTNT 7G7 antibody immobilized on an SPR biosensor surface (Andersson et al., 2010), (7) the binding rate coefficient ratio, k_2/k_1 as a function of the fractal dimension ratio, D_{f2}/D_{f1} for cTNT lF11 in solution to cTNT 7G7 immobilized on an SPR biosensor surface (Andersson et al., 2010).

Other predictive examples for the rate coefficients include (1) the increase in the binding rate coefficient, k (2) with an increase in the fractal dimension, D_{fi} for the binding of thrombin or in a mix with other components to APT-2 spot immobilized to an SPR chip surface. The binding rate coefficient, k_a is extremely sensitive to the degree of heterogeneity that exists on the surface since it exhibits an order of dependence greater than minus 15 and one-half (equal to -15.6) order of dependence on the fractal dimension, D_{fl}, and (b) the binding rate coefficient, k_2 exhibits an order of dependence of slightly more than 13 and one-half (equal to −13.52) on the fractal dimension, D_{f2}. These are some of the examples of predictive relations presented in this chapter.

As indicated previously in the preceding chapters, the fractal dimension values provide a quantitative indicator of the degree of heterogeneity present on the biosensor surface in the binding and in the dissociation phase. Perhaps, this may be due to the different mechanisms involved in these two phases. Binding and dissociation rate coefficient values along with affinity values are provided wherever possible. The fractal dimension value for the binding and the dissociation phase is not a typical value that may be directly manipulated, but is estimated from the equations presented in the chapter.

In a general sense, fractal models are fascinating since they provide a fresh outlook to the reactions occurring on the surface, and perhaps provide an avenue on how the surface and eventually the reactions may be manipulated. Newer avenues are definitely required to analyze and to help detect disease biomarkers which are often present in very dilute concentrations in solution. Either more and more sensitive biosensor techniques are required to detect these biomarkers in solution or the technique should be modified appropriately so as to enhance the eventual signal that can be measured. The sooner one is able to help detect these biomarkers for certain diseases accurately, the better is the prognosis of the disease. Needless to say, the detection of these biomarkers (and subsequently the onset of the diseases) is more and more difficult during the early stages. The example of the early diagnosis of cancer clearly demonstrates the distinct advantage of detecting this nefarious disease, and hopefully it leads to a better prognosis and quality of life. It is with this goal that the chapters in this book are written with the hope that they will contribute to the eventual better treatment of the various insidious diseases.

REFERENCES

Aklan, F., 2004. iNOS-mediated nitric oxide production and its regulation. Life Sciences 75, 639−653.

Amos, A.F., Mccarty, D.J., Zimmet, P., 1997. The rising global burden of diabetes and its complications: estimates and projections to the year 2010. Diabetic Medicine 14 (12 Suppl. 5), S7−S85.

Andersson, U., Hzagi, J., Oldgren, J., 2008. Background of the level of troponin is a powerful predictor of cardiaovascular events and mortality. European Heart Journal 29, 2252−2258.

Andersson, H., Steel, D., Asp, J., Dahlenborg, K., Jonsson, M., Jeppsson, A., Lindahl, A., Kagedal, S., Sartipy, P., Mandenius, C.F., October 1, 2010. Assaying cardiac biomarkers for toxicity using biosensing and cardiomyocytes derived from human embryonic stem cells. Journal of Biotechnology 150 (1), 175−181.

Aquino, D., Bizzi, A., Grisolim, Garavaglia B., Bruzzone, M.G., Nerdoui, N., et al., 2004. Radiology 252, 165−172.

Battaglia, T.M., Masson, J.F., Sierks, M.R., Beaudon, S.P., Rogers, J., Foster, et al., 2005. Quantification of cytokines involved in wound healing using surface plasmon resonance. Analytical Chemistry 77, 7016−7023.

Bernard, S., Enayati, A., Redwood, L., Roger, H., Bnstock, T., 2001. Autism: a novel form of mercury poisoning. Medical Hypothesis 56, 462–471.

Caspi, O., Itzhaki, I., Arbel, G., Kehat, I., Gepstein, A., Huber, I., Satin, J., Gepstein, L., 2009. in-vitro electrophysiological drug testing using human embryonic stem cell derived cardiomyocytes. Stem Cells Development 18, 161–171.

Cherif, B., Roget, C.L., Villers, R., Calemczuk, R., Leroncer, Y.V., Marche, P.N., Livache, T., Villiers, M.B., 2006. Clinically related protein-peptide interactions monitored in real time on novel peptide chips by surface plasmon resonance imaging. Clinical Chemistry 52, 255–262.

Chen, S., Yuan, R., Chai, Y., Min, L., Liw, Xuy, 2009. Electrochimica Acta 7242–7247.

Chang, C.W., Chen, S.F., Chang, C.L., Deng, C.K., Huang, J.J., Lei, T.F., 2008. IEEE Electron Devices Letters 29, 474–476.

Cho, H., Yun, C.W., Park, W.K., Kong, J.Y., Kim, K.S., Park, Y., Lee, S., Kim, B.K., 2004. Modulation of the activity of pro-inflammatory enzymes COX-2, and iNOS by chrysin derivatives. Pharmacological Research 49, 37–43.

Chou, T.H., Chuang, C.Y., Wu, C.M., July 2010. Quantification of interleukin-6 in cell culture medium using surface plasmon resonance biosensors. Cytokine 51 (1), 107–111.

Coleman, J.W., 2001. Nitric oxide in immunity inflammation. International Immunopharmacology 1, 1397–1408.

Daniel, M.C., Anstruc, D., 2004. Chemical Reviews 104, 293–334.

Deckert, F., Legay, F., 2000. Development and validation of an IL-6 immuno-receptor assay based on surface plasmon resonance. Journal of Pharmaceutical and Biomedical Analysis 23, 403–412.

Disis, M.L., Calenoff, E., McLaughlin, G., Muphy, A.E., Chen, W., Groner, B., Jeschike, M., Lydon, E., McGlynn, E., Livingston, R.B., Moe, R., Cheever, M.A., 1994. Existent T-cell and antibody immunity to Her-2 Neu patients with breast cancer. Cancer Research 54, 16–20.

Disis, M.L., Pupa, S.L., Grawlow, J.R., Dittadi, R., Menard, S., Cheever, M.A., 1997. Journal of Clinical Oncology 15, 3363–3367.

Dolci, A., Dominici, R., Cardimale, D., Sandri, M.T., Panteghini, M., 2008. Biochemical markers for prediction of chemotherapy-induced cardiotoxicity: systematic review of the literature and recommendation for use. American Journal of Clinical Pathology 130, 688–695.

Draghic, S., Chatterjee, M., Tainsky, M.A., 2005. Epitomics: serum screening for the early detection of cancer on microarrays using complex panels of tumor antigens. Expert Review of Molecular Diagnostics 5, 735–743.

Du, D., Zou, Z., Shin, Y., Wang, J., Wu, H., Engelhard, M.H., Liu, J., Aksay, I.A., Lin, Y., 2010. Analytical Chemistry 82, 2989–2995.

Englebiene, P., van Hoonacker, A., Verhas, M., 2003. Surface plasmon resonance: principles, methods, and applications in biomedical sciences. Spectroscopic—An International Journal 17, 255–273.

Franco, L., Doria, E., Belazzoni, E., Benini, A., Bassi, C., 1994. Increased expression of inducible nitric oxide synthase and cyclooxygenase-2 in pancreatic cancer. Prostaglandins Lipid Mediator 73, 915–916.

Gabay, C., Kushner, I.N., 1999. Acute-phase proteins and other systemic responses to inflammation. New England Journal of Medicine 340, 448–454.

Gadducci, A., Ferdeghini, M., Buttita, F., Coslo, S., Fanucchi, A., Annicchiaric, C., Gagetti, G., Bevlacqua, G., Genazzani, A.R., 1999. Assessment of the prognostic relevance of serum anti-p53 antibodies in epithelial ovarian cancer. Gynecologic Oncology 72, 76–81.

Gao, Z.Q., Agarwal, A., Trigg, A.D., Singh, N., Fang, C., Tung, C.H., Fan, Y., Buddharaju, K.D., Kong, J.M., 2007. Analytical Chemistry 79, 3291–3297.

Green, R.J., Frazier, R.A., Shakesheff, K.M., Davies, M.C., Roberts, C.J., Tendler, S.J.B., 2000. Surface plasmon resonance analysis of dynamic biological interactions with biomaterials. Biomaterials 21, 1823–1835.

Havlin, S., 1989. Molecular diffusion and reaction: In the Fractal Approach to Heterogeneous Chemistry: Surfaces, Colloids, Polymers. In: Avnir, D. (Ed.). Wiley, New York, pp. 251–269.

Havlin, S., Ben-Avraham, D., 1987. Diffusion in disordered media. Advances in Physics 36, 695–798.

Hayes, D.A., Leonard, L.M., Kennedy, R.O., 2007. Trends in Biotechnology 25, 125–131.

Hellgott, D.C., May, L.T., Stroeger, Z., Tamm, I., Sehgal, P.B., 1987. Bacterial lipopolysaccharide (endotoxin) enhances expression and secretion of beta-2 interferon by human fibroblasts. Journal of Experimental Medicine 166, 1300–1309.

Ho, J.A.C., Lin, Y., Wang, L.S., Hang, K.C., Chou, P.T., 2009. Analytical Chemistry 81, 1340–1346.

Homola, J., 2003. Present and future of surface plasmon resonance biosensors. Analytical and Bioanalytical Chemistry 377, 528–539.

Homola, J., Yee, S.S., Gauglitz, G., 1999. Surface plasmon resonance sensors: review. Sensors & Actuators: B Chemical 54, 3–15.

Hong, D.S., Angelo, L.S., Kurzock, R., 2007. Interleukin-6 and its receptor in cancer: implications for translational therapeutics. Cancer 110, 1911–1928.

Huang, K.J., Niu, D.J., Xie, W.Z., Wang, W., 2010. Analytica Chimica Acta 659, 102–108.

Jacobs, I.J., Menon, U., 2004. Progress and challenges in screening for early detection of ovarian cancer. Molecular & Cellular Proteomics 3, 355–366.

Juncker, D., Schmid, H., Drechster, U., Wolf, H., Wolf, M., Michel, B., de Rooij, N., Delmarche, E., 2002. Autonomous microfluidic capillary system. Analytical Chemistry 74, 139–6144.

Kanazawa, S., Morikawa, S., Harada, T., Yasuhira, K., Matsubara, Y., Ikeda, S., Soeda, S., Fujimura, M., 1982. Carcinoembryonic antigen producing cultured-cell lines enable detection of autoantibodies in sera from patients with gastrointestinal cancer. Cancer 50, 1775–1782.

Key, G., Schreiber, A., Feldbrugge, R., McNeil, C.J., Jergensen, P., Pelsers, N.M.A.L., 1999a. Multicellular evaluation of an amperometric immunoassay for plasma fatty acid binding protein. An early marker for acute myocardial infarction. Clincal Biochemistry 32, 229–231.

Key, G., Schreiber, A., Faldbrugge, R., McNeil, C.J., Jorgensen, P., Pelsers, M.A.L., Glantz, J.F.C., Spener, F., 1999b. Multicenter evaluation of an amperometric immunosensor for plasma fatty acid –binding protein: an early marker for acute myocardial infarction. Clinical Biochemistry 32, 229–231.

King, H., Aubert, R.E., Herman, W.H., 1998. Diabetes Care 21 (9), 1414–1431.

Kishimoto, T., 1989. The biology of interleukin-6. Blood 74, 1–10.

Ladd, J., Lu, H., Taylor, A.D., Goodbell, V., Diala, M.L., Jiang, S., April 1 , 2009. Direct detection of carcinoembryonic autoantibodies in clinical human serum samples using a surface plasmon sensor. Colloids and Surfaces B, Interfaces 70 (1), 1–6.

Lechpammer, M., Lukic, J., Lechpammer, S., Kovacevic, D., Loda, M., Kusic, Z., 2004. Humoral immune response to p53 correlates with clinical course in colorectal cancer patients during adjuvant chemotherapy. International Journal of Colorectal Disease 19, 114–120.

Lukaszewicz, M., Mroczko, B., Szmikowski, M., 2007. Clinical significance of interleukin-6 (IL-6) as a prognostic factor of cancer disease. Polish Archives of Internal Medicine 117, 247–251.

Malic, L., Veres, T., Tabrizian, M., 2011. Nanostructured digital microfluidics for enhanced surface plasmon resonance imaging. Biosensors & Bioelectronics 26, 2053–2059.

Mantzila, A.G., Maipa, V., Prodromidis, M.I., 2008. Analytical Chemistry 80, 1169–1175.

McCord, J., Nowak, R.M., McCullough, et al., 2001. Ninety-minute exclusion by use of quantitative point-of-care testing of myoglobin and troponin I. Circulation 104, 1483–1488.

Melosh, N.A., Boukai, A., Diana, F., Geradot, B., Badolato, A., Petrof, P.M., Heath, J.R., 2003. Science 300, 112–115.

Nakata, T., Hashimoto, A., Hase, M., Tsuchihashi, K., Tsuchhihashi, K., Shimamoto, K., 2003. Human heart fatty acid-binding protein as an early diagnostic and prognostic marker in acute coronary syndrome. Cardiology 99, 96–104.

Nathan, C., Xie, Q.W., 1994. Nitric-oxide synthases: roles, tolls, and controls. Cell 78, 915–918.

Njagi, J., Erlichman, J.S., Aston, J.W., Leiter, J.C., Andtescues, 2010. A sensitive electrochemical sensor based on chitosan and electropolymerized Meldola blue for monitoring NO in brain slices. Sensors & Actuators B 143, 673–680.

Noah, N.M., Alam, S., Sadik, O.A., June 15, 2011. Detection of inducible nitric oxide synthase using a suite of electrochemical fluorescence, and surface plasmon resonance biosensors. Analytical Chemistry 413 (2), 157–163.

Noah, N.M., Mwilu, S.K., Sadik, O.A., Fatah, A.A., Arcilesi, R.D., July 2011. Clinica Chimica Heta 412 (11–12), 1391–1398.

Okuno, J., Maeshi, K., Kerman, K., Takamura, Y., Matsumoto, K., Tamiya, E., 2007. Biosensors & Bioelectronics 22, 2377–2381.

Park, I., Li, Z., Pisano, A.P., Williams, R.S., 2010. Nanotech 21, 1–9.

Panini, N.V., Messina, G.A., Salinas, E., Fernandez, Reba, J., 2008. Biosensors & Bioelectronics 23, 1145–1151.

Ragmir, N.S., Yang, Y., Zacharias, M., 2010. Small 16, 1705–1722.

Ramakrishnan, A., Sadana, A., 2001. A single-fractal analysis of cellular analyte-receptor binding kinetics using biosensors. Biosystems 59, 35–51.

Ridker, P.M., Rifai, N., Rose, L., et al., 2002. Comparison of C-reactive and low-density lipoprotein cholesterol levels in the prediction of first cardiovascular events. New England Journal of Medicine 347, 1557–2156.

Rifai, N., Ridker, P.M., 2001. High-sensitivity C-reactive protein: a novel and promising marker of coronary disease. Clinical Chemistry 47, 403–411.

Sadana, A., 2001. A fractal analysis for the evaluation of hybridization kinetics in biosensors. Journal of Colloid and Interface Science 151 (1), 166–177.

Sadana, A., 2005. Fractal Binding and Disseriation Kinetics for Different Biosensors Applications. Elsevier, Amsterdam.

Sahin, U., Tureo, O., Schmitt, B., Cochlovius, B., Johannes, T., Schmitts, R., Stenner, F., Luo, G.R., Schubert, I., Pirendschuh, M., 1995. Human neoplasma elicit multiple specific immune responses in the autologous host. Proceedings of the National Academy of Sciences of the United States of America 92, 11810–11813.

Salami, S.S., Schmidt, F., Laxman, B., Regan, M.M., Rickman, D.S., Schern, D., Buet, G., Siddiqui, J., Tomlins, S.A., Wei, J.T., Chinnaiyan, A.M., Rubin, M.A., Sanda, M.A., Sanda, M.G., July 2013. Combining urinary detection of TMPRSS2: ERG and PCA with serum PSA to predict diagnosis of prostate cancer. Urology and Oncology 31 (5), 566–571.

Salini, A., Kavosi, B., Fathi, F., Hallaj, 2013. Highly sensitive immunosensing of prostate-specific antigen based on ionic liquid-carbon nanotubes modified electrode: application as cancer biomarker for prostate biopsies. Biosensor & Bioelectronics 42, 439–446.

Scanian, M.J., Gordan, J.D., Williamson, S., Stockert, E., Bander, N.H., Jorgeneel, V., Gore, A.O., Jager, E., Jager, A., Knuth, A., Chen, Y.T., Old, L.J., 1989. International Journal of Cancer 83, 456–464.

Steel, D., Hyliner, J., Bremer, S., 2009. Cardiomyocytes derived from human embryonic stem cells-characteristic and utility for drug discovery. Current Opinion in Drug Discovery Development 12, 133−140.

Squires, T.M., Messinger, R.J., Manalis, S.R., 2008. Nature Biotechnology 26, 417−426.

Tang, D., Ren, J., 2010. Analytical Chemistry 82, 1527−1534.

Tannous, M., Amin, R., Popoff, M.R., Forentine, C., Kowluru, A., 2001. Positive modulation by Ras of interleukin-1-beta-mediated nitric oxide generation in insulin-secreting clonal beta (HIT-T15) cells. Biochemical Pharmacological 62, 1459−1468.

Thaler, M., Metzger, J., Schrelegg, A., Denk, S., Glebner, A., Hauptmann, H., Luppa, P.B., 2005. Immunoassay for sex hormone-binding globulin in undiluted serum is influenced by high-molecular mass aggregate. Clinical Chemistry 51, 401−407.

Tosato, G., Jones, K.D., 1999. Interleukin-1 induces interleukin-6 production in peripheral blood monocytes. Blood 75, 1305−1310.

Wang, J., Liu, J.D., Lin, Y.H., 2006a. Small 1134−1138.

Wang, J., Liu, M.H., Engelhand, M.H., Lin, Y., 2006. Analytical Chemistry 78, 3974−6979.

Wang, L., Wang, K., Santra, S., Zhao, X., et al., February 1, 2008. Watching silica nanoparticles. Analytical Chemistry 647−654.

Wang, Y., Zhang, Z., Jain, V., Jinju, Y., Muellere, S., Sokolova, J., Liu, Z., Levon, K., Rigas, B., Rafailovicha, M.H., 8 April 2010. Potentiometric sensors based on surface molecular imprinting: Detection of cancer markers and viruses. sensors and actuators B: Chemical 149 (1), 381−387.

Wang, X.J., Yu, J.J., Sreekumar, A., Varambally, S., Shen, R.L., Giacherio, D., Mehra, R., Montie, J.E., Pienta, K.J., Sanda, M.G., Kantoff, P.W., Rubin, M.A., Wei, J.T., Ghosh, D., Chinnalyan, A.M., 2005. Autoantibody signatures in prostate cancer. New England Journal of Medicine 353, 1224−1235.

Wei, Q., Xiang, Z., He, J., Wang, G., Li, H., Qian, Z., Yang, M., 2010. Biosensors & Bioelectronics 26, 627−631.

Weinhold, B., Ruther, U., 1997. Interleukin-6-dependent and independent regulation of the human C-reactive protein gene. Biochemical Journal 327, 425−429.

Wolf, M., Juncker, D., Michel, B., Hunziker, F., Delmarche, E., May 15, 2004. Simultaneous detection of C-reactive protein and other cardiac markers in human plasmon using micro-mosaic immunoassays and self-regulating microfluidics networks. Biosensors and Bio-electronics 19 (10), 1193−1202.

Wu, J., Fu, Z., Yan, F., Ju, H., 2010. Trends in Analytical Chemistry 26, 627−631.

Zhong, L., Peng, X.J., Hidalgo, G.E., Doherty, D.E., Stromberg, A.J., Hirschowitz, E.A., 2003. Antibodies to HSP70 and HSP90 in serum in non-small cell lung cancer patients. Cancer Detection and Prevention 27, 285−290.

Zhou, Y., Yu, R., Chai, R., Hong, C., 2009. Journal of Electroanalytical Chemistry 628, 90−96.

FURTHER READING

Adams, J.E., Bodor, G.S., Davilla-Roman, V.G., et al., 1993a. Cardiac troponin I: a marker with high specificity for cardiac injury. Circulation 81, 101−108.

Adams 3rd, J., Abendschein, D., Jaffe, A., 1993b. Biochemical markers of myocardial injury, is MB creatine kinase the choice for the I990s? Circulation 88, 750−763.

Adhikari, B., Majumdar, S., 2004. Polymers in sensor applications. Progress in Polymer Science 29, 699−766.

Almazan, G., Liu, H.N., Khorchid, A., Sundararajan, S., Martinez-Bermudez, A.K., Chemtob, S., 2000. Exposure of developing oligodendrocytes to cadmium causes HSP72 introduction, free radical generation, reduction in glutathione levels, and cell death. Free Radicals in Biology and Medicine 29, 858–869.

Ameen, C., Strehl, R., Bjorquist, P., Lindahl, A., Hyliner, J., Sartipy, P., 2008. Human embryonic stem cells: current technologies and emerging industrial applications. Clinical Reviews in Oncology and Hematology 85, 54–80.

Apple, F.S., Jesse, R.L., Newby, L.K., Wu, A.H.B., Christensen, R.H., Cannon, C.P., Francis, G., Jesse, R., Morrow, D.A., Ravkilde, J., et al., 2007. National academy of clinical biochemistry and IFCC committee for standardization of markers of cardiac damage laboratory medicine practice guidelines: analytical issues for biochemical markers of acute coronary syndromes. Clinical Chemistry 53, 547–551.

Balasubramanian, K., 2010. Challenges in the use of ID nanostructures for on-chip biosensing and diagnostics. Biosensors & Bioelectronics 26, 1195–1204.

Barbas, J., Ellis, W.R., Santhanagopalan, V., Blaszczynski, M., Winge, D.R., 1992. Conversion in the peptides coating cadmium sulfide crystallites in Candida glabrata. Journal of Inorganic Biochemistry 48, 95–105.

Beard, K.M., Shangari, N., Wu, B., O'Brein, P.J., 2003. Metabolism, not autooxidation, plays a role in alpha-oxoaldehyde- and reducing sugar-induced erythrocyte GSH depletion: relevance for diabetes mellitus. Molecular Cellular Biochemistry 252, 331–338.

Bernard, A., Renault, J.P., Michel, l, Bosshard, H.R., Delmarche, E., 2003. Microcontact printing of proteins. Advances in Materials 12, 1067–1070.

Borque, L., Bellod, L., Rus, A., et al., 2000. Development and validation of an automated and ultrasensitive immunoturbidimetric assay for C-reactive protein. Clinical Chemistry 46, 1839–1842.

Boushey, C.J., Beresford, S.A.A., Omenn, G.S., Moulsky, A.G., 1995. A quantitative assessment of plasma homocysteine as a risk factor for vascular disease. Probable benefits of increasing folic acid intake. JAMA 274, 1049–1057.

Bradford, T.J., Wang, X.J., Chinnalyan, A.M., 2006. Cancer immunomics: using autoantibody signatures in the early detection of prostate cancer. Urologic Oncology-Seminars and Original Investigations 24, 237–242.

Canlan, M.J., Chen, Y.T., Williamson, B., Gore, A.O., Stockert, J.D., Gordan, J.D., Tureci, O., Sahin, U., Pfeundschuh, M., Old, L.J., 1998. Characterization of human colon cancer antigens recognized by autologous antibodies. International Journal of Cancer 76, 652–658.

Carlo, M.D., Loeser, R.F., 2003. Increased oxidative stress with aging reduces chondrocyte survival: correlation with intracellular glutathione levels. Arthritis Rheumatism 48, 3419–3430.

Caulum, M.M., Murphy, B.M., Ramsay, L.M., Henry, C.S., 2007. Detection of cardiac biomarkers using micellar electrokinetic chromatography and a cleavable tag immunoassay. Analytical Chemistry 79, 5249–5256.

Chapman, C., Murray, A., Chakrabarti, J., Thorpe, C., Woolston, C., Sahin, A., Barynes, A., Robertson, A., 2007. Autoantibodies in breast cancer: their use in early diagnosis. Annals of Oncology.

Chen, K., Li, B., Chen, Y., 2011. Nano Today 6, 131–154.

Chua, J.H., Chee, R.E., Agarwal, A., Wong, S.M., Zhang, G.J., 2009. Label-free electrical detection of cardiac biomarker with complementary metal-oxide semiconductor-compatible silicon nanowire sensor arrays. Analytical Chemistry 81, 6266–6271.

Clark-Taylor, T., Clark-Taylor, B.E., 2004. Is autism a disorder of fatty acid metabolism? Possible dysfunction of mitochondrial β oxidation by long chain acyl-CoA dehydrogenase. Medical Hypothesis 62, 970–975.

Clarkson, T.W., 1992. Mercury: major issues in environmental health. Enviromental Health Perspectives 100, 31–38.

Curreli, M., Rui, Z., Ishikawa, F.N., Hsiao-Kang, C., Coye, R.J., Chongwu, Z., Thompson, M.E., 2008. Real-time, label-free detection of biological entities using nanowire-based FETs. IEEE Transactions Nantechnology 7, 651–667.

Dameron, C.T., Reese, R.N., Mehra, R.K., Kortan, A.R., Carroll, P.J., Steigerwld, M.L., Brus, L.E., Winge, D.R., 1989. Biosynthesis of cadmium sulfide quantum semiconductor crystallites. Nature 338, 596–597.

Darain, F., Yager, P., Gan, K.L., Tjin, S.C., 2009. on-chip detection of myoglobin based on fluorescence. Biosensors & Bioelectronics 24, 1744–1750.

Darain, F., Shim, S.P.Y., 2003. Biosensors & Bioelectronics 18, 733–780.

Dewitt, D.L., 1981. Prostaglandin endoperoxide synthase- regulation of enzyme expression. Biochemical Biophysica Acta 1083, 121–134.

DeWitt, D.L., el-Harith, E.A., Kraemer, S.A., et al., 1991. The aspirin and heme-binding sites of ovine and murine gene prostaglandin endoperoxide synthases. Journal of Biological Chemistry 266, 14511–14518.

DiMasi, J.A., Hansen, R.W., Grabowski, H.G., 2003. The price of innovation: new estimates of drug development costs. Journal of Health Economics 22 (2), 151–185.

Disis, M.L., Pupa, S.M., Gralow, J.R., Dittadi, R., Manard, M.A., Cheever, M.A. High-time HER-2/neu protein-specific antibody can be detected in patients with early-stage breast cancer. Journal of Clinical Oncology 15, 3363–3367.

Draine, B.T., Flatau, P.J., 1994. Discrete –dipole approximation for scattering calculations. Journal of Optical Society American Association 11, 1491–1499.

Droge, W., 2002. Free radicals in the physiological control of cell function. Physiological Reviews 82, 47–95.

Duffy, D.C., McDonald, J.C., Schueller, O.J.A., Whitesides, G.M., 1998. Rapid prototyping of microfluidic systems in poly(dimethylsiloxane). Analytical Chemistry 70, 4974–4984.

Dufner, D., Previs, S.F., 2003. Measuring in vivo metabolism using heavy water. Current Opinion in Clinical Nutrition and Metabolic Care 6 (5), 511–517.

Dutra, R.F., Kubota, L.T., 2007. An SPR immunosensor for human cardiac troponin T using specific binding avidin to biotin at carboxymethyldextran-modified gold chip. Clinical Chimica Acta 376, 114–120.

Duyk, G., 2003. Attrition and translation. Science 302 (5645), 603–605.

Du, D., Wang, J., Smith, J.N., Timchalk, C., Lin, Y., 2011. Biosensors & Bioelectronics 26, 3857–3863.

ESC/ACC, Consensus Document, 2000. Myocardial infarction redefined: a consensus document of the Joint European Society of Cardiology/American College of Cardiology committee for the redefinition of myocardial infarction. European Heart Journal 21, 1502–1513.

Ellman, G.L., Courtney, K.D., Andres Jr., V., Fathestone, R.M., 1961. Biochemical Pharmacology 7, 89–95.

Etievent, J.P., Chocron, S., Toubin, G., Taberlet, C., Aolwan, K., Clement, F., Cordier, A., Schipman, N., Kantelip, J.P., 1995. Use of cardiac troponin I as a marker of perioperative myocardial Ischemia. Annals of Thoracic Surgery 59, 1192–1194.

Fanara, P., Turner, S., Busch, R., Killion, S., Awada, M., Turner, H., Mahaut, A., LaPrada, K.L., Stark, J.M., Hellerstein, M.K. In vivo measurement of microtubule dynamics using stable isotope labeling with heavy water: effect of taxanes. Journal of Biological Chemistry 279(48), 49940–49947

Fletcher, B.S., Lim, R.W., Varmuth, B.C., Kuzbu, R.A., Koshi, R.A., Herschman, H.R., 1990. Structure and expression of Cs21-a primary response gene induced by growth-factors and tumor promoters. Journal of Biological Chemistry 266, 5192–5196.

Le Fanu, J. (Ed.), 2000. The Rise and Fall of Modern Medicine. Caroll and Garf Publishers Inc, New York, NY, USA.

Gerard, M., Chaubey, A., Malhotra, B.D., 2002. Application of conducting polymers to biosensors. Biosensors & Bioelectronics 17, 345–359.

Graham, Im, Daly, L., Refsum, H.M., Robinson, K., Brattstrom, L.E., Ueland, P.M., Palma-Reis, Rj, et al., 1997. Plasma homocysteine as a risk factor for vascular disease: the European concerted project. JAMA 277, 1775–1781.

Heeger, P.S., Heeger, A.J., 1999. Making sense of polymer-based biosensors. Proceedings of the National Academy of Science of the United States of America 96, 12219–12221.

Hellerstein, M.K., Murphy, E., 2004. Stable isotope-mass spectrometric measurements in molecular fluxes in vivo, emerging applications in drug development. Current Opinion Molecular Therapy 8 (3), 249–264.

Hellerstein, M.K., 2004. New stable isotope-mass spectrometric techniques for measuring fluxes through intact metabolic pathways in mammalian systems: introduction of moving pictures into functional genomics and biochemical phenotyping. Metabolic Engineering 6 (1), 85–100.

Hla, T., Nelson, K., 1992. Human cyclooxygenase-2 DNA. Proceedings of the National Academy of Science 88, 7384–7386.

Jacobsen, D.W., 2000. Hyperhomocysteinemia and oxidative stress: time for a reality check? Arteriosclerosis Thrombosis and Vascular Biology 20, 1182–1184.

James, S.J., Metnyk, S., Jernigan, S., Cleves, M.A., Halsted, C.H., Wong, D.J., Cutler, P., Bock, K., Boris, M., Bradstreet, J.J., Baker, S.M., Gaylor, D.W., 2006. Metabolic endophenotype and related genotypes are associated with oxidative stress in children with autism. American Journal of Medical Genetics B Neuropsychiatric Genetics 141B, 947–956.

Kapoor, V., Singh, A.K., Dey, S., Sharma, S.C., Das, S.N., 2010. Circulating cyclooxygenase-2 in patients with tobacco-related intraoral squamous cell carcinoma and evaluation of its peptide inhibitors and potential antitumor agent. Journal of Cancer Research and Clinical Oncology 136, 1785–1804.

Katz, E., Willner, I., 2004. ChemPhysChem 5, 1084–1104.

Kurita, R., Yokota, Y., Sato, Y., Mizutani, F., Niwa, O., 2006. On-chip enzyme immunoassay of a cardiac marker using a microfluidic device combined with a portable surface plasmon system. Analytical Chemistry 78, 5525–5531.

Lee, I., Park, H., Park, S., Kim, M.J., Yun, M., 2008. Highly reproducible single polyaniline nanowire using electrophoresis method. Nano 3, 75–82.

Liedberg, B., Nylander, C., Lunstrom, I., 1983. Surface plasmon resonance for gas detection and biosensing. Sensing and Actuators 4, 299–304.

Lin, T.W., Hsieh, P.J., Lin, C.L., Fang, Y.Y., Yang, J.X., Tsai, C.C., Chaing, P.L., Pan, C.Y., Chen, Y.T., 2010. Label-free detection of protein-protein interactions using a calmodulin–modified nanowire transistor. Proceedings of the National Academy of Sciences of the United States of America 107, 1047–1952.

Lowe, C.R., 2011. The future: biomarkers, biosensors, neuroinformatics, and neuropsychiatry. International Review of Neurobiology 101, 375–405.

Lukosz, W., 1991. Principles and sensitivities of integrated optical and surface plasmon sensors for direct affinity sensing and immunosensing. Biosensors and Bioelectronics. 6, 215–225.

Malhotra, B.D., Chaubey, A., Singh, S.P., 2006. Prospects of conducting polymers in biosensors. Analytical Chimica Acta 578, 59–74.

Magariiti, M.G., Tsakalof, A.K., Tsatsakis, A.M., 2007. Therapeutic Drug Monitoring 29, 150–183.

Masson, J.F., Obando, L., Beaudoin, S., Booksh, K., 2004. Sensitive and real-time fiber-optic-based surface plasmon resonance sensors for myoglobin and cardiac troponin I. Talanta 62, 865–870.

Nair, P.R., Alam, M.A., 1988. Design considerations of silicon nanowire biosensors. IEEE Transactions Electron Devices 54, 45–56.

Needleman, P., Isakson, P.C., 1997. The discovery and function of COX-2. Journal of Rheumatology 24, 6–8.

Noort, D., Fidder, A., van der Schans, M.J., Hulst, A.G., 2006. Analytical Chemistry 78, 6640–6644.

Polhuijs, M., Langenberg, J.P., Benschhop, H.P., 2002. Toxicology and Applied Pharmacology 184, 116–126.

Plowman, T.E., Durstchi, J.D., Christensen, D.A., Herron, J.N., Reichert, W.M., 1999. Analytical Chemistry 71, 4344–4352.

Pope, C.N., 1997. Journal of Toxicology and Environmental Health 161–181. Part 2.

Roberts, W.L., Moulton, L., Law, T.C., et al., 2001. Evaluation of nine automated high-sensitivity C-reactive protein methods: implications for clinical and epidemiological applications. Clinical Chemistry 47 (Part 2), 418–425.

Schoenheimer, R., 2000. The Dynamic State of Body Constituents. Hafner Publishers Co, New York, NY, USA.

Seibert, K., Zhang, K., Leahy, K., et al., 1994. Pharmacological and biochemical demonstration of the role of cyclooxygenase 2 in inflammation and pain. Proceedings of the National Academy of Sciences of the United States of America 91, 12013–12017.

Smith, W.L., DeWitt, D.L., Garavillo, R.M., 1981. Cyclooxygenases: structural, cellular, and molecular biology. Annual Review Biochemistry 69, 145–182.

Somvanshi, R.K., Kumar, A., Kant, S., et al., 2007. Surface plasmon resonance studies and biochemical evaluation of a potent peptide inhibitor against cyclooxygenase-2 as an anti-inflammatory agent. Biochemical Biophysical Research Communication 36, 37–42.

Sporty, S.J.L.S., Lemiire, S.W., Jakubowski, E.M., Renner, J.A., Evans, R.A., Williams, R.F., Schmidt, J.G., van der Schans, M.J., Noort, D., Johnson, R.C., 2010. Analytical Chemistry 82, 6593–6600.

Story, P., Hooper, M.J., Astheimer, L.B., Buttemer, W.A., 2011. Environmental Toxicology and Chemistry 30, 1163–1169.

Tastsakis, A.M., Barbounis, M.G., Kavalakis, M., Kokkinakis, M., Terzi, I., Tzatzarakis, M.N., 2010. Joural of Chromatography B 878, 1246–1252.

Tolani, S., Craig, M., DeLong, R., Ghosh, K., Wanekaya, A., 2009. Towards biosensors based on conducting polymer nanowires. Analytical and Bioanalytical Chemistry 393, 1225–1231.

Turner, S.M., Murphy, E.J., Neese, R.A., Anstric, F., Thomas, T., Agarwal, A., Go, C., Hellerstein, M.R., 2003. Measurement of TG synthesis and turnover in vivo by $^{1}H_2O$ incorporation into the glycerol moiety and application of MIDA. American Journal of Physiological Endocrinological Metabolism 265 (4), E790–E803.

Tang, W.H.W., Francis, G.S., Morrow, D.A., Newby, L.K., Cannon, C.P., Jesse, R.L., Storrow, A.B., Christensen, R.H., Apple, F.S., Ravkilode, et al., 2007. National academy of clinical biochemistry laboratory medicine guidelines. Clinical utilization of cardiac biomarker testing in heart failure. Circulation E99–E109.

Waterlow, J.C., Garick, P.J., Millward, D.J., 1978. Protein Turnover in Mammalian Tissues and in the Whole Body. North-Holland Publishing Co., Amsterdam, The Netherlands, pp. 301–325.

Wessels, D., Barr, D.B., Mendola, P., 2003. Environmental Health Perspectives 111, 1939–1946.

Wild, S., Roglic, G., Green, A., Sicree, R., Kung, H., May 2004. Global prevalence of diabetes: estimates for the year 2000 and projections for 2030. Diabetes Care 27 (5), 1047–1053.

Wolfe, R.R., 1992. Radioactive and Stable Isotope Tracers in Biomedicine: Principles and Practice of Kinetic Analysis. Wiley-Liss, New York, NY, USA.

Worek, F., Masi, U., Kiderten, D., Diepoki, C., Eyer, P., 1999. Clinica Chimica Acta 288, 73–90.

Yun, M., Myung, N.V., Vaquez, R.P., Lee, C., Menke, E., Penner, R.M., 2004. Electrochemically grown wires for individually addressable sensor arrays. Nano Letters 4, 419–422.

Zhang, K., Mai, Q., Guan, G., Liu, B., Wang, Z., Zhang, Z., 2010. Analytical Chemistry 82, 9579–9586.

Chapter 11

Detection of Biomarkers for Different Diseases on Biosensor Surfaces Part II

11.1 INTRODUCTION

The previous chapters have described the detection of different biomarkers for different diseases. Each chapter analyzed the different biomarker for a particular disease. In this chapter we analyze the binding and dissociation (if applicable) kinetics of other biomarkers on biosensor surfaces. Some of the examples analyzed include the following:

1. Binding to and dissociation of different aptamer beacon modifications of interferon (IFN)-gamma in solution using fluorescence resonance energy transfer (FRET) and immobilized on an avidin-coated surface (Tuleuova et al., 2010).
2. Binding and dissociation of Glutathione S-transferase fused to the N-terminus of a protein (GST-N) protein in Phosphate Buffered Saline (PBS) to a localized surface plasmon resonance coupled fluorescence (LSPCF) biosensor (Huang et al., 2009).
3. Binding of a cytochrome c mutant to an amperometric superoxide biosensor (Wegerich et al., 2009).
4. Binding of different concentrations (in micromoles) of carbonic anhydrase-II (CA-II) in solution to immobilized 4-(2-aminoethyl)-benzene sulfonamide (ABS) using signal-locking surface plasmon resonance (SPR) (Williams et al., 2010).
5. Binding to and dissociation from a microfluidic platform of 50 μM glycerol secreted from differentiated (murine 3T3) adipocytes (Clark et al., 2010).
6. Binding of different concentrations of C-reactive protein (CRP) to a new sandwich-type assay design using a label-free detection method.

Some of the other biomarker studies that have appeared in the recent literature or have been presented at conferences include the following:

1. Alzheimer: a new theory (Agnvall, 2010).
2. Dual-Enzyme Biosensor for detection of organophosphorous compounds using organophosphorous Hydrolase and Horseradish Peroxide (Sahin et al., 2010).

Biomarkers and Biosensors. http://dx.doi.org/10.1016/B978-0-444-53794-2.00011-2

3. Multichannel mass organic analyzer and microfluidic networks for the automated in situ microchip electrophoretic analysis of organic biomarkers (Benhabib et a1., 2010).
4. Hybrid magnetic-plasmonic nanoparticles for biomarkers (Hirt et al., 2010).
5. Engineered Knottin peptides: a new class of agents for noninvasive molecular imaging of tumor biomarkers (Apte and Graves, 2010).
6. Identifying secreted biomarkers for murine evasion in cellular models of cancer (Kinke, 2010).
7. SPR biosensor for parallelized detection of protein biomarkers in diluted blood plasma (Pilarik et al., 2010).
8. Biomarkers in drug discovery and development: from target identification through drug marketing (Colburn and Keefe, 2000).
9. Validation of analytic methods for biomarkers used in drug development (Anonymous, 2006).
10. Electrochemical biosensors: toward point-of-care diagnostics (Wang, 2006).
11. Biosensors for biomarkers in medical diagnostics (Mancini and Tombelll, 2008).
12. Point-of-care biosensor systems for cancer diagnostics/prognostics (Sofer et al., 2006).
13. The demonstration of the immunochemical biomarkers in methyl methacrylate-embedded plucked human hair follicles (Anonymous, 2007).
14. Surface plasmon resonance biosensor based on Vroman effect: toward cancer biomarker detection (Choi and Chase, 2009).
15. Biogenic nanoporous silica-based sensor for enhanced electrochemical detection of cardiovascular biomarker proteins (Lin et al., 2010).
16. Nanomonitor: a miniature electronic biosensor for glycan biomarker detection (Nagaraj et al., 2010).
17. Multifunction dendrimer-template antibody presentation on biosensor surfaces for improved biomarker detection (Han et al., 2010).
18. Rapid and sensitive detection of protein biomarker using a portable fluorescence biosensor based on quantum dots and a lateral flow strip (Li et al., 2009).
20. A biomarker concept for assessment of insulin resistance, beta-cell function, and chronic system inflammation in type 2 diabetes mellitus (Pfutzner et al., 2008).
21. Multifunctional Au nanoparticle dendrimer-based surface plasmon resonance biosensor and its application for improved insulin detection (Frasconi et al., 2010).

Recently published reports are also available that describe in detail the different aspects of biomarkers and their applications in a clinical setting and the collaborative efforts that are required for their successful development

(LaRia, 2010). For example, they include the highlights of key technologies that are required for the development of imaging biomarkers. More importantly, case studies are presented of individual imaging biomarkers. Finally, the future of imaging biomarkers is also presented.

We now use fractal analysis to analyze the binding and dissociation kinetics of some of the different biomarkers available in the open literature. The examples were selected at random, with no particular bias toward analyzing a particular biomarker, or a class of biomarkers.

11.2 THEORY

Havlin (1987) has reviewed and analyzed the diffusion of reactants toward fractal surfaces. The details of the theory and the equations involved for the binding and the dissociation phases for analyte—receptor binding are available (Sadana, 2001). The details are not repeated here, except that the equations are given to permit an easier reading. These equations have been applied to other biosensor systems (Sadana, 2001; Ramakrishnan and Sadana, 2001; Sadana, 2005). For most applications, a single- or a dual-fractal analysis is often adequate to describe the binding and the dissociation kinetics. Peculiarities in the values of the binding and the dissociation rate coefficients, as well as in the values of the fractal dimensions with regard to the dilute analyte systems being analyzed will be carefully noted, if applicable.

In this chapter we analyze the binding and dissociation kinetics (if applicable) of (1) IFN-gamma as a function of aptamer variants and inclusion of spacer in addition to spacer (Tuleuova et al., 2010), (2) GST-N protein in PBS and GST-N protein in 10-fold diluted serum to an LPSCF fiber-optic biosensor (Huang et al., 2009), (3) cytochrome c mutant to a superoxide biosensor (Wegerich et al., 2009), (4) CA-II to an ABS ligand on an SPR biosensor surface (Williams et al., 2009), (5) glycerol secretion from differentiated (murine 3T3-L1) adipocytes to a microfluidic platform for fluorescence-based assay (Clark et al., 2010), and (6) different concentrations of CRP in solution to a sandwich-type assay using a label-free detection method, reflectometric interference spectroscopy (Albrecht et al., 2010).

11.2.1 Single-Fractal Analysis

11.2.1.1 Binding Rate Coefficient

Havlin (1989) indicates that the diffusion of a particle (analyte [Ag]) from a homogeneous solution to a solid surface (e.g., receptor [Ab]-coated surface) on which it reacts to form a product (analyte—receptor complex (Ab.Ag)) is given by

$$(\mathrm{Ab}\cdot\mathrm{Ag}) \approx \begin{cases} t^{(3-D_{f,\mathrm{bind}})/2} = t^{p} & t < t_c \\ t^{1/2} & t > t_c \end{cases} \tag{11.1}$$

Here $D_{f,bind}$ or D_f (used later on in the chapter) is the fractal dimension of the surface during the binding step. t_c is the crossover value. Havlin (1989) indicates that the crossover value may be determined by $r_c^2 \sim t_c$. Above the characteristic length, r_c, the self-similarity of the surface is lost and the surface may be considered homogeneous. Above time t_c the surface may be considered homogeneous, since the self-similarity property disappears, and "regular" diffusion is now present. For a homogeneous surface where $D_f = 2$, and when only diffusional limitations are present, $p = ½$ as it should be. Another way of looking at the $p = ½$ case (where $D_{f,bind} = 2$) is that the analyte in solution views the fractal object, in our case, the receptor-coated biosensor surface, from a "large distance." In essence, in the association process, the diffusion of the analyte from the solution to the receptor surface creates a depletion layer of width $(Đt)^{½}$ where Đ is the diffusion constant. This gives rise to the fractal power law (Analyte.Receptor) $\sim t^{(3-D_{f,bind})/2}$. For the present analysis, t_c is arbitrarily chosen and we assume that the value of the t_c is not reached. One may consider the approach as an intermediate "heuristic" approach that may be used in the future to develop an autonomous (and not time-dependent) model for diffusion-controlled kinetics.

11.2.1.2 Dissociation Rate Coefficient

The diffusion of the dissociated particle (receptor [Ab] or analyte [Ag]) from the solid surface (e.g., analyte [Ag]—receptor [Ab] complex-coated surface) into solution may be given, as a first approximation by

$$(\mathrm{Ab}\cdot\mathrm{Ag}) \approx -t^{(3-D_{f,diss})/2} = -t^{p} \quad (t > t_{diss}) \qquad (11.2)$$

Here $D_{f,diss}$ is the fractal dimension of the surface for the dissociation step. This corresponds to the highest concentration of the analyte—receptor complex on the surface. Henceforth, its concentration only decreases. The dissociation kinetics may be analyzed in a manner "similar" to the binding kinetics.

11.2.2 Dual-Fractal Analysis

11.2.2.1 Binding Rate Coefficient

Sometimes, the binding curve exhibits complexities and two parameters (k, D_f) are not sufficient to adequately describe the binding kinetics. This is further corroborated by low values of r^2 factor (goodness of fit). In that case, one resorts to a dual-fractal analysis (four parameters; k_1, k_2, D_{f1}, and D_{f2}) to adequately describe the binding kinetics. The single-fractal analysis presented above is thus extended to include two fractal dimensions. At present, the time ($t = t_1$) at which the "first" fractal dimension "changes" to the "second" fractal dimension is arbitrary and empirical. For the most part, it is dictated by the data analyzed and experience gained by handling a single-fractal analysis. A smoother curve is obtained in the "transition" region, if care is taken to select the correct number of points for the two regions. In this case, the product (antibody—antigen; or analyte—receptor complex, Ab.Ag or analyte.receptor) is given by

$$(\mathrm{Ab \cdot Ag}) \approx \begin{cases} t^{(3-D_{f1,bind})/2} = t^{p1} & (t < t_1) \\ t^{(3-D_{f2,bind})/2} = t^{p2} & (t_1 < t < t_2) = t_c \\ t^{1/2} & (t > t_c) \end{cases} \tag{11.3}$$

In some cases, as mentioned above, a triple-fractal analysis with six parameters (k_1, k_2, k_3, D_{f1}, D_{f2}, and D_{f3}) may be required to adequately model the binding kinetics. This is when the binding curve exhibits convolutions and complexities in its shape due perhaps to the very dilute nature of the analyte (in some of the cases to be presented) or for some other reasons. Also, in some cases, a dual-fractal analysis may be required to describe the dissociation kinetics.

11.3 RESULTS

Tuleuova et al. (2010) have developed an aptamer beacon for the detection of IFN-gamma, which is an important inflammatory cytokine. Boehm et al. (1997) indicate that it is secreted by immune cells in response to various pathogens. Tuleuova et al. (2010) indicate that the levels of this protein provide important information with regard to infectious diseases and the ability of the body to regulate an immune response. Panteleo and Koup (2004) indicate that there is vigorous production of IFN-gamma in human immunodeficiency virus-infected patients. Tuleuova et al. (2010) indicate that previous antibody-based detection techniques for IFN-gamma were very time consuming. Jayasena (1999) indicates that aptamer-based affinity strategies are coming into prominence. Ellington and Szostak (1990) indicate that aptamers are single-stranded DNA or RNA oligonucleotides that have been selected to bind to target analytes with high specificity and affinity. Aptamers have an advantage over antibodies since they are more robust; thus, aptamer-based biosensors can be regenerated and used over and over again. Furthermore, Balamurugan et al. (2008), Kirby et al. (2004), Nutiu (2005), and Luzi et al. (2003) indicate that aptamers are amenable to modification due to their simplicity and robustness. Tuleuova et al. (2010) indicate that FRET may be used to convert aptamers into real-time biosensors (Urata et al., 2007; Babendure et al., 2003).

Romangani et al. (1986) and Karlsson et al. (2003) indicate that IFN-gamma is an important immune response marker. Thus, Tuleuova et al. (2010) have designed a novel immune response marker to detect IFN-gamma. The DNA aptamer was biotinylated and immobilized on an SPR sensing surface by avidin—biotin interactions. The SPR biosensor was used to analyze the influence of biotinylation, fluorophore attachment, and spacer incorporation on the ability of the aptamer to bind to IFN-gamma.

IFN-gamma is a type II cytokine, and is critical for innate and adaptive immunity against viral and intracellular bacterial infections and for tumor control. Incorrect IFN-gamma expression is associated with a number of autoinflammatory and autoimmune diseases. It plays an important role in immunostimulatory and immunoregulatory effects. The IFN-gamma monomer consists of six alpha-helices and an extended unfolded chain in the C-terminal region (Ealick et al., 1991; Thiel et al., 2000).

Tuleuova et al. (2010) analyzed the influence of aptamer modification on the binding and dissociation of IFN-gamma in solution. Figure 11.1(a) shows the binding and dissociation of 100 nM IFN-gamma in solution to the modified aptamer 3′B (5′-GGG GTT GGT TGT GTT GGG TGT TGT GT-Biotin-3′; sequence with modification) beacon. A single-fractal analysis is adequate to describe the binding and the dissociation kinetics. The values of (1) the binding rate coefficient, k, and the fractal dimension for dissociation, D_f, for binding and (2) the dissociation rate coefficient, k_d, and the fractal dimension for dissociation, D_{fd} for a single-fractal analysis are given in Tables 11.1(a) and (b). Tuleuova et al. (2010) indicate that the highest level of the cytokine (IFN-gamma) binding was observed for the modified aptamer 3′B.

Tuleuova et al. (2010) also included a polyethylene glycol spacer between the aptamer and the biotin so that the nucleotides could be more accessible to the target analyte. Figure 11.1(b) shows the binding of 100 nM IFN-gamma to a 3′BS (5′-GGG GTT GGT TGT GTT GGG TGT TGT GT-C12-Biotin-3′)

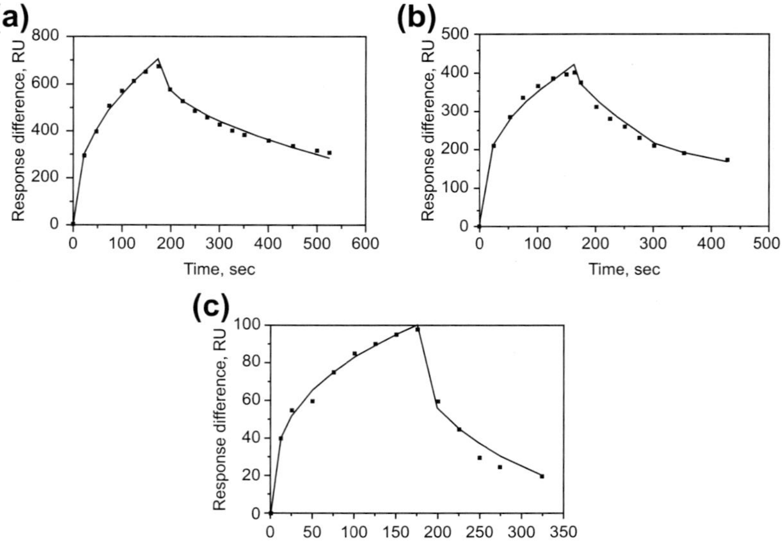

FIGURE 11.1 Binding and dissociation of IFN-gamma in solution to different aptamer beacon modifications for fluorescence resonance energy transfer (FRET) and immobilized on an avidin-coated surface (Tuleuova et al., 2010): (a) B*; (b) 3′BS, and (c) 5′BS.

TABLE 11.1a Binding and Dissociation Rate Coefficients for Interferon (IFN)-Gamma as a Function of (a) Aptamer Variants and (b) Inclusion of Spacer in Addition to Biotin (BS) (Tuleuova et al., 2010)

Analyte/ Receptor	Aptamer Variant	k	k_1	k_2	k_d
IFN-gamma/ aptamer beacon	3′B	75.790 ± 2.690	NA	NA	20.3111 ± 1.127
IFN-gamma/ aptamer beacon	3′BS	72.231 ± 3.407	NA	NA	3.934 ± 0.809
IFN-gamma/ aptamer beacon	5″BS	17.437 ± 0.880	NA	NA	13.682 ± 1.415
IFN-gamma/ aptamer + BS	Aptamer-B	161.99 ± 12.68	95.496 ± 7.926	656.49 ± 1.13	23.388 ± 2.514
IFN-gamma/ aptamer + BS	F-aptamer-B	166.65 ± 19.34	104.54 ± 7.65	509.19 ± 0.262	36.330 ± 1.289

TABLE 11.1b Fractal Dimensions for the Binding and the Dissociation Phases for Interferon (IFN)-gamma as a Function of (a) Aptamer Variants and (b) Inclusion of Spacer in Addition to Biotin (BS) (Tuleuova et al., 2010)

Analyte/ Receptor	Aptamer variant	D_f	D_{f1}	D_{f2}	D_{fd}
IFN-gamma/ aptamer beacon	3′B	2.1364 ± 0.0416	NA	NA	1.9914 ± 0.0414
IFN-gamma/ aptamer beacon	3′BS	2.3058 ± 0.0560	NA	NA	1.4332 ± 0.1402
IFN-gamma/ aptamer beacon	5′BS	2.3238 ± 0.04372	NA	NA	2.3082 ± 0.1082
IFN-gamma/ aptamer + BS	Aptamer-B	2.3854 ± 0.102	2.0568 ± 0.1167	2.9747 ± 0.00824	2.1812 ± 0.07782
IFN-gamma/ aptamer + BS	F-aptamer-B	2.4808 ± 0.08956	2.1906 ± 0.1034	2.9531 ± 0.00247	2.9358 ± 0.02646

modified aptamer beacon. In this case, in addition to the biotin (B) a spacer was included (BS). Once again, a single-fractal analysis is adequate to describe the binding and the dissociation kinetics. The values of (1) the binding rate coefficient, k, and the fractal dimension, D_f, and (2) the dissociation rate coefficient, k_d, and the fractal dimension for dissociation, D_{fd}, for a single-fractal analysis are given in Tables 11.1(a) and (b). It is of interest to note that as one goes from the modified aptamer beacon 3′B to the modified aptamer beacon 3′BS the binding rate coefficient, k, decreases by 4% from a value of k = 75.79 to k = 72.231, and the fractal dimension, D_f, increases by a factor of 1.079 from a value of $D_f = 2.1364$ to $D_f = 2.3058$. In this case, changes in the fractal dimension (degree of heterogeneity on the biosensor surface) and in the binding rate coefficient are in opposite directions (Tables 11.1 (a) and (b)).

Figure 11.1(c) shows the binding of 100 nM IFN-gamma in solution to the 5′BS(5′-Biotin-C12-GGG GTT GGT TGT GTT GGG TGT TGT Gt-3′) modified aptamer beacon. Once again, a single-fractal analysis is adequate to describe the binding and the dissociation kinetics. The values of (a) the binding rate coefficient, k, and the fractal dimension, D_f, and (b) the dissociation rate coefficient, k_d, and the fractal dimension for dissociation, D_{fd}, are given in Tables 11.1(a) and (b). Note that in this case there is an increase in the fractal dimension, D_f (the highest when compared with the B′ and B'S aptamer modifications), and a decrease in the binding rate coefficient, k (the lowest when compared with B′ and B'S aptamer modifications). In this case, there is a substantial decrease in the binding rate coefficient, k. The decrease is higher than a factor of four.

Figure 11.2(a) and Tables 11.1(a) and (b) show the increase in the binding rate coefficient, k, for a single-fractal analysis with an increase in the fractal dimension, D_f. For the data shown in Figure 11.2(a), the binding rate coefficient, k, is given by

$$k = (1.552 \pm 1.342)\,(D_f)^{3.01 \pm 1.81} \qquad (11.4a)$$

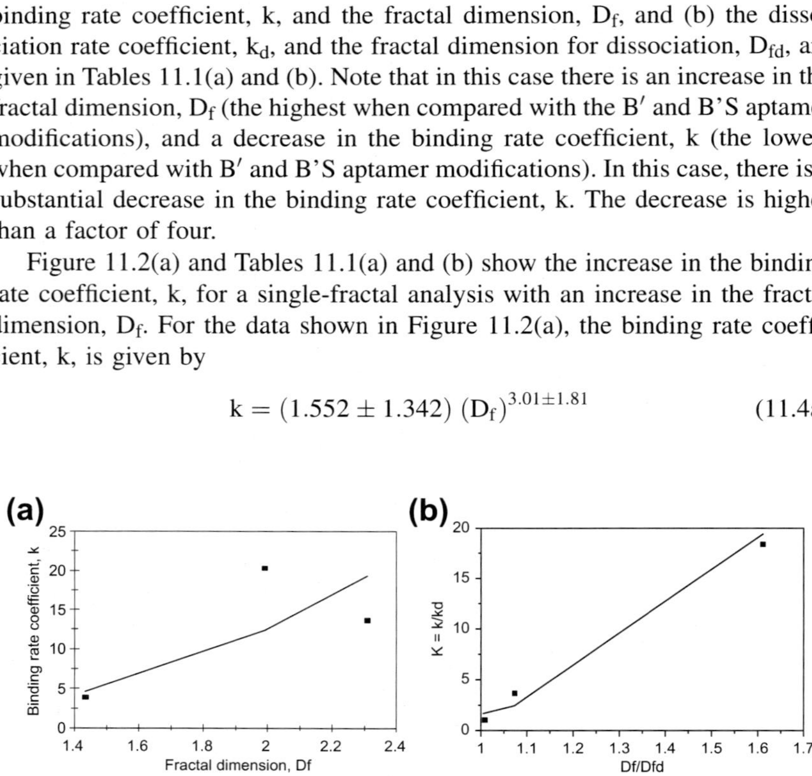

FIGURE 11.2 (a) Increase in the binding rate coefficient, k, with an increase in the fractal dimension, D_f, for a single-fractal analysis. (b) Increase in the affinity, K (=k/k_d), with an increase in the fractal dimension ratio, D_f/D_{fd}, for a single-fractal analysis.

The fit is reasonable. Only three data points are available. The availability of more data points would lead to a more reliable fit. The binding rate coefficient, k, exhibits close to a third (equal to 3.01) order of dependence on the fractal dimension, D_f, that exists on the biosensor surface. This indicates that the binding rate coefficient, k, is sensitive to the fractal dimension, D_f, or the degree of heterogeneity that exists on the biosensor surface (Tables 11.1(a) and (b)).

Figure 11.2(b) and Table 11.1(a) and (b) show the increase in the affinity, K ($=k/k_d$), for a single-fractal analysis with an increase in the fractal dimension ratio, D_f/D_{fd}. For the data shown in Figure 11.2(b), the affinity, K, is given by

$$K\,(=k/k_d) = (1.740 \pm 1.282)\,(D_f/D_{fd})^{5.064 \pm 1.498} \tag{11.4b}$$

The fit is very good. Only three data points are available. The availability of more data points would lead to a more reliable fit. The affinity, K (k/k_d), is very sensitive to the fractal dimension ratio (D_f/D_{fd}) as it exhibits close to a fifth (equal to 5.064) order of dependence on the fractal dimension ratio (D_f/D_{fd}). This is a very convenient way of manipulating the affinity, K, by changing the heterogeneity of the biosensor surface. Some ingenuity may be required here, since a change in the degree of heterogeneity on the biosensor surface would change both the binding as well as the dissociation rate coefficients.

Figure 11.3(a) shows the binding of the IFN-gamma to the biotin + aptamer B variant (Tuleuova et al., 2010). Once again, a dual-fractal analysis is required to describe the binding kinetics. A single-fractal analysis is adequate to describe the dissociation kinetics. The values of (1) the binding rate coefficient, k, and the fractal dimension, D_f, for a single-fractal analysis, (2) the binding rate coefficients, k_1 and k_2, and the fractal dimensions, D_{f1} and D_{f2}, for a dual-fractal analysis, and (3) the dissociation rate coefficient, k_d, and the fractal dimension for dissociation, D_{fd}, for a single-fractal analysis are given in Tables 11.1(a) and (b). In this case, the affinity values K_1 ($=k_1/k_d$) and K_2 ($=k_2/k_d$) are 4.083 and 28.07, respectively (Tables 11.1(a) and (b)).

Note that for dual-fractal analysis for the binding phase, an increase in the fractal dimension by 44.6% from a value of $D_{f1} = 2.0568$ to $D_{f2} = 2.947$ leads to an increase in the binding rate coefficient by a factor of 6.87 from a value of $k_1 = 95.496$ to $k_2 = 656.49$.

Tuleuova et al. (2010) also investigated the influence of a fluorophore (F) in addition to the biotin for the modified aptamer B during the binding of IFN-gamma. This was one of the aptamer variants. Figure 11.3(b) shows that a dual-fractal analysis is required to describe the binding kinetics. A single-fractal analysis is adequate to describe the dissociation kinetics. The values of (1) the binding rate coefficient, k, and the fractal dimension, D_f, for a single-fractal analysis, (2) the binding rate coefficients, k_1 and k_2, and the fractal dimensions, D_{f1} and D_{f2}, for a dual-fractal analysis, and (3) the dissociation rate coefficient, k_d, and the fractal dimension for dissociation, D_{fd}, for a single-fractal analysis are given in Tables 11.1(a) and (b). The affinity values K_1 ($=k_1/k_d$) and K_2 ($=k_2/k_d$) are 2.87 and 14.02, respectively.

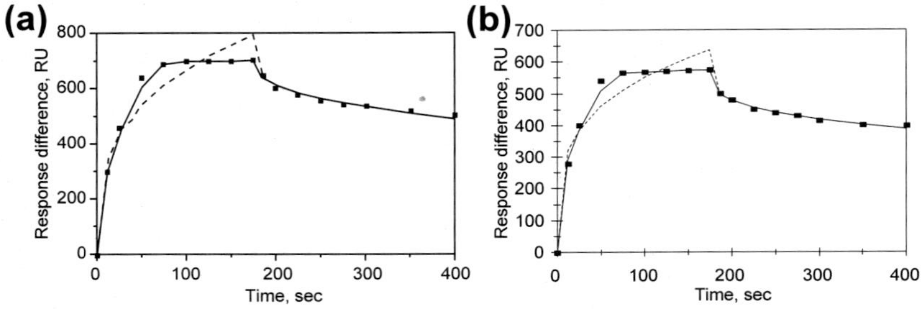

FIGURE 11.3 Binding and dissociation of (interferon) IFN-gamma in solution to aptamer-B (Tuleuova et al., 2010): (a) Aptamer + BS; (b) aptamer + BS + spacer.

Note that for dual-fractal analysis for the binding phase, an increase in the fractal dimension by 34.8% from a value of $D_{f1} = 2.1906$ to $D_{f2} = 2.9531$ leads to an increase in the binding rate coefficient by a factor of 4.87 from a value of $k_1 = 104.54$ to $k_2 = 509.19$ (Tables 11.1(a) and (b)).

Also, note that on comparing the affinity values K_1 and K_2 when the fluorophore is used and not used, the K_1 value is slightly lower and the K_2 value is significantly higher (by about 28%), respectively.

Huang et al. (2009) have developed a localized LSPCF fiber-optic biosensor for the detection of severe acute respiratory syndrome (SARS) coronavirus nucleocapsid protein in human serum. These authors indicate that SARS is a highly infectious disease. Drosten et al. (2004) indicate that SARS results in death in a large portion of patients. The SARS coronavirus (SARS-CoV) causes SARS, and is detectable in the respiratory secretions of patients after infection (Foucher et al., 2003). Wang et al. (2005) emphasize that SARS is highly contagious and exhibits the potential of becoming a large-scale future epidemic if effective therapeutic drugs are not discovered. Huang et al. (2004) and Che et al. (2005) emphasize the need for a rapid, sensitive, specific, and an accurate diagnostic method so that specific patients may be correctly assessed.

Huang et al. (2009) indicate that there are methods available to detect SARS. However, present methods such as reverse transcriptase polymerase chain reaction are not sensitive enough, and also require a specific laboratory with expertise in molecular diagnostics to confirm SARS in the acute phase (Fujimoto et al., 2008; Drosten et al., 2004).

Huang et al. (2009) indicate that gold nanoparticles (GNPs) have been introduced into biosensing (Manso et al., 2008; Cui et al., 2008). These GNPs possess special properties such as localized surface plasmons. Huang et al. (2009) have developed a novel fiber-optic biosensor where the property of LSPCF has been combined with the sandwich immunoassay. Huang et al. (2009) have used their LSPCF fiber-optic biosensor to detect SARS-CoV protein in diluted serum to a limit of 1 pg/ml. This according to these authors exhibits the potential for the early detection of clinical SARS-CoV infection.

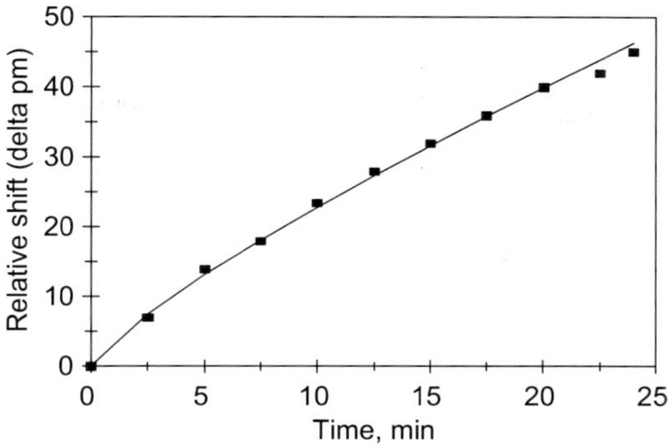

FIGURE 11.4 Binding of 1000 pg/ml GST-N protein in PBS to an LSPCF (localized surface plasmon coupled fluorescence) biosensor (Huang et al., 2009).

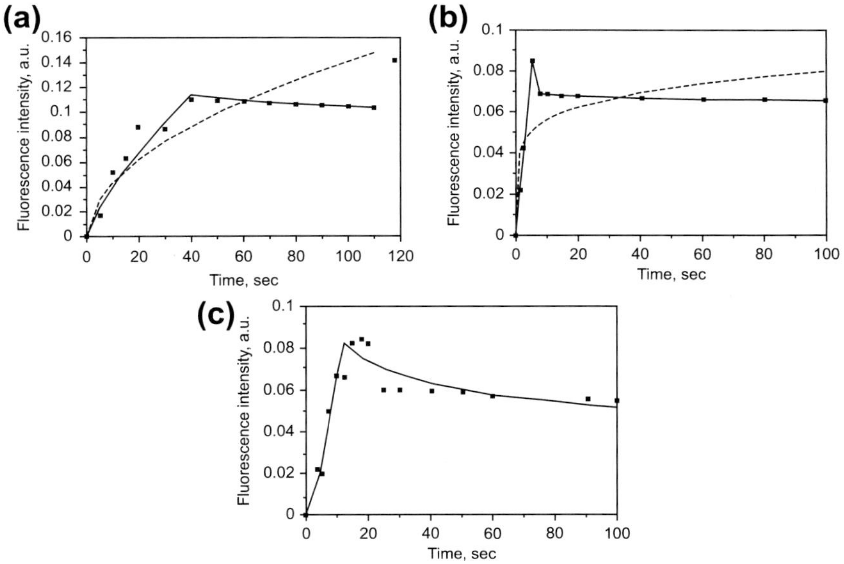

FIGURE 11.5 Binding of different concentrations of GST-N (in picomoles per milliliter) in 10-fold diluted human serum (Huang et al., 2009): (a) 1000; (b) 100; (c) 10.

Figure 11.4 shows the binding of 1000 pg/ml GST-N protein in solution to the LSPCF biosensor (Huang et al., 2009). A single-fractal analysis is required to describe the binding kinetics. The values of (a) the binding rate coefficient, k and the fractal dimension, D_f for a single-fractal analysis are given in Tables 11.2(a) and (b).

TABLE 11.2a **Binding Rate Coefficients for (a) GST-N Protein in PBS and (b) GST-N Protein in 10-Fold Diluted Human Serum to an LSPCF Fiber-Optic Biosensor (Huang et al., 2009)**

Analyte in Solution/Receptor on Surface	k	k_1	k_2
1000 pg/ml GST-N protein in PBS/LSPCF probe	0.00728 ± 0.00241	NA	NA
1000 pg/ml GST-N protein in 10-fold diluted human serum/LSPCF probe	0.01344 ± 0.00598	0.007192 ± 0.00205	0.1615 ± 0.0001
100 pg/ml GST-N protein in 10-fold diluted human serum/LSPCF probe	0.03951 ± 0.01435	0.01768 ± 0.0007	0.07217 ± 0.00045
10 pg/ml GST-N protein in 10-fold diluted human serum/LSPCF probe	0.02934 ± 0.01363	0.002714 ± 0.000670	0.1436 ± 0.00145

LSPCF, localized surface plasmon resonance coupled fluorescence.

TABLE 11.2b **Fractal Dimensions in the Binding Phase for (a) GST-N Protein in PBS and (b) GST-N Protein in 10-Fold Diluted Human Serum to an LSPCF Fiber-Optic Biosensor (Huang et al., 2009)**

Analyte in Solution/Receptor on Surface	D_f	D_{f1}	D_{f2}
1000 pg/ml GST-N protein in PBS/LSPCF probe	1.6224 ± 0.1154	NA	NA
1000 pg/ml GST-N protein in 10-fold diluted human serum/LSPCF probe	1.9782 ± 0.1912	1.4938 ± 0.2552	3−0.0060
100 pg/ml GST-N protein in 10-fold diluted human serum/LSPCF probe	2.9632 ± 0.1365	1.0502 ± 0.00941	3−0.00484
10 pg/ml GST-N protein in 10-fold diluted human serum/LSPCF probe	1.5912 ± 0.2054	0.2082 ± 0.6216	3−0.4444

LSPCF, localized surface plasmon resonance coupled fluorescence.

Huang et al. (2009) prepared GST-N protein samples in 10-fold diluted human serum. This allowed them to test their LSPCF biosensor in clinical samples. They measured the temporal fluorescence intensity of the biomolecular interaction between the LSPCF probes and the GST-N protein. They did this for the 10–1000 pg/ml rages GST-N protein in solution.

Figure 11.5(a) shows the binding of 1000 pg/ml GST-N protein in 10-fold diluted human serum in solution. A dual-fractal analysis is required to adequately describe the binding kinetics. The values of (a) the binding rate coefficient, k and the fractal dimension, D_f for a single fractal analysis, and (b) the binding rate coefficients, k_1 and k_2, and the fractal dimensions, D_{f1} and D_{f2} for a dual-fractal analysis are given in Tables 11.2 (a) and (b). For a dual-fractal analysis, an increase in the fractal dimension by a factor of 2.00 from a value of D_{f1} equal to 1.4938 to D_{f2} equal to 3.0 leads to an increase in the binding rate coefficient by a factor of 22.45 from a value of k_1 equal to 0.007192 to k_2 equal to 0.1615. Once again, an increase in the fractal dimension or the degree of heterogeneity on the LSPCF biosensor surface leads to an increase in the binding rate coefficient.

Figure 11.5(b) shows the binding of 100 pg/ml GST-N protein in 10-fold diluted human serum in solution. Once again, a dual-fractal analysis is required to adequately describe the binding kinetics. The values of (a) the binding rate coefficient, k and the fractal dimension, D_f for a single fractal analysis, and (b) the binding rate coefficients, k_1 and k_2, and the fractal dimensions, D_{f1} and D_{f2} for a dual-fractal analysis are given in Tables 11.2(a) and (b). For a dual-fractal analysis, an increase in the fractal dimension by a factor of 2.86 from a value of D_{f1} equal to 1.0502 to D_{f2} equal to 3.0 leads to an increase in the binding rate coefficient by a factor of 4.08 from a value of k_1 equal to 0.01768 to k_2 equal to 0.07217. Once again, an increase in the fractal dimension or the degree of heterogeneity on the LSPCF biosensor surface leads to an increase in the binding rate coefficient.

Figure 11.5(c) shows the binding of 10 pg/ml GST-N protein in 10-fold diluted human serum in solution. A dual-fractal analysis is required to adequately describe the binding kinetics. The values of (a) the binding rate coefficient, k and the fractal dimension, D_f for a single fractal analysis, and (b) the binding rate coefficients, k_1 and k_2, and the fractal dimensions, D_{f1} and D_{f2} for a dual-fractal analysis are given in Tables 11.2(a) and (b) once again, for a dual-fractal analysis, an increase in the fractal dimension by a factor of 14.4 from a value of D_{f1} equal to 0.2082 to D_{f2} equal to 3.0 leads to an increase in the binding rate coefficient by a factor of 52.9 from a value of k_1 equal to 0.002714 to k_2 equal to 0.1436. Once again, an increase in the fractal dimension or the degree of heterogeneity on the LSPCF biosensor surface leads to an increase in the binding rate coefficient.

Tables 11.2(a) and (b) show for a dual-fractal analysis the increase in the fractal dimension, D_f with an increase in the GST-N protein concentration in solution in the 10−1000 pg/ml range. Figure not shown. For the data shown in Table 11.2 the fractal dimension, D_f is given by:

$$D_{f1} = (0.09843 \pm 0.0794)\ [GST - N]^{0.420 \pm 0.164} \tag{11.5}$$

The fit is reasonable. Only three data points are available. The availability of more data points would lead to a more reliable fit. The fractal dimension, D_{f1} exhibits less than one-half (equal to 0.420) order of dependence on the GST-N protein concentration in solution in the 10−1000 pg/ml range. This indicates that the fractal dimension, D_{f1} exhibits a mild dependence on the GST-N protein concentration in solution.

The superoxide anion radical is present in several pathophysiological situations, such as sepsis (Valko et al., 2007; Vaklko et al., 2006). Electrochemical biosensors can detect this short-lived species (Lisdat, 2005; Prieto-Simon et al., 2008). Wegerich et al. (2009) indicate that the redox protein cytochrome c is used as a recognition element. These authors indicate that superoxide dismutase (SOD) biosensors used for the detection of the superoxide anion often lack the reproducibility due to immobilization problems. However, cyt c-based superoxide biosensors are more stable and may be used in *in vivo* applications (Buttemeyer et al., 2002; Scheller et al., 1999). In this case, the heme protein is reduced by the superoxide, followed by reduction by an electrode.

Wegerich et al. (2009) indicate that short-chain modified gold electrodes exhibit a highly efficient communication between cyt c and the electrode (Frew and Hill, 1988; Hinnen et al., 1983; Nahir et al., 1994; Taniguchi et al., 1982). They have been used for cyt c based superoxide sensors. Wegerich et al. (2009) analyzed the effect of introducing positive charges (lysines) in human cytochrome c on the redox properties and reaction rates of cyt c with superoxide radicals.

These authors claim that the eleven mutants analyzed were modified for structural integrity as well as axial coordination of the heme ion. Their results indicate that four mutants exhibited a higher reaction rate with the radical as compared with the wild type. These mutants were then used for the construction of the superoxide biosensors.

Figure 11.6 shows the binding of cyt c in solution to the superoxide biosensor (Wegerich et al., 2009). A single-fractal analysis is adequate to describe the binding kinetics. The binding rate coefficient, k and the fractal dimension, D_f, for a single-fractal analysis are 0.2109 ± 0.0302, and 1.234 ± 0.1806, respectively. These results are also shown in Tables 11.3(a) and (b). A single-fractal analysis is also adequate to describe the dissociation kinetics. The values of the dissociation rate coefficient, k_d and the fractal dimension, D_{fd} are 0.00116 ± 0.00015 and 0.3666 and 0.14702, respectively. In this case the affinity, K ($=k/k_d$) is equal to 181.81.

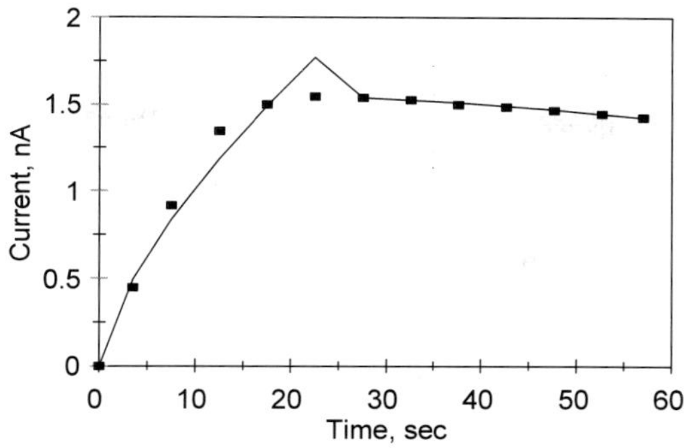

FIGURE 11.6 Binding of a cytochrome c mutant to an amperometric superoxide biosensor (Wegerich et al., 2009).

TABLE 11.3a Binding and Dissociation Rate Coefficients for (a) Cytochrome c Mutant to a Superoxide Biosensor (Wegerich et al., 2009) and (b) Binding of Carbonic Anhydrase-II (CA-II) to a 4-(2-Aminoethyl)-Benzene Sulfonamide (ABS) Ligand on an Surface Plasmon Resonance (SPR) Biosensor Surface (Williams et al., 2009)

Analyte in Solution/Receptor on surface	k	k_1	k_2	k_d	Reference
Cytochrome c mutant/superoxide biosensor	0.2109 ± 0.0302	NA	NA	0.00116 ± 0.00015	Wegerich et al. (2009)
2.3 μM CA-II/4-(2-aminoethyl-benzene sulfonamide (ABS) ligand	40.359 ± 5.604	20.817 ± 1.416	120.16 ± 0.0283	NA	Williams et al. (2010)
5.3 μM CA-II/4-(2-aminoethyl-benzene sulfonamide (ABS) ligand	116.76 ± 7.77	83.718 ± 4.847	152.87 ± 0.46	NA	Williams et al. (2010)
11.0 μM CA-II/4-(2-aminoethyl-benzene sulfonamide (ABS) ligand	98.584 ± 9.468	39.916 ± 8.403	238.66 ± 0.351	NA	Williams et al. (2010)

TABLE 11.3b Fractal Dimensions for the Binding and the Dissociation Phases for (a) Cytochrome c Mutant to a Superoxide Biosensor (Wegerich et al., 2009), and (b) Binding of Carbonic Anhydrase-II (CA-II) to a 4-(2-Aminoethyl)-Benzene Sulfonamide (ABS) Ligand on an (Surface Plasmon Resonance) SPR Biosensor Surface (Williams et al., 2009)

Analyte in Solution/Receptor on Surface	D_f	D_{f1}	D_{f2}	D_{fd}	References
Cytochrome c mutant/ superoxide biosensor	1.2334 ± 0.1806	NA	NA	0.3666 ± 0.1470	Wegerich et al. (2009)
2.3 μM CA-II/4-(2-aminoethyl-benzene sulfonamide (ABS) ligand	2.5114 ± 0.0802	2.1364 ± 0.0785	2.4698 ± 0.00412	NA	Williams et al. (2010)
5.3 μM CA-II/4-(2-aminoethyl-benzene sulfonamide (ABS) ligand	1.1336 ± 0410	2.6146 ± 0.0820	2.9294 ± 0.00692	NA	Williams et al. (2010)
11.0 μM CA-II/4-(2-aminoethyl-benzene sulfonamide (ABS) ligand	1.205 ± 0.075	1.9670 ± 0.2786	2.9807 ± 0.0019	NA	Williams et al. (2010)

Williams et al. (2010) have recently analyzed low noise detection of biomolecular interactions with signal-locking surface plasmon resonance. Surface plasmon resonance is a popular technique to analyze biomolecular interactions at a surface, especially since it is label-free. These authors indicate that the SPR technique is subject to the influence of noise and drift disturbances since that limits the minimum detectable mass change. The SPR technique uses the step response of the biomolecular interactions occurring on the biosensor surface. The technique proposed by Williams et al. (2010) measures the biomolecular interactions over a very narrow frequency range. This locks the measured response to a very specific narrow band signal. The authors used their technique to analyze the binding kinetics of carbonic anhydrase-II (CA-II) and immobilized 4-(2-aminoethyl)-benzenesulfonamide (ABS) to a SPR surface.

Carbonic anhydrases are a family of enzymes that catalyze the rapid conversion of carbon dioxide and water to bicarbonate and protons. These anhydrases are classified as metalloenzymes since the active site of most carbonic anhydrases contains a zinc ion. The primary function of this enzyme in animals is to maintain the acid-base balance in blood and other tissues, and to help transport carbon dioxide out of tissues. Carbon anhydrase II is a novel

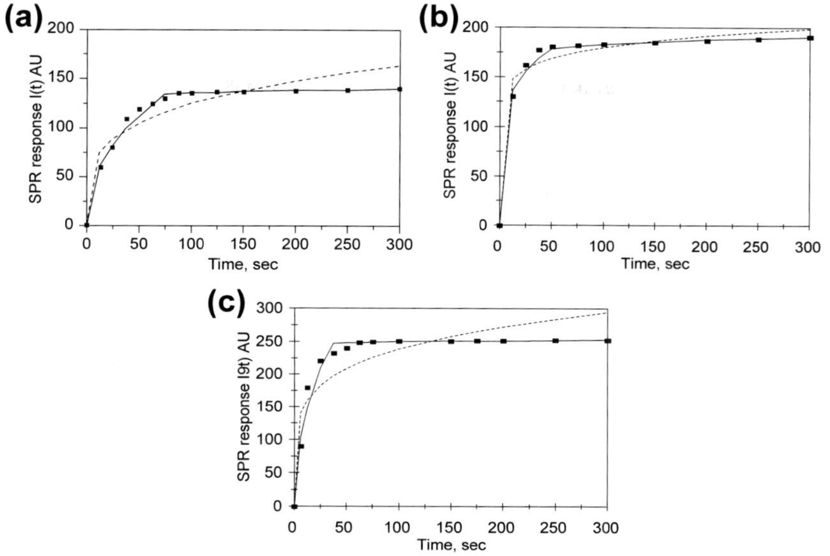

FIGURE 11.7 Binding of different concentrations (in micromoles) of carbonic anhydrase-II (CA-II) analyte in solution to immobilized 4-(2-aminoethyl)-benzene sulfonamide (ABS) using signal-locking surface plasmon resonance (Williams et al., 2010): (a) 2.3; (b) 5.3; (c) 11.0.

biomarker for gasterointestinal stomal tumors (Parkkila et al., 2010). These authors indicate that various carbonic anhydrase (CA) isoenzymes have been identified as potential targets against different cancers. They further indicate that high CA-II expression is associated with a better disease specific survival rate than low or no expression.

Figure 11.7(a) shows the binding of 2.3μM CA-II anhydrase in solution to a 4-(2-amino ethyl)-benzene sulfonamide (ABS) ligand on a SPR biosensor surface (Williams et al., 2009). A dual-fractal analysis is required to adequately describe the binding kinetics. The values of (a) the binding rate coefficient, k and the fractal dimension, D_f for a single fractal analysis, and (b) the binding rate coefficients, k_1 and k_2, and the fractal dimensions, D_{f1} and D_{f2} for a dual-fractal analysis are given in Tables 11.3(a) and (b). It is of interest to note that as the fractal dimension increase by a factor of 1.156 from a value of D_{f1} equal to 2.1364 to D_{f2} equal to 2.4698, the binding rate coefficient increases by a factor of 5.95 from a value of k_1 equal to 20.817 to k_2 equal to 120.16.

Figure 11.7(b) shows the binding of 5.3μM CA-II anhydrase in solution to a 4-(2-amino ethyl)-benzene sulfonamide (ABS) ligand on a SPR biosensor surface (Williams et al., 2009). Once again, a dual-fractal analysis is required to adequately describe the binding kinetics. The values of (a) the binding rate coefficient, k and the fractal dimension, D_f for a single fractal analysis, and (b) the binding rate coefficients, k_1 and k_2, and the fractal dimensions, D_{f1} and D_{f2} for a dual-fractal analysts are given in Tables 11.3(a) and (b). It is of interest to note that as the fractal dimension increase by a factor of 1.12 from a value of D_{f1} equal to 2.6146 to D_{f2} equal to 2.9294, the binding rate coefficient increases by a factor of 1.826 from a value of k_1 equal to 83.718 to k_2 equal to 152.87.

Figure 11.7(c) shows the binding of 11.0μM CA-II anhydrase in solution to a 4-(2-amino ethyl)-benzene sulfonamide (ABS) ligand on a SPR biosensor surface (Williams et al., 2009). Once again, a dual- fractal analysis is required to adequately describe the binding kinetics. The values of (a) the binding rate coefficient, k and the fractal dimension, D_f for a single fractal analysis, and (b) the binding rate coefficients, k_1 and k_2, and the fractal dimensions, D_{f1} and D_{f2} for a dual-fractal analysis are given in Tables 11.3(a) and (b) It is of interest to note that as the fractal dimension increase by a factor of 1.515 from a value of D_{f1} equal to 1.967 to D_{f2} equal to 2.9807, the binding rate coefficient increases by a factor of 5.98 from a value of k_1 equal to 39.916 to k_2 equal to 238.66.

Figure 11.8(a) and Table 11.3(a) show the increase in the binding rate coefficient, k_2 with an increase in the CA-II concentation in solution in the 2.3−11μM range. For the data shown in Figure 11.8(a), the binding rate coefficient, k_2, is given by:

$$k_2 = (80.452 \pm 8.66)\,[\mathrm{CA-II}]^{0.435\pm0.092} \tag{11.6a}$$

The fit is good. Only three data points are available. The availability of more data points would lead to a more reliable fit. The binding rate coefficient, k_2 exhibits less than one-half (equal to 0.435) order of dependence on the CA-II concentration in solution in the 2.3−11.0 μM range. This indicates that the binding rate coefficient, k_2 is only mildly sensitive to the CA-II concentration in solution.

Figure 11.8(b) and Table 11.3(a) show the increase in the fractal dimension D_{f2}, p with an increase in the CA-II concentration in solution in the 2.3−11 μM range. For the data shown in Figure 11.8(b), the fractal dimension, D_{f2} is given by:

$$D_{f2} = (2.279 \pm 0.130)[\mathrm{CA-II}]^{0.122\pm0.050} \tag{11.6b}$$

The fit is good. Only three data points are available. The availability of more data points would lead to a more reliable fit. The fractal dimension, D_{f2} exhibits a very mild dependence (equal to 0.122; close to zero order) on the CA-II concentration in solution in the 2.3−11.0 μM range. This indicates that the fractal dimension, D_{f2} is only mildly sensitive to the CA-II concentration in solution. The fractal dimension is based on a log scale, and even small changes in the fractal dimension indicate a reasonable change in the degree of heterogeneity on the biosensor surface.

Figure 11.8(c) and Tables 11.3(a) and (b) show the increase in the binding rate coefficient, k_2 with an increase in the fractal dimension, D_{f2}. For the data shown in Figure 11.8(b), the binding rate coefficient, k_2 is given by:

$$k_2 = (106.50 \pm 15.12)D_{f2}^{2.011\pm0.561} \tag{11.6c}$$

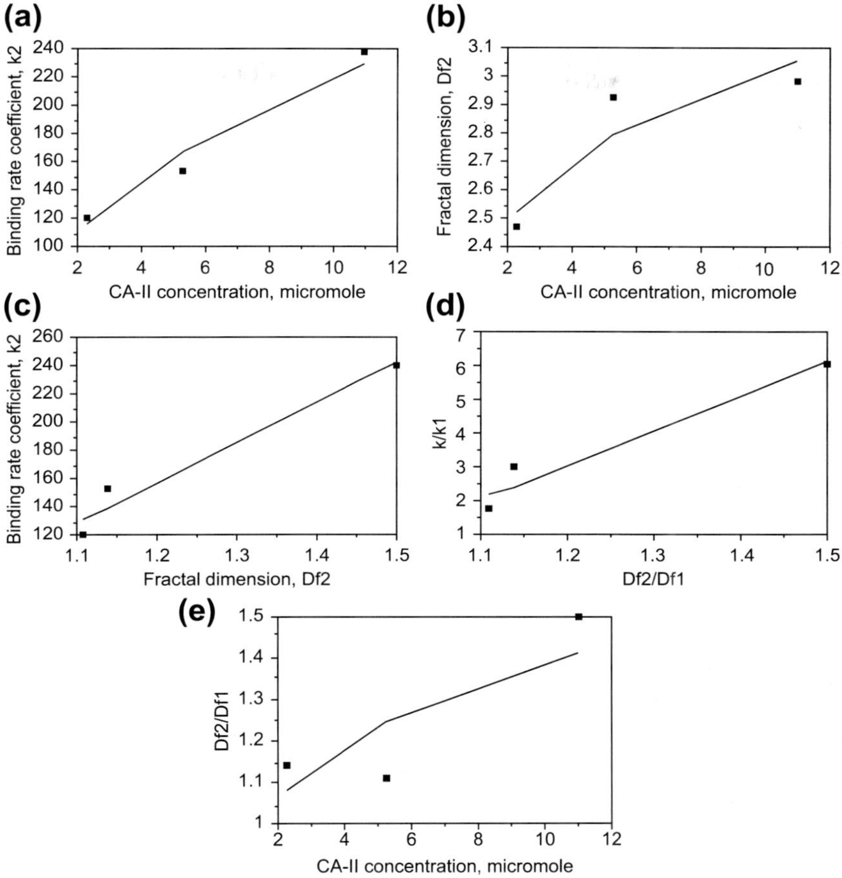

FIGURE 11.8 (a) Increase in the binding rate coefficient, for a dual-fractal analysis with an increase in the carbonic anhydrase II (CA-II) concentration (in micromoles) in solution. (b) Increase in the fractal dimension, D_f, with an increase in the carbonic anhydrase-II (CA-II) concentration (CA-II) (in micromoles) in solution. (c) Increase in the binding rate coefficient, k_2, with an increase in the fractal dimension, D_{f2}. (d) Increase in the binding rate coefficient ratio, k_2/k_1, with an increase in the fractal dimension ratio, D_{f2}/D_{f1}. (e) Increase in the fractal dimension ratio, D_{f2}/D_{f1}, with an increase in the carbonic anhydrase-II (CA-II) concentration (in micromoles) in solution.

The fit is good. Only three data points are available. The availability of mora data points would lead to a more reliable fit. The binding rate coefficient, k_2 exhibits close to a second (equal to 2.011) order of dependence on the fractal dimension, D_{f2} This indicates that the binding rate coefficient, k_2 is sensitive to the fractal dimension, D_{f2} or the degree of heterogeneity on the biosensor surface.

Figure 11.8(d) and Tables 11.3(a) and (b) show the increase in the binding rate coefficient ratio, k_2/k_1 with an increase in the fractal dimension ratio D_{f2}/D_{f1}. For the data in Figure 11.8(d) the binding rate coefficient ration k_2/k_1 is given by:

$$k_2/k_1 = (1.549 \pm 0.530]D_{f2}/D_{f1}^{3.386 \pm 1.249} \quad (11.6d)$$

The fit is good. Only three data points are available. The availability of more data points would lead to a more reliable fit. The binding rate coefficient ratio, k_2/k_1 exhibits an order of dependence between three and three and one half (equal to 3.386) on the ratio of the fractal dimensions, D_{f2}/D_{f1}. This indicates that the binding rate coefficient ratio is very sensitive to the fractal dimension ratio

Figure 11.8(e) shows the increase in the ratio of the binding rate coefficient, D_{f2}/D_{f1} with an increase in the CA-II concentration in solution. The ratio, D_{f2}/D_{f1} is only mildly dependent on the CA-II concentration in solution.

Clark et al. (2009) have recently developed a continuous-flow enzyme assay on a microfluidic chip for monitoring glycerol secretion from cultivated adipocytes. These authors indicate that different studies on using chips to monitor cellular secretion have appeared in the literature (Cheng et al., 2006; Lau et al., 2006; El-All et al., 2006; Kim et al., 2007; Meyvantsso et al., 2008; Urbanskl et al., 2008). Clark et al. (2009) indicate that physiological studies need to maintain cells or tissues in a controlled environment as one detects their physical, electrical and mechanical properties. These authors indicate micriofluidics facilitates such situations, since they permit creation of highly controlled cell-compatible environments along with measurement and cell maniplation methods.

Clark et al. (2009) emphasize that the prevalence of obesity-related disorders underscores the need to adipocyte physiology. Adipocytes store and release energy. Adipocytes store energy as triacylglycerol by lipolysls. Ths supplies energy for tissues and organs. Getty et al. (2006) and Getty-Kaushik et al., (2005a,b) indicate that the measurement of glycerol is used to determine the function and physiological state of adipocytes. Clark et al. (2009) have developed a dual-chip microfluidic system for culturing adipocytes and then monitoring the glycerol using a continuous fluorescent enzyme assay after a perfusion step. The authors used their system to demonstrate transient increases in glycerol secretion during exposure of the cells to isoproterenol, a β-adrenergic agonist. These adrenergic agonists act on receptors.

Beta receptors are specific molecules found in the body which receive and process signals for the nervous system and various hormones. These beta receptors are located at many places in the body, but are found in high numbers in the heart and blood vessels. Here they increase blood pressure when stimulated. Thus, they are attractive targets for high blood pressure treatment.

Clark et al. (2009) analyzed the glycerol secretion data from differentiated adipocytes and response to isoproterenol treatment. Figure 11.9(a) shows the binding of glycerol secretion from differentiated (murine 3T3-L1) adipocytes by a continuous-flow enzyme assay on a microfluidic chip. A dual-fractal

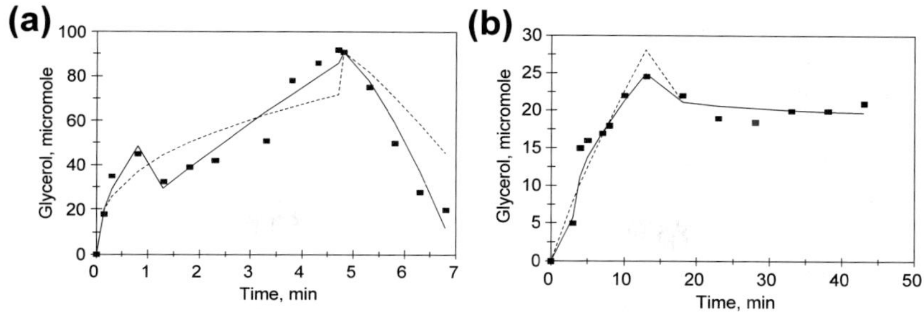

FIGURE 11.9 (a) Binding and dissociation of 50 μM glycerol secreted from differentiated (murine 3T3-L1) adipocytes to a microfluidic platform (Clark et al., 2010): (b) same as in; (a) but isoproterenol added.

analysis is required to model the binding and the dissociation kinetics. The values of (a) the binding rate coefficient, k and the fractal dimension, D_f for a single-fractal analysis, (b) the binding rate coefficients, k_1 and k_2 and the fractal dimensions, D_{f1} and D_{f2} for a dual-fractal analysis, (c) the dissociation rate coefficient, k_d and the fractal dimension, D_{fd} for a single-fractal analysis, and (d) the dissociation rate coefficients, k_{d1} and k_{d2} and the fractal dimensions for the dissociation phase, D_{fd1} and D_{fd2} for a dual-fractal analysis are given in Tables 11.4(a) and (b).

Figure 11.9(b) shows the binding of glycerol secretion from differentiated (murine 3T3-L1) adipocytesin the presence of iso proterenol by a continuous-flow enzyme assay on a microfluidic chip. A dual-fractal analysis is required to model

TABLE 11.4a Binding and Dissociation Rate Coefficients for Glycerol Secretion from (a) Differentiated (Murine 3T3-L1) Adipocytes to a Microfluidic Platform for Fluorescence-Based Enzyme Assay and (b) Influence of Isoproterenol on Online Enzyme Assay for Glycerol (Clark et al., 2010)

Analyte/ Receptor	k	k_1	k_2	k_d	k_{d1}	k_{d2}
Glycerol from differentiated (murine 3T3-L1) adipocytes/ microfluidic platform	40.25 ± 12.11	54.74 ± 14.29	23.64 ± 3.445	28.689 ± 8.239	36.862 ± 1.187	0.6187 ± 0.8260
Influence of isoproterenol on 50 μL glycerol/ online enzyme assay	3.113 ± 0.0934	0.4301 ± 0.1872	5.061 ± 0.175	2.4653 ± 0.8981	NA	NA

TABLE 11.4b Fractal Dimensions for the Binding and the Dissociation Phases for Glycerol Secretion from (a) Differentiated (Murine 3T3-L1) Adipocytes to a Microfluidic Platform for Fluorescence-Based Enzyme Assay and (b) Influence of Isoproterenol on Online Enzyme Assay for Glycerol (Clark et al., 2010)

Analyte/ Receptor	D_f	D_{f1}	D_{f2}	D_{fd}	D_{fd1}	D_{fd2}
Glycerol from differentiated (murine 3T3-L1) adipocytes/ microfluidic platform	2.065 ± 0.635	1.9440 ± 0.390	1.3358 ± 0.2304	0.288 ± 0.188	0+ 0.0388	0
Influence of isoproterenol on 50 μl glycerol/ online enzyme assay	1.7176 ± 0.4736	0 + 1.9446	1.7570 ± 0.1452	2.6102 ± 0.4192	NA	NA

the binding and the dissociation kinetics. The values of (a) the binding rate coefficient, k and the fractal dimension, D_f for a single-fractal analysis, (b) the binding rate coefficients, $k_1$and k_2 and the fractal dimensions, D_{f1} and D_{f2} for a dual-fractal analysis, and (c) the dissociation rate coefficient, k_d and the fractal dimension, D_{fd} for a single-fractal analysis are given in Tables 11.4(a) and (b).

Albrecht et al. (2010) have recently presented a new assay design for clinical diagnostics based on alternative recognition. These indicate that the assay format has an important impact in the practical handling as well in the sensitivity of the testing results. Jaras et al. (2007) indicate that for clinical diagnostics the sandwich assay format is frequently used due to (a) its lower limits of detection compared to other formats, (b) and reliable analysis of the different parameters. Albrecht et al. (2010) further indicate that a drawback of the sandwich assay format is the need for immobilization of the capture antibody on the surface. This often results in a significant loss in binding activity. Also, there is no guarantee that the binding sites on the antibodies immobilized on the surface are oriented in the 'correct' direction. This hinders the biosensor performance parameters such as sensitivity, loss in function, and stability of the sensor surface. In essence, these authors indicate that the recognition element on the sensor surface needs to exhibit a high affinity and specificity towards the antigen (analyte) in solution on being immobilized on the biosensor surface.

Albrecht et al. (2010) have presented an immunoassay set-up that uses a small and stable peptide sequence as the immobilized recognition element (receptor) (Baltzer, 2007). Albrecht et al. (2010) indicate that their recognition elements are small helix loop-helix motifs. These recognition elements contain natural binders of the target analyte. Furthermore, these motifs are easily accessible. Also, these authors indicate concerted modifications made for immobilization at the artificial helices do not affect binding properties.

Albrecht et al. (2010) have presented a new sandwich-type assay for the detection of C-reactive protein (CRP). They used a tailored binder as the capture element on the sensor surface, and an antibody as a detection element. C-reactive protein is a protein found in blood. Its levels rise in response to inflammation. Thompson et al. (1999) indicate that its physiological role is to bind to phosphocholine expressed on the surface of dead or dying cells in order to activate the complement system via the C1Q complex.

CRP is a general marker for inflammation and infection. It can be used as a very rough proxy for heart disease risk. Lloyd-Jones et al. (2006) emphasize that since many factors are responsible for CRP level elevations, thus it is not a very specific prognostic indicator. Also the patients with elevated basal levels of CRP are at an increased risk of diabetes (Pradhan et al., 2001), hypertension and cardiovascular disease.

Figure 11.10(a) shows the binding and dissociation of 5.45×10^{-8} M CRP in solution to the new sandwich assay design that contains a high affinity polypeptide scaffold as the immobilized capture element and an antibody for

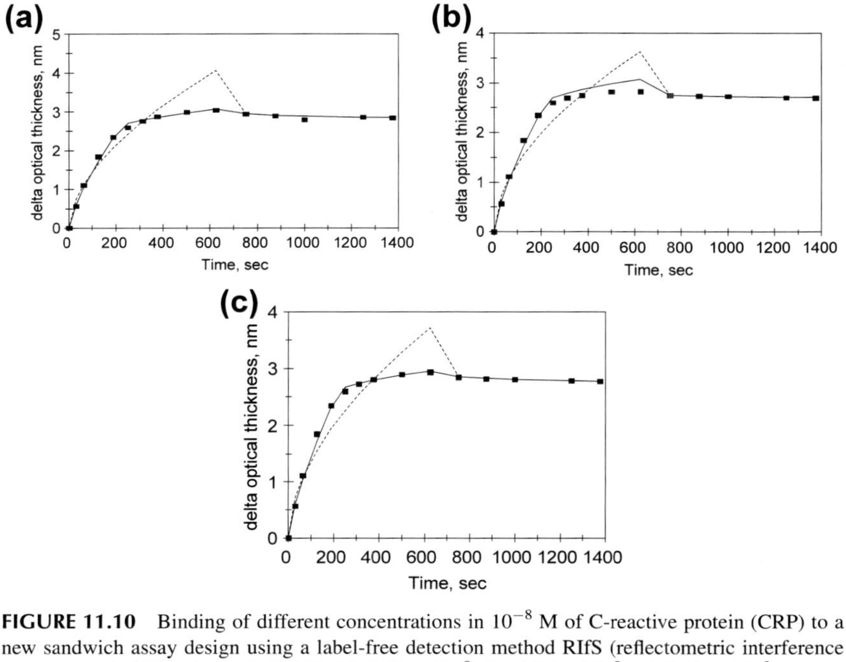

FIGURE 11.10 Binding of different concentrations in 10^{-8} M of C-reactive protein (CRP) to a new sandwich assay design using a label-free detection method RIfS (reflectometric interference spectroscopy) (Albrecht et al., 2010): (a) 5.45×10^{-8}; (b) 4.36×10^{-8}; (c) 2.18×10^{-8}.

detection (Albrecht et al., 2010). As mentioned above a biosensor based on reflectometric interference spectroscopy (RIfS) was used. A dual-fractal analysis is required to describe the binding kinetics. A single-fractal analysis is adequate to describe the dissociation kinetics. The values of (a) the binding rate coefficient k and the fractal dimension, D_f for a single-fractal analysis, (b) the binding rate coefficients, k_1 and k_2 and the fractal dimensions, D_{f1} and D_{f2} for a dual-fractal analysis, and (c) the dissociation rate coefficient, k_d and the fractal dimension, D_{fd} for a single-fractal analysis are given in Tables 11.5(a) and (b).

It is of interest to note that for a dual-fractal analysis, an increase in the fractal dimension by a factor of 1.78 from a value of D_{f1} equal to 1.5344 to D_{f2} equal to 2.723 leads to an increase in the binding rate coefficient by a factor of

TABLE 11.5a Binding and Dissociation Rate Coefficients for Different Concentrations of C-Reactive Protein (CRP) in Solution to a Sandwich-Type Assay Using a Label-free Detection Method, RIfS (Reflectometric Interference Spectroscopy) (Albrecht et al., 2010)

Analyte in Solution/ Receptor on Surface	k	k_1	k_2	k_d
5.48×10^{-8} M CRP/fluorescently labeled antibody	0.1088 ± 0.0211	0.04991 ± 0.0050	1.2621 ± 0.0124	0.0139 ± 0.00081
4.36×10^{-8} M CRP/fluorescently labeled antibody	0.1148 ± 0.0226	0.04911 ± 0.0050	1.4889 ± 0.0102	0.02078 ± 0.00068
2.18×10^{-8} M CRP/fluorescently labeled antibody	0.1206 ± 0.0252	0.04911 ± 0.0050	1.8137 ± 0.0117	0.0207 ± 0.0009

TABLE 11.5b Fractal Dimensions for the Binding and the Dissociation Phases for Different Concentrations of C-Reactive Protein (CRP) in Solution to a Sandwich Type Assay Using a Label-Free Detection Method, RIfS (Reflectometric Interference Spectroscopy) (Albrecht et al., 2010)

Analyte in Solution/ Receptor on Surface	D_f	D_{f1}	D_{f2}	D_{fd}
5.48×10^{-8} M CRP/ fluorescently labeled antibody	1.8950 ± 0.1210	1.5344 ± 0.0133	2.723 ± 0.03704	2.1488 ± 0.08472
4.36×10^{-8} M CRP/ fluorescently labeled antibody	1.92 ± 0.1288	1.5344 ± 0.1133	2.7872 ± 0.04378	02.3818 ± 0.0444
2.18×10^{-8} M CRP/ fluorescently labeled antibody	1.9430 ± 0.1133	1.5344 ± 0.1133	2.8601 ± 0.02534	2.4560 ± 0.05786

25.29 from a value of k_1 equal to 0.04991 to k_2 equal to 1.2621. Once again, an increase in the degree of heterogeneity on the biosensor surface leads to an increase in the binding rate coefficient. In this case, the affinity, K_1 $(=k_1/k_d)$ and $K_2(=k_2/k_d)$ are 3.59 and 90.8, respectively.

Figure 11.10(b) shows the binding and dissociation of 4.36×10^{-8} M CRP in solution to the new sandwich assay design (Albrecht et al., 2010). A dual-fractal analysis is once again required to describe the binding kinetics. A single-fractal analysis is adequate to describe the dissociation kinetics. The values of (a) the binding rats coefficient k and the fractal dimension, D_f for a single-fractal analysis, (b) the binding rate coefficients, k_1 and k_2 and the fractal dimensions, D_{f1} and D_{f2} for a dual-fractal analysis, and (c) the dissociation rate coefficient, k_d and the fractal dimension, D_{fd} for a single-fractal analysis are given in Tables 11.5(a) and (b).

It is of interest to note that for a dual-fractal analysis, an increase in the fractal dimension by a factor of 1.82 from a value of D_{f1} equal to 1.5344 to D_{f2} equal to 2.7872 leads to an increase in the binding rate coefficient by a factor of 30.32 from a value of k_1 equal to 0.04991 to k_2 equal to 1.4889. Once again, an increase in the degree of heterogeneity on the biosensor surface leads to an increase in the binding rate coefficient. In this case, the affinity, $K_1(=k_1/k_d)$ and $K_2(=k_2/k_d)$ are 2.40 and 71.6, respectively.

Figure 11.10(c) shows the binding and dissociation of 2.18×10^{-8} M CRP in solution to the new sandwich assay design (Albrecht et al., 2010). A dual-fractal analysis is once again required to describe the binding kinetics. A single-fractal analysis is adequate to describe the binding kinetics. The values of (a) the binding rate coefficient k and the fractal dimension, D_f for a single-fractal analysis, (b) the binding rate coefficients, k_1 and k_2 and the fractal dimensions, D_{f1} and D_{f2} for a dual-fractal analysis, and (c) the dissociation rate coefficient, k_d and the fractal dimension, D_{fd} for a single-fractal analysis are given in Tables 11.5(a) and (b)

It is of interest to note that for a dual-fractal analysis, an increase in the fractal dimension by a factor of 1.86 from a value of D_{f1} equal to 1.5344 to D_{f2} equal to 2.8601 leads to an increase in the binding rate coefficient by a factor of 36.93 from a value of k_1 equal to 0.04991 to k_2 equal to 1.8137. Once again, an increase in the degree of heterogeneity on the biosensor surface leads to an increase in the binding rate coefficient. In this case, the affinity, K_1 $(=k_1/k_d)$ and $K_2(=k_2/k_d)$ are 2.37 and 87.62, respectively.

Figure 11.11(a) and Table 11.5(a) show for a dual-fractal analysis the decrease in the binding rate coefficient, k_2 with an increase in the CRP concentration in solution in the $2.18{-}5.48 \times 10^{-8}$ M range. For the data shown in Figure 13.11(a) the binding rate coefficient, k_2 is given by:

$$k_2 = (2.444 \pm 0.148)[\mathrm{CRP}]^{-0.368 \pm 0.086} \qquad (11.7a)$$

The fit is good. Only three data points are available. The availability of more data points would lead to a more reliable fit. The binding rate coefficient, k_2

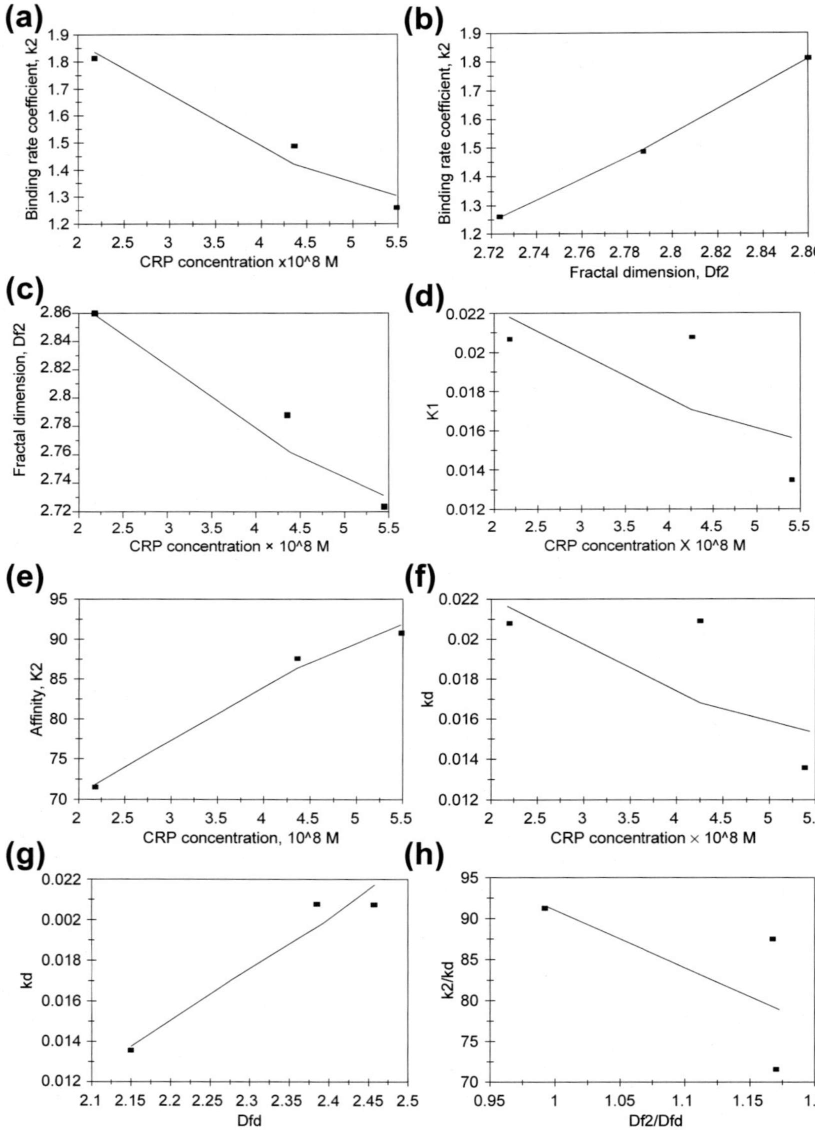

FIGURE 11.11 (a) Decrease in the binding rate coefficient, k_2, with an increase in the CRP concentration in solution. (b) Increase in the binding rate coefficient, k_2, with an increase in the fractal dimension, D_{f2}. (c) Decrease in the fractal dimension, D_{f2}, with an increase in the CRP concentration in solution. (d) Decrease in the dissociation rate coefficient, k_d, with an increase in the CRP concentration in solution. (e) Increase in the dissociation rate coefficient, k_d, with an increase in the fractal dimension for dissociation, D_{fd}. (f) Decrease in the affinity, K_2 ($=k_2/k_d$), with an increase in the fractal dimension ratio, D_{f2}/D_{fd}. (g) Increase in the dissociation rate coefficient, k_d with an increase in the fractal dimension, D_{fd}. (h) Decrease in the affinity, K_2 with an increase in D_{f2}/D_{fd}.

decreases with an increase in the CRP concentration in solution in the $2.18{-}5.45\times10^{-8}$ M concentration range, and exhibits less than a negative one half (equal to -0.368) order of dependence on the CRP concentration in solution.

Figure 11.11(b) and Table 11.5(a) show for a dual-fractal analysis the increase in the binding rate coefficient, k_2 with an increase in the fractal dimension, D_{f2}. For the data shown in Figure 11.11(b) the binding rate coefficient, k_2 is given by:

$$k_2 = (0.000732 \pm 0.000003) D_{f2}^{7.435 \pm 0.135} \tag{11.7b}$$

The fit is very good. Only three data points are available. The availability of more data points would lead to a more reliable fit. The binding rate coefficient, k_2 exhibits close to a seven and one-half (equal to 7.435) order of dependence on the fractal dimension, D_{f2} on the biosensor surface. This indicates that the binding fate coefficient, k_2 is very sensitive to the fractal dimension or the degree of heterogeneity on the biosensor surface.

Figue 11.11(c) show for a dual-fractal analysis the decrease in the fractal dimension, D_{f2} with an increase in the CRP concentration in solution in the $2.18{-}5.45 \times 10^{-8}$ M range. For the data shown in Figure 11.11(c) the fractal dimension, D_{f2} is given by:

$$D_{f2} = (2.977 \pm 0.026)\ \left[CRP \times 10^{-8}\right]^{-0.0494 \pm 0.01289} \tag{11.7c}$$

The fit is good. Only three data points are available. The availability of more data points would lead to a more reliable fit. The fractal dimension, exhibits a very mild dependence on the CRP concentration in the $2.18{-}5.45\times10^{-8}$ M concentration range.

Figure 11.11(d) shows the increase in the affinity K_1 ($=k_1/k_d$) with an increase in the CRP concentration in solution in the $2.18{-}5.45 \times 10^{-8}$ M range. For the data shown in Figure 11.11(d), the affinity, K_1 is given by:

$$K_1 = (k_1/k_d) = (1.724 \pm 0.456)\ [CRP]^{0.349 \pm 0.346} \tag{11.7d}$$

The fit is reasonable. Only three data points are available. The availability of more data points would lead to a more reliable fit. The availability of more data points would lead to a more reliable fit. The affinity, K_1 exhibits only a mild, less than one-half (equal to 0.349) order of dependence on the CRP concentration in solution in the $2.18{-}5.45 \times 10^{-8}$ M range.

Figure 11.11(e) shows the increase in the affinity K_2 ($=k_2/k_d$) with an increase in the CRP concentration in solution in the $2.18{-}5.45 \times 10^{-8}$ M range. For the data shown in Figure 11.11(e), the affinity, K_2 is given by:

$$K_2 = (k_2/k_d) = (58.422 \pm 1.076)\ [CRP]^{0.266 \pm 00269} \tag{11.7e}$$

The fit is very good. Only three data points are available. The availability of more data points would lead to a more reliable fit. The availability of more data points would lead to a more reliable fit. The affinity, K_2 exhibits only a mild, less than one-half (equal to 0.266) order of dependence on the CRP concentration in solution in the $2.18{-}5.45 \times 10^{-8}$ M range.

Figure 11.11(f) shows the decrease in the dissociation rate coefficient, k_d with an increase in the CRP concentration in solution in the 2.18–5.45 M × 10^{-8} M range. For the data shown in Figure 11:11(f), the dissociation rate coefficient k_d is given by:

$$k_d = (0.00290 \pm 0.0083)\ [CRP]^{-0.367-0.378} \tag{11.7f}$$

The fit is poor. There is scatter in the data. This is reflected in the error in the estimated value of the order of dependence of k_d on the CRP concentration in solution. Only the negative sign is applicable since the dissociation rate coefficient, k_d decreases with an increase in the CRP concentration in solution.

Figure 11.11(g) and Table 11.5(a) and (b) show the increase in the dissociation rate coefficient, k_d with an increase in the fractal dimension for dissociation, D_{fd}. For the data shown in Figure 11.11(g), the dissociation rate coefficient, k_d is given by:

$$k_d = (0.000979 \pm 0.000080)D_{fd}^{3.449 \pm 0.8030} \tag{11.7g}$$

The fit is good. Only three data points are available. The availability of more data points would lead to a more reliable fit. The dissociation rate coefficient, k_d exhibits close to a three and one-half (equal to 3.449) order of dependence on the fractal dimension in the dissociation phase, D_{fd}. This indicates that the dissociation rate coefficient, k_d is very sensitive to the degree of heterogeneity that exists on the biosensor surface in the dissociation phase.

Figure 11.11(h) shows the decrease in the affinity, K_2 (=k_2/k_d) with an increase in the ratio of the fractal dimensions, (D_{f2}/D_{fd}). For the data shown in Figure 11.11(h) the affinity, K_2 is given by:

$$K_2 = (k_2/k_d) = (90.76 \pm 13.61)\ (D_{f2}/D_{fd})^{-0.890-1.038.} \tag{11.7h}$$

The fit is poor. Only three data points are available. The availability of more data points would lead to a more reliable fit. The poor fit is expressed as the error in the power to which the ration of the fractal dimensions is raised. Only the negative power is applicable since the affinity, K_2 decreases with an increase in the fractal dimension, ratio, D_{f2}/D_{fd}.

Tang et al. (2010) have recently developed an integrated automatic electrochemical Immunosensor array for the detection of five hepatitis virus antigens: hepatitis A virus (HAV), hepatitis B virus (HBV), hepatitis C virus (HCV), hepatitis D virus (HDV), and hepatitis E virus (HEV) (Alavian and Ballantian, 2008). Tang et al. (2010) further indicate that hepatitis viruses are one of the leading causes of mortality (Bilora et al., 2009). Thus, an early diagnosis for hepatitis B viruses is critical. Tang et al. (2010) further emphasize that the simultaneous determination of multiple virus antigens is helpful in clinical diagnosis since the patient usually suffers from multiple virus antigens (Cornberg et al., 2008; Gitlin, 1997).

Tang et al. (2010) emphasize that potentiometric assays are high-throughput systems, are label-free, exhibit low assay cost, and their simplicity permits miniaturization as well as signal quantification (Wu et al., 2007; Tang et al., 2007). Thus, Tang et al., (2010) have developed their electrochemical immunosensor array for the simultaneous determination of five-type hepatitis virus antigens in five minutes.

The binding of 200 ng/ml of HAV to the Immunosensor array (Tang et al., 2010) may be modeled by a dual-fractal analysis. Figure not shown. The values of the binding rate coefficient, k and the fractal dimension, D_f for a single-fractal analysis ar 10.252 ± 1.313 and 2.4166 ± 0.2352, respectively. For a dual-fractal analysis, (a) the binding rate coefficients, k_1 and k_2 are 10.30 ± 0.818, and 20.493 ± 0.862, respectively, and (b) the fractal dimensions, D_{f1} and D_{f2} are 0.902 ± 0.147, and 2.417± 0.235, respectively.

Note that as the fractal dimension increases by a factor of 2.68 from a value of D_{f1} equal to 0.902 to D_{f2} equal to 2.417, the binding rate coefficient increases by a factor of 1.99 from a value of k_1 equal to 10.30 to k_2 equal to 20.493. Once again, an increase in the degree of heterogeneity or the fractal dimension on the immunosensor array surface leads to an increase in the binding rate coefficient.

13.4 CONCLUSIONS

A fractal analysis is used to analyze the binding and dissociation (if applicable) kinetics of biomarkers to different biosensor surfaces. Both single- and a dual-fractal analyses are used to analyze the binding and the dissociation kinetics. The dual-fractal analysis is used only if the single-fractal analysis does not provide an adequate fit.

For the binding and dissociation of IFN-gamma in solution to the aptamer modification (Tuleuova et al., 2010), and for a single-fractal analysis, the (1) binding rate coefficient, k, exhibits close to a third (equal to 3.01) order of dependence on the fractal dimension or the degree of heterogeneity that exists on the biosensor surface and (2) the affinity, K (=k/kd), exhibits close to a fifth (equal to 5.064) order of dependence on the ratio of fractal dimensions, D_f/D_{fd}. This indicates that both the binding rate coefficient, k, and the affinity, K, are very sensitive to the nature or the degree of heterogeneity that exists on the biosensor surface.

For the binding of different concentrations of CA-II anhydrase in solution (1) the binding rate coefficient, k, exhibits a mild (equal to 0.435) order of dependence on the CA-II anhydrase concentration in solution, (2) the binding rate coefficient, k, exhibits close to a second (equal to 2.011) order of dependence on the fractal dimension, D_f, that exists on the biosensor surface, and (3) the ratio of the binding rate coefficients, k_2/k_1, exhibits higher than a third (equal to 3.386) order of dependence on the ratio of fractal dimensions, D_{f2}/D_{f1}.

The relationships presented above are typical of the ones presented for the biomarkers for the other diseases analyzed and presented in this chapter. They provide a means by which these rate coefficients or affinities may be manipulated in desired directions in order to improve the different biosensor performance parameters. The more sensitive a biosensor is for a specific biomarker for a particular disease the earlier it may be detected. Needless to say the early detection of biomarkers for different diseases should lead to a better prognosis. Surely, as expected, there is considerable effort and resources being spent in this direction, and correctly so.

REFERENCES

Agnvall, E., 2010. Alzheimer's: a new theory. American Association of Retired Persons Bulletin 51 (7), 10–11.

Alavian, S.M., Fallantian, F., 2008. Hepatitis Monthly 8, 51–59.

Albrecht, C., Fechner, P., Honcharenko, L., Baltzer, L., Gauglitz, G., 2010. A new assay design for clinical diagnostics based on alternate recognition elements. Biosensors & Bloelectronics 25, 2302–2308.

Anonymous, 1 October 2006. Validation of analytic methods for biomarkers used in drug development. Clinical Cancer Research 14, 5967–5976.

Anonymous, 1 December 2007. The demonstration of the immunochemical biomarkers in methyl methacrylate-embedded plucked human hair follicle. Toxicological and Pathology 35, 962–967.

Anonymous, 1 May 2010. Transcriptional analysis of an E2F gene signature as a biomarker of activity of the cyclin-dependent kinase inhibitor PHA-793887 in tumor and skin biopsies from a phase I clinical study. Molecular Cancer Therapeutics 9, 1265–1273.

Apte, S., Graves, E.R., 7–12 November 2010. Engineered Knottin peptides: a new class of agents for non-invasive molecular imaging of tumor biomarkers. In: Annual American Chemical Engineers Meeting, Salt Lake City.

Babendure, J.R., Adams, S.R., Tsien, R.Y., 2003. Journal of the American Chemical Society 125, 14716–14717.

Balamurugan, S., Obubuafo, A., Sofer, S.A., Spivak, D.A., 2008. Analytical and Bioanalytical Chemistry 390, 1009–1021.

Baltzer, L., 2007. Polypeptide conjugate binders for protein recognition. Creative Chemical Sensor Systems 277, 89–106.

Benhahib, M., Stocktoin, A., Mathies, R.A., 7–12 November 2010. Multichannel mass organic analyzer (MoMOA) microfluidic networks for the automated *in situ* microchip electrophoretic analyzers of organic biomarkers. In: Annual American Chemical Engineers meeting, Salt Lake City.

Benhabib, M., Chiesl, T.N., Stockton, A.M., Scherer, J.R., Mathies, R.A., 2010. Multichannel capillary electrophoresis microdevice and instrumentation for in situ planetary analysis of organic molecules and biomarkers. Analytical Chemistry 82 (b), 2372–2379.

Boehm, U., Klamp, T., Groot, M., Howard, J.C., 1997. Annual Reviews in Immunology 15, 749–795.

Bilora, F., Campagnolo, L., Rinaldi, R., Rossaro, Arzenton, M., Petrobvelli F, F., 2009. Angiology 59, 717–720.

Buttemeyer, R., Phillip, A.W., Mall, J.W., Ge, B.X., Scheller, F.W., Lisdat, F., 2002. Microsurgery 22, 108–113.

Che, X.Y., Qui, L.W., Pan, Y.X., et al., 2005. Journal of Clinical Microbiology 42, 549–553.

Cheng, W., Klauke, N., Sedgwick, Smith, G.L., Cooper, J.M., 2006. Lab Chip 6, 1424–1431.

Choi, S., Chase, J., 10–12 June 2009. Surface plasmon resonance biosensor based on Vroman effect; towards cancer biomarker detection. In: 2009 IEEE 15th international Mixed Signals, Sensors and Systems, Test Workshop, Scottsdale, Arizona, pp. 1–3.

Clark, A.M., Sousa, K.M., Kennedy, R.T., 15 March 2009. Continuous-flow enzyme assay on a microfluidic chip for monitoring glycerol secretion from cultured adipocytes. Analytical Chemistry 81 (6), 2350–2356.

Clark, A.M., Sousa, K.M., Jennings, C., MacDougald, O.A., Kennedy, R.T., 2010. Continuous flow enzyme assay on a microfluidic chip for monitoring glycerol secretion from cultured adipocytes. Analytical Chemistry 82, 6025–6031.

Colburn, W.A., Keefe, D.L., 5 September 2010. Biomarkers in drug discovery and development: from target identification through drug marketing. http://jcb.sagepub.com/content/43/4/329.

Colburn, W.H., Keefe, D.L., June 2000. Bioavailability and bioequivalence average population and/or individual. Journal of Clinical Pharmacology 40 (6), 559–560.

Cornberg, M., Protzer, U., Dollinger, M.M., Petersen, J., Wedemeyer, H., Berg, T., Jilg, W., Manns, M.P., 2008. Journal of Viral Hepatitlis 15, 1–21.

Cropek, D.M., Banta, S.A., 7-12 November 2010. On chip electrochemical detection of biomarkers for detection of water borne toxins. In: Annual American Chemical Engineers Meeting, Salt Lake City.

Cui, R.J., Huang, H.P., Yin, Z.Z., et al., 2008. Biosensors & Bioelectronics 23, 1666–1673.

Dehghan, A., Kardys, I., De Maat, M.P., Uitterlinden, A.G., Sijbrands, E.J., Bootsma, A.H., Stijnen, T., Hoffman, A., et al., 2001. Genetic variation, C-reactive protein levels, and incidence of diabetes. Diabetes 56 (3), 872.

Dixit, C.K., Vashist, S.K., O'Neill, F.T., O'Reilly, B., MacCraith, B.D., O'kennedy, R.O., 2010. Development of a high sensitivity rapid sandwich ELISA procedure and its comparison with the conventional approach. Analytical Chemistry 82, 7049–7062.

Drosten, C., Doerr, H.W., Lim, W., et al., 2004. Emerging Infectious Diseases 10, 2200–2203.

Ealick, S.E., Cook, W.J., Vijay-Kumar, S., et al., 1991. Three-dimensional structure of recombinant human interferon-gamma. Science 252, 698–702.

El-Ail, J., Sorger, P.K., Jensen, K.F., 2006. Nature 442, 403–411.

Ellington, A.D., Szostak, J.W., 1990. Nature 346, 818–822.

Foucher, R.A.M., Kuiken, T., Schutten, M., et al., 2003. Nature 423, 240.

Frasconi, M., Tortollni, C., Botre, F., Mazzel, F., 2010. Multifunctional Au nanoparticle dendrimer-based surface plasmon resonance biosensor and its application for improved insulin detection. Analytical Chemistry 82, 7335–7342.

Frow, J.E., Hill, A.O., 1988. European Journal of Biochemistry 172, 261–269.

Fujimoto, K., Chan, K.H., Takeda, K., et al., 2008. Journal of Clinical Microbiology 46, 302–310.

Getty, L., Corkey, B., 2005. Diabetes 54, 629–637.

Getty-Kaushik, L., Richard, A.M.T., Corkey, B.E., 2005. Diabetes 54, 629–637.

Getty-Kaushik, L., Richard, M., Corkey, B.E., 2005. Obesity Research 13, 2058–2065.

Getty-Kaushik, L., Song, D.H., Boylan, M.D., Corkey, B.E., Wolfe, M.M., July 2006. Glucose dependent Insulinotropic polypeptide modulates adipocyte Lipolysis and Reesterfication. Obesity 14 (7), 1124–1131.

Gitlin, N., 1997. Clinical Chemsitry 43, 1500–1506.

Han, H.J., Rangaramanujam, M.K., Wang, S., Mao, G., Kusanovic, J.P., Romero, R., February 2010. Multifunctional dendrimer-templated antibody presentation on biosensor surfaces for improved biomarker detection. Plastics Engineering 20 (3), 408–421.

Havlin, S., 1989. Molecular diffusion and reaction: In the Fractal Approach to Heterogeneous Chemistry: Surfaces, Colloids, Polymers. In: Avnir, D. (Ed.). Wiley, New York, pp. 251–269.

Havlin, S., Ben-Avraham, D., 1987. Diffusion in disordered media. Advances in Physics 36, 695–798.

Hinnen, C., Parsons, R., Niki, K., 1983. Electroanalytical Chemistry 147, 329–337.

Hirt, A.M., Tuleki, A., Oratsinis, S.E., 7–12 November 2010. Hybrid magnetic-plasmonic nanoparticles for biomarkers. In: Annual American Institute of Chemical Engineers Meeting, Salt Lake City.

Huang, D., Chen, C., Strom, C.M., Bender, R.A., November 2004. High-throughput gene sequencing assay development for hereditary non-polyposis cancer. Clinical Colorectal Cancer 4 (4), 275–279.

Huang, J.C., Chang, Y.F., Chen, K.H., Su, L.C., Lee, C.W., Chen, C.C., Chen, Y.M.A., Chou, C., 2009. Detection of severe acute respiratory syndrome (SARS) coronavirus nucleocapsid protein in human serum using a localized surface plasmon coupled fluorescence fiber-optic biosensors. Biosensors & Bioelectronics 25, 320–325.

Jaras, K., Ressine, A., Nilsson, E., Malm, J., Marko-Varga, G., Lilja, H., Laurell, T., 2007. Reverse-phase versus sandwich antibody microarray, technical comparison from a clinical perspective. Analytical Chemistry 79 (15), 5817–5825.

Jayasena, S.D., 1999. Clinical Chemistry 458, 1628–1650.

Karlsson, A.C.J.N., Martin, S.R., Younger, B.M., Bredt, L., Epling, R., Ronquillo, A.V., Deeks, S.C., McCune, J.M., Nixon, D.F., Sinclair, A.E., 2003. Journal of Immunological Methods 283, 141–153.

Kim, L., Toh, Y.C., Voldman, J., Yu, H., 2007. Lab Chip 7, 681–694.

Kinke, D.J., 7–12 November 2010. Identifying secreted biomarkers for immune evasion in cellular models of cancer. In: 2010 Annual American Institute of Chemical Engineers meeting, Salt Lake City.

Kirby, R., Cho, E.J., Gehrke, B., Bayer, T., Park, Y.S., Neikirk, D.P., McDevitt, J.T., Ellington, A.D., 2004. Analytical Chemistry 76, 4066.

LaRia, R., 7 September 2010. Advances in imaging biomarkers, innovative technologies, applications in R & D and clinical practice, Informatics, and regulatory requirements. Cambridge Health Institute, personal communication, (email).

Lisdat, F., 2005. In: Grimes, C., Pishko, M. (Eds.), Encyclopedia of Sensors. American Scientific Publishers.

Lau, A.Y., Hung, P.J., Wu, A.R., Lee, L.P., 2006. Lab Chip 6, 1510–1515.

Li, T., Dong, S.J., Wang, E., 2009. Analytical Chemistry 81, 2144–2149.

Lin, K.C., Kunduru, V., Bothara, M., Rege, K., Prasad, S., Ramakrishna, B.L., 2010. Biogenic nanoporous silica-based sensor for enhanced electrochemical detection of cardiovascular biomarker proteins. Biosensors & Bioelectronics 25 (10), 2336–2342.

Lloyd-Jones, D.M., Liu, K., Tran, L., Greenland, P., June 2006. Narrative reviews: assessment of C-reactive protein in risk prediction for cardiovascular disease. Annals of Internal Medicine 145 (1), 35–42.

Luzi, E., Minunni, M., Tombelli, Mascini, M., 2003. True Trends in Analytical Chemistry 22, 810–818.

Mancini, M., Tombelli, S., 2008. Biosensors for biomarkers in medical diagnostics. Biomarkers 13 (7), 637–657.

Manso, J., Mena, M.L., Yanez-Sedeno, P., et al., 2008. Analytical Biochemistry 375, 345–353.

Meyvantsson, L., Beebe, D.J., 2008. Annual Reviews of Analytical Chemistry 1, 423–449.

Nagaraj, V.J., Athal, S., Eaton, S., Bothara, M., Wiktor, P., Prasad, S., 2010. Nanomedicine (london) 5 (3), 369–378.

Nahir, T.M., Ciark, R.A., Bowden, E.F., 1994. Analytical Chemistry 66, 2595–2598.

Nutiu, R., Li, Y., 2005. Methods 37, 16–25.

Parkkila, S., Casota, J., Fletcher, J.A., Ou, W.B., Kivela, A.J., Nuorva, K., Parkkila, J., Ollikainen, J., Sly, W.S., Waheed, A., Pastorekova, J., Isole, J., Miettinin, 2010. Carboni anhydrase II, A novel biomarker for gastrointestinal stomal tumors. Modern Pathology 23, 743–750.

Panteleo, G., Koup, R.A., 2004. Nature Medicine 10, 806–810.

Pfutzner, A., Weber, M.M., Forst, T., 2008. A biomarker concenpt for assessment of insulin resistance, beta-cell function and chronic systemic inflammation in type 2 diabetes mellitus. Clinical Laboratory 54 (11-12), 485–490.

Pilarik, M., Bochkoa, M., Homola, J., 7–12 November 2010. Surface plasmon resonance biosensor for parallelized detection of protein biomarkers in diluted blood plasma. In: Annual American Institute of Chemical Engineers Meeting, Salt Lake City.

Pradhan, A.D., Manson, J.E., Rifai, N., Buring, J.E., Ridker, P.M., 2001. C-reactive protein, interleukin 6, and risk of developing type 2 diabetes mellitus. Journal of the American Medical Association 286 (3), 327–334.

Prieto-Simon, B., Cortina, M., Campas, M., Calas-Blanchard, G., 2008. Sensors & Actuators 129, 458–466.

Ramakrishnan, A., Sadana, A., 2001. A single-fractal analysis of cellular analyte-receptor binding kinetics using biosensors. Biosystems 59, 35–51.

Romagnani, S., Giudizi, G.M., Almerigogna, F., Biagiotti, R., Alessi, A., Mingari, C., Liang, C.M., Moretta, L., Ricci, M., 1986. Analysis of the role of interferon-gamma, interleukin 2 and a third factor distinct from interferon- gamma and interleukin 2 in human B cell proliferation. Evidence that they act at different times after B cell activation. European Journal of Immunology 16 (6), 623–629.

Sadana, A., 2001. A fractal analysis for the evaluation of hybridization kinetics in biosensors. Journal of Colloid and Interface Science 151 (1), 166–177.

Sadana, A., 2005. Fractal Binding and Disseriation Kinetics for Different Biosensors Applications. Elsevier, Amsterdam.

Sahin, A., Banta, S., West, A.C., Cropek, D.M., 2010. A dual enzyme electrochemical assay for the detection of organophosphorus compounds using organophosphorus hydrolase and horseradish peroxidase. In: 2010 Annual American Chemical Engineers Meeting, Salt Lake City, Italy, Paper 203c.

Scheller, W., Jin, W., Ehrentreich-Forster, E., Ge, B.X., Lisdat, F., Butmeyer, R., Wolienberger, U., Scheller, 1999. Electroanalysis 11, 703–706.

Sofer, S.A., Brown, K., Ellington, A., Frazier, B., Garcla-Manero, G., Gau, V., Gutman, S.L., 2006. Point-of-eare biosensor system for cancer diagnostics/prognostics. Biosensors & Bioelectronics 21 (10), 1932–1942.

Tang, D., Tang, J., Su, B., ren, J., Chen, G., 2010. Simultaneous determination of five-type hepatitis virus antigens in 5 min using an integrated automatic electrochemuical immunosensor array. Biosensors & Bioelectronics 25, 1658–1662.

Tang, D.P., Yuan, R., Chai, Y., 2007. Clinical Chemistry 53, 1323–1329.

Taniguchi, I., Toyosawa, K., Yamaguchi, H., Yasukouchi, K.J., 1982. Electroanalytical Chemistry 140, 187–193.

Thiel, D.J., Du, M.H., Walter, R.L., D'Arcy, A., Chene, C., Fountoulakis, M., Garoth, G., Winkler, F.K., Ealick, S.E., September 2000. Observation of an unexpected third receptor molecule in the crystal structure of human interferon-gamma receptor complex. Strucure 8 (9), 927–936.

Thompson, D., Pepys, M.B., Wood, S.P., 1999. The physiological structure of human C-reactive protein and its complex with phosphocholine. Stucture 7 (2), 169–1177.

Tuleuova, N., Jones, C.N., Yan, J., Ramanculov, E., Yokobayashi, Y., Revzin, A., 2010. Development of an aptamer beacon for detection of interferon-gamma. Analytical Chemistry 32, 1851–1857.

Urata, H., Nomura, K., Wada, S., Akagi, M., 2007. Biochemical and Biophysical Research Communications 360, 459–463.

Urbanski, J.P., Johnson, M.T., Craig, D.D., Potter, D.I., Gardner, D.K., Thorsen, T., 2008. Analytical Chemistry 80, 6500–6507.

Valko, M., Leibritz, D., Moncol, J., Cronin, M.T.D., Mazur, M., Terser, J., 2007. International Journal of Biochemistry and Cellular Biology 39, 44–84.

Valko, M., Rhodes, C.J., Moncol, J., Ozakovic, M., Mazur, M., 2006. Chemical-Biological Interactions 160, 1–40.

Wang, W.K., Chen, S.Y., Liu, I.J., 2004. Emerging Infectious Diseases 10, 1213–1219.

Wang, J., Chen, W., 2005. SARS associated coronavirus transmitted from Human to Pig. Emerging Infectious Diseases 11, 446–448.

Wang, J., 15 April 2006. Electrochemical biosensors; towards point-of care cancer diagnostics. Biosensors and Bioelectronics 21 (10), 1887–1892.

Wegerich, F., Turano, P., Allegrozzi, M., Mohwald, H., Lisdat, F., 2009. Cytochrome c mutants for superoxide biosensors. Analytical Chemistry 61, 2976–2984.

Williams, L.D., Ghosh, T., Mastrangelo, C.H., 2010. Low noise detection of biomolecular interactions with signal locking surface plasmon resonance. Analytical Chemistry 62, 6025–6031.

Williams, L.D., Ghosh, T., Mastrangelo, C.H., 2009. Low noise detection of biomolecular interactions with signal-locking surface plasmon resonance. Journal of Physical Chemistry B.

Wu, J., Yan, F., Tang, J., Zhai, C., Ju, H., 2007. Clinical Chemistry 53, 1495–1602.

Chapter 12

A Fractal Analysis of Biomarkers for Different Diseases on Biosensor Surfaces

12.1 INTRODUCTION

The detection of biomarkers for different diseases at presently is a very active area of investigation, and this explains the increasing interest both at the industrial level and in academic surroundings. Essentially, this is a multidisciplinary area of investigation, and thus, groups of researchers are almost necessary for accomplishing future significant progress in this area. This, of course, is more easily accomplished at the industry level rather than at the university level. At the industrial level, very significant amount of resources can be brought to bear on a particular problem. However, even at the university level, useful collaborations can be forged between different groups at several universities, especially when seeking funding from government and other sources of funding. Nevertheless, the commercialization of these biosensors is, of course, restricted to a large extent to big and deep-pocketed companies.

There has been an abundance of meetings dealing with the different aspects of diagnostics, emphasizing the increasing importance in this area. Some of the more recent meetings include the following:

1. Introduction to the FDA and the Medical Device Approval Process-Comprehensive Monitoring for Medical Devices, July 30–August 1, 2013. This course provides an overview of medical device development process and the role the clinical research associate (CRA) plays in managing and monitoring medical device studies (Barnett In-Person Seminars Kathy@clin-ed.com, 2013).
2. The 5th Annual Ubiquitin Drug Discovery and Diagnostic Conference (July 22–24, 2013) (Luthra@ubiquitinconference.com, 2013). The workshop discusses novel insights into cardiovascular biology and obesity. Novel drug targets for anticancer approaches are presented. Inflammatory response and metabolic diseases are discussed.
3. Commercialization of Molecular Diagnostics, August 20–22, 2013 (kc@healthtech.com, 2013). The meeting discusses the commercial applications of gene patents and how to bring a novel companion diagnostic

Biomarkers and Biosensors. http://dx.doi.org/10.1016/B978-0-444-53794-2.00012-4

to the market. It also provides information on strategic assessment for market access decisions.

In this chapter, we analyze the kinetics of biomarkers for the different diseases that have recently appeared in the literature. They include the following:

1. Detection of human hepcidin-25 by surface plasmon resonance (SPR) biosensor (Scarano et al., 2013). Hepcidin-25 regulates iron in the blood.
2. Detection of vitamin D in the blood (Carluccia et al., 2013). Vitamin D is a key factor involved in avoiding skeletal damage and a variety of pathological conditions. Vitamin D deficiency is a condition of importance from the health point of view. Hypovitaminosis may be considered as a key factor in many diseases.
3. Detection of neutravidin by anti-lipoprotein immobilized on an SPR biosensor surface (Sciacca et al., 2013).
4. Detection of cardiac myoglobin (cMb) in serum using a gold nanoparticle (AuNP)-enhanced SPR (Gnedenko et al., 2013).
5. Direct detection of interferon-gamma in diluted blood plasma by an SPR biosensor (Šípová et al., 2012). Interferon-gamma (hIFNγ) is a cytokine that is critical for innate and adaptive immunity and for control of tumor growth. High hIFNγ expression may be associated with a number of autoimmune diseases (Šípová et al., 2012). Šípová et al. (2012) have used albumin-binding protein (ABD) as the binder to an SPR biosensor surface so that they may detect hIFNγ directly by the SPR biosensor.

12.2 THEORY

Havlin (1987) has reviewed and analyzed the diffusion of reactants toward fractal surfaces. The details of the theory and the equations involved for the binding and the dissociation phases for analyte–receptor binding are available (Sadana, 2001). The details are not repeated here, except that just the equations are given to permit an easier reading. These equations have been applied to other biosensor systems (Sadana, 2001; Ramakrishnan and Sadana, 2001; Sadana, 2005). For most applications, a single-fractal or a dual-fractal analysis is often adequate to describe the binding and the dissociation kinetics. Peculiarities in the values of the binding and the dissociation rate coefficients as well as in those of the fractal dimensions with regard to the dilute analyte systems being analyzed will be carefully noted, if applicable.

12.2.1 Single-Fractal Analysis

12.2.1.1 Binding Rate Coefficient

Havlin (1989) indicates that the diffusion of a particle (analyte (Ag)) from a homogeneous solution to a solid surface (e.g., receptor (Ab)-coated surface) on which it reacts to form a product (analyte–receptor complex, (Ab.Ag)) is given by

$$(\mathrm{Ab\cdot Ag}) \approx \begin{cases} t^{(3-D_{\mathrm{f,bind}})/2} = t^{p} & t < t_{\mathrm{c}} \\ t^{1/2} & t > t_{\mathrm{c}} \end{cases} \tag{12.1}$$

Here $D_{\mathrm{f,bind}}$ or D_{f} is the fractal dimension of the surface during the binding step and t_{c} is the crossover value. Havlin (1989) indicates that the crossover value may be determined by $r_{\mathrm{c}}2 \sim t_{\mathrm{c}}$. Above the characteristic length, r_{c}, the self-similarity of the surface is lost and the surface may be considered homogeneous. Above time t_{c}, the surface may be considered homogeneous, since the self-similarity property disappears, and "regular" diffusion is now present. For a homogeneous surface where $D_{\mathrm{f}} = 2$ and when only diffusional limitations are present, $p = ½$ as it should be. Another way of looking at the $p = ½$ case (where $D_{\mathrm{f,bind}} = 2$) is that the analyte in solution views the fractal object, in our case, the receptor-coated biosensor surface, from a "large distance." In essence, in the association process, the diffusion of the analyte from the solution to the receptor surface creates a depletion layer of width $(Đt)^{½}$, where $Đ$ is the diffusion constant. This gives rise to the fractal power law: (analyte.receptor) $\sim t^{(3-D_{\mathrm{f,bind}})/2)}$. For the present analysis, t_{c} is arbitrarily chosen and we assume that the value of the t_{c} is not reached. One may consider the approach as an intermediate "heuristic" approach that may be used in the future to develop an autonomous (and not time-dependent) model for diffusion-controlled kinetics.

12.2.1.2 Dissociation Rate Coefficient

The diffusion of the dissociated particle (receptor (Ab) or analyte (Ag)) from the solid surface (e.g., (analyte (Ag)–receptor (Ab)) complex-coated surface) into solution may be given as a first approximation by

$$(\mathrm{Ab\cdot Ag}) \approx -t^{(3-D_{\mathrm{f,diss}})/2} = -t^{p}(t > t_{\mathrm{diss}}) \tag{12.2}$$

Here $D_{\mathrm{f,diss}}$ is the fractal dimension of the surface for the dissociation step. This corresponds to the highest concentration of the analyte–receptor complex on the surface. Henceforth, its concentration only decreases. The dissociation kinetics may be analyzed in a manner "similar" to the binding kinetics.

12.2.2 Dual-Fractal Analysis

12.2.2.1 Binding Rate Coefficient

Sometimes, the binding curve exhibits complexities, and two parameters (k and D_{f}) are not sufficient to adequately describe the binding kinetics. This is further corroborated by low values of r^2 factor (goodness-of-fit). In that case, one resorts to a dual-fractal analysis (four parameters: k_1, k_2, D_{f1}, and D_{f2}) to adequately describe the binding kinetics. The single-fractal analysis presented above is thus extended to include two fractal dimensions. At present, the time ($t = t_1$) at which the "first" fractal dimension "changes" to the "second" fractal dimension is arbitrary and empirical. For the most part, it is dictated by the data analyzed and experience gained by handling a single-fractal analysis. A smoother curve is obtained in the "transition" region if care is taken to select the correct number of points for the two regions. In this case, the product (antibody–antigen or analyte–receptor complex, Ab.Ag or analyte.receptor) is given by

$$(\mathrm{Ab}\cdot\mathrm{Ag}) \approx \begin{cases} t^{(3-D_{\mathrm{f1,bind}})/2} = t^{p1} & (t < t_1) \\ t^{(3-D_{\mathrm{f2,bind}})} = t^{p2} & (t_1 < t < t_2) = t_c \\ t^{1/2} & (t > t_c) \end{cases} \tag{12.3}$$

In some cases, as mentioned above, a triple-fractal analysis with six parameters (k_1, k_2, k_3, D_{f1}, D_{f2}, and D_{f3}) may be required to adequately model the binding kinetics. This is when the binding curve exhibits convolutions and complexities in its shape due to perhaps the very dilute nature of the analyte (in some of the cases to be presented) or for some other reasons. Also, in some cases, a dual-fractal analysis may be required to describe the dissociation kinetics.

12.3 RESULTS

Scarano et al. (2013) have developed an SPR detection method for human hepcidin-25. According to these authors, human hepcidin-25 is a key hormone in the regulation of iron in the blood. Altered levels can lead to inflammatory events and iron dysfunctions. Hepcidin has also attracted the antidoping agencies for its role as a possible biomarker for erythropoietin (EPO) blood doping. Scarano et al. (2013) indicate that mature hepcidin is a cysteine-rich peptide hormone of 25 amino acids secreted by the liver in response to iron loading and inflammation. Increased hepcidin production leads to hypoferremic and anemia or inflammation diseases (Bartnikas, 2102; Ganz, 2011). The role played by hepcidin is that it binds to the iron transporter, ferroportin, and causes its degradation. That is how it performs its regulatory function (de Domenico et al., 2008; Nemeth et al., 2004; Kemna et al., 2008). As indicated above, it has also attracted the attention of antidoping agencies because of its possible role as an indirect marker for EPO blood doping (Krool et al., 2009; Laine et al., 2012; Piperno et al., 2009; Robach et al., 2009). Nevertheless, in spite of importance of hepcidin concentration determination, there is a lack of validated, fast, and low-cost assays (Scarano et al., 2013).

Schwarz et al. (2011) isolated a novel monoclonal antibody for ELISA (enzyme-linked immunosorbent assay)-based measurements that permitted the quantitative determination of serum hepcidin in the 0.9–140 ng/ml range.

Other methods for hepcidin determination include surface-enhanced laser desorption/ionization time-of-flight mass spectrometry (Altamura et al., 2009; Bozzzini et al., 2008; Castagna et al., 2010; Kenna et al., 2005), matrix-assisted ion desorption/time-of-flight mass spectrometry (Anderson et al., 2010, 2011), and liquid chromatography tandem mass spectrometry techniques (Bansal et al., 2009; Li et al., 2009; Mansion et al., 2011). These methods permit the semiquantitative or qualitative hepcidin determination. Krool et al. (2009) indicate that there is still a need for a reliable and reproducible method for hepcidin determination.

Scarano et al. (2013) present a method of hepcidin detection based on affinity-based biosensors using an SPR biosensor. They present two different methods using two different affinity receptors and compared their relative performance in a biosensor approach. They paid particular attention to the management of hepcidin, as this facilitated reproducibility with regard to working with standard samples. Finally, they indicate that their biosensor approach should find application in the clinical and antidoping areas.

Figure 12.1(a) shows the binding and dissociation of 100 ppb hepcidin-25 in solution to anti-hepcidin-25 immobilized on an SPR biosensor surface (Scarano et al., 2013). Figure 12.1(a) shows that the binding and the dissociation are adequately described by a first-order fractal kinetics. The values of the binding rate coefficient, k, and the fractal dimension, D_f, for a single-fractal analysis and dissociation rate coefficient, k_d, and the fractal dimension for dissociation, D_{fd}, are given in Table 12.1a and b. The affinity, K $(=k/k_d)$, is equal to 0.122. The ratio of fractal dimensions, D_f/D_{fd}, is equal to 0.674.

Figure 12.1(b) shows the binding and dissociation of 10 ppb hepcidin-25 in solution to anti-hepcidin-25 immobilized on an SPR biosensor surface (Scarano et al., 2013). Once again, Figure 12.1(b) shows that the binding and the dissociation are adequately described by a first-order fractal kinetics. The values of the binding rate coefficient, k, and the fractal dimension, D_f, for a single-fractal analysis and the dissociation rate coefficient, k_d, and the fractal dimension for dissociation, D_{fd}, are given in Table 12.1a and b. The affinity, K $(=k/k_d)$, is equal to 0.467. The ratio of fractal dimensions, D_f/D_{fd}, is equal to 0.857. It is of interest to note that even though there are only two data points available, the increase in the affinity values is in the same direction as the ratio of the fractal dimension values, D_f/D_{fd}.

Carluccia et al. (2013) indicate that vitamin D deficiency is important for health concerns (Holick et al., 2005; Dusso et al., 2005; Malabanan et al., 1998; Chapuy et al., 1997). Vitamin D is a factor involved in avoiding skeletal damage and a variety of pathological conditions (Lips, 2001; Holick, 2006). Hypovitaminosis D is considered as a key element in the pathogenesis of many multifactor diseases (Holick, 2006; Bischoff-Ferrari, 2006; Schwalfenberg, 2007). Heaney et al. (2003) indicate that the values of 25-OH vitamin D (25OHD) between 20 and 32 ng/ml are considered as vitamin deficiency.

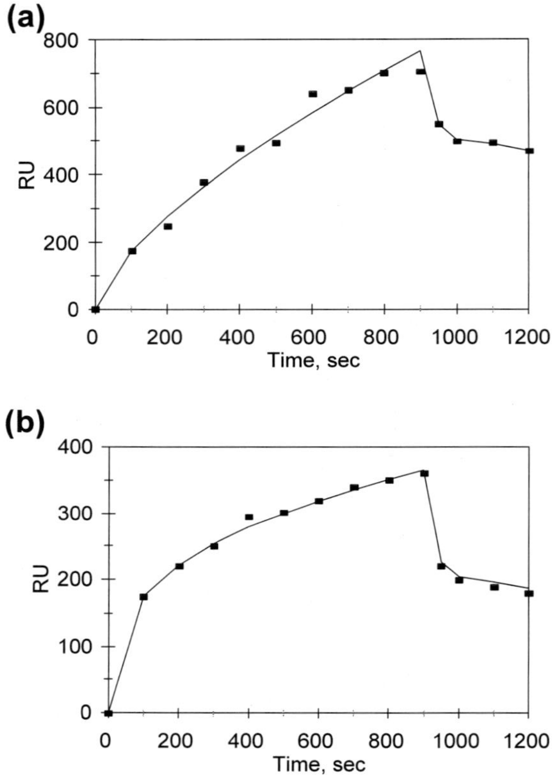

FIGURE 12.1 Binding of different concentrations of hepcidin in solution to anti-hepcidin immobilized on a surface plasmon resonance biosensor surface: (a) 100 ppb (b) 10 ppb. *Scarano et al. (2013).*

TABLE 12.1a Binding and Dissociation of Hepcidin in Solution to Anti-Hepcidin Immobilized on a Surface Plasmon Resonance Biosensor Surface Scarano et al. (2013)

Analyte in Solution/Receptor on Surface	k	k_d	k/k_d
100 ppb hepcidin/anti-hepcidin in PBS	7.6649 ± 0.572	63.477 ± 1.366	0.120
10 ppb hepcidin/anti-hepcidin in PBS	37.654 ± 0.872	80.615 ± 1.016	0.467

TABLE 12.1b Fractal Dimensions for Binding and Dissociation of Hepcidin in Solution to Anti-Hepcidin Immobilized on a Surface Plasmon Resonance Biosensor Surface Scarano et al. (2013)

Analyte in Solution/Receptor on Surface	D_f	D_{fd}	D_f/D_{fd}
100 ppb hepcidin/anti-hepcidin in PBS	1.6454 ± 0.708	2.441 ± 0.032	0.674
10 ppb hepcidin/anti-hepcidin in PBS	2.3304 ± 0.2250	2.7202 ± 0.2250	0.857

Dawson-Hughes et al. (2005) indicate that the levels greater than 32 ng/ml are sufficient to attain optimal absorption of calcium and the incidence of fractures is low (Dawson-Hughes et al., 2005).

Roth et al. (2008) and Lensmeyer et al. (2006) indicate that HPLC-MS has the highest sensitivity for detecting vitamin D (Adamec et al., 2011). Netzel et al. (2011) and Stepman et al. (2011) indicate that this technique is able to determine concentrations in the range of femtomoles. However, it has a very high instrumentation cost and thus is not suitable for routine use.

Carluccia et al. (2013) indicate that the biosensor method presented is the first attempt to detect vitamin D concentrations. They used two different transduction methods. They have used the SPR method and the screen-printed electrode method to determine vitamin D concentrations. Figure 12.2(a) shows the binding and dissociation of 50 μg/ml 25OHD to an Ab-25OHD-modified surface in the presence of 0.01 M PBS, pH 7.4, and 0.1 M NaCl. The binding kinetics is described by a single-fractal analysis. The dissociation kinetics is described by a dual-fractal analysis. The values of (1) the binding rate coefficient, k, and the fractal dimension, D_f, for a single-fractal analysis, (2) the dissociation rate coefficient, k_d, and the fractal dimension for dissociation, D_{fd}, for a single-fractal analysis, and (3) the dissociation rate coefficients, k_{d1} and k_{d2}, and the fractal dimensions for dissociation, D_{fd1} and D_{fd2}, for a dual-fractal analysis are given in Table 12.2a and b.

Figure 12.2(b) shows the binding and dissociation of 20 μg/ml 25OHD to an Ab-25OHD-modified surface in the presence of 0.01 m PBS, pH 7.4, and 0.1 M NaCl. Once again, the binding kinetics is described by a single-fractal analysis. The dissociation kinetics is described by a dual-fractal analysis. The values of (1) the binding rate coefficient, k, and the fractal dimension, D_f, for a single-fractal analysis, (2) the dissociation rate coefficient, k_d, and the fractal dimension for dissociation, D_{fd}, for a single-fractal analysis, are given in Table 12.2a and b.

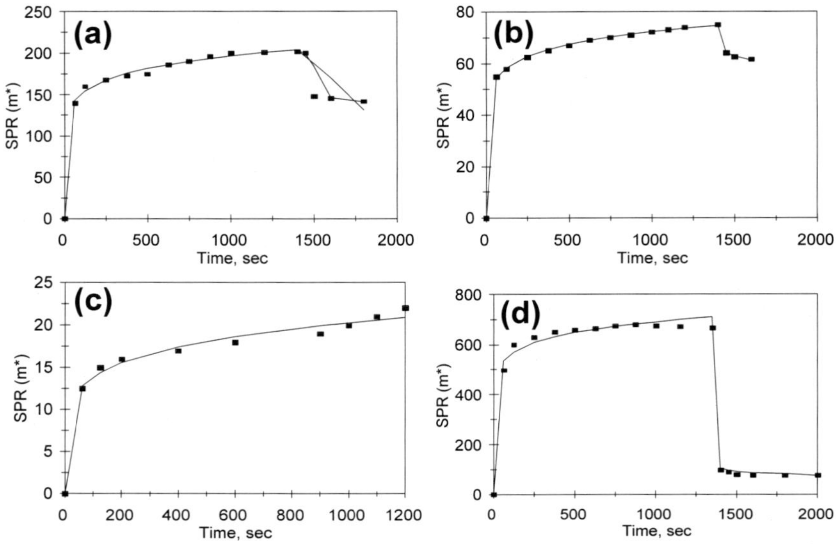

FIGURE 12.2 (a) Binding and dissociation of 50 μg/ml 25OHD in solution to Ab-25OHD-modified surface plasmon resonance (SPR) surface (b) Binding and dissociation of 20 μg/ml 25OHD in solution to Ab-25OHD-modified SPR surface (c) Binding of 10 μg/ml 25OHD in solution to Ab-25OHD-modified SPR surface (d) Binding and dissociation of 25OHD bound to AgNP particles in solution to Ab-25OHD-modified SPR surface. *Carluccia et al. (2013).*

TABLE 12.2a Binding and Dissociation Rate Coefficients for (1) different Concentrations of 25OHD in Solution to an Ab-25OHD-Modified Surface and (2) binding and Dissociation Rate Coefficients for 25OHD Bound to AgNP (Silver Nanoparticles) to an Ab-25OHD-Modified Surface Carluccia et al. (2013)

Analyte in Solution/ Receptor on Surface	k	k_d	k_{d1}	k_{d2}
50 μg/ml 25OHD/ Ab-25OHD-modified surface	88.34 ± 1.34	0.07288 ± 0.1509	0.00084 ± 0.000790	27.693 ± 0.565
20 μg/ml 25OHD/ Ab-25OHD-modified surface	35.745 ± 0.193	5.4447 ± 0.150	NA	NA
10 μg/ml 25OHD/ Ab-25OHD-modified surface	6.518 ± 0.258	NA	NA	NA
25OHD attached to AuNP/ Ab-25OHD-modified surface	368.63 ± 13.04	523.28 ± 5.86	NA	NA

TABLE 12.2b Fractal Dimensions for the binding and the Dissociation Phase for (1) different Concentrations of 25OHD in Solution to an Ab-25OHD-Modified Surface and (2) for the binding and Dissociation Rate Coefficients for 25OHD Bound to AgNP (Silver Nanoparticles) to an Ab-25OHD-Modified Surface Carluccia et al. (2013)

Analyte in Solution/ Receptor on Surface	D_f	D_{fd}	D_{fd1}	D_{fd2}
50 μg/ml 25OHD/ Ab-25OHD-modified surface	2.7688 ± 0.0130	0.7044 ± 1.1124	0 + 2.714	2.7366 ± 0.05146
20 μg/ml 25OHD/ Ab-25OHD-modified surface	2.7066 ± 0.00384	0.3536 ± 0.0244	NA	NA
10 μg/ml 25OHD/ Ab-25OHD-modified surface	2.6720 ± 0.02569	NA	NA	NA
25OHD attached to AgNP/ Ab-25OHD-modified surface	0.1816 ± 0.02674	2.9592 ± 0.0188	NA	NA

Figure 12.2(c) shows the binding and dissociation of 10 μg/ml 25OHD to an Ab-25OHD-modified surface in the presence of 0.01 M PBS, pH 7.4, and 0.1 M NaCl. In this case, the binding kinetics is described by a single-fractal analysis. The dissociation was negligible at this low concentration of 25OHD in solution. The values of the binding rate coefficient, k, and the fractal dimension, D_f, for a single-fractal analysis are given in Table 12.2a and b.

Figure 12.2(d) shows the binding and dissociation of 25OHD bound to AgNP (silver nanoparticles) in solution to an Ab-25OHD-modified surface in the presence of 0.01 M PBS, pH 7.4, and 0.1 M NaCl. In this case, the binding and the dissociation kinetics are described by a single-fractal analysis. The values of (1) the binding rate coefficient, k, and the fractal dimension, D_f, for a single-fractal analysis and (2) the dissociation rate coefficient, k_d, and the fractal dimension for dissociation, D_{fd}, are given in Table 12.2a and b.

Figure 12.3(a) shows the increase in the binding rate coefficient, k, with an increase in the 25OHD concentration in solution in micrograms per milliliter. For the data shown in Figure 12.3(a), the binding rate coefficient, k, is given by

$$k = (0.210 \pm 0.126)[25\text{OHD}]^{1.586 \pm 0.413} \quad (12.4a)$$

The fit is good. Only three data points are available. The availability of more data points would lead to a more reliable fit. The binding rate coefficient, k, for a single-fractal analysis exhibits slightly more that one and one-half (equal to 1.586) order of dependence on the 25OHD concentration in solution.

Figure 12.3(b) shows the increase in the fractal dimension, D_f, with an increase in the 25OHD concentration in solution in micrograms per

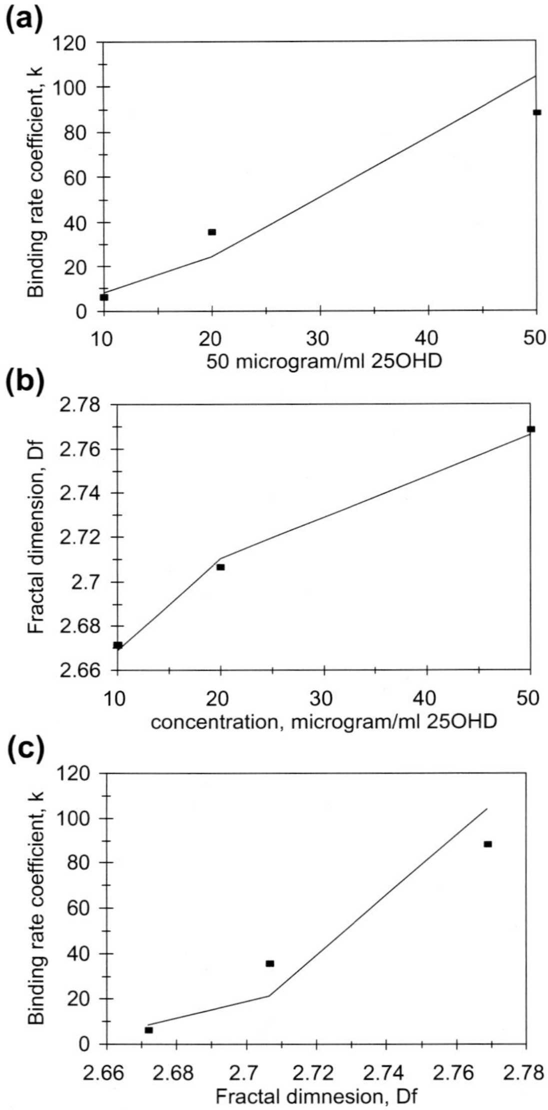

FIGURE 12.3 (a) Increase in the binding rate coefficient, k, for a single-fractal analysis with an increase in the 25OHD concentration in solution (b) Increase in the fractal dimension, D_f with an increase in the 25OHD concentration in solution. (c) Increase in the binding rate coefficient, k, with an increase in the fractal dimension, D_f.

milliliter. For the data shown in Figure 12.3(b), the fractal dimension, D_f, is given by

$$D_f = (0.2.536 \pm 0.005)[25OHD]^{0.0222 \pm 0.0017} \quad (12.4b)$$

The fit is good. Only three data points are available. The availability of more data points would lead to a more reliable fit. The fractal dimension, D_f, for a single-fractal analysis exhibits only a very slight order of dependence on the 25OHD concentration in solution.

Figure 12.3(c) shows the increase in the binding rate coefficient, k, with an increase in the fractal dimension, D_f. For the data shown in Figure 12.3(c), the binding rate coefficient, k, is given by

$$k = \left(02.12E - 29 \pm 1.79E - 29\right) D_f^{69.40 \pm 24.03} \quad (12.4c)$$

The fit is good. Only three data points are available. The availability of more data points would lead to a more reliable fit. The binding rate is extremely sensitive to the fractal dimension, D_f, or the degree of heterogeneity that exists on the surface, as is noted by the close to 70th (equal to 69.4) order of dependence exhibited.

Sciacca et al. (2013) have used radiative SPR for the detection of apolipoprotein in medical diagnostics application. They state that surface plasmons are electromagnetic waves at the interface of two media and have permittivity with two opposite directions (Dostalek et al., 2001). These surface plasmons can be excited by using a prism-based configuration (Kretschmann and Otto configuration) (Fan et al., 2008), by a waveguide approach (Salamon et al., 1997) and a fiber-based architecture (Jorgenson et al., 1993; Leung et al., 2007). The evanescent wave is exploited in these methods at the metal–dielectric interface (Liedberg et al., 1993). The SPR method is sensitive to changes in the refractive index of the dielectric—this yields a powerful dielectric method (Abbas et al., 2011; Kampranis et al., 1999; Laviolle et al., 2012; Lowe et al., 1998; Malmqvist et al., 1999; Markey et al., 1999).

The SPR technique is useful for the detection of large molecules. However, it can also be used for the detection of several small virus-like particles, for example, influenza A virus (thousands of kilodaltons), which is easily detected by a radiative SPR platform (Francois et al., 2011; Strandh et al., 1998).

Sciacca et al. (2013) explored the role of surface functionalization and noted that it ensures the specificity of the sensor improves the detection limit and enables the detection of small molecules. These authors also demonstrated that the detection limit can be reduced if the antibodies are properly oriented, enabling the detection of apolipoprotein E (apoE). This is a low-molecular-weight protein ($\sim$39 kDa) and is a major determinant in lipid transport. It plays a critical role in arteriosclerosis and other diseases such as Alzheimer's disease (Corder et al., 1994; Chen et al., 2011). This is also a potential biomarker for gastric cancer (Sakashita et al., 2008; Oue et al., 2004).

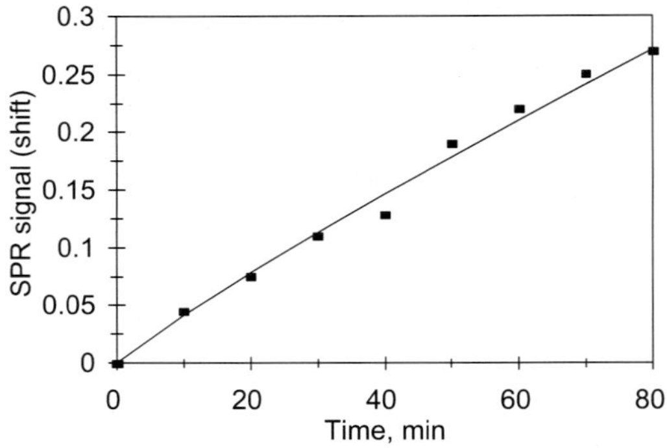

FIGURE 12.4 Binding of apoE lipoprotein in solution to apoE lipoprotein immobilized on a surface plasmon resonance (SPR) biosensor surface. *Sciacca et al. (2013).*

Figure 12.4 shows the binding of neutravidin in solution to anti-apolipoprotein (apoE) immobilized on an SPR biosensor surface (Sciacca et al., 2013). The binding kinetics is described by a single-fractal analysis. The values of the binding rate coefficient, k, and the fractal dimension, D_f, are given in Table 12.3.

Figure 12.5(a) shows the binding of 90 ng/ml apoE to anti-apoE immobilized to an SPR surface (Sciacca et al., 2013). A single-fractal analysis is adequate to describe the binding kinetics. The values of the binding rate coefficient, k, and the fractal dimension, D_f, are given in Table 12.4a and b.

Figure 12.5(b) shows the binding of 30 ng/ml apoE to anti-apoE immobilized to an SPR surface (Sciacca et al., 2013). A single-fractal analysis is adequate to describe the binding kinetics. The values of the binding rate coefficient, k, and the fractal dimension, D_f, are given in Table 12.4a and b.

Figure 12.5(c) shows the binding of 20 ng/ml apoE to anti-apoE immobilized to an SPR surface (Sciacca et al., 2013). A dual-fractal analysis is required to adequately describe the binding kinetics. The values of (1) the binding rate coefficient, k, and the fractal dimension, D_f, for a single-fractal analysis and (2) the binding rate coefficients, k_1 and k_2, and the fractal dimensions, D_{f1} and D_{f2}, for a dual-fractal analysis are given in Table 12.4a and b.

Figure 12.6 shows the shift in the SPR signal over time upon exposure of the neutravidin to the biotinylated surface (Sciacca et al., 2013). A dual-fractal analysis is required to adequately describe the binding kinetics. The values of (1) the binding rate coefficient, k, and the fractal dimension, D_f, for a single-fractal analysis and (2) the binding rate coefficients, k_1 and k_2, and the fractal dimensions, D_{f1} and D_{f2}, are given in Table 12.5a and b.

Figure 12.7 shows the binding of 30 μg/ml human apoE in solution on a biotin–neutravidin complex immobilized on an SPR biosensor surface (Sciacca et al., 2013). A single-fractal analysis is required to describe the

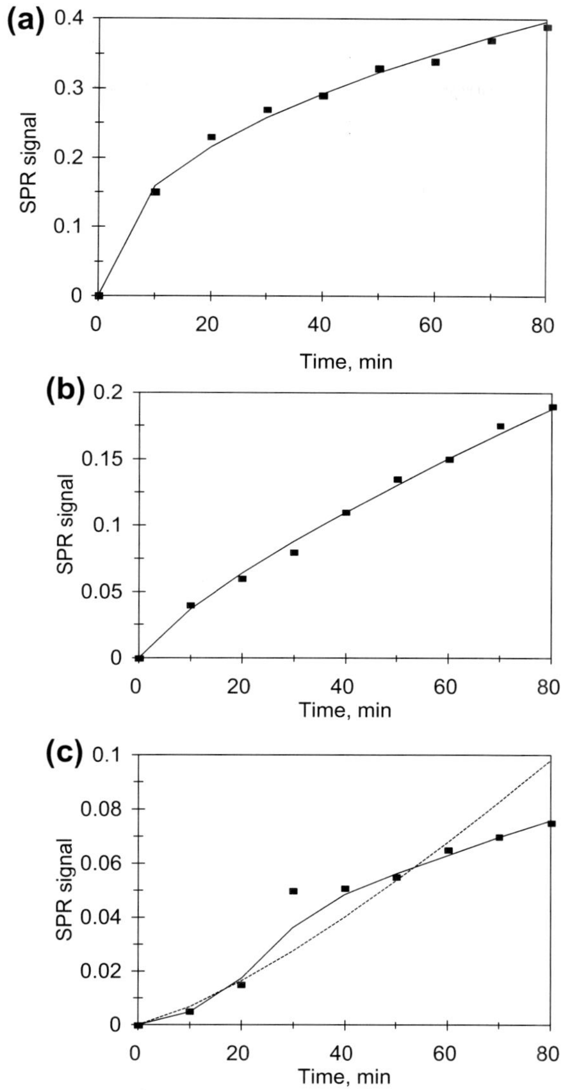

FIGURE 12.5 Binding of different concentrations (in ng/ml) of apolipoprotein E in solution to anti-lipoprotein immobilized to an SPR biosensor surface: (a) 90 (b) 30 (c) 20. *Sciacca et al. (2013).*

binding kinetics. The values of the binding rate coefficient, k, and the fractal dimension, D_f, are given in Table 12.6.

Gnedenko et al. (2013) have developed a highly sensitive reverse sandwich immunoassay for the detection of human cMb in serum using a gold

TABLE 12.3 Binding Rate Coefficient and the Fractal Dimension for Neutravidin in Solution to Anti-Apolipoprotein E (apoE) Bound to a Surface Plasmon Resonance (SPR) Biosensor Surface Sciacca et al. (2013)

Analyte in Solution/Receptor on SPR Surface	k	D_f
apoE/anti-apoE	0.005373 ± 0.000406	1.2094 ± 0.0824

TABLE 12.4a Binding Rate Coefficients for Different Concentrations of Apolipoprotein E (apoE) in Solution to Anti-apoE Immobilized to a Surface Plasmon Resonance (SPR) Biosensor Surface (Sciacca et al., 2013)

Analyte in Solution/ Receptor on Surface	k	k_1	k_2
90 ng/ml apoE/anti-apoE immobilized to an SPR surface	0.0578 ± 0.0027	NA	NA
30 ng/ml apoE/anti-apoE immobilized to an SPR surface	0.00618 ± 0.00037	NA	NA
20 ng/ml apoE/anti-apoE immobilized to an SPR surface	0.00034 ± 0.00013	(7.67 ± 2.53) E-05	0.0044 ± 0.00012

TABLE 12.4b Fractal Dimensions for Different Concentrations of Apolipoprotein E (apoE) in Solution to Anti-apoE Immobilized to a Surface Plasmon Resonance (SPR) Biosensor Surface (Sciacca et al., 2013)

Analyte in Solution/ Receptor on Surface	D_f	D_{f1}	D_{f2}
90 ng/ml apoE/anti-apoE immobilized to an SPR surface	2.1212 ± 0.04696	NA	NA
30 ng/ml apoE/anti-apoE immobilized to an SPR surface	1.5492 ± 0.0616	NA	NA
20 ng/ml apoE/anti-apoE immobilized to an SPR surface	0.434 ± 0.341	0 + 0.538	0.0044 ± 0.00012

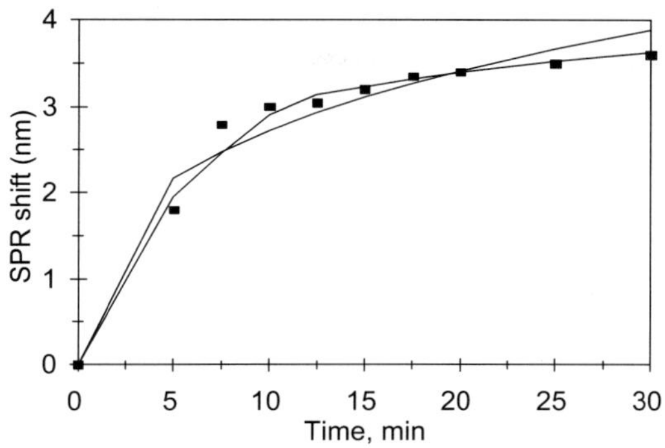

FIGURE 12.6 Typical real-time measured surface plasmon resonance (SPR) response for the neutravidin immobilization from a 400-nM solution. *Sciacca et al. (2013).*

TABLE 12.5a Binding Rate Coefficients for the Binding during Exposure of the Neutravidin in Solution to the Biotinylated Surface Sciacca et al. (2013)

Analyte in Solution/ Receptor on Surface	k	k_1	k_2
Neutravidin/biotinylated surface	1.2865 ± 0.1363	0.7698 ± 0.1009	2.0499 ± 0.0225

TABLE 12.5b Fractal Dimensions for the Binding during the Exposure of the Neutravidin in Solution to the Biotinylated Surface Sciacca et al. (2013)

Analyte in Solution/ Receptor on Surface	D_f	D_{f1}	D_{f2}
Neutravidin/biotinylated surface	1.2865 ± 0.1363	0.7698 ± 0.1009	2.0499 ± 0.0225

nanoparticle (AuNP)-enhanced SPR. These authors state that myoglobin is a heme-containing oxygen-binding protein, which is present in the cytoplasm of cardiac and skeletal muscle cells. Myoglobin is rapidly released into the blood when muscle cells are damaged. This occurs during acute myocardial infarction (AMI). Kemp et al. (2004) indicate that after AMI, cMb is elevated above the normal range and may be correlated with the extent of myocardial damage (Christenson and Azzazy, 1998). Lee and Goldman (1986), de Winter et al. (2000), and Chin et al. (1999) indicate that current detection methods for the biomarkers of AMI in serum have low accuracy. Thus, better methods are

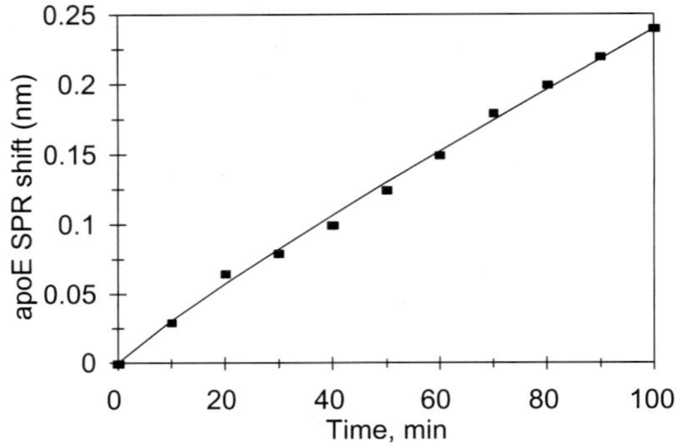

FIGURE 12.7 Surface plasmon resonance (SPR) response for 30 μg/ml human apoE sensing for biotin-neutravidin complex on an SPR surface. *Sciacca et al. (2013).*

TABLE 12.6 Binding Rate Coefficient, k, and Fractal Dimension, D_f, of Binding of 30 μg/ml Human apoE on Biotin–Neutravidin Complex Immobilized on a Surface Plasmon Resonance (SPR) Surface Sciacca et al. (2013)

Analyte in Solution/Receptor on Surface	k	D_f
30 μg/ml human apoE/biotin–neutravidin complex on an SPR surface	0.004051 ± 0.000226	1.2284 ± 0.04954

required. The SPR, as usual, is a suitable alternative (Homola et al., 2005). Gnedenko et al. (2013) indicate that it has better accuracy and reproducibility than ELISA (Avaramis et al., 2009; Lung et al., 2003; Valsocherova et al., 2009; Nechansky et al., 2010).

The higher sensitivity of the SPR biosensor is important to detect the cMb, as it has a low concentration in the circulating blood. Different approaches for cMb detection amplification have been presented in the literature (Goh et al., 2003; Liu et al., 2004; Cao et al., 2006; Seydack, 2005; Mitchell and Lowe, 2009; Luan et al., 2011; Besselink et al., 2004; Huang et al., 2007; Kim et al., 1995; Cao and Sim, 2007; Tang et al., 2007). AuNP particles enhance the SPR signal, and this method has been applied for the detection of different analytes (Johne et al., 1995; Huang et al., 2005; Pieper-Furst et al., 2005). Gnedenko et al. (2013) have developed a highly sensitive reverse sandwich immunoassay for human cMb using an AuNP-enhanced SPR biosensor.

Figure 12.8(a) shows the binding and dissociation of free cMb in solution to immobilized Mab1. A single-fractal analysis is required to adequately

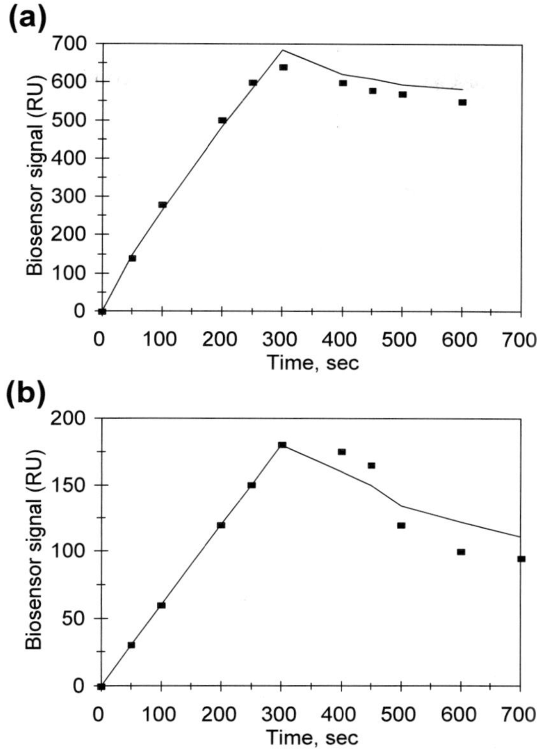

FIGURE 12.8 Binding and dissociation of (a) free cMb in solution to immobilized Mab1 on an SPR surface and (b) complex cMb–Mab2 to immobilized Mab1 on an SPR surface.

describe the binding and dissociation kinetics (Gnedenko et al., 2013). The values of the binding and dissociation rate coefficients, k and k_d, and the fractal dimension, D_f and D_{fd} for a single-fractal analysis are given in Table 12.7a and b.

TABLE 12.7a Binding and Dissociation rate coefficients of (1) Free cMb in Solution to Immobilized Mab1 and (2) complex cMb–Mab2 in Solution to Immobilized Mab1 Gnedenko et al. (2013)

Analyte in Solution/Receptor on Surface	k	k_d
Free cMb/immobilized Mab1	4.996 ± 0.3125	1.9895 ± 0.1542
Complex cMb–Mab2/immobilized Mab1	0.6 ± 0.0	1.9895 ± 0.1543

TABLE 12.7b Fractal Dimensions for the Binding and Dissociation of (1) Free cMb in Solution to Immobilized Mab1 and (2) Complex cMb–Mab2 in Solution to Immobilized Mab1 Gnedenko et al. (2013)

Analyte in Solution/Receptor on Surface	D_f	D_{fd}
Free cMb/immobilized Mab1	1.2744 ± 0.0818	1815 ± 0.1093
Complex cMb–Mab2/immobilized Mab1	1.0 ± 7.32E-15	18150 ± 0.1092

Figure 12.8(b) shows the binding and dissociation of complex cMb–Mab2 in solution to immobilized Mab1. A single-fractal analysis is required to adequately describe the binding and dissociation kinetics (Gnedenko et al., 2013). The values of the binding and dissociation rate coefficients, k and k_d, and the fractal dimension, D_f, for a single-fractal analysis are given in Table 12.7a and b.

Šípová et al. (2012) have developed a surface plasmon biosensor based on engineered proteins—the direct detection of interferon-gamma in diluted blood plasma. This interferon-gamma (hIFNγ) is a cytokine that is critical for innate and adaptive immunity and for control of tumor growth (Schroder et al., 2004; Hu et al., 2009). Certain autoimmune disorders also lead to a higher expression of hIFNγ. It may be detected using antibody-based sandwich immunoassay (Dheda et al., 2009). In simulated heparinized blood samples, hIFNγ can attain levels of hundreds of picograms per milliliter (Bartalesi et al., 2009). In tuberculosis, the levels of hIFNγ in blood samples are induced by specific tuberculosis antigens (Veerapathran et al., 2008). However, established methods of detection are time consuming and provide little information regarding the kinetics of the antigen–antibody interaction. This incidentally is a primary objective of the book: to provide an understanding of the basic kinetics involved in the antigen–antibody reaction, or in a more general sense, the analyte–receptor reaction.

As indicated in the different chapters throughout the book, the SPR is a useful biosensor for detecting analytes in solution (Fan et al., 2008). These include proteins, peptides, nucleic acids, bacteria, and viruses (Homola, 2008). Šípová et al. (2012) indicate that cytokines have been detected by biosensors (Battaglia et al., 2005; Chou et al., 2010). The SPR biosensor has been used to detect hIFNγ (Stigler et al., 2005; Stybayeva et al., 2010). A limit of detection of 16 nM (250 ng/ml) was achieved by Stigler et al. (2005) in the detection of hIFNγ.

Recently, Šípová et al. (2012) indicate that strategies involving aptamers have emerged as alternatives to antibody-based assays. They are cost-effective, stable, and show a high degree of specificity (van Dorst et al., 2010; Sefah et al., 2009; Tuleuva et al., 2010).

Šípová et al. (2012) used albumin-binding domain as the binder to an SPR biosensor surface so that they may detect hIFNγ directly by the SPR biosensor. They used two immobilization methods—the direct covalent coupling and streptavidin–biotin reaction.

Figure 12.9 shows the binding and dissociation of 100 nM hIFNγ in solution to ABD-Avi-Tag binders immobilized via a covalent coupling to an SPR biosensor surface (Šípová et al., 2012). A dual-fractal analysis is required to describe the binding kinetics. A single-fractal analysis is required to adequately describe the dissociation kinetics. The values of (1) the binding rate coefficient, k, and the fractal dimension, D_f, for a single-fractal analysis, (2) the binding rate coefficients, k_1 and k_2, and the fractal dimensions, D_{f1} and D_{f2}, for a dual-fractal analysis, and (3) the dissociation rate coefficient, k_d, and the fractal dimension for dissociation, D_{fd}, are given in Table 12.8a and b. It is of interest to note that for a dual-fractal analysis as the fractal dimension decreases by a factor of 3.09 from a value of 1.5348 to a value of 0.496, the binding rate coefficient decreases by a factor of 2.69 from a value of $k_1 = 0.01428$ to $k_2 = 0.005307$. The changes in the fractal dimension or the degree of heterogeneity on the surface are in the same direction.

Figure 12.10 shows the binding and dissociation of 100 nM hIFNγ in solution to ABD-ToIA fusion proteins immobilized via a covalent coupling to an SPR biosensor surface (Šípová et al., 2012). A dual-fractal analysis is required to describe the binding kinetics. A single-fractal analysis is required to adequately describe the dissociation kinetics. The values of (1) the binding rate coefficient, k, and the fractal dimension, D_f, for a single-fractal analysis, (2) the binding rate coefficients, k_1 and k_2, and the fractal dimensions, D_{f1} and D_{f2}, for a dual-fractal analysis, and (3) the dissociation rate coefficient, k_d, and

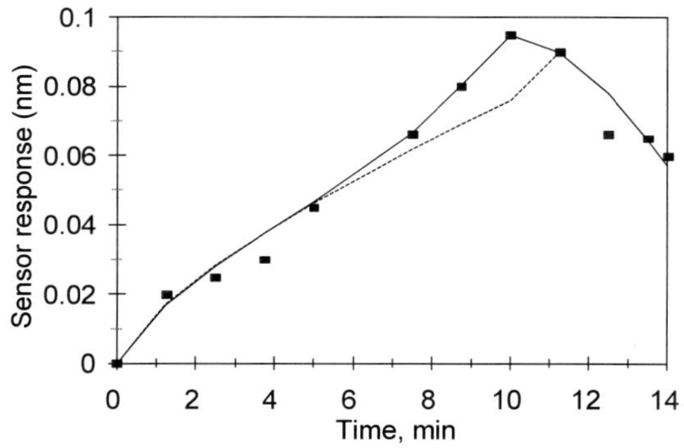

FIGURE 12.9 Binding and dissociation of 100 nM hIFNγ in solution to ABD-Avi-Tag binders immobilized via a covalent coupling to a surface plasmon resonance biosensor surface. *Šípová et al. (2012).*

TABLE 12.8a **Binding and Dissociation Rate Coefficients for hIFNγ in Solution to ABD0-TolA Immobilized Using Covalent Chemistry on a Surface Plasmon Resonance (SPR) Biosensor Surface Šípová et al. (2012)**

Analyte in Solution/ Receptor on Surface	k	k_1	k_2	k_d
500 nM hIFNγ/ABD0-TolA immobilized using covalent chemistry on an SPR surface	3.219 ± 0.734	3.045 ± 0.630	5.662 ± 0.011	0.5286 ± 0.0618
50 nM hIFNγ/ABD0-TolA immobilized using covalent chemistry on an SPR surface	0.04483 ± 0.00658	NA	NA	0.07946 ± 0.00859
20 nM hIFNγ/ABD0-TolA immobilized using covalent chemistry on an SPR surface	0.1289 ± 0.0300	0.1055 ± 0.0172	1.2917 ± 0.0331	NA
10 nM hIFNγ/ABD0-TolA immobilized using covalent chemistry on an SPR surface 10	0.0527 ± 0.0064	0.04 ± 0	0.2150 ± 0.007	NA

TABLE 12.8b **Fractal Dimensions for the Binding and Dissociation Rate Coefficients for hIFNγ in Solution to ABD0-TolA Immobilized Using Covalent Chemistry on a Surface Plasmon Resonance (SPR) Biosensor Surface Šípová et al. (2012)**

Analyte in Solution/ Receptor on Surface	D_f	D_{f1}	D_{f2}	D_{fd}
500 nM hIFNγ/ABD0-TolA immobilized using covalent chemistry on an SPR surface	2.220 ± 0.1500	1.7238 ± 0.2666	2.8486 ± 0.00878	1.6004 ± 0.0618
50 nM hIFNγ/ABD0-TolA immobilized using covalent chemistry on an SPR surface	0 + 0.2156	NA	NA	0.786 ± 0.1212
20 nM hIFNγ/ABD0-TolA immobilized using covalent chemistry on an SPR surface	1.0130 ± 0.1524	0.770 ± 0.2192	2.396 ± 0.09246	NA
10 nM hIFNγ/ABD0-TolA immobilized using covalent chemistry on an SPR surface 10	1.3416 ± 0.1126	1.0 ± 3.02E-16	2.3424 ± 0.0198	NA

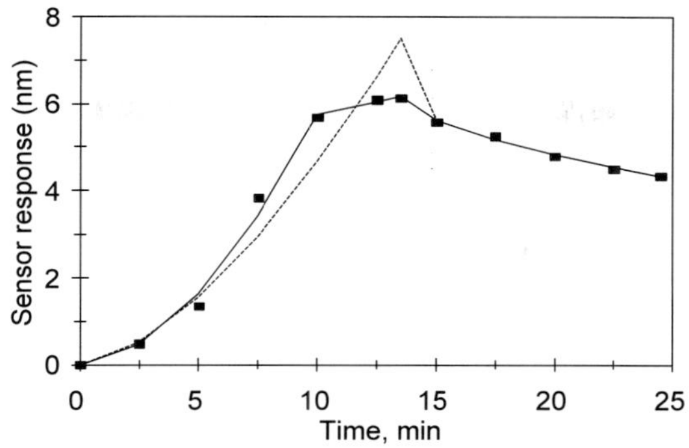

FIGURE 12.10 Binding and dissociation of 100 nM hIFNγ in solution to ABD-ToIA fusion proteins immobilized via a covalent coupling to an SPR biosensor surface. *Šípová et al. (2012).*

the fractal dimension for dissociation, D_{fd}, are given in Table 12.9a and b. It is of interest to note that for a dual-fractal analysis as the fractal dimension increases, the binding rate coefficient increases. The changes in the fractal dimension or the degree of heterogeneity on the surface are in the same direction.

Figure 12.11(a) shows the binding and the dissociation of 500 nM hIFNγ in solution to ABD0-ToIA immobilized using covalent chemistry to an SPR biosensor surface (Šípová et al., 2012). A dual-fractal analysis is required to adequately describe the binding kinetics. The dissociation is described by a

TABLE 12.9a Binding and Dissociation Rate Coefficients for (1) 100 nM hIFNγ in Solution to ABD-Avi-Tag Binders Immobilized via a Covalent Coupling to an SPR Biosensor Surface (Šípová et al., 2012) and (2) 100 nM hIFNγ in Solution to ABD-ToIA Fusion Proteins Immobilized via a Covalent Coupling to an SPR Biosensor Surface Šípová et al. (2012)

Analyte in Solution/ Receptor on Surface	k	k_1	k_2	k_d
100 nM hIFNγ/ABD-Avi-Tag binders immobilized via coupling to an SPR surface	0.01465 ± 0.00312	0.01428 ± 0.002664	0.005307 ± 0.000020	0.00364 ± 0.00051
100 nM hIFNγ in solution to ABD-ToIA fusion proteins immobilized via a covalent coupling to an SPR biosensor surface	0.1220 ± 0.0280	0.0852 ± 0.0238	3.105 ± 0.028	0.4128 ± 0.0231

TABLE 12.9b Fractal Dimensions for the Binding and Dissociation for (1) 100 nM hIFNγ in Solution to ABD-Avi-Tag Binders Immobilized via a Covalent Coupling to a SPR Biosensor Surface (Šípová et al., 2012) and (2) 100 nM hIFNγ in Solution to ABD-TolA Fusion Proteins Immobilized via a Covalent Coupling to an SPR Biosensor Surface Šípová et al. (2012)

Analyte in Solution/ Receptor on Surface	D_f	D_{f1}	D_{f2}	D_{fd}
100 nM hIFNγ/ABD-Avi-Tag binders immobilized via coupling to an SPR surface	1.4300 ± 0.1760	0.01428 ± 0.002664	0.005307 ± 0.000020	0.0 + 0.2896
100 nM hIFNγ in solution to ABD-TolA fusion proteins immobilized via a covalent coupling to an SPR biosensor surface	1.5348 ± 0.2096	1.5348 ± 0.2096	0.005307 ± 0.000020	0.0 + 0.2896

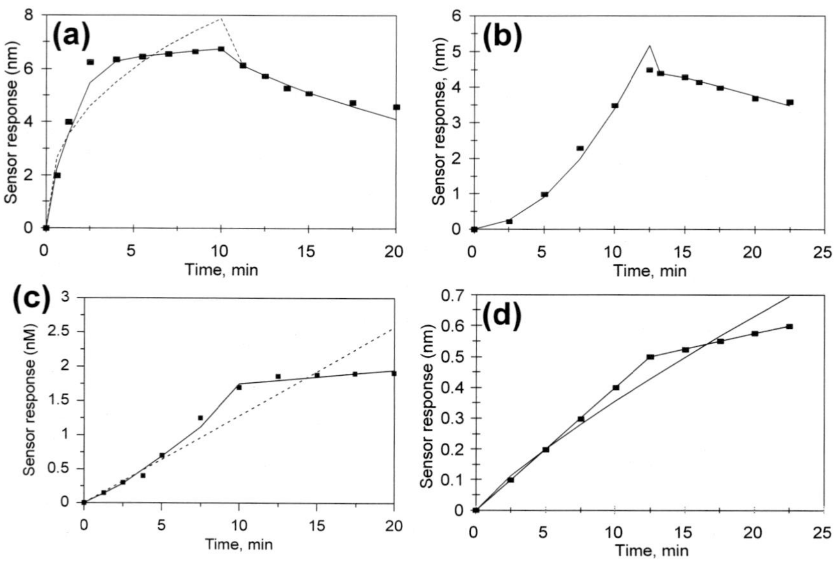

FIGURE 12.11 Binding of different concentrations of hIFNγ (in nM) in solution to ABD0-TolA immobilized using covalent chemistry to a surface plasmon resonance biosensor surface: (a) 500 nm (b) 50 nm (c) 20 nm (d) 10 nm. *Šípová et al. (2012).*

single-fractal analysis. The values of (1) the binding rate coefficient, k, and the fractal dimension, D_f, for a single-fractal analysis, (2) the binding rate coefficients, k_1 and k_2, and the fractal dimensions, D_{f1} and D_{f2}, for a dual-fractal analysis, and (3) the dissociation rate coefficient, k_d, and the fractal dimension for dissociation, D_{fd}, are given in Table 12.9a,b.

It is of interest to note that as the fractal dimension increases by a factor of 1.65 from a value of $D_{f1} = 1.7238$ to $D_{f2} = 2.8486$, the binding rate coefficient increases by a factor of 1.86 from a value of $k_1 = 3.045$ to $k_2 = 5.662$. An increase in the heterogeneity on the biosensor surface leads to an increase in the binding rate coefficient.

Figure 12.11(b) shows the binding and the dissociation of 50 nM hIFNγ in solution to ABD0-ToIA immobilized using covalent chemistry to an SPR biosensor surface (Šípová et al., 2012). A single-fractal analysis is adequate to describe the binding and the dissociation kinetics. The values of (1) the binding rate coefficient, k, and the fractal dimension, D_f, for a single-fractal analysis and (2) the dissociation rate coefficient, k_d, and the fractal dimension for dissociation, D_{fd}, are given in Table 12.9a,b.

Figure 12.11(c) shows the binding and the dissociation of 20 nM hIFNγ in solution to ABD0-ToIA immobilized using covalent chemistry to an SPR biosensor surface (Šípová et al., 2012). A dual-fractal analysis is required to adequately describe the binding kinetics. The dissociation is described by a single-fractal analysis. The values of (1) the binding rate coefficient, k, and the fractal dimension, D_f, for a single-fractal analysis, (2) the binding rate coefficients, k_1 and k_2, and the fractal dimensions, D_{f1} and D_{f2}, for a dual-fractal analysis, and (3) the dissociation rate coefficient, k_d, and the fractal dimension for dissociation, D_{fd}, are given in Table 12.9a,b.

It is of interest to note that as the fractal dimension increases by a factor of 3.56 from a value of $D_{f1} = 0.770$ to $D_{f2} = 2.396$, the binding rate coefficient increases by a factor of 2.80 from a value of $k_1 = 0.1055$ to $k_2 = 1.2917$.

Figure 12.11(d) shows the binding and the dissociation of 10 nM hIFNγ in solution to ABD0-ToIA immobilized using covalent chemistry to an SPR biosensor surface (Šípová et al., 2012). A single-fractal analysis is adequate to describe the binding and the dissociation kinetics. The values of (1) the binding rate coefficient, k, and the fractal dimension, D_f, for a single-fractal analysis and (2) the dissociation rate coefficient, k_d, and the fractal dimension for dissociation, D_{fd}, are given in Table 12.9a,b.

Figure 12.12(a) shows the increase in the binding rate coefficient k_1 with an increase in the hIFNγ concentration in nanomolarity in solution. For the data shown in Figure 12.12(a), the binding rate coefficient k_1 is given by

$$k_1 = (0.0036 \pm 0.0006)\,[\text{hIFN}]^{1.0879 \pm 0.0524} \tag{12.5a}$$

The fit is very good. Only three data points are available. The availability of more data points would lead to a more reliable fit. The binding

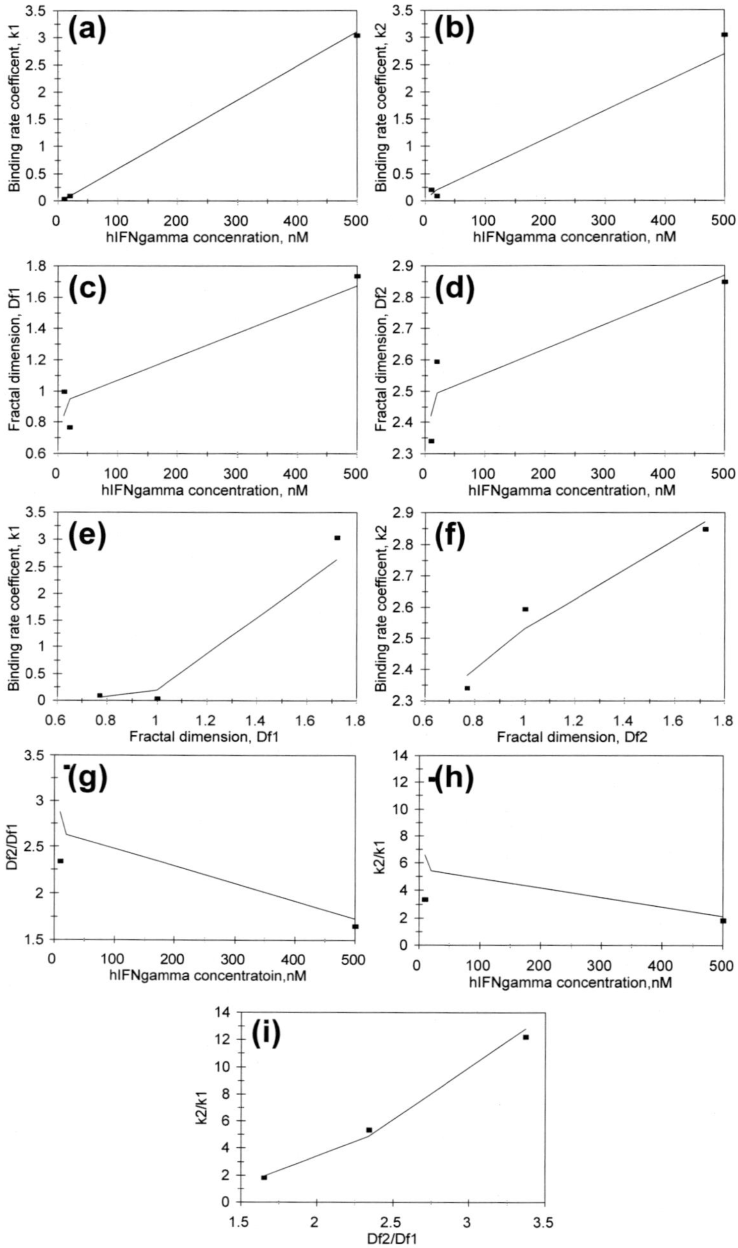

FIGURE 12.12 (a) Increase in the binding rate coefficient k_1 with an increase in the hIFNγ concentration in solution. (b) Increase in the binding rate coefficient k_2 with an increase in the hIFNγ concentration in solution. (c) Increase in the fractal dimension D_{f1} with an increase in the hIFNγ concentration in solution. (d) Increase in the fractal dimension D_{f2} with an increase in the hIFNγ concentration in solution. (e) Increase in the binding rate coefficient k_1 with an increase in the fractal dimension D_{f1}. (f) Increase in the binding rate coefficient k_2 with an increase in the fractal dimension D_{f2}. (g) Increase in the fractal dimension ratio, D_{f2}/D_{f1}, with an increase in the hIFNγ concentration in solution. (h) Increase in the binding rate coefficient ratio, k_2/k_1, with an increase in the hIFNγ concentration in solution. (i) Increase in the binding rate coefficient ratio, k_2/k_1, with an increase in the fractal dimension ratio, D_{f2}/D_{f1}.

rate coefficient k_1 exhibits slightly higher than a first (equal to 1.0879) order of dependence on the hIFNγ concentration in solution. This indicates that the binding rate coefficient k_1 is sensitive to the hIFNγ concentration in solution.

Figure 12.12(b) shows the increase in the binding rate coefficient k_2 with an increase in the hIFNγ concentration in nanomolarity in solution. For the data shown in Figure 12.12(b), the binding rate coefficient k_2 is given by

$$k_2 = (0.01966 + 0.02889)\,[hINF\gamma]^{0792 \pm 0.306} \qquad (12.5b)$$

The fit is very good. Only three data points are available. The availability of more data points would lead to a more reliable fit. The binding rate coefficient k_2 exhibits slightly an order of dependence between one-half and first (equal to 0.792) on the hIFNγ concentration in solution. This indicates that the binding rate coefficient k_2 is sensitive to the hIFNγ concentration in solution.

Figure 12.12(c) shows the increase in the fractal dimension D_{f1} with an increase in the hIFNγ concentration in nanomolarity in solution. For the data shown in Figure 12.12(c), the fractal dimension D_{f1} is given by

$$D_{f1} = (0.5608 \pm 0.1378)\,[hIFN\gamma]^{0.176 \pm 0.093} \qquad (12.5c)$$

The fit is very good. Only three data points are available. The availability of more data points would lead to a more reliable fit. The fractal dimension D_{f1} exhibits a slight dependence (equal to 0.176 order) on the hIFNγ concentration in solution. This indicates that the binding rate coefficient D_{f1} is not sensitive to the hIFNγ concentration in solution.

Figure 12.12(d) shows the increase in the fractal dimension D_{f2} with an increase in the hIFNγ concentration in nanomolarity in solution. For the data shown in Figure 12.12(d), the fractal dimension D_{f2} is given by

$$D_{f2} = (2.19 \pm 0.117)\,[hIFN\gamma]^{0.0434 \pm 0.0175} \qquad (12.5d)$$

The fit is very good. Only three data points are available. The availability of more data points would lead to a more reliable fit. The fractal dimension D_{f2} exhibits close to a zero-order dependence (equal to 0.043) on the hIFNγ concentration in solution. This indicates that the fractal dimension D_{f2} is not at all sensitive to the hIFNγ concentration in solution.

Figure 12.12(e) shows the increase in the binding rate coefficient k_1 with an increase in the fractal dimension D_{f1}. For the data shown in Figure 12.12(e), the binding rate coefficient k_1 is given by

$$k_1 = (2.532 \pm 0.080) D_{f1}^{0.2325 \pm 0.0541} \qquad (12.5e)$$

The fit is very good. Only three data points are available. The availability of more data points would lead to a more reliable fit. The binding rate coefficient k_1 is only mildly sensitive, as expected, to the fractal dimension D_{f1} or the degree of heterogeneity that exists on the biosensor surface, as it exhibits an order of dependence between zero and one-half on the fractal dimension that exists on the biosensor surface.

Figure 12.12(f) shows the increase in the binding rate coefficient k_2 with an increase in the fractal dimension D_{f2}. For the data shown in Figure 12.12(f), the binding rate coefficient k_2 is given by

$$k_2 = \left(0.1499 + 0.6322\right)D_{f2}^{4.758\pm2.848} \tag{12.5f}$$

The fit is very good. Only three data points are available. The availability of more data points would lead to a more reliable fit. The binding rate coefficient k_2 is very sensitive, as expected, to the fractal dimension D_{f2} or the degree of heterogeneity that exists on the biosensor surface, as it exhibits an order of dependence between four and one-half and five (equal to 4.758) on the fractal dimension that exists on the biosensor surface.

Figure 12.12(g) shows the decrease in the fractal dimension ratio, D_{f2}/D_{f1}, with an increase in the hIFNγ concentration in solution. For the data shown in Figure 12.12(g), the fractal dimension ratio, D_{f2}/D_{f1}, is given by

$$D_{f2}/D_{f1} = (3.880 \pm 0.493)\,[\text{hIFN}\gamma]^{-0.130\pm0.110} \tag{12.5g}$$

The fit is good. Only three data points are available. The availability of more data points would lead to a more reliable fit. The fractal dimension ratio, D_{f2}/D_{f1}, shows a very slight negative order of dependence (equal to -0.130) on the hIFNγ concentration in solution. This shows that the ratio is almost independent of the hIFNγ concentration in solution.

Figure 12.12(h) shows the decrease in the binding rate coefficient ratio, k_2/k_1, with an increase in the hIFNγ concentration in solution. For the data shown in Figure 12.12(h), the binding rate coefficient ratio, k_2/k_1, is given by

$$k_2/k_1 = (12.79 + 14.372)\,[\text{hIFN}\gamma]^{-0.2870\pm0.361} \tag{12.5h}$$

The fit is reasonable. Only three data points are available. The availability of more data points would lead to a more reliable fit. The binding rate coefficient ratio, k_2/k_1, shows a very slight negative dependence (equal to -0.2870) on the hIFNγ concentration in solution. This shows that the ratio is almost independent of the hIFNγ concentration in solution.

Figure 12.12(i) shows the increase in the binding rate coefficient ratio, k_2/k_1, with an increase in the fractal dimension ratio, D_{f2}/D_{f1}. For the data shown in Figure 12.12(i), the fractal dimension ratio, k_2/k_1, is given by

$$k_2/k_1 = (0.5143 + 0.0626)[D_{f2}/D_{f1}]^{2.646 \pm 0.2282} \quad (12.5i)$$

The fit is very good. Only three data points are available. The availability of more data points would lead to a more reliable fit. The binding rate coefficient ratio, k_2/k_1, is sensitive to the fractal dimension ratio, D_{f2}/D_{f1}, as it exhibits an order of dependence between two and one-half and three (equal to 2.646) on the fractal dimension ratio or the ratio of heterogeneities that exist on the biosensor surface.

Figure 12.13(a) shows the binding of 50 nM hIFNγ + 2% depleted human plasma in solution to ABD20-ToIA immobilized using covalent chemistry to a SPR biosensor surface. A single-fractal analysis is adequate to describe the binding kinetics. The values of the binding rate coefficient, k, and the fractal dimension, D_f, for a single-fractal analysis are given in Table 12.10.

Figure 12.13(b) shows the binding of 20 nM hIFNγ + 2% depleted human plasma in solution to ABD20-ToIA immobilized using covalent chemistry to an SPR biosensor surface. A single-fractal analysis is adequate to describe the binding kinetics. The values of the binding rate coefficient, k, and the fractal dimension, D_f, for a single-fractal analysis are given in Table 12.10.

Figure 12.13(c) shows the binding of 10 nM hIFNγ + 2% depleted human plasma in solution to ABD20-ToIA immobilized using covalent chemistry to an SPR biosensor surface. A single-fractal analysis is adequate to describe the binding kinetics. The values of the binding rate coefficient, k, and the fractal dimension, D_f, for a single-fractal analysis are given in Table 12.10.

Figure 12.14(a) shows the increase in the binding rate coefficient, k, with an increase in the hIFNγ concentration in nanomolarity in solution. For the data shown in Figure 12.14(a), the binding rate coefficient, k, is given by

$$k = (2.279 \pm 0.071)[\text{hIFN}\gamma]^{0.0657 \pm 0.0270} \quad (12.6a)$$

The fit is good. Only three data points are available. The availability of more data points would lead to a more reliable fit. The binding rate coefficient, k, exhibits close to a zero-order dependence (equal to 0.0657) on the hIFNγ concentration in solution. This indicates that the binding rate coefficient, k_1, is insensitive to the hIFNγ concentration in solution in the range presented.

Figure 12.14(b) shows the decrease in the binding rate coefficient, k, with an increase in the fractal dimension, D_f. For the data shown in Figure 12.14(b), the binding rate coefficient, k, is given by

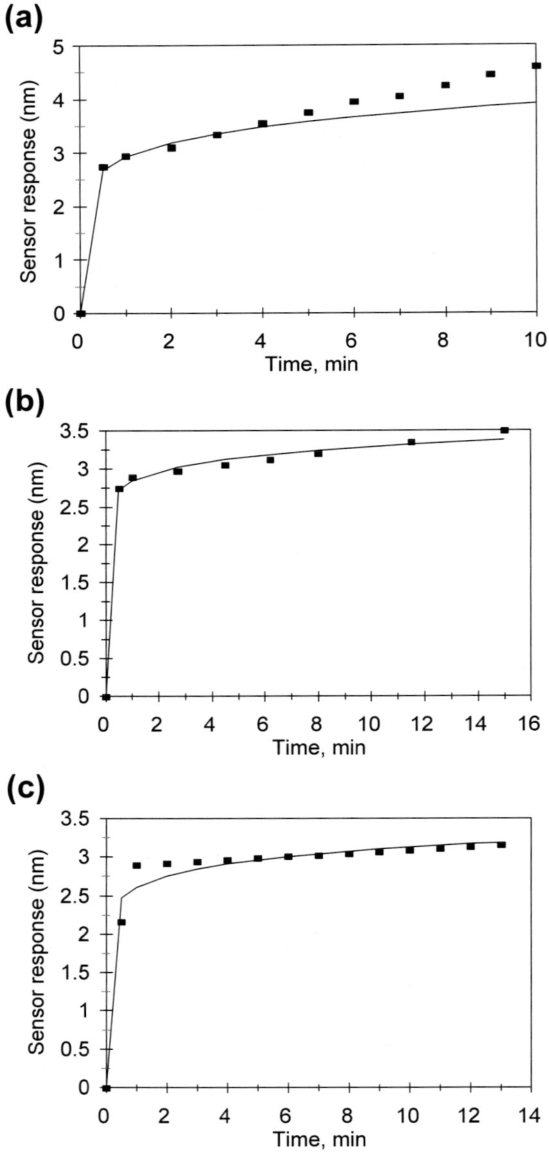

FIGURE 12.13 Binding of different concentrations (in nM) of hIFNγ + 2% depleted human plasma in solution to ABD20-ToIA immobilized using covalent chemistry to an SPR biosensor surface: (a) 50 nM (b) 20 nM (c) 10 nM.

TABLE 12.10 Binding Rate Coefficient, *k*, and the Fractal Dimension, D_f, for Different Concentrations of hIFNγ + 2% Depleted Human Plasma in Solution to ABD20-TolA Immobilized by Covalent Chemistry to a Surface Plasmon Resonance Biosensor Surface Šípová et al. (2012)

Analyte in Solution/Receptor on Surface	k	D_f
50 nM hIFNγ + 2% depleted human plasma/ABD20-TolA	2.916 ± 0.133	2.6546 ± 0.0294
20 nM hIFNγ + 2% depleted human plasma/ABD20-TolA	2.846 ± 0.071	2.8740 ± 0.0512
10 nM hIFNγ + 2% depleted human plasma/ABD20-TolA	2.614 ± 0.143	2.8447 ± 0.02772

$$k = (6.144 \pm 0.420) D_f^{-0.77-1.083} \tag{12.6b}$$

The fit is poor. Only three data points are available. The availability of more data points would lead to a more reliable fit. Since the curve exhibits a decreasing trend, only the negative power of dependence is shown. The binding rate coefficient, *k*, exhibits a negative power dependence between one-half and first (equal to −0.77) order of dependence on the fractal dimension, D_f, or the degree of heterogeneity on the biosensor surface.

Figure 12.14(c) shows the decrease in the fractal dimension, D_f, with an increase in the hIFNγ concentration in nanomolarity in solution. For the data shown in Figure 12.14(c), the fractal dimension, D_f, is given by

$$D_f = (3.205 \pm 0.196) \, [\text{hIFN}\gamma]^{-0.45 \pm 0.028} \tag{12.6c}$$

The fit is reasonable. Only three data points are available. The availability of more data points would lead to a more reliable fit. The fractal dimension, D_f, exhibits close to a negative one-half order of dependence on the hIFNγ in the concentration range presented. This indicates that the fractal dimension, D_f, is only mildly sensitive to the hIFNγ concentration in solution.

Figure 12.15 shows the binding and dissociation of 100 nM hIFNγ in solution to ABD20-TolA immobilized on an SPR biosensor surface (Šípová et al., 2012). A dual-fractal analysis is required to adequately describe the binding kinetics. The dissociation is described by a single-fractal analysis. The values of (1) the binding rate coefficient, *k*, and the fractal dimension, D_f, for a single-fractal analysis, (2) the binding rate coefficients, k_1 and k_2, and the fractal dimensions, D_{f1} and D_{f2}, and (3) the dissociation rate coefficient, k_d, and the fractal dimension for dissociation, D_{fd}, are given in Table 12.11a and b.

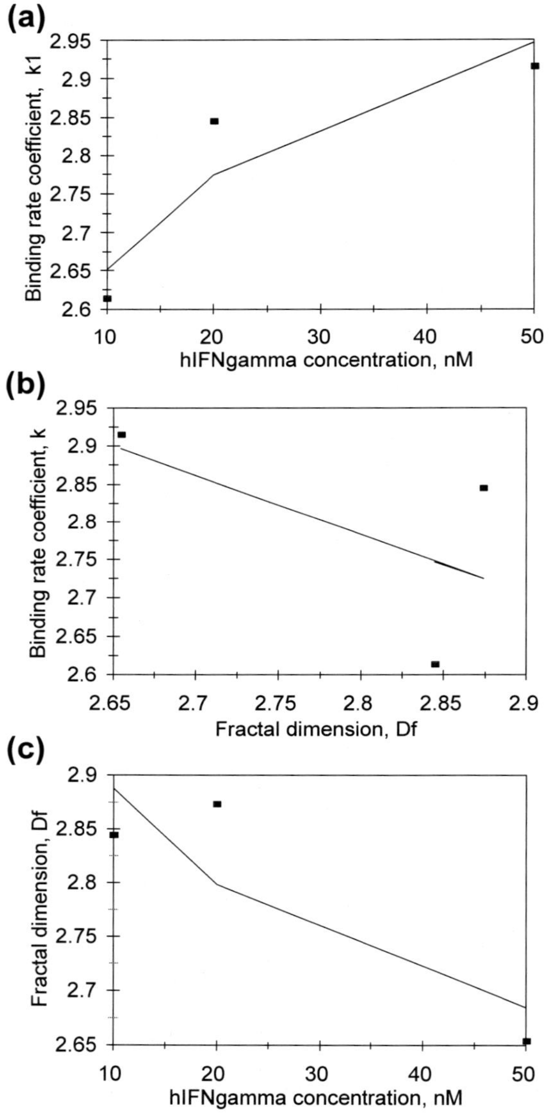

FIGURE 12.14 (a) Increase in the binding rate coefficient, k, with an increase in the hIFNγ concentration (in nM) in solution. (b) Decrease in the binding rate coefficient, k, with an increase in the fractal dimension, D_f. (c) Decrease in the fractal dimension, D_f, with an increase in the hIFNγ concentration (in nM) in solution.

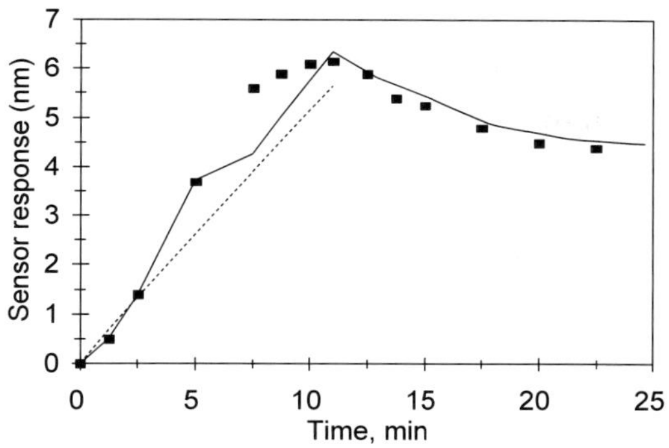

FIGURE 12.15 Binding of 100 nM hIFNγ in solution to ABD20-ToIA immobilized on a surface plasmon resonance biosensor surface (Šípová et al., 2012).

TABLE 12.11a Binding and Dissociation Rate Coefficients for hIFNγ in Solution to ABD20-ToIA Immobilized on a Surface Plasmon Resonance (SPR) Biosensor Surface (Šípová et al., 2012)

Analyte in Solution/ Receptor on the Surface	k	k_1	k_2	k_d
100 nM hIFNγ in solution/ ABD20-ToIA on an SPR surface	0.5535 ± 0.1769	0.3657 ± 0.0088	3.953 ± 0.029	0.3876 ± 0.0194

TABLE 12.11b Fractal Dimensions for Binding and Dissociation for hIFNγ in Solution to ABD20-ToIA Immobilized on a Surface Plasmon Resonance (SPR) Biosensor Surface (Šípová et al., 2012)

Analyte in Solution/ Receptor on the Surface	D_f	D_{f1}	D_{f2}	D_{fd}
100 nM hIFNγ in solution/ ABD20-ToIA on an SPR surface	1.0626 ± 0.2284	0.1126 ± 0.04810	2.6286 ± 0.09118	1.7202 ± 0.05604

12.4 CONCLUSIONS

In this chapter we present the kinetic analysis of the biomarkers for the different diseases that have recently appeared in the literature. As we have done throughout the book the binding and the dissociation kinetics is modeled first by a single-fractal analysis, and then by a dual-fractal analysis, if the single-fractal analysis was not sufficient.

The kinetics for the different biomarkers analyzed include: ((a) detection of human hepicidin-25 by the SPR biosensor (Scarrano et al., 2013). It regulates the iron in the blood (b) the detection of vitamin D in the blood (Carluccia et al., 2013). Vitamin D is a key factor involved in avoiding skeletal damage and a variety of pathological conditions, (c) detection of neutravidin by anti-lipoprotein immobilized on a SPR sensor surface (Sciacca et al., 2013), (d) detection cardiac myoglobin (cMb) in serum using a good nanoparticle (AuNP)-enhanced SPR (Gnedenko et al., 2013), (e) direct detection of interferon-gamma in diluted blood plasma by a SPR biosensor (Šípová et al., 2012), Interferongamma (hlFNγ) is a cytokine that is critical for innate and adaptive immunity and for control of tumor growth. (Šípová et al. (2012) have used albumin binding protein (ABD) as the binder to the SPR biosensor surface so that they may detect hlFNγ directly by an SPR biosensor.

Predictive relations are developed for the kinetics of binding and dissociation (if applicable) for the different biomarkers analyzed. For example, for the binding of vitamin D (25OHD) to a AB-25OHD modified surface, the binding rate coefficient, k exhibits a slightly more that one and one-half (equal to 1.582) order of dependence on the 25OHD concentration in solution. Also, the rate coefficient, k is extremely sensitive to the fractal dimension, D_f or the degree of heterogeneity on the biosensor surface since it exhibits a 69.4 order of dependence on D_f.

For the binding of HlFNγ onto the ABD/TolA immobilized on the biosensor surface, the binding rate coefficient, k_1 exhibits slightly more than a first (equal to 1.087)-order of dependence on the hlFNγ in solution. Similarly, for a dual-fractal analysis, the binding rate coefficient, k_2 exhibits less than a first (equal to 0.792) order of dependence on the hlFNγ in solution. Also, the binding rate coefficient, k_1 exhibits close to a zero (equal to 0.2515) order of dependence on the fractal dimension, D_{f1} or the degree of heterogeneity on the biosensor surface. Similarly, the binding rate coefficient, k_2 exhibits close to a fifth (equal to 4.258) order of dependence on D_{f2} or the degree of heterogeneity on the biosensor surface. This indicates that the binding rate coefficient, k_2 is extremely sensitive to the degree of heterogeneity or the fractal dimension on the biosensor surface in the second phase.

Similarly, predictive relations are developed for (a) k_2/k_1 vs hlFNγ in solution,(b) k_2/k_1 vs D_{f2}/D_{f1}, (c) the binding rate coefficient, k vs hlFNγ in solution, (d) the binding rate coefficient, k as a function of the fractal dimension or the degree of heterogeneity on the biosensor surface. These predictive relations provide insights into the influence of these parameters on the kinetics of binding and dissociation. These insights are critical in the different areas of applications. More often than not, the influence of diffusion and heterogeneity on the biosensor surface is neglected. Once again, as is amply demonstrated though out the different chapters in the book the degree of heterogeneity significantly influences, in general, the binding and dissociation kinetics on the biosensor surface.

Finally, a better understanding of all possible parameters that influence the kinetics of binding and dissociation of different analyte-receptor systems on biosensor surfaces is critical. These last chapters have attempted to analyze the kinetics of binding and dissociation of biomarkers for the different diseases that have recently appeared in the literature. The kinetic analysis is one possible avenue to obtain physical into these reactions occurring on the biosensor surfaces. It behooves practicing biosensorists to obtain these physical insights into these biomarkers for the different diseases. These physical insights, no matter how insignificant should help understand diseases in which they are involved in; which will and should lead to a better quality of life in general.

REFERENCES

Abbas, A., Linman, M.J., Cheng, Q.A., 2011. New trends in instrumental design for surface plasmon-based biosensors. Biosensors and Bioelectronics 26, 1815−1824.

Adamec, J., Jannasch, A., Huang, J., Hohman, E., Fleet, J.C., Peacock, M., Ferruzi, M.G., Martin, B., Weaver, C.M., 2011. Journal of Separation Science 34, 11−20.

Altamura, S., Kiss, J., Blattman, C., Gills, W., Muckenthaler, M.U., 2009. Biochimie 91, 1335−1338.

Anderson, D.S., Heeneym, M.M., Roth, U., Menzel, C., Fleming, M.D., Steen, H., 2010. Analytical Chemistry 82, 1551−1555.

Anderson, D.S., Kirchner, M., Kellogg, M., Kalish, L.A., Jeong, J.Y., Vanasse, G., 2011. Analytical Chemistry 83, 8357−8362.

Avaramis, V.I., Avramis, E.V., Hunter, W., Long, M.C., 2009. Anticancer Research 29, 299−302.

Bansal, S.S., Halkel, J.M., Fusova, J., Bomford, A., Simpson, R.J., Vasavda, N., Thein, S.L., Hider, R.C., 2009. Rapid Communications in Mass Spectrometry 23, 1531−1542.

Barnett In-Person Seminars, Kathy@clin-ed.com, 2013. Introduction to the FDA and the Medical Device Approach Process—Comprehensive Monitoring for Medical Decisions. July 30−August 1.

Bartnikas, T.B., Fleming, M.D., 2012. Hemojuvelin is essential for transferria-dependent and -independent hepcidia expression in mice. Haemotologia 97, 189−192.

Bartalesi, F., Vicidomini, S., Goletti, D., Fiorelli, C., Fiori, G., Melchiorre, D., et al., 2009. QuantiFERON-TB Gold and the TST are both useful for latent tuberculosis interferon infection screening in autoimmune diseases. European Respiratory Journal 33, 586−593.

Besselink, G.A., Kooyman, R.P., Van, O.S.P.J., Engbers, E.H., Schasfoort, R.B., 2004. Analytical Biochemistry 333, 165−173.

Battaglia, T.M., Masson, M.R., Sierks, M.R., Beaudoin, S.P., Rogers, J., Foster, K.N., et al., 2005. Quantification of cytokines involved in wound healing using surface plasmon resonance. Analytical Chemistry 77, 7016−7023.

Bischoff-Ferrari, H.A., Giovannucci, E., Willet, W.C., Dietrich, T., Dawson-Hughes, B., 2006. American Journal of Clinical Nutrition 84, 1253.

Bozzzini, C., Campostrini, N., Trombini, P., Nemeth, E., Castagna, A., Tenuti, I., Corrocher, R., Camaschella, C., Ganz, T., Olivieri, O., Piperno, A., Girelli, D., 2008. Blood Cells, Molecules and Diseases 40, 347−352.

Cao, C., Kim, J.P., Kim, B.W., Chae, H., Yoon, H.C., Yang, S.S., Sim, S.J., 2006. Biosensors & Bioelectronics 21, 2106–2113.

Cao, C., Sim, S.J., 2007. Biosensors & Bioelectronics 22, 1874–1880.

Carluccia, L., Gabriele, F., Tortolinia, C., Di Fusco Massimo, Romagnolic, E., Minisolac, S., Mazzeia, F., February 15, 2013. Several approaches for vitamin D determination by surface plasmon resonance and electrochemical affinity biosensors. Biosensors and Bioelectronics 40 (1), 350–355.

Castagna, A., Camposttrini, N., Zaninotto, F., Girelli, D., 2010. Journal of Proteomics 13, 527–536.

Chapuy, M.C., Prezosi, P., Maamer, M., 1997. Osteoporosis International 7, 439–443.

Christenson, R.H., Azzazy, H.M., 1998. Clinical Chemistry 44, 1855–1864.

Chen, J.L., Li, Q.Q., Wang, J.J., 2011. Topology of human apolipoprotein E3 uniquely regulates its diverse biological functions. Proceedings of the National Academy Sciences USA 108, 14813–14818.

Chin, A., Chan, W.K., Cheng, S.H., Leung, C.K., Choi, C.H., 1999. QJM, An International Journal of Medicine 92, 711–718.

Chiu, A., Chan, W.K., Cheng, S.H., Leung, C.K., Choi, C.H., 1999. QJM 92, 711–718.

Chou, E.H., Chuang, C.Y., Wu, C.M., 2010. Quantification of interleukin-6 in cell culture medium using surface plasmon resonance biosensors. Cytokine 51, 107–111.

Corder, E.H., Saunders, A.M., Risch, N.J., Strittmtter, W.J., Schmechel, D.E., Gaskell, P.C., et al., 1994. Protective effect of apolipoproteins-E type-2 allele for late-onset Alzheimers disease. Nature Genetics 7 (2), 180–184.

Dawson-Hughes, B., Heaney, R.P., Holick, M.F., 2005. Osteoporosis International 16, 713–716.

De Deminico, I., Nemeth, E., Nelson, J., Phillips, R., Ajioka, R., Kay, M.S., Kushner, J., Ganz, T., Ward, D., Kaplan, J., 2008. Journal of Clinical Investigation 8, 146–156.

Dheda, K., Smit, K.V., Badri, M., Pai, M., 2009. T-cell interferon-gamma release assays for the rapid immunodiagnosis of tuberculosis: clinical utility in high-burden vs. low-burden settings. Current Opinion in Pulmonary Medicine 15, 188–200.

Dostalek, J., Ctyroky, J., Homola, J., Brynda, E., Skalsky, M., Nevkvindova, P., et al., 2001. Surface plasmon resonance based on integrated optical waveguide. Sensors & Actuators B–Chemical 76, 6–12.

Dusso, A.S., Brown, A.J., Slatopolsky, E., 2005. American Journal of Physiology Renal Physiology 289, F8–F28.

Van Dorst, B., Mehta, J., Bekaet, K., Rouah-Martin, E., De Coen, W., Dubruel, et al., 2010. Recent advances in recognition elements of food and environmental biosensors; a review. Biosensors and Bioelectronics 26, 1178–1194.

Fan, X.D., White, J.M., Shopova, S.I., Zhu, H.Y., Suter, J.D., Sun, Y., 2008. Sensitive optical biosensors for unlabeled targets: a review. Analytica Chimica Acta 620, 8–26.

Francois, A., Boehm, J., Oh, S.Y., Kok, T., Momro, T.M., 2011. Collection mode surface plasmon fiber sensors: a new biosensing platform. Biosensors & Bioelectronics 26 (7), 3154–3159.

Ganz, T., 2011. Blood 117, 4425–4433.

Goh, J.B., Tam, P.L., Loo, T.W., Goh, M.C., 2003. Analytical Biochemistry 313, 262–266.

Gnedenko, O.V., Mezentsev, Y.V., Molnar, A.A., Lisitsa, A.V., Ivanov, A.S., Archakov, A.I., 2013. Highly sensitive detection of human cardiac myoglobin using a reverse sandwich immunoassay with a gold nanoparticle-enhanced surface plasmon resonance biosensor. Analytca Chimica Acta 759 (8), 105–109.

Havlin, S., 1989. Molecular diffusion and reaction: In the Fractal Approach to Heterogeneous Chemistry: Surfaces, Colloids, Polymers. In: Avnir, D. (Ed.). Wiley, New York, pp. 251–269.

Havlin, S., Ben-Avraham, D., 1987. Diffusion in disordered media. Advances in Physics 36, 695–798.

Heaney, R.P., Dowell, M.S., Hale, C.A., Bendich, A., 2003. Journal of the American College of Nutrition 142–146.

Holick, M.F., Siris, E.S., Binkley, N., Beard, M.K., Khan, A., Katzer, J.T., Petruschke, R.A., Cehn, E., de Papp, A.E., 2005. Journal of Clinical Endocrinology and Metabolism 90, 3215–3224.

Holick, M.F., 2006. Mayo Clinic Proceedings Mayo Clinic 81, 353–373.

Homola, J., 2008. Surface plasmon resonance sensors for detection of chemical and biological species. Chemical Reviews 108, 462–493.

Homola, J., Vaisocherova, H., Dostalek, J., Pillarik, M., 2005. Methods 37, 26–36.

Hu, X., Ivashkiv, L.B., 2009. Cross-regulation of signaling pathways by interferon-gamma: implications for immune responses and autoimmune diseases. Immunity 31, 539–550.

Huang, L., Reekmans, G., Saerens, D., Friedt, J.M., Frederix, F., Francis, L., Muyldermans, S., Campbell, A., Van Hoof, C., 2005. Biosensors & Bioelectronics 21, 483–490.

Huang, Y.Y., Hsu, H.Y., Huang, C.J., 2007. Biosensors and Bioelectronics 22, 960–965.

Johne, B., Hansen, K., mork, E., Holtlund, J., 1995. Journal of Immunological Methods 183, 167–174.

Jorgenson, R.C., Yee, S.S., Johmston, K.S., Compton, B.J., 1993. A novel surface plasmon resonance based fiber optic sensor applied to biochemical sensing. Proceedings of Fiber Optic sensors in Analytical Diagnostics 1866, 35–46.

Kampranis, S.C., Gormley, N.A., Tranter, R., Orphanides, G., Maxwell, A., 1999. Probing the binding of coumarins and cyclothialidines to DNA gyrase. Biochemistry 38, 1967–1976.

Kemna, E., Tjalsma, H., Laakkers, C., Nemeth, E., Wilems, H., Swinkels, D., 2008. Haematologica 93, 90–97.

Kenna, E., Tjalsma, H., Laarakkers, C., Nemeth, E., Willems, H., Swinkels, D., 2005. Blood 106, 3268–3270.

Kemp, M., Donovan, J., Higham, H., Hooper, J., 2004. British Journal of Anaesthesiology 93, 63–73.

Kim, M.G., Shin, Y.B., Jung, J.M., Ro, H.S., Chung, B.H., 1995. Journal of Immunological Methods 183, 167–174.

Krool, J.J.C., Kollaraki, V., Marinon, M., Vassil Kopoulos, T.P., Vavourakis, E., Tsochatzis, E., Pangalis, G.A., et al., 2009. Public Library of Science One 4e, 4581.

Laine, F., Laviolle, B., Ropert, M., Bougen, G., Morcet, J., Hamon, C., Massart, C., Westermann, M., Deugnier, Loreal, O., 2012. European Journal of Applied Physiology 112, 1391–1397.

Laviolle, B., Ropert, M., Bougen, G., Moncet, J., Hamon, C., Massart, M., Westermann, M., Deugnier, Y., Loreal, O., 2012. European Journal of Applied Physiology 112, 1391–1397.

Leung, A., Shankar, P.M., Mutharasan, R., 2007. A review of fiber-optic biosensors. Sensors and Actuators B: Chemical 125, 688–703.

Lee, T.H., Goldman, L., 1986. Annals of Internal Medicine 105, 221–233.

Liedberg, B., Lundstrom, J., Stenberg, E., 1993. Principles of biosensing with an extended coupling matrix and surface-plasmon resonance. Sensors and Actuators B: Chemical 11 (1–3), 63–72.

Lips, P., 2001. Endocrine Reviews 22, 474–501.

Liu, X., Sun, Y., Song, Y., Zhang, Q., Tian, Y., Bi, S., Zhng, H., 2004. Analytical Biochemistry 333, 99–104.

Li, H., Rose, M.J., Tran, L., Zhang, J., Miranda, L.P., James, C.A., Sasu, B.J., 2009. Journal of Pharmacological and Toxicological Methods 59, 171–180.

Luthra@ubitquinconference.com, 2013. The 5th Annual Ubiquitin Drug Discovery and Diagnostics Conference, July 2–24, 2013.

Lensmeyer, G.L., Wiebe, D.A., Binkley, N., Drezner, M.K., 2006. Clinical Chemistry 52, 1120–1126.

Lowe, P.A., Clark, T.J., Davies, R.J., Edwards, P.R., Kinning, T., Yeung, D., 1998. New approaches for the analysis of molecular recognition using the IAsys evanescent wave biosensor. Journal of Molecular Recognition 11, 194–199.

Luan, Q., Zhou, K., Tan, H., Yang, D., Yao, X., 2011. Biosensors & Bioelectronics 26, 2473–2477.

Lung, F.D., Chen, H.Y., Lin, H.T., 2003. Protein Peptide Letters 10, 313–319.

Malabanan, A., Veronikia, I.E., Holick, M.F., 1998. Lancet 351, 805–806.

Malmqvist, M., 1999. BIACORE: an affinity biosensor system for characterization of biomolecular interactions. Biochemical Society Transactions 27, 335–340.

Mansion, F., Chiap, P., Houbart, V., Crommen, J., Servais, A.C., Fillet, M., 2011. Journal of Separation Science 34, 1820–1827.

Markey, F., 1999. What is SPR anyway? BIA Journal 6, 14–17.

Mitchell, J.S., Lowe, T.E., 2009. Biosensors & Bioelectronics 24, 2177–2183.

Nemeth, E., Tuttle, M., Powelson, J., Vaughn, M., Donovan, A., ard, D., Ganz, T., Kaplan, J., 2004. Science 306, 2090–2093.

Netzel, B.C., Cradic, K.W., Bro, E.T., Girtman, A.B., Cyr, R.C., Singh, R.J., Grebe, S.K.G., 2011. Clinical Chemistry 57, 431–440.

Nechansky, A., 2010. Journal of Pharmaceutical and Biomedical Analysis 51, 252–254.

Oue, N., Hamai, Y., Mitani, Y., Matsumura, S., Oshimo, Aung, P.P., et al., 2004. Gene expression profile of gastric carcinoma: identification of genes and tags potentially involved in invasion, metastasis and carcinogenesis by serial analysis of gene expression. Cancer Research 64, 2397–2405.

Pieper_Furst, U., Stocklein, W.F.M., Warsinke, A., 2005. Analytica Chimica Acta 550, 69–76.

Piperno, A., Mariani, R., Trombini, P., Girelli, D., 2009. World Journal of Gastroenterology 15, 538–555.

Ramakrishnan, A., Sadana, A., 2011. A single-fractal analysis of cellular analyte-receptor binding kinetics using biosensors. Biosystems 59, 35–51.

Robach, P., Recalcali, S., Gelfi, C., Aachmaa-Anderson, N.J., Thomsen, J.J., Norgaard, A.M., Alberghini, A., Campostrini, N., Castagna, A., Vigano, A., Santambrogio, P., Kemppf, T., Wollert, K.C., Mouttereau, S., Ludby, C., Cairo, G., 2009. Blood 113, 6707–6715.

Roth, H.J., Schmid-Gayk, H., Weber, H., Niederau, C., 2008. Annals of Clinical Biochemistry 45, 153–159.

Sadana, A., 2001. A fractal analysis for the evaluation of hybridization kinetics in biosensors. Journal of Colloid and Interface Science 151(1),166–177.

Sadana, A., 2005. Fractal Binding and Disseriation Kinetics for Different Biosensors Applications. Elsevier, Amsterdam.

Sakashita, K., Tanaka, F., Zhang, X., Mimori, K., Kamohara, Y., Inouue, H., et al., 2008. Clinical significance of ApoE expression in human gastric cancer. Oncology Reports 20 (8), 1313–1319.

Salamon, Z., Macleod, H.A., Tollin, G., 1997. Coupled plasmon waveguide resonators: a new spectroscopic tool for probing proteolipid film structure and properties. Biophysical Journal 73, 2791–2797.

Scarano, S., Vestri, A., Emini, M.L., Minummi, M., 15 Feb 2013. Biosensors and Bioelectronics 40 (1), 135−140.

Scarano, S., Mascinia, M., Turner, A.P.F., Minunnia, M., 2010. Surface plasmon resonance imaging for affinity-based biosensors. Biosensors and Bioelectronics 25, 957−966.

Schroder, K., Hertzog, P.J., Ravasi, T., Hume, D.A., 2004. Interferon-gamma: an overview of signals, mechanisms and functions. Journal of Leukocyte Biology 75, 163−189.

Sciacca, B., Francois, A., Klinger-Hoffmann, M., Brazatti, J., Penno, M., Hoffmann, P., Monso, T.M., May 2013. Nanomedicine: Nanotechnology. Biology and Medicine 9 (4), 550−557.

Sefah, K., Phillips, J.A., Xiong, X.L., Meng, E., van Simaeys, D., Chen, H., et al., 2009. Nucleic acid aptamers for biosensors and bio-analytical applications. Analyst 134, 1765−1775.

Seydack, M., 2005. Biosensors & Bioelectronics 20, 2454−2469.

Schwalfenberg, G., 2007. Canadian Family Physician Medcin de Familie Canadien 53, 841−854.

Stepman, H.C.M., Vanderroost, A., Van Uylfangle, K., Thienpont, L.M., 2011. Clinical Chemistry 57, 441−448.

Schwarz, P., Strnad, P., van Figura, G., Janetzko, A., Krayenbuhl, P., Adler, G., Klaksiz, H., 2011. Journal of Gastroenterology 46, 648−658.

Šípová, H., Ševců, V., Kuchař, M., Ahmad, J.N., Mikulecký, P., Osička, R., Homola, J., November 2012. Surface plasmon resonance biosensor based on engineered proteins for direct detection of interferon-gamma in diluted blood plasma. Sensors and Actuators B, Chemical 174, 306−311.

Stigler, E.C.A., de Jong, G.J., van Bennekom, W.P., 2005. An improved coating for the isolation and quantification of interferon-gamma in spiked plasma using surface plasmon resonance (SPR). Biosensors and Bioelectronics 21, 474−482.

Strandh, M., Persson, B., Roos, H., Ohlson, S., 1998. Studies of interactions with weak interactions with weak affinities and low-molecular-weight compounds using surface plasmon resonance technology. Journal of Molecular Recognition 11, 188−190.

Stybayeva, G., Karova, M., Ramanculov, E., Simonian, A.L., Revzin, A., 2010. Detecting interferon-gamma release from human CD4 T-cells using surface plasmon resonance. Colloids & Surfaces B 80, 231−235.

kc@healthtech.com, 2013. Commercialization of Molecular Diagnostics 2013. August 20−22, 2013.

Tang, H., Wang, Q., Xie, Q., Zhang, Y., Tan, L., Yao, S., 2007. Biosensors & Bioelectronics 23, 668−674.

Tuleuva, N., Jones, C.N., Ramaculov, J.V.E., Yokobayshi, Y., Revzin, A., 2010. Development of an aptamer beacon for the detection of interferon-gamma. Analytical Chemistry 82, 1851−1857.

Valsocherova, H., Faca, V.M., Taylor, A.D., Hanash, Jiang, S., 2009. Biosensors & Bioelectronics 24, 2143−2148.

Veerapathran, A., Joshi, K., Goswami, K., Dogra, S., Moodie, E.E., Reddy, M.V., et al., 2008. T-cell assays for tuberculosis infection: deriving cut-offs for conversions using reproducibility data. PLoS One 3, e1850.

De Winter, J., Lijmer, J.G., Koster, R.W., Hoek, F.J., Sanders, G.T., 2000. Annals of Emerging Medicine 35, 113−120.

Zeder −Lutz, G., Neurath, A.R., van Regenmortel, M.H., 1999. Kinetics of interaction between 3-hydoxyphthaloyl-beta-lactoglobulin and CD4 molecules. Biologicals 27, 29−34.

Chapter 13

A Fractal Analysis of the Detection of Biomarkers for Different Diseases on Biosensor Surfaces

13.1 INTRODUCTION

Biosensors may be used to detect different biomarkers for different diseases. In this chapter we analyze the binding and dissociation (if applicable) kinetics of different biomarkers irrespective of any particular disease, unlike the previous chapters where we concentrated on particular diseases. This chapter demonstrates the versatility of biosensors, which have been used to detect a wide variety of biomarkers. The kinetic data for this chapter for the biosensors were selected at random with no particular attention to any diseases, as mentioned above. The analysis is similar to the fractal analysis presented for the detection of biomarkers for the different diseases presented in earlier chapters. Sometimes, the kinetics of the same biomarker is analyzed by different authors, for example, lysozyme.

13.2 THEORY

13.2.1 Single-Fractal Analysis

13.2.1.1 Binding Rate Coefficient

Havlin (1989) indicates that the diffusion of a particle (analyte [Ag]) from a homogeneous solution to a solid surface (e.g., receptor [Ab]-coated surface) on which it reacts to form a product (analyte−receptor complex; (Ab.Ag)) is given by:

$$(\mathrm{Ab \cdot Ag}) \approx \begin{cases} t^{(3-D_{f,\mathrm{bind}})/2} = t^p & t < t_c \\ t^{1/2} & t > t_c \end{cases} \tag{13.1}$$

Here $D_{f,bind}$ or D_f is the fractal dimension of the surface during the binding step and t_c is the crossover value. Havlin (1989) indicates that the crossover

Biomarkers and Biosensors. http://dx.doi.org/10.1016/B978-0-444-53794-2.00013-6

value may be determined by $r_c^2 \sim t_c$. Above the characteristic length, r_c, the self-similarity of the surface is lost and the surface may be considered homogeneous. Above time, t_c of the surface may be considered homogeneous, since the self-similarity property disappears, and "regular" diffusion is now present. For a homogeneous surface where D_f is equal to 2 and when only diffusional limitations are present, p = ½ as it should be. Another way of looking at the p = ½ case (where $D_{f,bind}$ is equal to 2) is that the analyte in solution views the fractal object, in our case, the receptor-coated biosensor surface, from a "large distance." In essence, in the association process, the diffusion of the analyte from the solution to the receptor surface creates a depletion layer of width $(Đt)^{½}$, where Đ is the diffusion constant. This gives rise to the fractal power law, $(\text{analyte}\cdot\text{receptor}) \sim t^{(3-D_{f,bind})/2}$. For the present analysis, t_c is arbitrarily chosen and we assume that the value of the t_c is not reached. One may consider the approach as an intermediate "heuristic" approach that may be used in the future to develop an autonomous (and not time-dependent) model for diffusion-controlled kinetics.

13.2.1.2 Dissociation Rate Coefficient

The diffusion of the dissociated particle (receptor [Ab] or analyte [Ag]) from the solid surface (e.g., analyte [Ag]—receptor [Ab] complex-coated surface) into solution may be given, as a first approximation, by:

$$(\text{Ab}\cdot\text{Ag}) \approx -t^{(3-D_{f,diss})/2} = -t^p (t > t_{diss}) \tag{13.2}$$

Here $D_{f,diss}$ is the fractal dimension of the surface for the dissociation step. This corresponds to the highest concentration of the analyte—receptor complex on the surface. Henceforth, its concentration only decreases. The dissociation kinetics may be analyzed in a manner "similar" to the binding kinetics.

13.2.2 Dual-Fractal Analysis

13.2.2.1 Binding Rate Coefficient

Sometimes, the binding curve exhibits complexities and two parameters (k and D_f) are not sufficient to adequately describe the binding kinetics. This is further corroborated by low values of r^2 factor (goodness-of-fit). In that case, one resorts to a dual-fractal analysis (four parameters: k_1, k_2, D_{f1}, and D_{f2}) to adequately describe the binding kinetics. The single-fractal analysis presented above is thus extended to include two fractal dimensions. At present, the time ($t = t_1$) at which the "first" fractal dimension "changes" to the "second" fractal dimension is arbitrary and empirical. For the most part, it is dictated by the data analyzed and experience gained by handling a single-fractal analysis. A smoother curve is obtained in the "transition" region, if care is taken to select the correct number of points for the two regions. In this case, the product (antibody—antigen or analyte—receptor complex, Ab.Ag or analyte.receptor) is given by:

$$(Ab \cdot Ag) \approx \begin{cases} t^{(3-D_{f1,bind})/2} = t^{p1} & (t < t_1) \\ t^{(3-D_{f2,bind})/2} = t^{p2} & (t_1 < t < t_2) = t_c \\ t^{1/2} & (t > t_c) \end{cases} \tag{13.3}$$

In some cases, as mentioned above, a triple-fractal analysis with six parameters (k_1, k_2, k_3, D_{f1}, D_{f2}, and D_{f3}) may be required to adequately model the binding kinetics. This is when the binding curve exhibits convolutions and complexities in its shape due to perhaps the very dilute nature of the analyte (in some of the cases to be presented) or for some other reasons. Also, in some cases, a dual-fractal analysis may be required to describe the dissociation kinetics.

13.3 RESULTS

Matavelli et al. (2010) have developed a simple and low-cost potentimetric biosensor for pharmaceutical and in vivo adrenaline determination. Adrenaline is a catecholamine, and is an important neurotransmitter. Adrenaline is a potent agonist of cardiac α- and β-receptors. It has been suggested that local cardiac adrenaline storage and synthesis does play an important role in regulating cardiac function (Kawada et al., 1998). For example, restoring the cardiac rhythm in cardiac patients (Goodman and Gilman, 2006). Thus, the importance of adrenaline determinations, in vivo in physiological studies. Large quantities of adrenaline are released during emotional states and conditions of physical or emotional stress, for example, hypoglycemia (Robertson et al., 2003), exercise (Watt et al., 2001), acute pain (Wortsman, 2002), and other situations involving mental excitement. The analytical determinations of catecholamines in biological matrices require analytical procedures capable of low detection limits since they occur in low concentrations (nmol/L). Different instrumental techniques have been employed, for example, some of the techniques employed are fluorimetry (Guo et al., 2002), spectrophotometry (Medina et al., 2000), cyclic voltametry (Liu et al., 2005), Flame Ionization Absorption (FIA) with fluorometric detection (Torres et al., 1998), electrochemiluniesence (Jie et al., 2009), potentiometric (Moreira et al., 2005), and amperometric (Felix et al., 2006).

Matavelli et al. (2010) studied the influence of 5.50×10^{-5} M mol/L adrenaline in solution to polyphenol oxides (PPO) immobilized on a polyethylene tube filled with carbon paste. Figure 13.1(a) shows the binding and dissociation of run #1 of adrenaline in solution. A first-order fractal analysis is adequate to the binding and dissociation kinetics. The values of (1) the binding rate coefficient, k, and the fractal dimension, D_f, for a single-fractal and analysis, and (2) the dissociation rate coefficient, k_d, and the fractal dimension

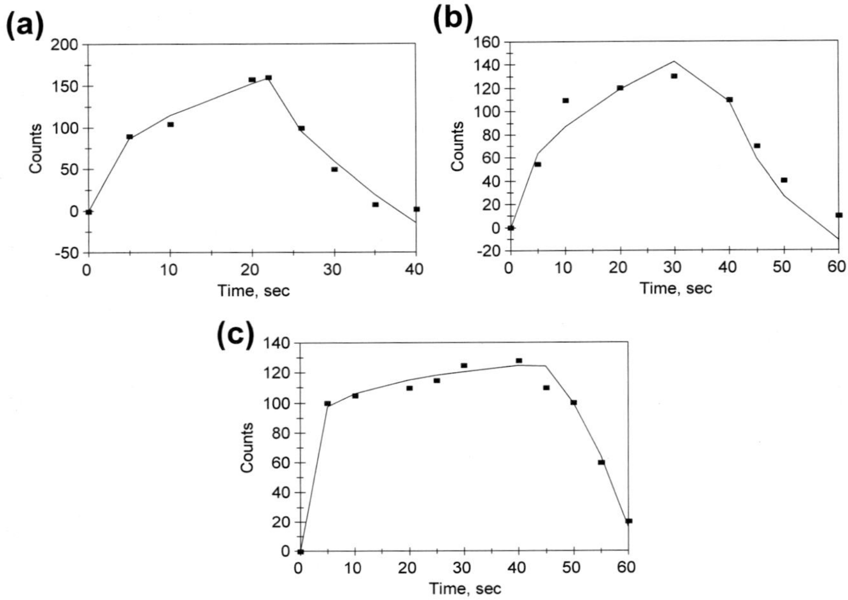

FIGURE 13.1 Binding and dissociation of 8×10^{-5} M of adrenaline in phosphate buffer pH 7 (Matavelli et al., 2010). Influence of consecutive runs: (a) Run #1 (b) Run #2 (c) Run #3.

TABLE 13.1 Binding Rate Coefficients and Fractal Dimensions for Adrenaline in Solution to Polyphenol Oxidase (PPO) Immobilized on a Polyethylene Tube Filled with Carbon Paste. Influence of Repeat Runs (a–c) 5×10^{-5} Mol/L Adrenaline in Solution, (d,e) Rat Blood Sample (Matavelli et al., 2010)

Analyte in Solution/ Receptor on Surface	k	k_d	D_f	D_{fd}
(a) 5×10^{-5} mol/L adrenaline; run#1/PPO	44.31 ± 13.33	25.207 ± 2.979	2.1746 ± 0.03374	1.6610 ± 0.2034
(b) 5×10^{-5} mol/L adrenaline; run#2/PPO	30.687 ± 7.063	54.56 ± 1.050	2.0546 ± 0.3020	0
(c) 5×10^{-5} mol/L adrenaline; run#3/PPO	80.818 ± 2.973	0.3129 ± 0.0258	2.7640 ± 0.0416	0
(d) Rat blood sample	66.187 ± 4.719	50.266 ± 0.504	2.5680 ± 0.09296	2.4826 ± 0.0914
(e) Rat blood sample	62.396 ± 7.729	2.413 ± 0.102	2.418 ± 0.1014	0.9330 ± 0.08154

for dissociation, D_{fd} for a single-fractal analysis are given in Table 13.1. The affinity, K, value is 1.758.

Figure 13.1(b) shows the binding and dissociation of 5.5×10^{-5} mol/L adrenaline in solution to immobilized PPO. This is run #2 of the sequence of three runs. Once again, a single-fractal analysis is required to describe the binding and dissociation kinetics. (1) The binding rate coefficient, k, and the fractal dimension, D_f, for a single-fractal analysis, and (2) the dissociation rate coefficient, k_d, and the fractal dimension, D_{fd}, for a single-fractal analysis are given in Table 13.1. In this case, the affinity, K, value is 0.56.

Figure 13.1(c) shows the binding and dissociation of 5.5×10^{-5} mol/L adrenaline in solution to immobilized PPO. This is run #3 of the sequence of three runs. Once again, a single-fractal analysis is required to describe the binding and dissociation kinetics. (1) The binding rate coefficient, k, and the fractal dimension, D_f, for a single-fractal analysis, and (2) the dissociation rate coefficient, k_d, and the fractal dimension, D_{fd}, for a single-fractal analysis are given in Table 13.1. In this case, the affinity, K, value is 250.29.

Figure 13.2 shows the increase in the binding rate coefficient, k, with an increase in the fractal dimension, D_f, for a single-fractal analysis. For the data shown in Figure 13.2, the binding rate coefficient, k, is given by:

$$k = (3.364 \pm 0.650) D_f^{3.144 \pm 0.830} \quad (13.4)$$

The fit is good. Only three data points are available. The availability of more data points would lead to a more reliable fit. The binding rate coefficient, k, exhibits a slightly higher than third order (equal to 3.144) of dependence on the fractal dimension, D_f, or the degree of heterogeneity that exists in the biosensor surface.

Also, Matavelli et al. (2010) analyzed the binding and dissociation of adrenaline in rat blood samples to immobilized PPO. Figure 13.3(a) shows the binding and dissociation of run #1 rat blood sample. The binding and

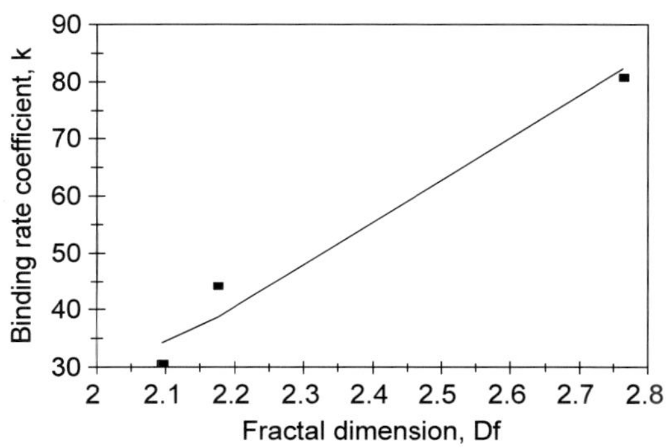

FIGURE 13.2 Increase in the binding rate coefficient, k, for a single-fractal analysis with an increase in the fractal dimension, D_f.

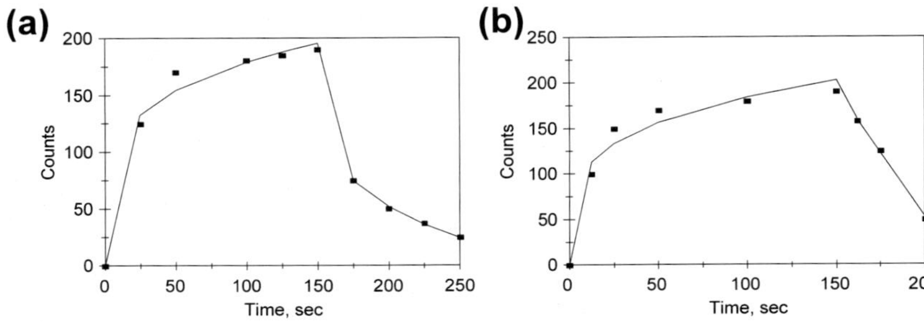

FIGURE 13.3 Binding and dissociation of adrenaline in rat blood sample (Matavelli et al., 2010). Influence of consecutive runs: (a) Run #1 (b) Run # 2.

dissociation kinetics are adequately described by a single-fractal analysis. The values of (1) the binding rate coefficient, k, and the fractal dimension, D_f, for single-fractal analysis, and (2) the values of dissociation rate coefficient, k_d, and the fractal dimension for dissociation, D_{fd}, for a single-fractal analysis are given in Table 13.1. In this case, the affinity, K, value is 1.316.

Figure 13.3(b) shows the binding and dissociation of run #2 (consecutive) rat blood sample to immobilized PPO. The binding and dissociation kinetics are adequately described by a single-fractal analysis. The values of (1) the binding rate coefficient, k, and the fractal dimension, D_f, for single-fractal analysis, and (2) the values of dissociation rate coefficient, k_d, and the fractal dimension for dissociation, D_{fd}, for a single-fractal analysis are given in Table 13.1. In this case, the affinity, K, value is 25.86.

Stobiecka and Hepel (2011) have developed a label-less electrochemical sensor for the immunodetection of glutathione (GSH). These authors state that a range of biosensors has been developed using the principle of molecular recognition for the detection of different diseases, utilizing antibodies, DNA, and receptors in sensory films. For example, these authors indicate that these principles have been used to detect the various biomarkers for different diseases (Mirkin et al., 1996; Stobiecka and Hepel, 2010a; Uehara et al., 2007, and Zhang et al., 2002). Hepel and Stobiecka (2011) more recently developed a biosensor to detect a biomarker for oxidative stress. These authors indicate that oxidative stress has been implicated in a variety of diseases including the aging process (Noble et al., 2004) and contributes to the development of autism in children (Amaral et al., 2008). The analysis of biomarkers of oxidative stress, such as GSH, glutathione disulfide, 3-nitrotyrosine, and homocysteine is a key factor for the preventive treatment of oxidative stress (Mahadik et al., 2001; Reddy and Bradley, 2004; Stobiecka et al., 2009, 2010a,b).

Stobiecka and Hepel (2011) indicated that surface-assisted laser desorption/ionization mass spectrometry has been proposed for the analysis of GSH in cell lysates (Cheng et al., 2010). A reliable method is based on fluorescence enhancement of dyes highly specific to GSH monochlorobimane (Kameric et al., 2000), and *o*-phthalaladehyde (Puri and Rostoski, 1988). Other techniques have also been employed (Stobiecka et al., 2010a,b; Stobiecka and Hepel, 2011).

Stobiecka and Hepel (2011) indicate that the pioneering work for the detection of GSH by biosensors was done by Gordon and Swalen (1977). In the present analysis, Stobiecka and Hepel (2011) investigated the effect of potential barriers introduced into the IgG-based sensing films for the detection of GSH. These authors indicate that the influence of electrostatic interactions in designing sensory films has been well established. This includes layer-by-layer deposition of oppositely charged polyelectrolytes. However, these authors emphasize that the effect of electric field on antibody–antigen interactions and sensory response has not been investigated.

Figure 13.4(a) shows the binding of 1.25 m MSGH in solution to a QC/Au/ANT/Ab piezoimmunsensor (Stobiecka and Hepel, 2011). A single-fractal analysis is adequate to describe the binding kinetics. The values of the binding rate coefficient, k, and the fractal dimension, D_f, for a single-fractal analysis are given in Table 13.2.

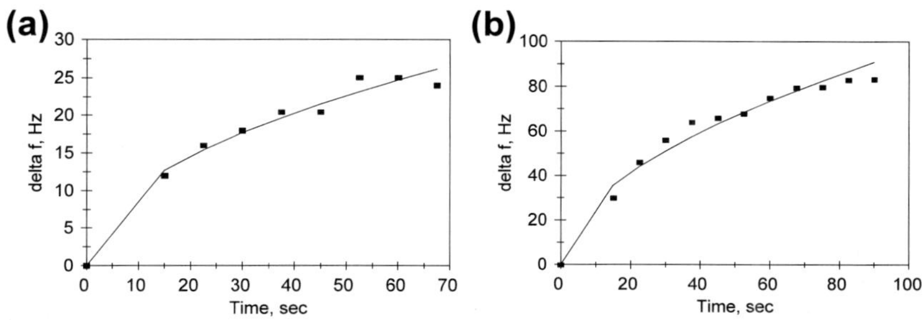

FIGURE 13.4 (a) Binding of 1.25 mM Mitochondrial Glutathione (MGSH) to a QC/Au/ANT/Ab piezoimmunosensor (Stobiecka and Hepel, 2011). (b) Binding of 0.95 nM AuNP@GSH to a QC/Au/ANT/Ab piezoimmunosensor.

TABLE 13.2 Binding Rate Coefficients and Fractal Dimensions for (a) a QC/Au/ANT/Ab After the Addition of 1.25 m MGSH (b) 0.95 nM AuNP@gSH (Stobiecka and Hepel, 2011)

Analyte in solution/Receptor on surface	k	D_f
Qc/Au/ANT/Ab	3.445 ± 0.214	2.0382 ± 0.08846
0.95 nM Au NP@ GSH	8.5148 ± 0.6472	1.9476 ± 0.0932

Figure 13.4(b) shows the binding of 0.95 nM NP@GSH to AC/Au/ANT/Ab piezoimmunsensor (Stobiecka and Hepel, 2011). Once again, a single-fractal analysis is adequate to describe the binding kinetics. The values of the binding rate coefficient, k, and the fractal dimension, D_f, for a single-fractal analysis are given in Table 13.2.

Glutamate is one of the important neurotransmitters in the central nervous system (CNS) and plays an important role in brain functioning and cognitive processes (Oldenziel et al., 2006a,b). It also plays a role in mediating energy transfer (Tong et al., 2007; Wilson and Johnson, 2008). Kualgina et al. (1999) indicate that it is important to monitor glutamate levels as it is heavily implicated in diseases such as schizophrenia, Parkinson's disease, and epilepsy. Glutamate has been detected by spectrophotometric, fluorimetric, and chromatographic methods (Jeffries et al., 1998) but they compared these methods, to electrochemical enzymatic biosensors, which are superior when used in conjunction with the microdialysis technique (Mao and Yamamoto, 2000; Niwa et al., 1996). The microdialysis sampling technique due to its short analytical tie, high sensitivity, and specificity with real-time measurements is a suitable method to provide near real-time measurements (Hogant and Lunte, 1994).

Yu et al. (2010) indicate that metal nanoparticles (NPs) due to their unusual electronic properties and high surface/volume ratio are suitable materials, for enzyme biosensors (Tian et al., 2006; Daniel and Anstruc, 2004). Incorporation of functional groups in room temperature ionic liquids (RTILs) is becoming increasingly important since they exhibit increasing enzyme stability and improved catalyst retention (Fei et al., 2007). The incorporation of functional groups in RTILs has been made them more flexible in order to tune them to a specific application.

Yu et al. (2010) have constructed a third-generation glutamate biosensor based on {$C_3(OH)_2$mim/BF_4]-AuPt film, which was constructed with glutamate oxidase as the recognition element. The authors tested their glutamate biosensor with respect to sensitivity, stability, selectivity, linear range, and detection limit.

Figure 13.5(a) shows the binding and dissociation of 15 nmol/L rhodamine derivative solutions + UO_x + uric acid to a chemiluminescent paper microfluidic biosensor based on an enzymatic reaction (Yu et al., 2010). A dual-fractal analysis is required to adequately describe the binding kinetics. The values of (1) the binding rate coefficient, k, and the fractal dimension for binding, D_f, and (2) the binding rate coefficients, k_1 and k_2, and the fractal dimensions, D_{f1} and D_{f2}, for a dual-fractal analysis are given in Table 13.3(a) and (b). An increase in the fractal dimension by a factor of 1.5 from a value of D_{f1} equal to 2 to D_{f2} equal to 3 leads to an increase in the binding rate coefficient by factor of 280.6.

Figure 13.5(b) shows the binding and dissociation of 20 nmol/L rhodamine derivative solutions + UO_x + uric acid to a chemiluminescent paper

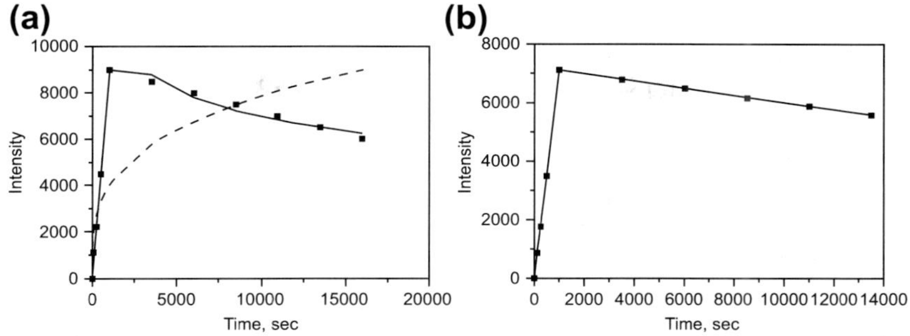

FIGURE 13.5 (a) Chemiluminescent intensity (binding) of 15 mmol/L rhodamine derivative solutions + UO_x + uric acid (Yu et al., 2010). (b) Chemiluminescent intensity (binding) of 20 mmol/L rhodamine derivative solutions + UO_x + uric acid (Yu et al., 2010).

microfluidic biosensor based on an enzymatic reaction (Yu et al., 2010). The binding and dissociation kinetics is adequately described by a single-fractal analysis. The values of (1) the binding rate coefficient, k, and the fractal dimension for binding, D_f, and (2) the dissociation rate coefficient, k_{1d}, and the fractal dimensions, D_{fd}, for a single-fractal analysis are given in Table 13.3(a) and (b). In this case, the affinity, $K = k/k_d$, is 57.0.

The binding and dissociation of glutamate in the striatum of brain dialysate of anesthetic rats to the GlutaOx-[C_3 $(OH)_2$mim][BF_4]-Au-Pt-Nafion-modified electrode has been analyzed (Yu et al., 2010). These authors studied the influence of two repeat runs. This is run #1. Figure is not shown. The binding kinetics is adequately described by a dual-fractal analysis. The dissociation kinetics is adequately described by a single-fractal analysis. The values of

TABLE 13.3a Binding Rate and Dissociation Rate Coefficients for the Binding and the Dissociation Phase for (a) Rhodamine Derivative Solution + UO_x + Uric Acid in Solution (15 mmol/L), and (b) Rhodaimine Derivative Solution + UO_x + Uric Acid (20 mmol/L) (Yu et al., 2010)

Analyte in Solution/ Receptor on Surface	k	k_1	k_2	k_d
Rhodamine derivative solutions + UO_x + uric acid (15 mmol/L)	537.39 ± 310.56	8.997 ± 0.0	56506.20 ± 2629.82	NA
Rhodamine derivative solutions + UO_x + uric acid (20 mmol/L)	6.7762 ± 0.0351	NA	NA	0.1188 ± 0.0012

TABLE 13.3b Fractal Dimensions for the Binding and the Dissociation Phase for (a) Rhodamine Derivative Solution + UO_x + Uric Acid in Solution (15 mmol/L), and (b) rhodamine Derivative Solution + UO_x + Uric Acid (20 mmol/L) (Yu et al., 2010)

Analyte in solution/receptor on surface	D_f	D_{f1}	D_{f2}	D_{fd}
Rhodamine derivative solutions + UO_x + uric acid (15 mmol/L)	2.4176 ± 0.1706	2.0 ± 2.4 E−15	3.0 ± 0.0	NA
Rhodamine derivative solutions + UO_x + uric acid (20 mmol/L)	0.0351	NA	NA	0.9868 ± 0.0070

(1) the binding rate coefficient, k, and the fractal dimension, D_f, for a single-fractal analysis, (2) the binding rate coefficients, k_1 and k_2, and the fractal dimensions, D_{f1} and D_{f2}, for a dual-fractal analysis, and (3) the dissociation rate, k_d, and the fractal dimension for dissociation, D_{fd}, for a single-fractal are given in Table 13.3(c) and (d).

It is of interest to note that for a dual-fractal analysis, an increase in the fractal dimension by a factor of 1.826 from a value of D_{f1} equal to 1.4396 to

TABLE 13.3c Binding and Dissociation Rate Coefficients for glutamate in the Striatum of Brain Dialystae of Anesthetic Rats to the GlutaOx-[$C_3(OH)_2$mim(BF_4]-Au/Pt-Nafion-Modified Membrane Electrode. Influence of Repeated Runs (Yu et al., 2010)

Run#	Analyte in Solution/ Receptor on Surface	k	k_1	k_2	k_d
Run #1	Glutamate/GlutaOx [$C_3(OH)_2$mim] [BF_4]-Nafion modified electrode	0.0501 ± 0.0113	0.01586 ± 0.00166	0.5777 ± 0.0219	0.01760 ± 0.00235
Run #2	Glutamate/GlutaOx [$C_3(OH)_2$mim] [BF_4]-Nafion modified electrode	0.01763 ± 0.0032	NA	NA	0.005764 ± 0.000073

TABLE 13.3d Fractal Dimensions for the Binding and Dissociation Phase for Glutamate in the Striatum of Brain Dialysate of Anesthetic Rats to the GlutaOx-[$C_3(OH)_2$mim(BF_4]-Au/Pt-Nafion-Modified Membrane Electrode. Influence of Repeated Runs (Yu et al., 2010)

Run#	Analyte in Solution/Receptor on Surface	D_f	D_{f1}	D_{f2}	D_{fd}
Run#1	Glutamate/GlutaOx [$C_3(OH)_2$mim] [BF_4]-Nafion-modified electrode	1.8872 ± 0.1414	1.4396 ± 0.1235	2.6286 ± 0.1008	1.2868 ± 0.2022
Run#2	Glutamate/GlutaOx [$C_3(OH)_2$mim] [BF_4]-Nafion-modified electrode	1.6106 ± 0.0986	NA	NA	1.0968 ± 0.01727

D_{f2} equal to 2.6286 leads to an increase in the binding rate coefficient by a factor of 36.42 from a value of k_1 equal to 0.01586 to k_2 equal to 0.5777. An increase in the degree of heterogeneity on the Nafion membrane surface leads to an increase in the binding rate coefficient. The affinities, $K_1 = k_1/k_d$ and $K_2 = k_2/k_d$, are 0.935 and 32.82, respectively.

The binding and dissociation of glutamate in the striatum of brain dialysate of anesthetic rats to the GlutaOx-[$C_3(OH)_2$mim][BF_4]-Au-Pt-Nafion-modified electrode (Yu et al., 2010) was analyzed. This is run #2. Figure is not shown. The binding and dissociation kinetics is adequately described by a single-fractal analysis. The values of (1) the binding rate coefficient, k, and the fractal dimension, D_f, for a single-fractal analysis, (2) the dissociation rate, k_d, and the fractal dimension for dissociation, D_{fd}, for a single-fractal are given in Table 13.3(c) and (d).

Nakamoto et al. (2010) indicate that disease biomarkers have been used for clinical diagnosis due to their homeostatic property and provide a quick response to the severity of a symptom. Price (2001) indicates that since blood sampling in invasive urinary sampling is frequently used for daily monitoring. Wang et al. (2009a,b) point out that diagnosis is often difficult from urinary samples alone. This is due to the fact that the urinary concentration fluctuates. However, Narayanan and Appleton (1980) indicate that same amount of creatinine is excreted everyday.

Nakamoto et al. (2008) had used a surface plasmon resonance (SPR) biosensor to detect the biomarker, transferrin, and the calibration marker creatinine. The SPR method has attracted a lot of attention in the Point-of-Care-Testing (POCT) format since it provides real-time and label-free measurement (Homola et al., 1999; Miura et al., 2003). In the present analysis Nakamoto et al. (2010) describe a one-chip biosensor for transferrin and creatinine. They also investigated the efficiency with which interfering substances in a sample solution can be eliminated by using enzymes that could be decompose these interferences.

Figure 13.6(a) shows the binding of 10 times diluted creatinine in a urine sample to a trienzyme layer of creatinine, creatinine oxidase, and sucrosine oxidase immobilized on an SPR biosensor surface (Nakamoto et al., 2008). A first-order fractal analysis is adequate to describe the binding kinetics. The values of the binding rate coefficient, k, and the fractal dimension, D_f, are given in Table 13.4.

Figure 13.6(b) shows the binding of 10 times diluted transferrin in a urine sample to an antitransferrin immobilized on an SPR biosensor surface (Nakamoto et al., 2008). A first-order fractal analysis is adequate to describe the binding kinetics. The values of the binding rate coefficient, k, and the fractal dimension, D_f, are given in Table 13.4a and b.

Figure 13.6(c) shows the binding of the interference substance, uric acid to uricase immobilized on an SPR biosensor surface (Nakamoto et al., 2008). A first-order fractal analysis is adequate to describe the binding kinetics and

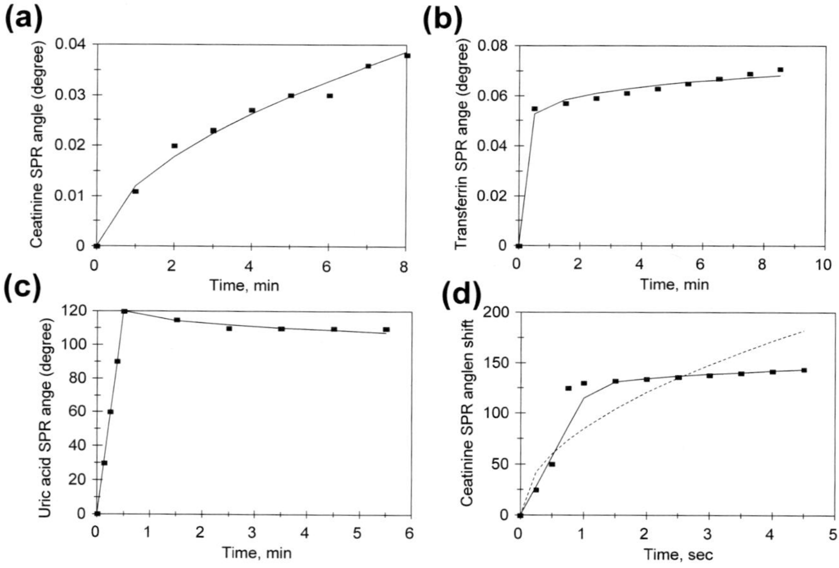

FIGURE 13.6 (a) Binding of creatinine 10 times diluted in a urine sample to an electrochemical surface plasmon resonance biosensor. (b) Binding of transferrin 10 times diluted in urine sample to an electrochemical surface plasmon resonance biosensor (Nakamoto and Kurita (2010)). (c) Binding of 100 μM uric acid (inference substance) to an electrochemical biosensor (d) Binding of 100 μM creatine (interference substance) to an electrochemical biosensor (Nakamoto and Kurita, 2010).

TABLE 13.4a Binding Rate and Dissociation Rate Coefficients for (a) Creatinine in Urine Sample (10 Times Diluted), (b) Transferrin in Urine Sample (10 Times Diluted), (c) Interference Substance 100 μM Uric Acid, and (d) Interference Substance 100 μM Creatinine (Nakamoto et al., 2010)

Analyte in Solution/Receptor on Surface	k	k_1	k_2	k_d
Creatinine in urine sample (10 times diluted)/trienzyme layer of creatinine, creatinine and sucrosine oxidase immobilized on an SPR biosensor surface	0.0128 ± 0.00093	NA	NA	NA
Transferrin in urine sample (10 times diluted)/ antitransferrin antibody immobilized on SPR biosensor surface	0.05616 ± 0.00181	NA	NA	NA
100 μM uric acid/uricase immobilized on SPR biosensor surface	240 ± 0	NA	NA	5.8051 ± 1.1488
100 μM creatinine/enzyme trilayer of creatinine, and creatinine and sucrosine oxidase on an SPR biosensor surface	84.68 ± 32.46	115.07 ± 39.51	127.08 ± 0.63	NA

SPR, surface plasmon resonance.

TABLE 13.4b Fractal Dimensions for the Binding and Dissociation Phase for (a) Creatinine in Urine Sample (10 Times Diluted), (b) transferrin In Urine Sample (10 Times Diluted), (c) Interference Substance 100 μM uric acid, and (d) Interference Substance 100 μM Creatinine (Nakamoto et al., 2010)

Analyte in Solution/Receptor on Surface	D_f	D_{f1}	D_{f2}	D_{fd}
Creatinine in urine sample (10 times diluted)/ trienzyme layer of creatinine, creatinine and sucrosine oxidase immobilized on an SPR biosensor surface	1.8846 ± 0.0796	NA	NA	NA
Transferin in urine sample (10 times diluted)/anti-transferrin antibody immobilized on SPR biosensor surface	2.8198 ± 0.02147	NA	NA	NA
100 μM uric acid/uricase immobilized on SPR biosensor surface	2.0 ± 0	NA	NA	2.0342 ± 0.2840
100 μM uric creatinine/enzyme tri-layer of creatinine, and sucrosine oxidase on a SPR biosensor surface	1.9830 ± 0.2192	0.762 ± 0.046	0.9422 ± 0.4304	NA

the dissociation kinetics. The values of (1) the binding rate coefficient, k, and the fractal dimension, D_f, and (2) the dissociation rate coefficient, k_d, and the fractal dimension for dissociation, D_{fd}, for a single-fractal analysis are given in Table 13.4a and b.

Figure 13.6(d) shows the binding of the interference substance, uric acid to uricase immobilized on an SPR biosensor surface (Nakamoto et al., 2008). A dual-fractal analysis is required to describe the binding kinetics. The values of (1) the binding rate coefficient, k, and the fractal dimension, D_f, and (2) the binding rate coefficient, k_1 and k_2, and the fractal dimensions, D_{f1} and D_{f2}, for a dual-fractal analysis are given in Table 13.4a and b.

Lysozyme (E.C.3.2.1.7) is an enzyme that catalyzes the binding of the β1-4 bond in peptidoglycan residue of bacterial cell walls between N-acetylmuramic acid and N-acetylglucosamine (Sener et al., 2010). It may be considered as the body's own antibiotic. Due to its small size and simple molecular structure it has been used in developing new detection methods. For example, the determination of lytic activity by *Micrococcus lysodelkticus* by ELISA is often used (Liao et al., 2001). However, Vidal et al. (2005) and van Hengel (2007) indicate that high cost, low shelf-life, and cross-reactivity limit the use of this method for the detection of lysozyme.

Sener et al. (2010) indicate that lysozyme can serve as a biomarker for some health problems. For example, the lysozme concentration increased in the cerebrospinal fluid in meningitis patients (Porstmann et al., 1989). Lysozyme is shown to increase in serum and in urine in case of leukemia patients (Pascual et al., 1973) and in kidney problems (Hospacsy et al., 1978). Serra et al. (2002) indicate that lysozyme may also be a new prognostic factor in patients with breast cancer. Finally, antibodies against citrullinated variants of lysozyme have been found in rheumatoid arthritis patients (Ireland et al., 2006).

Wu and Syu (2006) indicate that the quartz crystal microbalance (QCM) due to its high selectivity, low cost, portability, stability, and simplicity is used often for the detection of different analytes. In the QCM biosensor there is an increased mass on the surface due to the binding reaction, which results in a decrease in the oscillating frequency (Biosensor, Q.C.M., 2014). Sener et al. (2010) indicate that QCM biosensors have been used to detect analytes as clinical targets (Chou et al., 2002), and enviromental contaminants (Feng et al., 2007).

Sener et al. (2010) indicate that many of the problems in, for example, conventional imprinting processes for the detection of lysozyme by biosensors have been alleviated by nanotechnology principles. In order to solve these types of problems nanotechnology provided a novel approach, which is imprinting into NPs. These authors indicate that using nanotechnology a more homogeneously distributed imprinted cavity is obtained on the surface or near the nanoparticles (Hayden et al., 2006; Bonin et al., 2007). Sener et al. (2010) have prepared a QCM nanosensor using lysozyme-imprinted NPs.

Sener et al. (2010) studied the influence of repeat runs on the binding of lysozyme to a QCM biosensor. Figure 13.7(a) shows the binding of lysozyme

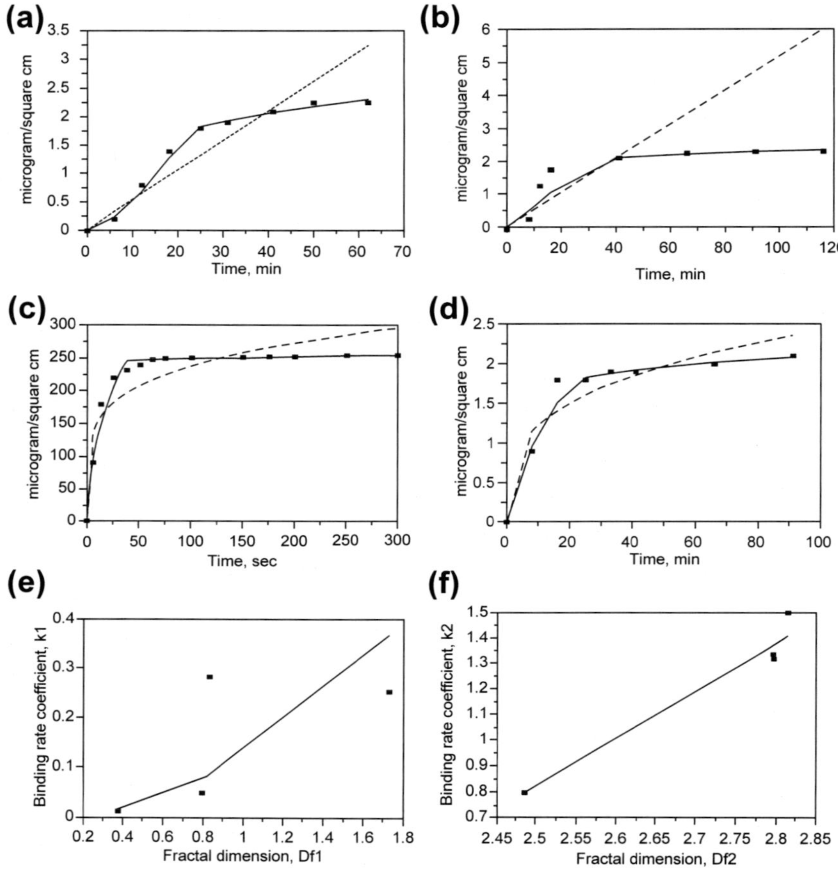

FIGURE 13.7 Binding of lysozyme to imprinted QCM biosensor. Influence of repeat runs (Sener et al., 2010): (a) Run #1 (b) Run # 2 (c) Run #3 (d) Run #4 (e) Increase in the binding rate coefficient, k, with an increase in D_{f1} (f) Increase in the binding rate coefficient, k_2, with an increase in the fractal dimension, D_{f2}.

in run #1. A dual-fractal analysis is required to adequately describe the binding kinetics. The values of (1) the binding rate coefficient, k, and the fractal dimension, D_f, for a single-fractal anaysis, and (2) the binding rate coefficients, k_1 and k_2, and the fractal dimensions, D_{f1} and D_{f2}, for a dual-fractal analysis are given in Table 13.5.

Figure 13.7(b) shows the binding of lysozyme in run #2 (Sener et al., 2010). A dual-fractal analysis is, once again, required to adequately describe the binding kinetics. The values of (1) the binding rate coefficient, k, and the fractal dimension, D_f, for a single-fractal anaysis, and (2) the binding rate coefficients, k_1 and k_2, and the fractal dimensions, D_{f1} and D_{f2}, for a dual-fractal analysis are given in Table 13.5. It is of interest to note that as the

TABLE 13.5 Binding Rate Coefficients and Fractal Dimensions for Lysozyme to a QCM Biosensor (Sener et al., 2010). Influence of Consecutive Runs

Analyte in Solution/ Receptor on Surface	D_f	D_{f1}	D_{f2}	k	k_1	k_2
Run#1	1.0312 ± 0.3702	0 + 0.382	2.4862 ± 0.1231	0.0562 ± 0.025	0.01337 ± 0.00302	0.7992 ± 0.0442
Run#2	1.8216 ± 0.4350	0.7958 ± 1.2162	2.8123 ± 0.0403	0.1847 ± 0.1397	0.04928 ± 0.05316	1.4963 ± 0.0237
Run#3	1.8158 ± 0.2214	0.8150 ± 0.1683	2.7942 ± 0.0089	0.1600 ± 0.0408	0.2820 ± 0.00982	1.332 ± 0.005
Run#4	2.4190 ± 0.1745	1.725 ± 0.548	2.7960 ± 0.0344	0.6309 ± 0.1232	0.2538 ± 0.0625	1.3157 ± 0.0182

QCM, quartz crystal microbalance.

fractal dimension increases by a factor of 3.53 from a value of D_{f1} equal to 0.7958 to D_{f2} equal to 2.8123, the binding rate coefficient increases by a factor of 30.36 from a value of k_1 equal to 0.04928 to k_2 equal to 1.4963.

Figure 13.7(c) shows the binding of lysozyme in run #3 (Sener et al., 2010). A dual-fractal analysis is, once again, required to adequately describe the binding kinetics. The values of (1) the binding rate coefficient, k, and the fractal dimension, D_f, for a single-fractal analysis, and (2) the binding rate coefficients, k_1 and k_2, and the fractal dimensions, D_{f1} and D_{f2}, for a dual-fractal analysis are given in Table 13.5. It is of interest to note that as the fractal dimension increases by a factor of 3.43 from a value of D_{f1} equal to 0.8150 to D_{f2} equal to 2.7942, the binding rate coefficient increases by a factor of 4.723 from a value of k_1 equal to 0.2820 to k_2 equal to 1.332.

Figure 13.7(d) shows the binding of lysozyme in run #4 (Sener et al., 2010). A dual-fractal analysis is, once again, required to adequately describe the binding kinetics. The values of (1) the binding rate coefficient, k, and the fractal dimension, D_f, for a single-fractal analysis, and (2) the binding rate coefficients, k_1 and k_2, and the fractal dimensions, D_{f1} and D_{f2}, for a dual-fractal analysis are given in Table 13.5. It is of interest to note that as the fractal dimension increases by a factor of 1.62 from a value of D_{f1} equal to 1.725 to D_{f2} equal to 2.7960, the binding rate coefficient increases by a factor of 5.184 from a value of k_1 equal to 0.2538 to k_2 equal to 1.3157.

Figure 13.7(e) shows the increase in the binding rate coefficient, k_1, with an increase in the lysozyme concentration in solution in microgram per milliliter. For the data shown in Figure 13.7(a), the binding rate coefficient, k_1, is given by:

$$k_1 = (0.009862 + 0.011750)\ [\text{lysozyme}]^{0.7908\ \pm\ 0.2406} \qquad (13.5)$$

The fit is good. Only three data points are available. The availability of more data points would lead to a more reliable fit. The binding rate coefficient, k_1, exhibits an order of dependence between one half and first (equal to 0.7908) order on the lysozyme concentration in solution.

Figure 13.7(f) shows the increase in the binding rate coefficient, k_2, with an increase in the lysozyme concentration in solution in microgram per milliliter. For the data shown in Figure 13.7(b), the binding rate coefficient, k_2, is given by:

$$k_2 = (0.4778 \pm 0.1860)\ [\text{lysozyme}]^{0.3134\ \pm\ 0.1008} \qquad (13.6)$$

The fit is good. Only three data points are available. The availability of more data points would lead to a more reliable fit. The binding rate coefficient, k_2, exhibits an order of dependence less than one-half (equal to 0.3134) order on the lysozyme concentration in solution.

Figure 13.8(a) shows the binding of 3 μg/mL lysozyme in solution to a QCM biosensor. A dual-fractal analysis is required to adequately describe the binding kinetics. The values of (1) the binding rate coefficient, k, and the fractal dimension, D_f, for a single-fractal analysis, and (2) the binding rate coefficients, k_1 and k_2, and the fractal dimensions, D_{f1} and D_{f2}, for a

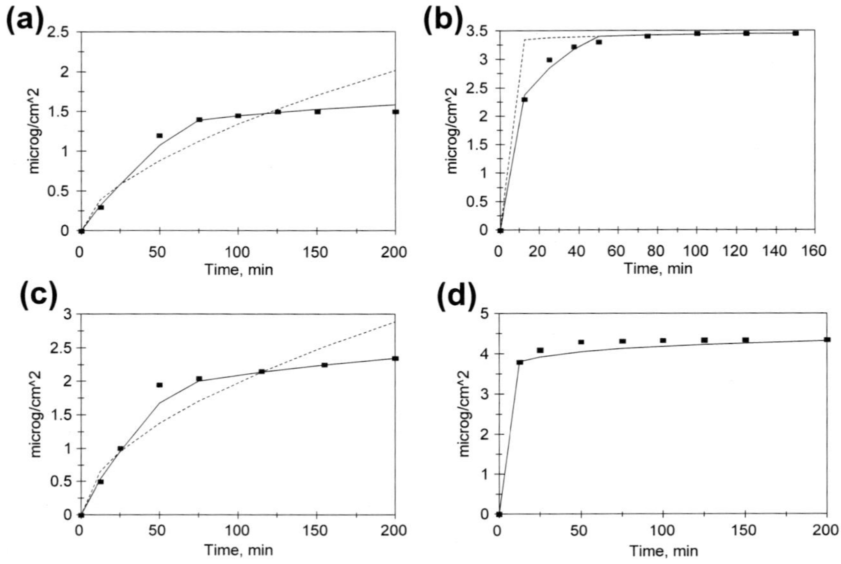

FIGURE 13.8 Binding of lysozyme in microgram per milliliter in solution to a QCM biosensor (Sener et al., 2010): (a) 3 0 (b) 25 (c) 300 (d) 1500.

dual-fractal analysis are given in Table 13.6a. It is of interest to note that as the fractal dimension increases by a factor of 2.235 from D_{f1} equal to 1.2220 to D_{f2} equal to 2.7316, the binding rate coefficient increases by a factor of 23.4 from k_1 equal to 0.03321 to k_2 equal to 0.7796. An increase in the fractal dimension or the degree of heterogeneity on the biosensor surface leads to an increase in the binding rate coefficient.

Figure 13.8(b) shows the binding of 25 μg/mL lysozyme in solution to a QCM biosensor. A dual-fractal analysis is required to adequately describe the binding kinetics. The values of (1) the binding rate coefficient, k, and the fractal dimension, D_f, for a single-fractal analysis, and (2) the binding rate coefficients, k_1 and k_2, and the fractal dimensions, D_{f1} and D_{f2}, for a dual-fractal analysis are given in Table 13.6a,b. It is of interest to note that as the fractal dimension increases by a factor of 1.20 from D_{f1} equal to 2.4652 to D_{f2} equal to 2.9743, the binding rate coefficient increases by a factor of 2.68 from

TABLE 13.6a **Binding Rate Coefficients for Lysozyme to a QCM Biosensor (Sener et al., 2010). Influence of Different Concentrations (μg/ml)**

Analyte in solution, μg/ml	k	k_1	k_2
3	0.0857 ± 0.0208	0.03321 ± 0.00388	0.7796 ± 0.01115
25	0.1694 ± 0.0457	1.2050 ± 0.0673	3.2343 ± 0.0168
300	1.356 ± 0.2346	0.06632 ± 0.01112	1.0031 ± 0.0021
1500	2.9087 ± 0.01740	NA	NA

QCM, quartz crystal microbalance.

TABLE 13.6b **Fractal Dimensions for the Binding of Lysozyme to a QCM Biosensor (Sener et al., 2010). Influence of Different Concentrations (μg/ml)**

Analyte in solution, μg/ml	D_f	D_{f1}	D_{f2}
3	1.8072 ± 0.1734	1.2220 ± 0.1616	2.7316 ± 0.05806
25	1.3452 ± 0.1278	2.4652 ± 0.09848	2.9743 ± 0.0136
300	1.9298 ± 0.1942	1.3478 ± 0.2268	2.6790 ± 0.0105
1500	0 ± 0.0770	NA	NA

QCM, quartz crystal microbalance.

k_1 equal to 1.2050 to k_2 equal to 3.2343. Once again, an increase in the fractal dimension or the degree of heterogeneity on the biosensor surface leads to an increase in the binding rate coefficient.

Figure 13.8(c) shows the binding of 300 μg/mL lysozyme in solution to a QCM biosensor. A dual-fractal analysis is required to adequately describe the binding kinetics. The values of (1) the binding rate coefficient, k, and the fractal dimension, D_f, for a single-fractal analysis, and (2) the binding rate coefficients, k_1 and k_2, and the fractal dimensions, D_{f1} and D_{f2}, for a dual-fractal analysis are given in Table 13.6a,b. It is of interest to note that as the fractal dimension increases by a factor of 1.98 from D_{f1} equal to 1.3478 to D_{f2} equal to 2.6790, the binding rate coefficient increases by a factor of 15.12 from k_1 equal to 0.06632 to k_2 equal to 1.0031. Once again, an increase in the fractal dimension or the degree of heterogeneity on the biosensor surface leads to an increase in the binding rate coefficient.

Figure 13.8(d) shows the binding of 1500 μg/mL lysozyme in solution toa QCM biosensor. A single-fractal analysis is required to adequately describe the binding kinetics. The values of (1) the binding rate coefficient, k, and the fractal dimension, D_f, for a single-fractal analysis, and (2) the binding rate coefficients, k_1 and k_2, and the fractal dimensions, D_{f1} and D_{f2}, for a dual-fractal analysis are given in Table 13.6a,b.

Figure 13.9(a) shows the increase in the binding rate coefficient, k_1, with an increase in the lysozyme concentration. For the data shown in Figure 13.9(a), the binding rate coefficient, k_1, is given by:

$$k_1 = \left(0.013358 \pm 0.00205\right) [\text{lysozyme}]^{0.5.00 \pm 0.2667} \tag{13.7}$$

The fit is good. Only three data points are available. The availability of more data points would lead to a more reliable fit. The binding rate coefficient, k_1, is one-half it exhibits a order dependence on D_{f1}.

Figure 13.9(b) shows the increase in the binding rate coefficient, k_2, with an increase in the lysozyme concentration. For the data shown in Figure 13.9(b), the binding rate coefficient, k_2, is given by:

$$k_2 = (2.5\text{ E} - 06 \pm 1.1\text{ E} - 06)\ [\text{lysozyme}]^{12.858 \pm 4.585} \tag{13.8}$$

The fit is good. Only three data points are available. The availability of more data points would lead to a more reliable fit. The binding rate coefficient, k_2, is extremely sensitive to the lysozyme concentration in this case since it exhibits close to a 13th order dependence on the lysozyme concentration. This is unusual.

Figure 13.9(c) shows the increase in binding coefficient, k_1, for a dual-fractal analysis, with an increase in the fractal dimension, D_{f1}. For the data shown in Figure 13.7(e), the binding rate coefficient, k_1, is given by:

$$k_1 = \left(0.1254 + 0.2071\right) D_{f1}^{1.977 \pm 0.9164} \tag{13.9a}$$

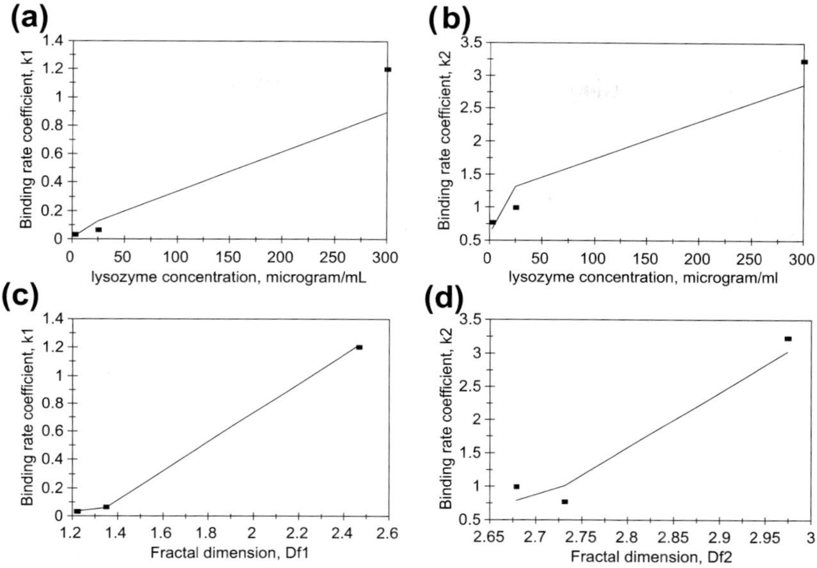

FIGURE 13.9 (a) Increase in the binding rate coefficient, k_1 with an increase in the lysozyme concentration (b) Increase in the binding rate coefficient, k_2 with an increase in the lysozome concentration. (c) Increase in the binding rate coefficient, k_1, with an increase in the fractal dimension, D_{f1}. (d) Increase in the binding rate coefficient, k_2, with an increase in the fractal dimension, D_{f2}.

The fit is quite good. Only four data points are available. The availability of more data points would lead to a more reliable fit. The binding rate coefficient, k_1, exhibits close to a second (equal to 1.977) order of dependence on the fractal dimension, D_{f1}, or the degree of heterogeneity that exists on the biosensor surface. No table is provided with the data.

Figure 13.9(d) shows the increase in binding coefficient, k_2, for a dual-fractal analysis, with an increase in the fractal dimension, D_{f2}. For the data shown in Figure 13.7(f), the binding rate coefficient, k_2, is given by:

$$k_2 = \left(0.0119 + 0.00068\right)D_{f2}^{4.611 \pm 0.5343} \tag{13.9b}$$

The fit is quite good. Only four data points are available. The availability of more data points would lead to a more reliable fit. The binding rate coefficient, k_2, exhibits close to a four and one-half (equal to 4.611) order of dependence on the fractal dimension, D_{f1}, or the degree of heterogeneity that exists on the biosensor surface. The binding rate coefficient, k_2, is very sensitive to the degree of heterogeneity or fractal dimension that exists on the biosensor surface.

Yu et al. (2010) have recently developed a $[C_3(OH)_2mim][BF_4]$-Au/Pt biosensor for glutamate sensing. These authors indicate that glutamate plays a significant part in brain functioning since it plays an important excitatory neurotransmitter role in the CNS. These include cognitive processes and in the development of and plasticity of the CNS (Oldenziel et al., 2006a,b). Also, it is responsible for mediating energy transfer and insulin−glucose homostasis (Tong et al., 2007; Wilson and Johnson, 2008). Kulagina et al. (1999) indicate that glutamate is heavily involved in neurological disorders such as schizophrenia, Parkinson's disease, and epilepsy. Precise monitoring of glutamate levels in essential to provide a fundamental understanding of these diseases. Pasco et al. (1999) indicate that compared with spectrometric and chromatographic methods of analysis, electrochemical enzymatic sensing is a suitable approach for the detection of glutamate (Mao and Yamamoto, 2000; Niwa et al., 1996). Hansen et al. (1999) indicate that due to a short analytical time, high sensitivity, specificity, and the ability to provide real-time measurements, the microdialysis method is a suitable approach to provide both temporal and chemical information.

Metal NPs are suitable for use in electrochemical biosensors since they possess high surface/volume ratios and increased photochemical activity (Daniel and Anstruc, 2004). RTILs are suitable solvents for nanostructured materials, and are being increasingly used to incorporate functional groups, due to increased catalyst stability and improved catalyst retention (Fei et al., 2007). The incorporation of different functional groups in the RTILs has made these RTILs more flexible to tune their physiochemical properties to a specific application (Zheng et al., 2007). Yu et al. (2010) have developed a third-generation glutamate biosensor using $[C_3(OH)_2$ mim][BF_4]-Au/Pt film with glucose oxidase as the recognition element.

Figure 13.10(a) shows the amperometric response to a GlutaOx-$(C_3(OH)_{2mim}(BF)_4$-Au/Pt-Nafion biosensor toward a 2 μM glutamate solution (Yu et al., 2010). A single-fractal analysis is required to adequately describe the binding and the dissociation kinetics. The values of (1) the binding rate coefficient, k, and the fractal dimension, D_f, for a single-fractal analysis, and (2) the dissociation rate coefficient, k_d, and the fractal dimension for dissociation, D_{fd}, for a single-fractal analysis are given in Table 13.7. The affinity, $K = k/k_d$, value is 0.729.

Figure 13.10(b) shows the amperometric response to a GlutaOx-$(C_3(OH)_{2mim}(BF)_4$-Au/Pt-Nafion biosensor toward a 5 μM glutamate solution (Yu et al., 2010). A single-fractal analysis is required to adequately describe the binding and the dissociation kinetics. The values of (1) the binding rate coefficient, k, and the fractal dimension, D_f, for a single-fractal analysis, and (2) the dissociation rate coefficient, k_d, and the fractal dimension for dissociation, D_{fd}, for a single-fractal analysis are given in Table 13.7. The affinity, $K = k/k_d$, value is 0.699.

Figure 13.10(c) shows the amperometric response to GlutaOx-$(C_3(OH)_{2mim}(BF)_4$-Au/Pt-Nafion biosensor toward a 10 μM glutamate solution (Yu et al., 2010). A single-fractal analysis is required to adequately describe the binding and the dissociation kinetics. The values of (1) the binding rate coefficient,

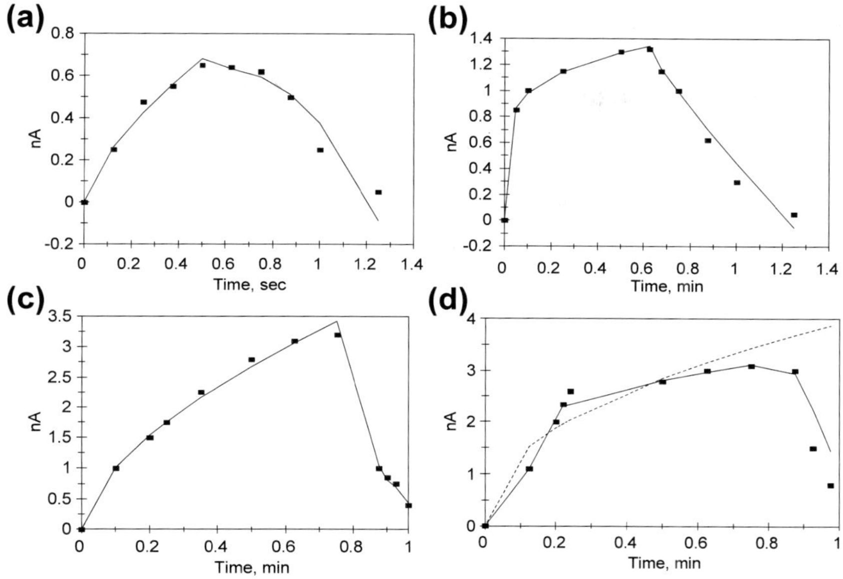

FIGURE 13.10 Amperometric response of a GlutaOX-($C_3(OH)_2$mim)[BF_4}-Au/Pt) Nafion biosensor toward different glutamate solutions (in μM) (Yu et al., 2010): (a) 2 (b) 5 (c) 10 (d) 20.

k, and the fractal dimension, D_f, for a single-fractal analysis, and (2) the dissociation rate coefficient, k_d, and the fractal dimension for dissociation, D_{fd}, for a single-fractal analysis are given in Table 13.7. The affinity, K = 1.005.

Figure 13.10(d) shows the amperometric response to a GlutaOx-($C_3(OH)_{2mim}(BF)_4$-Au/Pt-Nafion biosensor toward a 20 μM glutamate solution (Yu et al., 2010). A dual-fractal analysis is required to adequately describe the

TABLE 13.7 Binding of Different Concentrations of Glutamate in Solution to a GlutaOx-($C_3(OH)_{2mim}(BF)_4$ Au/Pt-Nafion Biosensor (Yu et al., 2010)

Analyte in Solution/Receptor Surface	k	k_d	D_f	D_{fd}
2 μM glutamate/Nafion biosensor	1.091 ± 0.109	1.4959 ± 0.895	1.628 ± 0.1838	0
5 μM glutamate/Nafion biosensor	1.434 ± 0.030	2.0801 ± 0.2602	2.6542 ± 0.0190	2.3220 ± 0.120
10 μM glutamate/Nafion biosensor	4.0672 ± 0.157	4.0458 ± 0.0979	1.7894 ± 0.04930	2.4446 ± 0.0766

SPR, surface plasmon resonance.

binding kinetics and the dissociation kinetics by a single-fractal analysis. The values are not shown for the binding and dissociation rate coefficients as well as the fractual dimension values.

Figure 13.11(a) shows the decrease in the binding rate coefficient, k, with an increase in the fractal dimension, D_f. For the data shown in Figure 13.11(a), the binding rate coefficient, k, is given by:

$$k = \left(6.511 + 5.084\right) D_f^{-0.3683-2.6588} \quad (13.10a)$$

There is scatter in the data. Only three data points are available. The availability of more data points would lead to a more reliable fit. For the data shown in Figure 13.11(a), the binding rate coefficient, k, exhibits an order of dependence less than one-half (equal to −0.368) order of dependence on the fractal dimension, D_f.

Figure 13.11(b) shows the increase in the dissociation rate coefficient, k_d, with an increase in the fractal dimension in the dissociation phase, D_{fd}. For the data shown in Figure 13.11(b), the dissociation rate coefficient, k_d, is given by:

$$k_d = \left(1.879 \pm 0.256\right) D_{fd}^{0.716 \pm 0.119} \quad (13.10b)$$

The fit is good. Only four data points are available. The availability of more data points would lead to a more reliable fit. The dissociation rate coefficient, k_d, exhibits less than first (equal to 0.716) order of dependence on the fractal dimension in the dissociation phase, D_{fd}.

Graphene is a NP that has exhibited promising applications in electrochemical biosensors due to its large specific area, good conductivity, and biocompatabilty (Li et al., 2008, 2009a,b; Tan et al., 2010). Electrochemical immunosensors are highly sensitive, easy to fabricate, and have a low cost compared to other immunological methods (Stoeva et al., 2006; Darain et al., 2003), Furthermore,

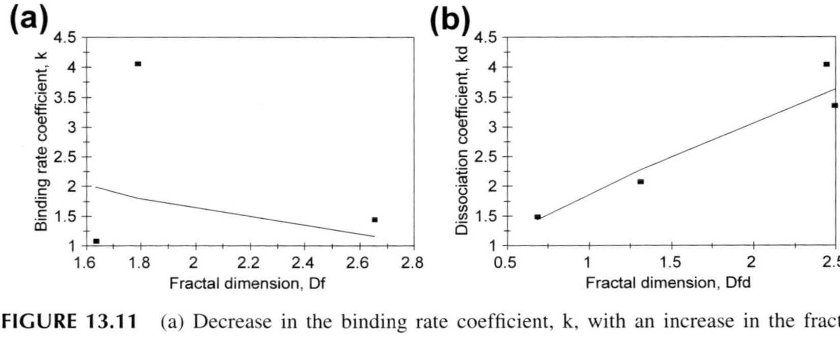

FIGURE 13.11 (a) Decrease in the binding rate coefficient, k, with an increase in the fractal dimension, D_f. (b) Increase in the dissociation rate coefficient, k_d, with an increase in fractal dimension, D_{fd}.

Wei et al (2011) indicate that in the design of electrochemical sensors, the immobilization of antibodies is a critical step (Sanchez et al., 2008).

Recently, HRP (horse radish peroxidase) has been immobilized on different surfaces for electrochemical biosensing. These include chitosan (Shi and Ma, 2010), titanium dioxide (Zhang et al., 2010), silica (Qu et al., 2008), and mesoporous silica nanoparticles (MSN) (Liu et al., 2007). Wei et al. (2011) indicate that human chorionic gonadotrophin (HCG) is a glycoprotein produced in pregnancy. HCG is also produced by some kinds of tumor and plays an important role as a tumor biomarker in gestational trophobastic disease (Liu et al., 2005).

Yang et al. (2010) have previously used aminofunctional MSN as a label, for immunosensors to detect HCG. They have employed an ultrasensitive immunosensor to detect HCG employing a new type of label based on HRP and Pt nanoparticle-modified Health-Care Sensor Networks (HSN).

Pillarik et al. (2010) have developed an SPR biosensor for detecting the cancer biomarker, HCG. They indicate that the optical biosensors can rapidly and sensitively detect chemical and biological substances in the fields of food safety, environmental monitoring, drug development, and medical diagnostics (Homola, 2005). They indicate that protein biomarkers for diseases can readily be detected in body fluids such as blood plasma. The cancer biomarker HCG can be found in healthy individuals at levels of 1 ng/ml in blood plasma, whereas in patients suffering from trophoplastic cancer, epithelial or ovarian cancer, or breast cancer these levels can rise up to microgram per milliliter.

Pillarik et al. (2010) indicate that future medical diagnostics are expected to detect a panel of biomarkers (Choolani et al., 2001; Feng et al., 2009). The detection of biomarkers by the SPR biosensor poses challenges (Pillarik et al., 2010). For example, these authors indicate that in the analysis of complex substances there is nonspecific adsorption of nontarget molecules on the sensing surface. This effect can be minimized by employing nonfouling surfaces in low-throughput SPR biosensors. In the present report, Pillarik et al. (2010) have developed a high-throughput SPR biosensor for the screening of protein biomarkers in blood samples. These authors have detected the protein biomarker, HCG.

Figure 13.12(a) shows the binding of 1000 μg/mL ALCAM in solution to anti-ALCAM immobilized on an SPR biosensor chip surface. A single-fractal analysis is adequate to describe the binding kinetics. The values of the binding rate coefficient, k, and the fractal dimension, D_f, for a single-fractal analysis are given in Table 13.8.

Figure 13.12(b) shows the binding of 500 μg/mL ALCAM in solution to anti-ALCAM immobilized on an SPR biosensor chip surface. A single-fractal analysis is adequate to describe the binding kinetics. The values of the binding rate coefficient, k, and the fractal dimension, D_f, for a single-fractal analysis are given in Table 13.8.

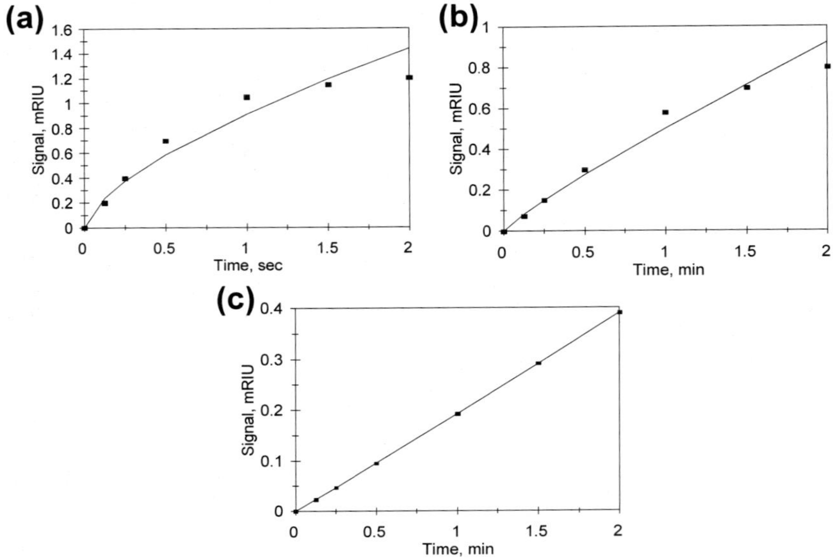

FIGURE 13.12 Binding of different concentrations of ALCAM (in μg/mL) to anti-ALCAM antibody immobilized on the SPR surface: (a) 1000 (b) 500 (c) 200 (Pillarik et al. (2010)).

Figure 13.12(c) shows the binding of 200 μg/mL ALCAM in solution to anti-ALCAM immobilized on an SPR biosensor chip surface. A single-fractal analysis is adequate to describe the binding kinetics. The values of the binding rate coefficient, k, and the fractal dimension, D_f, for a single-fractal analysis are given in Table 13.8.

TABLE 13.8 Binding Rate Coefficients and Fractal Dimensions for Different Concentrations of ALCAM to Anti-ALCAM Antibody Immobilized on an SPR Biosensor Chip Surface (Pillarik et al., 2010)

Analyte in Solution/Receptor on Surface	k	D_f
1000 μg/mL/anti-ALCAM antibody immobilized on SPR chip surface	0.9157 ± 0.1730	1.7030 ± 0.1435
500 μg/mL/anti-ALCAM antibody immobilized on SPR chip surface	0.5014 ± 0.0630	1.2532 ± 0.0932
200 μg/mL/anti-ALCAM antibody immobilized on SPR chip surface	0.1928 ± 0.010	0.9700 ± 0.04570

SPR, surface plasmon resonance.

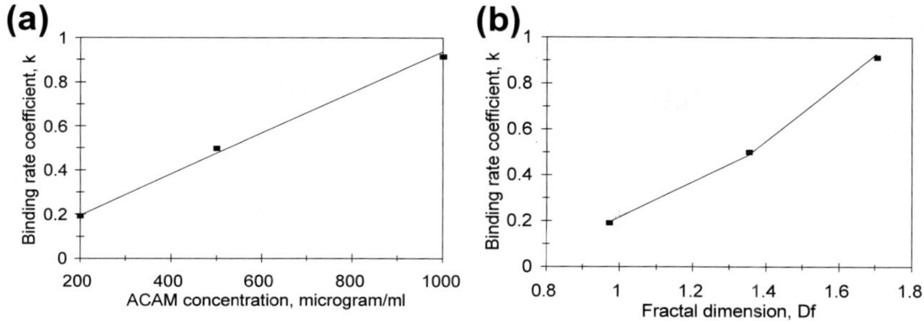

FIGURE 13.13 (a) Increase in the binding rate coefficient, k, with an increase in the ACCAM concentration in solution (in μg/mL). (b) Increase in the binding rate coefficient, k, with an increase in the fractal dimension, D_f.

Figure 13.13(a) shows the increase in the binding rate coefficient, k, for a single-fractal analysis with an increase in the ALCAM concentration in solution. For the data shown in Figure 13.13(a), the binding rate coefficient, k, is given by:

$$k = (0.00112 \pm 0.000076)\,[\text{ALCAM}]^{0.9746 \,\pm\, 0.0497} \tag{13.11}$$

The fit is quite good. Only three data points are available. The availability of more data points would lead to a more reliable fit. The binding rate coefficient, k, exhibits close to a first (equal to 0.9746) order of dependence on the ALCAM concentration in microgram per milliliter in solution.

Figure 13.13(b) shows the increase in the binding rate coefficient, k, for a single-fractal analysis with an increase in the fractal dimension, D_f. For the data shown in Figure 13.13(b), the binding rate coefficient, k, is given by:

$$k = \left(0.2099 \pm 0.0073\right) D_f^{2.790 \,\pm\, 0.0747} \tag{13.12}$$

The fit is good. Only three data points are available. The availability of more data points would lead to a more reliable fit. The binding rate coefficient, k, exhibits an order of dependence between two and one half and three (equal to 2.790) order of dependence on the fractal dimension or the degree of heterogeneity that exists on the biosensor surface.

Su et al. (2007) have recently analyzed the binding of angiotensin-converting enzyme (ACE) with lisinopril using a 35 mHz QCM and SPR. These authors indicate that some enzymes are important drug targets for developing drugs against various diseases such as cancer, AIDS, diabetes, and hypertension (Cushman and Ondetti, 1994; Johnson et al., 2002; Noble et al., 2004). ACE plays an important role in homestatic mechanisms of mammals for regulating blood pressure (Erdos, 1976; Jimsheena and Gowda, 2008; Peach, 1977). Brown and Vaughan (1998); and Mancini et al. (1996) indicate that ACE inhibitors are effective in the treatment of hypertension and congestive heart failure. Lisinopril is a carboxylic third-generation ACEI for hypertension treatment (Busssien et al., 1985).

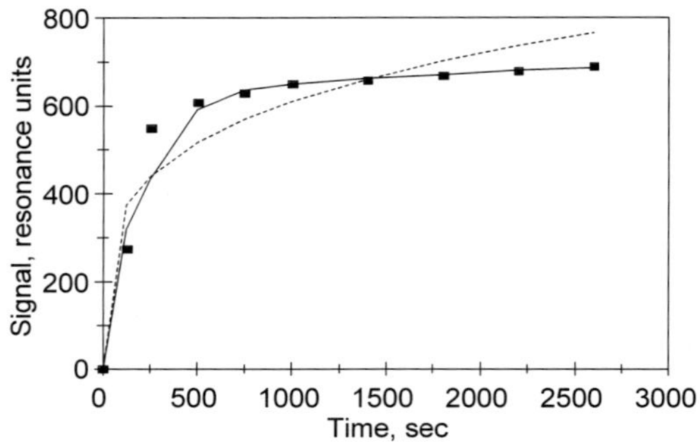

FIGURE 13.14 Binding of ACE in solution to ACEI/C12SH/Au immobilized on an SPR surface (Su et al., 2007).

Su et al. (2007) suggest that QCM is a powerful quantitative tool to investigate various modified substances (Xie et al., 1999). Su et al. (2007) have developed a highly sensitive QCM at 35 mHx fundamental frequency. This was used to make the binding process of lisinopril (ACEI) to ACE adsorbed at a modified electrode.

Figure 13.14 shows the binding of ACE in solution to ACEI to C12SH/Au immobilized on an SPR biosensor surface. A dual-fractal analysis is required to adequately describe the binding kinetics. The values of (1) the binding rate coefficient, k, and the fractal dimension, D_f, for a single-fractal analysis, and (2) the binding rate coefficients, k_1 and k_2, and the fractal dimensions, D_{f1} and D_{f2}, for a dual-fractal analysis are given in Table 13.9(a) and (b).

Njagi et al. (2010) have recently developed an amperometric biosensor for the in vivo detection of dopamine using an enzyme-based carbon fiber microbiosensor. They used tyrosinase immobilized in a biocompatible matrix that consisted of a biopolymer, chitosan, and ceria-based metal oxides deposited on the surface of a carbon fiber microelectrode. These authors state that dopamine is a catecholamine neurotransmitter that is present in the CNS. It modulates several aspects of brain circuitry. It plays a critical role in learning and memory (Robinson et al., 2003; Hefco et al., 2003). Njagi et al. (2010)

TABLE 13.9a Binding Rate Coefficients for 2.5 μmol/L ACE in PBS Solution to C12SH/Au Immobilized on an SPR Biosensor Surface (Su et al., 2007)

Analyte in Solution/Receptor on Surface	k	k_1	k_2
2.5 μmol/LACE in PBS/C12SH/Au immobilized on surface	119.52 ± 9.054	37.741 ± 9.054	422.21 ± 1.52

ACE, angiotensin-converting enzyme; SPR, surface plasmon resonance.

TABLE 13.9b Fractal Dimensions for 2.5 μmol/L ACE in PBS Solution to C12SH/Au Immobilized on an SPR Biosensor Surface (Su et al., 2007)

Analyte in Solution/Receptor on Surface	D_f	D_{f1}	D_{f2}
2.5 μmol/LACE in PBS/C12SH/Au immobilized on surface	2.5258 ± 0.1171	2.1148 ± 0.3148	2.8760 ± 0.00946

ACE, angiotensin-converting enzyme; SPR, surface plasmon resonance.

further indicate low levels of dopamine have been implicated in Parkinson's disease and schizophrenia (Swamy and Vinton, 2007; Wightman et al., 1988).

Dopamine is measured by, for example, electrochemical techniques (Mecker et al., 2008). Dopamine is electrochemically active and can be measured directly using an electrode held at an appropriate potential. Njagi et al. (2010) indicate that dopamine has been measured using amperometry (Yavich and Tilhonen, 2000), differential pulse voltametry (Valentini et al., 2007), and scan cyclic voltammetry (Cheer et al., 2007).

Njagi et al. (2010) do point out that direct electrochemical detection of dopamine has limitations, however, due to electrode passivation. Njagi et al. (2010) suggest that to enhance electrode sensitivity and restrict the detection of interferring agents one needs to lower operating potential difference on using an enzyme-cycling approach.

Njagi et al. (2010) indicate that traditionally tyrosinase catalyzes the conversion of phenol to their respective quinine derivatives (Andreescu and Sadik, 2004). Tyrosinase biosensors for the detection of dopamine exhibit reduced sensitivity (Zhou et al., 2007; Njagi et al., 2008). Njagi et al. (2010) emphasize the need for strategies to enhance the sensitivity of tyrosinase-based biosensors for the detection of dopamine. They have used a tyrosinase-based biosensor for the detection of dopamine using a ceria-based biocomposite. They recently report (Njagi et al., 2010) on the development of an implantable tyrosinase-based biosensor for dopamine that permits real-time measurements on dopamine in the brain of anesthesized rats.

Figure 13.15 shows the binding of (amperometric response) dopamine in solution to a tyrosinase biosensor in the presence of ceria-based metal oxide immobilized on a carbon fiber electrode. A single-fractal analysis is adequate to describe the binding kinetics. The values of the binding rate coefficient, k, and the fractal dimension, D_f, are given in Table 13.10.

The binding of (amperometric response) dopamine in solution to a tyrosinase biosensor in the absence of ceria-based metal oxide immobilized on a carbon fiber electrode was also analyzed. Figure not shown. A single-fractal

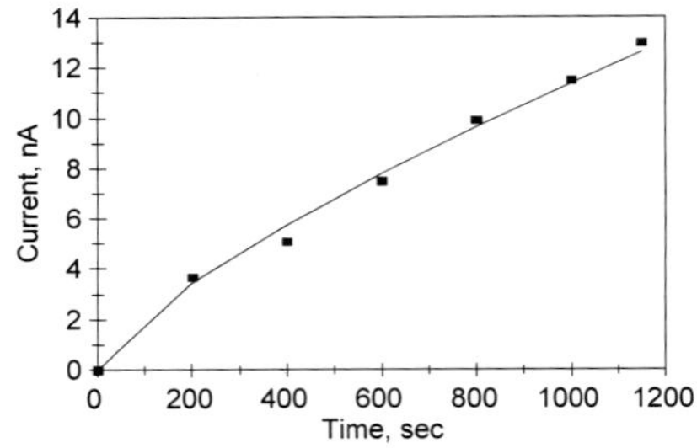

FIGURE 13.15 Binding of dopamine in solution to the tyrosinase biosensor (Njagi et al., 2010) ceria-based metal oxides used.

analysis is adequate to describe the binding kinetics. The values of the binding rate coefficient, k, and the fractal dimension, D_f, are given in Table 13.10. It is of interest to note that when one compares the binding of dopamine to the tyrosinase biosensor when ceria-based metal oxides are used and when they are not used, the fractal dimension and the binding rate coefficient increase when ceria-based metal oxides are used and when they are not used.

Ching et al. (2011) have recently developed a biosensor for the precise determination of homocysteine. Homocysteiene is a risk factor in coronary heart disease. They indicate that cardiovascular disease (CVD) is a challenging global public health problem. WHO estimates that 16.7 milion people die every year of CVD and this is over 20% of the total deaths annually. Furthermore, Ross (1993) indicates that CVD is responsible for 50% of mortality in developed countries. In Taiwan, heart disease held the second position in all deaths in 2007 (Department of Health, Executive Yuan, 2004).

Ching et al. (2011) indicate that plasma homocysteine bears a relationship for the risk of CVD (Homocysteine Studies Collaboration, 2002; Elkelboom

TABLE 13.10 Binding Rate Coefficients and Fractal Dimensions for Dopamine in Solution to a Tyrosinase Biosensor (Njagi et al., 2010)

Analyte in Solution/Receptor on Surface	k	D_f
Dopamine/tyrosinase, biocompatible matrix, biopolymer, chitosan, and ceria-based metal oxide immobilized on a carbon fiber microelectrode	0.06658 ± 0.00525	1.5124 ± 0.1038
Dopamine/tyrosinase, biocompatible matrix, biopolymer, chitosan, immobilized on a carbon fiber microelectrode	0.029 ± 0.005	1.1984 ± 0.1794

et al., 1999; Splaver et al., 2004). Ubbink et al. (1991) and von Eckardstein et al. (1994) indicate that the degree of atherosclerosis bears a positive correlation with plasma homocysteine levels. Furthermore, Cramer (1998) indicates that plasma homocysteine levels correlate better than cholesterol levels with increased risk of arteriosclerosis. Several studies have reported the detection of homocysteiene by biosensors (Gong et al., 2004; Zhang et al., 2007; Alcam et al., 2007; Acharya et al., 2008; Agula et al., 2007). However, none of these have used biosensors in an array format. Biosensors used in an array format have a higher signal to noise ratio, and Ching et al. (2011) have developed a biosensor in an array format for the accurate determination of homocysteine.

Figure 13.16 shows the binding of 100 μM homocysteine in solution to D-amino acid oxidase immobilized on screen-printed carbon electrode with silver—silver chloride (Au/AuCl) as a reference electrode (Ching et al., 2011). A single-fractal analysis is adequate to describe the binding kinetics. The values of the binding rate coefficient, k, and the fractal dimension, D_f, are given in Table 13.11.

Abdelwahab et al. (2010) have developed a selective nitric oxide (NO) nanocomposite biosensor based on direct electron transfer of microperoxidase (MP). The selective NO biosensor was developed by immobilizing MP onto an multi-walled carbon nanotubes (MWCNT)-poly-5.2′.5′.2″/terthiophene-3′-carboxylic CID (PTTCA) nanocomposite catalase (CAS) and superoxide dismutase (SOD) was co-immobilized on the probe to protect it from interferents, H_2O_2 and O_2^- during the detection of NO. NO is a free radical molecule generated in biological systems by nitric oxide synthases. Parker (1996) indicates that it is an important messenger molecule that is involved in many physiological and pathological processes and is both detrimental and beneficial.

A deficiency in NO plays a significant role in Parkinson's disease and Alzheimer's disease. An increase in NO plays a role in cancer (Bredt, 1999;

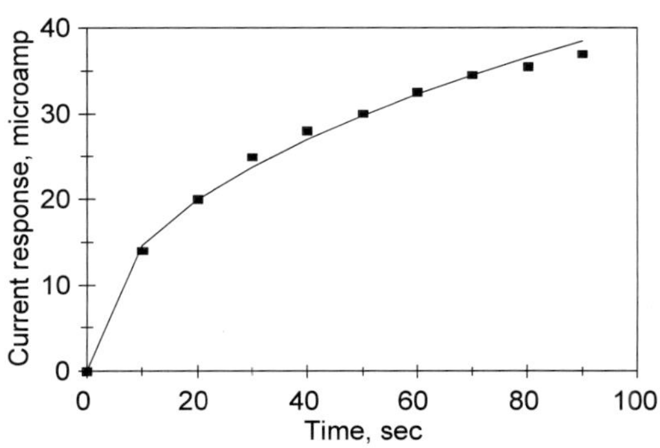

FIGURE 13.16 Binding (amperometric response) of 100 μM homocysteine to an HBA (homocysteine biosensor array) (Ching et al., 2011).

TABLE 13.11 Binding (Amperometric Response) Rate Coefficient, k and Fractal Dimension, D_f for 100 μM Homocysteine in Solution to a D-Amino Acid Oxidase HBA (Homocysteine Biosensor Array) (Ching et al., 2011)

Analyte in Solution/Receptor on Surface	k	D_f
100 μM homocysteine/D-amino acid oxidase immobilized on HBA	5.335 ± 0.193	2.1220 ± 0.03482

Christopherson and Bredt, 1997). The quantification of NO is difficult due to its spontaneous chemical reactivity (Wink and Mitcheel, 1998). Thus, various methods of NO detection have been proposed (Taha, 2003). Electrochemical techniques exhibit the most promise (Isik et al., 2004; Villamena and Zweier, 2004; de Groot et al., 2005; Koh et al., 2008). Also, NO can be detected via electrooxidation (Friedmann et al., 1996) or by electroreduction (Maskus et al., 1996). Interferences make the electrochemical detection of NO challenging (Wu et al., 2002; Wang et al., 2005a,b; Du et al., 2008).

Conducting polymers are useful material for their use in biosensor due to their conducting nature and biocompatibility (Rahman et al., 2008). Carbon nanotubes also exhibit promise for use in biosensors (Iijima, 1991; Ajayan, 1999). Abdelwahab et al. (2010) have developed a highly sensitive NO detection biosensor by covalently immobilizing three enzymes, MP, CAS, and SOD on to a PTTCA nanocomposite layer.

Figure 13.17 shows the binding of NO in solution to MP immobilized on to a PTTCA nanocomposite biosensor (Abdelwahab et al., 2010). A single-fractal analysis is adequate to describe the binding kinetics. The values of the binding rate coefficient, k, and the fractal dimension, D_f, are given in Table 13.12.

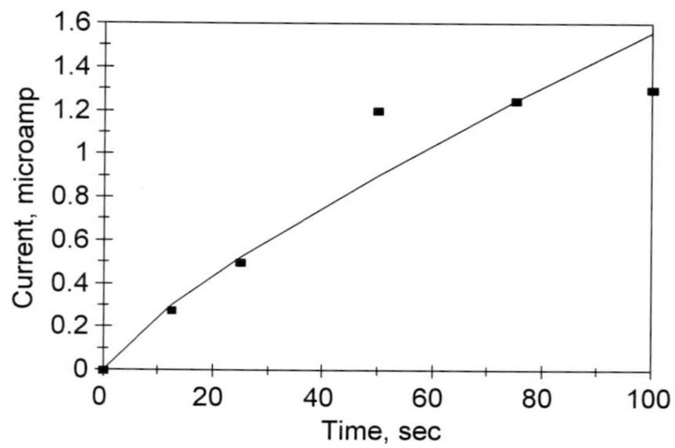

FIGURE 13.17 Binding of NO to a selective nanocomposite biosensor based on direct electron transfer of microperoxidase (Abdelwahab et al., 2010).

TABLE 13.12 Binding Rate Coefficient, k and the Fractal Dimension, D_f for Nitric Oxide (NO) in Solution to Microperoxidase (MP) Immobilized onto an MWCNT-Poly-5.2′.5.2″-Tetriphiophehe-3′-Carbxylic Acid (PTTCA) Nanocomposite Biosensors (Abdelwahab et al., 2010)

Analyte in Solution/Receptor on Surface	k	D_f
NO in solution/MP on nanocomposite biosensor	0.04055 ± 0.00907	1.4148 ± 0.2386

Jing et al. (2010) indicate that molecular imprinting polymers (MIPs) have received a lot of attention due to their specific recognition properties (Tan et al., 2008). The imprinting of macromolecules like proteins still possess some problems, which include unfavorable kinetics of adsorption and desorption, heterogeneity of receptor sites, and denaturation of the macromolecules (Bereli et al., 2008; Ouyang et al., 2008; Jin and Tang, 2009; Flavin and Resmine, 2009). Surface polymerization has been proposed to prepare the MIP films on a solid support to alleviate the problems (Jin and Tang, 2009).

Magnetic nanoparticles have shown superior characterisitics compared to the MIPs. They are of small size, have a high surface-to-volume ratio, exhibit high-binding characteristics to the biomolecules. This has attracted a lot of attention in the areas of bioseparation and detection of biological molecules, chemical diagnostics, and targeted drug therapy (Gupta and Gupta, 2005; Gu et al., 2006; Gao et al., 2009). Combining the magnetic separation with molecular imprinting leads to advantages that promote the selective recognition of the analytes along with effective separation (Li et al., 2009a,b; Wang et al., 2009a,b).

Jing et al. (2010) indicate that lysozyme (*N*-acetylmuramide glycohydrolase) is considered as a self-defense enzyme, which is produced in serum, mucus, and many organs of vertebrates. It is a common index in the diagnosis of different diseases including renal disease and leukemia. Guo et al. (2007) and Zheng et al. (2007) indicate that lysozyme is also used as a drug for the treatment of ulcers and infections.

Jing et al. (2010) have synthesized magnetic molecularly imprinted NPs for the recognition of lysozyme. Fe_3O_4 NPs were initially synthesized. Then they were coated with thin MIP films. A sensitive chemiluminescence method was developed to determine lysozyme in human serum samples.

Figure 13.18(a) shows the binding of MIP to lysozyme (0.2 mol/L, pH 6.2) containing 0.02 mol/L NaCl (Jing et al., 2010). A single-fractal analysis is adequate to describe the binding kinetics. The values of the binding rate coefficient, k, and the fractal dimension, D_f, for a single-fractal analysis are given in Table 13.13.

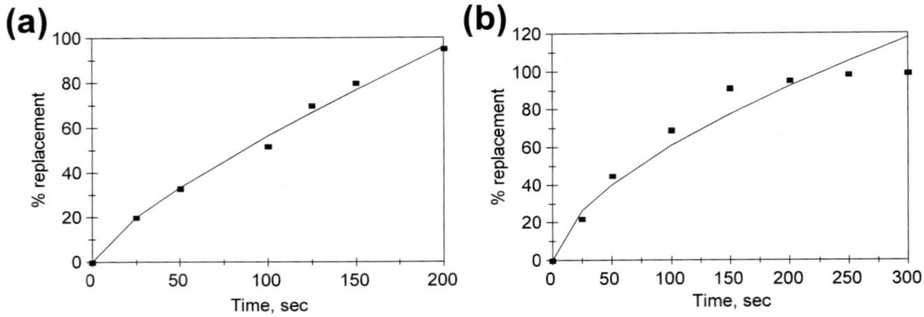

FIGURE 13.18 Binding of (a) MIP and (b) NIP to lysozyme (0.2 mol/L; pH 6.2) (Jing et al., 2010).

TABLE 13.13 Binding Rate Coefficient, k and the Fractal Dimension, D_f for (a) MIP and (b) NIP to Lysozyme (0.2 mol/L; pH 6.2) Containing 0.02 Mol/L

Analyte in Solution/Receptor on Surface (Repeat Runs)	k	D_f
Lysozyme/MIP	1.6866 ± 0.0901	1.4750 ± 0.06008
Lysozyme/NIP	3.8119 ± 0.5418	1.7972 ± 0.1420

Figure 13.18(b) shows the binding of NIP to lysozyme (0.2 mol/L, pH 6.2 containing 0.02 mol/L NaCl (repeat run)). A single-fractal analysis is adequate to describe the binding kinetics. The values of the binding rate coefficient, k, and the fractal dimension, D_f, for a single-fractal analysis are given in Table 13.13.

Huang et al. (2011) have developed a microcantilever (MCL) biosensor modified by a specific peptide for the selective detection of trimethylamine (TMA). According to these authors TMA is a gas at room temperature and has an irritating and ammonia-like odor. Siminhoff et al. (2004) indicate that in the field of medical diagnostics the breath of patients is a sign of viremic disease. Traditionally, TMA may be detected by instruments such as gas chromatography (GC) (Zaitsu et al., 2008), high-performance liqud chromatography (HPLC) (Hyotylainen et al., 2001), and mass spectrometry (Ampuero et al., 2002). These techniques are nonportable, expensive, and complex (Huang et al. (2011)). Huang et al., 2011 emphasize the need for field analysis of TMA samples.

Yang et al. (2010) have reported on a MCL-based sensor for the detection of TMA. Huang et al. (2011) have immobilized a recognition molecule for TMA, a peptide with the sequence of Cys-Pro-Ser-Ala-Asu-Asu-Ser-Thr-Val-Lys-Glu-Gly-Cys-Glu immobilized on a MCL. Huang et al. (2011) state that it is a fragment of a natural receptor protein, which can selectively bind TMA.

Figure 13.19a show the binding of TMA to a modified specific peptide immobilized on a MCL surface. A single-fractal analysis is adequate to describe the binding kinetics. The values of the binding rate coefficient, k, and the fractal dimension, D_f, are given in Table 13.14(a).

Huang et al. (2011) analyzed the influence of replicate (two) runs on the bending response of a peptide-modified MCL on the injection of 800 ppm TMA in 0.1M phosphate buffer. Figure 13.19(b) (first run) shows that a single-fractal analysis is adequate to describe the binding and the dissociation kinetics. The values of (1) the binding rate coefficient, k, and the fractal dimension, D_f, and (2) the dissociation rate coefficient, k_d, and the fractal dimension for the dissociation phase, D_{fd}, are given in Table 13.14b. The affinity, K ($=k/k_d$), is equal to 1.3×10^5.

Figure 13.19(c) (second run) shows that a single-fractal analysis is adequate to describe the binding and the dissociation kinetics. The values of (1) the binding rate coefficient, k, and the fractal dimension, D_f, and (2) the dissociation rate coefficient, k_d, and the fractal dimension for the dissociation phase, D_{fd}, are given in Table 13.14(b). The affinity, K ($=k/k_d$), is equal to 2.6×10^{13}.

Wang et al. (2010) have recently developed an SPR biosensor with core/shell Fe_3O_4/Au nanocomposites modified by 3-mercaptopropionic acid (MPA). These are used as solid support for the goat antihuman IgM, which was

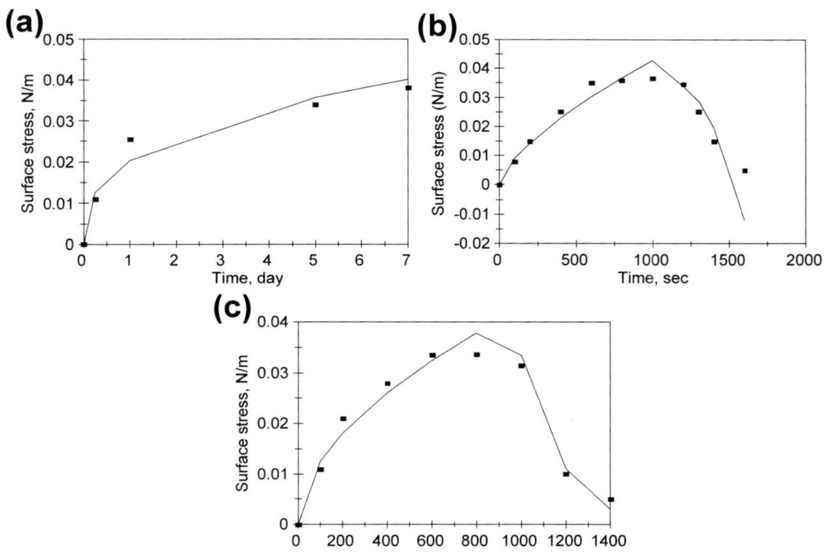

FIGURE 13.19 (a) Binding of trimethylamine (TMA) to a modified specific peptide immobilized on a microcantilever surface (Huang et al., 2011). (b) and (c) Replicate runs—binding response as a function of time for a peptide-modified microcantilever on injection of 800 ppm of TMA in phosphate buffer (Huang et al., 2011).

TABLE 13.14a **Binding Rate Coefficient, k, and Fractal Dimension, D_f, for the Selective Detection of Trimethylamine (TMA) by a Modified Cantilever Surface (Huang et al., 2011)**

Analyte in Solution/Receptor on Surface	k	D_f
800 ppm TMA in 0.1 M phosphate/synthetic peptide on microcantilever surface	0.0203 ± 0.0043	2.3014 ± 0.1432

TABLE 13.14b **Binding and Dissociation Rate Coefficients and Corresponding Fractal Dimensions for Replicate (Two) Runs for 800 ppm TMA in 0.01 M Phosphate Buffer (Huang et al., 2011). Bending Response as a Function of Time for a Peptide-Modified MCL**

Analyte in Solution/ Receptor on Surface	k	k_d	D_f	D_{fd}
800 pm TMA/Peptide-modified MCL (run #1)	0.000377 ± 0.000053	2.9 E −09 ± 1.5E-05	1.6294 ± 0.1231	0 ± 1.096
800 ppm TMA/Peptide-modified MCL (run#2)	0.001226 ± 0.001066	4.7E-17 + 17.7E-17	1.9318 ± 0.1651	0

TMA, Trimethylamine; MCL, microcantilever.

immobilized on the SPR surface. This, according to these authors, not only simplified the procedure for immobilizing but facilitated in the regeneration process of the sensing membrane. These authors state that magnetic nanomaterials have applications in the field of medical diagnostics such as cancer (Jun et al., 2005). Magnetic nanomaterials may be used with metal or metal oxides, which lead to surface modification. For example, Gu et al. (2004) deposited amorphous CdS on the surface of Fe/Pt NPs to form a metastable core/shell structure.

Au is considered a suitable material to coat magnetic nanoparticles. Xu et al. (2007) have indicated that a synthesis of Au-coated magnetic nanoparticle displays multifunctional properties, which are useful for diagnostic and medical applications. IgM is a polypeptide chain-rich I disulfide structure. These IgM antibodies are good in clumping microorganisms for eventual elimination from the body (Wang et al., 2010). Thus, their presence in sera is an indicator of infections. Wang et al. (2010) indicate that the detection of IgM is important in disease testing.

Wang et al. (2010) analyzed the binding of antihuman IgM to Fe_3O_4/Au nanocomposites. Figure 13.20a. A dual-fractal analysis is required to describe the binding kinetics. The values of (1) the binding rate coefficient, k, and the fractal dimension, D_f, for a single-fractal analysis, and (2) the binding rate coefficients, k_1 and k_2, and the fractal dimensions, D_{f1} and D_{f2}, for a dual-fractal analysis are given in Table 13.15(a) and (b). It is of interest to note that by a factor of 2.6 from D_{f1} equal to 1.1008 to D_{f2} equal to 2.881, binding rate coefficient increases by a factor of 39.16 from a value of k_1 equal to 0.4728 to k_2 equal to18.516.

Once again, Wang et al. (2010) analyzed the binding of antihuman IgM to an MPA-modified sensing membrane (Figure 13.20b). A single-fractal analysis is adequate to describe the binding kinetics. The values of the binding rate coefficients and the fractal dimensions are given in Table 13.15(a) and (b).

Figure 13.21 shows the binding of PSA in solution to anti-PSA-coated NPs (Sabatte et al., 2007). A single-fractal analysis is adequate to describe the binding kinetics. The values of the binding rate coefficient, k, and the fractal dimension, D_f, are given in Table 13.16.

TABLE 13.15a Binding Rate Coefficients for Antihuman IgM to (a) Fe_3O_4/Au Nanocomposite and (b) MPA-Modified Monolayer on Sensing Membrane (Wang et al., 2010)

Analyte in Solution/Receptor on Surface	k	k_1	k_2
Antihuman IgM/Fe_3O_4/Au nanocomposite	1.6217 ± 0.5900	0.4728 ± 0.1850	18.516 ± 0.057
Antihuman IgM/MPA-modified sensing membrane	2.3605 ± 0.1697	NA	NA

MPA, 3-mercaptopropionic acid.

TABLE 13.15b Fractal Dimensions for Binding of Antihuman IgM to (a) Fe_3O_4/Au Nanocomposite and (b) MPA-Modified Monolayer on Sensing Membrane (Wang et al., 2010)

Analyte in Solution/Receptor on Surface	D_f	D_{f1}	D_{f2}
Antihuman IgM/Fe_3O_4/Au nanocomposite	1.8158 ± 0.3050	1.1008 ± 0.5192	2.8881 ± 0.01309
Antihuman IgM/MPA-modified sensing membrane	2.4618 ± 0.3050	NA	NA

MPA, 3-mercaptopropionic acid.

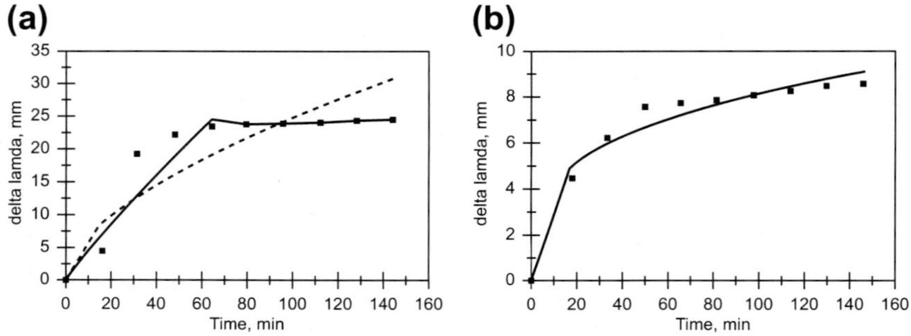

FIGURE 13.20 (a) Binding of Fe_3O_4/Au nanocomposite to antihuman IgM. (b) Binding of MPA-modified monolayer to antihuman IgM on sensing membrane (Wang et al., 2010).

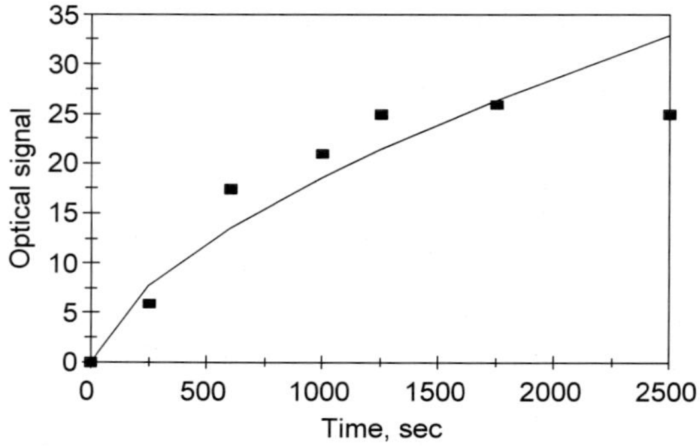

FIGURE 13.21 Binding of PSA in solution to anti-PSA coated on magnetic beads.

Wang et al. (2010) have recently developed an immunochromatographic electrochemical biosensor (IEB) for the rapid detection of metabolite biomarker of exposure to organophosphorous insecticides. Wang et al. (2010) indicate that chloropyrifos (CPF), is chloropyrifos-methyl (CPF-m), a class of chemical compounds that has been used in agricultural production (Jiang et al., 2008; Singh et al., 2010; Choi et al., 2006). Recently, it has been linked to accidents with organophosphorous compounds (Caldas et al., 2008).

TABLE 13.16 Binding Rate Coefficient and Fractal Dimension for PSA in Solution to Anti-PSA Coated on Magnetic Beads (Sabatte et al., 2007)

Analyte in Solution/Receptor on Surface	k	D_f
PSA/anti-PSA coated on magnetic beads	0.2434 ± 0.0696	1.7456 ± 0.2734

Organophophorous insecticides or their actual metabolites inhibit the activity of the enzyme, acetylcholinesterase. This enzyme is invoved in the CNS or the peripheral nervous system (Willie et al., 2010; Thompson et al., 2010).

Wang et al. (2010) indicate that the thionophosphate insecticide like CPF metabolizes to form TCP (3,5,6-trichloro-2-pyridinol) (Timchalk et al., 2007; Karnatki et al., 1976; Chanda et al.,1977; Pond et al., 1988). Lee et al. (2010) indicate that the primary metabolite is TCP when exposed to CPF.

Wang et al. (2010) indicate that TCP has been detected by different instrumental techniques: GC (Randhawa et al., 2007), gas chromatography—mass spectrometry (Daiz and Barello, 2006), HPLC (Abu-Qare and Abou-Donia, 2001a,b), and liquid chromatography—positive ion electrospray tandem mass spectrometry (Rain a and Sun, 2008). However, Wang et al. (2010) indicate that these are expensive methods and are time consuming. IEBs are alternate sensitve and inexpensive methods for detection (Liu and Lin, 2007; Lin et al., 2008; Blazkova et al., 2011). Wang et al. (2010) have presented a novel design of the IEB device for the sensitive detection of the metabolite, TCP.

Figure 13.22 shows the binding of TCP in solution to (Hyper Text Catching Protocol) HTCP-HRP precoated on a glass fiberIEB (Wang et al., 2010). A dual-fractal analysis is required to adequately describe the binding kinetics. The values of (1) the binding rate coefficient, k, and the fractal dimension, D_f, for a single-fractal analysis, and the (2) the binding rate coefficients, k_1.and k_2, and the fractal dimensions, D_{f1} and D_{f2}, for a dual-fractal analysis are given in Table 13.17(a) and (b).

Wang et al. (2010) have recently developed a potentiometric sensor based on surface molecular imprinting for the detection of cancer biomarkers and viruses. Molecular imprinting permits the detection of a wide range of substances. These authors have applied surface molecular imprinting using self-assembled monolayers for the sensing of cancer biomarkers and other proteins. Most

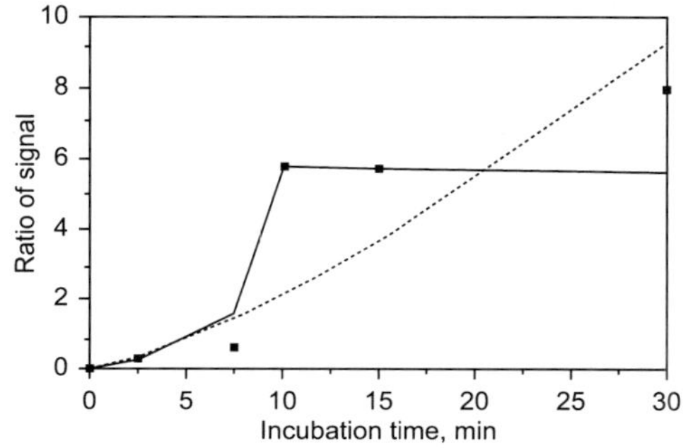

FIGURE 13.22 Binding of trichloropyridinol (TCP) to HTCP-HRP preloaded on a glass fiber immunochromatographic electrochemical biosensor (IEB) (Wang et al., 2010).

TABLE 13.17a Binding Rate Coefficients for Trichloropyrdinol to HTCP-HRP Preloaded on a Glass Fiber Immunochromatgraphic Biosensor (IEB) (Wang et al., 2010)

Analyte in Solution/Receptor on Surface	k	k_1	k_2
TCP/HTCP-HRP on glass fiber	0.09485 ± 0.1398	0.0505 ± 0.2242	6.091 ± 0.381

TABLE 13.17b Fractal Dimensions for Trichloropyrdinol to HTCP-HRP Preloaded on a Glass Fiber Immunochromatgraphic Biosensor (IEB) (Wang et al., 2010)

Analyte in Solution/Receptor on Surface	D_f	D_{f1}	D_{f2}
TCP/HTCP-HRP on glass fiber	0.3038 + 0.9396	1 + 1.2366	0 + 0.0158

cancer biomarkers are single proteins or protein-based molecular assemblies (Wu et al., 2007). Modern high-throughput methods have accelerated biomarker discovery (Sofer et al., 2006). Wang et al. (2010) state that current detection methods are time consuming and often suffer from cross-reactivity of the detecting antibody (Zheng et al., 2005; Wilson and Nie, 2006).

Molecular imprinting is a useful alternate method to detect a wide range of substances (Pauling, 1940; Sellergren, 2001; Piletsky et al., 2001; Cram, 1988; Bartsch and Maeda, 1994; Mosbach, 1994; Sellergren and Shea, 1994; Haupt and Mosbach, 2000; Vlakis 1993; Wulff, 1995). The molecular imprinting method is limited to the detection of small molecules (Zhou et al., 2005; Alzenburg et al., 1999; Mirsky et al., 1999; Li and Husson, 2006). Wang et al. (2010) indicate that 2-D imprinting procedures are required (Du et al., 2005; Das et al., 2003; Turner et al., 2006; Bossi et al., 2007). Wang et al. (2010) present a method to successfully detect carcinoembryonic antigen and poliovirus.

Figure 13.23 shows the binding of poliovirus in solution to the poliovirus biosensor (Wang et al., 2010). A single-fractal analysis is adequate to describe the binding kinetics. The values of the binding rate coefficient, k, and the fractal dimension, D_f, for a single-fractal analysis are given in Table 13.18.

Noah et al. (2011) analyzed the binding of different concentrations of COX-2 enzyme to the antibody immobilized in a sandwich assay format. Figure 13.24(a) shows the binding kinetics of 7.46e-3 mg/ml of COX-2 to the

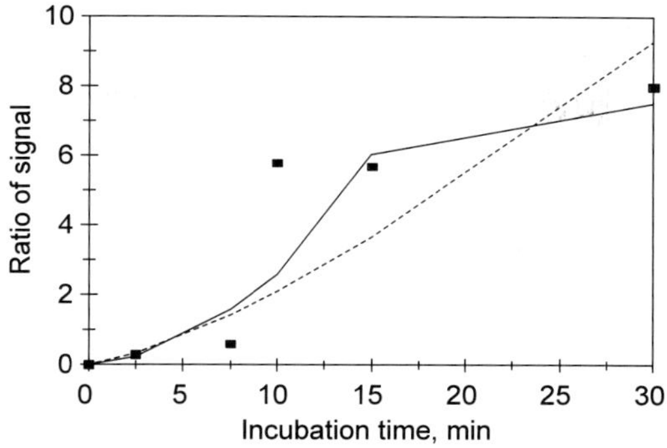

FIGURE 13.23 Binding of poliovirus to the poliovirus sensor (Wang et al., 2010).

TABLE 13.18 Binding Rate Coefficient and Fractal Dimension for Poliovirus to the Poliovirus Biosensor (Wang et al., 2010)

Analyte in Solution/Receptor on Surface	k	D_f
Polivirus/poliovirus biosensor	0.2765 ± 0.0712	0.6086 ± 0.2730

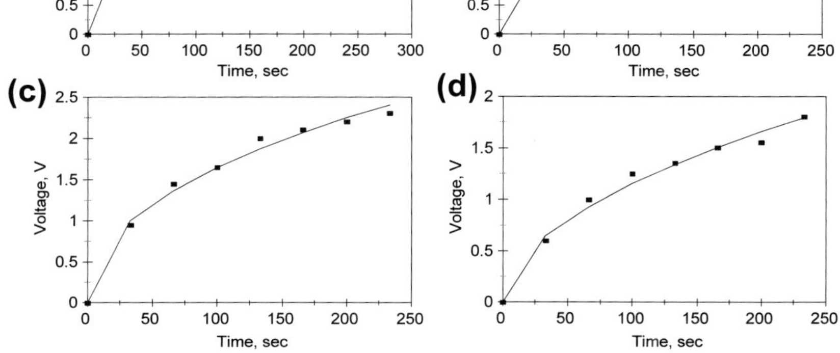

FIGURE 13.24 Binding of different concentrations of COX-2 enzyme and rabbit anti-COX-2 polyclonal antibody in solution to the goat anti-COX antibody (Noah et al., 2011): (a) 7.46e-3 ng/m (b) 7.46e-2 ng/ml, (c) 7.46e-1 ng/ml (d) 7.47e-0 ng/ml.

antibody immobilized on the sensor surface. A single-fractal analysis is adequate to describe the binding kinetics. The values of the binding rate coefficient, k, and the fractal dimension, D_f, for a single-fractal analysis are given in Table 13.19.

Figure 13.24(b) shows the binding kinetics of 7.46e-2 mg/ml of COX-2 to the antibody immobilized on the sensor surface (Noah et al., 2011). A single-fractal analysis is adequate to describe the binding kinetics. The values of the binding rate coefficient, k, and the fractal dimension, D_f, for a single-fractal analysis are given in Table 13.19.

Figure 13.24(c) shows the binding kinetics of 7.46e-1 mg/ml of COX-2 to the antibody immobilized on the sensor surface (Noah et al., 2011). A single-fractal analysis is adequate to describe the binding kinetics. The values of the binding rate coefficient, k, and the fractal dimension, D_f, for a single-fractal analysis are given in Table 13.19.

Figure 13.24(d) shows the binding kinetics of 7.46e-0 mg/ml of COX-2 to the antibody immobilized on the sensor surface (Noah et al., 2011). A single-fractal analysis is adequate to describe the binding kinetics. The values of the binding rate coefficient, k, and the fractal dimension, D_f, for a single-fractal analysis are given in Table 13.19.

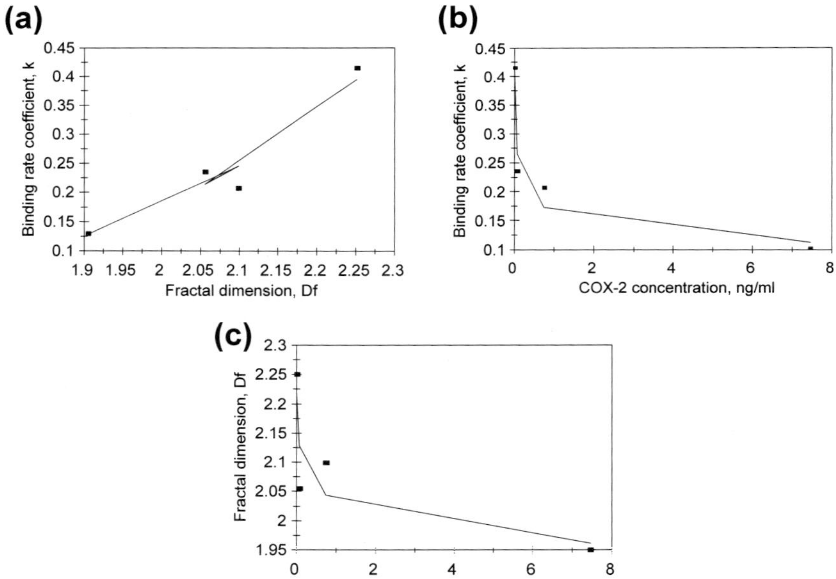

FIGURE 13.25 (a) Increase in the binding rate coefficient, k, with an increase in the fractal dimension. (b) Decrease in the binding rate coefficient, k, with an increase in the COX-2 concentration, in mg/ml. (c) Increase in the fractal dimension, D_f, with an increase in the COX-2 concentration, in mg/ml.

TABLE 13.19 Binding Rate Coefficients and Fractal Dimensions for Different Concentrations of COX-2 Enzyme to Antibodies in a Sandwich Immunoassay (Noah et al., 2011)

Analyte in Solution/Receptor on Surface	k	D_f
7.46e-3-ng/ml/antibody immobilized in sandwich immunosensor format	0.4160 ± 0.0033	2.2510 ± 0.009964
7.46e-2-ng/ml/antibody immobilized in sandwich immunosensor format	0.2362 ± 0.0085	2.0560 ± 0.0418
7.46e-1-ng/ml/antibody immobilized in sandwich immunosensor format	0.2076 ± 0.0110	2.0994 ± 0.0612
7.46e-0-ng/ml/antibody immobilized in sandwich immunosensor format	0.1032 ± 0.0070	1.9506 ± 0.07858

Figure 13.25(a) (interchange a and b) and Table 13.19 show the increase in the binding rate coefficient, k, with an increase in the fractal dimension, D_f for a single-fractal analysis. For the data shown in Figure 13.25(a), the binding rate coefficient, k, is given by:

$$k = (0.001644 \pm 0.000743)\, D_f^{6.754 \pm 1.222}$$

The fit is good. Only four data points are available. The availability of more data points would lead to a more reliable fit. The binding rate coefficient, k, for a single-fractal analysis exhibits a very strong (equal to 6.754) order of dependence on the fractal dimension or the degree of heterogeneity that exists on the biosensor surface.

Figure 13.25(b) and Table 13.19 show the decrease in the binding rate coefficient, k, with an increase in the COX-2 concentration in solution. For the data shown in Figure 13.27(b), the binding rate coefficient, k, is given by:

$$k = (0.1634 \pm 0.0295)\, [COX - 2]^{-0.187 \pm 0.032}$$

The fit is good. Only four data points are available. The availability of more data points would lead to a more reliable fit. The binding rate coefficient, k, for a single-fractal analysis exhibits only a mild order of dependence (equal to −0.187) on the COX-2 concentration in solution.

Figure 13.25(c) and Table 13.19 show the decrease in the fractal dimension, D_f, with an increase in the COX-2 concentration in solution. For the data shown in Figure 13.27(c), the fractal dimension, D_f, is given by:

$$D_f = (2.033 \pm 0.069)\, [COX - 2]^{-0.075 \pm 0.0064}$$

The fit is good. Only four data points are available. The availability of more data points would lead to a more reliable fit. The fractal dimension, D_f for a single-fractal analysis exhibits only a mild order of dependence (equal to −0.075) on the COX-2 concentration in solution.

Figure 13.26(a) shows the binding of 0.0364 ng/ml of COX-2 enzyme and rabbit anti-COX-2 polyclonal antibody in solution to the goat anti-COX-2 antibody immobilized on a sensor surface by a competitive reaction (Noah et al., 2011). A single-fractal analysis is adequate to describe the binding kinetics. The values of the binding rate coefficient, k, and the fractal dimension, D_f, are given in Table 13.20.

Figure 13.26(b) shows the binding of 0.364 ng/ml of COX-2 enzyme and rabbit anti-COX-2 polyclonal antibody in solution to the goat anti-COX-2 antibody immobilized on a sensor surface by a competitive reaction (Noah et al., 2011). A single-fractal analysis is adequate to describe the binding kinetics. The values of the binding rate coefficient, k, and the fractal dimension, D_f, are given in Table 13.20.

Figure 13.26(c) shows the binding of 3.64 ng/ml of COX-2 enzyme and rabbit anti-COX-2 polyclonal antibody in solution to the goat anti-COX-2 antibody immobilized on a sensor surface by a competitive reaction (Noah et al., 2011). A single-fractal analysis is adequate to describe the binding kinetics. The values of the binding rate coefficient, k, and the fractal dimension, D_f, are given in Table 13.20.

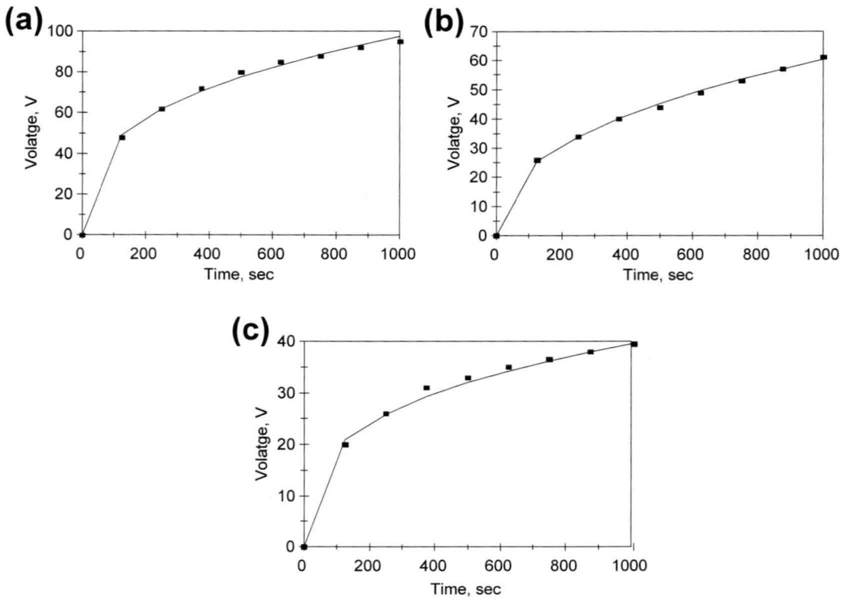

FIGURE 13.26 Binding different concentrations of COX-2 enzyme and rabbit and anti-COC-2 polyclonal antibody in solution to the goat anti-COX antibody (Noah et al., 2011): (a) 0.0364 ng/ml (b) 0.364 ng/ml (c) 3.64 ng/ml.

TABLE 13.20 Binding Rate Coefficients and Fractal Dimensions for Different Concentrations of the COX-2 Enzyme Plus Rabbit Anti-COX-2 Polyclonal Antibody to the Antigoat-Polyclonal Antibody Immobilized on the Sensor Surface (Noah et al., 2011)

Analyte in Solution, ng/mL	k	D_f
0.0364	1.2627 ± 0.0345	2.2024 ± 0.02894
0.364	4.8005 ± 0.1848	2.3890 ± 0.0362
3.64	3.480 ± 0.064	2.1744 ± 0.0888
36.4	10.031 ± 0.232	2.3416 ± 0.0246

The binding of 36.4 ng/ml of COX-2 enzyme and rabbit anti-COX-2 polyclonal antibody in solution to the goat anti-COX-2 antibody immobilized on a sensor surface by a competitive reaction (Noah et al., 2011). The binding is similar to the above figures, hence it is not shown. A single-fractal analysis is adequate to describe the binding kinetics. The values of the binding rate coefficient, k, and the fractal dimension, D_f, are given in Table 13.20.

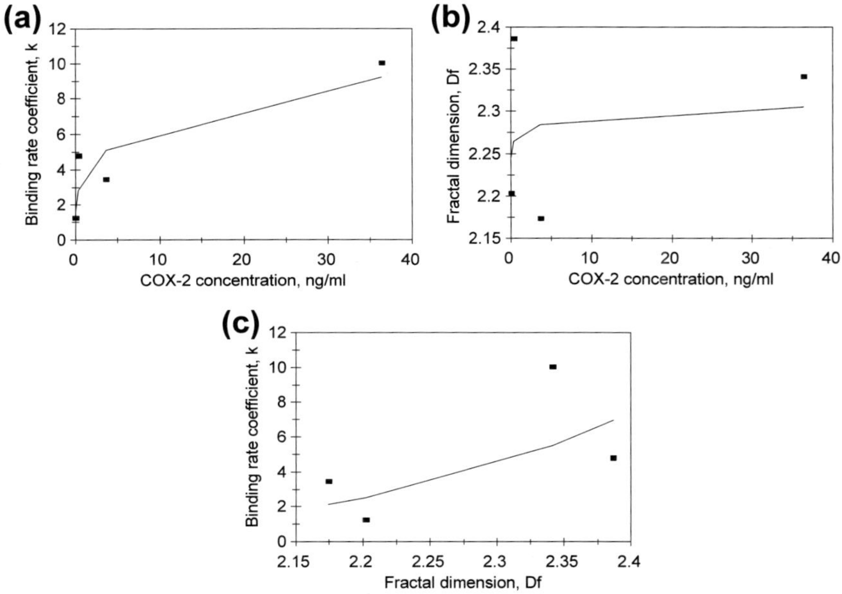

FIGURE 13.27 (a) Increase in the binding rate coefficient, k with an increase in the COX-2 enzyme concentration in solution. (b) Increase in the fractal dimension, D_f with an increase in the COX-2 enzyme concentration in solution. (c) Increase in the binding rate coefficient, k with an increase in the fractal dimension, D_f.

Figure 13.27(a) shows the increase in the binding rate coefficient, k, for a single-fractal analysis with an increase in the COX-2 enzyme concentration in solution in a competitive reaction. For the data shown in Figure 13.27(a), the binding rate coefficient, k, is given by:

$$k = (3.678 \pm 2.2328)\,[COX - 2]^{0.256\,+\,0.952}$$

There is scatter in the data. Only four data points are available. The availability of more data points would lead to a more reliable fit. The binding rate coefficient, k, exhibits less than one-half (equal to 0.256) order of dependence on the COX-2 enzyme concentration in solution.

Figure 13.27(b) shows the slight increase in the fractal dimension, D_f, for a single-fractal analysis with an increase in the COX-2 enzyme concentration in solution in a competitive reaction. For the data shown in Figure 13.27(b), the fractal dimension, D_f, is given by:

$$D_f = (2.273 \pm 0.126)\,[COX - 2]^{0.003838\,+\,0.0104}$$

The fit is good. Only four data points are available. The availability of more data points would lead to a more reliable fit. The fractal dimension, D_f, exhibits close to a zero (equal to 0.003838) order of dependence on the COX-2 enzyme concentration in solution.

Figure 13.27(c) shows the increase in the binding rate coefficient, k, for a single-fractal analysis with an increase in the fractal dimension, D_f. For the data shown in Figure 13.27(c), the binding rate coefficient, k, is given by:

$$k = \left(0.000116 \pm 0.000137\right) D_f^{12.65\,\pm\,9.887}$$

The fit is good. Only four data points are available. The availability of more data points would lead to a more reliable fit. The binding rate coefficient, k, is very sensitive to the fractal dimension, D_f, of the degree of heterogeneity that exists on the surface as it exhibits more than a twelve and a half (equal to 12.65) order of dependence on the fractal dimension or the degree of heterogeneity that exists on the biosensor surface.

13.4 CONCLUSIONS

A fractal analysis is presented for the binding and dissociation of different biomarkers for different diseases on biosensor surfaces. The biomarkers analyzed are those that have not been discussed in any of the previous biosensor chapters and have not been characterized with respect to any particular diseases. A single-fractal analysis parameters (k and D_f) is initially used to analyze the binding and dissociation (if applicable) kinetics. Only if a suitable fit is not obtained then we present a dual-fractal analysis to model the data using four parameters (k_1, k_2, D_{f1}, and D_{f2}).

Some of the kinetics for the biomarkers of different analyzed include (1) the binding of adrenaline in solution to polyphenol oxidase (PPO) immobilized on a polyethylene tube filled with carbon paste (Matavelli et al., 2010), (2) a label-free electrochemical sensor for the immunodetection of GSH (Stobiecka and Hepel, 2011), (3) a glutamate biosensor using glutamate oxidase as a recognition element (4) the detection of the biomarker transferring and the calibraton marker, creatine by an SPR biosensor (Nakamoto et al., 2008), (5) detection of lysozyme by a QCM biosensor including the influence of repeat runs (Sener et al., 2010), (6) a $\{C_3(OH)_{2mim}]$ $[BF_4]$-Au/Pt biosensor for glutamate sensing, (7) an SPR biosensor for the detection of the cancer biomarker, HCG, (8) binding of ACE to ACEI using a QCM (Su et al., 2007), (9) an amperometric biosensor for the in vivo detection of doapamine using an enzyme-based carbon fiber microbiosensor (Njagi et al., 2010), (10) a biosensor for the precise determination of homocysteine, (11) a selective NO nanocomposite-based biosensor based on direct electron transfer of MP. A deficiency of NO plays a significant role in Parkinson's disease and Alzheimer's disease, (12) binding of MIP to lysozyme, (13) a MCL biosensor modified by a specific peptide for the selective detection of TMA (Huang et al.,. 2011), (14) binding of antihuman IgM to Fe_3O_4/Au nanocomposite (Wang et al., 2010). The biosensor was modified by MPA, (15) binding of Prostate-Specific Antigen (PSA) in solution to anti-PSA-coated NPs (16) binding of trichloropyridinol (TCP) in solution to HTCP-HRP precoated on a glass fiber IEB (Wang et al., 2010), (17) binding of poliovirus to the poliovirus biosensor and, (18) binding of different concentrations of COX-2 enzyme for the goat anti-COX antibody immobilized in a sandwich assay format (Noah et al., 2011). As indicated above, a variety of biomarkers that are exhibited and maybe detected to indicate the presence of different diseases that are predicted. The biomarkers were selected at random form in the literature.

Both single- and double-fractal analysis were used to model the binding and dissociation (wherever applicable) kinetics. This chapter unlike the previous chapters, which was concentrated on the detection of biomarkers for specific diseases, shows the versatility of biosensors to be able to detect a wide range of biomarkers. It behooves the people working with biosensors to improve the detection of different biomarkers for the different diseases, so that the prevalance of different diseases is detected as early as is possible (for example, cancer) and the prognosis is significantly improved permitting the application of medical protocols as soon as is possible.

REFERENCES

Abdelwahab, A.A., Koh, W.C.A., Noh, H.B., Shim, Y.B., November 15, 2010. A selective nitric oxide nanocomposite biosensor based on direct electron transfer of microperoxidase: removal of inerferences by co-immobilized enzymes. Biosensors and Bioelectronics 26 (3), 1080–1086.

Abu-Qare, A.W., Abou–Donia, M.B., 2001b. DEET(N,N-diethyl-m-toulamide) alone and in combination with permethrin increased urinary excretion of 6-beta-hydroxycortisol in rats, a marker of hepatic CYP3A induction. Journal of Toxicology ,Environmental Health A 64 (5), 373–384.

Abu-Qave, A.W., Abou-Donia, M.B., 2001a. Combined exposure to DEET(N,N-diethyl-m-toulamide) and permethrin–induced release of rat brain, mitochondrial cytochromec. Journal of Toxicology and Environmental Health A 63 (4), 243–252.

Acharya, G., Chang, C.L., Holland, D.P., Thompson, D.H., Savran, C.A., 2008. Rapid detection of S-adenosyl homocysteine using self-assembled optical diffraction gratings. Angewante Chemistry International Edition 47, 1051–1053.

Agula, L.O., Pena-Farfala, C., Yanez-Sedenoa, P., Pingarron, J.M., 2007. Electrochemical determination of homocysteine at a gold nanoparticle-modified electrode. Talanta 74, 412–420.

Ajayan, P.M., 1999. Chemical Reviews 99, 1787–1800.

Alacam, S., Timur, S., Telefoncu, A., 2007. A novel biosensor based on I-homocysteine desulfhydrase enzyme immobilized in eggshell membrane. Journal of Molcular Catalysis B—Enzymes 49, 55–60.

Alzenberg, J., Black, A.J., Whitesides, G.M., 1999. Contol of crystal nucleation by patterned self-assembled monolayers. Nature 398, 495–498.

Amaral, D.G., Schumann, C.M., Nordahl, C.W., 2008. Neuroanatomy of autism. Trends in Neuroscience 31 (3), 137–145.

Ampuero, S., Zeisger, T., Gustafsson, V., Lunden, A., Bosset, J.O., 2002. European Food Research and Technology 214, 163–167.

Andreescu, S., Sadik, O.A., 2004. Analytical Chemistry 76, 552–560.

Bartsch, R.A., Maeda, M., 1994. Molecualr imprinting. Trends in Biochemical Sciences 19, 9–14.

Berelli, N., Andac, M., Baydemir, G., Say, R., Galaev, I.Y., Denzili, A., 2008. Journal of Chromatography A 1190, 18–26.

Biosensor, Q.C.M., 2014. Biological Science Instruments. www.Bio-Logic.Info.

Blazkova, M., Javurkova, B., Fukal, L., Rauch, P., 15 February 2011. Immunochromatographic strip test for detection of genus cronobacter. Biosensor and Bioelectronics 26 (6), 2828–2834.

Bonini, F., Piletsky, S., Turner, A.P.F., Speghini, A., Bossi, A., 2007. Biosensors & Boelectronics 22, 1131–1137.

Bossi, A., Bonini, F., Turner, A.P.F., Piletsky, J.A., January 2007. Molecularly imprinted polymers for the recognition of proteins, the state of art. Biosensor and Bioelectronics 22 (6), 1131–1137.

Bredt, D.S., 1999. Free Radical Research 31, 577–596.

Brown, N.J., Vaughan, D.E., 1998. Circulation 97, 1411–1420.

Bussien, J.P., Waeber, B., Nussberger, J., Gomez, H.J., Brunner, H.R., 1985. Current Therapy Research 37, 342–351.

Caldas, E.D., Rebelo, F.M., Helliodoro, V.O., Magajhaes, A.F., Rebelo, R.M., 2008. Clinical Toxicology 46 (10), 1058–1063.

Chanda, S.M., Mortenson, S.R., Mser, V.C., Padilla, S., 1977. Fundamentals in Applied Toxicology 38 (2), 148–157.

Chang, C.K., Lin, Y.W., Chen, W.T., Chang, H.T., 2010. Biology and Medicine 10, 1008.

Cheer, J.F., Wassum, K.M., Sombers, L.A., Heien, M.L.A.V., Arlangen, J.L., Aragona, B.J., Phillips, P.E.M., Wightman, R.M., 2007. Neuroscience 27, 791–795.

Chen, W., Cai, S., Ren, Q.Q., Wen, W., Zhao, Y.D., 2012. Recent advantages in electrochemical sensing for hydrogen peroxide analyst 137, 49–58.

Ching, C.T.S., Chon, T.R., Sun, T.P., et al., February 2011. Simultaneuos, non invasive and transdermal extraction of urea and homocysteine by nervous lonophoros. International Journal of nanomedicine 6, 417–423.

Choi, K., Joo, H., Rose, R.L., Hodgson, E., 2006. Journal of Biochemical Molecular Toxicology 20 (6), 279–291.

Choolani, M., Narsimhan, K., Kolla, V., Hahns, S., 2001. Expert Reviews in Proteonomics 6 (1), 87–101.
Chou, S., Hsu, W., Hwang, J., Chen, C., 2002. Analytical Chemistry 453 (2), 181–189.
Christopherson, K.S., Bredt, D.S., 1997. Journal of Clinical Investigation 100, 2424–2429.
Cram, D.J., 1988. The design of molecular hosts, guests, and their complexes (nobel lecture). Angewandte Chemie-Internationa Edition in English 27, 1009–1020.
Cramer, D.A., 1998. Homocysteine vs. cholesterol: competing views, or unifying. Explanation of arterioschlerotic cardiovascular disease. Laboratory Medicine 29, 410–417.
Cushman, D.W., Ondetti, M.A., 1999. Nature Medicne 5, 1110–1112.
Daniel, M.C., Astruc, D., 2004. Chemical Reviews 104, 293–346.
Darain, F., Park, S.U., Shim, Y.B., 2003. Disposable amperometric immunosensor system for rabbit IgG using a conducting polymer modified screen-printed electrode. Biosensors & Bioelectronics 18, 773.
Department of Health, Executive Yuan, R.O.C. (Taiwan), 2007. Health Statistical Reports. Available. http://www.doh.gov/tw/CHT2006/DisplayStatisticFile.aspx?d=69275&s=1.
Diaz-Cruz, A.S., Barcelo, D., 2006. Journal of Chromatography 1132 (1–2), 21–27.
Das, K., Panelle, J., Rotello, V.M., 2003. Selective picomolar detection of hexachlorobenzene in water using a quartz crystal coated with a molecularly imprinted polymer thin film. Langmuir 19, 3921–3925.
Du, X.Z., Hlady, V., Britt, D., 2005. Langmuir monolayer approaches to protein recognition through molecular imprinting. Biosensors & Bioelectronics 20, 2053–2060.
Du, F., Huang, W., Shi, Y., Wang, Z., Cheng, J., 2008. Biosensors & Bioelectronics 24, 415–421.
Erdos, E.G., 1976. America Journal of Medicine 60, 749–759.
Von Eckardstein., Malinow, M.R., Upson, B., Heinrich, J., Schulte, H., Schiönfeld, R., Kohler, E., Assman, G., 1994. Effect of age lipoproteins and hemostatic parameters on the role of homocysteinemia as a candiovascular risk factor in men. Journal of American heart Association, Anteriosclerosis Thrombosis and Vascular Biology 14, 460–464.
Elkelboom, J.W., Lonn, E., Geneest Jr., J., Hankey, G., Yusuf, S., 1999. Homocysteine and cardiovascular disease: a critical review of the epidemiologic evidence. Annals Internal Medicine 131, 363–375.
Fei, Z.F., Zhao, D.B., Pierraccini, D., Ang, W.H., Geldbach, T.J., Scopelliti, R., Chiappe, C., Dyson, P.J., 2007. Organometallics 26, 1588–1598.
Felix, E.S., Yamashita, M., Angnes, L., 2006. Biosensors & Bioelectronics 21, 2283–2289.
Feng, K., Li, J., Jiang, J., Shen, G., Yu, R., 2007. Biosensors & Bioelectronics 22 (8), 1651–1657.
Feng, B., Yu, F., Zheng, M.H., 2009. Advances in Clinical Chemistry 47, 45–57.
Flavin, K., Resmini, M., 2009. Analytical and Bioanalytical Chemistry 393, 437–444.
Friedemann, M.N., Robinson, S.W., Gerhardt, G.A., 1996. Analytical Chemistry 68, 2621–2628.
De Groot, M.T., Merkx, M., Wonders, A.H., Koper, M.T.M., 2005. Journal of The American Chemical Society 127, 7579–7586.
Gao, J.H., Gu, H.W., Xu, B., 2009. Accounts of Chemical Research 42 (8), 1097–1107.
Goodman, L.S., Gilman, A., 2006. As basesfarmacologicas da therapeutic, eleventh ed. McGraw-Hill, Interamericana do brasil, Rio de Janeirio.
Gordon, J.G., Swalen, J.D., 1977. Optics communication 22, 374–376.
Gong, K., Dong, Y., Xiong, S., Chen, Y., Mao, L., 2004. Novel electrochemical method for sensitive determination of homocysteine with carbon nanotube-based electrodes. Biosensors & Bioelectronics 20, 253–259.
Gu, H., Xu, K., Xu, B., 2006. Chemical Communication 9, 941–949.

Gu, H.W., Zheng, R.K., Zhang, X.X., Xu, B., 2004. Journal of The American Chemical Society 126, 5664–5665.

Guo, J.Z., Guo, F.Z., Hong, C.Y., 2002. Analytica Chimica Acta 463, 257–263.

Guo, T.K., Zhao, X.H., Xie, X.D., Chen, Z.H., Zhou, C.S., Wei, L.L., Zhang, J., 2007. Journal of Intenal Medicine Research 35, 353–360.

Gupta, A.K., Gupta, M., 2005. Biomaterials 26, 3995–4201.

Hansen, D.K., Davies, M.L., Lunte, S.M., Lunte, C.E., 1999. Pharmacokinetic and Metabolism Studies using Microdialysis sampling. Journal of Pharmacetical Science 88 (1), 14–27.

Haupt, K., Mosbach, K., 2000. Molecular imprinted polymers and their use in biomimetic biosensors. Chemical Reviews 100 (7), 2495–2504.

Havlin, S., 1989. Molecular diffusion and reaction. In: Avnin, D. (Ed.), In The Fractal Approach to Heterogeneous Chemistry: Surfaces, Colloids, Polymers. Wiley, New York, pp. 251–269.

Hayden, O., et al., 2006. Quartz crystal microbalance based nanosensor for lysosyme. Nature Materials 352–356.

Hepel, M., Stobiecka, M., 2011. Comparative kinetic model of fluorescence enhancement in selective binding of monochlorobimane to glutathione. Journal of Photochemistry and Photobiology A: Chemistry 225, 72–80.

Hefco, V., Yamada, K., Hefco, A., Hritcu, L., Tiron, A., Nabeshima, T., 2003. European Journal of Pharmacology 475, 55–60.

Hogant, B.L., Lunte, S.M., 1994. Analytical Chemistry 66, 596–602.

Homola, J., 2005. Chemical Reviews 108 (2), 462–493.

Homola, J., Yee, S.S., Gauglitz, G., 1999. Sensors & Actuators B: Chemical 54 (1–2), 3–15.

Horpacsy, G., Zinsmeyer, J., Schroder, K., Mebel, M., 1978. Clinical Chemistry 24, 74–79.

Huang, X., Li, M., Xu, X., Chen, H., Ji, H.F., Zhu, S., December 15, 2011. Microcantilevers modified by specific peptide for selective detection of trimetylamine. Biosensors and Bioelectronics 30 (1), 140–144.

Hyotylainen, T., Savloa, N., Lehtonen, P., Riekotta, M.L., 2001. Analyst 126, 2124–2127.

Van Hengel, A.J., 2007. Analytical and Bioanalytical Chemistry 389, 111–118.

Iijima, S., 1991. Nature 354, 56–58.

Ireland, J., Herzog, J., Unanue, E.R., 2006. Journal of Immunology 177, 1421–1425.

Isik, S., Etienne, M., Oni, J., Biochi, A., Reiter, S., Schuhmann, W., 2004. Analytical Chemistry 76, 6389–6394.

Jeffries, C., Chen, L., Carpita, N.C., Reiter, W.D., Wilson, R.H., McGann, M.C., November 1998. A rapid method to screen cell-wall mutants using discriminant analysis of Fourier transform infrared spectra. The Plant Journal 16 (3), 385–392.

Jiang, X.S., Li, D.Y., Xu, X., Ying, Y.B., Li, Y.B., Ye, Z.Z., Wang, J.P., 2008. Biosensors & Bioelectronics 23 (11), 1577–1587.

Jie, G.L.L., Chen, C., Xuan, J., Zhu, J.J., 15 July 2009. Enhanced electrochemiluminescence of CdSe quantom dots composited with CNTs and PDDA for sensitive immunoassay. Biosensors and Bioelectronics 24 (11), 3352–3358.

Jimsheena, V.K., Gowda, L.R., 2008. Analytical Chemistry 81, 9388–9394.

Jin, G.Y., Tang, Y.W., 2009. Microchimica Acta 165, 143–149.

Jing, T., Du, H., Dai, Q., Xia, H., Niu, J., Hao, Q., Mei, S., Zhou, Y., October 15, 2010. Magnetic molecularly imprinted nanoparticles for recognition of lysozyme. Biosensors and Bioeletronics 26 (2), 301–306.

Johnson, T.O., Ermolieff, J., Jirousek, M.R., 2002. Natural Reviews of Drug Discovery 1, 696–709.

Jun, Y.W., Huh, Y.M., Choi, JS. Lee, J.H., Song, H.T., Kim, S., Yoon, S., Kim, K.S., Shin, J.S., Sun, J.S., Cheon, J., 2005. Journal of American Chemical society 127, 5732–5733.

Karnatki, T., Leelin, M.C.M., Belcher, D.H., Neal, R.A., 1976. Drug Metabolism and Disposal 4 (2), 180–189.

Kameric, I., Lyon, A., Paterson, P.G., Jyurlink, B.H., 2000. Monochlorobimane Fluoremetric method to measure tissue glutathione. Analytical Biochemistry 286, 35–37.

Kawada, T., Yamasaki, T., Akiyama, T., Sato, T., Shishido, M., Sugimachi, M., Inagaki, I., Alexander, J., Sunagawa, K., 1998. Journal of Chromatograph B 714, 375–378.

Kualgina, R.V., Shankar, L., Michael, A., 1999. Analytical Chemistry 71, 5093–5100.

Koh, W.C.A., Rhaman, M.A., Choe, E.S., Lee, D.K., Shim, Y.B., 2008. Analytical Chemistry 23, 1374–1381.

Kron, O.S., Park, S.J., Jang, J., June 2010. A high-performance VEGF aptamer functionalized polypyrrole nanotube biosensor. Biomaterials 31 (13), 4470–4477.

Lee, S., Poet, T.S., Smith, J.N., Busby-Hjerpe, A.L., Timchalk, C., 2010. Chemical and Biological Interactions 184 (3), 449–457.

Li, J., Guo, S.J., Zhai, Y.M., Wang, E., 2009a. Nafion-graphene nanocomposite film as enhanced sensing platform for ultrasensitive determination of cadmium. Electrocommunication Communications 11, 1085.

Li, L., He, X.W., Chen, L.X., Zhang, Y.K., 2009b. Chemistry Asian Journal 4, 286–293.

Li, X., Husson, S.M., September 15 2006. Adsorbption of dansylated amino acids on molecularly imprinted surfaces: a surface plasmon resonance study. Biosensors and Bioelectronics 22 (13), 336–348.

Li, Muller MB., Gitle, S., Kaner, R.B., Wallace, G.G., 2008. Processable aqueous dispersions of graphene nanosheets. Nature Nanotechnology 3, 101.

Liao, Y.H., Brown, M.B., Martin, G.P., 2001. Journal of Pharmacy and Pharmacology 53, 549–554.

Lin, Y.Y., Wang J., Lui, G., Wu, H., Wai, C.M., Lin, Y., 15 June 2008. A nanoparticle label/immunochromatographic electrochemical biosensor for rapid and sensitive detection of prostate specific antigen. Biosensors and Bioelectronics 23 (11), 1659–1665.

Liu, A., Honma, I., Zhou, H., 2005. Biosensors and Bioelectronics 21, 809–816.

Liu, G., Lin, Y.Y., Wang, J., Wu, H., Wai, C.M., Lin, Y., 2007. Analytical Chemistry 79 (20), 7644–7653.

Liu, G., Lin, Y., December 2007. Nanomaterial label in electrochemical immuno sensor and immunoassays. Talanta 74 (3), 308–317.

Lu, F., Wang, K.H., Lin, Y.H., 2005. Analyst 130, 1513–1517.

Mahadik, S.P., Evans, D., Lai, H., April 2001. Oxidetive stress and role of antioxident and CO_3 essential fatty acid suplementation in schizophrenia. Progress in Neuro Psychopharmacology and Biological Psychiatry 26 (3), 463–493.

Mao, L.Q., Yamamoto, K., 2000. Talanta 51, 187–195.

Mancini, G.B.J., Henry, G.C., Macaya, C., O'Neill, B.J., Pucilo, A.L., Carere, R.G., Wargovich, T.J., Mudra, H., Luscher, T.F., Klibaner, M.I., Haber, H.E., Uprichard, A.C.G., Pepine, C.J., Pitt, B., 1996. Circulation 94, 258–265.

Matavelli, L.R.V., Antunes N., de. J., Ribeiro, M., Brigagao, P.L., de Magalhaes, C.S., Wisniewski, C., Lucas, P.O., 15 October 2010. Evaluation of a simple and low cost potentiometric biosensor for pharmaceutical and in vivo adrenaline determination. Biosensors & Bioelectronics 26 (2), 798–802.

Maskus, M., Pariente, F., Wu, Q., Toffanin, A., Shaplegih, J.P., Abruna, H.D., 1996. Analytical Chemistry 68, 3128–3134.

Mecker, L.C., Martin, R.S., 2008. Analytical Chemistry 80, 9257–9264.
Medina, A.R., De Cordova, M.L.F., Diaz, A.M., 2000. Mikrochimica Acta 134, 101–105.
Merkozi, A., December 2006. Electrochemical biosensing with nanoparticles. Federation of the European Biological society, FEBS 274 (21), 310–317.
Mirkin, C.A., Letzinger, R.L., Mucic, R.C., Storhoff, J.J., 1996. Nature 382, 607–609.
Mirsky, V.M., Hirsch, T., Piletsky, S.A., Wolfbeis, O.S., 1999. A spreader-bar approach to molecular architecture: formation of stable artificial chemoreceptors. Angewandte Chemie-International Edition 38, 1108–1110.
Miura, N., Sasaki, M., Gobi, K.V., Katoka, C., Shoyama, Y., 2003. Biosensors & Bioelectronics 18 (7), 953–959.
Moreira, L.N., Maghahaes, C.S., Luccas, P.O., 2005. Infarma 16, 59–62.
Mosbach, K., 1994. Molecular imprinting. Trends in Biochemical Science 19 (1), 9–14.
Nakamoto, K., Kurita, R., 2010. One-chip biosensor for simultaneous disease marker/calibration substance measurement in human urine by electrochemical surface plasmon resonance method. Biosensors and Bioelectronics 26 (4), 1536–1542.
Nakamoto, K., Kurita, R., Sekioka, N., Niwa, O., 2008. Chemical Letters 37, 698–699.
Narayanan, S., Appleton, H.D., 1980. Clinical Chemistry 26 (8), 1119–1126.
Niwa, O., Torimitsu, K., Morita, M., Osborne, P., Yamamoto, K., 1996. Analytical Chemistry 68, 1865–1870.
Niwa, O., Honuchi, T., Torimistm, K., 1997. Continous Monitoring of L-Glutamate released from Cultured nerve cells by an online sensor coupled with micro sampling. Biosensors and Bioelctronics 12 (4), 311–319.
Noah, N.M., Mwilu, S.K., Sadik, O.A., Fatah, A.A., Arcilesi, R.D., July 2011. Immunosensor for quantifying cylooxygenase 2 pain biomakers. Clinica Chemica Acta International Journal of Clinical Chemistry 412 (15–16), 1391–1398.
Noble, M.E.M., Endicot, J.A., Johnson, L.N., 2004. Science 303, 1800–1805.
Nobel, J. McC., Canoll, P., Honig, L.S., 24, August 2005. Brain tumor- Associated Dementia. Aging Knowledge and Environment 34, 2805–2814 p dn 2 JAMA, December 26, 2006.
Njagi, J., Chernov, M.M., Leiter, J.G., Andreescu, S., 2010. Amperometric detection of dopamine in vivo with an enzyme based carbon fiber microbiosensor. Analytical Chemistry 82, 989–996.
Njagi, J., Ispas, C., Andresescu, S., 2008. Analytical Chemistry 80, 7266–7274.
Oldenziel, W.H., Dijkstra, G., Cremers, T.I.F.H., Westerink, B.H.C., 2006a. Brain Research 1118, 34–42.
Oldenziel, W.H., Dijkstra, G., Cremers, T.I.F.H., Westerink, B.H.C., 2006b. Analytical Chemistry 78, 3366–3378.
Ouyang, R.Z., Lei, J.P., Ju, H.X., 2008. Chemical Communications 5761–5763.
Parker, L. (Ed.), 1996. Nitric Oxide, Part B; Physiological and Pathological Processes Methods in Enzymology, vol. 269. Academic Press, San Diego, California.
Pasco, N., Jeffries, C., Davies, Q., Downard, A.J., Roddick-Langiotta, A.D., Groton, L., 6 March 1999. Characterization of a thermophilic dehydrogenase biosensor for a merometric determination of L-glutamate by flow injection analysis. Biosensors and Bioelectronics 14 (2), 171–186.
Pascual, R., Gee, J., Finch, S., 1973. New England Journal of Medicine 289, 1074.
Pauling, L., 1940. A theory of the structure and process of formation of antibodies. Journal of the American Chemical Society 62, 2643–2657.
Peach, M.J., 1977. Physiological Reviews 57, 313–370.
Piletsky, S.A., Alcock, S., Turner, A.P.F., 2001. Molecular imprinting: at the edge of the third millennium. Trends in Biotechnology 19, 9–12.

Pillarik, M., Bockova, M., Homola, J., December 1, 2010. Surface plasmon resonance biosensor for parallelized detection of protein biomarkers in diluted blood plasma. Biosensors and Bioelectronics 26 (2), 1656–1661.

Pond, A.L., Chambers, H.W., Coyne, C.P., Chambers, J.E., 1998. Journal of Pharmacology Experimental Therapy 286 (3), 1404–1411.

Porstman, B., Jung, K., Schmechta, H., Evers, U., 1989. Clinical Biochemistry 22, 349–355.

Price, C.P., 2001. British Medical Journal 322, 1285–1288.

Puri, R.N., Roskoski, R., 1988. Analytical Biochemistry 173, 26–32.

Qu, F.L., Shi, A.W., Yang, M.H., Shen, G.L., Yu, R.Q., 22 February 2008. Amperometric H_2O_2 biosensor based on poly-thionine nanowire/HRP/nano-Au modified glassified carbon electrode. Sensors and Actuators B: Chemical 129 (21), 769–783.

Rahman, M.A., Noh, H.B., Shim, Y.B., 2008. Analytical Chemistry 80, 8020–8027.

Raina, R., Sun, L., 2008. Journal of Environmental Science Health Part B—Pestides Food Contaminants and Agricultural Wastes 43 (4), 323–332.

Randhawa, M.A., Anjum, F.M., Ahmed, A., Randhawa, M.S., 2007. Food Chemistry 103 (3), 1016–1023.

Reddy, S., Bradley, J., 2004. Annals New York Academy of Sciences 1037, 199–202.

Robertson, R.P., Sutherland, D.E., Seaquist, E.R., Nanz, K.J., 2003. Diabetes 1689-1694.

Robinson, D.L., Venton, B.J., Heien, M.L.A., Wightman, R.M., 2003. Clinical Chemistry 49, 1763–1773.

Ross, R., 1993. The pathogenesis of atherosclerosis: a perspective for the 1990s. Nature 32, 801–809.

Sabatte, J., Ceballos, A., Raiden, S., Vermenlen, M., Nahmod, K., Maggini, J., Salamone, G., Salamone, H., Amigorena, S., Yeffner, J., 2007. Human seminal plasma abrogates the capture and transmission of human immunodeficiency virus type I to CD4+ T cells mediated by DC-SIGN. Journal of Virology 81, 13723–13734.

Sanchez, S., Roldan, M., Perez, M., Fabregas, E., 2008. Toward a fast, easy, and versatile immobilization of bimolecules into carbon nanaotubes/polysulfone-based biosensors for the detection of HCG hormone. Analytical Chemistry 80, 6508.

Sellergen, B., Shea, K.J., 1994. Trends in Polymer Science 2, p. 166.

Sellergren, E., 2001. Molecularly Imprinted Polymers, Man-made Mimics of Antibodies and Their Application in Analytical Chemistry. Elsevier, New York.

Sener, G., Ozgur, E., Yilmaz, E., Uzun, L., Say, R., Denizli, A., October 15, 2010. Biosensors and Bioelectronics 26 (2), 815–821.

Serra, C., Vizoso, F., Alonso, L., Rodriguez, J.C., Gonzalez, L.O., Fernandez, M., Lamelas, M.L., Sanchez, L.M., Garcia-Muniz, J.L., Baltasar, A., Medrano, J., 2002. Breast Cancer Research 4, R16.

Shi, W., Zhanfang, Ma, November 2010. Amperometric glucose biosensor based on a triangular silver nanoprisms/chitosan composite film is immobilization matrix. Biosensors and Bioelectronics 26 (3), 1098–1103.

Siminhoff, M.L., Burke, J.F., Saukkonon, J.J., Ordinario, A.T., Doty, R., Dunn, S., 2004. New England Journal of Medicine 297 (9), 2478–2481.

Singh, J.N., Wang, J., Lin, Y.H., Timchalk, C., 2010. Toxicology Science 113 (2), 315–325.

Sofer, S., et al., 2006. The T cell receptor influences V beta element usage in response to myoglobin. Journal of experimental Medicine 174 (1), 83–92.

Splaver, A., Lamas, G.A., Hennekens, C.H., 2004. Homocysteine and cardiovascular diseases: biological mechanisms, observational epidemiology, and the need for randomized trials. American Heart Journal 148, 34–40.

Stobiecka M., Coopersmith K., and Helpel M., 16–20 August, 2009, Resonance Light Scattering Spectroscopy of Triot and Biomolecule–Mediated Assembly of gold nanoparticles, American Chemical Society National Meeting, (Washington DC, USA).

Stobiecka, M., Hepel, M., 15 April 2011. Effect of buried potential barrier in label-free electrochemical immunodetection of glutathione and glutathione-capped gold nanoparticles. Biosensors and Bioelectronics 26 (8), 3524–3530.

Stobiecka, M., Coopersmith, K., Hepel, M., 2010a. Journal of Colloid and Interface Science 350, 168–177.

Stobiecka, M., Deeb, J., Hepel, M., 2010b. Biophysical Chemistry 146, 98–107.

Stoeva, S.I., Lee, J., Smith, J.E., Rosen, S.T., Mirkin, C.A., 2006. Multiplexed detection of protein cancer markers with biobarcoded nanoparticle probes. Journal of the American Chemical Society 26, 8378.

Su, Y.Y., Chen, H., Wang, Z.M., Lev, Y., 2007. Applied Spectroscopic Reviews 42, 139.

Swamy, B.E.K., Venton, B.J., 2007. Analyst 132, 876–881.

The Homocysteine Studies Collaboration, 2002. Homocysteine and risk of ischemic heart disease and stroke: a meta-analysis. Journal of the American Medical Association 288, 2015–2022.

Taha, Z.H., 2003. Talanta 61, 3–10.

Tan, C.J., Chua, H.G., Ker, K.H., Tong, Y.W., 2008. Analytical Chemistry 80, 683–692.

Tan, L., Zhou, K.G., Zhang, Y.H., Wang, H.X., Wang, X.D., Guo, X.D., Guo, Y.F., Zhang, H.L., 2010. Nanomolar detection of dopamine in the presence of ascorbic acid at β-cyclodextrin/graphene nanocompositde platform. Electrochemistry Communications 12, 557.

Thompson, C.M., Prins, J.M., George, K.M., 2010. Health Perspectives 118 (1), 11–19.

Timchalk, C., Campbell, J.A., Liu, G.D., Lin, Y.H., Kousba, A.A., 2007. Toxicology Applications and Phamacology 219 (2–3), 217–225.

Tian, Y., Liu, H.Q., Zhao, G.H., Tatsuma, T., 2006. Journal of Physical Chemistry B 110, 23478–23481.

Tong, Q., Ye, C., McCrimmon, R.J., Dhillon, H., Choi, B., Kramer, M.D., Yu, J., Yang, Z., Christinasen, L.M., Lee, C.E., Choi, C.S., Zigman, J.M., Shulman, G.I., Sherwin, R.S., Elmqusit, J.K., Lowell, B.B., 2007. Cell Metabolism 5, 383–393.

Torres, C., Romero, M.A., Catatyud, M.J., 1998. Mikrochimica Acta 128, 187–190.

Turner, N.W., Jeans, C.W., Brain, K.R., Allender, C.J., Hlady, V., Britt, D.W., 2006. Frorm 3rd to 2nd: a review of the molecular imprinting of proteins. Biotechnology Progress 22, 1474–1489.

Ubbink, J.B., Vermaak, W.J.H., Bennett, Becker, P.J., van Staden, D.A., Bassbort, 1991. The prevalence of homocysteinemia and hypercholoesteroiemia in angiographically defined coronary heart disease. Klin Wochenschr 69, 527–534.

Uehara, A., Tekada, H., March 2007. Functional TLRs and NODs in human gingival fibroblasts. Journal of Dental Research 86 (3), 249–254.

Valentini, F., Palleschi, G., Lopez, E., Orlanducci, S., Tambum, E., Terachova, M., April 2007. Electroalaysis 19 (7–8), 859–869.

Vidal, M.L., Gautron, J., Nys, Y., 2005. Journal of Agricultural and Food Chemistry 53, 2379–2385.

Villamena, F.A., Zweier, J.L., June 2004. Antioxidants & Redox signaling 6 (3), 619–629.

Vlakis, G., February 1993. Drug assay using antibody mimics made by molecular imprinting. Nature 361, 645–647.

Wang, Y., Ma, X., Wen, Y., Xing, Y., Zhang, Z., Yang, H., July 2010. Direct Electro Chemistry and bioelectrocatalysis of horseradish peroxide based on the gold nano–seeds dotted Tio_2, nanocomposite. Biosensors and Bioelectronics 25 (11), 2442–2446.

Wang, J., Sun, Y., Wang, L., Zhu, X., Tian, Y., Zhang, H., December 1, 2010. Surface plasmon resonance biosensor based on Fe_3O_4/Au nanocomposites. Colloids and Surfaces B: Biointerfaces 81 (2), 600–606.

Wang, X.Y., Wang, M.J., Zou, X., Xu, X., Lei, R., Li, K.A., Li, N., 2009a. Analytical Bioanalytical Chemistry 395 (7), 2397–2403.

Wang, X., Wang, L.Y., He, X.W., Zhagn, Y.K., Chen, L.X., 2009b. Talanta 78, 327–332.

Wang, M., Zhao, F., Liu, Y., Dong, S., 2005a. Biosensors & Bioelectronics 21, 159–166.

Wang, Y.Z., Li, Q., Hu, S.S., 2005b. Bioelectrochemistry 65, 135–142.

Watt, M.J., Howlett, K.F., Febbaraio, M.A., Spriet, L.L., Hargreaves, J., 2001. Physiology 534, 269–278.

Wei, Q., Du, B., Han, D., Cai, Y., Zhao, Y., Xin, X., Li, H., Yang, M., March 31, 2011. Multifunctional mesoporous silica nanoparticles as sensitive labels for immunoassay of human chorionic gonadotropin. Sensors and Actuators B: Chemical 153 (1), 256–260.

Wightman, R.M., May, L.J., Michael, A.C., 1988. Analytical Chemsitry 60, 769A–779A.

Wilson, G.S., Johnson, M.A., 2008. In Vivo electric chemistry, what can we learnt about living systems? Chemical Reviews 108, 2254–2584.

Wilson, M.S., Nie, W.Y., 2006. Multiplexed measurement of seven tumor markers using an electrochemical protein chip. Analytical Chemistry 78, 6476–6483.

Willie, T., Thiermann, H., Worek, F., 2010. Toxicology In vitro 24 (3), 1026–1031.

Willis S.L., Tennstedt S.L., Marsiske M., Ball K., Elics J., Keopke K.M., Mornis J.N., Rebok E.W., Unvergaqt F.W., Stoddard A.M., and Wright, Long-term effects of cognitive training on Everyday, Funcional Outcomes of Older Adults(Active Study Group).

Wink, D.A., Mitchell, J.B., 1998. Free Radicals in Biology and Medicine 25, 434–456.

Wortsman, J., 2002. Endocrinology Metabolism Clinical North America 31, 79–84.

Wu, F.H., Zhao, G.C., Wei, X.W., 2002. Electrochemistry Communications 4, 690–694.

Wu, A.H., Syu, M.J., 2006. Biosensors & Bioelectronics 21, 2345–2353.

Wu, J., Fu, Z.F., Yan, F., Ju, Hx., 2007. Biomedical and clinical applications of immunoassays and immnosensors for tumor markers. TRAC-Trends in Analytical chemistry 26, 679–688.

Wulff, G., September 15 1995. Molecular imprinting in cross linked materials with the aid of moleclar templates A Way Towards Artificial Antibodies. Angewandte Chemie English edition 24 (17), 1812–1832.

Xie, Q., Zhang, Y., Xu, M., Li, Z., Yuan, Y., Yao, S., 1999. Electroanalytical Chemistry 478, 1–8.

Xu, Z.C., Hou, Y.I., Sun, S.R., 2007. Journal of the American Chemical Society 129, 8698–8699.

Yang, M.H., Li, H., Javadi, A., Gong, S.Q., 2010. Multifunctional mesoporous silica nanoparticles as labels for the preparation of ultrasensitive electrochemical immunosensors. Biomaterials 31, 3281.

Yavich, L., Tihonen, J., 2000. Neuroscience Letters 293, 41–44.

Yu, J.H., Ge, L., Dai, P., Ge, S.G., Liu, S.Q., 2010. Biosensors & Bioelectronics 25, 2065–2070.

Zaitsu, K., Katagi, M., Kamata, H., Kamata, T., Shima, N., Miki, A., Iwamura, T., Tsuchihashi, H., 2008. Journal of Mass Spectrometry 43, 528–534.

Zhang, S.Q., Tsiaras, W.G., Anaki, T., Wen, G., Minichiello, L., Klein, R., Neel, B.G., June 2002. Molecular Cell Biology 22 (12), 4062–4072.

Zhang, H., Li, Z.P., Zhang, J., Zhang, S.G., Zhu, L.Y., Yang, J., Zhang, X.P., Deng, Y.Q., 2007. Journal of Physical Chemistry B 111, 2864–2872.

Zhang, Z., Wang, Y., Ma, X., Wen, Y., Xing, Y., Yang, H., 15 July 2010. Direct electrochemistry and bioelectrocatalysis of horseradish peroxidase based on gold nano-seeds dotted TiO_2 nanocomposite. Biosensors and Bioelectronics 25 (11), 2442–2446.

Zheng, G.F., Patolsky, F., Cui, Y., Wang, W.U., Lieber, C.M., 2005. Multiplexed electrical detection of cancer markers with nanowire sensor arrays. Nature-Biotechnology 23, 1294–1301.

Zheng, W.B., Tian, C., Chen, X.H., 2007. Immunological Letters 113, 107–116.

Zhou, Y.L., Tian, R.H., Zhi, J.F., 2007. Biosensors & Bioelectronics 22, 822–828.

Zhou, Y.X., Yu, B., Levon, K., 2005. Potentiometric sensor for dipicolonic acid. Biosensors & Bioelectronics 20, 1851–1855.

Chapter 14

Biosensor Economics and Manufacturing

14.1 INTRODUCTION

Biosensor economics and manufacturing cost of biosensors are presented in this chapter. There is very little information available on these topics in the open literature. Thus, it is worthwhile to present information together in one place. Most of the information gathered and presented here is from whatever has recently appeared on the Internet. There is a question of reliability here. This chapter on economics is critical, since some of the terms such as markets for biosensor manufacturing processes, costs, and competition are mentioned here. They are, however, not discussed in any detail, but some of these features are outlined so as to provide one with some sort of perspective. This chapter is the capstone for the entire book, because without it, one is not able to understand the full picture. Similarly, the kinetics of binding and dissociation are presented which helps one to understand the complete picture. Both the kinetics and economics are not presented in any detail elsewhere, and if they are, the kinetics are generally presented for a specific analyte. Besides, if one is to understand the different aspects involved in the economics process for a particular biosensor, then one presumably draws some valuable insights into one's process for biosensor manufacture. Of course, one would have to tailor make or modify some of the comments or conclusions to better suit one's process. It is hoped that this chapter provides some leads and thoughts into what is involved in the so-called biosensor processes, and how one can gain from this knowledge. Surely, this should just serve as a starting point in this very critical endeavor.

One should not get the impression that since it is the last chapter of the book, it is not important or that it has very little influence on the field of biosensors as a whole. This could not be more further from the truth. For the universities, this may be partially correct, as they are very much interested in the research aspects of biosensors, and in general do not pay much attention or only scant attention to the economics of the process. What their mindset is that if they will be able to detect to particular analyte of interest—cost is not a major concern here. In an industrial setting, however, the "bottom line" is a primary concern, and whatever moves the company makes, or changes therein have to keep the economics in mind.

Biomarkers and Biosensors. http://dx.doi.org/10.1016/B978-0-444-53794-2.00014-8

Companies also have to keep a very close eye on their competition as well as the market projections for biosensors in the biosensor field, at least for the coming 3–4 years. If the market is right and is expected to grow, surely one can expect companies to invest more and more in particular types of biosensors. Thus, market projections have a certain value here, and they can give the company a "leg up" on its competition in that it can invest quickly in a particular area of biosensor application. Of course, these projections should be as accurate as possible, as the companies will rely on this. Misinformation or wrong information in a battle scenario will be devastating. In one place, you have a cut-throat competition, whereas in the other, actual lives are involved, and much depends on the decisions made by the leaders of a company or in battle field conditions. Extreme care needs to be taken, and all possible avenues should be considered before these "critical" decisions are made. Therefore, market projections are presented via the different reports that have been published in the open literature. As expected, these reports are expensive (thousands of dollars), and rather have a short life where the information presented therein is of significant value.

Data on a few companies and start-up companies are also presented in order to provide one with a perspective as to what does it take to run a successful company. Besides, one gets an idea of how long does it take for a company to be profitable, and one needs to interact with investment companies, private investors, etc. One has to raise capital to run a company, and many companies fail in that they have not given this a lot of thought, or they expected the company to be profitable sooner. Other more vexing problems may be involved. Finally, one should not downplay the importance of research (whether in-house or contract research), and how it may influence the economics and subsequently the market of the biosensor development process. For example, the development of a successful noninvasive method for diabetic monitoring would, needless to say, be very significant in influencing the biosensor market in this area. No wonder the companies involved in biosensor development in this area of sugar-level measurement are spending tremendous amounts of money to obtain a noninvasive biosensor.

14.2 BIOSENSOR COST

Leander (2013) indicates that students at Arizona State University (ASU) have created a low-cost biosensor to detect contaminated water in developing countries. They state that diarrheal disease represents the second leading cause of death of children under 5 years of age. According to WHO (World Health Organization), approximately 1.5 million children are killed as a result of diarrheal disease every year.

Nazirizadeh et al. (2010) have developed a low-cost biosensor using photonic crystals embedded between crossed polarizers. The authors indicate that there is a strong need for low-cost biosensors to enable rapid, on-site analysis of biological, biomedical, or chemical analysis. They propose a platform for label-free optical biosensors based on applying the analyte onto a surface-functionalized photonic crystal slab and performing a transmission measurement with two crossed polarizers. Their method also provides for efficient background suppression.

Professor Birch of the University of Luton, UK, has commented on the barriers to commercialization of biosensors. He indicates that blood glucose and pregnancy are large markets, and users of biosensors will pay economic prices for biosensors that may be used in these applications.

Liu et al. (2012) indicate that due to the desire to (1) decrease the cost of health care and (2) shift some of the analytical tests from centralized facilities to "front line" physicians and nurses who need to obtain accurate information more quickly about the health of a patient, there is subsequently an increasing demand for low-cost biosensors.

The Center for Computational Science at Tulane University (2013) indicates that an article on biosensors reports that immunosensors are low-cost platforms for disease diagnosis, population screening, and environmental monitoring since they provide real-time information about the presence of biological and chemical agents. The authors indicate that biosensors represent an important component in the fight against terrorism.

In a lecture given on September 2012 entitled "The Potential of an Integrated Biosensor," Purvis (2012) in addressing the technology barriers for point-of-care (POC) analysis indicates the cost, ease of use, robustness, and integration of biosensors. He emphasizes that the commercially available biosensor has reduced complexity for the worker. The cartridge-based desktop is a portable device. The Vantix VT System is a complete system for POC analysis, and real-time and accurate results are obtained within 2−10 min. Besides, these results are as accurate as obtained in a regional laboratory.

The Vantix potentiometric biosensor detects a peroxidase substrate reaction by converting it to a change in potential (mV). Simple, low-cost electronic circuits are used. The sensor detects changes in the electric open circuit potential. This is generated by biochemical reaction at the surface. Furthermore, the biosensor is manufactured by a low-cost manufacturing process, is easily adaptable to new applications, and permits multiplexing, which allows the simultaneous measurement of multiple analytes.

The author emphasizes that the biosensor is robust, since the surface area does not come into play. It simplifies the manufacturing process. The authors emphasize the use of standard electronic industrial materials, and volume manufacturing techniques yield a low cost per sensor production. It is instructive to provide some typical assay times for different compounds:

1. Glucose oxidase/peroxidase assay and anti-HbA1c immunosensor, approximately 5 min
2. Total cholesterol, approximately 2 min

The author emphasizes that the Vantix VT biosensor enables all possible assays requested by physicians.

Furthermore, the Vantix platform is able to detect different analytes that rival the existing POC techniques offered by different companies. In essence, the test is performed within 10 min, the results are received, and the patient is informed. The therapy is begun immediately.

BLU biosensors (2013) indicates that deaths from waterborne diseases can be prevented. They have developed low-cost, rapid, and effective biosensor. BLU indicates that 1.5 million children of developing countries die from diarrheal disease every year. Cost-effective biosensors to detect these pathogens (such as *Escherichia coli* and *Salmonella*) are not available in these countries, and also there is a lack of trained personnel. Rapid response is unavailable for a wide range of pathogens, since the biosensors are not customizable. BLU has sought to address these problems by creating a cost-effective and user-friendly biosensor that rapidly detects these pathogens in developing countries.

BLU has designed a protein-based biosensor that can be cheaply produced, purified with lab-cultured microbes, and then sent to wherever they are needed. BLU emphasizes that no other company has produced a biosensor that is robust enough to be used in rural communities. The advantages of their biosensor are: cheap and easy distribution, adjustable modifications to detect a wide range of pathogens, and quick and easy visible detection. Their device is simple. If it changes color, then the people need to take precautions like boiling the water. If it does not change color, then the water is safe to drink. Finally, BLU adds that their device provides peace of mind in the sense that it lets the residents know about the safety of the water they drink and for other purposes. They do plan to a pilot test for their technology in Guatemala.

BLU emphasizes that it will save the governments in different countries from the unnecessary cost required to treat "clean" water. Diarrheal diseases might contribute to about 60% of deaths in Guatemala. It is planned to reduce these deaths by 5% in the first year of distribution. This saves 400 children. They have also planned to test the water in 200 households in the rural highlands of Guatemala.

In an article entitled "Ancient Color-Shifting Goblet Inspires Nanoplasmonic Biosensor," Drake (2013) indicates that an ancient Roman cup that changes color in different lighting is the inspiration for a new nanoplasmonic biosensor. In essence, the tiny sensor changes color when target molecules bind to it. This is because of the optical properties of the materials it is made from. The researchers indicate that it is a low-cost alternative to conventional techniques used to study DNA, proteins, etc. The array changes color when target molecules are detected. The authors claim that their biosensor could cost less than US $10 who has optimized the array. As a consequence, experiments done by their device would be much less expensive when compared to the half million dollar devices with similar functions. The developers of the device indicate that it brings the optical physics inherent to the color of the iconic Lycurgus Cup down to a nanoscale. Their array is filled with Lycurgus Cup, and they are so small that literally each cup can hold a single virus particle. As the substances are introduced, they bind to the array. The optical refractive index is changed and different colors are illuminated. These can be easily observed by the eye or a cell phone camera. This is simpler than the other technologies wherein molecules must first be labeled with fluorescent tags, etc. Liu (one of the researchers) adds that the color change represents protein or DNA binding. The authors emphasize that the results are also quantitative. If more protein is in the solution, the color change will be more intense, whereas when more protein is not present, then the color change will not be that intense.

Pharmaco-Kinesis Corporation (PKC) (2013) has developed the PKC biosensor. This is an impedance-based affinity detection platform. It indicates that the electrochemical impedance spectroscopy (EIS) is a sensitive technique that is well suited to help characterize the biological interactions occurring on the surface. This is for the interactions occurring at the interface of conductive surfaces and liquid solutions such as bacterial growth media, cerebrospinal fluid (CSF), serum, and plasma (Gomez et al., 2002; Klosgen et al., 2011; K'Owino and Sadik, 2005; Lisdat and Schafer, 2008; Munoz-Berbel et al., 2008). They indicate that their biosensor is better than present-day technologies such as PCR and ELISA by combining high sensitivity, cost effectiveness, label-free rapid response, and real-time monitoring of samples. Their biosensor may be used to detect (1) pathogenic bacteria in foods and beverages to make sure that they are safe to consume and (2) fluctuating levels of vascular endothelial growth factor proteins in CSF of brain cancer patients. This may then be used to optimize chemotherapeutic drugs delivered to a tumor-affected area.

Biosensor has been developed of high reliability, lower-cost, and electrochemical behavior to meet the critical needs of millions of people. It is derived from two distinct platforms explained below. This proprietary position is applicable for diabetic disposable enzymatic sensor market. There is market for indwelling sensors. The author indicates that the goal is to replace the rapidly aging current strip technology. The market is large, as there are 20 million diabetic patients in the United States alone, with approximately 250 million worldwide.

The two platforms are

1. Conductive composite management (CCM) technology.

   This platform is owned entirely by Pepex. This is a market-ready technology that will make its debut in Trio™ blood glucose system. The CCM system combines a whole new series of sensor architectures. Besides, the CCM platform is easy to manufacture.

2. The wired enzyme is a big leap forward in medical diagnostics (Dx).
   It was discovered by Professor Adam Heller. The biosensor is being developed for commercial use by Pepex and Abbott Diabetic Care. Basically, it includes the electrical connection of enzymes (for example, glucose or lactate oxidase) with the electrodes via redox polymers to form amperometric biosensors. Electrons are shuttled from the enzyme to a meter whenever the electrode encounters with the desired molecule. The electrical current produced in the body, which is recorded by the meter readings, is directly proportional to the enzyme concentration.

The company emphasizes that the wired enzyme chemistry gives a significant edge to Pepex over the currently available biosensor in medical devices that are designed specifically for continuous or trend monitoring. Finally, Pepex states that it has leveraged Dr Heller's chemistry to fabricate biosensors for use in detecting glucose, lactate trend monitors, and sensors in drug delivery systems. This is of course combined with their proprietary CCM sensors.

Andersson (2013) indicates that researchers at the University of Leeds, UK, have developed a biosensor that can detect adenovirus in a noninvasive manner. This virus is responsible for quite a few illnesses, including the common cold and gastroenteritis. This biosensor can detect the virus and also identify the strain. It also has the capability to detect the number of virus particles.

Antibodies are attached to the electrical sensor, and the sensor's electrical charges are measured in the presence of adenovirus. Conventional testing of the adenovirus is slow and complicated. Their new technique not only helps diagnose patients quicker, but at a lower cost.

The adenovirus can have serious consequences if it afflicts children with an immature or compromised immunity, and even patients with HIV.

14.3 BIOSENSOR MARKETS

Azonano (2010) indicates that the advent of microfabrication techniques has led to the miniaturization of biosensors, for example, the use of nanosized electronic components. These authors emphasize that though the medical applications market offers ample opportunities, there are hindrances to commercialization. These include (1) high cost, (2) availability of effective alternate technologies, (3) stability, (4) sensitivity, (5) response time, and (6) quality assurance. Many new players are, however, entering the market.

Azonano (2010) indicates that just over two-thirds of the medical market for biosensors is in the United States and in the Europe. This was true in the year 2008. By 2012, the projections for the Asia-Pacific area are about US \$800 million. For medical applications, the majority of biosensors (about 90%) are used in blood gas analyzers, electrolyte analyzers, glucose meters, and metabolite analyzers.

Azonano (2010) emphasizes that diabetes mellitus (DM) has reached epidemic proportions and its management creates a very strong demand for glucose meters. Biosensor development has advanced in recent years with the inclusion of noninvasive monitoring (a very significant improvement that leads to better compliance) and wireless technologies. By 2012, the estimated market for glucose meters is US $1.28 billion. Water quality is increasingly important in European countries like France, Germany, Sweden, and Spain. Azonano (2010) estimates that for water quality testing in Germany, the biosensor market is expected to reach approximately US $33 million. Azonano (2010) indicates that there are two types of players in the biosensor market. These include the companies that develop biosensor-based devices and biosensor technology developers. Finally, Azonano (2010) indicates that the major market dynamics, trends, and competition are presented in the report entitled "Sensors in Medical Diagnostics: A Global Strategic Business Report."

Gaspar and Azonano (2013) in an article entitled "Biosensors Market" indicates that the biosensor market is categorized as a growth market. Research universities and the different industries are creating biosensors that are more precise, sensitive, noninvasive, and energy efficient. According to him, this has resulted in an enormous growth in the areas of health, environment, and nutrition. He emphasizes that the biosensor area is interdisciplinary in its nature and requires large investments for its development.

He provides a graph with percentage revenues in markets such as security, environment, domestic biodefense, and Dx. He indicates that till 2016, there is a continually growing trend. For example, Gaspar and Azonano (2013) indicates that the world biosensor revenues in 2009 for the following application areas were: (1) POC 47.9%, home diagnostics 19.2%, environmental 12.6%, research laboratories 11.2%, process industries 6.8%, and biodefence 2.6%. The author indicates that presently biosensors have nearly 50 applications and the number keeps on increasing.

As far as market revenues are concerned, the global market for biosensors is projected to grow to €14 billion over the next 7 years (2009–2016). He estimates that the biosensor market will increase by 12–14% every year. The author emphasizes that the POC application for biosensors will remain the largest market share in 2016. Over the next 4–5 years, the biosensor applications in the areas of environment and biodefense are expected to surge. The author emphasizes that over the last 6–7 years, biosensor applications have increased by about 50%. He expects more innovations in biosensor devices in the coming years. The author emphasizes that accuracy is a very important aspect in biosensor development.

Research and Markets (2013) has offered the latest title in their offerings on Biosensors Markets in their report entitled "Research and Markets: Biosensors in Medical Diagnostics: Global Strategic Business Report 2012." Their report analyzes the medical diagnostics in million US dollars by following the different product segments. This includes medical biosensors, glucose biosensors, environmental biosensors, and other biosensors. The report analyzes the demand in different countries or areas. They provide an analysis for the years 2010 through 2017. The countries or areas include United States, Canada, Asia-Pacific, Canada, Japan, Europe, and the rest of the world. Besides, a 6-year analysis is provided for these markets. Furthermore, their report includes the names of the major companies involved that make biosensor-based devices. They include Abbott Point of Care Inc., Siemens Healthcare Diagnostics Inc., LifeScan Inc., Medtronic Diabetes, Hoffman La Roche, AgaMatrix Inc., M-Biotech, and Cranfield Health to name a few.

The report provides a global market overview and analysis, product overview, application areas, and product launches. It does include a section on competitive scenario. It also provides the number of players who make these biosensors involved in the different countries (competitive landscape). For example, and as there are 64 companies involved in biosensors in the United States, whereas in Asia-Pacific (excluding Japan) and in the Middle East, there are 4 and 1, respectively.

Business Wire, London (2013) in a report entitled "Biosensors—A Global Market Overview" published in March 2012 indicates that the global market for biosensors in 2012 was estimated to be US $8.5 billion, and was estimated to roughly double (US $16.8 billion) by 2018. This represents a compound annual growth rate (CAGR) of approximately 12 %. The United States as expected is the single largest user with an estimated US $2.6 billion, followed by Europe. The report estimates that the Asia-Pacific region is expected to have the largest CAGR (11%) from 2008 to 2018. Primarily, the market is being driven by health care concerns and the increase in affordability regarding health care. Newer techniques, technologies, microfluidics, and noninvasive technologies are driving R & D activities that would lead to better, more sensitive, and reliable biosensors. It is a 232-page report (US $4050) and is expensive by university standards, though industries involved in this highly competitive area would find it a reasonable investment.

Though as mentioned above, the report may be expensive, it does provide the different principles of detection that include photometric, electrochemical, and ion-channel switch. It also includes a section on the categories of biosensors that includes optical biosensors, resonant biosensors, amperometric biosensors, and thermal detection biosensors. Finally, the report does analyze the application of biosensors in the POC testing, home diagnostics, environmental monitoring, process industry, and in the ever-increasing demand in the strategic security and biodefense areas.

Thusu (2013) in a Frost and Sullivan report entitled "Strong Growth Predicted for Biosensors Market" published in October 1, 2010 indicates that the biosensors market is expected to grow in the areas of health care, industry, environment, and monitoring. Also, it is expected to find increasing applications in the security and biodefense markets. The author emphasizes that biosensors have penetrated newer and diverse areas. They are becoming more noninvasive, are smaller, more affordable, and more sensitive. Advances in nanotechnology are expected to play an increasing role in the effectiveness of these biosensors in newer as well as the traditional areas of applications. Also, more and more applications are based on a single platform, for example, the testing for diabetes and cholesterol.

Thusu (2013) indicates that though most of the biosensors are under patent protection, their market penetration is often limited by the resources of the patent company. Furthermore, there still remain a lot of concerns with regard to commercialization. These include sensitivity, readout times vary greatly from one biosensor to another. Sometimes, these readout times can be greater than 20 s. A critical factor is the biological molecules used for detection purposes. The life span of these biomolecules is limited. There is always a problem associated with shelf life and long-term stability. Miniaturization and manufacturing are also difficult, and expensive, so much so that some biosensors are too expensive for commercial production. However, in the areas of drug discovery, advances in the different areas of biosensor development are bound to decrease and alleviate some of these limitations.

Thusu (2013) does indicate that despite some of the disadvantages outlined above, the biosensor is a low-cost alternative for many areas, especially optical biosensors. The further development of optical fiber technology will lead to biosensors that are more and more affordable suitable for high volume and mass use. Drug discovery is an important area of biosensor applications. Needless to say, "Big Pharma" has "deep pockets," and is in a position to effectively invest in biosensor technologies.

Thusu (2013) highlights some of the areas where more headway needs to be made. These include development of biosensors capable of detecting multiple analytes and monitoring. This would be useful in following and monitoring the progress of diseases such as systemic lupus erythematosus, development of integrated biosensor platforms, availability of wireless options, and the development of a self-configuring biosensor.

Finally, Thusu (2013) indicates the total biosensor market: percent revenue for each area of application. This author projected for 2009 and 2016. For the year 2009, the values are as follows: environmental 12.6%, biodefense 2.6%, research laboratories 11.2%, home diagnostics 19.2%, and POC 47.9%. As expected, the POC section has the largest percentage. For 2016, the values are as follows: environmental 14.3%, biodefense 3.3%, research laboratories 10.7%, home diagnostics 2.2%, and POC 44.9%. Once again, the projected largest percentage is for POC. In fact, for the 7 years (2009–2016) projected, there has been insignificant change in the relative percentages.

Prolog (2013) in a recent report by biosensor markets indicates that the market is expected to grow by 9.6% to reach US $16.8 billion by 2018. This, the author indicates, is due to standardization of equipment, and test processes in the areas of drug discovery, biodetection, environmental monitoring, and narcotic detection.

PR Newswire, in an article entitled "Biosensors—A Global Market Overview," indicates that the biosensor market is on a growth curve with increasing applications in a variety of areas. PR Newswire indicates that the technological developments will continue to provide a healthy trend for the growth of biosensors. Furthermore, the continued R & D efforts in biosensors will permit the effective use of biosensors in increasing areas.

In a recent Reuters report indicates that Professor Anthony Turner started research in the area of biosensors about 30 years ago, and the biosensor market was worth only US $5 million. Nowadays, Turner indicates that the biosensor market is worth a billion dollars. About 6000 research papers in the area of biosensors are published every year. Indicates that companies have been quick to realize the applications of biosensors to diabetes, cancer, and other insidious diseases. Indicates that the biosensor market is still dominated by diabetes and the detection of sugar levels (85%). However, this market is bound to expand to other areas such as environmental monitoring. The challenge for sugar level measurement is to make the biosensor noninvasive, and companies such as Siemens, among others, are working furiously in this area. Finally, other areas of interest include detection of "marker gases" to detect variety of diseases. The advantages of this technique are the noninvasive and user-friendly detection systems based on nothing but a quick breath test. Finally, Wickham (2013) indicates that the early and speedy detection, for example, could be crucial in the detection of diseases where the health networks are not that robust.

An article entitled "Commercially Available Biosensors" (Biosensors manufacturing cost, commercially available biosensors, 2013) indicates that for medical diagnostics, 90% of biosensors are for glucose monitors, blood gas analyzers, and electrolyte or metabolite analyzers. The authors indicate that half of all biosensors produced are for glucose monitoring. For 2012, the sales for these glucose monitors was projected to be US $1.28 billion.

Lifescan (Universal Biosensors, 2013) has launched One Touch Verio Sensor for use by diabetics patients. It uses Universal Biosensor's innovative opposing electrode technology. Lifescan is a Johnson & Johnson company. The sensor is available in Europe and Australia. The sensor was initially introduced in the market in January 2010, and further launches were expected in 2011. Universal Products (2013) indicates that the market has reacted positively to the sensor because of its novel electrochemical approach.

Universal Products (2013) indicates that POC testing for glucose first appeared in the 1980s. These tests allowed the patients with diabetes to take care of their insulin medication. It is estimated that 16–18 million disposable sensor strips using electrochemical cells are used every year. This remains as the largest market in the POC, and is valued at up to US $9.8 billion per year.

At 50—60 cents per strip, this is a large market. So, many companies do not charge for the actual sensor. This is based on the "shaving blade razor" model, where the companies make money on the shaving blades used and not on the actual razor itself. The same may be said for the table-top photocopy machine where the companies make their money on the cartridges and not on the machines themselves. Other examples, where this is effectively used, can be found in practice.

Universal Biosensors, following its success for the measurement of glucose levels, has adapted its sensor to measure prothrombin. This test measures and monitors the blood-thinning anticoagulant, warfarin. The purpose of the test is to maintain patients on a safe and effective dose. It should be noted that if the patients take too much warfarin, then they are at a risk of serious bleeding, whereas if they take too little warfarin, then they are at a risk of thrombosis.

Universal Biosensors (2013) suggests that the immunoassay technique represents an important technique for the measurement of biomarkers for the different diseases. It accounts for approximately 20% of the US $38 billion per year in vitro diagnostics (IVD) market. Universal Biosensors concludes by adding that many of these technologies can be easily transferred to POC diagnostics. Finally, Universal Biosensors also concludes that it is in the process of (immunoassay) detecting for D-dimer, which is a biomarker for thrombosis. Apparently, Universal Biosensors has come out with biosensor platform (business niche), and have plans to exploit it to the hilt.

Innovative Biosensors Inc. is a diagnostic company that sells instruments for the analysis of diseases. It is a high-valued activity since the results of testing are primarily responsible health care decisions. They claim that Dx is a growing market with a mid-single growth. The market for Dx is just over US $30 billion, and even the submarket for Dx outside the body is growing.

For example, the CAGR for the IVD is roughly 7%. It was estimated that it would reach US $9.3 billion by 2012. The company emphasizes that by identifying the protein biomarkers for the different diseases, rapid diagnosis testing has the potential to help significantly in clinical prognosis. Furthermore, the company estimates that the detection of infectious diseases is one of the largest submarkets for Dx. It is estimated to have a CAGR of 8.1%, and is expected to grow to US $12.4 billion by 2013. The company claims that there is significant increasing interest in detecting drug-resistant tuberculosis, influenza, and encephalitis. Finally, the company claims that cardiac testing and the diagnosis for cardiovascular diseases are becoming increasing important in managing patients and health care costs. The company emphasizes that the detection of protein biomarkers will play a very significant role in the early detection of diseases, and help in improving the sensitivity and specificity of these diagnostic tests.

Cranfield University in the United Kingdom is scheduled to offer three short courses entitled "Biosensors—From Fundamentals to Manufacturing Technology and Key Market Drivers." This course is to be offered in three countries including Cranfield University, UK, April 29–30, 2013; Barcelona, Spain, March 4, 2013; and Bangalore, India, May 30–31, 2013. Cranfield University is one of the leading universities on Biosensors, with the pioneering work performed by Professor Anthony Turner.

The course intends to provide an introduction to the biosensor market with a clear description of the market drivers and the ongoing R & D activities. It will also provide an overview on current manufacturing techniques. Finally, it will examine market trends and explore possible future technological trends and advances. For example, the course will analyze the future development drivers in the area of POC tests, and the commercial opportunities therein. It will also describe the design of disposable electrode biosensor strips for commercial manufacture of blood glucose and other biomedical biosensors.

News Medical (2013) in a recent article dated March 15, 2013 indicates that biosensors are characterized by slower pace of commercialization cost, availability of effective technologies, sensitivity, and quality assurance. Nevertheless, the field of biosensors is to a large extent untapped, and many new players are entering the field. Their study analyzes the trends in the field of biosensors. Some of these trends and limitations, opportunities, etc., are briefly outlined below.

The challenges ahead for the commercialization of biosensors are described. Regulations are hindering the growth of biosensors. They anticipate that painless or noninvasive technologies will spur growth in glucose determination. They indicate that there is intense competition. There is a market for home glucose monitors, and as indicated above, for noninvasive products. Finally, the authors provide a brief overview of the corporate developments that are occurring worldwide. For example, (1) Medtronic receives FDA approval clearance one one-touch, (2) Mexico inventors develop integrated optical biosensor, (3) new biosensor detects avian influenza virus, (4) Abbott introduces FreeStyle Lite, and (5) Medtronic introduces guardian real-time system management. Thus, there is a lot of emphasis on improving the biosensor system for different applications as a whole. This is only bound to increase with the increase in health problems, diabetes, etc., and the ease of obtaining quantitative results for different analytes of interest using biosensors.

Frost and Sullivan (2013) in a recent article acknowledge that PKC biosensor is expected to become the standard in pathogen testing. Based on an analysis of the biosensor market, Frost and Sullivan awarded them the 2013 global Frost and Sullivan award for new product innovation. Apparently, the company has set a new standard in the biosensor market with its highly sensitive and highly innovative biosensor. The biosensor uses EIS to detect pathogens such as *E. coli* 0157:H7 and the SARS virus. Their biosensor is highly sensitive, accurate, and results in cost savings. Furthermore, PKC indicates that it is portable and reusable. There is also a shorter, less complex testing process than other biosensors in the market. This reflects on PKC expertise in this area.

The PKC biosensor requires minimal human intervention, and provides for real-time pathogen detection. An additional advantage is that it operates, according to the company, with repeated use for a period of 5 years or so. The PKC biosensor uses a container with an electrochemical sensor attached to the base. It uses a liquid mixer and a microprocessor that records the biosensor readings.

The cost advantages of this biosensor are

1. Lateral flow method requires 8 h for sample preparation, whereas the PKC biosensor requires only half the time, i.e., 4 h.
2. PKC technology requires a consumable input of only bacteria culture medium, which makes the solution more cost effective.
3. Additional expensive equipment and the need to train personnel/technicians are avoided due to automated signal generation. Only an initial investment for a PKC instrument is required instead of recurring investments needed for the test solution. In summary, the PKC instrument lowers the cost of pathogen testing to almost one-half that of lateral flow technology.

The F&S award recognizes the value-added feature/benefits of the product and increases return on investment (ROI) it offers to customers. This leads to increasing customer acquisition and overall market penetration potential.

Nanofolio (2013), in an article entitled "Insulating Film-Based Biosensors," indicates that scientists at the University of Leeds have been working on a range of low-cost biosensors. This is based on the impedimetric measurement technique. The first step is the seating on the electrode surface with a mixed self-assembled monolayer. The antibody is attached to this monolayer. When the electrode is bathed in a solution containing the analyte of interest, the analyte binds to the sensing biomolecule. This is concentrated by binding reaction at the surface. Very small amounts can affect the biosensor characteristics since even a single molecule has a profound influence. For example, a single bound molecule has a significant effect on the impedance characteristics.

The sensor can be used to detect myoglobulin, which is released in increasing amounts following myocardial infarction (heart attack), as the muscle deteriorates. Apparently, free myoglobulin in the blood stream is thought to be an indicator of this myoglobulin immediately released following a heart attack, with increasing amounts being released as the muscle deteriorates. Finally, the author indicates that biosensor design is very flexible, and different antibodies may be bound to the surface, thus increasing the range of possible applications for the biosensor.

The goal of Professor Nader Pourmand's lab is to develop new tools and technologies using biological and electrical principles to detect and study genes and proteins. These tools should be able to lower the cost and increase the speed of the analysis. The primary focus is to develop new technologies and to validate these novel technologies by implementing them in important biological systems. They are especially interested in those technologies that directly convert a biological component into an electrical signal. This avoids the expense and time usually involved in biological amplification of the signal and the use of special readout material such as, for example, fluorescent dyes. The device could be battery powered, handheld, and inexpensive. The biosensor technologies under development should lead to either speed, accuracy, throughput, or cost when compared to the existing technologies.

14.3.1 Biosensor Manufacturing Cost

Biosensor manufacturing cost, commercially available biosensors (2013), indicates that due to the large development cost, manufacturing of devices in areas will be specialized, and concentrated in which they receive the most response from the market. Miniaturization has significantly reduced the cost of fabrication of these sensors, and makes these sensors more marketable. These authors further indicate that R & D of biosensors will focus on the creation of newer sensors and the miniaturization of newer biosensors. The authors emphasize that because of the high cost of manufacture of biosensors, miniaturization allows these biosensors to be made with less material, energy, and effort. They emphasize that new research helps keep the companies and universities ahead of this quickly changing field, and these companies have to be nimble and should adapt to newer technologies such as nanotechnology along with the many benefits therein.

Diagnostic Biosensors (Diagnostic Biosensors, Press Release, 2013) located in Minneapolis, Minnesota, USA, is setting the standard for biosensors, and has contributed to global MEMS (microfluidics manufacturing standard). The authors did indicate that in 2011 (September), there were no generally accepted standards relating to high-density microfluidics interfaces. They further indicated that details were required to design interfaces between fluidic routing cards, electrofluidic MEMS devices, and circuit boards. A minimal set of parameters had to be specified in order to standardize the design of a component or interface. Their Senior Manager, Mr Paul Trio stated "SEMI MS9-061 was developed to reduce redundant design time, enable more rapid miniaturization of MEMS systems, reduce system complexity, and most importantly reduce the cost by standardizing interfaces." He emphasizes that their standard will promote the development of MEMS microfluidics manufacturing and commercialization. The setting of this new standard will not only allow Diagnostic Biosensors to further develop their core sensor and fluidics products, but will also permit the company to collaborate with other companies in the MEMS-microfluidics marketplace.

OJ-Bio, in an article entitled "Mass Manufacture of Diagnostic Devices" (2013), indicates that JRC (Japan Radio Corporation) is one of its parent companies, and has over a hundred years experience in wireless communications. JRC has mass manufacturing capacity and an extensive range of patents in wireless, integrated circuits design, and in surface acoustic wave (SAW) technology. They have manufactured SAW devices for cell phones and GPS systems for over 30 years.

OJ-Bio has used this expertise and turned this technology for state-of-the-art biosensor technology by using SAW chips coated with appropriate receptors (for example, disease-specific biocapture surfaces). They have used the knowledge for OJ-Bio's other parent company, Orla Protein Technologies. Antibodies are immobilized in the appropriate orientation on these chips and react with the disease pathogens. This results in a phase angle shift of the SAW wave passing across the chip. This is then translated into an electrical signal. Their biosensor is able to detect pathogens from serum, urea, and saliva samples.

The authors claim that their biosensor has excellent sensitivity and a wide working range and uses direct measurement protocols. Their biosensors, OJ-Bio claims, are suitable for POC biosensors for a wide range of diseases. They are at present collaborating with a number of possible partners for commercial applications to detect a wide range of diseases.

JRC is able to manufacture millions of SAW devices every year using the techniques common in the electronics industry. Thus, OJ-Bio is able to mass manufacture biosensor devices.

Another company that is interested in developing SAW technology is ASR&D Corporation (2013). In an article entitled "Solving Problems with Acoustic Wave Technology," it indicates that it is actively engaged in developing tools for POC diagnostics based on the company's patented acoustic wave array affinity wave biosensor technology. The intent of the biosensor is to provide a diagnostic system using a reusable handheld reader. This is capable of push button operation for the automated analysis of the samples. It uses cost-effective microfluidic disposable cartridges that have been functionalized to help identify multiple clinical targets relevant to infectious diseases. The authors indicate that their chip incorporates microfluidic channels and nanostructured biologically active binding films in order to obtain rapid, detection of multiple infectious agents. The results are obtained within 20 min, and the authors claim that even untrained personnel can operate their biosensor.

Besides being able to detect a wide range of infectious diseases such as cancer markers, viruses, bacteria, proteins, and nucleic acids, their biosensor can also detect STDs (sexually transmitted diseases). These STDs are a major cause of global health problems and cause a lot of acute illness in women and infertility. The company states that its cost-effective biosensor is well-suited to meet this unmet need of detecting STDs.

The company indicates that their biosensor is a more sensitive quartz crystal microbalance (QCM) biosensor. Furthermore, their biosensor is being developed with volume manufacturing in mind. They emphasize that the problem associated with the widespread commercial distribution of these SAW biosensors is complex packaging and fluid handling and this has been a constraint. The company also emphasizes that the ASR&D biosensor eliminates the problem associated with complex packaging and fluid handling by incorporating multiple channels on a chip. Their novel device, the company claims, will allow for mass production and utilization of standard wafer processing and device packaging techniques. Their biosensor in combination with low-cost plastic microfluidic cartridges for sample handling and sensor/reader interface permits low-cost and high-volume disposable cartridge production.

Dey and Goswami (2011) have recently reviewed quantum nanoscale electronics device fabrication. They state that the development of optical biomolecular devices is a new move towards the revolution of nano-bioelectronics. The authors do point out that the emergence from research laboratory to the marketplace has been slow. A major problem for the realistic mass production of biosensors is the cost factor. There are problems associated with integrated biosensor systems that offer automatic monitoring systems. The size of the market also has an impact on the type of biosensor specified, as some are more amenable to mass production than others. The authors emphasize that optical sensing techniques offer advantages particularly when used in an integrated scheme. The authors also point out that the technology of integrated optics allows the integration of several active and passive optical components onto the same substrate. This permits the flexible development of compact sensing devices with the additional possibility of fabrication of multiple sensors on a single chip.

Epocal Corporation (2013) in an article entitled "Biosensors-on-Flex" indicates that it converts flex circuits into biosensor arrays in a single continuous in-line manufacturing process. In their process, the raw material to finished diagnostic card is ready for sale in minutes. Epocal indicates that the flex circuits are adapted from industry-standard smart card modules manufacturers on 35-mm tape-on-reel format. It indicates that the simple design and manufacturing process reduces the otherwise complex process of sensor manufacturing process and even the current competitor technologies. Rotors, discs, and cartridges containing manufactured chips are thereby eliminated.

They emphasize that their 35-mm module carrier results in the lowest cost per function of any sensor technology. Epocal adds that it has developed proprietary membrane and reagent formulation technology for application in blood test formats that include electrolytes, dissolved gases, hematocrit, and metabolites.

Wangmaung et al. (2013) have developed a biosensor-based molecular differential diagnosis of α-thalassemia (Southeast Asia deletion). This is a genetic hematologic disease which in the homozygous form can cause either death in utero or shortly after birth. The authors claim that it is absolutely necessary to accurately identify high-risk heterozygous couples. They have developed a QCM to detect this disease and have used a silver electrode presently, which replaced the previously used gold electrode. This reduces not only the production cost but also the analysis time. The diagnostic potency of the silver electrode was evaluated. Finally, the authors add that their silver thalassemia QCM was specific, sensitive, rapid, cheap, and field applicable.

Innocentive Challenge (2013) in an article entitled "Processes for DNA Biosensor Manufacture" has initiated an electronic request for partners (eRFP) challenge. The intent was to obtain a written proposal for establishing a collaborative partnership.

The seeker has assets of over US $2 billion. Partners are requested for the development and scale-up of processes for the manufacture of DNA biosensors for medical applications. Solvers with expertise in this area and resources to help optimize consistent manufacture of functionalized solid surfaces are encouraged to apply. Innocentive is a global innovation market place where creative minds solve some of the important problems for cash awards upto US $1 million.

LasX (2013) in an article entitled "Biosensor Components—Biosensor Test Strip Manufacture" indicates that laser biosensor components manufacturing yields less expensive, more precise, and more accurate test strips. Over 150 million people suffer from DM, not counting the millions of children not affected as yet. Every year, the patients need to use billions of disposable blood glucose test strips to be able to continuously monitor the blood glucose levels.

LasX indicates that lasers are key to the development and manufacture of next generation biosensor devices. Lasers permit thin film conductive coatings to create precise electrodes. These electrodes allow for more accurate results and permit the patients monitor the disease better.

TopSens Biosensors (2003) in an article entitled "Manufacturing Process" describe the five steps that are involved in the manufacturing process. These include:

1. screen printing;
2. first vision test;
3. first assembly;
4. reagents deposition (if required);
5. second assembly;
6. second vision test; and
7. packaging (if required).

Screen printing: The process is carried out by automatic, sophisticated, and technologically advanced machines. There is a pressurized environment and special filtration systems are used to eliminate impurities.

Vision test: First accurate vision test is done on every sensor with special high-precision machines. Only sensors with suitable requirements will proceed to assembly.

Assembly: Assembly and die-cut process before and after the deposition of reagents takes place for specific systems. They are protected to assure precision and accuracy.

Reagent deposition: A high level of experience about the deposition of the reagents in its own production process, utilizing exclusive systems and highly innovative technologies. This leads to high precision and reliability.

Packaging: Intermediate sensors that are without reagents are usually supplied in multiple sheets produced according to the customer's requirements.

The Gwent Group (2013) in a recent article stated that development of biosensors has been rapid in recent years. There have been key developments in the areas of integration of sensor systems, miniaturization of smaller systems, cheaper components, and mass production.

The group also states that Gwent Electronic Materials Ltd (GEM) and Applied Enzyme Technology (AET) are leaders in ink manufacturing and biosensor development. They manufacture products according to customer needs as well as standard biosensors. The Gwent group emphasizes that its sister company, AET specializes in protein stabilization of many enzymes. The authors emphasize that there is synergy between the two companies: the expertise of AET in enzyme stabilization and the skills of GEM in the design and production of specialist sensor materials. The company emphasizes that it has produced biosensors for diagnostic purposes, environmental sensors, and electrochemical sensors.

The company indicates that they use screen printing for nonbiological base transducers. However, materials can also be supplied for flexographic or gravue printing. They sell three types of sensor materials at GEM:

1. thermoplastic polymeric materials;
2. high-temperature materials; and
3. thermosetting polymeric materials.

Finally, the Gwent group indicates that the two main areas where their materials may be used are in single shot disposable sensors (detection of blood glucose) and reusable systems for blood gas analysis. The major measurement technique used is with electrochemical biosensors.

Zeta Corporation (2013) is the leading supplier or research equipment, for developing and the production of diagnostic strips biosensors and biochips. Using its core technologies, it is expanding worldwide. They are one of the leaders in R & D of dispensing technology applicable for biosensors, biochips, and microarrays. They are collaborating with government agencies and universities on this. The company says that it is always trying to simplify the technology and make it more customer-friendly. This assists in enhancing production efficiency and reduces the production cost. Finally, it states that its motto is "maximizing customer growth" and future development of technologies will bear this in mind.

IMICROQ (2013) (Integrated microsystems for quality of life) indicates that it has applied its knowledge for the treatment of serigraphy as a biosensor and microsystem fabrication. They state that they have considerable expertise in this area with regard to R & D. They have successfully integrated the structural and functional needs of the device. They also emphasize that their sensors are of low cost and easy to operate using electrochemical methods to allow for various applications.

Furthermore, the IMICROQ platform is low-cost engineered and easy to implement in mass production with low fabrication costs. This is a distinct advantage. IMICROQ indicates that serigraphy (also known as screen printing) has been around for years and used in mass-producing two-dimensional reproductions of a masked pattern. The authors indicate that the technique is widely used in microelectronics or as a cheap electrode material for biosensors.

IMICROQ has used their technology to develop a device to detect pathogens in food and environmental samples. These kits made by them allow for magnetic separation of the pathogens from a sample. This allows for detection of the pathogens by electrochemical methods within the microsystem.

14.3.2 Biosensor Start-Up Companies

In this chapter on biosensor economics, it is of interest to note how much effort and capital it takes to start up a biosensor company. Also, more importantly to note what is ROI and the time it takes to make the start-up turn to a profit. Surely, more companies fail or run out of money or capital before they turn a profit. More often then not, if a small company has successfully built a prototype, then in all probability will be bought out by one of the larger "deep pocket" biosensor companies. Perhaps, as time goes by, there will be just a handful of established biosensor companies, with a few smaller "boutique" type companies where presumably all, or most of, the relevant and innovative research with regard to biosensor research takes place. This is definitely not to say that the larger "Big Pharma" or the more established companies are not doing useful and relevant research with regard to getting a better hold or share of the biosensor market.

In order to provide a better perspective of biosensor start-up companies, two examples have been selected and analyzed briefly. They include:

1. University of Rochester Medical Center (URMC) startup funding to propel biosensor chips, and
2. heart disease start-up using "iCoaches" and biosensors to improve patient outcomes (Stone, 2013).
   a. URMC start-up funding to propel biosensor chips

A cardiologist at the University of Rochester wants to keep patients out of the hospital by using implantable biosensor chips by detecting health abnormalities before they really become a serious problem. Of course, the earlier you can detect possible heart problems the better will be the final prognosis. Biosensors are well-suited to do this task. Dr Spencer Rosero, MD indicates that the implanted living chip has the potential to revolutionize medical monitoring.

The company is called Physiologic Communications, LLC and it has received US $60,000 in funding to develop the technology from Excell Partners, Inc. Excell generally has a cap of US $250,000 on its seed investment. If other co-investors are involved, then the seed investment could be higher.

The company is integrating living cells with electronics to create a biological chip. The chip when it is implanted under the skin is able to detect physiological and chemical changes more quickly and accurately than traditional diagnostic tests. The chip is relatively small in size and not larger than a nickel. Also, it is expected to last about a minimum of 8 months. Also, Dr Rosero indicates that a patient with diabetes can constantly monitor the glucose level. If a change is detected, a signal is sent to a wireless device, and the patient can be alerted directly.

Patients with a history of heart failure can also be monitored by this device. Changes in protein levels can be detected. Doctors can alter the medication to correct the problem before it becomes serious and dangerous. Some of the patients who suffer from disorders such as congestive heart failure and life-threatening heart rhythms can avail of this device. Up until now, the company has been testing the individual biological and electronic components. One has to integrate these two types of components together, and the sensitivity of the device depends significantly on the interactions at the interface.

The company has built its first generation prototype. It is seeking further seed money and expects to generate a profit of revenue in about 3.5 to 4 years. Finally, Dr Rosero adds that biotech is a high-risk and long-term venture. Patience is essential.

b. Heart disease start-up using "iCoaches" and biosensors to improve patient outcomes

CardioVIP, a company dealing with cardiovascular disease has obtained US $2 million in Series B financing to create better outcomes for cardiovascular disease (MedCity News, 2013). CardioVIP has built a series of algorithms that combine the information from biosensors in the blood and diagnostic tests performed by primary care physicians. The company has plans to partner with the Cleveland Heart Clinic (one of the world's leaders in heart treatment care) to crunch the data and evaluate the a person's disease risk factor. This then leads to a tailored treatment of heart-related illnesses. This comes under the present day emphasis of "personalized care." The Vice-President of the company, Jamie Richter indicates the information maybe displayed on the web via a proprietary medical communication system that is available to both patients and physicians.

Richter has plans to expand the Houston-based company to the Philadelphia region. Cardiovascular disease risk is apparently broken down into five components: dyslipidemia, inflammation, diabetes and prediabetes, hypertension, and metabolic disease. The algorithm assists the physician by offering a set of treatment plans, which the physicians can tailor make the patient at hand. Also there are web coaches or "iCoaches" which may continuously assist the patient during his or her treatment. Finally, MedCity News (2013) estimates that in the United States there at present 19 million people who have risk factors associated with cardiovascular disease, and more importantly, an estimated 46 million who go undiagnosed, and are at a substantial risk of developing the disease.

nanoRETE (2013) indicates that the Global Food Protection Institute (GFPI) will make investments in two innovative start-up companies that have developed products that rapidly detect pathogens in food. The two companies selected were: Seattle Sensor Systems and nanoRETE. The President and CEO of GFPI indicates that one of the challenges in the food processing industry is field-based technology that is able to detect potential problems more rapidly. The investments to nanoRETE and Seattle Sensor Systems will help get these technologies to the market sooner.

nanoRETE (2013) indicates that it expects the Air Force to award its DOD SBIR Phase II proposal entitled "Development of a Field-Appropriate Biosensor for Detecting *Mycobacterium tuberculosis*." The proposal has been selected as one of the proposal that the Air Force will award. nanoRETE, Inc. is a Lansing, Michigan, USA company that is currently developing technologies that provide real-time detection of pathogens and toxins. It uses customized and proprietary nanoparticle biosensors. The company's biosensor can be used to detect single as well as multiple pathogens. The company emphasizes that it uses a simple device that can generate rapid screening in a cost-effective manner. The biosensor has applications in the food industry, military, and homeland security. The handheld device can generate screening results in about an hour. In contrast, current techniques require precious time and also sophisticated equipment and dedicated laboratory environment.

Michigan State University's biosystems and agricultural engineering professor, Dr Evangelyn Alocilja has developed a nanoparticle-based biosensor that is marketed by nanoRETE. nanoRETE is a spin-off company that will develop and commercialize this inexpensive biosensor to detect the disease causing organisms rapidly. Dr Alocilja indicates that its goal is to cut down the response time from 2 to 3 weeks to about an hour. This will fight the diseases and also help detect them quickly.

On January 25, 2012, Michigan State University has licensed a suite of technologies to detect a wide range of pathogens and pathogens to a Michigan-based start-up company, nanoRETE. The company was launched by the Michigan Accelerator Fund I (MAF-1). MAF-1 is focused on Michigan-based early stage life science and technology companies.

14.3.3 Biosensor Companies

Universal Biosensors is a biosensor company (Universal Biosensor, 2013), which specializes in medical diagnostics, and is focused on research, development, and manufacture of IVD products for consumer and professional POC use. They are using their electrochemical platform to expand into the blood glucose market, coagulation testing market, and other electrochemical-cell-based tests. The Chairman Mr Andrew Denver indicates that the year 2011 was a transitional year in which UBI moved from a technology start-up company to a company whose technology platform has been validated by the launch of its LifeScan OneTouch Verio blood glucose product. The chairman's letter does indicate that it has formed a strategic partnership with Siemens Healthcare Diagnostics, Inc. This should considerably assist in the development and commercialization of their device. They intend to market their product in Australia, Europe, and North America.

The CEO, Mr Paul Wright indicates that they have targeted their product to POC testing and have ended the year 2011 with a solid cash balance and growing product volumes. The intention is to expand beyond diabetes testing. The CEO indicates that the volatility exhibited during the early stages of their new product launch is typical, for example, the fluctuation of revenues. He indicates that there are exciting opportunities in the POC diagnostics field. He points out the two major requirements in the POC field: (1) rapid diagnosis is essential so that medical intervention is carried out if the need arises, (2) improved convenience that may be required for chronic ailments or for ongoing therapy. He adds that POC market is bound to exhibit double-digit growth in the future with the emergence of new, cost-effective technologies. Most diagnostic companies have realized this potential for the increase in market growth that is exhibited in this area. The company emphasizes that their electrochemical cell technology is a well-suited platform that could be adapted to different central laboratory tests to a POC format.

The company has been careful enough to protect their rights that include patent protection, trademark, and trade and secret laws, as well confidentiality agreements. They emphasize that their continued success depends on substantially protecting and maintaining their owned and licensed patents, patent applications, etc.

As far as revenue is concerned, the major part of the revenue is from LifeScan. In Australian dollars, the revenue income from products and services was approximately 12 million and 2.63 in the year 2011, 11.8 and 6.42 million in the year 2010, and 132,000 and 2.85 million in the year 2009, respectively. The percent income from LifeScan was 96%, 94%, and 96%, respectively. The company needs to diversify into other sources of income from products and is aware of this. According to the Annual Report published at the end of the year 2011, the company is seeking further collaborative arrangements and strategic alliances. As expected, and since the company has begun to move from the start-up phase to a company whose product has been validated, the R & D expenditures are a significant percentage of the revenues. For the years 2009, 2010, and 2011, the research expenditures are approximately 14.9, 6.48, and 9.81 million, Australian dollars, respectively.

Finally, as of March 2012, the company had 111 full-time employees, and they operate in a single geographic location, Australia. They are not as yet a multinational company, but presumably have plans to do so in the future.

It is instructive to get an indication of the stock price of a couple of companies. This, of course gives an indication of the idea and perspective of the company's earning potential and other entities. Biosensors International Group Ltd, Singapore completed an offering of US $300 million fixed-rate notes due in the year 2013. The offering was completed on January 23, 2013. Biosensors International Group Pvt Ltd, Singapore, a wholly owned subsidiary of the company issued US $300 million in principal amount of 4.875% fixed-rate notes due in the year 2017. The oversubscribed order book had a total value of US $1.25 billion. The notes were issued under the US $800 million multicurrency medium established by the issuer on January 4, 2013.

The company indicates that its US $300 million offering was 8 times oversubscribed. The company is "cash-rich" and Biosensors International may announce and acquisition of US $800 million as the company has more than sufficient funds for day-to-day operations in FY 2013. China Stock Exchange (CS) has listed its stock as OUTPERFORM, and is indicative that Biosensors International has emerged as a strong player in the region. Besides, it may benefit by China's push into the biomedical market.

Ft.com/markets data indicate that Biosensors International Group Ltd had 419.15 (SGD) in revenue with a net income of US $141.09 million. It has 312 employees. It is engaged in investment holding and licensing of medial technology. The company develops, manufactures, and commercializes medical devices. It is listed in medical equipment and supplies. The medical devices are used in interventional cardiology and intensive care products. It operates in three subsections:

1. interventional cardiology, such as drug-eluting stents;
2. critical care segments, such as intensive care and monitoring; and
3. licensing revenue segment, payments, and royalties for its products.

On October 2011, it acquired 50% interest in JW Medical Care Systems, Ltd.

OJ-Bio based in Newcastle, UK has obtained funding from the Biomedical Catalyst program toward its one-million-pounds project to develop a flu diagnostics device. The technology combines the biosensor materials with advanced electronics. The device is able to accurately detect flu and other respiratory conditions from patient-supplied samples. The low cost POC device obtains the results in minutes. It can successfully detect influenza A and B viruses and respiratory synctyial virus more quickly than current devices. OJ-Bio indicates that the latest round of funding will allow the company to develop a lab-based device into a functionally pilot device which is capable of carrying out large-scale clinical trials. Finally, OJ-Bio is a joint venture between the Newcastle-based biotechnology company Orla Protein Technologies, and JRC. JRC's expertise with wireless technology gives the detection device the flexibility to be wireless, allowing connectivity in their networks.

The Biomedical Catalyst program in the United Kingdom provides financial support for UK academics, and to small- and medium-sized companies operating in the life science sector.

14.4 CONCLUSIONS

Biosensor economics and manufacturing cost of biosensors is presented in this chapter. There is very little information available on these topics in the open literature. Thus, it is worthwhile to present the information together in one place. Most of the information presented is taken from the Internet. Thus, there is a reliability factor here. This chapter on economics is critical since it defines terms such as markets for biosensors, manufacturing costs, competition on a national and international level, etc. Besides, it does provide one with the changing dynamics in the biosensor area and it is not unusual to see companies collaborating on an international level in this area of biosensors.

One could presumably obtain the information presented in this chapter, and perhaps even in a clearer and more detailed fashion in reports that are available in the market from different sources. These reports (a few examples of the more recent ones are presented in this chapter), but these are expensive (typically, they cost about thousands of dollars). Universities, in general, would be hard-pressed to justify the cost of purchasing them. However, industries which have "deep pocket" should be able to buy these reports. These expensive reports are time-sensitive and begin to lose their value rather quickly. Thus, this type of economic information is presented collectively here in one place. Surely, industries involved in the biosensors, will have this information, but needless to say are not ready to part with this type of "confidential" and perhaps competitive information.

The chapter is divided into sections. We first consider the cost of a biosensor. Examples of different companies who are making these biosensors are presented. Only those companies where they claim to have a low-cost biosensor are presented. Examples of different detection techniques have been utilized to produce these biosensors are given. They have been selected at random from whatever is available in the literature. Most companies are trying to keep abreast of the newer technologies (like nanotechnology) as applied to biosensors. They are making all efforts to include and incorporate these newer technologies in the manufacture of these biosensors. This should help keep the cost of the biosensor down and help improve the different biosensor parameters (such as sensitivity, stability, residence time, lower detection limit, validity, reliability, etc.). This is essential in this highly competitive field. It makes no sense to develop a sensor with excellent performance parameters but which prices itself out of this competitive market.

In general, if one improves the biosensor performance of one parameter, one of the others (or quite a few) will deteriorate in its performance capability. Thus, it has to be optimized keeping this in mind. It seems that it is almost an art and not exactly science. The primary objective is the traditional "bottom line" that is keeping the cost of the functional biosensor down, while improving one or more of the performance parameters.

A brief overview of the biosensor markets is presented. There are expensive reports available in the open market, which deal with this issue in detail. Biosensors is a rapidly changing and dynamic area, and predictions of the market growth change from year to year. The major market for biosensors is in the area of medical diagnostics with blood glucose levels monitoring comprising a large fraction of the use of biosensors. But, due to the ease of biosensor use in different areas of application, there will be greater and greater applications in other areas such as environmental and biodefense applications. These trends are bound to take away some of the market share of biosensors from the traditional medical diagnostics field. The chapter does provide some of the reasons why biosensors have not been commercialized to a large extent in other areas, or find it difficult to penetrate in a market sense. Nevertheless, it is hoped that with future improvements in technology like nanotechnology, biosensors, in general will play a significant role in one's life. The applications of biosensors are apparently limitless, and this area is poised to expand substantially. The detection of biomarkers for the different diseases (the major area which this book deals) is one such area. There are other areas too, the number of which are bound to increase where biosensors may be effectively used. It is of interest to note that Cranfield University, a leading university in the areas of biosensors (apparently Professor Anthony Turner started and initiated the work on biosensors there) considers the biosensor *economics* and markets a very important area in which it ran a short course on Biosensors and markets in England, Spain, and India in the year 2013. Particular emphasis was to be placed on the commercialization of biosensors.

Manufacturing processes involved in biosensor production do play an important role in overall biosensor cost. Companies adapt their processes, if they can, to biosensor production. The development cost for manufacturing processes used in biosensor production is rather high. However, miniaturization allows these biosensors to be made with less energy, materials, and effort. Here is one of the avenues where nanotechnology can be of assistance. Standardization of the different components involved in manufacturing also leads to a better process. The different components involved in biosensor manufacture may be made by different methodologies. However, it seems that the use of lasers is estimated to play a bigger and bigger role in the cost-effective manufacture of biosensor components.

Investors are supplying the financial wherewithal to quite a few biosensor start-up companies. They do expect a good ROI from these start-up companies. The investors can be privately owned or state-run agencies (which would for example start-up capital for typically university-type projects). A couple of randomly selected start-up companies are presented. They do present novel ideas: basically in help improving patient care.

Finally, a couple of examples of biosensor companies are given. They are analyzed in some detail. Universal Biosensors in Australia is a good example wherein the company has recently transitioned from a start-up company to a company that has just begun to make a profit.

There is an urgent need to obtain more clear and detailed economic information in the open literature. Private companies who have the data will not part with it. The reports are very expensive and not everybody can attend the short courses that are offered which also have registration costs and other costs related to living expenses. It behooves the researchers working in the universities to start paying more attention to economies in biosensors and how that affects the advancements in this area.

REFERENCES

Andersson, C. Biosensor Technology Detects the Presence of Viruses, Medtech Insider. http://medtechinsider.com/archives/26982, downloaded March 20, 2013.

ASR&D Corporation. Solving Problems with Acoustic Wave Sensor Technology. http://www.asrdcorp.com/research/biosensor-research.html, downloaded March 8, 2013.

Azonano. Biosensor Market to Reach $6.1 Billion by 2012. http://www.azonano.com/news.aspx?newsID=8571, downloaded February 19, 2010.

Birch, B. Introduction to Biosensors, Barriers to Commercialization, University of Luton, UK, downloaded February 18, 2013.

BLU Biosensors. Dell Social Innovation Challenge, http://dellchallenge.org/projects/blu-biosensors, downloaded March 17, 2013.

Business Wire Research and Markets: Biosensors in Medical Diagnostics—Global Strategic Business Report, February 14, 2013. Research Markets, Dublin, Press Release. http://finance-yahoo.com/news/research-markets-biosensor-medical-diagnostics, downloaded February 19, 2013.

Business Wire, London Global Biosensors Market Review by Industry Experts Recently Published at Market Publishers.com, March 16, 2012. http://www.bus.newswire.com/news/home/20120316005408/en/global-biosensors-market, downloaded February 19, 2013.

Biosensor Manufacturing Cost, Commercially Available Biosensors. http://does/gogle.com/viewr?a=v&q=cche:Lw6_ds.d8_4J18J:faculty:uml.edu/xwang/16.54, downloaded March 8, 2013.

Biosensors International Group Ltd. http://markets.ft.com/research/markets/tearsheets/Business-profile?s=B20.SES.

Cranfield University. Short Course/CPD, Biosensors—From Fundamentals to Manufacturing Technology and Key Market Drivers. http://www.cranfield.ac.uk/health/shortcourses/page53015.html.

Diagnostic Biosensors, Press Release. Setting the Standard—Diagnostic Biosensors Contributes to Global MEMS Microfluidics Manufacturing Standard. www.DiagnosticBiosensors.com, downloaded March 8, 2013.

Dey, D., Goswami, T., 2011. Optical biosensors: a revolution towards quantum nanoscale electronics device fabrication. Journal of Biomedicine and Biotechnology 7. http://dx.doi.org/10.1155/2011/348218. Article ID 348218.

Drake, N. Ancient color-shifting goblet inspires nanoplasmonic biosensor. Science. News of your neurons. http://www.wiredscience/2013/02/nanoplasmonic-biosensor, posted February 25, 2013, downloaded March 17, 2013.

EPA. Extramural Research, MEMS Biosensor for In Situ Drinking Water Analysis. http://fpub.epa.gov/ncer_abstracts/index.cfm/fuseacton/display.abstract/13, downloaded March 24, 2013.

Epocal Corporation. Biosensors-on-Flex Technology. http://www.epocal.com/biosensors.htm, downloaded March 8, 2013.

Frost and Sullivan, 2013. Frost and Sullivan Acknowledges That PKC's Biosensors Are Expected to Become the New Industry Standard in Pathogen Testing, PKC Innovation Platform Is Highly Responsive Accurate, Fail-Safe, Cost-Effective Technology. http://finance.yahoo.com/news/frost-sullivan-acknowledges-pkcs-biosensors-170500808.ht, downloaded March 17, 2013.

Gomez, R., Bashir, R., Bhunia, A.K., 2002. Microscale electronic detection of bacterial metabolism. Sensors & Actuators 86 (2–3), 198–208.

Gwent Group. Advanced Material Systems, downloaded March 11, 2013.

Gaspar, C., Azonano. Biosensor Market-FL1. http://sites.google.com/site/biosensorsfl1/detector-cancro, downloaded March 8, 2013.

Innocentive Challenge. Processes for DNA Biosensor Manufacture. http://www.innocentive.com/ar/challenge/9844222, downloaded March 11, 2013.

IMICROQ. Integrated Microsytems for Quality of Life S.L., downloaded March 2013.

Innovative Biosensors, 2013. About Us, Markets. http://www.innovativebiosenors.com/ibi_markets.html. downloaded March 4, 2103.

Klosgen, B., Rumenapp, C., Gelich, B., 2011. Bioimpedance spectroscopy. In: Boob-Bamek, B., et al. (Eds.), BetaSys. Springe, New York, pp. 241–271.

K'Owino, I.O., Sadik, O.A., 2005. Impedance spectroscopy: a powerful tool for rapid biomolecular screening and cell culture monitoring. Electroanalysis 17 (23), 2101–2113.

LasX Industries Inc. Biosensor Components, Biosensor Test Strip Manufacture. http://www.lasx.com/biosensors_components.php.

Leander, S. Students Create Low-Cost Biosensor to Detect Contaminated Water in Developing Nations. http://asnews.asu.edu/20120906_water-biosensor, downloaded February 18, 2013.

Lisdat, F., Schafer, D., 2008. The use of electrochemical impedance spectroscopy for biosensing. Analytyical and Bioanalytical Chemistry 391 (5), 1555–1567.

Liu, Y., Matharu, Z., Howland, M.C., Revzin, A., Simonian, A.L., September 2012. Affinity and enzyme-based biosensors: recent advances and emerging applications in cell analysis and point-of-care testing. Analytical and Bioanalytical Chemistry 404 (4), 1181–1196. http://www.ncbc.nim.nih.gov/pubmed/22722742.

Maalouf, R., et al. Label-Free Detection of Bacteria by Electrochemical Impedance Spectroscopy MarketResearch.com, Market Research Estimated Biosensors Market at $16.8 Billion by 2018. http://www.marketwire.com/press-release/market-research-estimates-biosensors-market-at-, downloaded March 4, 2013.

MedCity News. Heart Disease Startup Using "iCoaches" and Biosensors to Improve Patient Outcomes, Digital Platforms. http://medcitynews.com/2012/09/cardiovascular-disease-startup using-i-coaches-to-improve, downloaded March 4, 2013.

Munoz-Berbel, X., Godino, N., Larzka, O., Baldrich, E., Munoz, F.X., del Campo, F.J., 2008. Impedance-based biosensors for pathogen detection. In: Zourob, M., Elwary, S., Turner, A. (Eds.), Principles of Bacterial Detection: Biosensors, Recognition Receptors and Microsystems. Springer, New York, pp. 341–376.

Nanofolio, 2013. Insulated Film-Based Biosensors. http://www.nanofolio.org/researcg/paper04.php, downloaded March 17, 2013.

nanoRETE. http://www.nanorete.om/In_the_News.html.

Nazirizadeh, Y., Bog, U., Sekula, S. com/releases_/medical_, Mappes, T., Lemmer, U., Gerken, M., 2010. Low-cost label-free biosensors using photonic crystals embedded between polarizers, *Optic Express* Optics Express 18, 19120–19128. http://www.Opticsinfobase.org/oe/abstract.cfm?uri=18-1819120, downloaded January 19, 2013.

News Medical. Trends in Global Market for Biosensors. http://www.news-medical.net/news/20100130/Trends-in-the-global-market-for-biosensors, downloaded March 15, 2013.

OJ-Bio. Mass Manufacture of Diagnostic Devices. http://www.oj-bio.com/technology-manufacturing.aspx, downloaded March 8, 2013.

Pharmaco-Kinesis Corporation. PKC Biosensor: An Impedance-Based Affinity Detection Platform. http://www.pharmaco-kinesis.com/tech_biosensor_tec.php, downloaded March 20, 2013.

Prolog, 2013. World Biosensor Market to Grow by 9.6% to Reach $16.8 Billion by 2018. http://www.prlog.org/11827989-worldbiosensor-market-to-grow-by 9.6-to-reach-16.8-billion, downloaded February 19, 2013.

PR Newswire Biosensors—A Global Market Overview, May 22, 2012, New York. http://www.prnewswire.com/news-releases/biosensors-a-global-market-overview-152556, downloaded February 19, 2013.

Purvis, D., September 26, 2012. The Potential of an Integrated Biosensors, Vantix. Healthcare Technologies, Life Science Leadership Series, Chesterfield Research Park. www.vantix.co.uk.

Stone, M. URMC Startup Funding to Propel Biosensor Chips, Rochester Business Journal, Daily Edition, Excell Partners Inc., "Providing Seed capital for Growth", http://www.excelny.com/index.php?mact=News, cntnt01, detail,&cntnto1 articleid=10& cht.

TopSens Biosensors. Manufacturing Process. http://www.topsens.it/engl_,manufacturing.html, downloaded March 11, 2011.

Thusu, R., Frost and Sullivan Sensors, October 1, 2010. Strong Growth Predicted for Biosensor Market. http://www.sensorsmag.com/speciality-markets/medical//strong-growth-predicted-biosensor, downloaded February 19, 2013.

Tulane University, Center for Computational Science, Research, Biosensors. http://tulane.edu.sse/ccs/research/index.cfm, downloaded January 13, 2013.

Universal Biosensors. Annual Report, Year Ended December 31, 2011.

Universal Biosensors, Products, First Product in Market, One Touch Verio Glucose Test. http://www.universal.biosensors.com/product.aspx, downloaded February 18, 2013.

Wangmaung, N., Promptmas, C., Chomean, S., Sanchomphu, C., Ittarat, W., February 1–7, 2013. Low cost biosensor-based molecular differential analysis of alpha-thalassemia (Southeast Asia deletion). Clinical Chemistry Laboratory Medicine 51 (6), 1199–1205.

Wu V.C.H., Low-Cost Nano-Biosensor to Detect Foodborne Pathogen That Causes Listeriosis, March 14, 2013, http://www.nanomagazine.co.uk/index.php?Option=com_content&view=article&id=2088:1.

Young, L. Strategic Business Insights, Biosensors. http://www.strategicbasicinsights.com/explored/bsshtml, downloaded March 4, 2013.

Zeta Corporation. Lateral Flow, Biosensor Production. http://zetacorporation.com/eng/company_1.html, downloaded March 15, 2013.

Index

Note: Page numbers followed by "f" and "t" indicate figures and tables respectively.

A

C

D

E

F

G

H

M

N

O

P

T

U

Edwards Brothers Malloy
Thorofare, NJ USA
December 12, 2014

[illegible]

[illegible]

[illegible] [illegible]

[illegible] [illegible]nology